D0845242

Coastal Upwelling

Coastal and Estuarine Sciences

A series developed to advance the knowledge of physical, chemical, and biological processes in coastal and estuarine regimes and their relevance to societal concerns.

1. Coastal Upwelling

Coastal and Estuarine Sciences 1

Coastal Upwelling

Francis A. Richards
Editor

American Geophysical Union
Washington, D.C.
1981

Published under the aegis of the Geophysical Monograph Board:
Robert Van der Voo, Chairman; Eric J. Essene, Donald W. Forsyth,
Joel S. Levine, William I. Rose, and William R. Winkler, members.

Coastal Upwelling

Library of Congress Cataloging in Publication Data

Main entry under title:

Coastal upwelling research, 1980.

(Coastal and estuarine sciences; 1)
Includes bibliographies.
1. Upwelling (Oceanography)—Congresses. 2. Coastal ecology—Congresses. I. Richards, Francis A. II. Series.
GC228.5.C6 551.47'01 81-14866
ISBN 0-87590-250-2 AACR2

Copyright 1981 by
American Geophysical Union
2000 Florida Avenue, N.W.
Washington, D.C. 20009

Figures, tables and short excerpts may be reprinted in scientific books and journals if the source is properly cited; all other rights reserved.

Printed in the United States of America.

CONTENTS

Models

Some Chemical Consequences of Upwelling

Upwelling, Primary Production, and Phytoplankton

Upwelling and Zooplankton

Upwelling and Macrofauna

The Nekton

The Benthos

Upwelling and Biological Systems

EDITOR'S PREFACE

This volume is one outcome of the IDOE International Symposium on Coastal Upwelling, but it is not a true symposium proceedings. Some of the papers in the book were not presented at the meeting, some papers and posters that were presented at the symposium are not in the book, and none of the discussions following oral presentations is included. All papers submitted to the editor were reviewed by anonymous referees, who recommended for or against inclusion and suggested changes. The editor wishes to thank the more than 100 scientists who helped improve the book by their expert reviews of the individual papers.

It has seemed best to include contributions to the symposium that deal with the history of the CUEA program and societal matters in an introductory section, and so the *Epilogue* delivered by Feenan Jennings is found at the end of this section rather than at the end of the book.

The editor wishes to thank the coveners of the symposium, Christopher N.K. Mooers and Theodore T. Packard, for their help and reinforcement of the editor's decisions. Richard T. Barber and Dolors Blasco helped the editor cast some of the papers into more idiomatic English, and Robert L. Smith supplied the drawings for Plates I, II, and III. Vicki Jones did the final typing.

Francis A. Richards

Department of Oceanography
University of Washington
Seattle, Washington 98195, U.S.A.

ORGANIZERS' REMARKS

In February, 1980, a week-long International Symposium on Coastal Upwelling was held in Los Angeles, California. It was co-sponsored by The American Geophysical Union, The American Society of Limnology and Oceanography, and the American Meteorological Society. The intent of the symposium was to summarize recent progress in the multi-disciplinary aspects of upwelling research, especially the progress made during the IDOE (International Decade Of Ocean Exploration) in the 1970's. Over 100 papers were presented, either orally or with posters. Participants came from five continents and all major disciplines of oceanography. The symposium was organized by several of the investigators in the multi-year, multi-institutional, multi-disciplinary IDOE program sponsored by the U.S. National Science Foundation that was called CUEA (Coastal Upwelling Ecosystems Analysis); the National Science Foundation funded many of the costs of the symposium as well as this volume. However, care was taken to ensure that CUEA investigators did not dominate the symposium and that the broadest possible participation was achieved. The symposium itself was organized in a multi-disciplinary fashion, with biological and physical papers mainly grouped together by scale: large scale, synoptic scale, mesoscale, and small scale. Approximately one-third of the papers presented at the symposium are published in this volume after peer review. They provide a representative cross-section of the papers presented at the symposium. The volume is augmented by some papers that were not presented at the symposium. The 1970's saw several attempts to study jointly physical and biological processes on the several scales of variability noted above. The general objective was to develop consistent physical and biological data sets and models of the coastal upwelling ecosystem. Though the models tended to be conceptual and statistical, substantial progress has been made in identifying causal relations between physical and biological fields and processes. By conducting somewhat similar studies in a variety of regions around the world, a basis was provided for comparison of similarities and differences.

Christopher N.K. Mooers

Department of Oceanography
U.S. Naval Postgraduate School
Monterey, California 93940, U.S.A.

Upwelling areas are important because they supply 50% of the world's sea food. Upwelling is studied to understand the mechanisms that lead to the generation of the rich fisheries and to develop a capacity to predict the effect on the fisheries of perturbations in these mechanisms. Upwelling itself is caused by a divergence of the surface flow field which, in turn, can be caused by wind, bottom topography, and several types of internal waves. In each instance, sub-surface, nutrient-rich water is added to the light-rich surface waters and stimulates the development of phytoplankton blooms. The blooms provide the food supply for the zooplanktonic, nektonic, and benthic components of the upwelling ecosystem.

The Coastal Upwelling Ecosystems Analysis (CUEA) Program was developed to study the environmental forcing functions and the physical, chemical, and biological aspects of the upwelling ecosystem. The CUEA group studied intensively four different upwelling regions: off Baja California during the MESCAL I and II expeditions in March and April, 1973; off Oregon during the CUE-2 expedition in July and August, 1972 (Plate II); off northwest Africa during the JOINT-I expedition in March and April, 1974 (Plate I); off Peru during the JOINT-II expedition from March to October, 1976 and March to May, 1977 (Plate III a, b). Many of the articles in this volume are based on the results these expeditions or similar ones made by other research groups to the same regions.

The IDOE International Symposium held at the University of Southern California from 4 to 8 February, 1980, was the fourth of a series of symposia and workshops that started with the working conference on the Analysis of Upwelling Systems held in Barcelona, Spain in March, 1970. The Barcelona meeting and the published proceedings (*Investigaciones Pesqueras, 35*, 1, 1971) were so successful that a second meeting was planned to be held in Marseille. At the Marseille meeting (May, 1973) there was increased participation from the community of physical oceanographers. The proceedings (*Tethys, 6*, 1-2, 1974) reflect their contribution as well as the general broadening of the multi-disciplinary base of upwelling research. The third meeting followed soon afterwards, in Kiel, West Germany, in September, 1975, and like the Marseille meeting, was strongly interdisciplinary. The published record appeared as the book *Upwelling Ecosystems*, edited by R. Boje and M. Tomczak, Springer-Verlag, 1978. In the present volume the organizers and the editor collaborated to preserve the multi-disciplinary flavor of the previous volumes.

Theodore T. Packard

Bigelow Laboratory for Ocean Sciences
West Boothbay Harbor, Maine 04575, U.S.A.

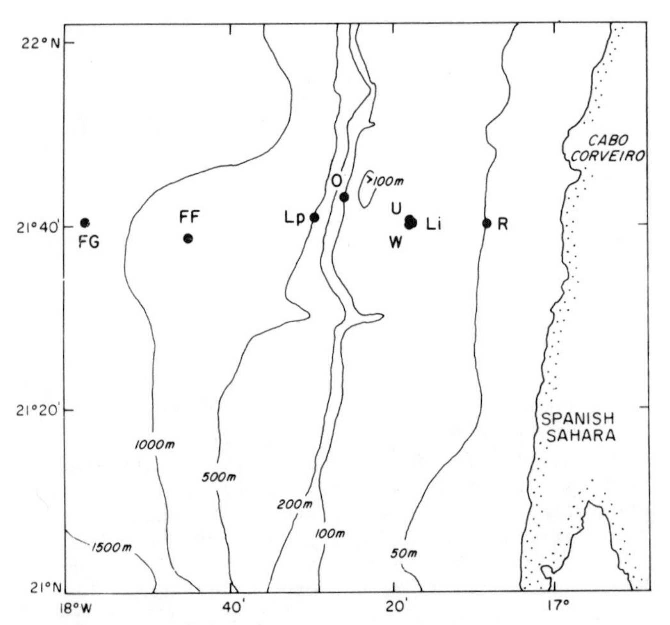

Plate I. Site and current-meter moorings during the JOINT I Expedition off northwest Africa, 1974. The letters designate mooring locations: FG = Foxglove; FF = Forest Fern; Lp = Lupine; O = Oregon Grape; U = Urbinia; W = Weed; L = Lisa; R = Rhododendron. Courtesy of R.L. Smith.

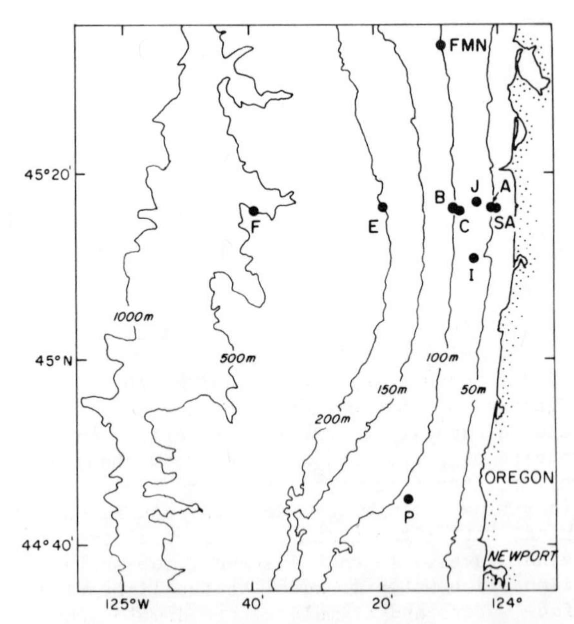

Plate II. Site of the CUE 2 observations and current meter moorings off the Oregon coast in July and August 1973. Courtesy of R.L. Smith.

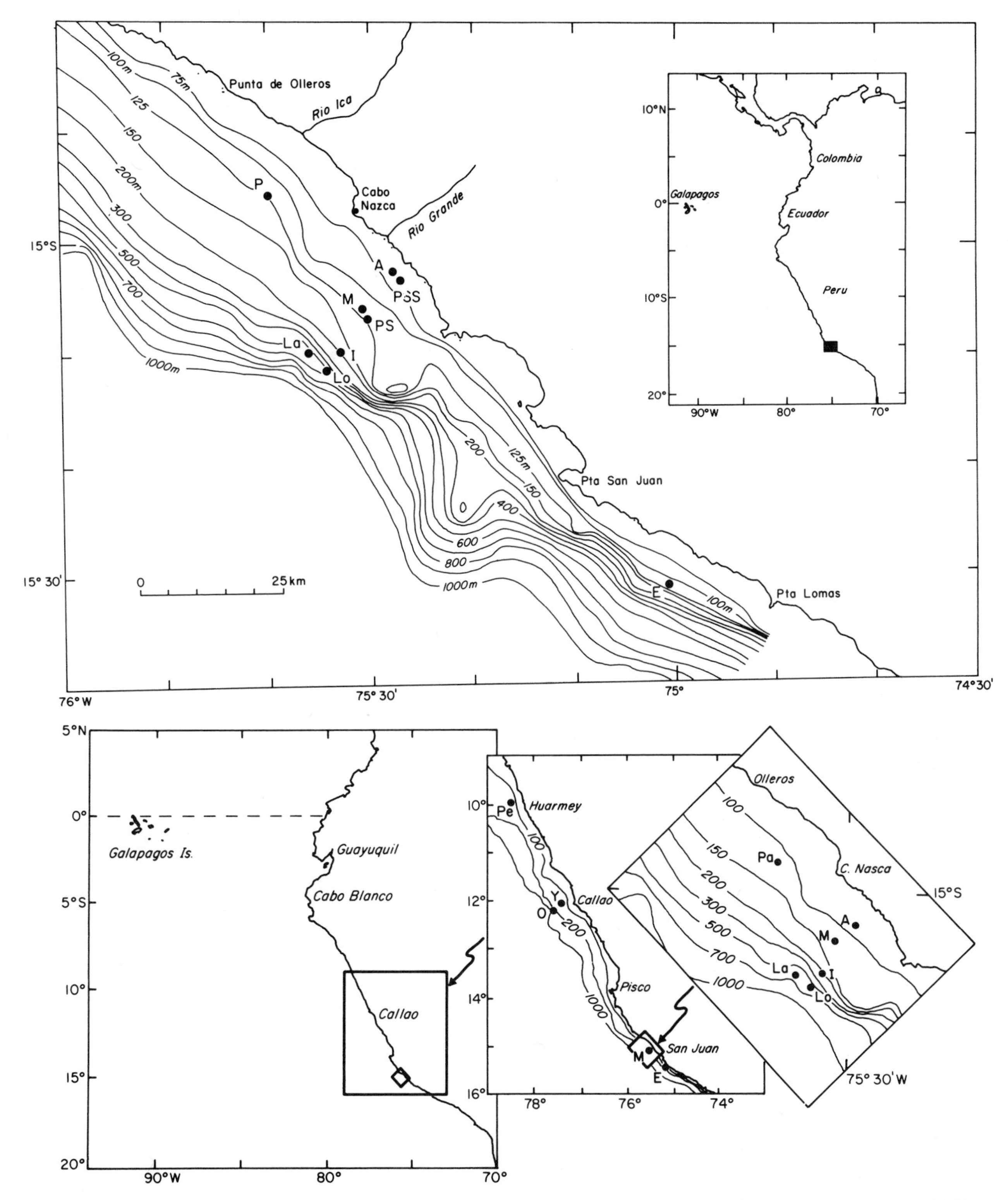

Plate III. a) Site and mooring locations during the JOINT II Expedition off Peru, March to May 1977. The C line extended along 215°T from the coast approximately through moorings A, PSS, M, PS, I, La, and Lo. The letters designate mooring locations: P = Parodia; A = Agave; PSS; PS; M = Mila; I = Ironwood; La = Lagarta; Lo = Lobivia; E = Euphorbia. b) Pe = Peyote; Y = Yucca Too; O = Opuntia. Courtesy of R.L. Smith.

AN UPWELLING MYTHOLOGY

Warren S. Wooster

Institute for Marine Studies, University of Washington,
Seattle, Washington, 98195, U.S.A.

The occurrence of this international symposium on coastal upwelling demonstrates the widespread interest in the subject. The upwelling literature has grown enormously in the last few years, in part a direct consequence of the considerable investment (more than $20 million in the United States) in the Coastal Upwelling Ecosystems Analysis (CUEA) project of the International Decade of Ocean Exploration (IDOE).

The history of sustained study of coastal upwelling goes back much further, at least to 1949 when the field work of the California Cooperative Oceanic Fishery Investigations (CalCOFI) began. The methods available then for studying upwelling were limited to classical water bottle and bathythermograph surveys repeated at monthly intervals throughout the year. The scales of variability were little appreciated in the early years of the program. For example, estimates of upwelling were attempted from the vertical displacement of isotherms from one month to the next. The first anchor stations then showed that changes over a few hours could be as great as during a whole month. Distributions in space were also found to be patchy and discontinuous. No methods were available then — or now — for direct measurement of the vertical component of velocity, and the methods available then — and now — for measuring the horizontal components were incompatible with such an extensive survey. While understanding of the process of upwelling did not advance rapidly during the first decades of CalCOFI, the climatological setting of the phenomenon in the region was usefully documented — to say nothing of the invaluable biological monitoring that also occurred.

Understanding of upwelling was accelerated by a number of factors. One of the most important was the establishment of a "school" of upwelling investigation at Oregon State University under the leadership of the late June Patullo, from whence issued many of today's most distinguished upwelling specialists. It was at Oregon State that modern velocity measurements were first applied systematically to the study of upwelling. Another factor was the development of theory, evolving from the classical paper of the late Kozo Yoshida (1967) and contributing to numerical models such as those flowing from the modelling school at Florida State University in Tallahassee, which could serve to guide the subsequent experimental and field work. Finally, one must acknowledge the impetus provided by the IDOE and its support for CUEA and the succession of expeditions to the west coasts of North and South America and North Africa. These interdisciplinary, international, and comparative studies have provided most of the observations on which the papers of this volume are based.

It is not surprising that the synthesis of present understanding of the upwelling process has not kept up with the explosion of research. For example, to my knowledge there is still no adequate treatment that can be used for teaching purposes. That treatment when written will presumably differ in significant ways from that contained in *The Oceans* (Sverdrup, H.V., M.W. Johnson, and R.H. Fleming, Prentice Hall, 1942, 1087 pp.) which is still the basis for most textbook discussions of the subject.

There is a popular image of upwelling, what I would call an upwelling mythology, that is slow to disappear, even in the face of scientific findings. Here, for example, is a statement about upwelling that might have been made 20 years ago without serious challenge and whose equivalent is often repeated today:

> Upwelling is a coastal process whereby cold, nutrient-rich bottom water is brought to the surface. Vertical speeds are of the order of 10^{-3} cm/s, and the vertical motion is caused by the local winds, as demonstrated by Ekman. Upwelling causes high productivity leading to large quantities of fish. Upwelling research should be supported so that fishermen can catch more fish.

If one looks carefully at this statement, there are several features that are open to question. It is still fair to say that upwelling is a process -- although one that is often poorly defined -- and that it involves vertical motion. It does occur in some coastal regions, especially on the eastern sides of the Atlantic and Pacific oceans, but it can also be found in the open ocean, for example along the equator. Although the term is sometimes applied to the very slow, ocean-wide upward compensation for intermediate and bottom water formation, it seems more useful to confine its use to the more concentrated motions within a few hundred meters of the surface. In this sense, the water is unlikely to come from the bottom except on the shelf, and it is not necessarily cold and nutrient rich. Thus it has been suggested that during *El Niño* the usual source water for coastal upwelling off Peru can be replaced by warm, oligotrophic waters, which are then brought to the surface.

Although a variety of estimates have shown vertical speeds of 10^{-3} cm/s, it is noteworthy that these are all continuity estimates from averages of horizontal motion or distributions of properties related thereto. Satisfactory direct measurements of the vertical motion have not, as noted before, yet been made. But it seems likely that the motion is not a widespread upward ooze but rather is concentrated in plumes where the speed may be one of two orders of magnitude more rapid than the conventional estimates.

The vertical motion is certainly in the simplest case driven by the local winds, and the accepted explanations derive from Ekman theory. Ekman himself did not make this application, but rather Thorade in 1909. Only in the last few years has it become apparent that distant events may have a profound effect on the local manifestations of upwelling through the action of long, equatorially or coastally trapped waves. Thus the current paradigm for explaining *El Niño* does not depend on variations in the local, coastal wind stress; one assumes that the observed phase differences between maximum equatorward wind stress and maximum upwelling along the southwestern coast of Africa have an analogous explanation.

As a rule, coastal (and equatorial) upwelling is associated with high primary production, and coastal upwelling regions are the locations of large concentrations of fish. But now there is evidence that upwelling is not always beneficial to fish production and in fact may have harmful effects on recruitment success. I refer here to the work of Lasker (1978) who found that the sudden onset of upwelling at the wrong time in the development of anchovy larvae could cause a most appropriate food, in this case small flagellates, to be replaced with larger, less desirable diatoms to the detriment of the anchovy year class concerned. I note also the recent findings of my colleague, Kevin Bailey, that hake larvae can be carried away from their coastal nurseries by the offshore transport associated with upwelling, again to the disadvantage of the particular year class.

I have always felt that upwelling research was well justified by the added understanding of our planet that it should produce. But proposals for such research nearly always refer to the potential benefit for fisheries. Let us look critically at what that benefit might be. It will presumably not involve increased production of fish in the well known upwelling areas such as those off California, Peru, Mauretania, or Namibia. That is, the research is unlikely to increase the rate of carbon fixation or the efficiency of energy transfer through the food web. Nor is the discovery of new upwelling areas likely to lead to the development of new fishing grounds. One might argue that on occasion a result of upwelling research might be the catching of less fish. That is, one might hope that a product of such research would be a better understanding of the relation between variations in upwelling location, timing, and intensity and the inter-annual variations in year class strength. One might then occasionally forecast poor recruitment that would require restriction of the fishery. I myself believe that the perfection of such prognoses will eventually pay the costs of upwelling research many times over.

Another justification claimed for upwelling research is its relation to improved understanding of climate. The ocean has its own climate, of course, and one of the most dramatic climatic events in the ocean is *El Niño* as it is observed in the equatorial and eastern tropical Pacific. As already discussed, it is probably changes of climatic scale in coastal regions that determine the inter-annual variations in year class strength of many organisms. But much of the current interest in climate concerns that of the atmosphere, modulated as it must be by conditions in the ocean. Here the major concern, at least initially, will be with determining the meridional heat flux, where it is the western boundary currents, for example the Gulf Stream, that presumably dominate the ocean's contribution. Remember, however, that Bjerknes (1966) showed that the equatorial regions, and notably the eastern tropical Pacific, are also regions of direct and consequential ocean-atmosphere interaction of large scale. The variations in the heat content of coastal waters on the eastern sides of oceans must also surely contribute to the determination of climate changes over the nearby continents.

To return to my upwelling mythology, I have not tried to construct a more correct statement about upwelling. There may well be errors in what I have already said that are more serious than those in my hypothetical statement. I have tried to illustrate that conventional wisdom on upwelling has fallen behind what has been learned in the last few years. The explosion in investigating this process has led to an enormous proliferation of ideas and facts. The job of synthesis and generalization that will make this body of knowledge available to more than the upwelling enthusiasts has yet to be done. My guess is that the time is now ripe for this effort and that the synthesis when produced will lead to yet another burst in our level of understanding of this important and fascinating process.

References

Bjerknes, J., A possible response of the atmospheric Hadley circulation to equatorial anomalies of ocean temperature, *Tellus, 18*, 820-828, 1966.

Lasker, R., The relation between oceanographic conditions and larval anchovy food in the California Current: identification of factors contributing to recruitment failure, *Rapports et Procés-Verbaux des Reunions, Conseil International pour l'Exploration de la Mer, 173*, 212-230, 1978.

Thorade, H., Über die kalifornische Meeresströmung, *Annalen der Hydrographie und maritimen Meteorologie, 37*, 63-76, 1909.

Yoshida, K., Circulation in the eastern tropical oceans with special reference to upwelling and undercurrent, *Japanese Journal of Geophysics, 4*, 1-75, 1967.

THE IMPORTANCE OF COASTAL UPWELLING RESEARCH FOR PERU

Felipe Ancieta

Instituto del Mar del Peru
Callao, Peru

In spite of constituting only one-tenth of one percent of the area of the world's oceans, the coastal upwelling regions are outstanding in terms of economic and scientific interest because their production accounts for approximately half of the world's fish harvest. It is well known that in the upwelling region off Peru there has developed the largest single species fishery in the world based on the Peruvian anchovy. The catch of this fishery increased spectacularly from its beginning in 1956 and reached 12 million metric tons by 1971. At that time it constituted 20% of the world's fish catch.

Aside from the fluctuations in abundance that can be attributed to the fishery, it is known that the abundance of the huge anchovy population varies with the intensity and duration of the coastal upwelling off Peru. Thus, added to overfishing, there are effects of the drastic oceanographic changes of 1972 - 1973 and the relatively minor changes of 1976. These variations in the upwelling process reduced the anchovy biomass from 20 million tons in the 1960's to scarcely 3 million tons in 1978. At the same time pelagic fishes other than anchovy experienced an unexpected increase. Species such as sardine, saury, horse mackerel and mackerel that were previously of minor importance in Peruvian fisheries became important economic considerations. The guano industry, another mainstay of the Peruvian economy, suffered a decline from ocean-climate changes such as the El Niño phenomena of 1925, 1940-1941, and 1957-1958.

The causal relations between upwelling, the high biological productivity, and the existence of exploited or potential fishery resources of exceptional magnitude were noticed a long time ago. An understanding of the upwelling process and its eventual forecast is an important component of fishery management. The need for such understanding has motivated resource managers to promote and encourage scientific research in the major upwelling areas of the world. In Peru, investigations into the processes of upwelling, mainly in respect to the anchovy, started systematically at the beginning of the 60's with the creation of the Instituto de Investigaciones de los Recursos Marinos, which in 1964 changed its name to Instituto del Mar del Peru (IMARPE). The upwelling research carried out by IMARPE has been expanded within the Institute in parallel to new findings in the field. Upwelling investigations at the ecosystem level by the Coastal Upwelling Ecosystems Analysis (CUEA) program became important for Peru with the execution of the JOINT-II expedition in the 15°S area of coast near San Juan. It is also important to mention the completion of a cooperation project between Peru and Canada, entitled "Study of the Peruvian Anchovy and its Ecosystem" (ICANE). ICANE combined the practical experience of IMARPE with the scientific resources of some Canadian institutions to study the cause of recent decreases of the anchovy population.

I must congratulate those responsible for CUEA for suggesting the subject of this paper, because, if I were asked to select a subject, I would, as a representative of IMARPE and indirectly of Peru, have chosen the same title: "The importance of research on coastal upwelling for Peru, with emphasis on the social importance." This is so for two important reasons, (1) the geographic situation of Peru and (2) its condition of being a "developing country". Concerning IMARPE we can transfer this concept to a "developing institution". Added to these facts there is a circumstance we cannot ignore, and it is that we are living in a decade full of possibilities never dreamed of before for wealth and transcendence. I am referring here to the International Decade of Ocean Exploration which coincides with the decade of the Law of the Sea. We recognize the social importance of upwelling research, because the debate on the new Law of the Sea has pivoted around the revindication of rights to the renewable resources of coastal areas as being exclusive rights of coastal nations, including the right of exploration and research. Only when this complex social, political, and economic climate is considered is it possible to describe the importance of upwelling research to Peru.

I shall consider upwelling investigations with

special reference to the CUEA results, which are of interest to my country and institution, in relation to the following questions:

(1) To what extent have the CUEA results been useful or could be useful for the rational management of our fisheries resources, and to what extent are those investigations, as well as the work to follow, of our concern?

(2) In terms of increase in scientific capacity, to what extent has IMARPE profited from CUEA collaboration?

These questions are pertinent, considering that (i) the marine resources of our country are dependent on coastal upwelling, (ii) marine resources in our country are an important fraction of the total wealth (20-30% of the income of foreign currency in the last 10 years) and (iii) as a developing country, we must see that any scientific exchange yields a profit in knowledge we need, as well as providing increased capacity to produce knowledge by ourselves.

The answer to the first question must be partial, because interpretation and synthesis of some of the results are still going on. The results are still being analyzed especially in relation to other investigations, and consitute a quarry which is still being very actively mined. For the time being, and assuming the inherent risks of a summary done by a non-specialist, I will mention the results, which to my judgement, are most important to Peru:

(a) The upwelling process is focused both in space and time, and the geographic distribution of the foci is more or less permanent. This was known before the CUEA investigations, but the work has contributed more powerful and detailed evidence than that previously available.

(b) It is possible to describe each upwelling center in terms of a characteristic structure, the "plume", and in terms of a pattern of historical evolution of the plume as it drifts towards the north and west.

(c) The upwelling of nutrients does not always culminate with the production of phytoplankton and the biomass of phytoplankton produced is not always in proportion to the intensity and concentration of the nutrients upwelled.

(d) Variations in the physical and biological upwelling processes are partially linked to local environmental changes, but also to the origin of the upwelled waters, which, in turn, is linked to remote processes taking place far from the coast of Peru.

(e) There is a close relationship between the distribution of enhanced primary production in the upwelling foci and anchovy concentrations. The results emphasize the variability in intensity and area of primary productivity and the coincidence of the centers of productivity with the maximum concentrations of anchovy.

The identification of local and remote processes that are driving upwelling as being the factors causing the variability in primary production establishes the design of research programs needed to elucidate the mechanisms of such variability. Future long term forecasts will require an understanding of these local and remote factors. The areas of investigation will be far from the coast of Peru to monitor distant processes and will also have to be carried out on our own coast over a larger area than that studied in JOINT-II to resolve the variability due to local factors and remote processes.

On the other hand, the knowledge of the local character of the upwelling process, of the general characteristics of the upwelling centers, and of their coincidence with the anchovy concentrations have been used by us to improve confidence limits of our acoustic estimates of the biomass of anchovy.

Answering the first question we can affirm that the CUEA results have contributed to improved management of our living resources. In a short term perspective they open a promising and clear path for a more precise estimate of the fish biomass and in the medium and long term, they allow better definition of the research required on the origin of upwelling variability to implement more complete and useful models for resource management. For example, there are good prospects based on the initial results obtained in ICANE that a better knowledge of upwelling variability could be used in models that predict the length of the favorable period for spawning and larval survival during the period. It should be noted that the ICANE advances would not have been possible without having the previous CUEA results, and above all, without having the direct collaboration of CUEA personnel during planning stages.

In the longer term, it can be stated that CUEA's investigations, as well as any others on the upwelling system off Peru, are of direct benefit to Peru and my institution as long as some of the results can be used to improve our estimates of fish resources. Any improvement in the accuracy of our stock estimates supports and enhances the fishery management policies of Peru.

With regard to the question on increased scientific capability, I will make a very short reference to the CUEA-IMARPE collaboration without entering into many details. After an agreement between CUEA and IMARPE in 1975 a group of Peruvian scientists took on a CUEA component as their responsibility for the 1976 and 1977 field work and for analysis and interpretation in the 1978 to 1980 period. In a general way, and in spite of the barrier of the language as well as other limitations, we have had some significant results with regard to:

(a) training at different levels for young scientists,
(b) increasing experience for both junior and senior scientists,
(c) improvement of the technology in different fields,
(d) improvement of knowledge of the upwelling process, and
(e) development of better communication between Peruvian and foreign scientists.

NASA's POTENTIAL REMOTE SENSING CAPABILITIES THAT COULD BE APPLIED TO UPWELLING STUDIES

R.R.P. Chase*

California Institute of Technology
Jet Propulsion Laboratory
Pasadena, California 91103 U.S.A.

Introduction

The National Aeronautics and Space Administration has been involved with aquatic remote sensing research for approximately 12 years. Sponsorship of the work resides within the Oceanic Processes Branch where the emphasis has evolved from site-specific water quality research to a well focused oceanographic research program that includes research within the coastal environment. During this evolutionary period, the instruments we use for coastal zone measurements have undergone substantial refinement and have been greatly augmented by new measurement techniques. Today, we feel that certain useful measurements can be made remotely which supplement *in situ* data, and consequently we look forward to the mid-1980's when we expect these data to be routinely available to the oceanographic community.

Techniques

Four fundamental measurement techniques form the basis for remote sensing instruments employed by NASA. These include both passive and active measurements made within either the visible (and infrared) or microwave portions of the electromagnetic spectrum.

Passive instruments operating within the visible portion of the spectrum utilize reflected sunlight as a source of information (imagery). Active measurements at visible wavelengths are relatively new, using laser technology. The instruments measure either back-scatter energy or induced fluorescence within the water column.

In the microwave portion of the spectrum, passive instruments capable of sampling at various frequencies detect differential microwave emissions from the sea surface while the active instruments emit and then measure the reflected microwave energy. Many different types of radar instruments comprise the active class and include those capable of "imaging" the sea surface as well as the more traditional instruments that measure the transit time between the emitted and returned pulse.

The physically unambiguous and quantitative geophysical interpretation of the remote sensing signals obtained through application of these techniques is often difficult and typically dictates the geophysical measurement capabilities and limitations of the instruments.

Capabilities and Limitations

Table 1 lists various remote sensing instruments according to measurement technique. Column 3 summarizes the geophysical parameters that can be derived; the precision of the basic measurement is given in Column 4. Although deployable on aircraft, most instruments are designed for space-borne usage; Column 5, therefore, reflects the instrument's space-borne resolution while Columns 6 and 7 list the most recent flight and next scheduled flight of each of the instruments.

Based upon the derivable geophysical parameters, all of the listed remote sensing instruments potentially could contribute to upwelling research. However, given the current state of development and understanding of instruments and their measurements together with their planned flight schedules, only a few of the listed instruments are likely to contribute to coastal upwelling research during this decade. Specifically, the Coastal Zone Color Scanner, Infrared Radiometer, Scatterometer, and Altimeter appear to offer the most potential.

From the color radiance data (CZCS), integrated chlorophyll *a* measurements can be derived. As currently configured, the measurements have a spatial resolution of approximately 800 m and maximum temporal sampling frequency of roughly one image a day for 4 of every 5 days at midlatitudes. The integration depth for these mea-

---

* Currently assigned to Oceanic Processes Branch (Code EBC-8) NASA Headquarters, Washington, D.C. 20546

TABLE 1. Instruments, Measurements, and Characteristics

Technique		Instrument	Primary Derivable Measurement	Measurement Precision	Resolution and Swath	Last Flight	Next Flight
Passive	Visible	Coastal Zone Color Scanner (CZCS)	Chlorophyll *a* and Phaeophytin	Factor of 2	800 m 1600 m	NIMBUS-7 (1978)	NOSS * (1986)
		Visual and Infrared Radiomater (VIRR)	Sea Surface Temperature	$\pm$ 1.5°C	3 km 1800 km	Operational NOAA* Instrument	
	Microwave	Scanning Multichannel Microwave Radiometer (SMMR)	Sea Surface Temperature	$\pm$ 1.5°C	50-100 km 900 km	NIMBUS-7 (1978)	NOSS* (1986)
Active	Visible	LIDAR	Chlorophyll *a*, Yellow Substance, Total Particulate	Undetermined	Airborne Instrument in Early Development Stages		
	Microwave	Altimeter (ALT)	Surface Topography	$\pm$ 7 cm	1.6 km 12 km	SEASAT (1978)	NOSS* TOPEX (1986)
		Scatterometer (SCAT)	Surface Winds	$\pm$ 2 m/s $\pm$ 20°C	50 km 1000 km	SEASAT (1978)	NOSS* (1986)
		Synthetic Aperture Radar (SAR)	Imagery	N/A	22.5 m 100 km	SEASAT (1978)	?

*The NOSS spacecraft referred to in this paper will not be deployed as originally planned. NASA is currently examinining the desirability and feasibility of deploying select instrumentation from that system (e.g. CZCS, SCAT) by alternate means.

surements is variable and depends upon the quantity of particulate matter present in the water column as well as its vertical distribution. Consequently, at a given location *in situ* measurements of downwelling and upwelling irradiance are frequently made to determine the exact depth of integration. An operational limitation posed by the color scanner is that it requires daytime cloud-free conditions for effective use. However, even with these constraints, the large space and time scales of upwelling phytoplankton blooms appear amenable to color radiance measurement.

Sea surface temperatures derivable from infrared imagery (VIRR) are equally promising in their potential contribution to coastal upwelling research. Although the precision of these measurements are $\pm$ 1.5°C the data are sufficiently accurate to locate the relatively large surface thermal gradients normally manifest in upwelling regions. Because these measurements are skin temperatures, the utility of the instrument is obviated when surface thermal signatures are obscured (either by clouds or an isothermal ocean surface).

Active microwave instruments (scatterometer and altimeter) will contribute primarily to the physical oceanographic aspects of upwelling research. Wind velocities derived from the scatterometer are distributed spatially at 50-km intervals and temporally every 3 days at mid-latitudes (SEASAT). Although the variance in these data is approximately $\pm$ 2 m/s, the threshold measurement is 4 m/s and extends to a maximum of approximately 25 m/s; appropriate spatial and temporal averaging can further reduce the uncertainty of the measurement. Altimeter data can be used to delineate cross-shelf and longshore pressure gradients and, although not fully developed, tidal analysis schemes that use the same data base are being devised. The current class of altimeters has a root mean square precision of 7 cm (SEASAT), which represents an order one error in coastal pressure gradient calculations. However, by properly filtering the data and utilizing differences of repetitive transects, the data are adequate to characterize the proper order of magnitude for these gradients. The next generation of altimeters (TOPEX) is being designed to achieve a precision of order 2 cm and should be a much more useful research tool.

Today, only two of the four cited data products of interest are available from spacecraft. Infrared radiometer-derived sea-surface temperature data are routinely available; these data are an operational data product produced by NOAA and can be obtained through NOAA/EDIS. Coastal zone color scanner data are currently being obtained by NASA's NIMBUS-7 research spacecraft. Although this data will be available through the NOAA/EDIS archive, ground processing problems have delayed the reduction and distribution of the data.

Future Prospects

Within the next decade, the Oceanic Processes Branch is planning the deployment of two oceanographic satellite observing systems, the National Oceanic Satellite System (NOSS) and the ocean dynamic Topography Experiment (TOPEX).

As currently configured, NOSS will carry a color scanner, scatterometer, and SEASAT-class altimeter, in addition to an advanced version of the scanning multichannel microwave radiometer. NOSS is a limited operational demonstration project sponsored jointly by NOAA, DOD, and NASA. Data derived from this system will, therefore, be available to the research community either in real time (with a suitable ground station) or through the NOAA/EDIS archive.

The dynamic Topography Experiment is being designed to address the oceans' general circulation. We anticipate that the satellite will carry the new high-precision altimeter in addition to some complementary instruments. In contrast to NOSS, TOPEX is a NASA research satellite program and as such, the Oceanic Processes Program is responsible for the mission from concept through final analyses of data and publication of the results. TOPEX data (surface topography, significant wave height, tidal analyses, wind speed, and short wavelength geoid) will be available to the NASA sponsored experiment teams and principal investigators, via the TOPEX data archive, by real time acquisition (not reduced) or through NOAA/EDIS.

Infrared radiometer measurements of sea-surface temperature will continue to be available through NOAA. With the advent of NOSS, these temperature measurements will be augmented by the microwave radiometer's all-day, all-weather capabilities.

Summary

Remote sensing measurements of ocean color, surface winds, sea surface temperature, and coastal pressure gradients show the most potential for contributing to coastal upwelling research. Currently, only color and surface temperature measurements are available from space, although during the latter part of this decade all of these measurements will be routinely available to the coastal upwelling research community.

THE COASTAL UPWELLING ECOSYSTEM ANALYSIS PROGRAM: SOCIAL IMPLICATIONS

James C. Kelley

San Francisco State University, San Francisco, California 94132, U.S.A.

The Program Committee has asked me, as I understand their request, to address the success of the Coastal Upwelling Ecosystems Analysis (CUEA) program in the larger social environment in which the program has been conducted. The question is, "if CUEA met its own goals, as they are perceived within the program, and met the goals of the National Science Foundation in supporting the program (n.b., these sets are not necessarily in one-to-one correspondence), has this success been accompanied by any evidence of social impact?" Reduced to terms more characteristic of the 20th Century Vulcans slaving over the fiery smithy of science policy in the U.S. Senate — "If CUEA got where it was going, does it make any difference?"

I am tempted to terminate my presentation here with the argument that the very existence of the question in the program is its own *prima facia* affirmation. The fact that a group of active research scientists is genuinely interested in the social impact of its efforts would, in some intellectual fraternities be greeted as novel, refreshing, socially significant, and quite likely a harbinger of declining rigor and integrity. Rather than yield to this temptation, I feel that CUEA itself -- to say nothing of the program committee -- would be better served if I were to develop a rationale for the somewhat aberrant behavior of the research group and then to attempt an extrapolation in the general direction of social impact.

In order to address the question somewhat more cogently, it will be helpful to subdivide it into the separate effects that CUEA has visited upon at least three levels of society it might have been expected to have changed: (1) its own fraternity of investigators, (2) the oceanographic research community in general, and (3) the larger society in which it was developed and supported. This artful dodge is necessitated by the fact that my comments on the latter effect are so sparce and hypothetical. It is too early to discern impacts on the larger society, but we may be able to extrapolate from the more proximate communities. The impacts of CUEA in the first two circles, on scientists within and outside the program, are tightly coupled. Those investigators within the program approached a large, complex problem in a manner which, though not unique, represented significant progress in experimental design and accomplishment in oceanography. The impact on the larger research community stems largely from the fact that CUEA scientists proved their design concepts and experimental strategies work. In order to understand these combined effects it will be useful to review a short prehistory of the program.

Prior to the mid-1950's, oceanographic research had been largely basic and essentially wholly non-synoptic. However, by 1954 the need was recognized for synoptic sea surface temperature maps in at least one major upwelling area, the California Current, as an aid to fishermen. In 1958, the Bureau of Commercial Fisheries' Pacific Oceanic Fishery Investigation -- predecessor of the Honolulu Biological Laboratory -- developed a series of 10-day mean sea surface temperature maps of the eastern North Pacific. These were followed by the establishment of the BCF's Tuna Resources Laboratory in San Diego in 1959 which continued the project. Today the synoptic maps in *Fishing Information* continue to provide data of at least equal significance to oceanographers, fisheries biologists, and the fishermen themselves. I recount this history to emphasize that the source of the pressure was outside the oceanographic research community, but it stimulated a quick response from some of its members. Certain others (presumably those of the enlightened whose mantra includes the mystic words "HMS *Challenger*") have yet to succomb to these temptations of the flesh — or at least of the fish.

It is, I believe, significant to the understanding of the history of the CUEA program to recognize the development, during the period of 1955 to 1965, of a group of synoptic oceanographers engaged in an activity more reminiscent, to some, of the "battle-zone" economics of their ports of call than the intellectual currency of the research laboratories. These researchers used the tools and time scales of meteorology to produce, from oceanographic data, products for the primary consumption of fishermen! That such eclectic behavior was tolerated at all by the orthodox oceanographic research community was probably attributable more to the self-congratulatory disdain of the *literati* than to their charity. The schism persists and the rent fabric of the discipline is still worn today with only occasional, strategically located patches attached for the sake of modesty.

During the last half of the decade, the combined effects of the 1968 Mansfield Amendment and the formation of the National Oceanic and Atmospheric Administration (NOAA) produced a major restructuring of the oceanographic research community. The primary characteristics of the restructuring have been the increasing bureaucratization of the federal agency research community and some tendency for the academic research community to circle the wagons of basic research in reaction. In 1970, Henry Stommel* warned of the tendency to "subordinate formulation of specific critical scientific problems to those that are less clearly defined but that on the surface appear to be politically attractive". At the time, Stommel was appealing for greater national attention to the fundamental problems of ocean dynamics -- instead of following the bureaucratic response to the Report of the Commission on Marine Science, Engineering, and Resources, which reflected so great an infatuation with the methods and tools that might be used in the process, that the deployment of large sensor arrays and instrument platforms was apparently being considered as an end in itself. The appeal first to identify a problem was apparently finally heard, at least in the office of the International Decade of Ocean Exploration (IDOE/NSF) and programs such as CUEA and MODE (the Mid-Ocean Dynamics Experiment) were developed in response. Unfortunately, while NSF was moving in the direction of designing programs to solve fundamental problems, the National Marine Fisheries Service (NMFS) retreated from its support of synoptic oceanography and the efforts of the U.S. Navy were largely redirected and its cooperation with other agencies diminished.

It was within the rather tumultuous research climate that characterized the early '70's that CUEA grew up. The program attempted to combine two goals, one basic and one "applied" in a unique way. In the first summary written for the NSF cover sheet, this schizophrenic theme is expressed in the words "... the primary goal of the program is to understand the upwelling process far enough in advance to be useful to man." I will not embarrass the author by an *a posteriori* identification, but *a priori*, with some years of funded research activity ahead, this statement seemed a most accurate, even glib, expression of the intent and the expectation of the research team. In context, however, the concept — that of combining an attempt at basic understanding with an expectation that this understanding would find expression as a useful and presumably predictive tool — elicited certain flagellatory inclinations in some parts of the research community. The criticisms seemed to hinge upon the notion that it was impossible to serve both ends equitably and, by extension, the suspicion that one or the other of them was included in the statement for, shall we say symmetry, but did not reflect serious intent on the part of the program.

In the context of the time, it is not surprising that some misunderstanding surrounded the early years of the CUEA program. Earlier programs that were intended to be "useful", such as the Bureau of Commercial Fisheries' synoptic temperature maps or the Fleet Numerical Weather Center's surface weather predictions, were explicitly pragmatic endeavors that did not pretend to membership in the fraternity of basic research. It was, simultaneously, characteristic of the basic research community to eschew persuit of "practical" ends. In the words of one of my prominent senior colleagues at the time, "if I understood the practical applications of a research problem, I wouldn't do it!" This comment, though perhaps extreme (and in this case uncharacteristically virginal), indicates at least the singular lack of neutrality with which CUEA was received.

There exists in the statement of the CUEA goal, however, the kernel of the "social" impact of the program — even if the impacted society is only the research community — and it is embedded in the concept of "useful understanding" as it was developed in this context. The development of our understanding of the ocean has been characterized by highly creative extrapolations of very precise measurements. Unfortunately, the extrapolation distance began to vary inversely with the error of the estimate. More precise analysis took more time, which in the traditional mode of oceanographic sampling, further separated the analyses in space, time, or both.

Prior to the formal initiation of the CUEA program, the significance of both horizontal and temporal variability in upwelling areas began to receive attention. Through the period during which formal CUEA activities were carried out, there was a series of experiments designed to reveal the spatial and temporal structures of important variables in upwelling areas. These include the traditional sea-surface temperature maps, including the significant and effective use of aircraft mapping, maps of nutrients and chlorophyll, and exciting new information from maps of the electron transport system (ETS) and carbon uptake in primary production. Within the water column the resolution of current velocities within the upper layer, the structure and significance of fronts and the three-dimensional structure of the entire regime led to a perception of the upwelling phenomenon and the processes by which it is maintained fundamentally

*Stommel, H., Future prospects for physical oceanography, *Science, 168,* 1531-1537, 1970.

different from earlier models. Perhaps the most significant effect of this spatial definition outside of the program will be seen as the link it will form between early synoptic mapping and the full utilization of remote sensing technology in the future. Had CUEA not developed this basic spatial understanding of the coastal zone it is not clear that the necessary momentum could have been maintained.

Significant technological developments were fostered by CUEA funding, which would not have developed without it. The development of a sophisticated and dedicated time-sharing computer system and the advances in modelling technology — especially in the improved graphics software — come to mind.

Most of the impacts of these developments are now being realized. The effect on the larger scientific community has been in providing strength to the faint of heart, courage to the reticent, and by helping to still some of the more strident voices of parochialism. Of course this is how any scientific field develops. While hardly of the status of one of Thomas Kuhn's scientific revolutions, the research of the last decade has certainly expanded and refined the paradigm.

Finally, and with some trepidation, I will address the impact of CUEA research on the society at large -- however embryonic that impact may be. One might attempt to recite a short litany of societal effects, but, if you will indulge a short — and probably amateurish — diversion into political science, I think the test of CUEA's social impact is perhaps better discussed in respect to the global statement of the problem which, by implication, CUEA was supposed to have addressed. That larger question is: "How do we effectively couple the results of scientific investigations to the development of environmental policy through the political process?"

First, let us consider the goals of the two areas of human activity:

Relevant goals of scientific investigation include:

1) Hypothesis testing
2) Fundamental understanding
3) Predictive capability.

Relevant goals of political process include:

1) Compromise among special interests
2) Protection of the public — securing the public good
3) Effective resource management — the public trust.

Between these sets of goals there exist some articulation problems caused by mismatches:

1) Time frames: The time frames of science are generations; those of politics are terms of office. Scientific progress is usually a function of the development of a new generation of scientists, which has an implicit time scale of perhaps 20 years. Political progress is a function of the electoral process, with a time scale of 2 to 8 years.

   The impact of this mismatch is that long-range solutions are inherently of less interest in the political process. A measure of potential political success is whether short term solutions result from the research. The responsibility for eliminating this mismatch is essentially political.

2) Nomenclature and scientific illiteracy in the public domain in general: The political process is naive with respect to science. Some of the naivete results from lack of familiarity with the jargon in science, but some is fundamental and conceptual.

   The impact of this mismatch is that the political process often focuses on an irrelevant or minor, but tractable point and misses the more subtle, but essential, underlying concept. Therefore, the two sides in the dialogue talk past one another. To the extent that the scientific community persists in obfuscatory jargon, the mismatch is exacerbated. Ultimately the responsibility for overcoming this mismatch lies with the scientific community.

3) Inconclusive scientific results lack appeal for politicians faced with the immediate and vocal realities of special interests: To a large degree this mismatch is characterized by the academic scientific approach vs. the practical engineering approach.

   The academic model prizes (and rewards) open-ended, open-minded investigation and discovery. Academicians are taught to avoid closure, the seductive appeal of finality, of the absolute, definitive conclusion.

   The engineering model is designed to result in a specific outcome. There are few results more final, absolute, or definitive than a dam, a power plant, or a freeway. Engineers are taught to reach closure efficiently, professionally, and creatively. Because the engineering model produces responsive and definitive results it is competitive in the political arena with the reality of special interests which can be measured as jobs, votes, dollars, etc. The academic model produces results that typically are not competitive because they appear, by comparison, to be diffuse, inconclusive, and confusing. The impact of this mismatch is that scientific results have little impact on political decision making because they become obscured by the strident tones of more immediate special interests. The responsibility for eliminating this mismatch is shared by the scientific and political communities. Scientists should be less reticent to discuss the implications of their research and politicians should be less demanding of concrete, practical, short-term results.

How successful has CUEA been at transcending the scientific-political mismatches and by extention, in affecting the political process?

1) On the first count, that of time scales, CUEA has been highly successful because it was fortunate to have as a subject phenomenon one that varies on a sufficiently short time scale that there exists a good impedence match with the political process.

2) On the second count, in the task of overcoming lay scientific illiteracy, CUEA has also been successful, especially in the early days. Toward the end, increasing emphasis on articulation of results through the scientific literature has shifted energy away from the task.

3) On the matter of definitiveness of results, the success of CUEA has depended upon the success of the modelling effort. In this regard the success has been greatest in physical modelling where, from a social perspective it is less critical, than in the biological modelling effort.

In sum, I believe that CUEA has had significant direct and indirect effects on the scientific community. It has demonstrated that multidisciplinary international programs are feasible and can provide significant scientific results. For many of the participants the program has proved that the divergent evolutionary pattern that characterizes the sciences and results in mutually exclusive vocabularies in physicists and biologists is reversible.

On the larger social scale the program has had reasonable success in affecting the world in which it lives. In some cases the impact was serendipitous. More often significant levels of courage, commitment, and vision helped it develop and, in the end, a lot of plain hard work and tenacity insured its survival.

In a time when the problem of communication and interaction between the public and the scientific community is of serious consequence, a situation described by James Burk in the terms "Never have so many understood so little about so much." The impact of CUEA has been significant and I expect it will persist.

THE COASTAL UPWELLING ECOSYSTEMS ANALYSIS PROGRAM
EPILOGUE

Feenan D. Jennings

Sea Grant Center for Marine Research, Texas A&M University,
College Station, Texas 77843, U.S.A.

Since 1972, support for studies of coastal upwelling has been provided by the International Decade of Ocean Exploration Program of the National Science Foundation. The organizers of this symposium were kind enough to invite me to give the epilogue to this week-long meeting, and especially to the CUEA Program because I had the responsibility for directing the IDOE for the first seven and a half years of its life. They were also intelligent enough to put me on the agenda at 5:00 p.m. of the final day, in case they don't agree with some of my remarks.

When the IDOE was established in 1969, the goals set forth by the Vice President did not include any direct mention of living resources. As it turned out, this was no accidental omission. The Office of Management and Budget had insisted that programs associated with fisheries or food from the sea be left out of the IDOE charter. They were negotiating with the Bureau of Commercial Fisheries about its programs and did not want any new living resources programs started in the federal government until the issues were resolved.

Our NSF-appointed IDOE Advisory Committee and the NAS-Ocean Sciences Board IDOE Advisory Committee both commented on the lack of a living resources component in IDOE and strongly recommended that corrective action be taken. In 1971, after discussions with the Office of Management and Budget we undertook support of research in living resources.

The program which evolved was the Coastal Upwelling Ecosystems Analysis Program (CUEA). It was an outgrowth of two ongoing projects. One was a physical oceanographic project to evaluate the theory of upwelling through a carefully designed field experiment off the Oregon Coast. This project, CUE-I, was being supported by the Environmental Forecasting component of IDOE. The other was the Upwelling Biome Project being supported by the International Biological Program of NSF. It involved chemical and biological oceanographers developing theoretical and numerical models of the upwelling ecological systems and evaluating the models through field experiments off Mexico, northwest Africa, and Peru.

In the five years covering 1972 to 1977 the CUEA scientists conducted six field experiments in four geographic areas. In 1972 and '73 the CUE-I and CUE-II field experiments in the physics of coastal upwelling were completed off the Oregon coast, and the MESCAL I and II field experiments in the chemistry and biology of coastal upwelling were completed off the Mexican coast. CUE-II also served as a pilot in which meteorologists, physicists, chemists, and biologists worked together on some preliminary field studies.

Having sharpened their skills in these separate experiments, the two groups joined forces in the spring of 1974 to conduct the JOINT-I field experiment off northwest Africa along with scientists from Federal Republic of Germany, Mauritania, France, German Democratic Republic, and Spain. And from March 1976 to May 1977 the CUEA team conducted the JOINT-II field experiment off the coast of Peru with the full involvement of Peruvian institutions and scientists.

The overall goal of the experiments was to attain scientific understanding of coastal upwelling sufficient for predictive use in fishery resources management. Research objectives associated with the goal were:

1. Description of the spatial and temporal scales of the circulation and the distribution of both nutrients and organisms in coastal upwelling regions.
2. The measurement of the physical, biological, and chemical processes and mechanisms causing and affecting upwelling, and the importance of those processes to the observed spatial and temporal variability.
3. Documentation of interactions between physical and biological processes that combine in coastal upwelling ecosystems to give such extraordinarily high biological production.
4. The development of a series of theoretical and simulation models for upwelling ecosystems that would provide a basis for predicting the response of upwelling ecosystems.

The final field experiment has been completed and the CUEA scientists are deeply involved in the analysis and interpretation of the data and in synthesizing the results. Many of the papers presented at this symposium are tangible evidence of those activities. To find out the extent of such tangible products, I asked the National Oceanographic Data Center, which is the manager for IDOE reports and data, for a computer print-out of the CUEA documentation in their files. As of February, 1980, they listed the following documents:

- Over 200 scientific papers from refereed journals, symposium volumes, and other publications.
- 65 technical reports, and
- 60 data reports.

These numbers tell us that the CUEA scientists have been hard at work but they don't shed any light on what we've learned nor what scientific advances have been made.

Richard Barber and John Costlow have recently written an article on the biological component of IDOE for a forthcoming issue of *Oceanus* magazine. In it, they have described in some detail the advances made by the CUEA program in our understanding of coastal upwelling ecosystems. I would summarize those advances as concisely as possible by saying

- CUEA has documented the importance of the periodicity of local wind intensity to circulation in coastal upwelling and to phytoplankton production and zooplankton complexity in each of the upwelling areas studied;
- The role of bottom topography and slope in circulation and its subsequent impact on productivity has been determined;
- The importance of longshore surface currents and the sub-surface countercurrent to biological production has been determined;
- The 1976 field experiment off Peru and the cooperative CUEA-NORPAX *El Niño* Watch clearly demonstrated the impact of large scale remotely driven events on local oceanographic conditions and its overriding influence on the biological character of the upwelling system.

With regard to the important question of how well CUEA has done in meeting its four research objectives, I would like to quote directly from Barber and Costlow:

> "The first two objectives, description of the variables that define the ecosystem and the quantification of the critical processes, are largely completed. The third objective, elucidation of the physical/biological processes that make upwelling uniquely productive, is on its way to realization but not nearly finished. Work on it will continue for several more years since each synthesis of the physical, chemical, and biological interrelationships suggests more interactions that need to be analyzed for in the CUEA data suite. Modelling accomplishments, especially in circulation have been substantial but ecosystem modelling has yielded less insight than expected when the fourth objective was defined."

This seems like a reasonable assessment of progress toward the initial objectives, and to me, it represents very impressive progress.

CUEA scientists also made impressive progress in learning to work as a team. It took quite a bit of time and effort for the meteorologists, physicists, chemists, and biologists to learn to work together. Each had to give up some of his autonomy, but they did it successfully and clearly demonstrated its value to the study of large, complex oceanographic phenomena.

In CUEA, effective working relationships were developed with the oceanographers from the coastal nations of upwelling areas and close bonds were established that will last far beyond this impressive international symposium.

A major contribution to further advances in our understanding of coastal upwelling and our prediction of it is represented by the data contained in the 200 scientific papers and in the technical reports and data reports already published. These data are particularly valuable because of their interdisciplinary character. They will serve as the basis for continued advances by those of you attending this symposium and by future research scientists as you, and they, build on the concepts, theories, and models developed from the data thus far.

When the organizers of this symposium first asked me to give the closing remarks, they referred to it as a benediction. Subsequently, they decided to call it an epilogue. According to Webster's dictionary, a benediction is "a short blessing with which public worship is concluded". In a sense, this symposium is a worship at the altar of scientific learning, and a benediction might not be inappropriate. However, there are those who may rightfully question my qualifications as a holy man, so rather than give a blessing, I will conclude these remarks with a few expressions of gratitude:

- As a science administrator associated with the program, I am grateful to the leaders of CUEA for setting forth a clear set of objectives in the beginning, and for designing and implementing a long-term research program that has made important advances toward those goals.
- Speaking on behalf of the oceanographic community I am grateful to all of the CUEA participants for:

1) demonstrating that biologists, chemists, physicists, meteorologists, and engineers can work together as a very effective team over a long time period to tackle the kinds of large complex problems we face in the oceans,

2) giving focus to a world-wide program in the study of coastal upwelling during the past 10 years,
3) the many scientific advances which have resulted from the CUEA program and the excellent data sets that will be so important to continued advances in our understanding of upwelling systems in the future.

- And finally I believe we should all be grateful for the privilege of having been able to take part in such a grand experiment, in spite of the problems each of us may have encountered as individuals.

If there are questions, I am certain there is sufficient expertise in the room to answer them. If there are none, go forth and continue your important scientific work.

ESTIMATES OF LARGE-SCALE ATMOSPHERIC PRESSURE VARIATIONS IN THE UPWELLING AREA OFF NORTHWEST AFRICA

Norbert Michelchen

Institute of Marine Research, Rostock-Warnemünde,
German Democratic Republic

Abstract. Long-time series of mean monthly air pressure, p_a, meridional gradient, $\partial p_a/\partial y$, and special wind data at eight northwest African weather stations are analyzed with regard to periodicities of some years. Both the common method of estimation of energy density and the statistical entropy method are used. Characteristic periods have been found in the range of 1/2, 1, 2, and approximately 5 to 6 years. The amplitudes of the signals decrease from north to south. These significant atmospheric variations could be expected to generate analogous large-scale variations in upwelling processes. A simple estimate shows that the mean air pressure gradient along the northwest African coast cannot produce a steady northward flow by the inverse barometric effect seawards of the shelf edge.

Introduction

The upwelling area off northwest Africa has been intensively studied the last few years, for instance during the CINECA program. It was pointed out by Tauber (1974) that anomalies in the seasonal variations of horizontal atmospheric pressure gradients generate anomalies in the northeast trade winds. The anomalies are primarily caused by shifts of the Azores high and the Sahara low both meridionally and especially in the zonal direction; the seasonal air pressure variations are secondary in their effects on the wind anomalies. The meridional migration of the northeast trade wind system during the course of the year is described for instance by Schemainda, Nehring, and Schulz (1975), who refer to strong deviations from the general state, e.g., in spring, 1971. It seems that anomalies of the large-scale trade wind field are closely connected with large-scale upwelling periodicities both in time and space (Tauber, 1974). Meteorological variations, such as those that cause years of severe drought in the Sahel region, also produce inter-annual variations in the adjacent oceans. Therefore, the estimation of periods greater than a half year is important in upwelling research (Codispoti, Dugdale, and Minas, 1979).

Autospectra estimates

Mean monthly air pressure values p_a at eight northwest African weather stations, published in World Weather Records together with general meteorological statistics for several stations, were used for estimating autospectra (Table 1). In spite of strong fluctuations, on an average a certain regularity is apparent. Air pressure maxima occur at all stations in January (for Banjul in December) with a secondary maximum in June or July. The pressure minima occur at the northern stations in August and at Dakar and Banjul in April. The secondary minimum appears in April at northern stations and in August at Dakar and Banjul. The average annual pressure variation decreases from north to south and the monthly average is more variable in winter than in summer. Statistical analyses reveal periodicities of 2 and 4.5 years during periods with anomalous air pressure relationships. The time series were further investigated using spectral estimates (k = degree of freedom). The stations, which are north of or immediately in the region of northeast trade winds, have the highest energy density in a one-year period. In transitional regions between trade winds and calms (Saint Louis, Dakar, Banjul) the energy density increases for the half-year period and becomes equal to or higher than the one-year peak. All spectra (except at Sidi Ifni) show peaks of energy density at lower frequencies, e.g., 24 months. Variability with this periodicity is well known from the upper atmosphere (Reed, 1962). Time series of 20 to 30 years show that approximately 5 to 6-, 2-, 1- and 1/2-year periods are significant (Fig. 1). The longer periods were confirmed by the maximum-entropy spectrum (Fig. 2). The length of the autocorrelation function was always chosen with N/3 (N = number of values). The periodic frequency variations of other meteorological values, e.g., the

TABLE 1. Northwest African Weather Stations and Air-pressure Time Series

Station	Latitude	Longitude	From	To	Number of Values	Notes
Casablanca-Anfa	33°34'N	07°40'W	Jan. 1924	Jan. 1972	576	
Sidi Ifni	29°22'N	10°11'W	Jan. 1951	Dec. 1960	120	
Villa Cisneros	23°42'N	15°52'W	Jan. 1951	Dec. 1960	120	
Nouadhibou	20°56'N	17°03'W	Jan. 1941	March 1978	435	except 1952
Nouakchott	18°07'N	15°56'W	Jan. 1941	March 1978	423	except 1952, 1953
Saint Louis	16°03'N	16°27'W	Jan. 1951	March 1978	327	
Dakar-Yoff	14°44'N	17°30'W	Jan. 1951	March 1978	327	
Banjul-Yundum	13°21'N	16°40'W	Aug. 1945	March 1978	392	

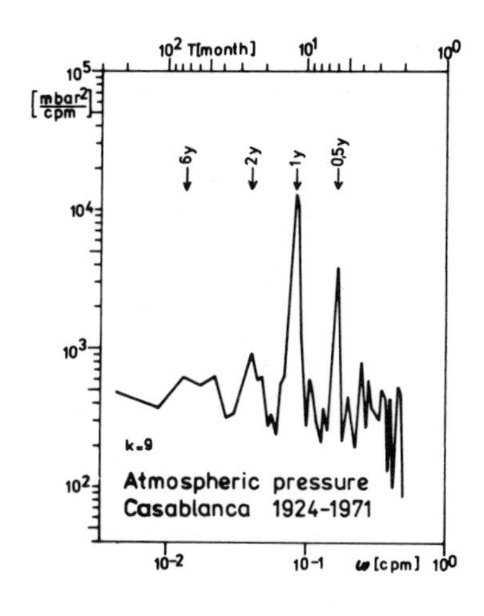

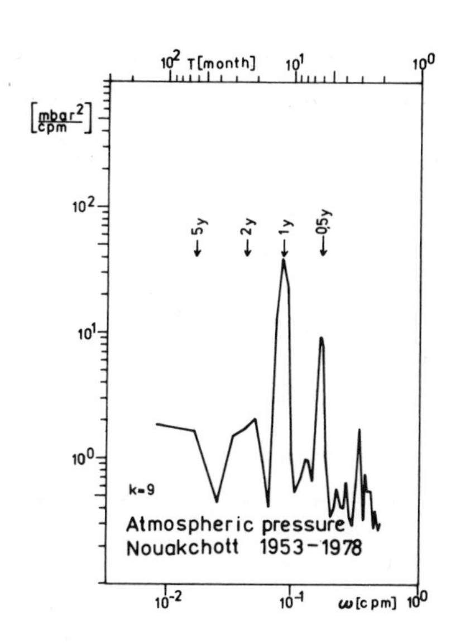

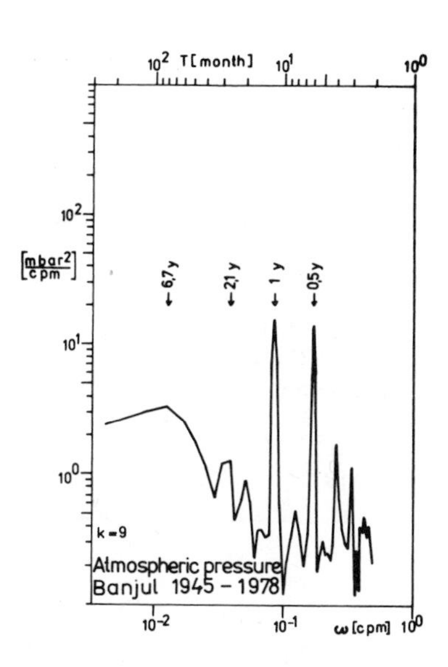

Fig. 1. Autospectra of atmospheric pressure at Casablanca, Nouakchott, and Banjul.

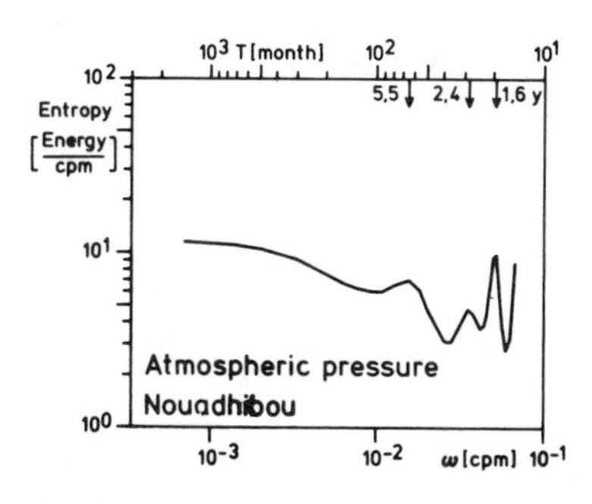

Fig. 2. Maximum-entropy spectrum of atmospheric pressure at Nouadhibou.

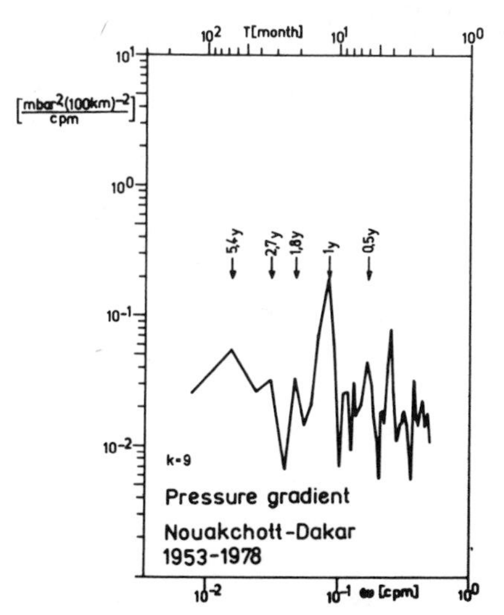

Fig. 3. Autospectrum of air-pressure gradient between Nouakchott and Dakar.

meridional pressure gradient, $\partial P_a/\partial y$, and measured wind (Saint Louis, Dakar) were determined from the same spectral estimates. The results also show periodicities of 1/2, 1, 2, and approximately 5 years (Figs 3, 4, 5). Klaus and Stein (1978) obtained similar results for temporal variations of European Grosswetterlagen. Table 2 shows the amplitudes of air pressure variations for different periods. They decrease from north to south and demonstrate the the first four stations are under the influence of a subtropical high only and that at the southern stations the transitional region dominates. It is to be expected that significant atmospheric variations generate analogous large-scale variations in the upwelling processes. Current measurements along the the continental slope frequently show a mean northward flow from the near-surface layer down to depths of 500 to 1000 m (Mittelstaedt, 1978). An estimate of the average large-scale meridional air pressure gradient between Casablanca and Dakar shows that the air pressure difference along the African coast cannot generate steady northward flow seawards of the shelf edge as a permanent part of the large-scale upwelling. In the last 28 years the maximum difference of mean monthly pressure values between Casablanca and Dakar was 16.7 mbar (February, 1977). This corresponds to a rise of sea level from north to south of approximately 16.7 cm in a distance of 2300 km from the inverse barometric effect. Hence a northern countercurrent of about 1.2 cm/s is possible when the mean monthly atmospheric pressure is at a maximum, as was the case in February, 1977. The mean differences between Casablanca and Dakar are 8.5 mbar in January and 3.7 mbar in July. This does not rule out the possibility that air pressure differences are responsible for mesoscale northward flow, especially in winter.

TABLE 2. Amplitudes (mbar) of Air-pressure Variations for Different Periods

	Period			
	5 y	2 y	1 y	0.5 y
Casablanca	0.33	0.28	1.07	0.58
Sidi Ifni	-	-	1.02	0.58
Villa Cisneros	-	0.28	1.11	0.52
Nouadhibou	-	0.18	0.98	0.37
Nouakchott	-	0.18	0.8	0.4
Saint Louis	-	0.12	0.3	0.37
Dakar	0.18	0.12	0.1	0.31
Banjul	0.16	0.1	0.36	0.33

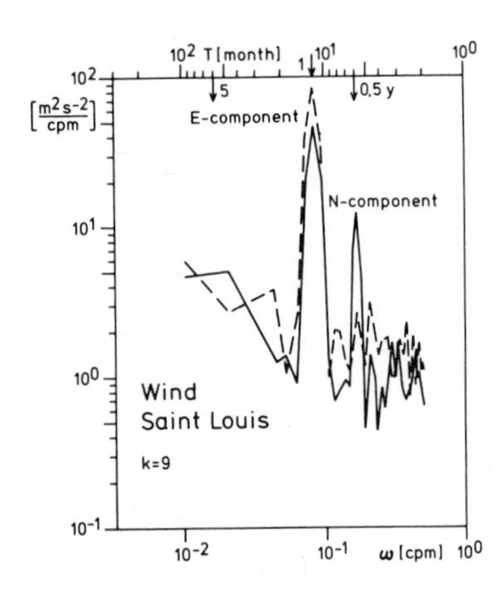

Fig. 4. Autospectra of wind at Saint Louis.

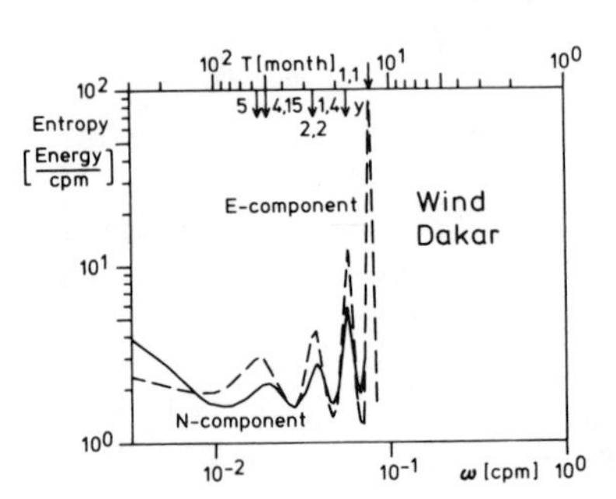

Fig. 5. Maximum-entropy spectra of wind at Dakar.

References

Codispoti, L.A., R.C. Dugdale, and H.J. Minas, A comparison of the nutrient regimes off N.W. Africa, Peru and Baja California, *CUEA Newsletter, 8(1)*, 2-53, 1979.

Klaus, D., and G. Stein, Temporal variations of the European Grosswetterlagen, *Geophysical Astrophysical Fluid Dynamics, 11*, 89-100, 1978.

Mittelstaedt, E., Physical oceanography of coastal upwelling regions with spectral references to northwest Africa, Symposium on the Canary Current: *Upwelling and Living Resources, 40(ii)*, 1-11, 1978.

Reed, R.J., On the cause of the 26-month periodicity in the equatorial stratospheric winds, *Meteorologische Abhandlungen der Freien Universität Berlin, 36*, 245-257, 1962.

Schemainda, R., D. Nehring, and S. Schulz, Ozeanologische Untersuchungen zum Produktionspotential der nordwest-afrikanischen Wasserauftriebsregion 1970 - 73, *Geodätische und Geophysikalische Veröffentlichungen, R. IV, 16*, 3-85, 1975.

Tauber, G.M., The trade wind circulation in the North Atlantic and its influence on thermal regime of the surface waters (in Russian), *Trudy GOIN/Leningrad, 120*, 42-69, 1974.

SEA-SURFACE TEMPERATURES AND WINDS DURING JOINT II: PART I. MEAN CONDITIONS

G.L. Moody

DET. 30, 2nd Weather Sq., Vandenberg A.F.B., California 93437

David W. Stuart

Department of Meteorology, Florida State University, Tallahassee, Florida 32306

A.I. Watson

National Hurricane Experimental Meteorology Laboratory, NOAA
Coral Gables, Florida 33416

M.M. Nanney

National Weather Service, NOAA, San Antonio, Texas 78209

Abstract. A research aircraft was used to map sea-surface temperatures (SST) and flight-level (152 m) winds over the JOINT II region on eight flights during March and April, 1976 and on 12 flights during March, April, and May, 1977. The individual flight maps were gridded and overall mean maps of SST and winds are presented for each period in 1976 and 1977.

The mean maps for the two years are similar. A definite plume of cold upwelled water was apparent south of Cabo Nazca. There was a wind maximum over the cold plume with the highest wind speeds 10 to 15 km offshore and a wind minimum north of Cabo Nazca over the warmer water. The marked along-shore variation of the wind was in part related to thermally-induced circulations up the Ica River Valley. The slightly warmer SST in 1977 are consistent with the weaker mean winds at 152 m.

Data from the two years suggest a possible useful linear relationship between the maximum speed of the flight-level winds and the value of minimum SST when the flight-level winds are greater than 13 m/s.

Introduction

During 1976 and 1977 the Coastal Upwelling Ecosystems Analysis (CUEA) Program conducted an upwelling study (JOINT II) along the Peruvian coast (near 15°S). For the periods of intensive study (March, April, May) a research aircraft was available to map the sea-surface temperatures (SST), flight-level (152 m) winds, and vertical structure of the atmosphere over the study area. This paper presents mean maps from the fine-resolution mapping flights flown each year. A companion paper (Stuart, 1980) presents temporal variability observed in 1977. Figs 1 and 2 show the flight tracks along with the location of coastal meteorological stations and buoys and topographic features. No meteorological buoy data were available during the aircraft flights in 1976, but the coastal stations 2-6 (Fig. 2) were functioning.

Data Reduction

The National Center for Atmospheric Research (NCAR) provided a completely instrumented research aircraft — a twin engine Beechcraft Queen Air. The SST data were obtained with a Barnes PRT-5 radiometer having a 2° field of view sensing in the spectral region 9.5 to 11.5 microns. Aircraft position was determined by an inertial navigation system (LTN-51 by Litton). After the field work, flight-level winds were computed using the NCAR computer from position, heading, temperature, and pressure data recorded on magnetic tape during each second of flight. The flight altitude (152 m) was determined by a radio altimeter (see Nanney, 1978 and Moody, 1979 for further details).

Each flight (8 in 1976, 12 in 1977) was flown between 1000 and 1600 LST always progressing from the southwest to the northwest corner of the pattern (Figs 1 and 2). The SST data were digitized in the aircraft and individual maps were plotted and subjectively analyzed the same day. SST val-

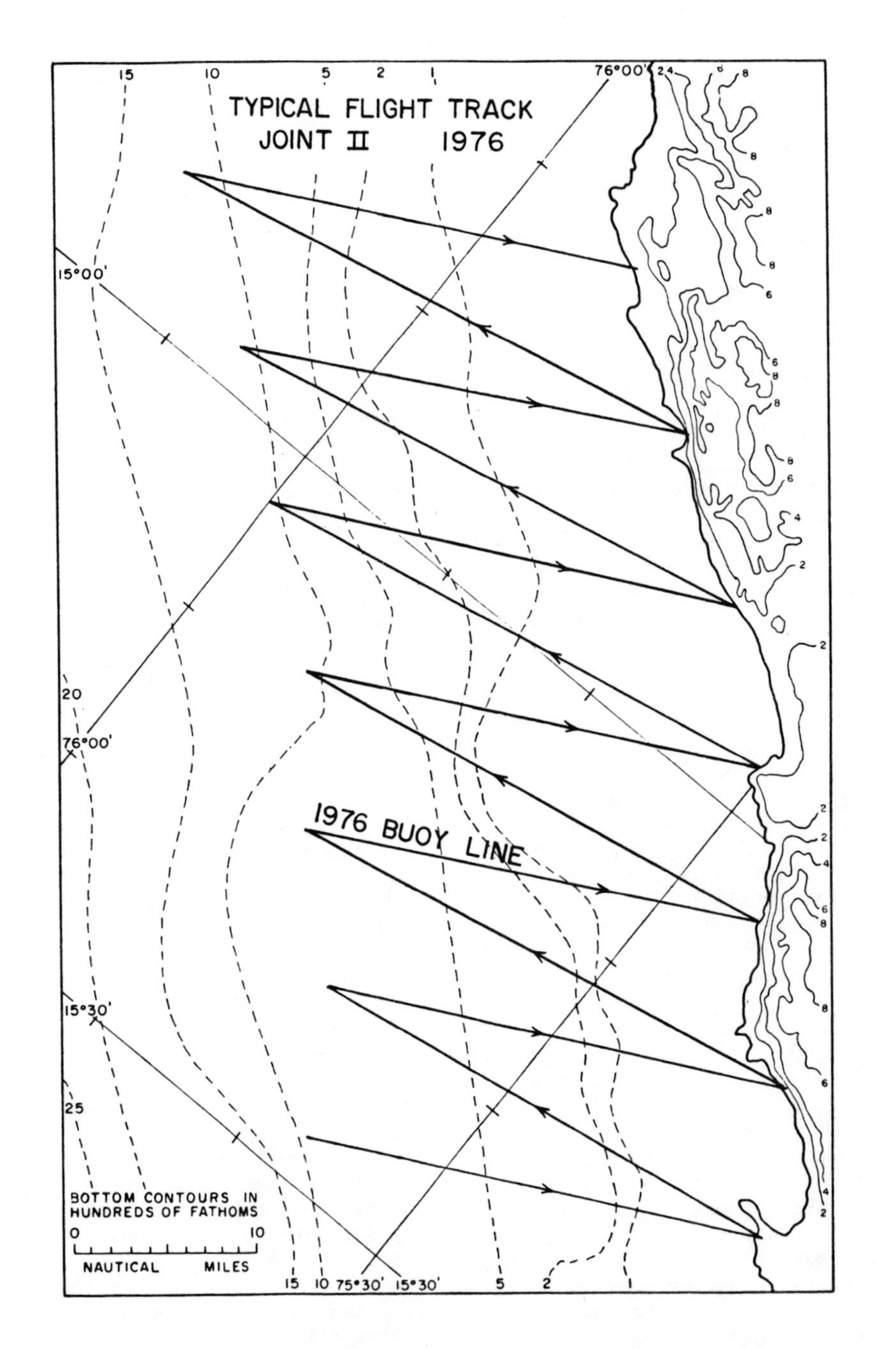

Fig. 1. Typical flight track during JOINT II, 1976.

ues were plotted for each minute along the track or approximately each 3.5 to 5.5 km (2-3 nautical miles). The wind data were also plotted for each minute using a 30-s running vectorial average of the inertially computed winds available each second.

To form a composite of the data a two-dimensional grid (2 x 2 nautical miles, Fig. 3) was overlain on each individual wind and SST map. From these grid point data, maps of the mean conditions for each of the two years were drawn (Figs 3-6). Besides the total wind, maps of the component parallel to the coast V_T (V-Tangent) are also presented (Figs 7 and 8).

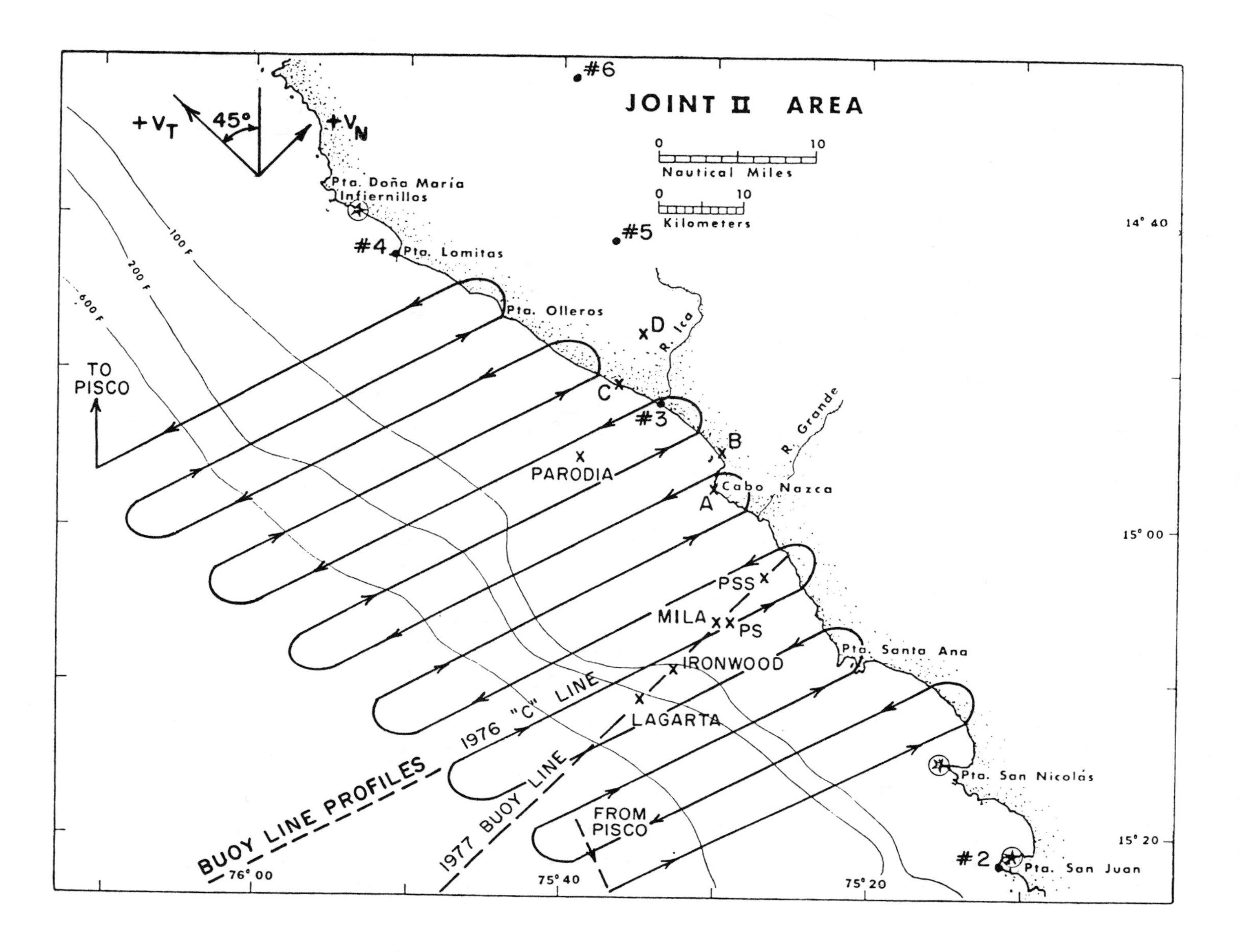

Fig. 2. The JOINT II region with locations of coastal stations, meteorological buoys, and typical flight track during 1977.

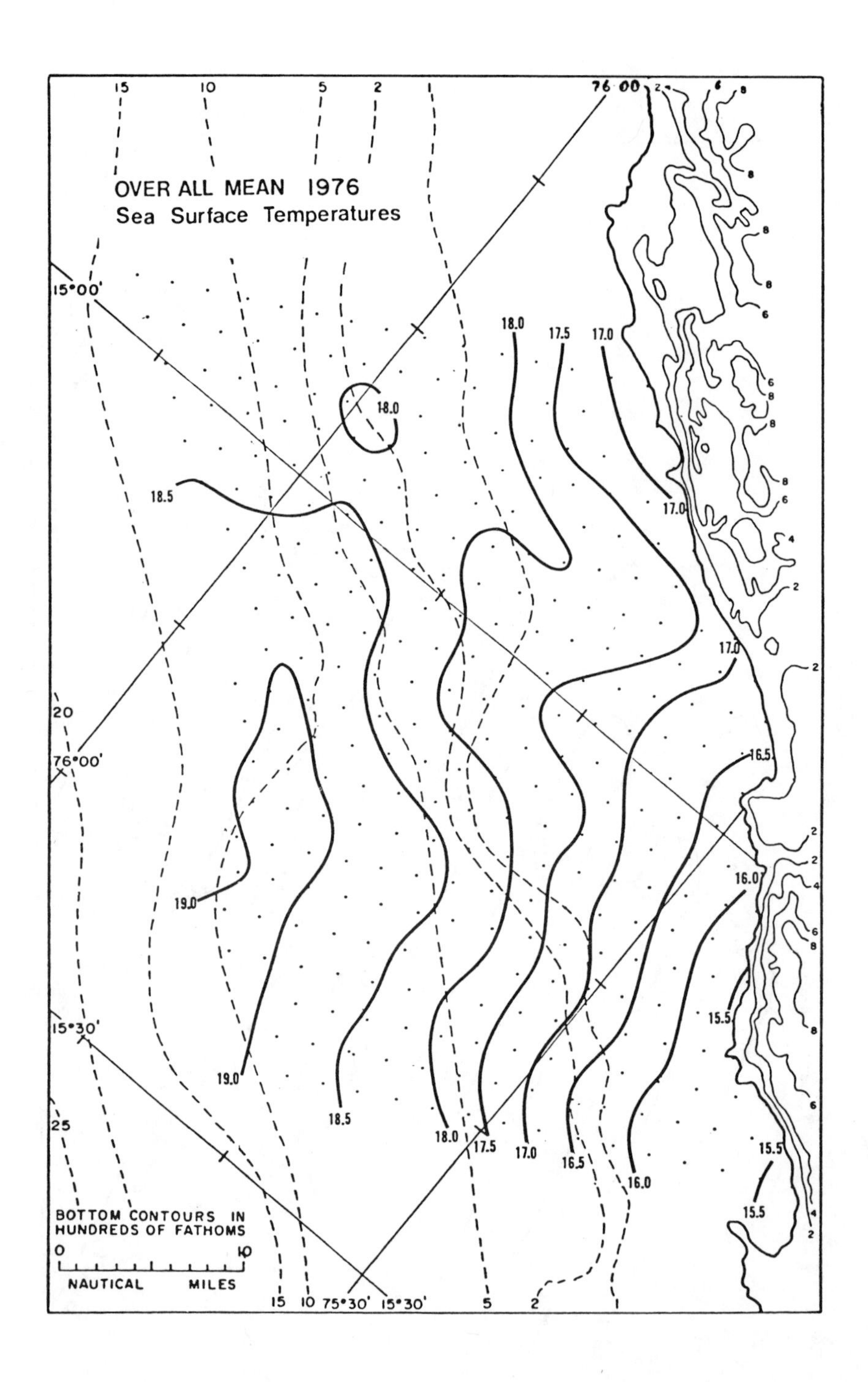

Fig. 3. The mean sea-surface temperature pattern as recorded by aircraft during March and April, 1976. Land contours in hundreds of meters.

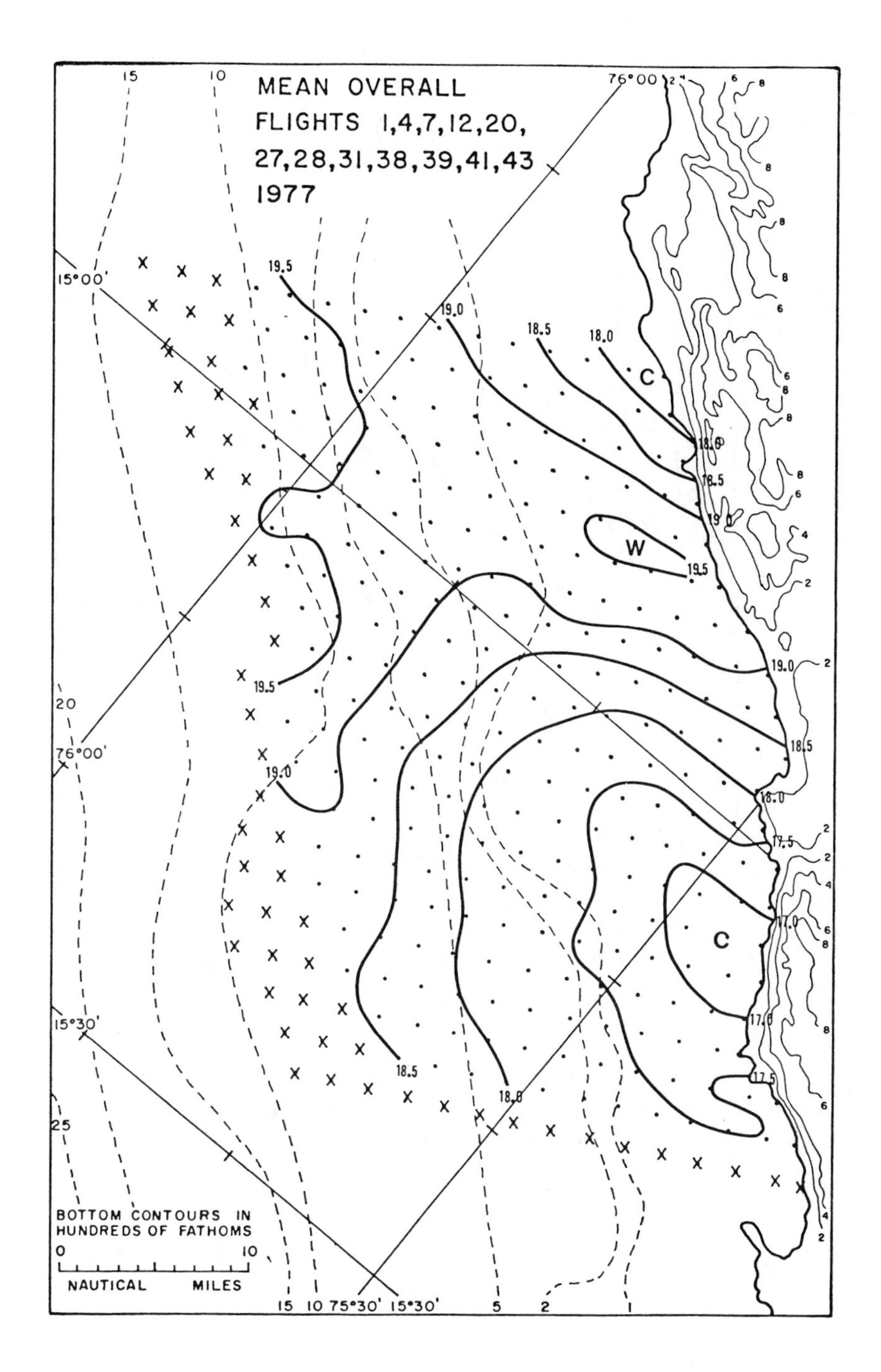

Fig. 4. The mean sea-surface temperature pattern as recorded by aircraft during March, April, and May, 1977. X indicates grid points with data for fewer than nine flights. Land contours as in Fig. 3.

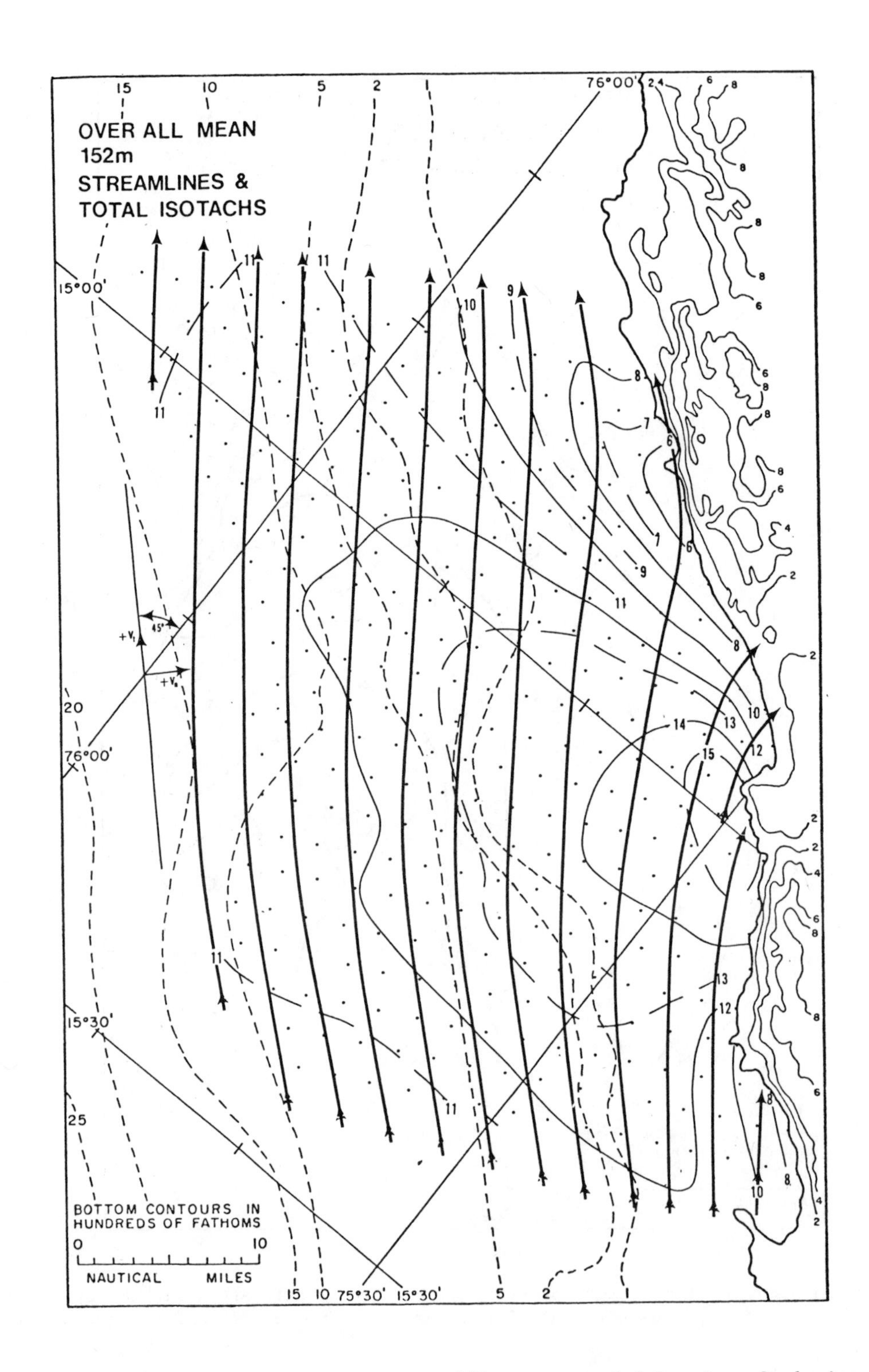

Fig. 5. The mean streamline and isotach patterns at 152 m as recorded by aircraft during March and April, 1976. Isotachs in m/s. Land contours as in Fig. 3.

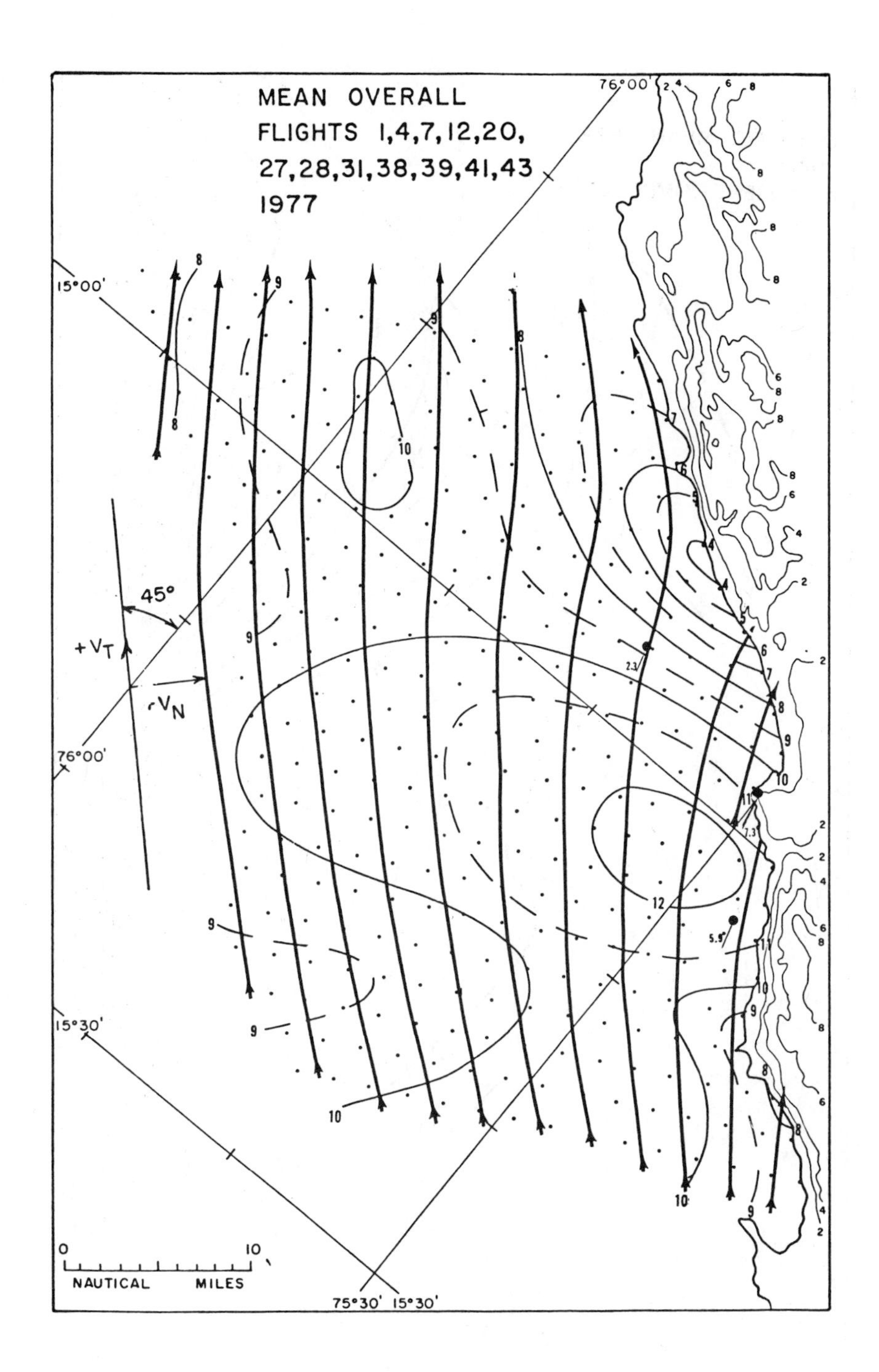

Fig. 6. The mean streamline and isotach patterns at 152 m as recorded by aircraft during March, April, and May, 1977. Isotachs in m/s. Mean winds (for flight time) at buoys and Cabo Nazca also shown in m/s. Land contours as in Fig. 3.

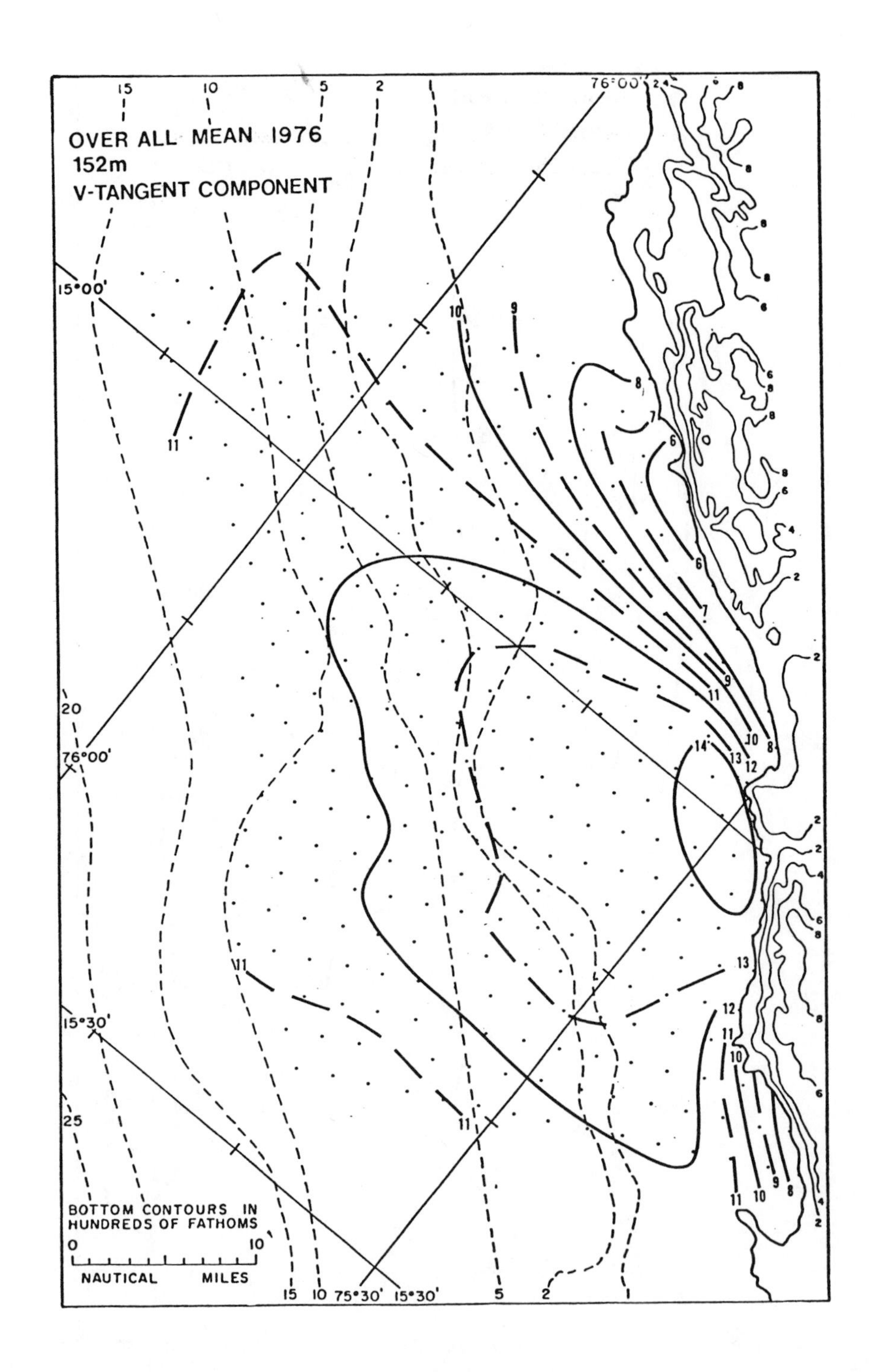

Fig. 7. The mean V_T component pattern at 152 m as recorded by aircraft during March and April, 1976. Isotachs in m/s. Land contours as in Fig. 3.

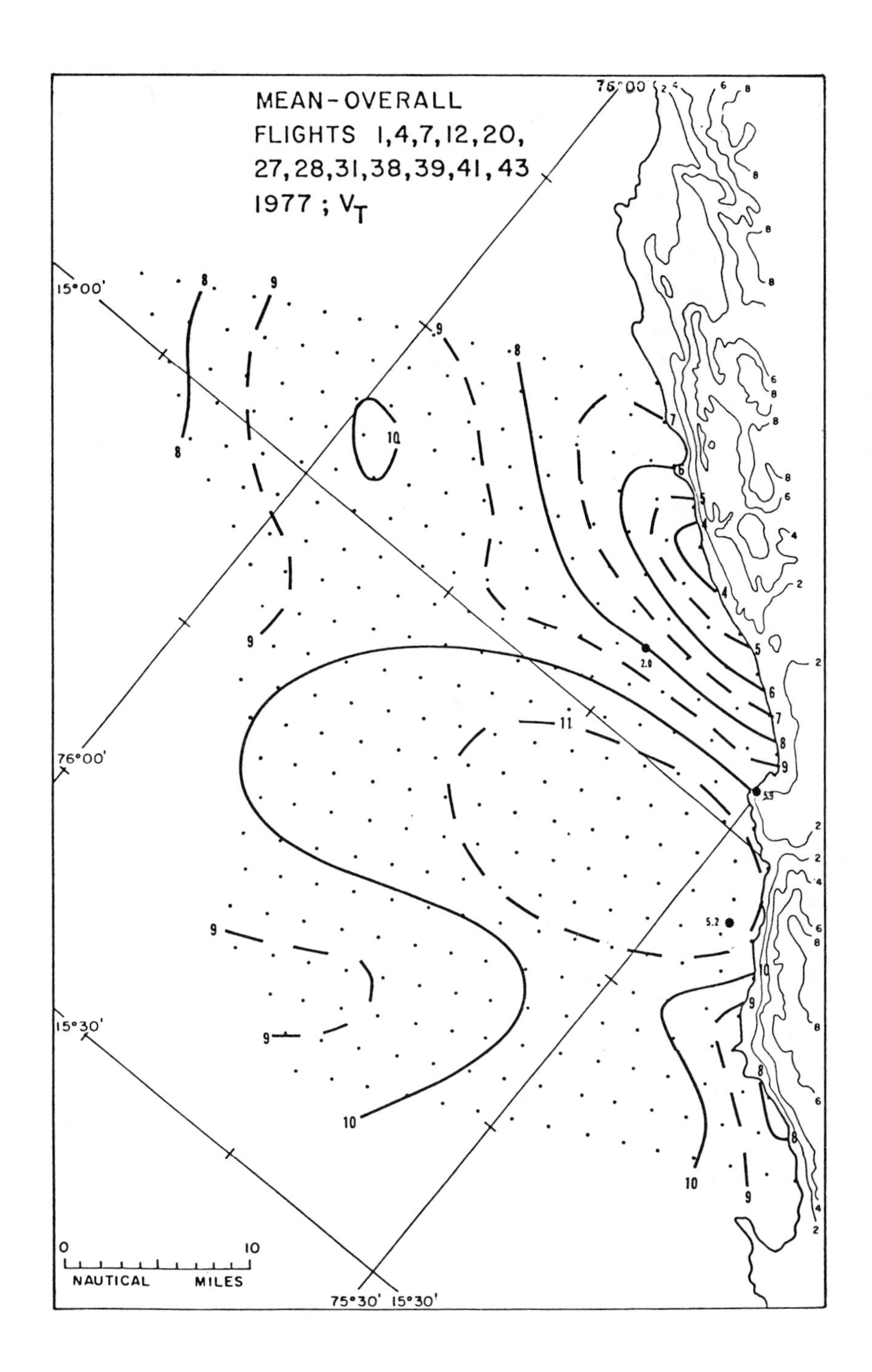

Fig. 8. The mean V_T component pattern at 152 m as recorded by aircraft during March, April, and May, 1977. Isotachs in m/s. Mean V_T (for flight time) at buoys and at Cabo Nazca also shown in m/s. Land contours as in Fig. 3.

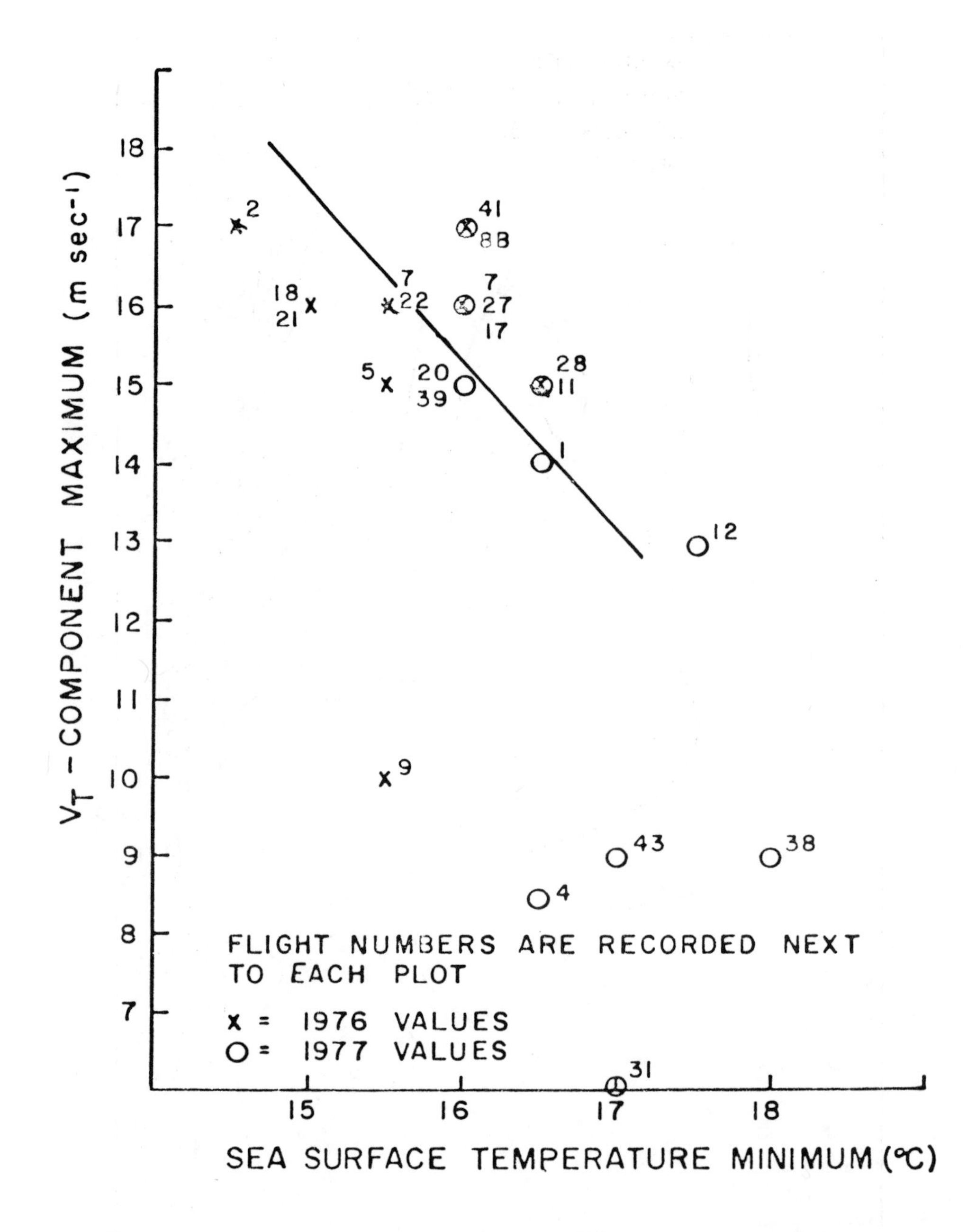

Fig. 9. Scatter diagram of V_T component maximum values at 152 m vs. associated sea-surface temperature minimum values from aircraft data during 1976 and 1977. The solid line is the regression line computed using all data for $V_T > 13$ m/s. The regression line is SST(°C) = 23.044 - 0.459V_T (m/s) and the correlation coefficient is r = -0.70.

Results

The mean maps for the two years are similar. A definite plume of cold, upwelled water is evident south of Cabo Nazca with its axis running west - northwestward from the coast. A wind maximum (at 152 m) is apparent over the cold plume with highest wind speeds some 10 to 15 km offshore. There is a wind minimum (at 152 m) north of Cabo Nazca over the warmer water. The marked alongshore variation of the wind is in part related to thermally-induced circulations up the Ica River Valley. The wind maximum (magnitude and location) is a result of complicated feedback between the normal land-sea breeze, localized channelling by the coastal bluffs south of Cabo Nazca, and the extra cool coastal waters associated with the upwelling. The slightly warmer SST in 1977 are consistent with the weaker mean winds at the 152-m level. However, the nature of the water and subsurface currents involved in the upwelling must be considered when comparing the values of the mean SST of the the two years.

Using data from the two years, a plot of the lowest SST vs. the maximum of the alongshore component of the wind (V_T) at 152 m was prepared. Fig. 9 suggests a linear relationship between the maximum speed of the flight-level winds and the value of the minimum SST, especially when the flight-level winds (at 152 m) are greater than 13 m/s. The regression line is SST(^{o}C) = 23.044 - 0.459 V_T (in m/s). With a correlation coefficient r = -0.07 a linear relationship accounts for 49% of the variance (see Stuart, Goodwin, and Duval, 1980, for comparison of 152-m winds with surface winds).

Acknowledgements. The National Center for Atmospheric Research (NCAR) provided the instruments for the surface meteorological stations as well as the research aircraft with its highly trained crew and support teams. Data processing was at NCAR and the Computing Center at Florida State University. We are indebted to the students who participated in the field phases of JOINT II, and we appreciate the technical assistance of D. Rudd and P. Mullinax. This is a contribution of the CUEA Program of the International Decade of Ocean Exploration (IDOE) funded by the National Science Foundation under Grants OCE 76-82831 and OCE 77-27735.

References

Moody, G.L., Aircraft derived low level winds and upwelling off the Peruvian coast during March, April, May, 1977, CUEA Technical Report 56, Department of Meteorology, Florida State University, Tallahassee, 110 pp. (Available from the National Technical Information Service, Accession No. PB 80-119571), 1979.

Nanney, M.M., The variability of sea surface temperature off the coast of Peru during March and April, 1976, CUEA Technical Report 42, Department of Meteorology, Florida State University, Tallahassee, 99 pp. (Available from the National Technical Information Service, Accession No. PB 287355/AS), 1978.

Stuart, D.W., Sea surface temperatures and winds during JOINT II: Part II - Temporal Variability, This Volume, 1980.

Stuart, D.W., R.J. Goodwin, and W.P. Duval, A comparison of surface winds and winds measured at 152 m during JOINT I 1974 and JOINT II 1977, This Volume, 1980.

SEA-SURFACE TEMPERATURES AND WINDS DURING JOINT II: PART II. TEMPORAL FLUCTUATIONS

David L. Stuart

Department of Meteorology, Florida State University, Tallahassee, Florida 32306

Abstract. The temporal and spatial fluctuations of a plume of cold surface seawater off Peru were investigated during March and early April, 1977. Maps and profiles of sea-surface temperature (SST) and aircraft flight level winds (152 m) are presented along with surface winds from coastal stations and buoys.

The daily average alongshore wind was always upwelling-favorable, ranging from 2.5 to 8.5 m/s. Such wind fluctuations are reflected in changes in the area of the surface plume and in the lowest temperatures. The area and temperature of the plume are correlated with the wind stress. The data support the contention that wind-induced upwelling is important in the region.

Introduction

During 1976 and 1977 the Coastal Upwelling Ecosystems Analysis (CUEA) group conducted the JOINT II expedition along the Peruvian coast. During March, April, and May of both years intensive observations were made near 15°S latitude. Besides strings of current meters, surface meteorological buoys, coastal stations, and research vessels, an instrumented Queen Air research aircraft from the National Center of Atmospheric Research (NCAR) mapped the sea-surface temperatures (SST) and flight-level winds (at 152 m) over the study region. The preceding paper by Moody, Stuart, Watson, and Nanney presents the mean SST and winds as determined by aircraft. This paper treats the temporal fluctuations focusing largely on data from the aircraft, most particularly the 1977 phase when the observational program was most intense.

Reference is made to the preceding Part I (especially Figs 2, 4, 6, and 8) for maps of the study region, mean SST, mean total wind (at 152 m) and the mean alongshore component (V_T) for the JOINT II 1977 study. As the coastline is oriented along 135-315°, all winds are resolved into alongshore (V_T-tangential) and onshore (V_N-normal) components. Part I showed that the mean conditions exhibit a distinct plume of cold upwelled water extending west-northwestward from the coast between Cabo Nazca and Pta. Santa Ana with warmer water to the north off the mouth of the Ica River. The aircraft winds (at 152 m) show a pronounced alongshore variation in the total wind speed as well as in V_T with the axis of strong wind generally aligned along the axis of cold water. On the whole the surface ocean currents (upper 20 m) are parallel to the coast and directed equatorward while below 20 m the ocean currents are directed poleward. Underwater canyons and ridges are important as they affect the deeper currents providing the cold water observed near the surface. This paper emphasizes the fluctuations of this plume of cold surface water and how the fluctuations are related to the wind. In the following section two examples from March, 1977 will be examined and in the summary section the results will be broadened by considering wind and SST data from all flights during the two years of JOINT II.

Examples

Fig. 1 (see Brink, Smith, and Halpern, 1978) provides time plots of the smoothed V_T component of the surface winds for two buoys (PA, PSS) and one coastal station (San Juan) during March, April, and May, 1977. Even at the surface the wind is weaker off the mouth of the Ica River (i.e., at PA). The V_T at the surface is relatively strong in the general vicinity of the plume of cold water, as it is at 152 m. The smoothed V_T wind rarely is negative (i.e., from the northwest), yet the upwelling-favorable wind (+ V_T) does have temporal fluctuations that occur nearly simultaneously over the whole study region. The dates of the flights that yielded maps of SST and wind data are indicated along the time axis of the PA curve in Fig. 1. A distinct SST plume was observed on each flight except 31 and 38, which would not be expected because the tangential wind and hence the Ekman transport was weak during those flights. Buoy line profile flights were made on most days between the mapping flights.

The two examples chosen are the periods 24, 25, 26, and 29, 30, 31 March. Each example is charac-

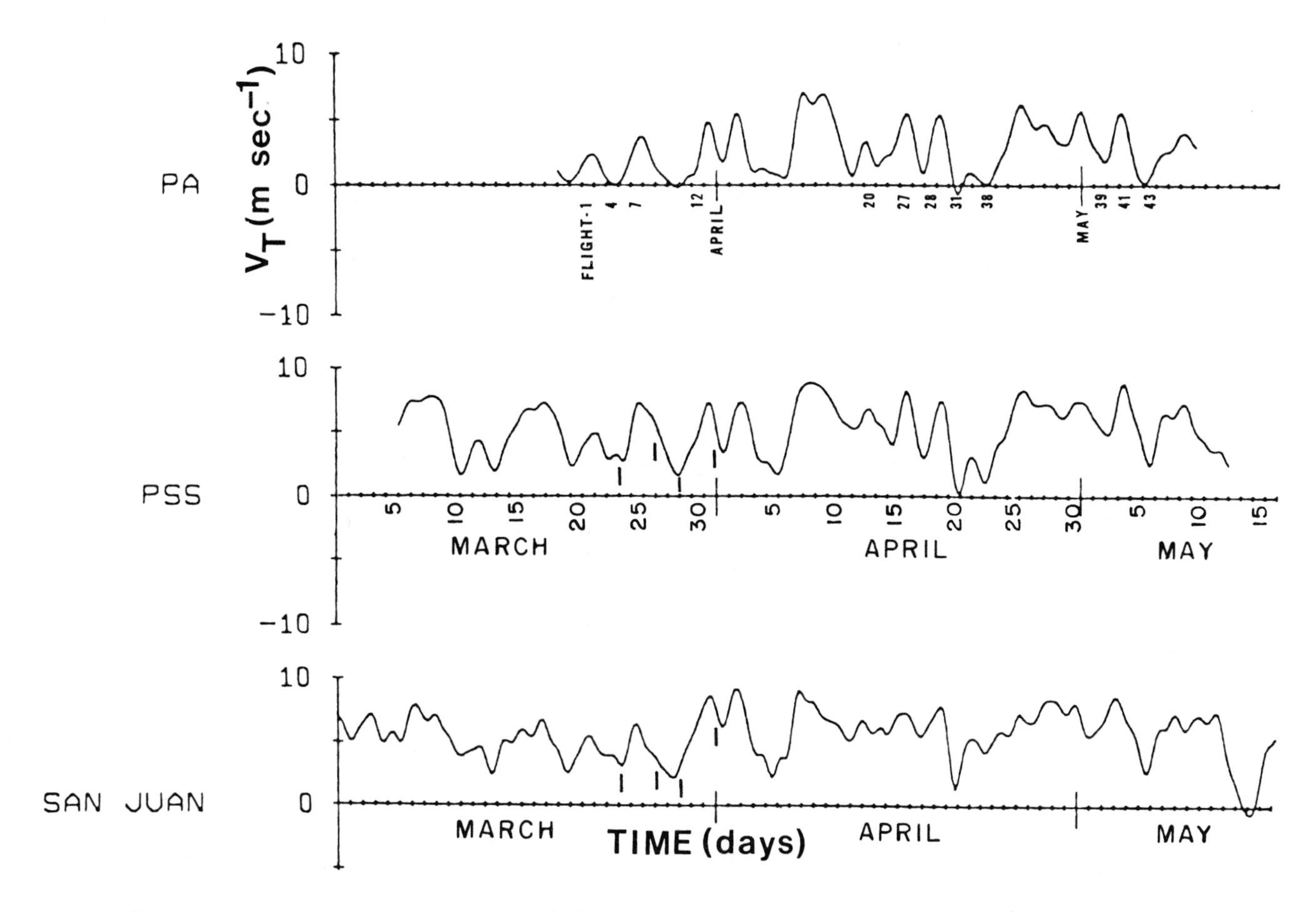

Fig. 1. Time plots of alongshore wind (V_T) during March, April, and May, 1977 for San Juan and two buoys (PA and PSS).

terized by increasing V_T, each day is characterized by profiles of V_T and SST along the buoy line, and maps of the SST on 25 and 30 March are available.

24, 25, and 26 March: Maps of SST and wind (at 152 m) on 25 March 1977 (Fig. 2) show a well defined plume (colder than the mean) with a sizeable area within the 16.5°C isotherm. Warmer water to the north is evident and the wind field well reflects the mean. The wind maximum was over 16 m/s. Profiles of SST and V_T (at 152 m) along the buoy line (Fig. 3) show the progressive lowering of the SST from 24 to 26 March (from 18.5 to 16.0°C) as the tangential speed increased from 9.5 m/s some 30 km offshore on 24 March to 15.5 m/s some 4 km offshore on the 25th. By the 26th the tangential winds had decreased somewhat but still exceeded 12 m/s between the coast and 25 km offshore. A marked increase in V_T favored upwelling between the coast and 20 km offshore (Rossby radius of deformation). The surface currents partially explain the transport of cool water further offshore (see Stevenson, Stuart, and Heburn, 1980).

29, 30, and 31 March: Maps of SST and wind on 30 March (Fig. 4) show a well defined plume but warmer than the mean with the coldest water (17.5°C) closer to Pta. Santa Ana. The warmer conditions were consistent with the wind changes as the V_T minimum occurred early on 29 March (see Fig. 1). The wind maximum was weaker, broader, and nearer Pta. Santa Ana. However, a 10 m/s variation alongshore was still observed. Profiles of SST and V_T along the buoy line on each day (Fig. 5) show the progressive lowering of the SST from 29 to 31 March (from 19.0 to 16.5°C) as the tangential speed increased from 13.0 to 15.5 m/s near 22 km offshore on 31 March. On 31 March the buoy line showed water colder than 16.5°C between the coast and 15 km offshore. Even though V_T had decreased at the coast, the wind maximum 23 km offshore by 30-31 March led to a positive wind stress curl and hence increasing Ekman transport and upwelling between the coast and 23 km.

Because there were so many observations in March, I shall attempt to extend the results of the above two examples to the period 15 March to 2 April, during which period 11 flights yielded SST profiles along the buoy line and of these,

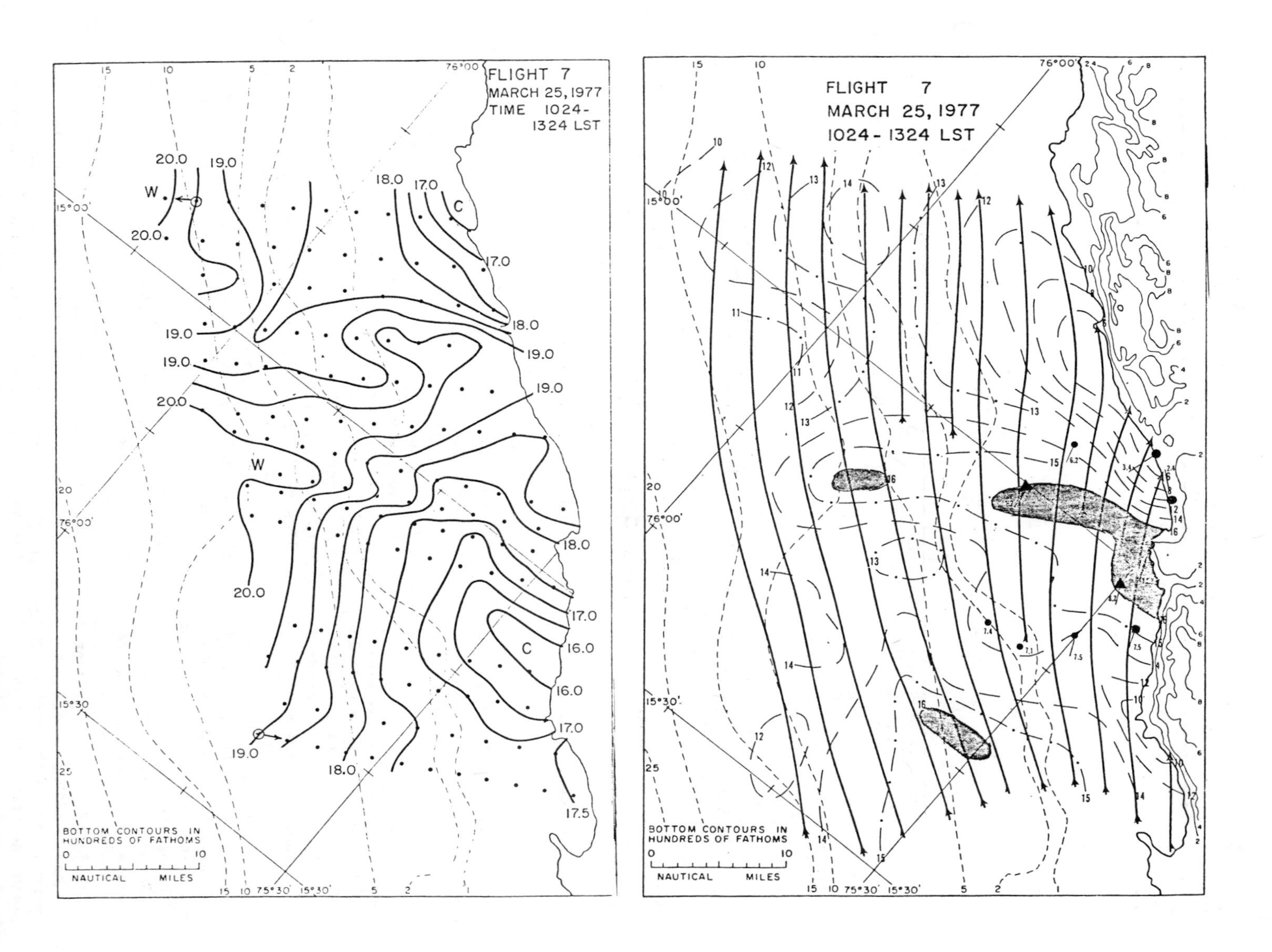

Fig. 2. Maps of sea-surface temperatures (°C) (left panel) and winds (at 152 m) (right panel) 25 March, 1977. Wind speeds (m/s) shown by isotachs (dashed lines) and wind direction by streamlines (solid lines with arrows). Surface winds (at buoys and coastal stations - dots - and at ships - triangles - on right panel) for appropriate flight time also shown in m/s. On left panel dots show 1-min positions of aircraft while circles with arrow indicate start (lower left) and end (upper left) points along track. (See Part I, Fig. 2 for identification of buoy and coastal station locations.) Land contours in hundreds of meters.

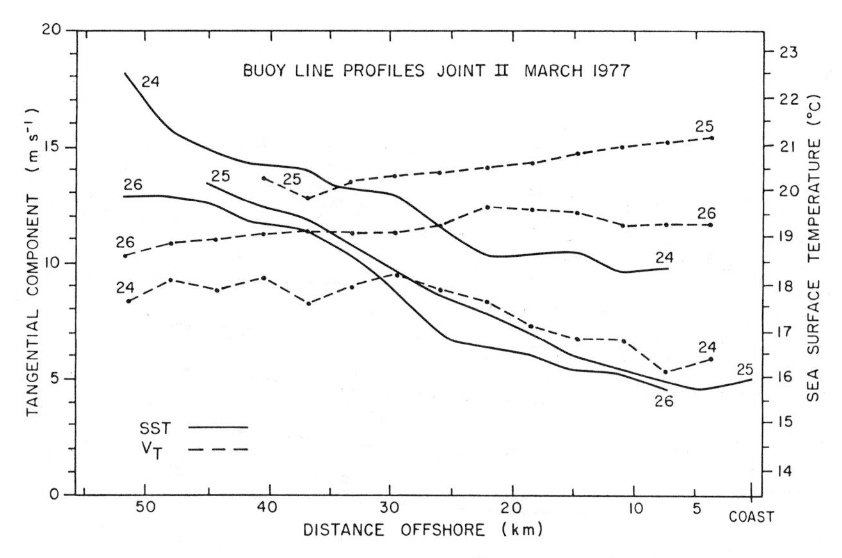

Fig. 3. Buoy line profiles of sea-surface temperature (^{o}C) and alongshore wind, V_T (m/s at 152 m), for 24, 25, and 26 March, 1977.

five also yielded maps of SST. Figure 6 shows a plot of the average alongshore stress, $\overline{\tau}_T$, for the 24-h period ending at 1200 GMT (0700 LST) each day (i.e., 3 to 4 h prior to flight time). The lowest SST (to the nearest 0.5^{o}C) observed on each flight (dashed line, Fig. 6) also shows some response to $\overline{\tau}_T$. Figure 6 also shows the areas within the 16.5, 17.0, and 17.5^{o}C isotherms vs. time. The areas could be determined only from mapping flights, so the curves are somewhat crude, but they do show the area of cold water responses to variations of the surface wind stress to be in agreement with Ekman transport ideas.

Summary

From the preceding section it is clear that the SST and the area of the plume fluctuated with the wind (stress), at least during the observations in March, 1977. Furthermore the fluctuations were observed even though V_T varied only in magnitude, there were no distinct reversals.

In an attempt to corroborate the above results, data from all flights in 1976 and 1977 that yielded maps of SST were correlated with the surface winds at San Juan. (PSS winds were not available in 1976.) Of the correlations only two are presented. The alongshore stress averaged over 24 h ($\overline{\tau}_T$) preceding each flight was correlated with the lowest SST as observed during each of the 23 flights that yielded a SST map (Fig. 7). The regression line is $T(^oC) = 17.1 - 1.74\ \overline{\tau}_T$ (in dynes/cm^2). The linear fit is significant at the 5% level but not at the 1% level. The standard deviation is 0.64^{o}C and the correlation coefficient is 0.47. Such a fit is somewhat surprising because the observed SST is surely a function of the depth of the thermocline, the nature of the upwelling water, surface stirring, wind stress curl, etc. However, the regression is not too useful because a tripling of the alongshore stress (from 0.3 to 0.9) lowers the lowest SST by only one degree (from 16.5 to 15.5^{o}C).

A slightly better correlation was obtained between the area within the 16.5^{o}C isotherm and the average alongshore stress ($\overline{\tau}_T$) for the 48 h prior to each flight (Fig. 7). The 16.5^{o}C isotherm most frequently intersected the coast to form a rather definite area of cold water. The regression equation is $A(\text{in km}^2) = -473 + 1512\overline{\tau}_T$ (in dynes/cm^2). The linear fit is significant at the 1% level, the standard deviation is 370 km^2, and the correlation coefficient is 0.62. The area within a rectangle one degree latitude by one degree longitude is approximately 120 x 10^2 km^2, hence during our studies the 16.5^{o}C isotherm bounds regions of at most 1/6 of a one-degree latitude - longitude square. The regression equation indicates the stress must average 1/3 of a dyne/cm^2 for two days for there to be any area within the 16.5^{o}C isotherm and the stress must average twice this (0.64 dynes/cm^2) for there to be an area of at least 500 km^2 (i.e., about 1/25 of a one-degree latitude - longitude square). The scatter also shows that the area within a given isotherm is determined by more than just the wind stress, but a knowledge of the wind at San Juan is useful for estimating the size of the plume defined by the 16.5^{o}C isotherm.

Acknowledgements. I wish to thank the Research Aircraft Facility of the National Center for Atmospheric Research for their expert support. The Field Observing Facility of NCAR provided the instruments for the surface weather stations. The Computing Center at Florida State University was used for processing the data and preparing the graphs. I am grateful to my graduate students

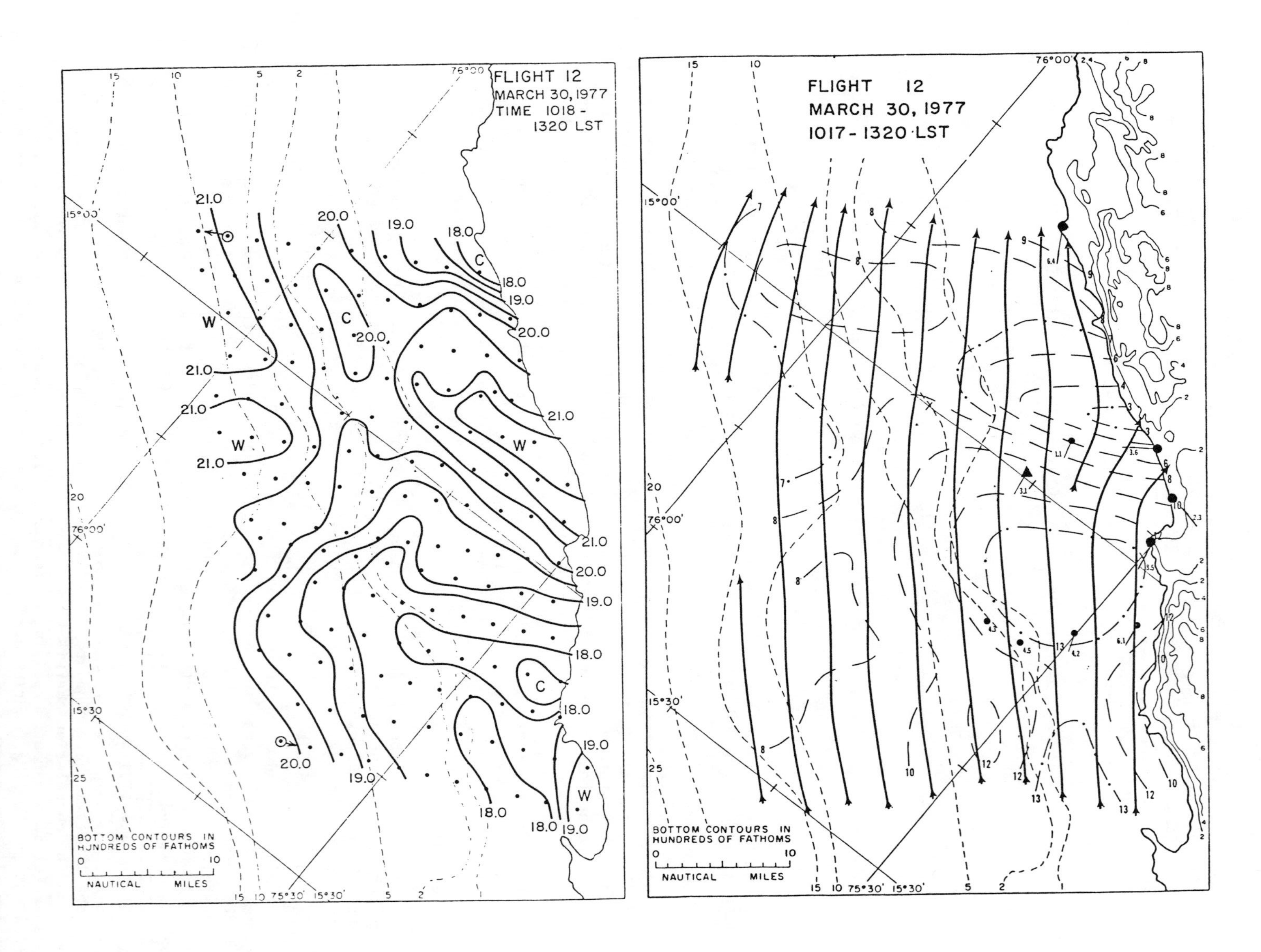

Fig. 4. Maps of sea-surface temperature (^{o}C) (left panel) and winds (at 152 m) (right panel) for 30 March, 1977. Conventions as in Fig. 2.

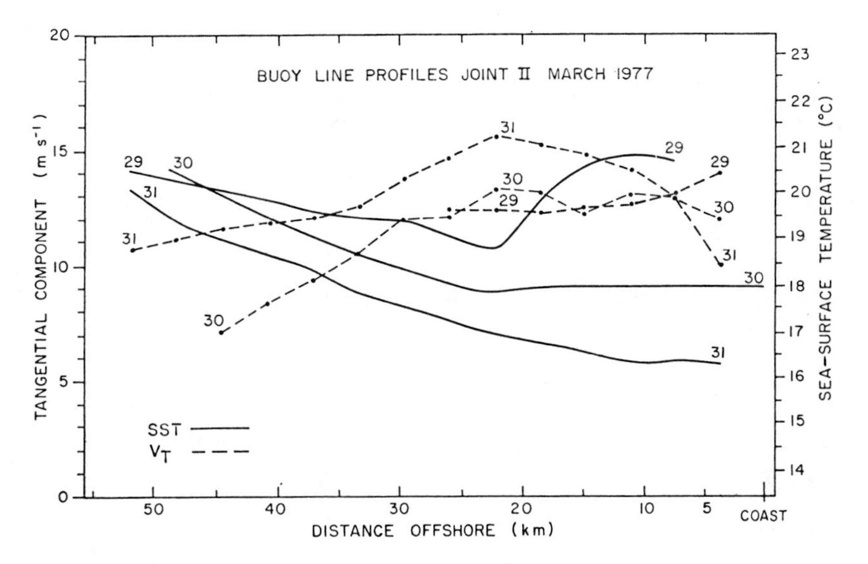

Fig. 5. Buoy line profiles of sea-surface temperature (°C) and alongshore wind, V_T (m/s at 152 m) for 29, 30, 31 March 1977.

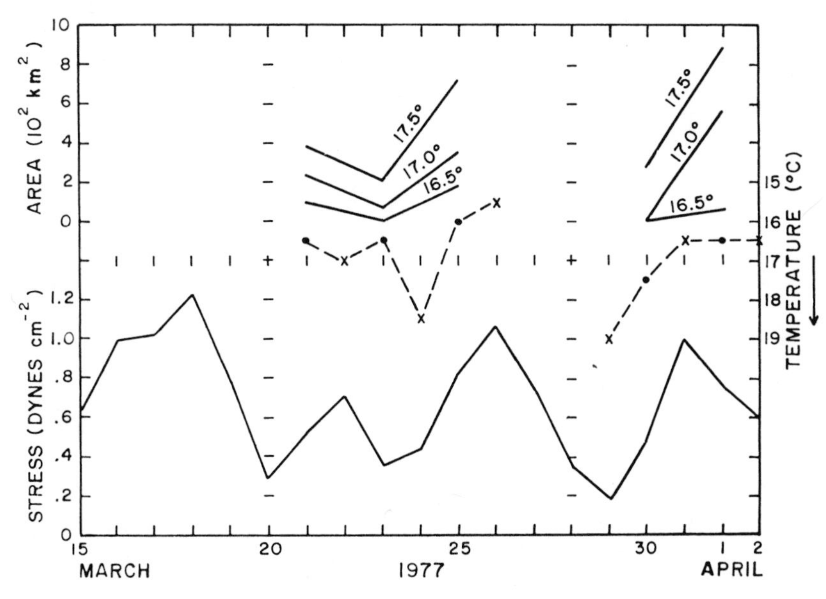

Fig. 6. Time plots of alongshore stress, lowest sea-surface temperature and area within 16.5°C isotherm. The stress is from buoy PSS data (Brink *et al.*, 1978) averaged over 24 h ending at 1200 GMT each day. The lowest SST, to the nearest 0.5°C, is from the aircraft flights. Dots refer to mapping flights and x refers to buoy line profile flights.

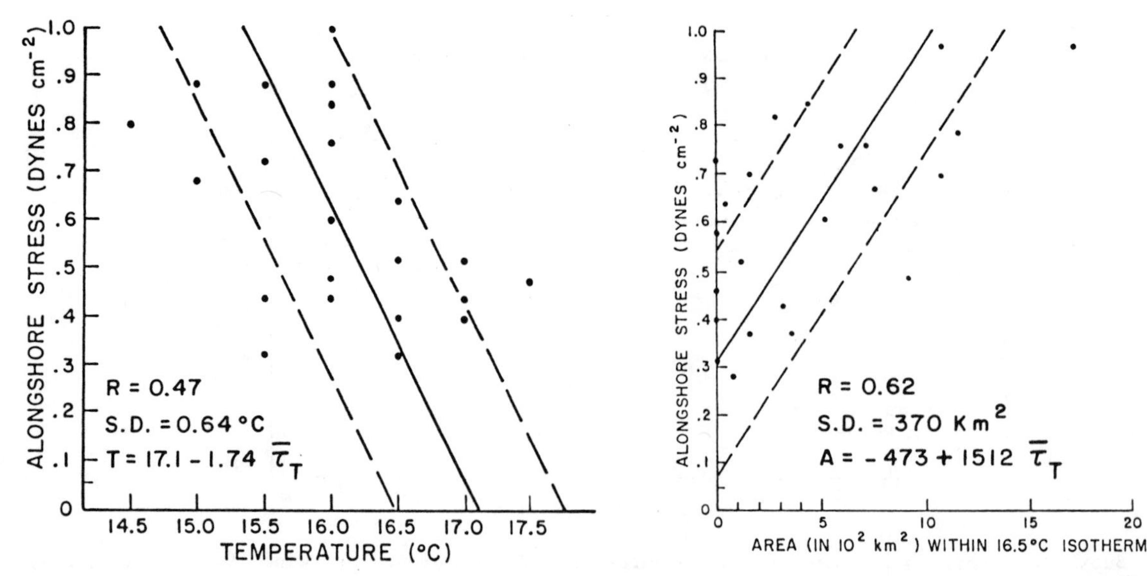

Fig. 7. Scatter diagrams of alongshore stress vs. lowest SST (left panel) and area within 16.5°C isotherm (right panel) for JOINT II, 1976 and 1977. Alongshore stress is for San Juan winds averaged over 24 h (left panel) and 48 h (right panel) prior to each flight.

who assisted in the field work and data processing, to Dewey Rudd for the drafting, and to Patti Mullinax for typing. This is a contribution to the CUEA Program, a program of the International Decade of Ocean Exploration (IDOE) funded by the National Science Foundation under Grant OCE 77-27735.

References

Brink, K.H., R.S. Smith, and D. Halpern, A compendium of time series measurements from moored instrumentation during the MAM 77 Phase of JOINT II, CUEA Technical Report 45, School of Oceanography, Oregon State University, Corvallis, 72 pp. (Available from the National Technical Information Service, Accession No. PB 293296/AS), 1978.

Moody, G.L., D.W. Stuart, A.I. Watson, and M.M. Nanney, Sea-surface temperatures and winds during JOINT II: Part I - Mean Conditions, *This Volume*, 1980.

Stevenson, M., D.W. Stuart, and D. Heburn, Short term variations observed in the circulation, heat content, and surface mixed layer of an upwelling plume off Cabo Nazca, Peru, *This Volume*, 1980.

A COMPARISON OF SURFACE WINDS AND WINDS MEASURED AT 152 m DURING JOINT I 1974 AND JOINT II 1977

David W. Stuart

Robert J. Goodwin

William P. Duval

Department of Meteorology, Florida State University, Tallahassee, Florida 32306

Abstract. During the JOINT I 1974 and JOINT II 1977 expeditions, winds were observed at the surface by meteorological buoys and at 152 m by a research aircraft. The surface and aircraft data were screened for proximity in space and time and data pairs were extracted if the aircraft flight track was within 2.8 km of the buoy location. Scatter diagrams were plotted for surface wind components vs. aircraft wind components.

An analysis of variance for the JOINT I data showed that a linear relation between the two levels explained about 84% of the variance for the component normal to shore and about 54% of the variance for the component parallel to shore. For the JOINT II data, about 76% of the variance was explained for the parallel component and virtually none of the variance was explained for the normal component.

Introduction

During the JOINT I 1974 and JOINT II 1977 field observations by the Coastal Upwelling Ecosystems Analysis (CUEA) group, an instrumented aircraft was used to measure winds at 152 m. This paper examines the relationship between those winds and winds measured at the surface by an array of meteorological buoys. In earlier works, Halpern (1978, 1979) and Krishnamurti and Krishnamurti (1978) examined the relationship between low-level cloud motion vectors and surface winds over open ocean regions. This study differs in that the study areas are coastal regions subject to pronounced orographic and diurnal effects.

Data Availability and Reduction

JOINT I 1974 was conducted off northwest Africa (Fig. 1) during February and March, 1974 while JOINT II 1977 was off Peru (Fig. 2) during March, April, and May, 1977. Both locations have pronounced coastal upwelling.

The wind field aloft for each region was measured by an instrumented research aircraft flying at a mean height of 152 m (500 ft) as determined by a radio altimeter (Sperry Rand model AA-220, overall accuracy of $\pm$ 10 m). The standard deviation of the height was 5.8 m (19 ft) and no observations were used if the flight level deviated by more than 18 m (59 ft) from the mean. The flight patterns and initial data processing were described by Duval (1977) and Moody (1979).

Two types of moored buoys provided surface wind measurements. Buoys Li(1974), PSS(1977), and PS(1977) had vector averaging wind recorders (VAWR) at 3 m constructed from AMF Model 610 vector averaging current meters. Their basic recording interval was 7.5 min. All other buoys had Aanderaa Model D124 wind recorders at 2.4 m. The Aanderaa recorders average the wind speed during the recording interval, but they yield the instantaneous wind direction at the end of the recording interval. The recording interval varied from 5 or 10 min (JOINT I) to 20 min (JOINT II). Intercomparison studies (Halpern and Smith, 1975) indicated that unfiltered winds from the two buoy systems should be used with caution. The JOINT II study used data mostly from the VAWR system (buoys PSS and PS) and the inclusion of data from the one Aanderaa system (buoy Pa) does not appear to have improved the results.

Whenever the aircraft flight track passed within 2.8 km (1.5 nm) of a surface buoy mooring, the aircraft and buoy wind measurements nearest in time and space were extracted as a data pair. Aircraft winds were available at 1-min intervals. The buoys with VAWR had averaged data at 15- (PSS and PS) and 30-min (Li) intervals. The buoys with Aanderaa recorders recorded winds at 20-min intervals for JOINT II (Pa, L, I), but the wind values for JOINT I (R1, R2, U1, U2) were available each 5 to 10 min even though they were vectorially averaged over 20 to 30 min before being used for a data pair. Therefore, the largest interval in space and time allowed for a data pair to be used was 2.8 km and 15 min.

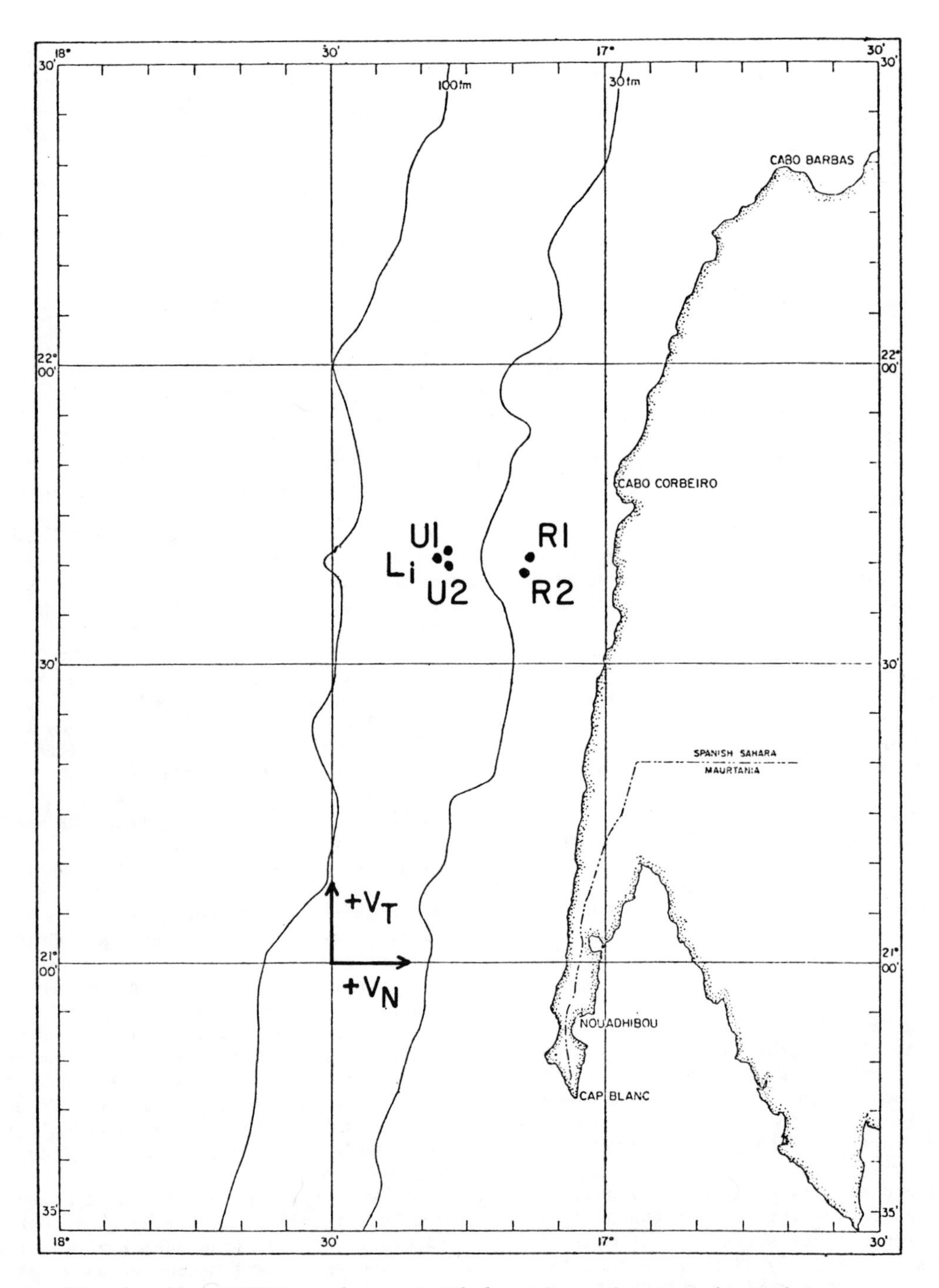

Fig. 1. The JOINT I study area with locations of meteorological buoys.

Winds were decomposed into a component parallel to the coast (V-tangent or V_T) and a component perpendicular to the coast (V-normal or V_N). For JOINT I, V_T was positive when the tangential component was blowing from the south and V_N was positive when the normal component was from the west (onshore)(Fig. 1). For the JOINT II region, V_T was positive when it was from the southeast and V_N was positive from the southwest (onshore) (Fig. 2).

Scatter diagrams were plotted for the total wind speed, and, if possible, for V_T and V_N for each buoy. Additional scatter diagrams were plotted for the sets of buoys that reported both speed and direction. For JOINT I these were the second Rhododendron mooring (R2), both Urbinia

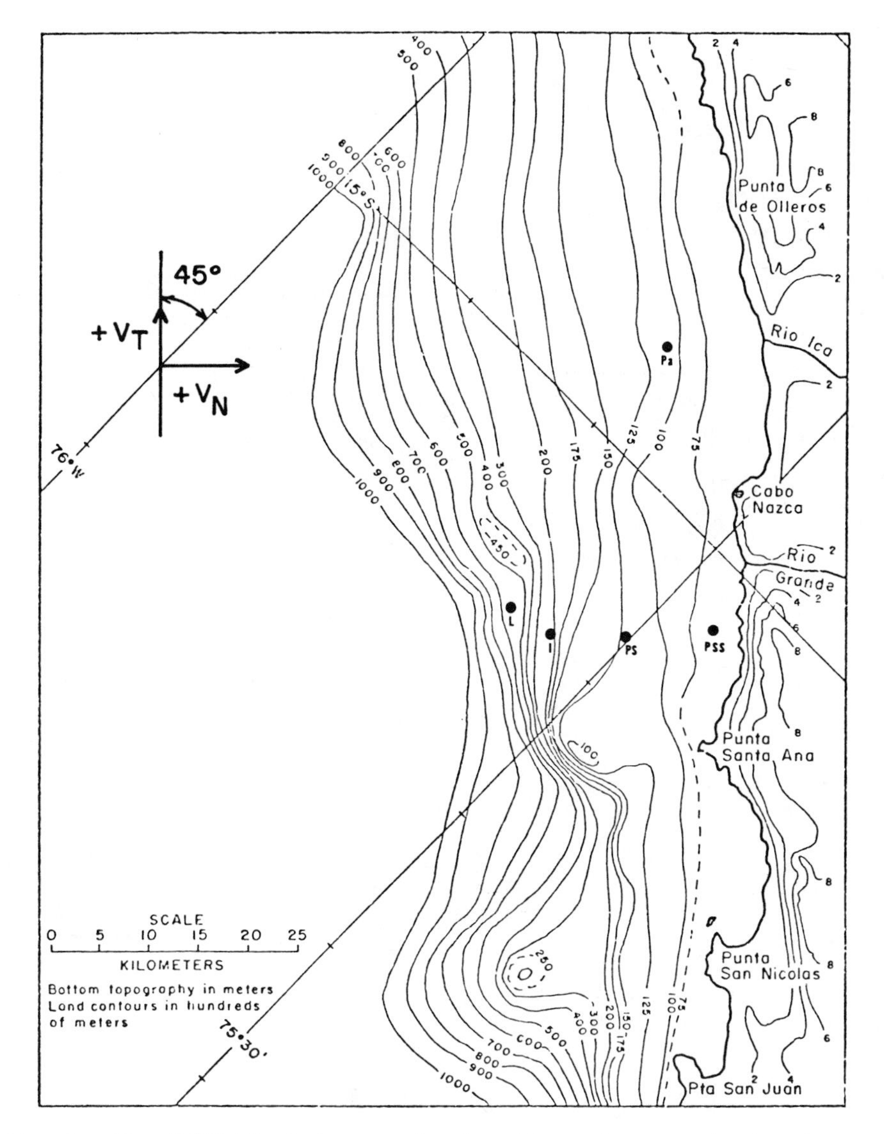

Fig. 2. The JOINT II study area with locations of meteorological buoys.

moorings (U1, U2), and Lisa (Li). For JOINT II they were PSS, PS, and Parodia (Pa). Buoys providing only wind speed were the first Rhododendron mooring (R1) from JOINT I and Ironwood (I) and Lagarta (L) from JOINT II. A corresponding series of analysis of variance was calculated. In the following we present results from only those buoys that recorded both speed and direction.

Results

Figs 3 and 4 are scatter diagrams of total wind speed recorded by the buoys versus total wind speed observed by the aircraft for JOINT I and JOINT II. Figs 5 and 6 show the regression lines for total wind speed, V_T and V_N, for each expedition. With the exception of V_N for JOINT II, each scatter diagram gives some indication of a linear relationship between the winds at the surface and at 152 m.

The results of the analyses of variance are given in Table 1. In addition to the slope and intercept of the regression lines, the table gives the correlation coefficients (r), the square of the correlation coefficient (r^2), the "f" ratio, and the corresponding critical values of f at the 95% (f_5) and 99% (f_1) confidence levels (from Panofsky and Brier, 1968).

The statistics support linear relationships in most cases. The linear relations between wind measured at the surface and at 152 m explain 54% of the variance for V_T and 84% of that for V_N during JOINT I. Also, 71% of the variance in total wind speed is explained and all three corresponding correlation coefficients are signifi-

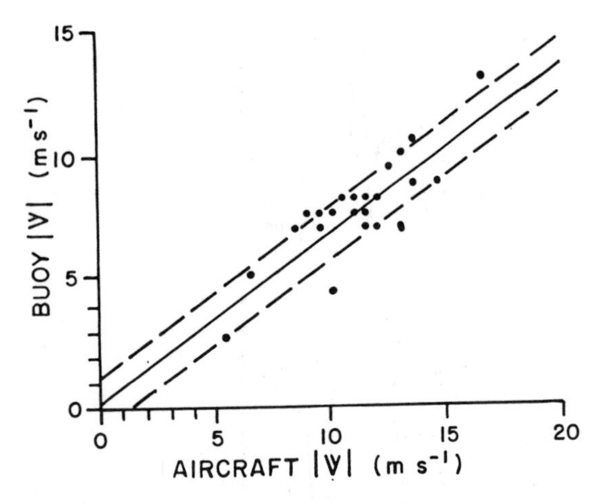

Fig. 3. Scatter diagram of buoy total wind speed ($|\underset{\sim}{V}|$) vs. aircraft total wind speed ($|\underset{\sim}{V}|$) for JOINT I. All buoys except R1 are included. Dashed lines are ± one standard deviation from regression line.

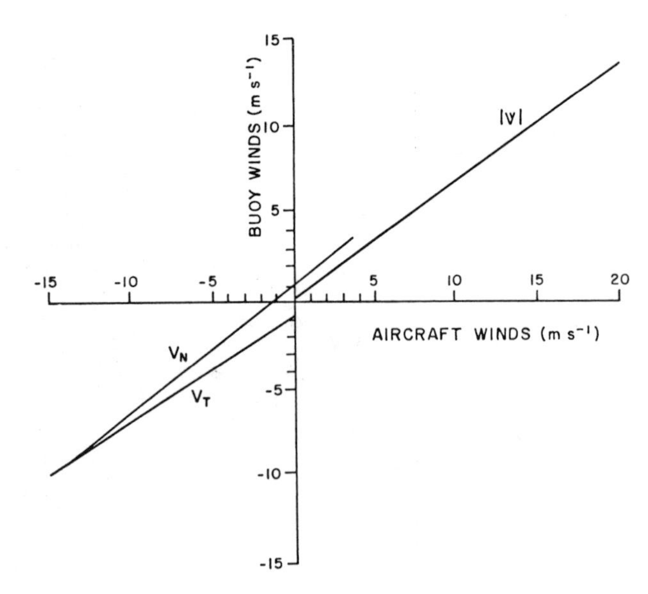

Fig. 5. Regression lines for buoy total wind speed ($|\underset{\sim}{V}|$) V-tangent (V_T) and V-normal (V_N) vs. aircraft ($|\underset{\sim}{V}|$), V_T and V_N for JOINT I. The equations of the regression lines are:

Buoy $|\underset{\sim}{V}|$ = 0.20 + 0.67 Aircraft $|\underset{\sim}{V}|$
Buoy V_T = -0.97 + 0.61 Aircraft V_T
Buoy V_N = 1.02 + 0.74 Aircraft V_N.

cant at the 99% confidence level. For JOINT II, the explained variances for V_T, V_N, and total speed are 76, 1, and 68%, respectively. For V_T and total speed, correlation coefficients are again significant at the 99% confidence level. However, a linear relation explains virtually none of the variance and the correlation coefficient is not significant for V_N.

Conclusions

The results suggest that it may be possible to estimate the surface wind field in coastal regions from aircraft measurements of the low-level wind field. Such estimates, however, must take into account regional differences. The JOINT I and JOINT II regions show significant differences. In the JOINT I region, more of the variance was explained by a linear relationship for the normal component than for the tangential component

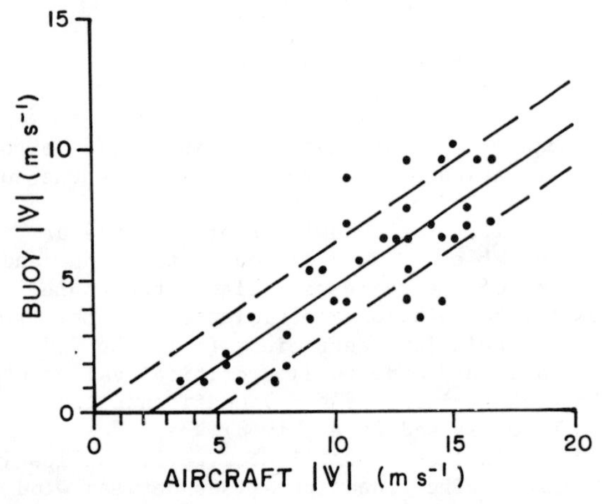

Fig. 4. Scatter diagram of buoy total wind speed ($|\underset{\sim}{V}|$) vs. aircraft total wind speed ($|\underset{\sim}{V}|$) for JOINT II. Includes PSS, PS, and Pa. Dashed lines are ± one standard deviation from the regression line.

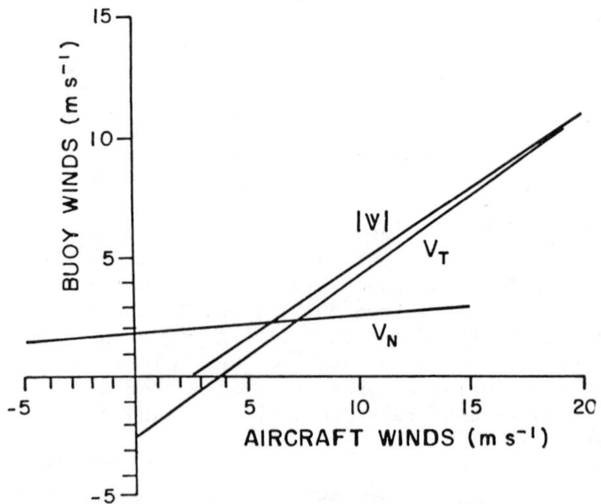

Fig. 6. Regression lines for buoy total wind speed ($|\underset{\sim}{V}|$) V-tangent (V_T) and V-normal (V_N) vs. aircraft ($|\underset{\sim}{V}|$), V_T and V_N for JOINT II. The equations of the regression lines are:

Buoy $|\underset{\sim}{V}|$ = -1.26 + 0.59 Aircraft $|\underset{\sim}{V}|$
Buoy V_T = -2.43 + 0.65 Aircraft V_T
Buoy V_N = 1.98 + 0.07 Aircraft V_N.

TABLE 1. Aircraft vs. Buoy Comparisons for JOINT I 1974 and JOINT II 1977.

Experiment	Intercept	Slope	r	r^2	f	f_5	f_1
			Total Speed				
JOINT I	0.20	0.67	0.84	0.71	57.63	4.26	7.82
JOINT II	-1.26	0.59	0.82	0.68	78.06	4.11	7.39
			V_T				
JOINT I	-0.97	0.61	0.74	0.54	28.19	4.26	7.82
JOINT II	-2.43	0.65	0.87	0.76	116.44	4.11	7.39
			V_N				
JOINT I	1.02	0.74	0.92	0.84	124.77	4.26	7.82
JOINT II	1.98	0.07	0.10	0.01	2.38	251.00	6286.00

JOINT I data consist of R2, U1, U2, and Li. JOINT II data consist of PSS, PS, and Pa. Figs 1 and 2 give buoy locations.

while the opposite was true for the JOINT II region. V_N was generally offshore during JOINT I and larger than during JOINT II, where V_N was primarily onshore. Thus, a sea-breeze circulation in the JOINT I region was manifested as a reduction in the magnitude of V_N. In the JOINT II region, however, the onshore component near the coast decreased with height and reversed near 960 mb (Wirfel, 1980). This reversal in the flow near 152 m contributes to the lack of a linear relationship for V_N in the JOINT II region because day-to-day fluctuations in the depth of the sea-breeze circulation tend to increase the scatter of the sample. These results are based on flights that took place during mid-morning and mid-afternoon during JOINT I and around mid-morning during JOINT II. The flights thus took place at times when developing sea breeze circulations could be expected to influence the correlations. Even so, linear relationships generally explain a large percentage of the variances in both regions during the daytime hours.

Acknowledgements. **Thanks are extended to R.L. Smith and David Halpern, who provided the buoy data, and to Dewey Rudd for technical assistance. We are indebted to the Research Aircraft Facility of the National Center for Atmospheric Research, which provided the aircraft support.**

This is a contribution to the CUEA Program, a program of the International Decade of Ocean Exploration (IDOE) sponsored by the National Science Foundation (NSF) under Grant OCE 76-82831.

References

Duval, W.P., Aircraft observed winds over a coastal upwelling region, CUEA Technical Report 29, Department of Meteorology, Florida State University, Tallahassee, 120 pp. (Available from National Technical Information Service, Accession No. PB267369/AS), 1977.

Halpern, D., Comparison of low-level cloud motion vectors and moored buoy winds, *Journal of Applied Meteorology, 17*, 1866-1871, 1978.

Halpern, D., Surface wind measurements and low-level cloud motion vectors near the intertropical convergence zone in the central Pacific Ocean from November 1977 to March 1978, *Monthly Weather Review, 107*, 1525-1534, 1979.

Halpern, D. and R.L. Smith, On near-surface wind observations, Part III, CUEA Newsletter 4(1), 2-3, (Available from Duke University Marine Laboratory, Beaufort, N.C.), 1975.

Krishnamurti, T.N. and R. Krishnamurti, Surface meteorology over the GATE A-scale, Technical Report 78-6, Departments of Meteorology and Oceanography, Florida State University, Tallahassee, 45 pp., 1978.

Moody, G.L., Aircraft derived low level winds and upwelling off the Peruvian coast during March, April, and May 1977, CUEA Technical Report 56, Department of Meteorology, Florida State University, Tallahassee, 110 pp. (Available from National Technical Information Service, Accession No. PB80-119571), 1979.

Panofsky, H.O. and G.W. Brier, *Some Applications of Statistics to Meteorology*, University Park, Pennsylvania, 224 pp., 1968.

Wirfel, W.P., The atmospheric structure over the Peruvian upwelling region, CUEA Technical Report 63, Department of Meteorology, Florida State University, Tallahassee, 134 pp. (Available from National Technical Information Service), 1980.

SOUTHERN CALIFORNIA SUMMER COASTAL UPWELLING

Clive E. Dorman

David P. Palmer

Department of Geological Sciences
San Diego State University
San Diego, California, U.S.A.

Abstract. The conditions for significant summer upwelling along the Southern California Bight are studied. Background information and previous studies are reviewed. Significant upwelling events occur twice a summer. They are simultaneous only from La Jolla to Balboa. Local winds occur at the same time and place as the significant upwelling events and force the summer upwelling.

Introduction

Various authors have discussed different conditions of summer upwelling in the Southern California Bight. In this paper, we review the conditions and introduce our conclusion about the cause of the upwelling. In particular, we focus on the conditions causing local upwelling over the shelf.

General Climatology

One of the conditions affecting upwelling is climatology. To get a general idea of the climatology of the entire California Bight area, we refer to Emery (1960) and Jones (1971). The major feature of the bight is a counter-clockwise gyre whose principal source of water is the California current. Reid, Roden, and Wylie (1958) made an extensive study of this southward-flowing current that brings cool, low-salinity water past Point Conception to 31 or 32°N. At this latitude, some of the California current turns east and then north. The northward current, called the Southern California Countercurrent, SCC, stays at least 20 km offshore. However, in the spring the northward-flowing countercurrent is replaced by the southward-flowing California current, which moves further to the east than during the rest of the year.

Inshore of the Southern California Countercurrent is another current. Tsuchiya (1980) defines "inshore" as within 20 km of the coast and notes that if the offshore current is weak, the inshore current is in the opposite direction. If the offshore current is strong, the inshore current is in the same direction. The seasonal march of the inshore current direction and speed is correlated with the southerly wind stress at 34°N, 120°W.

Currents in and around the Southern California Bight are largely wind-driven. Broad-scale features and large-scale variations of the currents are determined by the large-scale wind stress (Reid *et al.*, 1958). Munk (1950) related currents west of Catalina Island to the curl of the wind stress to explain the seasonal variation in the California Current and the SCC.

Besides the surface currents, there is a deep undercurrent that moves toward the north through the bight, details of which were reviewed by Jones (1971) and Tsuchiya (1976, 1980). The undercurrent forms to the south and brings high-salinity water into the bight at depths between 100 and 300 m. The main core of the poleward-flowing undercurrent stays within a few kilometers of the continental slope.

Surface currents determine the general pattern of surface properties in the bight. Isotherms and isohalines bend southward in the California current (Reid *et al.*, 1958) and northward in the California Countercurrent. A local temperature maximum appears in the surface temperature fields offshore between San Diego and Newport in the summer and through autumn (see Fig. 4 for a typical example). At the same time, a narrow band of cold water is isolated over the shelf south of Los Angeles (Jones, 1971). This water is presumed to be due to upwelling. Also, in a narrow band from about Newport to 100 km south of San Diego, Reid *et al.* (1958) indicated a surface salinity minimum in August.

As the California Current moves south along the central California coast, horizontal mixing causes the water in the upper tens of meters to increase in salinity. By the time the California current swings east, there is a shallow salinity minimum at the 100-m depth (Reid *et al.*, 1958). This minimum, which shoals to the east, is advected around the bight at 30 to 50 m (Reid and Worrall, 1964) and may be clearly seen in the hydrographic data from over the shelf just north of San Diego (Tsuchiya, 1976). Water from this salinity minimum is brought to the surface during upwelling, causing the surface salinity to decrease. This is unusual for the Southern California Bight because, further north, upwelling is associated with increased salinity.

Variations

Other variations in upwelling in the Southern California Bight have been observed. For example, on a climatic scale, there was a 5.6-cm increase in sea level at Scripps Institution of Oceanography (SIO), La Jolla, California, between the periods 1948-1957 and 1958-1969 (Namias and Huang, 1972). The change was due to a reduction in the northerly wind stress on the eastern Pacific that allowed more eastward-moving water with its higher temperatures to occupy this area. The higher temperatures increased the oceanic volume and raised the sea level.

Many authors have reported anomalous conditions in the bight lasting months to a year. The long record at Scripps has allowed authors to relate temperature variation to changes in the biology. Jones (1971) reviewed a number of such studies.

Monthly anomalies of water temperature and salinity are weakly coherent at stations along the California Bight (Roden, 1961). At Scripps, monthly anomalies of temperature and salinity tend to be in phase (mostly less than a 60° difference) at periods of less than one year. For these periods, cold water is associated with lower salinity. The association between cold water and lower salinity could be due to more cold, low-salinity water advected by the California current. The other possible cause is upwelling, because both cold and low-salinity water is just below the surface. The role of the two mechanisms cannot be further clarified by the coastal station data alone.

Another related variation in upwelling was observed by Enfield and Allen (1980), who did a spectral analysis of monthly mean sea level, temperature, and wind stress along eastern Pacific coastal stations, including San Diego. Consistent with other investigations, they found that annual and inter-annual variations in sea level at San Diego are correlated with other stations and with large-scale events in the South Pacific. The correlation length scale along the coast decreases as the time scale decreases for periods less than several months. This re-expresses the observations by Reid *et al.* (1958) that La Jolla is representative of open ocean conditions for periods of months but not necessarily so for time scales very much shorter.

Enfield and Allen (1980) also found evidence for planetary sea-level waves that propagate northward by San Diego. Such planetary waves should have some surface temperature signature although none was observed.

No significant correlation was found between the wind stress and either sea level or temperature on monthly scales. Enfield and Allen (1980) do qualify the latter results by noting that their wind stress estimates may not be completely representative of the actual conditions in the Southern California coastal area.

Stewart (1960) also analyzed water temperature and sea level in the Southern California Bight using monthly values. Stewart's graphs reveal that monthly mean temperature and sea level were actually out of phase from January to May, 1957 and in phase from June, 1957 through March, 1958 for La Jolla and Los Angeles. But Santa Monica and Port Hueneme temperatures and sea levels were in phase from January, 1957 through March, 1958. This suggests that the temperature and sea level relationships are more complicated south of Santa Monica than those at and north of Santa Monica. The suggestion might be tempered by the fact that no corrections in the sea level for changes in atmospheric pressure were made.

Roden (1972) computed the divergence of the Ekman transport off California based upon surface wind stress. The surface wind stress was constructed from monthly mean sea level atmospheric pressure derived from a 5° grid and refined to a 1° grid by spline fitting. Significant upwelling was predicted within about 200 km of the Southern California Bight coast for spring through summer with the strongest upwelling occurring in spring. Because of the agreement between the surface temperature distribution and the predicted upwelling pattern, Roden concluded that large-scale upwelling is largely determined by the divergence of the Ekman mass transport.

Reid *et al.* (1958) suggested that surface advection by the major currents, which also respond to the seasonal wind field, might also produce the observed seasonal surface temperature patterns. When using the surface pressure at 5° grid points, some of the eastern pressure points are over interior valleys or deserts separated from the coast by mountain ranges. As noted by Enfield and Allen (1980), surface winds inferred from large-scale pressure patterns in this area will probably differ significantly from the actual surface wind. Questions about surface advection and the suitability of the wind calculation leave the role of large-scale upwelling uncertain. Still, the arguments of other authors (Reid *et al.*, 1958; Jones, 1971) and Roden's calculation suggest that large-scale upwelling should have a significant effect in the California Bight.

Weekly averaged conditions at SIO have been presented by Tont (1975, 1976). He assumed that anomalously cold water at Scripps Pier indicates upwelling and found that sudden decreases in water temperature — which he defines as upwelling — coincide with large changes in the diatom biomass. He remarked that upwelling is also coincident with increased surface salinity, citing Reid *et*

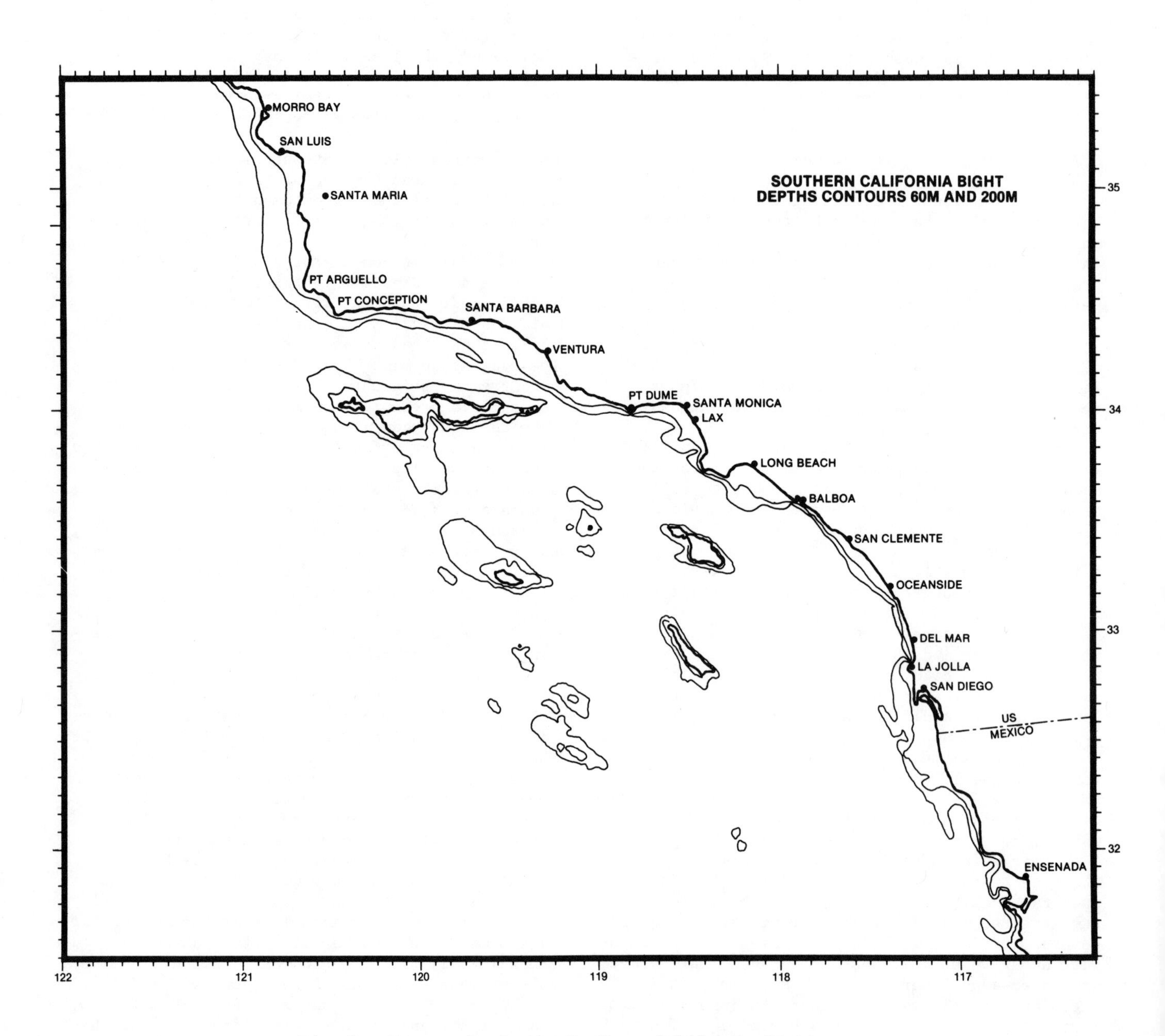

Fig. 1. Topography in the Southern California Bight.

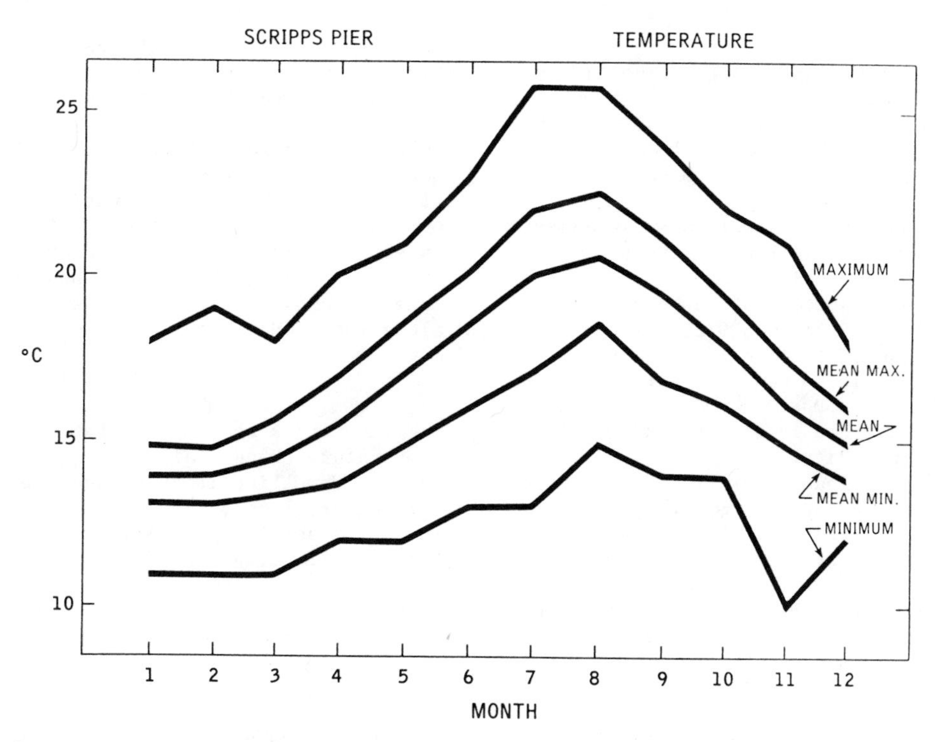

Fig. 2. Scripps Pier surface temperatures based upon the years 1945 through 1964.

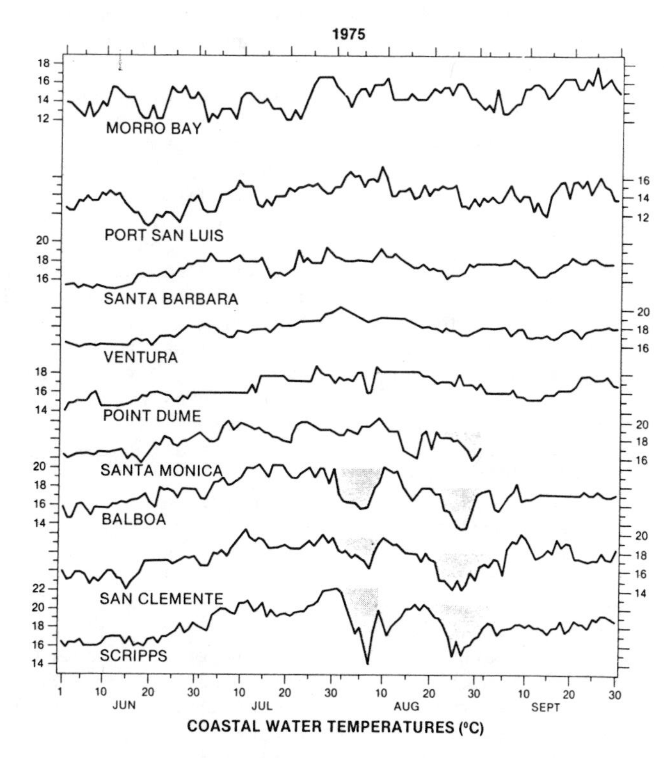

Fig. 3. Summer 1975 coastal water temperatures for stations along the Southern California coast. The shading highlights major upwelling events.

al. (1958) and Tont (1975).

For short-term, local upwelling, as noted earlier, continuous salinity records have shown that there is a salinity minimum at 30 to 50 m in the Southern California Bight, and local coastal upwelling results in a salinity decrease at the surface.

Tont tried to show that monthly mean salinity is related to Bakun's (1973) wind stress index. However, the winds he used were 90 miles off the coast and not representative of inshore winds. Reid *et al.* (1958) and Roden (1972) pointed out that winds increase significantly offshore.

Tont's conclusion is in conflict with Enfield and Allen (1980), who observed no significant relationship between the local wind and sea level anomalies at Scripps Pier.

Reid *et al.* (1958) found that the SCC variations account for the summer surface salinity increases that Tont noted. Tsuchiya (1980) observed that the significant increase in the SCC in the late spring is coincident with the wind stress. Therefore, Tont's correlation between the offshore wind stress and the surface salinity at Scripps is really more indicative of how the surface water responds to seasonal wind forcing.

Extensive observations of the thermal structure have been made at a tower in 18 m of water 1.6 km offshore between Scripps Pier and San Diego. This site is on a gently sloping shelf and should be representative of the conditions in the Scripps area. In a report on the seasonal thermocline cycle, Cairns and La Fond (1966) found that the sharpest gradients occur in late June and were as great as 1.27°C/m. The temperature variance is greatest in the late summer.

At the same tower, Cairns and La Fond (1966) found that simple wind-induced Ekman transport and tides account for all of the significant variation of the seasonal thermocline at periods from hours to months. They were able to formulate a simple linear relationship between the winds and the non-tidal component of the thermocline variation. The relation includes a 14- to 16-h response time of the ocean to the surface winds measured at the San Diego airport. The non-tidal thermocline motions include a diurnal oscillation and an oscillation at 4.5 days — both wind induced. In retrospect, no significant variations were observed in the four-month study between May through August, 1961 that were not locally correlated with the wind. There was no evidence that remotely forced planetary waves caused significant thermocline variations. This study suggests that the thermal structure off Scripps Pier is forced by the local wind.

A few bathythermograph studies have been made in and around the La Jolla to Oceanside area (Arthur, 1954, 1960; Leipper, 1955). Typical summer surface diurnal ranges are on the order of 1°C with extremes limited to less than 2.2°C. There is a strong vertical temperature gradient that is about 4°C over the upper 9 m at the end of Scripps Pier and sometimes is as much as 7°C. Diurnal temperature changes at the base of Scripps Pier (at 9-m depth) are of the order of 5°C.

Leipper concluded that variation away from the surface is largely due to tidal stirring. Arthur (1960) found that surface variations are coherent over 5 km but not over the 40-km separation between Oceanside and Scripps Pier. These studies show that tides cannot be responsible for coherence of temperature changes between stations along the bight.

Coastal Upwelling

We must look to other mechanisms to explain temperature changes along the Southern California Bight. Upwelling is an obvious mechanism, in particular, local, wind-driven upwelling. The review of upwelling by Smith (1968) provides the theoretical background for our discussion. The primary data to investigate upwelling are pier and beach station observations because there is a lack of detailed measurements over the shelf. The coastal stations also provide long and continuous records (Figs 2, 3). Location of stations and selected bathymetry are shown in Fig. 1.

Temperatures at Scripps Pier (Fig. 2) are typical of the Southern California coastal stations. The average temperature, the mean monthly maximum, mean minimum, and the absolute maximum and minimum are constructed for the years 1945 through 1964 (U.S. Dept. of Commerce, 1967). The range between the mean maximum and minimum is greatest in the summer.

Summer water temperature at Scripps Pier in an individual year is occasionally marked by sudden drops. The records suggest that these cold periods are significant events when they last more than six days and the maximum temperature drop is more than 3°C. It is relatively easy to use both temperature and time to establish unambiguously when an event has occurred, although the criteria do not indicate when smaller scale events have occurred.

The cold water events occur along the coast from San Diego to Balboa (see Fig. 4 for the 1975 case) within one day, so the typical 10 cm/s coastal current (Tsuchiya, 1980) could not have simply advected the cold water along the coast. The nearest surface waters of only somewhat lowered temperature were more than 100 km away.

Fig. 4 shows the typical temperature field observed on a California Cooperative Oceanic Fisheries Investigations hydrographic survey between June 24 and July 9, 1975. The shelf between San Diego and Balboa is isolated from any offshore upper layer cold water source. Therefore, simple onshore advection cannot account for the appearance of the cold water off the San Diego to Balboa section.

Cold water is less than 10 m below the surface (Fig. 5). Cold events tend to be accompanied by

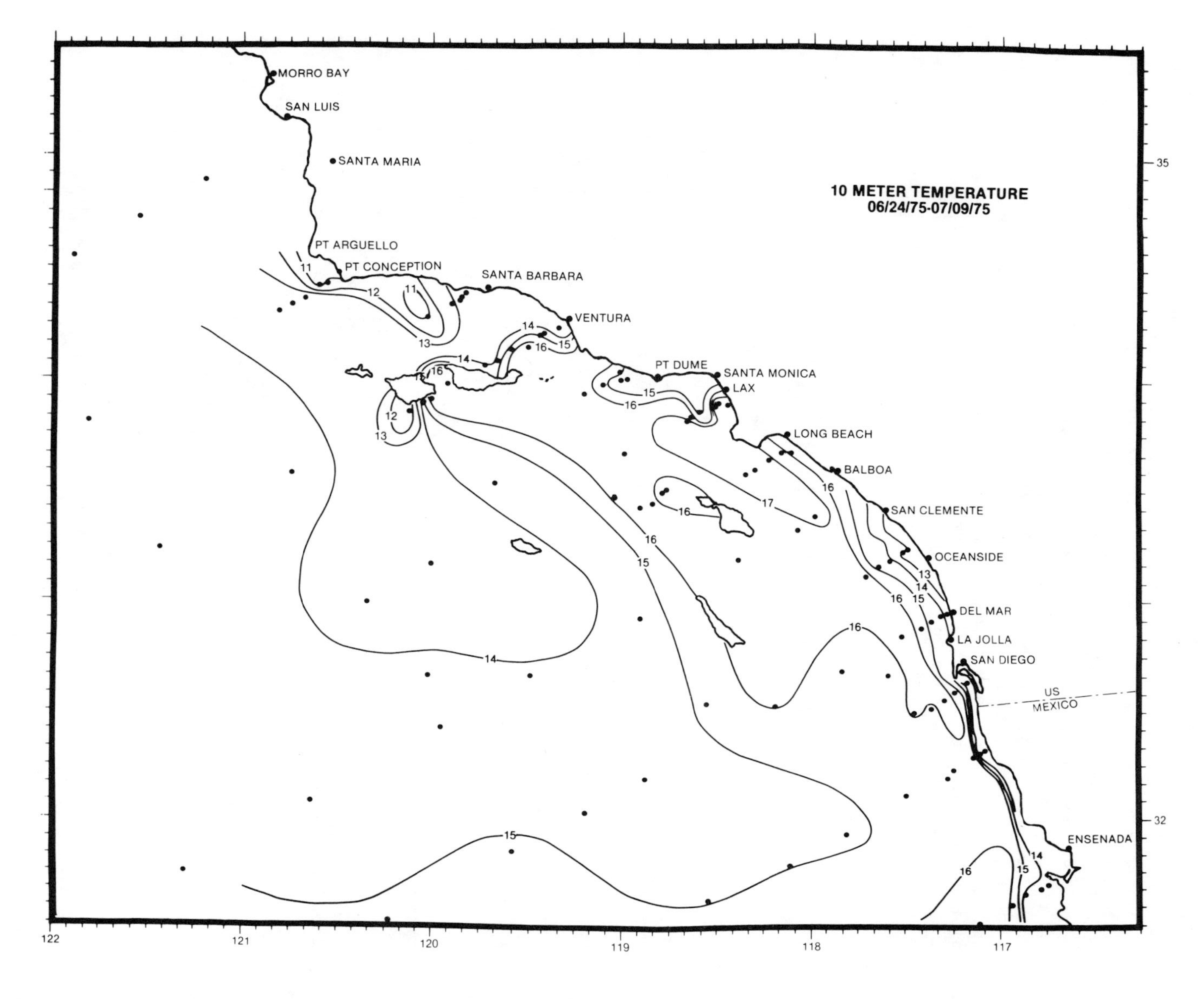

Fig. 4. Ten-meter water temperatures observed between June 24 and July 9, 1975. Lower temperatures over the shelf between San Diego and Balboa are typical of the summer, although the cold water did not reach the surface at this time.

sudden decreases in the surface salinity (0.1 to 0.2) at Scripps Pier. The source of this less saline water is the salinity minimum just below the surface (Fig. 6). Because water with the lower salinity and temperature is just below the surface and the lower temperatures could not have moved in on the surface, these major cold events are assumed to be upwelling.

Major upwelling events occur on an average of twice a summer (Table 1). The maximum temperature change is 4.7°C and the average duration is 11.4 days. Of 20 identified events, 16 were coherent and occurred simultaneously from La Jolla to Balboa, but only once was a similar temperature change seen at Santa Monica. Stations further upcoast do have significant cold events, but they are not coincident with those in the La Jolla to Balboa section.

There is some evidence that the latter events are locally wind generated by simple Ekman dynamics. That is, along coastal wind stress toward the south will cause water to be transported away from the coast and replaced by colder water from below (Smith, 1968). To test this idea, the San Diego airport winds are compared with the Scripps Pier temperature (Fig. 7); sustained higher winds to the south are coincident with significant upwelling. Examination of other years reveals a similar result: upwelling events are correlated with the local wind field.

The Los Angeles wind stress is much less than that at San Diego in terms of the average, variation, or direction (Fig. 8). There is no obvious increase of the wind at Los Angeles coincident with upwelling events further south. The upwelling events are usually limited to an area south

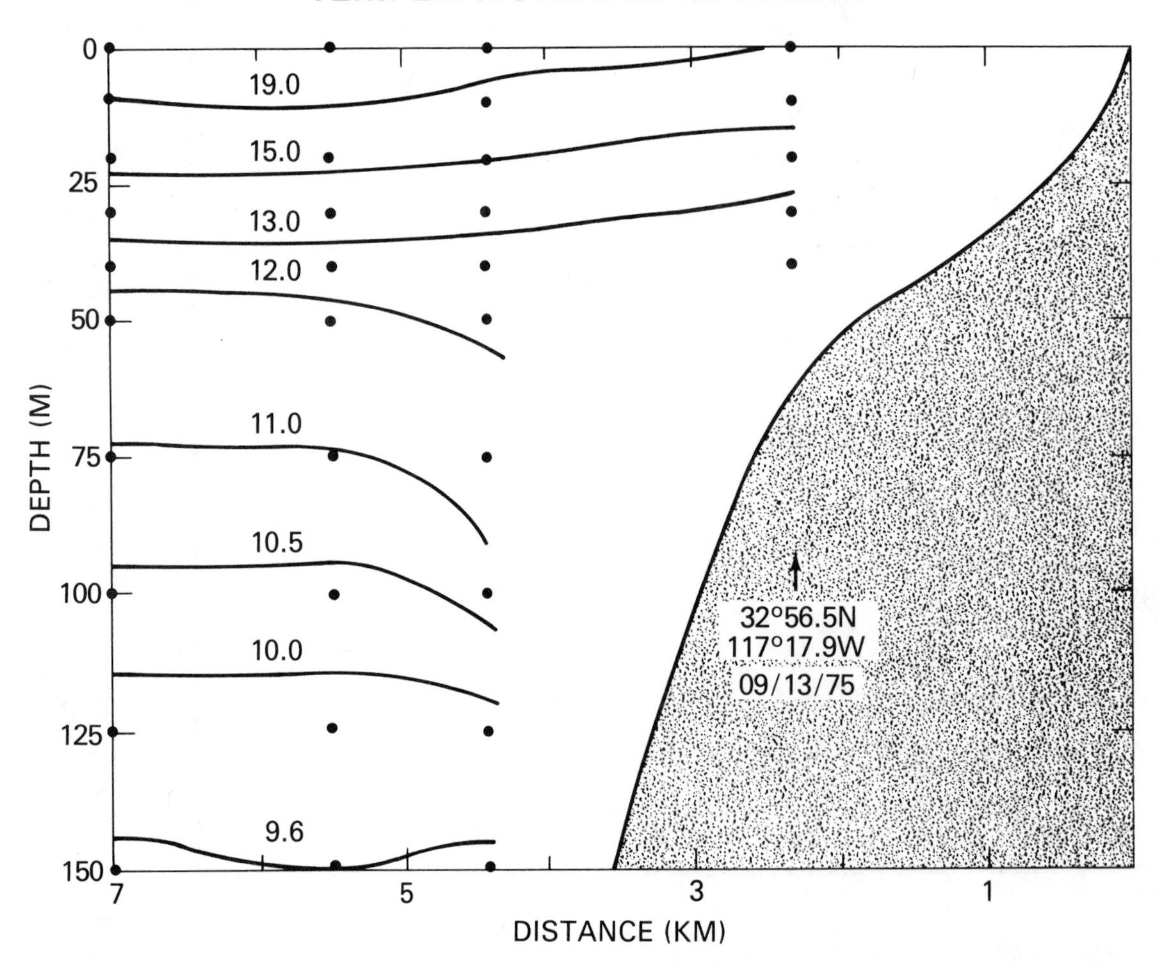

Fig. 5. Temperature section off Del Mar observed about two weeks after a major upwelling event.

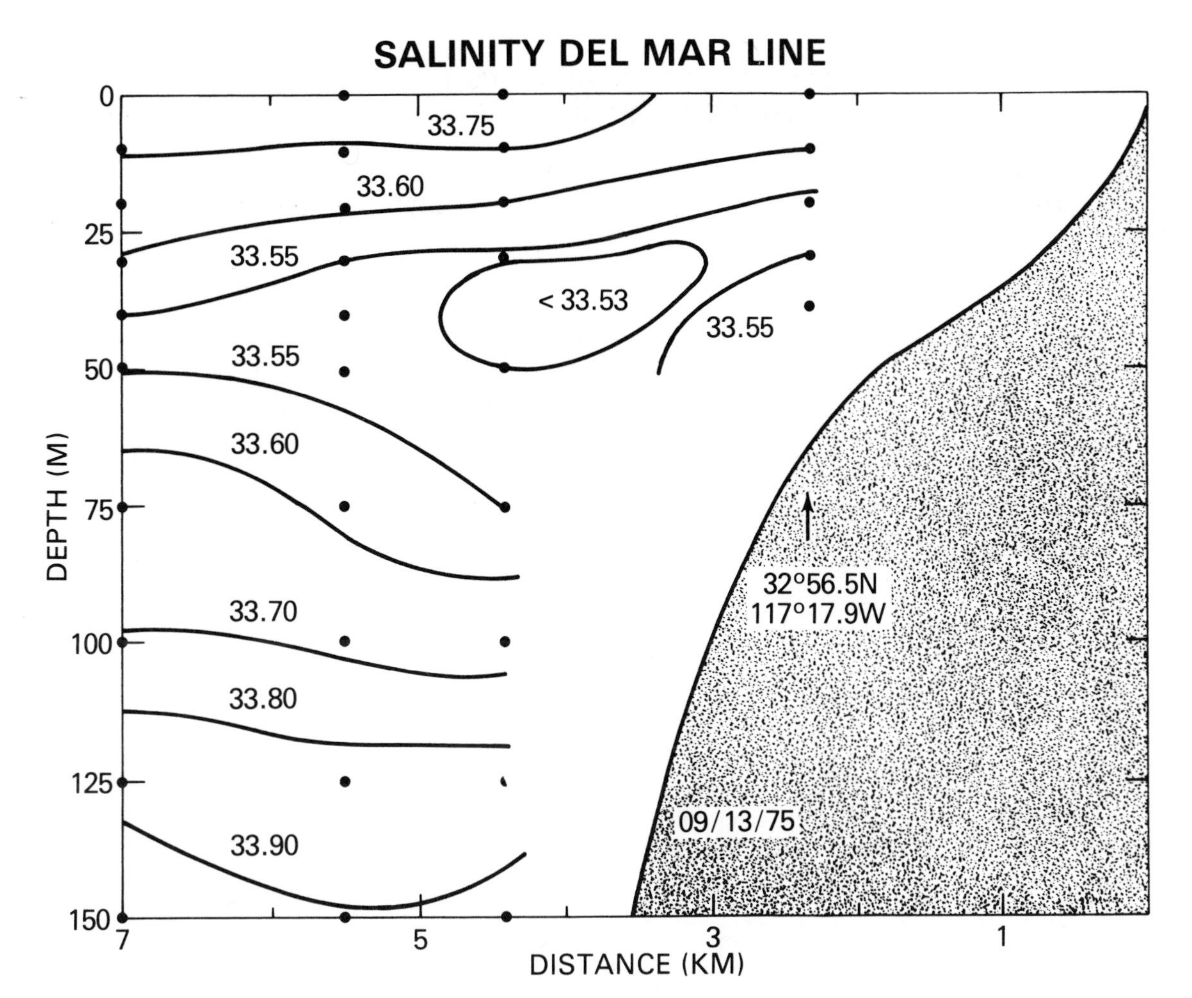

Fig. 6. Salinity section off Del Mar observed at the same time as the temperature data in Fig. 5. Note the salinity minimum around 50 m.

of Los Angeles because this area does not experience the local wind forcing that San Diego does.

Currents over the shelf are correlated with San Diego winds. Tareah Hendricks made current meter measurements off San Diego as part of the Southern California Coastal Water Research Project. His current meter was deployed off La Jolla at 41-m depth in 56 m of water from mid-July through September, 1976. An upwelling event during this period is evident in Fig. 9. Before upwelling, the along-coast wind stress is in phase with the near-bottom current; afterward, it is out of phase. Some transition in the phase relationship occurred at about the start of the upwelling. While it would be difficult to construct an unambiguous model of what occurred, there is a distinct relationship between the local winds and the currents during the event.

If the local winds cause upwelling, they should also cause downwelling. A downwelling signal is difficult to identify from the coastal station records because there is no sharp temperature change as there is with upwelling. But a downwelling case has been documented by Winant (1980). Winant had deployed current meters off Scripps Pier late in one summer when the decaying remnants of a hurricane passed over the coast just to the north. Using the San Diego airport winds, he showed how southerly winds forced the inshore currents and caused downwelling.

The horizontal surface temperature field for a downwelling event is available for a case in 1969. Fig. 10 shows the surface temperature field measured by a Coast Guard infrared radiometer flight on Aug. 15, 1975. For a few days before, the winds at San Diego were from the south, which should cause downwelling. In fact, temperatures at Scripps Pier rose abruptly from Aug. 11 through the 15th, but San Clemente had a cooling trend at the same time and Balboa and Santa Monica had steep temperature drops. By 15 August, the infrared measurements indicated the moderately cooler water over the shelf just north of La Jolla had been replaced by a surface temperature maximum.

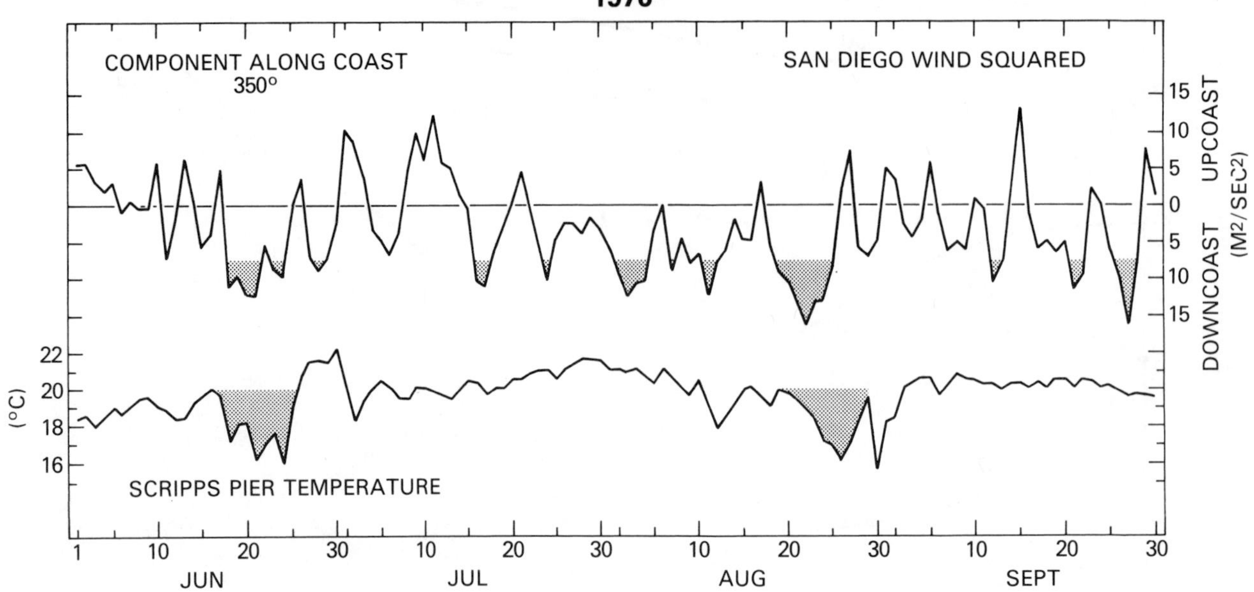

Fig. 7. A comparison of winds at the San Diego airport and surface temperatures at Scripps Pier. The wind is the daily averaged sum of the square of the along-coast component. A southerly wind is blowing toward the south. Shading on the temperature curve highlights a major upwelling event. Shading on the wind curve highlights estimated threshold speeds associated with major upwelling events.

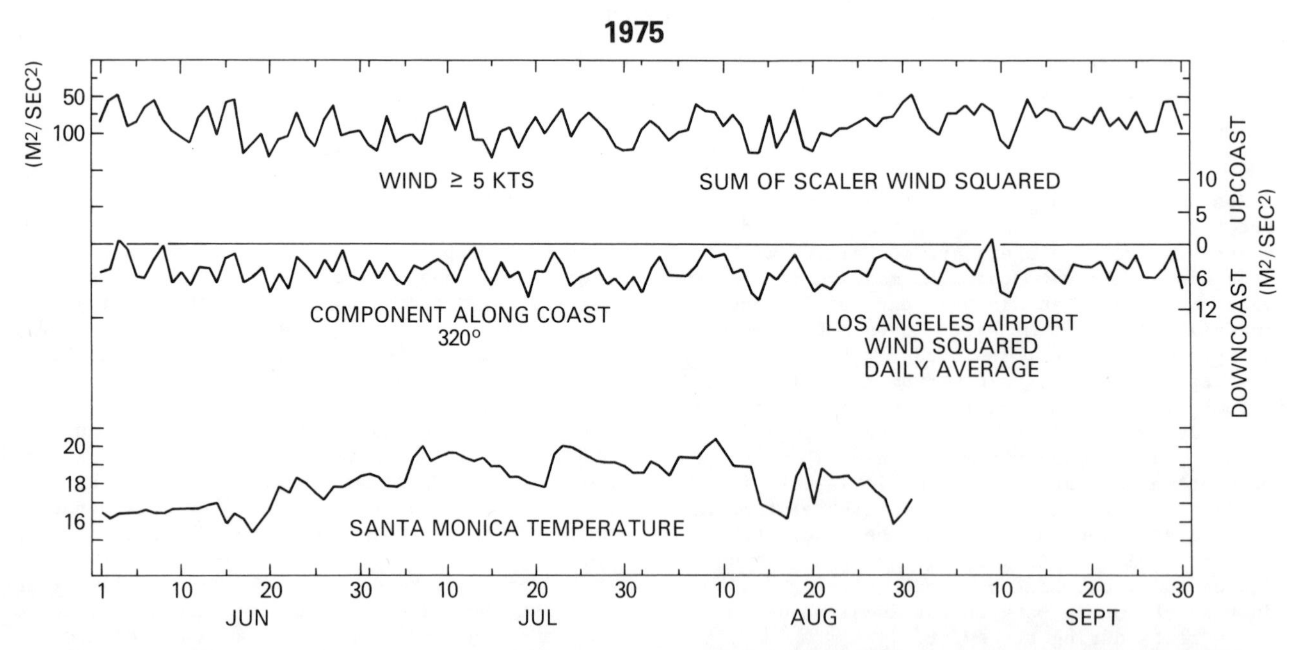

Fig. 8. Comparison of winds at the Los Angeles airport with the Santa Monica water temperature. The convention for winds is as in Fig. 7. In addition, the sum of the scaler wind squared is plotted in the upper panel to show that the total variability is small regardless of direction. In this example, the daily sum of the wind speed greater than or equal to five knots is plotted.

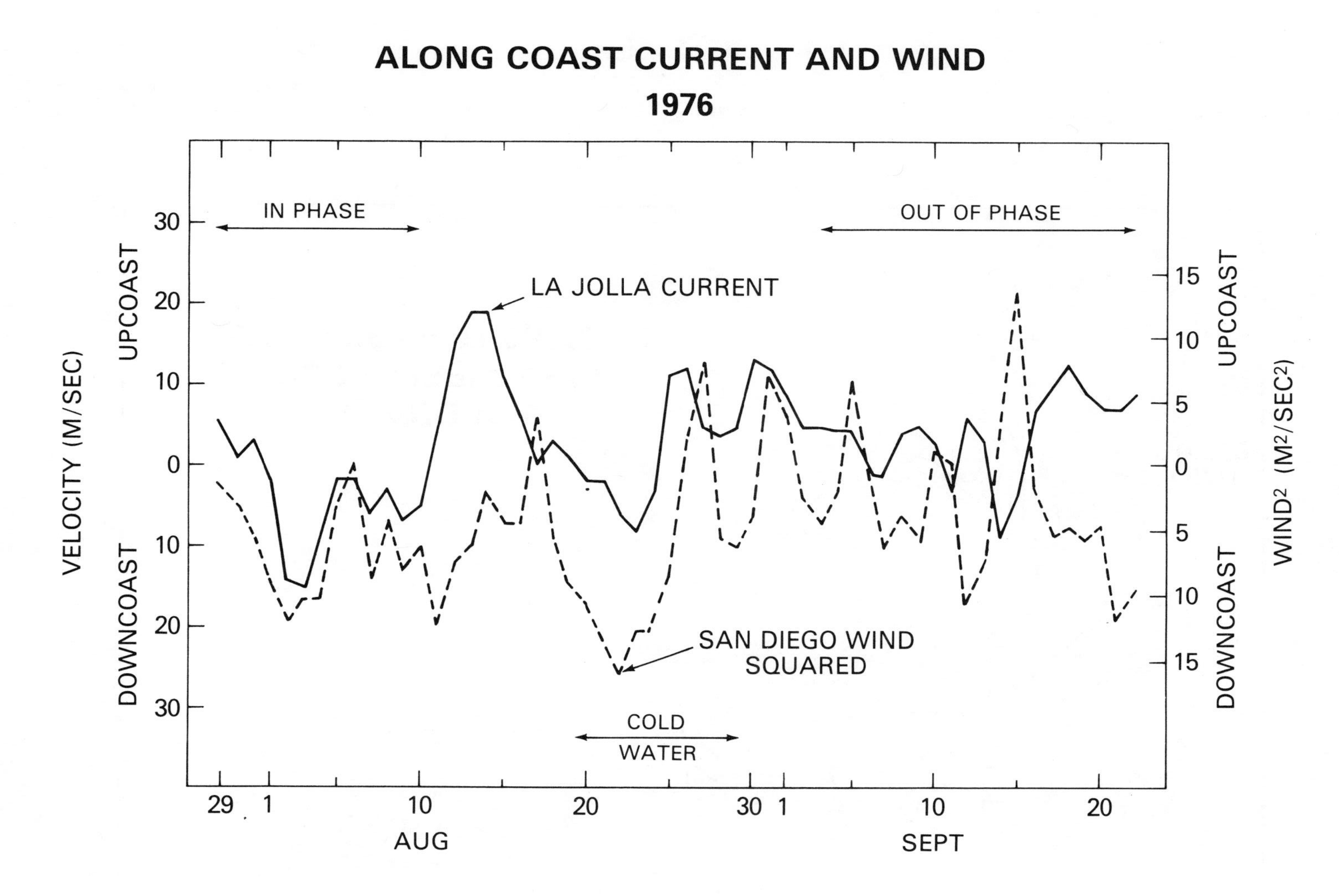

Fig. 9. Comparison of the daily mean-squared long-shore wind component at San Diego airport (as in Fig. 7) and the daily averaged along-shore current measured off Scripps Pier at 41-m depth in 56 m of water.

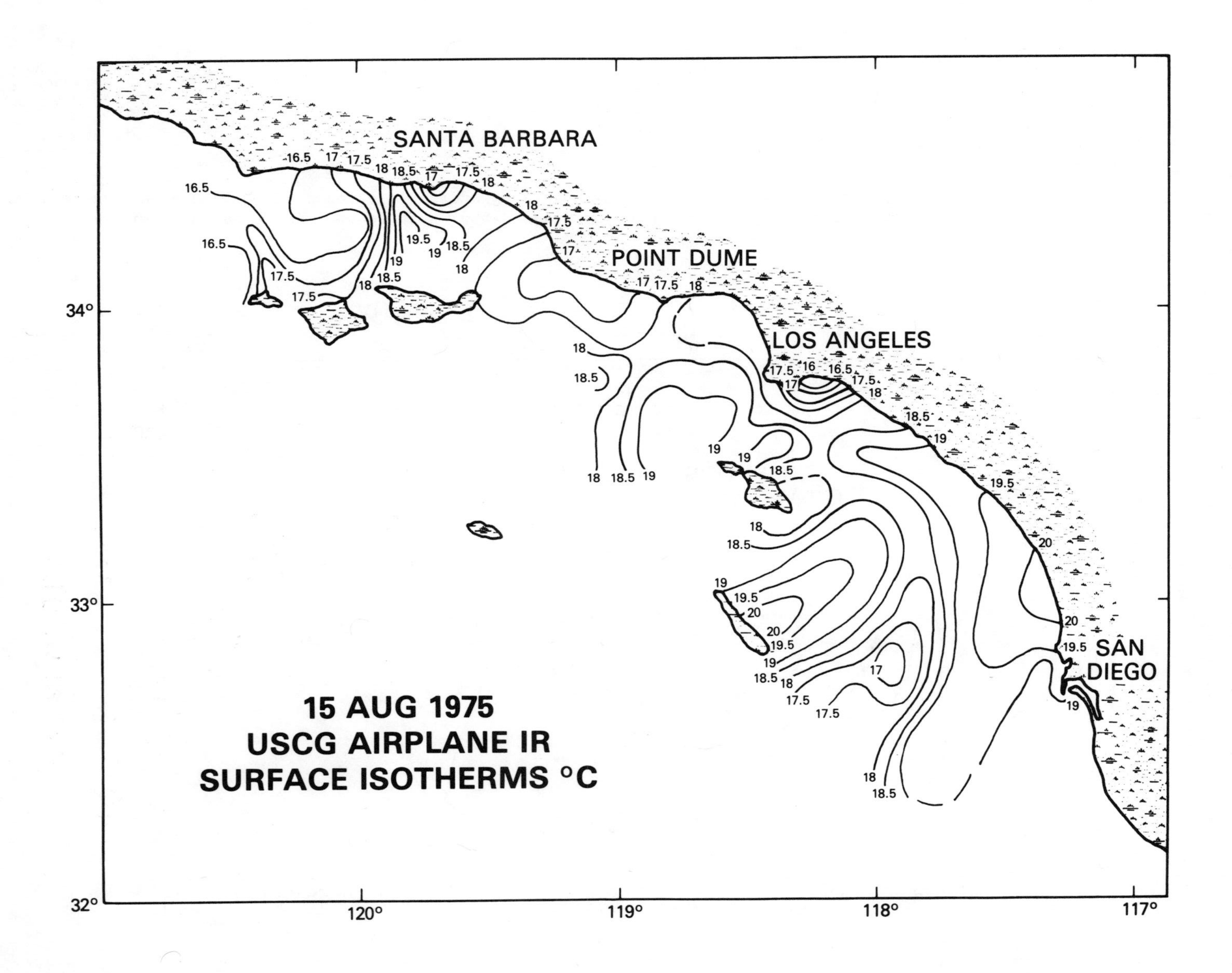

Fig. 10. Surface temperatures during a downwelling event. Temperatures were measured by an infrared radiometer on a United States Coast Guard airplane.

TABLE 1. Major Summer Upwelling Events at Scripps Pier from 1967 through 1976, Also noted is the extent of the upwelling along the California Bight.

Year	Date Begun	Maximum Temperature Drop (°C)	Duration (days)	Coastal Extent	Comments
1967	27 June	3.7	10	Scripps and San Clemente	Not at Oceanside
1968	30 July	5	14	?	
1969	9 July	6.5	24	Scripps-Balboa	
	10 August	4.3	21	"	
	3 September	4.8	10	"	
1970	8 July	5.9	11	"	
	29 July	5	6	"	
	11 August	4.8	10	"	
	29 August	5.3	5	Scripps-Oceanside	
	13 September	9.5	12	"	No record of other stations
1971	25 August	3.4	8	Scripps-Balboa	
	5 September	3.6	11	"	
1972	18 July	6	14	"	
	22 August	3.3	11	"	Very irregular
1973	3 June	3	7	"	
	31 August	3.9	9	"	Poor definition
1975	31 July	8	9	"	
	19 August	5.5	14	Scripps-Santa Monica	
1976	16 June	4	9	Scripps-Balboa	
	19 August	3.8	10	"	

The downwelling event appears to have been localized in the San Diego-Oceanside area while upwelling may have occurred in the Balboa-Long Beach area.

Discussion

There are two scales of upwelling that may occur in the California Bight. Roden (1972) calculated that large-scale upwelling, extending 200 km off the coast, occurs from spring through summer. However, his study leaves undefined the role of upwelling *vis-á-vis* advection. It would be difficult to distinguish these processes from other simultaneous processes.

Significant upwelling caused by long period waves does not seem likely. As the waves move upcoast, cold water should appear progressively through the entire Southern California Bight. However, significant upwelling is observed only south of Balboa. Long-period waves were not observed to cause significant changes in the temperature structure during an extended study at the Mission Beach tower.

Although Enfield and Allen (1980) observed planetary waves in the sea level at San Diego, and although they may cause small changes in the temperature structure, their signals are too small to be observed above the other variations in the shore stations and tower measurements.

The second scale of upwelling occurring in the bight is local upwelling. We have studied such upwelling for the summer period and find that it occurs about twice a summer. Local upwelling events occur simultaneously from San Diego to Balboa and are forced by local winds.

References

Arthur, R.S., Oscillations in sea temperatures at Scripps and Oceanside piers, *Deep-Sea Research*, *2*, 107-121, 1954.

Arthur, R.S., Variation in sea temperature off La Jolla, *Journal of Geophysical Research*, *65*, 4081-4086, 1960.

Bakun, A., Coastal upwelling indices, west coast of North America, Technical Report, National Marine Fisheries Service, SSRF-671, 103 pp., NOAA, Boulder, Colorado, 1973.

Cairns, J.L., and E.C. LaFond, Periodic motions of the seasonal thermocline along the Southern California coast, *Journal of Geophysical Research*, *71*, 3903-3915, 1966.

Emery, K.O., *The Sea off Southern California, A modern habitat of petroleum*, John Wiley, 366 pp., 1960.

Enfield, D.B., and J.S. Allen, On the structure and dynamics of monthly mean sea level anomalies along the Pacific coast of North and South America, *Journal of Physical Oceanography, 10,* 557-578, 1980.

Jones, J.H., General circulation and water characteristics in the Southern California Bight, Southern California Coastal Water Research Project, Los Angeles, 37 pp., 1971.

Leipper, D.F., Sea temperature variations associated with tidal currents in stratified shallow water over an irregular bottom, *Journal of Marine Research, 14,* 234-252, 1955.

Munk, W.H., On the wind-driven ocean circulation, *Journal of Meteorology, 7,* 79-83, 1950.

Namias, J. and J.C.K. Huang, Sea level at Southern California: a decadal fluctuation, *Science, 177,* 351-353, 1972.

Reid, J.L., Jr., and C.G. Worrall, Detailed measurements of a shallow salinity minimum in the thermocline, *Journal of Geophysical Research, 69,* 4767-4771, 1964.

Reid, J.L., Jr., G.I. Roden, and J.G. Wylie, Studies of the California current system, California Cooperative Oceanic Fisheries Investigations, Progress Report, 1 July, 1957 to 1 January, 1958, pp. 27-56, 1958.

Roden, G.I., Nonseasonal temperature and salinity variations along the west coast of the United States and Canada, California Cooperative Oceanic Fisheries Investigations Reports, 8, 95-119, 1961.

Roden, G.I., Large-scale upwelling off northwestern Mexico, *Journal of Physical Oceanography, 2,* 184-189, 1972.

Smith, R.L., Upwelling, *Oceanographic and Marine Biology, An Annual Review, 6,* London, Allen, and Unwin, pp. 11-46, 1968.

Stewart, H.B., Coastal water temperature and sea level -- California to Alaska, California Cooperative Oceanic Fisheries Investigations, Progress Report, 1 January, 1958 to 30 June, 1959, pp. 97-100, 1960.

Tont, S.A., The effect of upwelling on solar irradiance near the coast of Southern California, *Journal of Geophysical Research, 80,* 5031-5034, 1975.

Tont, S.A., Short-period climatic fluctuations: effects on diatom biomass, *Science, 194,* 942-944, 1976.

Tsuchiya, M., California undercurrent in the Southern California Bight, California Cooperative Oceanic Fisheries Investigations, Reports, 18, 155-158, 1976.

Tsuchiya, M., Inshore circulation in the Southern California Bight, 1974-1977, *Deep-Sea Research, 27A,* 99-118, 1980.

United States Department of Commerce, Temperature and densities for the west coast stations of North America, p. 57, 1967.

Winant, C.D., Downwelling over the Southern California shelf, *Journal of Physical Oceanography, 10,* 791-799, 1980.

UPWELLING: EFFECTS ON AIR TEMPERATURES AND SOLAR IRRADIANCE

Sargun A. Tont

Scripps Institution of Oceanography, A-020
University of California San Diego, La Jolla, California

Abstract. An analysis of 31 upwelling events monitored near the coast of southern California indicates that during a majority of the events air temperatures either decreased from their previous levels or their increase was less than expected. The percent of possible sunshine, which is indicative of solar irradiance reaching the ground, did not decrease from its previous level during nearly half of the events.

Introduction

Eastern boundary currents flowing from high latitudes combined with upwelling result in cold water masses along the western edge of the continents. The cold water masses, in turn, have a marked effect on the climate of the regions adjacent to the coast. The contribution of upwelling to climate has been summarized by Smith (1968): (1) cooling of the air masses passing over upwelled waters, (2) the formation of stratus clouds or fog, and (3) strengthening of the diurnal sea breeze. In a later review, Thompson (1978) concluded that the coastal upwelling "...also tends to produce the terrestrial coastal deserts evident of South America, and southwest Africa." It should be emphasized that neither of the authors considered upwelling as the sole cause of the processes referred to above, but as only one of the factors making up the climate of so called "upwelling regions". A careful examination of any climatic atlas (e.g. Climatic Atlas of the United States, 1968) verifies the conclusions reached by Smith (1968) and Thompson (1978).

Early investigators of upwelling along the California coast assumed upwelling to be a steady-state phenomenon that generally begins in March and continues uninterrupted until July (e.g., Sverdrup, 1942). Later studies off the California coast (Tont, 1976) and elsewhere (Huyer, 1976 and references cited therein) indicate that an upwelling season is characterized by a series of episodic events, the periods of which range from a few days to weeks. This paper is an attempt to use historical data to quantify the changes in the climatic variables (sea-surface temperature, air temperature, and solar irradiance) during each upwelling event and to evaluate the changes with respect to climatic normals and fluctuations. For a comprehensive study of climate-upwelling relationships off Baja California in which the longer time scales of upwelling are discussed, see Bakun and Nelson (1977). In the absence of detailed hydrographic cross sections, the criteria used here for the identification of an upwelling event is a quasi-uniform decrease in sea-surface temperature for a period of at least one week, accompanied by a diatom bloom. (The justification for such an approach can be found in Tont, 1976 and Kamykowski, 1973).

In addition to the reviews by Smith (1968) and Thompson (1978), there are numerous writings dealing with specific aspects of the upwelling-climate relationship. Some are those by Petterssen (1938), Sverdrup (1942), Neiburger, Edinger, and Bonner (1971), Namias (1959), Tont (1975), and Mickelson (1978). Locations discussed in the text are shown in Fig. 1.

Materials and Methods

All the atmospheric data used in this study, unless otherwise indicated, can be found in Local Climatological Data Reports, San Diego, California, 1920-1938. Copies of sea-surface temperatures (SST) recorded at Scripps Pier can be obtained from Francis Wilkes, Scripps Institution of Oceanography, La Jolla, California. Solar irradiance data are from the original data records of G.F. McEwen; some were published by McEwen (1938). The records are primarily used to show the seasonal cycle of solar irradiance during each upwelling event. During the period under study (1920-1938) they are fragmentary and incomplete. Instead, to evaluate the changes during each upwelling event I use the percent of possible sunshine (the amount of time the sun is not obscured by clouds divided by total daylight hours times 100), which is a good indicator of cloud cover (Bennett, 1965).

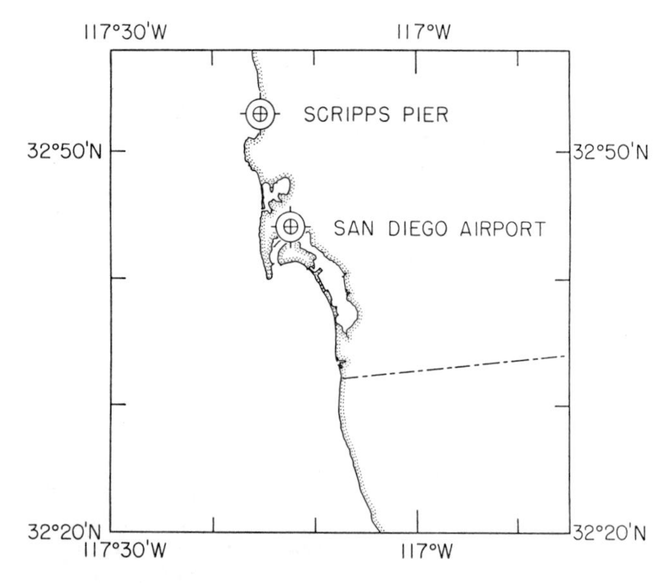

Fig. 1. Locations discussed in the text. Sea-surface temperatures (observed daily around 0700 h local time) were recorded at "Pier"; solar irradiance (integrated from hourly increments) was recorded at Scripps Institution of Oceanography, adjacent to the pier. Air temperatures and percent of possible sunshine data (both calculated from daily averages) were recorded at the San Diego Airport.

Results and Discussion

General Characteristics

The large scale northwesterly air flow, which effects the coastal regions of southern California, is maintained by a high pressure area that forms over the North Pacific Ocean and a low pressure area over the southwestern interior United States (southern Nevada and southern California deserts). (For long-term atmospheric pressure patterns for this region and elsewhere, see Namias, 1975). One of the predominant climatic features of the region is the trade-wind inversion or subtropical inversion which, according to Neiburger *et al.* (1971), is "produced by the sinking of air as it moves equatorward around the eastern end of the high-pressure areas (anticyclones) that remain over the subtropical oceans most of the time". Consequently, "fog, stratus, or stratocumulus are present much of the time along the California coast". The land breeze-sea breeze system also has a marked effect on the local climate. The sea breeze develops in the early morning hours and reaches its maximum velocity in the mid-afternoon, whereas the land breeze generally sets in about midnight. (For a detailed discussion of local winds, see Flohn, 1969). The net result of the processes on the climate of the area can be seen in Fig. 2. Increased cloud cover during spring and summer is reflected in a decrease of percent of possible sunshine (% SS), which in turn results in a significant decrease of solar irradiance (H) reaching the ground. This is a unique feature of the coast and not found in inland locations (see monthly maps of daily insolation for the United States prepared by Bennett, 1965, for comparisons).

Short-period Effects (days, weeks). Fig. 3 represents changes in air temperature during four upwelling events. During the first two air temperature was reasonably well correlated with sea-surface temperature, whereas in 1929 the correlation was weak, and most importantly, during the upwelling event of 1936, the air temperature actually increased. Tabulated averages (Table 1) of all the upwelling events examined show that the duration of upwelling was 2.64 weeks, resulting in a drop of sea-surface temperature of -1.59°C, equivalent to -0.76°C per week. Although air temperature drops on the average of -0.41°C during upwelling events, on 11 out of 31 occasions it increased, which is also indicated by the relatively large standard deviation (Table 1). However, such increases in air temperature do not mean that upwelling has no cooling effect on the air. It is possible that the increases were less than they would have been had upwelling not taken place. All the upwelling events examined were during the first part of the year, when air temperatures are expected to increase (Fig. 2). A more meaningful method to evaluate changes in both sea-surface and air temperature during an upwelling event is to normalize both with respect to their long-term average seasonal cycles (e.g., if air temperature increases by 1.5°C during a 3-week period, its rate of increase is 0.5° per week. If the expected increase in air temperature during the same period is 0.8°C per week (estimated from Fig. 1), then the normalized value is -0.3°C per week.) The same considerations apply to sea-surface temperatures. The normalization indicates that in 25 cases (80% of the total) air temperature either decreased or its increase was less than the long-term average increase for that seasonal segment. The seasonal variations in percent of possible sunshine and upwelling index are significantly correlated off the southern California coast (Tont, 1975). As can be seen in Table 1, in only 50% of the cases does the percent of possible sunshine increase during an upwelling event. (No normalization of the percent of possible sunshine was attempted because during spring and summer, when most of the upwelling events took place, the percent of possible sunshine is expected to decrease (Fig. 2), which would not compensate for increases observed during many upwelling events).

The results should not be interpreted as indicating that the presence of cold water masses in the region has no effect on the formation of stratus clouds or fog. The formation of the inversion layer, which creates optimum conditions for stratus or fog formation, is mainly due to large-scale atmospheric events, that is, the sink-

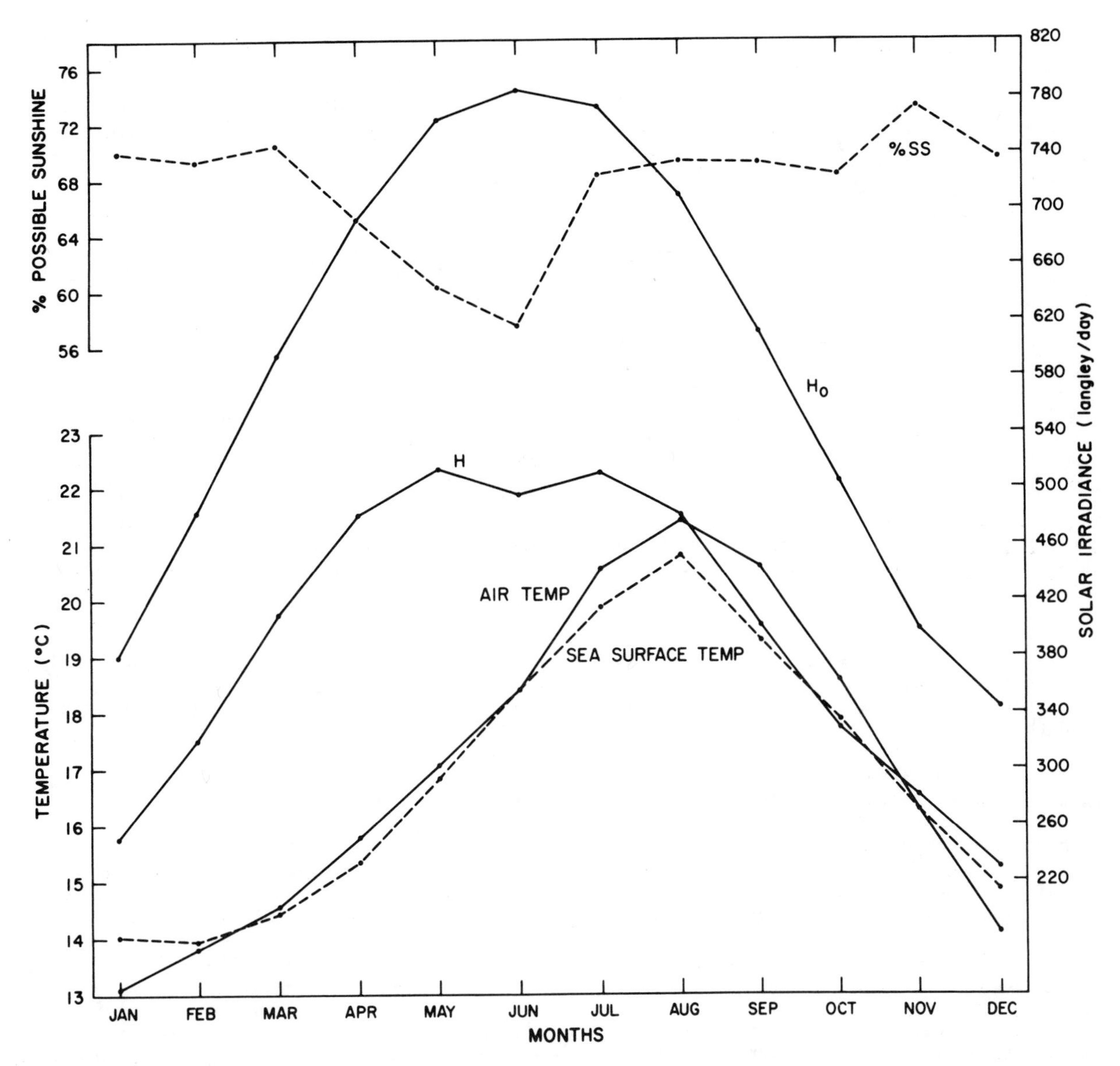

Fig. 2. 1920-1973 monthly means of calculated clear sky solar irradiance (H_o), percent of possible sunshine (%SS), air temperature, sea-surface temperature, and 1930-1951 monthly mean of solar irradiance reaching the ground (H) at locations shown in Fig. 1.

ing of air as it moves equatorward around the eastern end of the high-pressure area (Neiburger *et al.*, 1971). The relatively cool water masses brought down by the California Current, reinforced by upwelling, simply strengthen the inversion layer. As was pointed out by Petterssen (1938) "...the presence of cold water along the coast greatly influences the air masses in the formation of fog; but it is not satisfactory to say that there is fog in California because of the cold water along the coast".

There are several factors that, either individually or in combination, result in increases of both air temperature and percent of possible sunshine despite the presence of cold water masses along the coast (only a brief summary of our results will be given here). Stratus clouds typical of spring and summer along the California coast start forming around 1600 h (local standard time) and persist through the night and early morning hours; by mid-day most of the clouds are dissipated or "burned off". For example, in June the average percent of possible sunshine is 10% between 0500 and 0600 h (local standard time) but by 1300 h it increases to 90%; therefore any cooling effect by the stratus clouds during daylight may be compensated or overriden by warming during the night by what one may call the "lid effect".

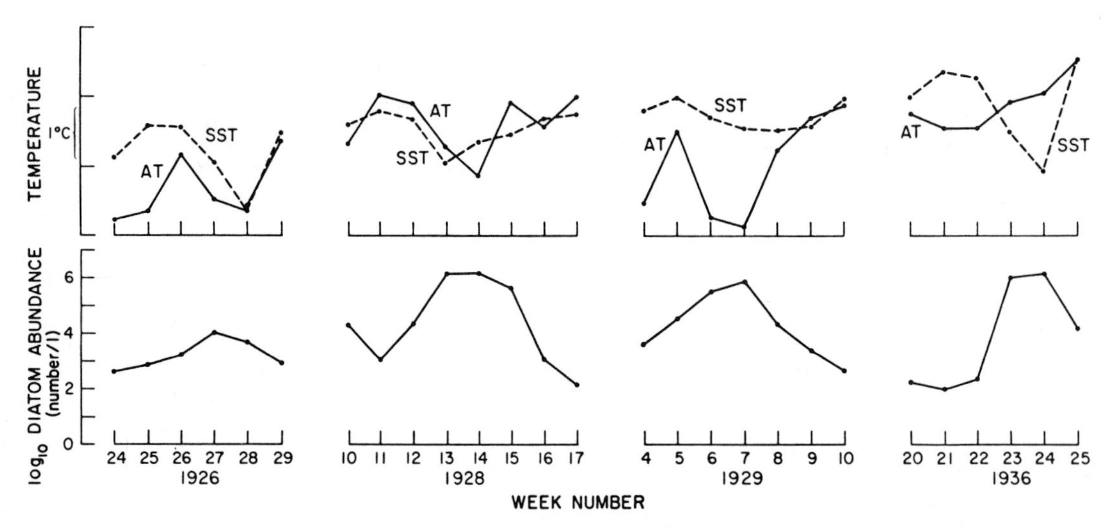

Fig. 3. Air temperature (AT), sea-surface temperature (SST), and diatom abundance (used in this study primarily for identification of upwelling events) changes during four upwelling events. Durations of upwelling events are calculated from the point where a nearly uniform decrease in SST was observed to the point where the SST is lowest. For example, the duration of upwelling was two weeks (between the 26th and 28th weeks) in 1926.

That is, the cloud cover acts as a lid by keeping in a part of the outgoing radiation. Optimum conditions for upwelling are created by northerly winds and the air characteristics of the regions where the winds originate should be transported (and mixed with other air masses on the way) to the regions of upwelling. Theoretically, a northeasterly wind originating in dry and warm regions inland (popularly called Santa Ana) could cause upwelling as strong as that caused by alongshore winds originating over cool and moist oceanic regions. Consequently the air temperature would, in a large degree, depend on the origin of the air masses passing over the region under study. In summary, however, one can conclude that upwelling, or events associated with it, has a marked effect on air temperature, because nearly 80% of the time air temperatures either decrease from their previous levels, or their increase is less than expected (Table 1). Considering the short periodicity (2.64 weeks) of the changes, the effects are on weather but not necessarily on climate.

Effects on Climatic Change

Both the spatial and temporal extent of sea-surface and air temperature anomalies (several hundred kilometers, months to years) indicate that they are the result of large-scale phenomena (e.g., Namias, 1959). The upwelling is confined to coastal regions, so upwelling can only act as a modifier of the large-scale climatic anomalies. In the absence of concurrent data from off the coast where the effects of upwelling are expected to be negligible, it is difficult to quantify the

TABLE 1. Averages of Various Changes Observed During 31 Upwelling Events, (1920-1938).

	1 Duration	2 ΔSST	3 ΔSST/ week	4 Normalized ΔSST/week	5 ΔAT	6 ΔAT/ week	7 Normalized ΔAT/week	8 % SS/ week
μ	+2.64	-1.59	-0.76	-0.92	-0.41	-0.06	-0.35	+0.60
σ	±1.52	±0.55	±0.44	±0.53	±1.72	±0.98	±0.99	±7.60
# increase		0	0	0	11	11	6	16
# decrease		31	31	31	20	20	25	15
N	31	31	31	31	31	31	31	31

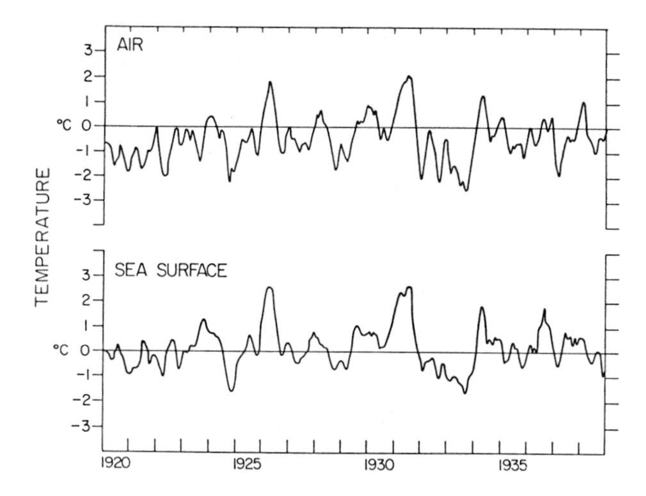

Fig. 4. Air-temperature and sea-surface temperature anomalies (based on 1920-1973 means) near the coast of southern California.

effect of upwelling on both sea-surface and air temperature anomalies. Another difficulty in ascertaining the effects of upwelling on climatic change is that both upwelling and large-scale processes are often linked. For example, an increase in the strength of the subtropical gyre would tend to lower sea-surface temperatures through increased coastal upwelling, southward advection, and direct cooling through ocean-atmosphere heat exchange. Consideration of monthly anomalies of air and sea-surface temperatures based on long-term means (1920-1938) yields some preliminary conclusions. On 21 occasions (68% of the total) air temperatures during the months in which the upwelling events took place show negative anomalies (Table 2). On 14 of the occasions, however, the air-temperature anomaly during the preceding months was also negative. Considering that air-temperature anomalies generally tend to persist for several months (Fig. 4), there is a good possibility that on those 14 occasions the monthly air-temperature anomalies would have been negative regardless of upwelling. On seven occasions, however, air-temperature anomalies that were positive the month before became negative during the month of upwelling (Table 2). Once (the upwelling event during February, 1923) a decrease of $-0.92^{o}C$ per week during the event could account for the resulting monthly anomaly of $-0.44^{o}C$ (not shown), but on other occasions the resultant monthly climatic anomalies were too large to be accounted for by the temperature decrease during upwelling events. The same considerations apply to sea-surface temperatures.

References

Climatic Atlas of the United States, Environmental Data Service, United States Department of Commerce, Washington, D.C., 80 pp., 1968.

Bakun, A. and C.S. Nelson, Climatology of upwelling related processes off Baja California, *California Cooperative Oceanic Fisheries Investigations, 19,* 107-127, 1977.

Bennett, I., Monthly maps of mean daily insolation for the United States, *Solar Energy, 9,* 145-158, 1965.

Flohn, J., Local wind systems, in: *General Climatology, Vol. 2,* H. Flohn (ed.), pp. 139-169, Elsevier Publishing Co., 1969.

TABLE 2. Comparison of Upwelling Events to Large-scale Climatic Fluctuations. Column 1 shows the number of those months in which upwelling events occurred and the climatic anomaly was positive; Column 2, when the climatic anomaly was negative; Column 3, the number of months when upwelling occurred and the climatic anomalies for both the month and the preceding month were negative. For further details see the text.

	1 # (+) anomalies during upwelling months	2 # (-) anomalies during upwelling months	3 # (-,-) anomalies during upwelling and preceding months	4 # (+,+) anomalies during upwelling and preceding months	5 # preceding month (+) to upwelling month (-)	6 # preceding month (-) to upwelling month (+)
Air temperature	10	21	11	8	7	2
Sea-surface temperatures	13	18	12	12	6	1

Huyer, A., A comparison of upwelling events in two locations: Oregon and northwest Africa, *Journal of Marine Research, 34,* 531-546, 1976.
Kamykowski, D., Some physical and chemical aspects of the phytoplankton ecology of La Jolla Bay, Ph.D. Thesis, University of California at San Diego, 174 pp., 1973.
Local Climatological Data, San Diego, California, Environmental Data Service, National Climatic Center, Asheville, N.C., 1920-1938.
McEwen, G.F., Some energy relations between the sea surface and the atmosphere, *Journal of Marine Research, 1,* 217, 1938.
Mickelson, M.J., Solar radiation in Peru during JOINT-II, A guide for modelers, Coastal Upwelling Ecosystems Analysis Technical Report 43, 25 pp., 1978.
Namias, J., Recent seasonal interactions between North Pacific waters and the overlying atmospheric circulation, *Journal of Geophysical Research, 64,* 631-646, 1959.
Namias, J., Northern hemisphere seasonal sea level pressure and anomaly charts, 1947-1974, *CalCOFI Atlas No. 22,* Scripps Institution of Oceanography, La Jolla, California, 1975.
Neiburger, M., J.G. Edinger, and W.D. Bonner, Understanding our atmospheric environment, W.H. Freeman and Company, San Francisco, 293 pp. 1971.
Petterssen, S., On the causes and forecasting of the California fog, *Bulletin of the American Meteorological Society, 19,* 49-55, 1938.
Smith, R.L., Upwelling, in: *Oceanography and Marine Biology Annual Review, Vol. 6,* H. Barnes (ed.), pp. 11-46, Allen and Unwin, London, 1968.
Sverdrup, H.U., *Oceanography for Meteorologists,* Prentice Hall, 246 pp., 1942.
Thompson, J.D., Ocean deserts and ocean oases, *Climatic Change, 1,* 205-230, 1978.
Tont, S.A., The effect of upwelling on solar irradiance near the coast of southern California, *Journal of Geophysical Research, 80,* 5031-5034, 1975.
Tont, S.A., Short-period climatic fluctuations: Effects on diatom biomass, *Science, 194,* 942-944, 1976.

COLD WATER INTRUSIONS AND UPWELLING NEAR CAPE CANAVERAL, FLORIDA

Thomas D. Leming

National Marine Fisheries Service
NSTL Station, Mississippi 39529, U.S.A.

Christopher N.K. Mooers

Department of Oceanography
U.S. Naval Postgraduate School
Monterey, California 93940, U.S.A.

Abstract. A pool of cold water (<18°C) was trapped on the bottom over the mid-shelf region near Cape Canaveral after an apparent offshore meander of the Florida Current in July, 1971. The cold pool was 80 km, 10 km, and 5 m in the alongshore, cross-shelf, and vertical directions, respectively. An XBT (Expendable Bathythermograph) transect taken 61 h later revealed that the integrity of the cold pool apparently had been destroyed by meandering of the Florida Current. The temperature depression associated with the cold water intrusion was also recorded at four bottom-mounted thermographs.

Cold bottom water upwelled to the surface in a narrow coastal band north of Cape Canaveral and was associated with upwelling-favorable winds. This assymmetry was probably due to the change in relative vorticity of the flow around Cape Canaveral.

Introduction

Anomalously low sea-surface temperatures occur during summer along the central and northern east coast of Florida (Green, 1944; Wagner, 1957). Summer low water temperatures are associated with depressed sea level and upwelling-favorable winds (Taylor and Stewart, 1959). The greatest temperature anomalies (6°C) have appeared between Daytona Beach and Canova Beach and the anomalies never have occurred north of Fernandina Beach and only rarely south of Canova Beach (Fig. 1).

Stone and Azarovitz (1968) noted an unusually cold patch of surface water off St. Lucie Inlet (Fig. 1) on 20 June, 1968. The temperature depression, compared to the previous year, was about 7°C. They speculated that the anomaly resulted from upwelling-favorable winds produced by tropical storms "Abby" and "Brenda" in their passage across South Florida during the first three weeks of June, 1968.

The possible effects of Florida Current meanders and eddies on the shelf circulation near Cape Canaveral were largely ignored in the previous studies. The coastal waters are relatively homogeneous, both horizontally and vertically, in temperature and salinity during the winter and spring (Bumpus, 1964). During August, however, intense vertical stratification, primarily temperature dependent, develops over the shelf. These conditions are similar to those over the continental shelf off Onslow Bay, North Carolina (Fig. 1) (Bumpus, 1955; Bumpus and Pierce, 1955; Webster, 1961; Blanton, 1971; Stefánsson, Atkinson, and Bumpus, 1971). These authors proposed that the circulation and hydrographic properties over the shelf off Onslow Bay may be dominated, especially during summer, by horizontal variations in the axial position of the Gulf Stream. Webster (1961) stated that Gulf Stream meanders off Onslow Bay with horizontal amplitudes of 10 km were apparently related to the onshore component of the wind stress.

Thermal Structure of Florida Current Meanders

During 1970 and 1971, the National Marine Fisheries Service conducted an investigation of variables (primarily temperature and currents) in the coastal waters near Cape Canaveral (Fig. 1). As a part of a larger set of observations (Leming, 1979), the July 1971 XBT survey data revealed an apparent onshore-offshore meander of the Florida Current. Over the mid-shelf zone (20 to 50 m) a pool of cold (<18°C) water was trapped along the bottom. The bottom temperature (Fig. 3) suggested that the pool extended along the coast for about 80 km. The average onshore-offshore width was about 10 km and the mean height above the bottom was about 5 m. For these dimensions, the total volume was about 2 x $10^9 m^3$. Large cut-off pools of cold water also occur off

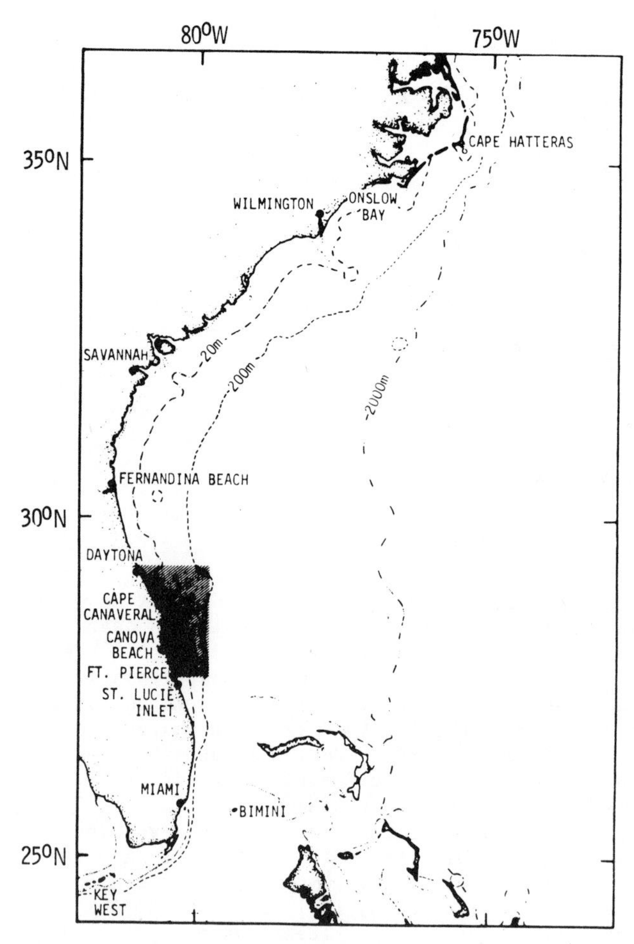

Fig. 1. Map of Florida Straits and South Atlantic Bight with locations referred to in text. The general study area is shaded.

the North Carolina coast (Blanton, 1971; Stefánsson *et al.*, 1971). Over the mid-shelf area in Raleigh, Onslow, and Long bays these cold pools were semi-permanent features during summer months. However, this does not appear to be typical off Cape Canaveral, as discussed below. More recent time series of temperature and currents in the coastal waters of the South Atlantic Bight (Atkinson, 1977; Blanton and Pietrafesa, 1978) and over the shelf and slope off Miami (Lee and Mooers, 1977) suggest that such cold pools are periodic or intermittent rather than permanent features.

The short-lived nature of the cold water pool was clearly evident off Cape Canaveral. During the July cruise, Transect 3 (see Fig. 2) was occupied first from 2235 EDT on 23 July to 0250 EDT on 24 July, and again from 1107 to 1508 EDT on 26 July, about 61 h later. During 23 to 24 July (Fig. 4), a pool of cold water remained in contact with the bottom as the 18°C isotherm had apparently retreated offshore. On 26 July (Fig. 5), the 18°C isotherm had returned over the shelf and the inshore position retreated somewhat, intersecting the bottom at about 25 m. The much steeper slope of the isotherms offshore from the shelf break for the 26 July transect suggests that the Florida Current had intensified and moved onshore from 23 and 24 July. Also, in water depths of 40 to 50 m, large vertical movement of the isotherms occurred between Stas 5 and 7 for the 23 to 24 July transect and 53 to 55 for the 26 July transect. For example, when the Florida Current had apparently moved offshore, the 20°C isotherm at Sta. 6 (Fig. 4) was about 10 m deep. After the onshore excursion, it was 25 m deep there (Fig. 5). This change reflected the general increase in the downward slope of the isotherms in the offshore direction. That is, when the Florida Current moved toward shore, warm water was diverted into the upper layers while colder water was introduced along the bottom. The upward bowing of the isotherms over the shelf break during the onshore movement (Fig. 5) suggests bottom-trapped upwelling at the shelf break. The winds during this entire time were generally southeasterly or southwesterly. The alongshore stress was always upwelling-favorable at about 0.1 to 0.2 dynes/cm^2 (Leming, 1979).

A comparison of surface (Fig. 6) with bottom (Fig. 3) temperatures suggests that any coastal upwelling that surfaces does so in a narrow coastal band. Although there was a large pool of cold water over the shelf in the lower layer, the only significant surface effect occurred in a band inshore and north of the cold pool (Fig. 6), a distance of 15 to 20 km from shore. The surface temperature in this band was between 23.8 and 26.0°C. In addition, a large temperature depression (5°C) existed in the surf temperature at Daytona Beach as well as at several bottom-mounted thermographs over the shelf (Leming, 1979). Theoretical (e.g., Allen, 1973) and observational (e.g., Smith, 1974; Walin, 1972) considerations suggest that the major baroclinic effects in upwelling zones are confined to a narrow coastal band defined by the internal Rossby radius of deformation. This value is given by $\delta_R = HN/f$, where H is a characteristic depth, N is the Brunt-Väisälä frequency, and f is the Coriolis parameter. A characteristic depth for this shelf area is 40 m. During 23 to 25 July, the mean vertical temperature gradient (dt/dz) was about 0.3°C/m. Assuming a constant salinity of 36.0×10^{-3}, the characteristic N was 3×10^{-2} rad/s. With a value of 7×10^{-5} rad/s for f, δ_R was about 17 km, which approximates the offshore distance to which the surface expression of the temperature depression was observed.

Although XBT transects were not repeated on other cruises during 1971, a suggestion of another mid-shelf, bottom-trapped cold pool (<15°C) was observed along Transect 3 during June 28-29, 1971 (Fig. 7). Although the bottom water was cooler than in July, the surf temperature depression at Daytona Beach was only about 2°C, perhaps

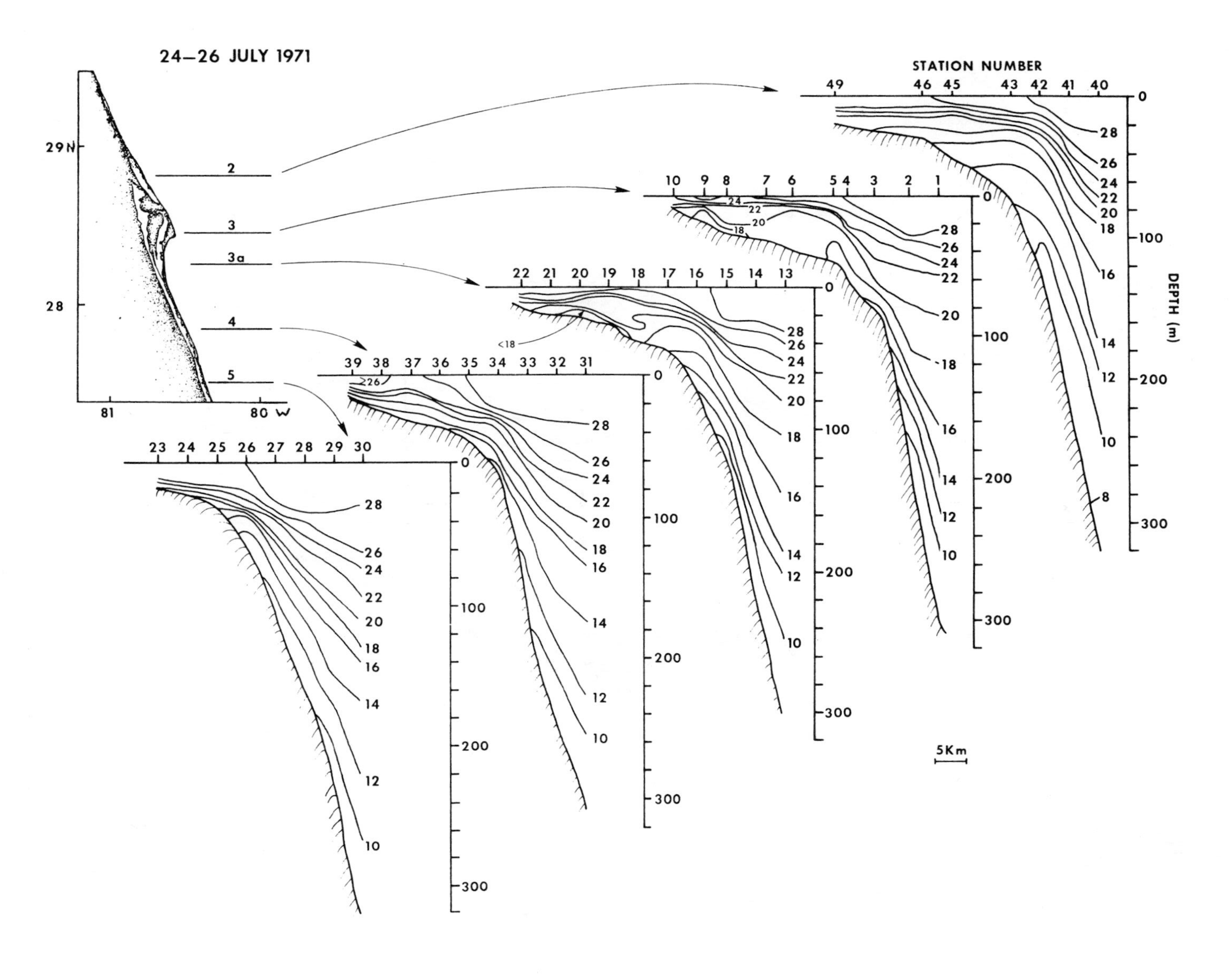

Fig. 2. Thermal structure over the continental shelf off Cape Canaveral from XBT survey, 24-26 July, 1971.

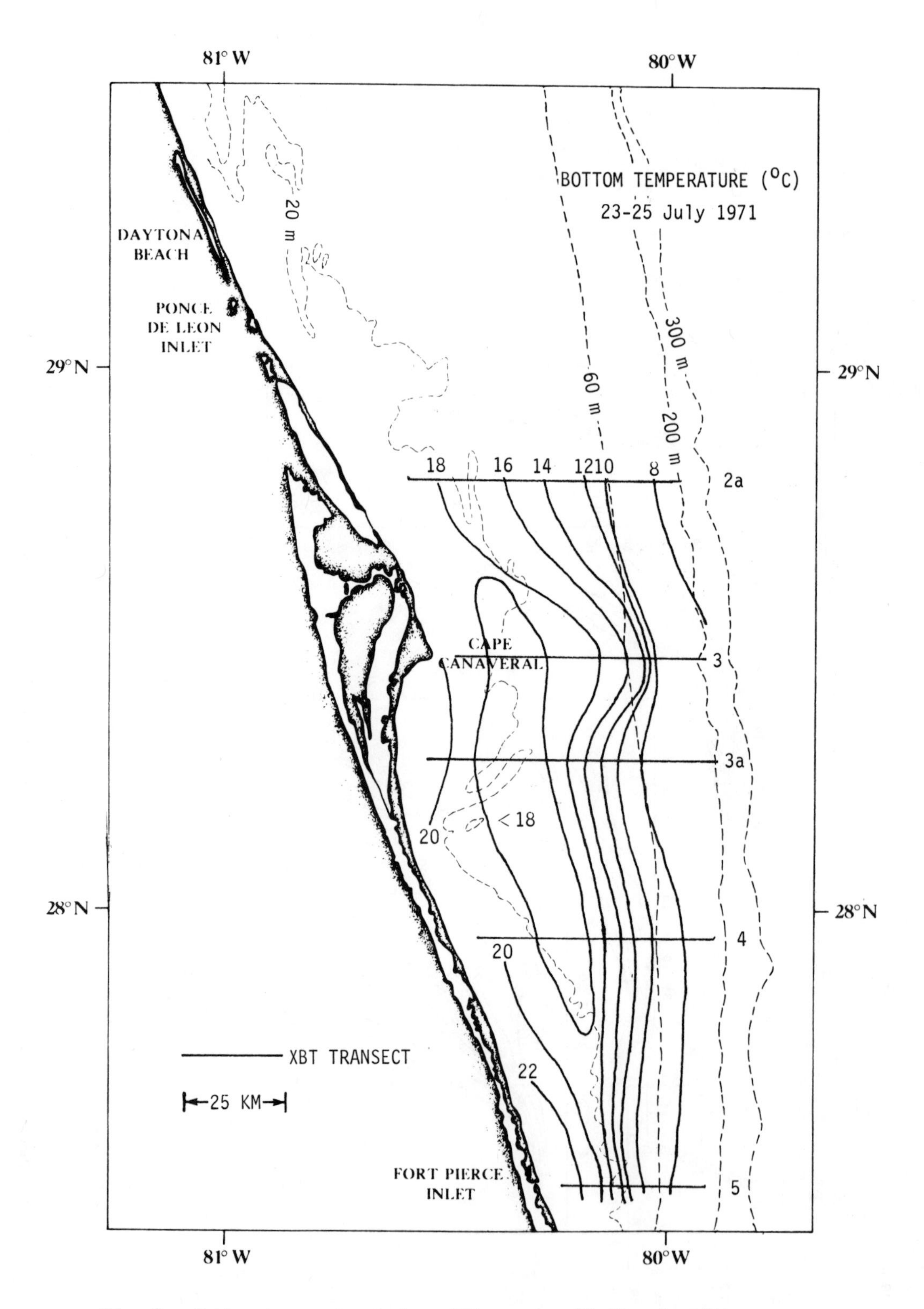

Fig. 3. Bottom temperatures from XBT survey, 24-26 July 1971.

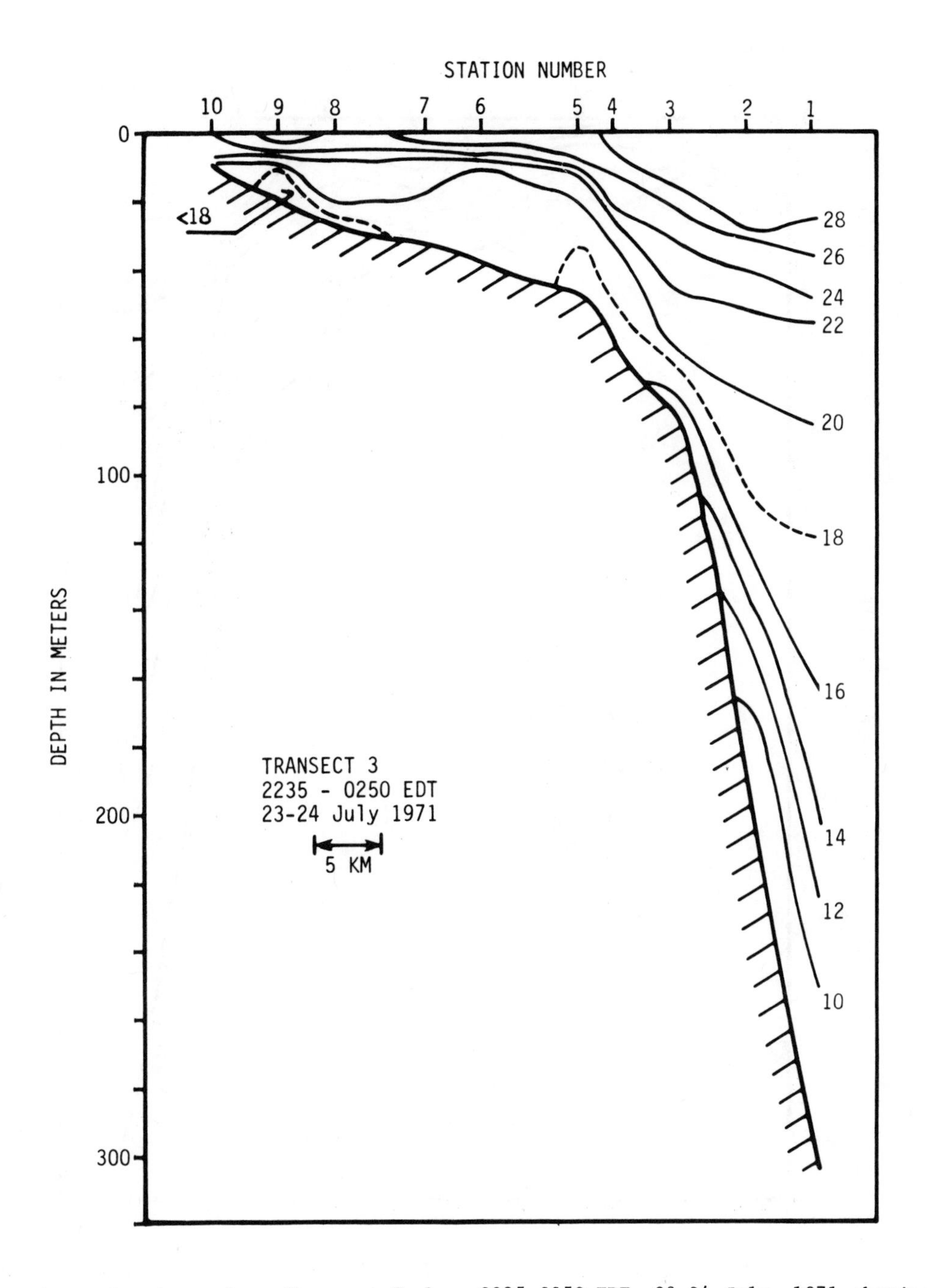

Fig. 4. Temperature structure along Transect 3 from 2235-0250 EDT, 23-24 July, 1971 showing the cut-off cold pool.

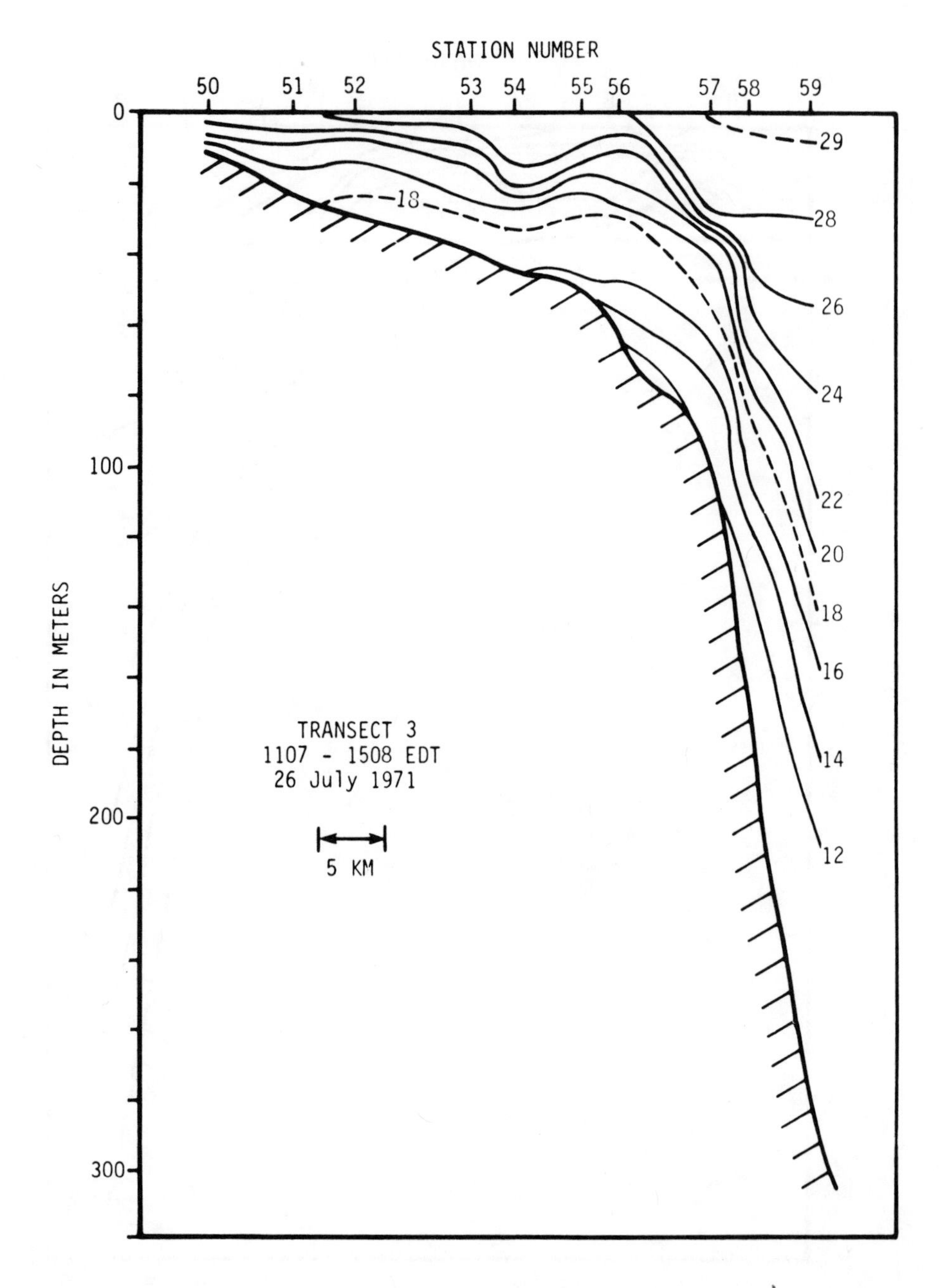

Fig. 5. Temperature structure along Transect 3 from 1107-1508 EDT, 26 July 1971, or about 61 h later than Fig. 4. Note that the cut-off cold pool has disappeared.

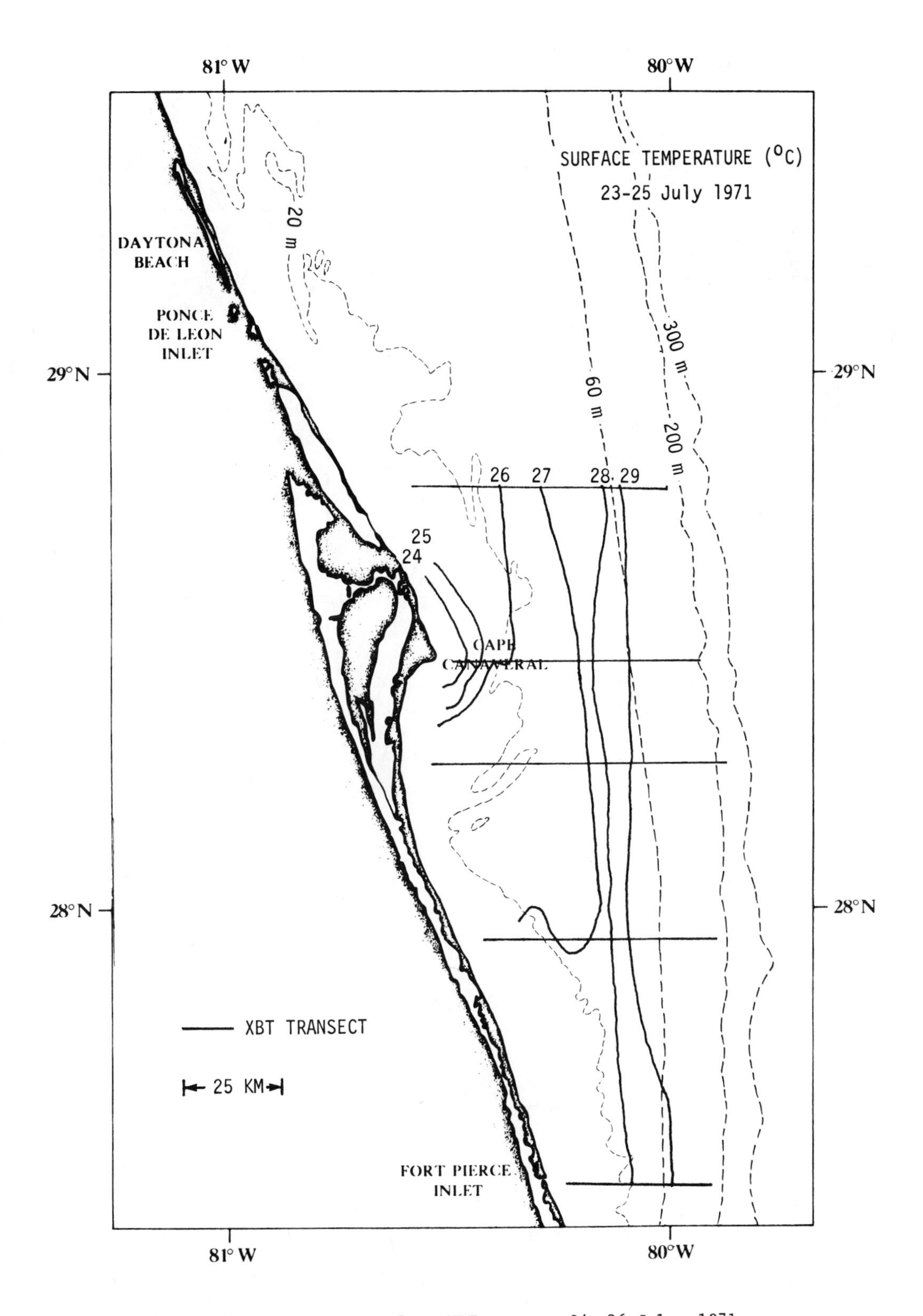

Fig. 6. Surface temperature from XBT survey, 24- 26 July, 1971.

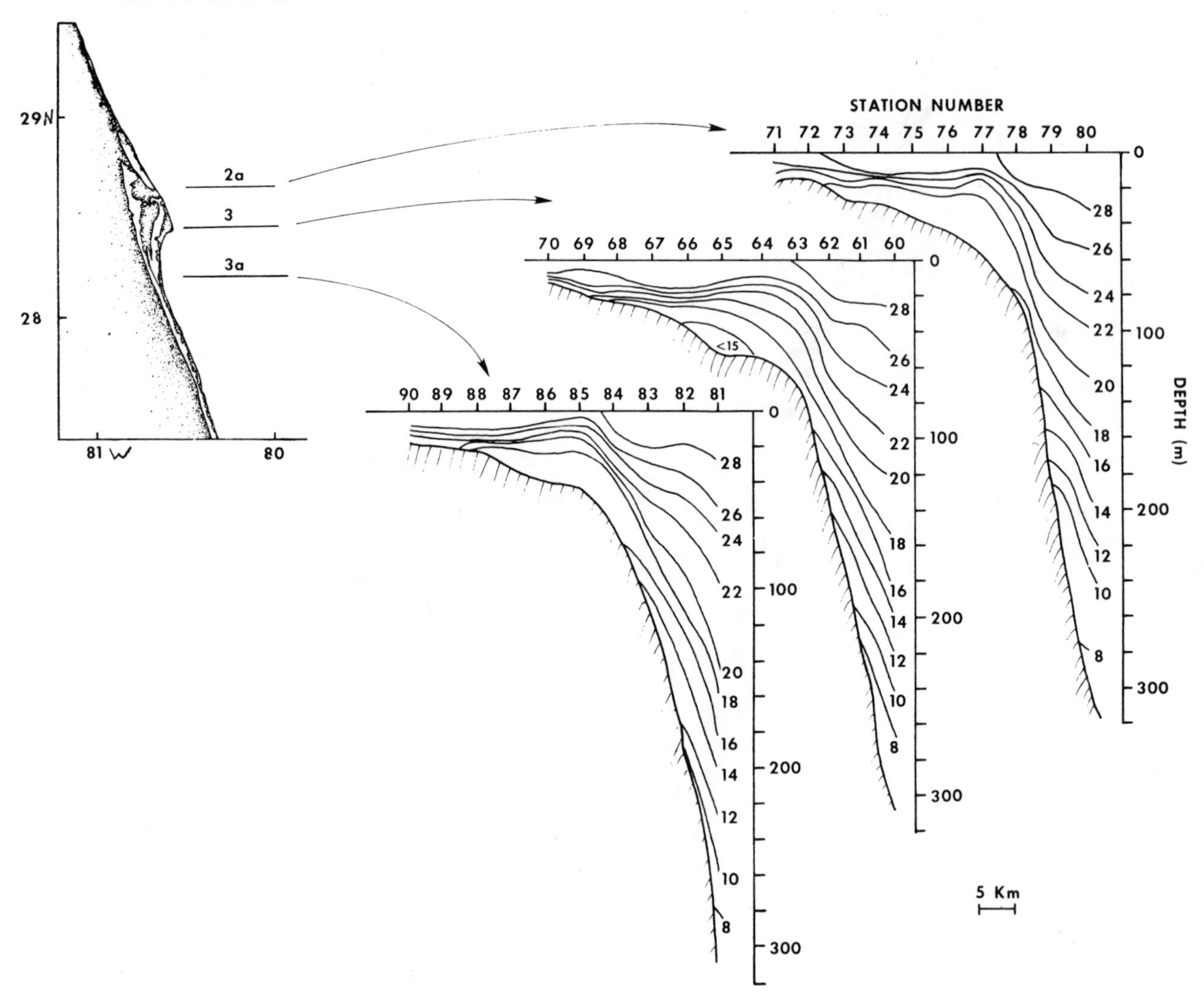

Fig. 7. Thermal structure over the continental shelf off Cape Canaveral from XBT survey, 28-29 June, 1971.

due to a stronger and more consistent onshore wind stress component during June than in July (Leming, 1979).

Effects of Cape Canaveral

The summer occurrences of cool coastal water anomalies are largely confined between Fernandina Beach and Canova Beach, Florida (see Fig. 1), (Taylor and Stewart, 1959). The greatest anomalies are located about 50 km on either side of Cape Canaveral, the only prominent cape in the South Atlantic Bight south of the Carolinas. The depression of surface temperature (for example, see Fig. 6) was confined to a narrow coastal band mainly north of Cape Canaveral. However, the depression of bottom temperature was confined to a narrow along-shelf band mainly south of the cape.

Manifestations of intensified upwelling are observed near capes along the California coast (e.g. Reid, Roden, and Wylie, 1958; Hubbs, 1960), in particular, upwelling effects are intensified south of the capes. To explain the intensification of upwelling equatorward of capes on the west coasts of continents, it is necessary to include the change of relative vorticity along a curved streamline as well as planetary vorticity in the estimation of horizontal divergence (Arthur, 1965). Analogous reasoning accounts for the intensification of upwelling poleward of capes on the east coasts of continents.

In summary, Cape Canaveral has a significant effect on coastal upwelling. Both northward and southward flow would contribute to divergence upwelling north of the cape and to convergence downwelling to the south. This is consistent with the surface temperature depression (Fig. 6), which was generally confined to a narrow coastal band north of Cape Canaveral. Evidence (Leming, 1979) suggests that some of the temperature and current fluctuations near Cape Canaveral are probably due to propagating disturbances such as continental shelf waves generated by winds to the north of Cape Canaveral. The effects of alongshore variations of the coastline and bottom topography on continental shelf waves (Allen, 1976) would also lead to intensification of coastal upwelling near Cape Canaveral.

References

Allen, J.S., Upwelling and coastal jets in a continuously stratified ocean, *Journal of Physical Oceanography, 3,* 245-257, 1973.

Allen, J.S., Continental shelf waves and alongshore variations in bottom topography and coastline, *Journal of Physical Oceanography, 6,* 864-878, 1976.

Arthur, R.S., On the calculation of vertical motion in eastern boundary currents from determinations of horizontal motion, *Journal of Geophysical Research, 70,* 2799-2803, 1965.

Atkinson, L.P., Modes of Gulf Stream intrusion into the South Atlantic Bight shelf waters, *Geophysical Research Letters, 4,* 583-586, 1977.

Blanton, J., Exchange of Gulf Stream water with North Carolina shelf water in Onslow Bay during stratified conditions, *Deep-Sea Research, 18,* 167-178, 1971.

Blanton, J., and L.J. Pietrafesa, Flushing of the continental shelf south of Cape Hatteras by the Gulf Stream, *Geophysical Research Letters, 5,* 495-598, 1978.

Bumpus, D.F., The circulation over the continental shelf south of Cape Hatteras, *Transactions of the American Geophysical Union, 36,* 601-611, 1955.

Bumpus, D.F., Report on non-tidal drift experiments off Cape Canaveral during 1962, Woods Hole Oceanographic Institution Reference Number 64-6, 15 pp., 1964.

Bumpus, D.F., and E.L. Pierce, The hydrography and the distribution of chaetognaths over the continental shelf off North Carolina, Papers in Marine Biology Oceanographic Supplement, *Deep-Sea Research, 3,* 92-109, 1955.

Green, C.K., Summer upwelling, northeast coast of Florida, *Science, 100,* 546-547, 1944.

Hubbs, C.L., The marine vertebrates of the outer coast, *Systematic Zoology, 9,* 134-147, 1960.

Lee, T.N., and C.N.K. Mooers, Near-bottom temperature and current variability over the Miami Slope and Terrace, *Bulletin of Marine Science, 27,* 758-775, 1977.

Leming, T.D., Observations of temperature, current, and wind variations off the central eastern coast of Florida during 1970 and 1971, NOAA Technical Memorandum, NMFS-SEFC-6, 197 pp. (Available from author: NMFS, NSTL Station, MS. 39529), 1979.

Reid, J.L., Jr., G.J. Roden, and J.G. Wylie, Studies of the California Current system, California Cooperative Fisheries Investigation Progress Report, 1 July 1956 to 1 January, 1958, 27-56 (Marine Research Commission, California Department of Fish and Game, Sacramento), 1958.

Smith, R.L., A description of current, wind, and sea-level variations during coastal upwelling off the Oregon coast, July - August 1972, *Journal of Geophysical Research, 79,* 435-443, 1974.

Stefânsson, U., L.P. Atkinson, and D.F. Bumpus, Hydrographic properties and circulation of the North Carolina shelf and slope waters, *Deep-Sea Research, 18,* 383-420, 1971.

Stone, R.B., and T.R. Azarovitz, An occurrence of unusually cold water off the Florida coast, *Underwater Naturalist, 5,* 14-17, 1968.

Taylor, C.B., and H.B. Stewart, Jr., Summer upwelling along the east coast of Florida, *Journal of Geophysical Research, 64,* 33-40, 1959.

Wagner, L.P., Note on the Daytona Beach cold water, Technical Report of Marine Laboratory, University of Miami, Ref. 57-9, 81-82, 1957.

Walin, G., Some observations of temperature fluctuations in the coastal region of the Baltic, *Tellus, 24,* 187-198, 1972.

Webster, F., A description of Gulf Stream meanders off Onslow Bay, *Deep-Sea Research, 8,* 130-143, 1961.

MESOSCALE UPWELLING VARIATIONS OFF THE WEST AFRICAN COAST

Eberhard Hagen

Institute of Marine Research, Rostock-Warnemünde,
German Democratic Republic

Abstract. Dispersion curves of the first three dynamical stable modes of free continental shelf waves (CSW) are numerically estimated for the coastal upwelling areas off West Sahara and Namibia. As a first approximation, depending upon the geometry of the shelf, the wave number-frequency combinations of CSW's with disappearing group velocity are regarded as one of several possibilities to generate characteristic mesoscale upwelling. Empirically approximated characteristic shelf profiles normal to the coast line have been determined. The typical mode-dependent structure of eigenfunctions is compared with spectra of time series and distributions of oceanographic fields.

Introduction

It is useful to divide coastal upwelling into a large scale steady state part and a mesoscale unsteady state part. For the steady state part of the near-shore current system off northwest Africa some essential dynamic aspects are diagrammed in Fig. 1. Corresponding conditions had been observed in other upwelling regions as, e.g., by Hart and Currie (1960) off South West Africa. This simple model assumes that mesoscale upwelling processes are linearly superimposed on the large scale dynamics of upwelling. The energy-rich mesoscale dynamics are among the important baroclinic processes connected with dynamics of long barotropic continental shelf waves (CSW), the periods of which are a few days and their wave lengths are a few hundred to thousand kilometers in the alongshore direction. The typical mode structure of these waves perpendicular to the coast line is mainly determined by the mean shelf geometry. The CSW's are marked by periodic anomalies of water level $h = h(x, y, t)$ along the meridionally oriented shore lines and with vertical amplitudes of some centimeters. According to Reid (1958) these waves travel from the equator to the poles. CSW's are probably induced by external force fields, for example by the alongshore component of wind stress and (or) by disturbances of air pressure not only in the nearshore zone but also further offshore as was investigated for example by Allen (1976a). The space and time structures of the potential energy of CSW's induce corresponding structures in the kinetic energy of barotropic currents. To obtain insight into the general mode structure of CSW's it is sufficient to start with the analysis of free CSW's in regions where the typical shelf profile is given, for example in main upwelling areas off northwest and South West Africa.

A right-handed Cartesian coordinate system (x, y, z) is commonly used with the x-axis aligned onshore-offshore (zonal), the y-axis aligned alongshore (meridional), and the z-axis vertical. Corresponding current components are (u, v, w) and time is t. The shelf bottom topography is $z = H(x)$ for the first approximation step, while $\partial H/\partial x >> \partial H/\partial y$ is really an accomplished assumption in coastal upwelling areas. According to Kundu, Allen, and Smith (1975) $H^{-1}\partial H/\partial x$ is a typical trapping scale of CSW's. Consequently the approximate variant of shelf profiles $H(x)$ is important in resolving analytically or numerically the basic equation for $h(x, y, t)$, as was done by Clarke and Louis (1975). The necessary boundary conditions are $w = \partial h/\partial t$ at $z = 0$ and $w = u\ \partial H/\partial x$ at $z = H(x)$. Also desired are $u = 0$ at $x = 0$ and $\lim_{x\to\infty} h(x) = 0$. With these boundary conditions, a barotropic modal structure is guaranteed for $h(x)$ because the system is barotropic to start with. When n is the mode number the required parameters of CSW's are the eigenfunctions $h(n)$ in the x direction, the frequency $\omega(n) = 2\pi/T(n)$ with period $T(n)$, and the alongshore wave number $m(n) = 2\pi/\lambda(n)$ with the wavelength $\lambda(n)$ in the y direction. It is evident that the structure of the dispersion relation

$$G_n(\omega, m) = 0 \qquad (1)$$

depends on the choice of the typical shelf profile $H(x)$. The dispersion relation of CSW's for a linear approximation of $H(x)$ has been given by Reid (1958), and for an exponential approximation of $H(x)$ by Efimov and Kulikov (1978). It is generally expected that the real characteristic conditions of G_n are found between the weak linear and the strong exponential approximation variant.

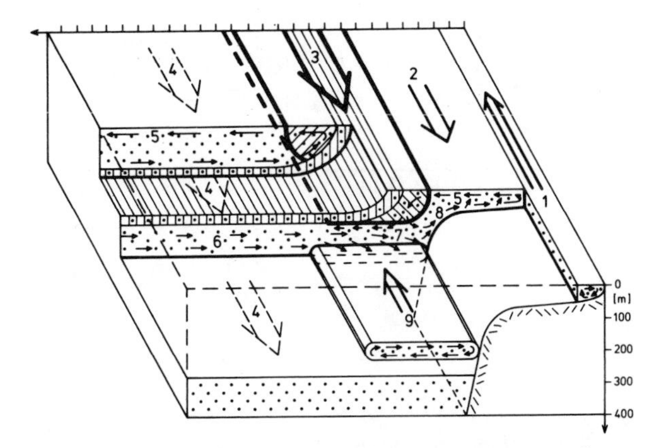

Fig. 1. Some essential aspects of the steady state current system in the coastal upwelling area off northwest Africa. 1. Zone of a weak near-shore countercurrent with sporadic changes in latitude and time of year as a result of a large-scale meridional pressure gradient with low pressure in the north and high pressure in the south. 2. Primary upwelling zone with wind induced currents between the main density front on the shelf and the coast line. 3. Main density front along the shelf edge as a result of pycnoclines upwarping to the surface. This front system is embedded in the strong geostrophic surface current parallel to the coast. 4. Weak branches of the general Atlantic circulation (the Canary Current). (Sometimes weak surface countercurrents alternate with flows to north or to south, as suggested by the dashed-line current arrows.) 5. The wind-generated offshore current induces downwelling at the frontal barrier. This zone is marked by intense downward momentum transport and vertical temperature inversions. 6. The onshore compensation current is the main source of coastal upwelling. 7. Downward branch of (6), a part of the divergence zone off the shelf slope. 8. Upward branch of (6), a part of the divergence zone off the shelf slope feeding into (2). 9. Undercurrent counter to the direction of the surface current with a thickness of some hundred meters.

For H(x) a simple power function approximation is therefore used with empirically determined coefficients. The representative shelf profile normal to the coast line is chosen as the arithmetic mean of at least 10 profiles in an alongshore scale of 150 nautical miles. These mean profiles are located where the error square approximation is at a minimum. Here L is the shelf width, H_a is the depth at x = 0 (roughly located along the shore at a distance of 12 nautical miles) while H_e is the depth of the open ocean. The coefficients for the three approximations are s for the linear variant (Lin.) given in equation (2), b for the exponential variant (Exp.) given in equation (3), and A, E for the empirical variant (Appr.) given in equation (4).

linear:

$$H = H_a \qquad x = 0$$
$$H = s \cdot x \qquad 0 < X < L$$
$$H = H_e \qquad L \leq x < \infty \tag{2}$$

exponential:

$$H = H_a e^{bx} \qquad 0 \leq x < L$$
$$H = H_a e^{bL} \qquad L \leq x < \infty \tag{3}$$

empirical:

$$H = H_a \qquad x = 0$$
$$H = H_a + A \cdot x^E \qquad 0 < x < L$$
$$H = H_e \qquad L \leq x < \infty \tag{4}$$

Results

Table 1 lists characteristic values of these three variants for the upwelling regions both off northwest Africa and off South West Africa (see also Fig. 2). By applying the Runge-Kutta proce-

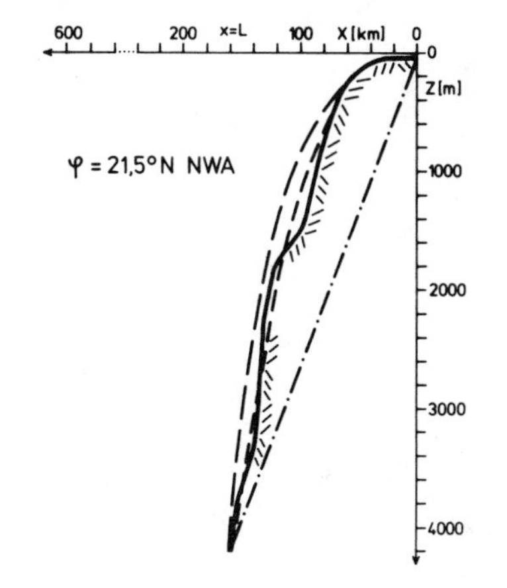

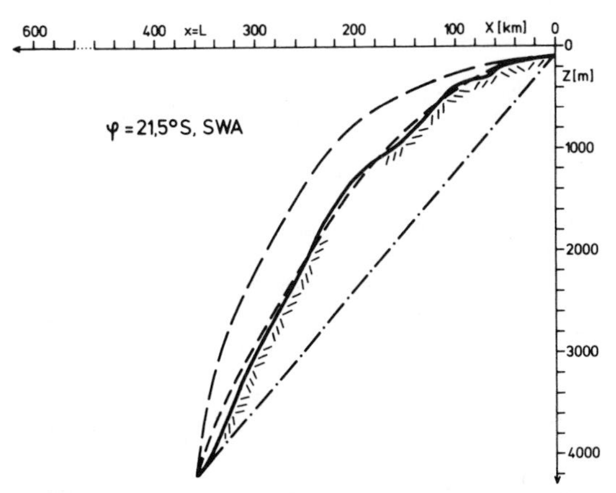

Fig. 2. Approximation variants of mean shelf profiles (full line) by parameters of Table 1 for a linear (—◦—◦), an exponential (— — —) and also for an empirical approximation (– – –) off the coasts of NWA and SWA.

TABLE 1. Typical Parameters of H(x) for Three Approximation Variants (Linear = Lin., Exponential = Exp., Power Function = Appr.), both off northwest Africa ($21^{o} < \lambda < 22^{o}30'$N) and off South West Africa ($19^{o} < \lambda < 22^{o}$S) (Further explanations in the text.).

$\phi = 21.5^{o}$ (N, S)	Shelf Geometry H_a(m)	H_e(m)	L(km)	Lin. $s \times 10^{-2}$	Exp. $b \times 10^{-2}$(km^{-1})	Appr. $A \times 10^{-2}$	E
NWA	50	4200	160	2.63	2.77	0.14	2.953
SWA	85	4200	360	1.14	1.08	3.75	1.967

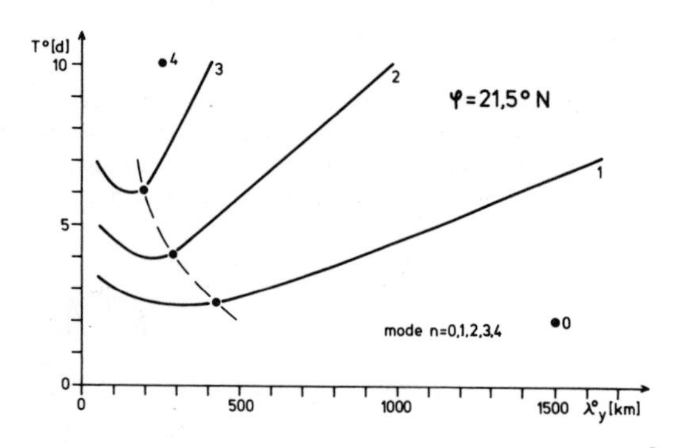

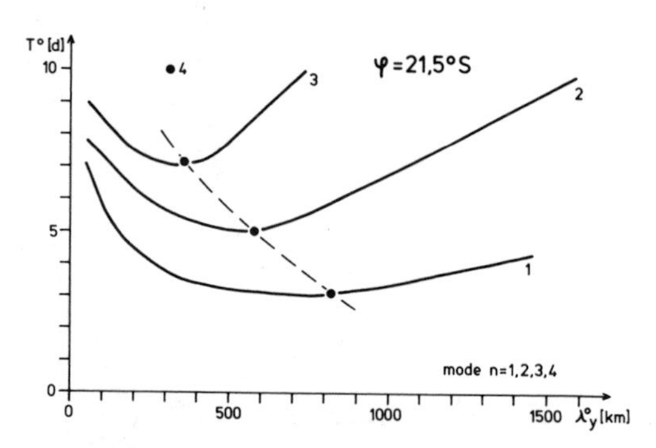

Fig. 3. Numerical estimations of dispersion curves of the first three mode numbers of free CSW's (The dotted line connects period-wave-length combinations with the zero group velocity.)

dure the differential equation of free CSW's has been solved numerically for constant frequency steps and variable wave number by Hagen (1979).

The resulting dispersion curves are given in Fig. 3 for the areas of investigation between 21 and $22^{o}30'$N off northwest Africa (NWA), and also between the latitude of 19 and 22° off South West Africa (SWA). It is a commonly accepted notion but not generally proven for waves that energy accumulates in the dispersion relationship under disappearing energy transport values, i.e., where the wave group velocity is zero. This condition

$$c_g = d\omega/dm = 0 \qquad (5)$$

is marked by index (^{o}).

Examples of such cases are quoted in Adams and Buchwald (1969) and in Wunsch and Gill (1976).

Experiments by Cutchin and Smith (1973) support this hypothesis for the upwelling region off the Oregon coast. From this the probability follows that, among other things, the mesoscale upwelling dynamics are essentially influenced by the wave length-period combinations of first low-mode numbers of CSW's, where $c_g = 0$. For example the linear approximation variant (2) yields the characteristic frequencies

$$\omega^{o}(n) = |f|/(2n+1) \qquad (6)$$

if f is the Coriolis frequency and n is the mode number. On the other hand the exponential variant (3) yields these frequencies with

$$\omega^{o}(n) = |f|/(n^2+1)^{\frac{1}{2}}. \qquad (7)$$

Mesoscale periods T^{o} (n) for different approximation variants are compared in Fig. 4. Typical periods and wave-lengths of variant (4) determined from the empirical coefficients given in Table 1 are listed in Table 2 for the first three low mode numbers. The time series of sea level by Meincke, Mittelstaedt, Huber, and Koltermann (1974) also support this hypothesis with the same typical periods off NWA as given in Table 2 (see also Fig. 6). Corresponding eigenfunctions of sea level $h^{o}(x)$ are plotted in Fig. 5 with a

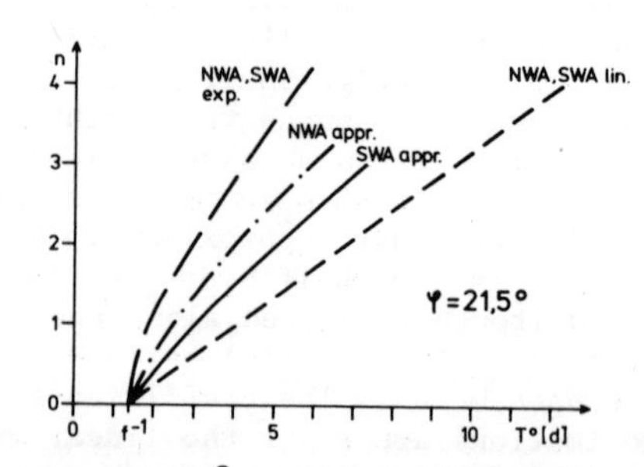

Fig. 4. Periods $T^{o}(n)$ of the modes n = 1, 2, 3 with $c_g = 0$ as estimated by empirical profile approximation (appr.), by linear (lin.), and by exponential profile approximation (exp.).

typical value of wave amplitude of 2 cm at the coast line, i.e., $h_a = 2$ cm.

Discussion

Near shore the potential energy maximum is at the low modes and decreases with increasing mode numbers. The eigenfunction is most intensively super-elevated with increasing n at the shelf edge

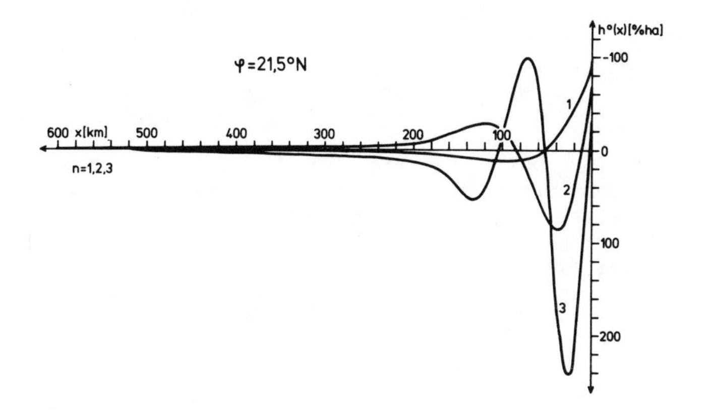

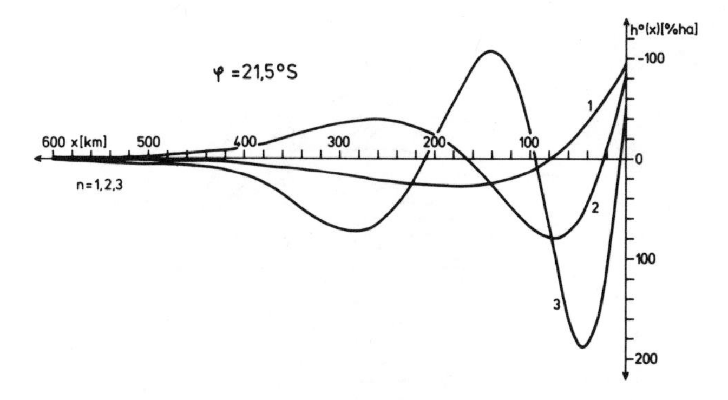

Fig. 5. Estimates of eigenfunctions of sea level $h^o(x)$ of the first three free CSW modes with the zero group velocity in percentage of sea level $h_a(x = 0)$ at the coast line for the areas off NWA and SWA.

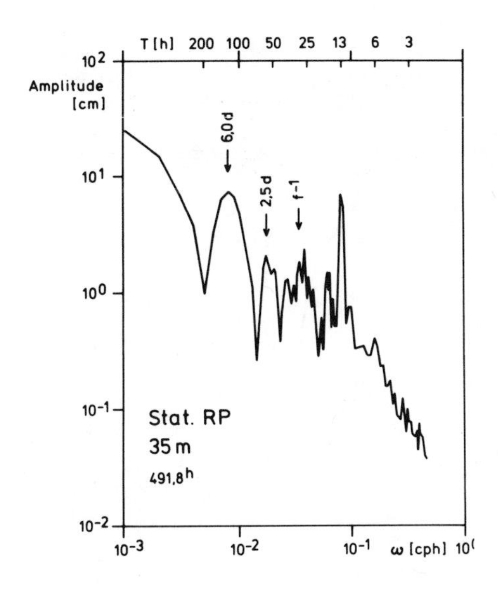

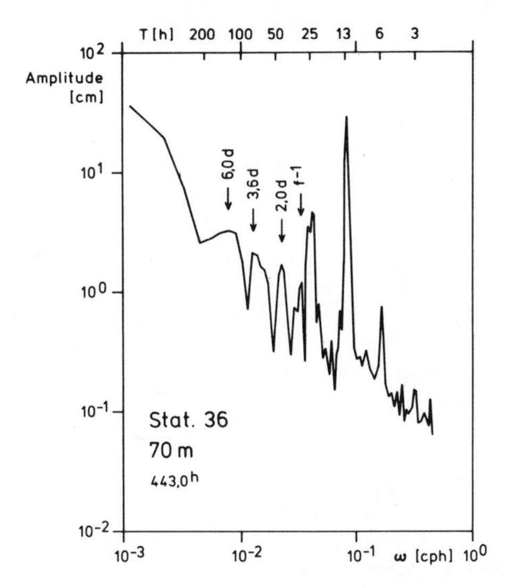

Fig. 6. Amplitude spectra of sea level in the shelf region off NWA near $\phi = 19^o 40'$N, $\Lambda = 17^o 05'$W (southwest of Banc d'Arguin) in springtime 1972 according to Meincke *et al.* (1975).

TABLE 2. Characteristic Periods (d), Wave Lengths (km), Phase Speeds (km/day) of the First Three Mode Numbers of Free CSW's with the Zero Group Velocity, Calculated with a Simple Empirical Profile Approximation off NWA and off SWA near Latitude 21.5°N, S.

$\phi = 21.5^o$ N, S	NWA			SWA		
n	T^o	λ^o	c^o	T^o	λ^o	c^o
1	2.5	440	176	3.1	850	274
2	4.0	280	70	5.0	600	120
3	6.0	200	33	7.2	350	49

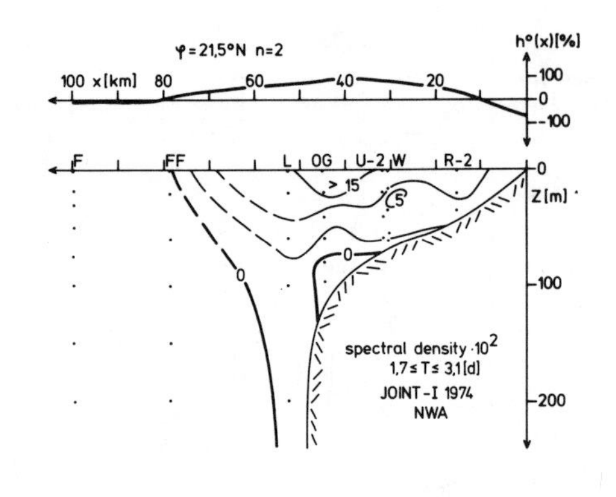

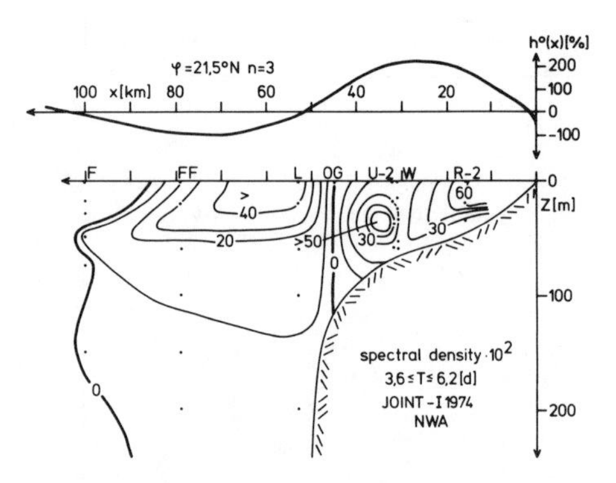

Fig. 7. Kinetic current energy (cm^2/s^2) for the period range between 6 and 2 days according to values by Pillsbury *et al.* (1974) in comparison to the eigenfunction $h^o(x)$ of sea level for the modes n = 2, 3 in percentage of $h_a(x = 0) = 2$ cm.

region. Under similar conditions a narrow shelf produces smaller mesoscale variations in time and space than a broad one, but more intensive with horizontal pressure gradients.

The values of phase speed mark the fact that this wave parameter is of the same order for $n \geq 3$ as the mean alongshore current components of the steady state current part. Therefore the mode numbers $n \geq 3$ are not independent of the mean current and non-linearities dominate. The linear theory of CSW's breaks down. As was pointed out by Allen (1976b) energy flux is given between the different modes as a result of the alongshore variations both of coastal configuration and of bottom topography. The dispersion relation is deformed by these effects. If the phase speed is shifted to such low values by non-linear interactions, dynamically unstable high modes are probable. An energy transfer is possible from these unstable modes of CSW's into barotropic eddies with a diameter of some nautical miles as pointed out by Mysak (1977). Such eddies are especially evident in the shelf-edge zone where the horizontal parameter gradients have a maximum. Therefore the surface frontal system outlined in Fig. 1

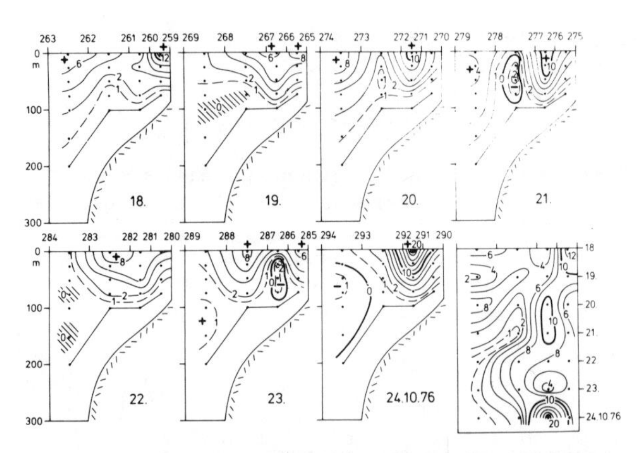

Fig. 8. Daily variations of the calculated geostrophic part of the alongshore component of current (cm/s , relative to the 400-dbar reference level) perpendicular to the coast line off Dune Point ($\phi = 20^oS$) in October 1976 by Wolf (personal communication). The variation of surface conditions with time are given in the lower right-hand panel. The station distance is 10 n.m.

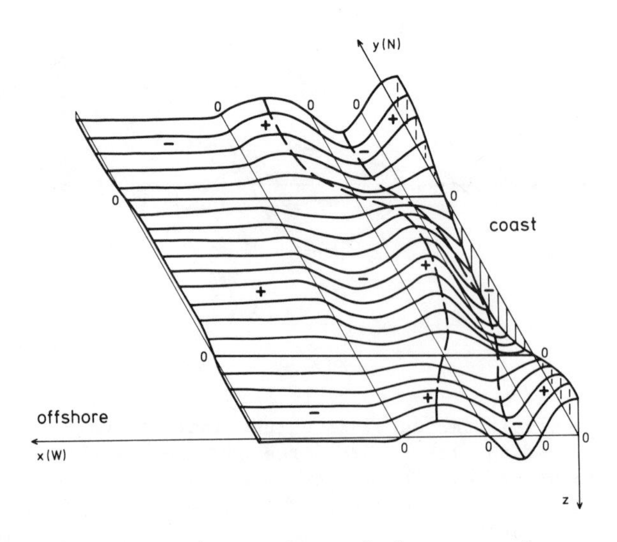

Fig. 9. Mesoscale patterns of the sea-surface elevation $h^o(x)$ for mode n = 3.

is primarily influenced by mesoscale dynamics. The observations in all upwelling areas clearly demonstrate that the mesoscale processes are essentially determined by baroclinic effects in the shelf-edge zone.

The modification of shelf-wave dispersion curves by continuous stratification (without a strong pycnocline over the shelf) has been investigated numerically by Huthnance (1978), from which work it follows that trapped baroclinic motions are important in the shelf-edge zone. On the other hand many mesoscale experiments show a typical barotropic structure in upwelling regions. There-

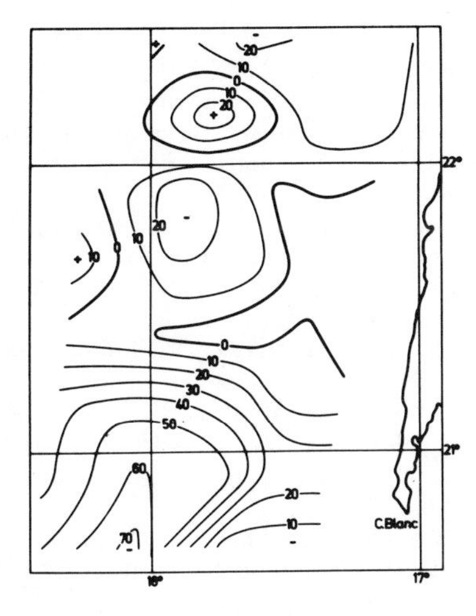

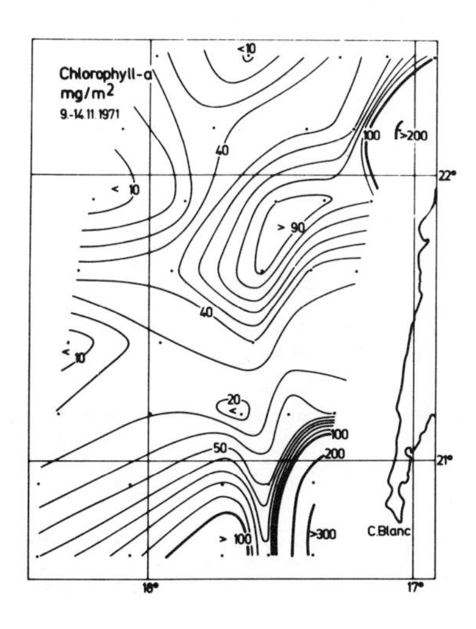

Fig. 10. Difference of the dynamic topography of the sea surface in dyn mm relative to the 500-dbar reference level (25.10./29.10.), (9.11./14.11.1971) and the chlorophyll *a* distribution between Cap Blanc and Cap Barbas according to values by Schulz *et al.* (1973).

fore, it is highly probable that barotropic CSW's determine essential mesoscale structures. The influence of CSW modes on the current fluctuations and corresponding period ranges during the JOINT-I observations are demonstrated in Fig. 7, drawn from current spectra by Pillsbury, Bottero, Still, and Mittelstaedt (1974). The graphical treatment was objectively limited by values lower than 500 cm^2/s^2, and lower values are arbitrarily assumed to be zero. Comparison of the structure of the eigenfunctions $h^0(x)$ for n = 2, 3 with the structure of kinetic current energy perpendicular to the coast demonstrates that the maxima of kinetic current energy are observed where a peak factor of potential water level energy is estimated. Further the energy level of mode n = 3 is twice the energy level of mode n = 2. This observation corresponds with both Figs 5 and 6. Thus we derive the hypothetical deduction that the barotropic dynamics sometimes stamp their structure on the baroclinic dynamics. An example is given in Fig. 8; the daily variations of geostropic alongshore current calculated by the thermal wind equation after Wolf (personal communication) reflect a typical structure normal to the coast, a structure influenced by CSW modes off SWA with a period range of 4 or 5 days at the sea surface (lower right-hand frame of Fig. 8). This period could correspond to the period of T^0 (n = 2) = 5d of Table 2. The current measurements by Hagen (1979) prove that on the shelf off Namibia this period for the alongshore component is independent of depth.

A second example of the influence of barotropic CSW dynamics on oceanographic field distributions is given for the NWA upwelling region. The surface topography generated by mode n = 3 is schematically presented in Fig. 9. This chequered structure of potential energy caused by variations of h^0(n = 3) with time is also observed in the difference representation of dynamical sea surface topography (mm) relative to a 500-dbar reference level according to values obtained by Schulz, Schemainda, and Nehring (1973).

Ship observations have been carried out from south to north in the directions of travel of CSW's. The mean meridional ship speed was roughly an integer of the phase velocity for mode n = 3. The zonal distances of this chequered difference structure harmonize with the $h^0(x)$-scales of mode n = 3 in Fig. 5. Their alongshore scale corresponds with the wavelength for mode n = 3 as given in Table 2. This situation is compared in Fig. 10 with the chequered chlorophyll *a* distribution between Cap Blanc and Cap Barbas according to values by Schulz *et al.* (1973) for the same observation period.

The examples support the importance of free barotropic CSW's with a permanent energy accumulation by $c_g = 0$ to their modification of mesoscale upwelling processes. On the other hand, it is clear that the mode number with energy accumulation changes in the meridional direction with the shelf profile as well as with the energy input into the CSW's during the annual course of the forcing function, for instance by the annual cycle of the northeast and southeast trade wind systems.

Conclusions

In the future it will be necessary to examine intensively such problems as:

1) The resonance action between wind and mode structure of forced CSW's,
2) The mutual energy exchange between the dynamic stable low modes,
3) The energy flux between the dynamic stable and dynamic unstable modes,

4) The role of unstable mode structures in respect to formation of energy-rich mesoscale eddies, and
5) The estimation of typical life cycles of mesoscale eddies with regard to a practical use for the ecosystem modelling in upwelling areas.

References

Adams, J.K., and V.T. Buchwald, The generation of continental shelf waves, *Journal of Fluid Mechanics, 35,* 815-826, 1969.

Allen, J.S., Some aspects of forced wave response of stratified coastal regions, *Journal of Physical Oceanography, 6,* 113-119, 1976a.

Allen, J.S., Continental shelf waves and alongshore variations in bottom topography and coastline, *Journal of Physical Oceanography, 6,* 864-878, 1976b.

Clarke, D.J., and J.P. Louis, Edge waves over an exponential continental shelf in a uniformly rotating ocean, *Deutsche Hydrographische Zeitschrift, 28,* 168-179, 1975.

Cutchin, D.L., and R.L. Smith, Continental shelf waves: low frequency variations in sea level and currents over the Oregon continental shelf, *Journal of Physical Oceanography, 3,* 73-82, 1973.

Efimov, V.V., and E.A. Kulikov, Adaptive wave number - frequency spectrum analysis of the trapped waves, *Atmospheric and Oceanic Physics, 14,* 748-756, 1978 (in Russian).

Hagen, E., Zur Dynamik charakteristischer variationen mit barotropem Charakter in mesoskalen ozeanologischen Feldverteilungen küstennaher Auftriebsgebiete, *Geodätische und Geophysikalische Veröffentlichungen, Reihe IV, Heft 29,* 71 pp., 1979.

Hart, T.J., and R.I. Currie, The Benguela Current, *Discovery Reports, 31,* 123-298, 1960.

Huthnance, J.M., On coastal trapped waves: analysis and numerical calculation by inverse iteration, *Journal of Physical Oceanography, 8,* 74-92, 1978.

Kundu, P.K., J.S. Allen, and R.L. Smith, Modal decomposition of the velocity field near the Oregon coast, *Journal of Physical Oceanography, 5,* 683-704, 1975.

Meincke, J., E. Mittelstaedt, K. Huber, and K.P. Koltermann, Strömung und Schichtung im Auftriebsgebiet vor Nordwest-Afrika, Deutsches Hydrographisches Institut, No. 2149/13, *Meereskundliche Beobachtungen und Ergebnisse, 41,* 117 pp., 1975.

Mysak, L.H., On the stability of the California Undercurrent off Vancouver Island, *Journal of Physical Oceanography, 7,* 904-917, 1977.

Pillsbury, R.D., J.S. Bottero, R.E. Still, and E. Mittelstaedt, Wind, currents and temperature off northwest Africa along 21°40'N during JOINT-I. A compilation of observations from moored current meters, Vol. 8, *Data Report 62,* 143 pp., Oregon State University, 1974.

Reid, R.O., Effect of Coriolis force on edge waves. I. Investigation of the normal modes, *Journal of Marine Research, 16,* 109-144, 1958.

Schulz, S., R. Schemainda, and D. Nehring, Beiträge der DDR zur Erforschung der küstennaben Wasserauftriebsprozesse im Ostteil des nördlichen Zentral-atlantiks, Teil III: Das ozeanographische Beobachtungsmaterial der Messfahrt vom 16. 9. - 17. 12. 1971, *Geodätische und Geophysikalische Veröffentlichungen, Reihe IV, Heft 10,* 65 S., 1973.

Wunsch, C. and E.A. Gill, Observations of equatorially trapped waves in Pacific sea level variations, *Deep-Sea Research, 23,* 371-390, 1976.

THE PROPAGATION OF TIDAL AND INERTIAL WAVES IN THE UPWELLING REGION OFF PERU

W.R. Johnson

College of Marine Studies, University of Delaware, Lewes, Delaware 19958

Abstract. Current meter observations made as part of the JOINT II study were analyzed for a 10-day period (March 19-29, 1977), which coincided with an upwelling wind event and the passage of a coastal trapped wave. The semi-diurnal motions had larger amplitude in the alongshore than in the onshore-offshore component by a ratio of approximately 3:1. The diurnal motions were largest in the surface layer (less than 15 m deep) on the inner shelf and were smaller elsewhere, consistent with wind forcing by the diurnal sea breeze. The inertial motions were highly variable over time in both amplitude and phase, but amplitudes as large as 10 cm/s persisted for a day in response to the change in the wind forcing. The inertial amplitude was largest near the surface at mid-shelf, but amplitudes were significant over the water column throughout the shelf region. The propagation of inertial energy through the slowly-varying mean flow is analyzed with ray diagrams. Regions of possibly enhanced mixing are identified.

Introduction

As part of the Coastal Upwelling Ecosystems Analysis program (CUEA), a field program, entitled JOINT II, was undertaken off Peru during 1976 and 1977. The purpose of the overall program was to investigate the relationship between the physical environment and the high biological productivity.

Tidal and inertial period motions were studied from the current meter observations acquired as part of the program. The current measurements consisted of 35 records from Oregon State University (OSU) and NOAA Pacific Marine Environmental Laboratory (PMEL) current meters and 33 time series from one University of Miami Cyclesonde (an unattended profiling current meter). Analyses were over a 10-day period, 0400 19 March to 1000 29 March, 1977. The positions of the current meter arrays are shown in Plate III, except for the Cyclesonde, which was moored on the 200-m isobath approximately 5 km north and west of the Ironwood mooring. The temperature, salinity, and density observations consisted of 11 C-line sections (the primary hydrographic section, near Cabo Nazca) from the R.V. *Iselin*, four from the R.V. *Melville*, and one from the R.V. *Cayuse* during the 10-day period. The current meter and wind data were rotated 45° counterclockwise from north to align with the general direction of the coastline (Brink, Halpern, and Smith, 1980). Thus, the velocity discussion is presented in onshore-offshore ($\underline{u}$, positive onshore) and alongshore ($\underline{v}$, positive equatorward) components. From these, the mean flow and density fields were calculated and least squares frequency and complex demodulation analyses were carried out.

Data

The wind stress was measured at the coast (Cabo Nazca) and at two buoy positions offshore (PSS, 4 km and PS, 12 km offshore) (Fig. 1). The wind stress was modulated by a diurnal sea breeze, which was strongest at the coast and decreased offshore. The winds were usually weak at night, increasing equatorward and onshore during the days. During a wind event on 25 and 26 March the wind maintained near-daytime strength during the nighttime periods.

The mean flow during the 10-day period was similar to the 3-month (March through May) mean flow observed by Enfield, Smith, and Huyer (1978) and Brink *et al.* (1980). The mean alongshore flow (Fig. 2) was equatorward near surface and near shore and poleward sub-surface and offshore. The core of the poleward undercurrent over the shelf was between 50 and 60 m deep, and its speed increased from 10 cm/s at mid-shelf to more than 20 cm/s at the shelf break. The mean onshore-offshore flow (Fig. 3) was offshore near the surface in a 5- to 10-m thick Ekman layer. Below the surface Ekman layer, over the mid and inner shelf, the flow was onshore between 30 and 80 m, with values as high as 3 cm/s. The flow reversed to offshore near the bottom over the mid and inner shelf, consistent in direction with a bottom Ekman layer associated with a poleward undercurrent near the bottom. At the shelf break and outer

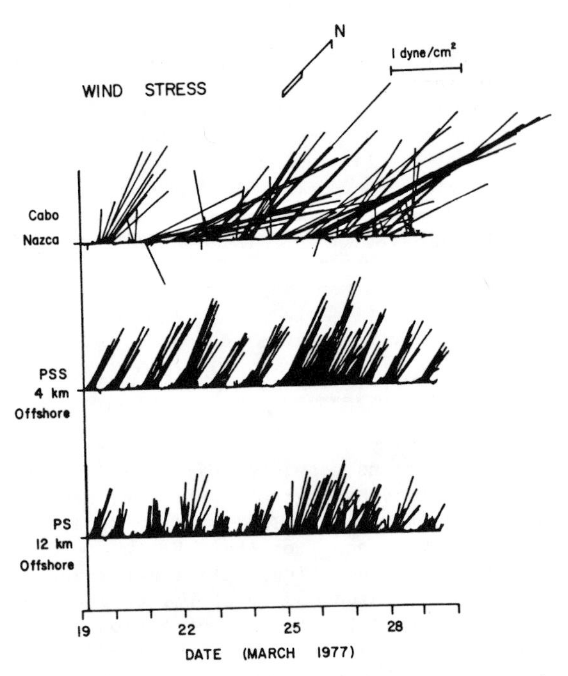

Fig. 1. Wind stress at the coast (Cabo Nacza) and at two buoy positions offshore (PSS and PS).

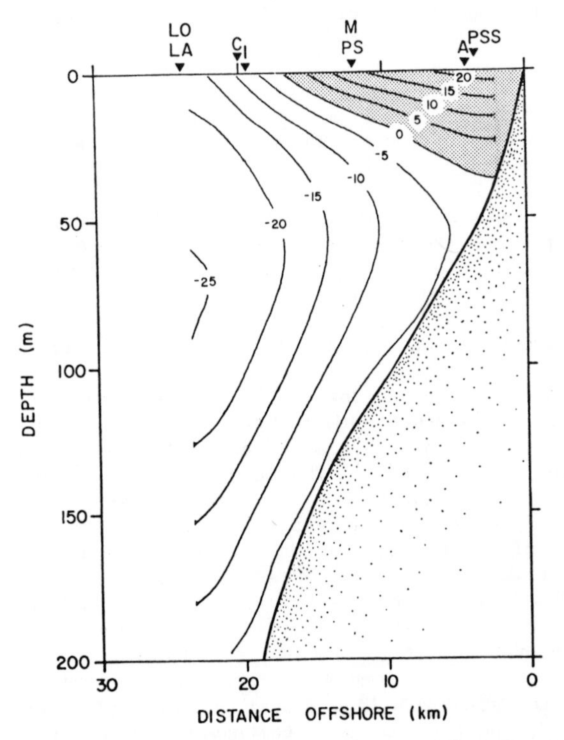

Fig. 2. Average alongshore velocity from the current meters for the 10-day period. The top scale indicates the positions of the current meters shown in Plate III. The shaded region denotes equatorward flow. Values are in cm/s.

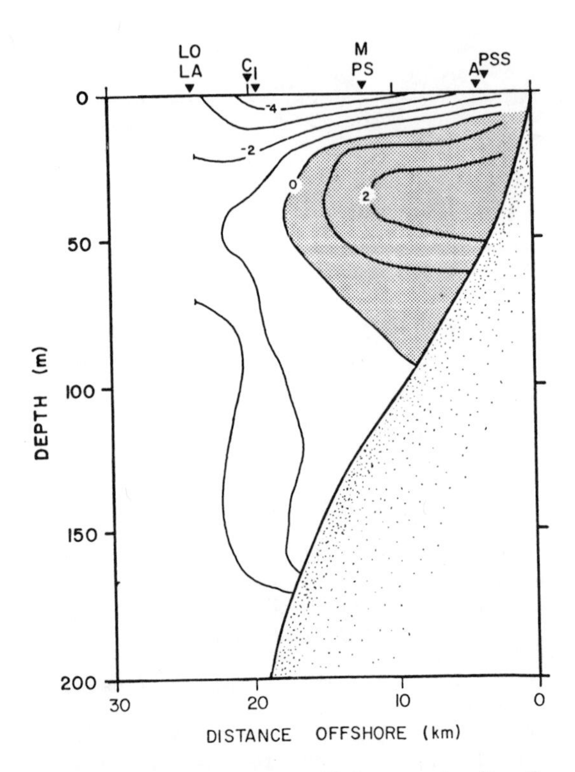

Fig. 3. Average onshore-offshore velocity from the current meters for the 10-day period. The shaded region denotes onshore flow. Values are in cm/s.

shelf (200 to 600 m), the flow was offshore throughout the water column, with values as large as 3 cm/s at 100 m. The mean onshore-offshore flow pattern indicated the possibility of alongshore convergence or divergence, because two-dimensional (x, z) mass balance was not achieved in all profiles. Brink *et al.* (1980) also found a mass divergence between the shelf break and mid-shelf and a mass convergence between mid-shelf and the coast in the 3-month (March through May) mean flow pattern.

The average density (*sigma-t*) cross section along the C line (Fig. 4) was calculated from 16 individual sections during the 10-day period. Above about 75 m, the isopycnals sloped upward toward the coast. Below, the isopycnals sloped downward toward the coast. The isopycnal slopes were consistent with a nearly geostrophically balanced poleward undercurrent in the interior. Due to the leveling of the isopycnals very near the coast (Fig. 4) in an observed strong shear zone (Fig. 2), the shear in the near-surface, inner shelf, alongshore equatorward flow appeared to be in part ageostrophic.

Analysis

Least squares frequency analysis of the current meter data was conducted for three frequencies,

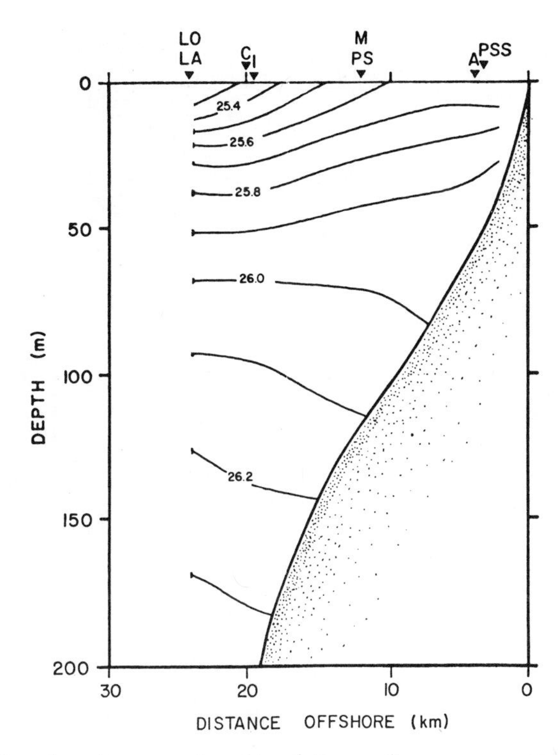

Fig. 4. Average density (*Sigma-t*) cross section along the C line.

semi-diurnal (M2, 1.93 cpd), diurnal (01, 0.93 cpd), and inertial (I, 0.53 cpd at $15^{o}30'$S). The results for these frequencies were representative of a band around the central frequency, rather than an astronomical tidal line. The alongshore semi-diurnal kinetic energy (ρ_o $v^2/2$, Fig. 5) dominated the onshore-offshore semi-diurnal kinetic energy (ρ_o $u^2/2$, Fig. 6), by about four to one. The alongshore semi-diurnal tidal signal had nearly uniform phase over the entire shelf, with all the values within 50 degrees of each other. The total diurnal kinetic energy ρ_o $(u^2 + v^2)/2$ was maximum near the surface near the coast (Fig. 7). A region of higher values extended along the bottom from the inner shelf to near the shelf break. The total inertial kinetic energy (Fig. 8) was maximum at the surface over the mid-shelf. A secondary maximum occurred at 26 m at Agave (inner shelf). A broad minimum in inertial kinetic energy occurred over the mid-shelf from about 25 to 75 m, with larger values below. The interpretation of the results is that the semi-diurnal tide appeared to be predominantly barotropic and an alongshore propagating wave, while both the diurnal and inertial motions were forced by the wind, resulting in maximum values near the surface.

Smith (1978) and Brink, Allen, and Smith (1978) found that the spectra of current fluctuations off Peru were dominated by energy with periods of 6 to 10 days. This 10-day case study represents a passage of one of these low-frequency waves. To study the event, the slowly-varying mean flow was calculated as uniform averages over 48-h overlapping segments, each separated by 12 h. (The results for one of the moorings are shown in Fig. 9.) In all the averaged time series, especially that of the alongshore flow, the occurrence of a low-frequency coastal trapped wave was apparent over the 10-day period (Brink *et al.*, 1980). The alongshore flow at all the moorings reached an equatorward maximum on March 22 or 23. At Mila (Fig. 9), a mid-shelf station, the wave passage resulted in equatorward flow throughout the water column. Also at mid-shelf there was strong onshore flow at 19 and 39 m during the equatorward maximum, with reversal to offshore at 19 m later in the period.

The time series were complex demodulated at the semi-diurnal, diurnal, and inertial frequencies. The complex demodulations were performed over the 48-h segments together with the slowly-varying mean flow calculation. The amplitude and phase of the semi-diurnal motion varied smoothly over the 10-day period, especially in the stronger alongshore component. The onshore-offshore component was more variable, especially in phase. The diurnal motion had larger fluctuations than the semi-diurnal motions in both components, but they were not obviously related to either the long coastal trapped wave or the wind forcing. The inertial motions responded to the wind forcing,

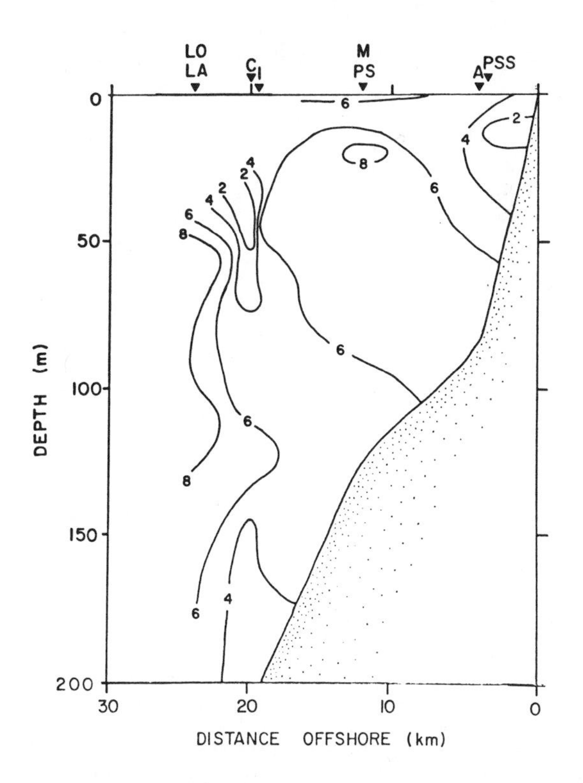

Fig. 5. Alongshore semi-diurnal kinetic energy (ρ_o $v^2/2$) in erg/cm^3.

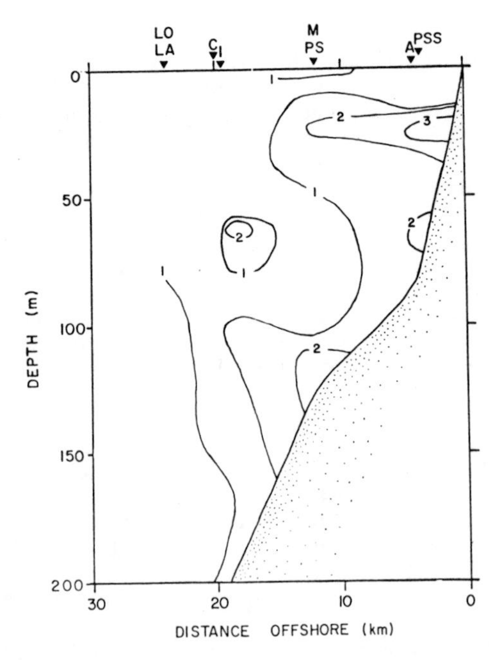

Fig. 6. Onshore-offshore semi-diurnal kinetic energy (ρ_o $u^2/2$) in erg/cm^3.

especially near the surface over the inner shelf (PSS and Agave). The maximum response at PSS in the onshore-offshore component was on 25 March, during the wind event (Fig. 10a). The alongshore inertial response at PSS was less distinct, with large amplitudes occurring on 22 and 23 March.

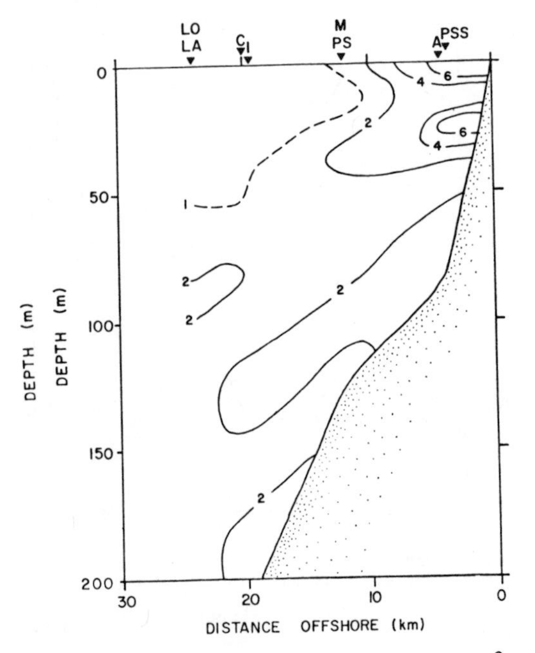

Fig. 7. Diurnal total kinetic energy $|\rho_o(u^2 + v^2/2)|$ in erg/cm^3.

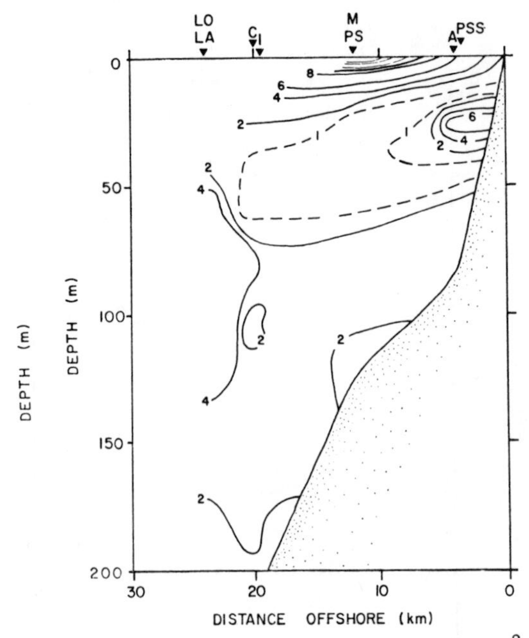

Fig. 8. Inertial total kinetic energy $|\rho_o(u^2 + v^2/2)|$ in erg/cm^3.

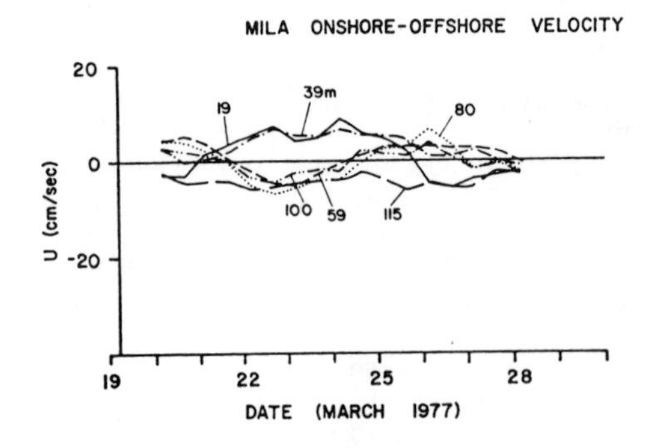

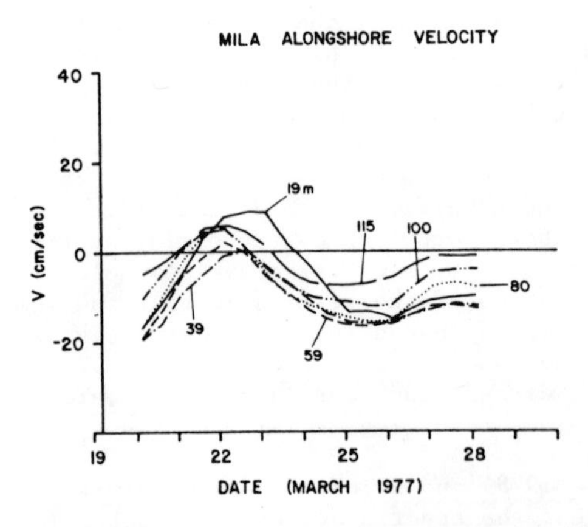

Fig. 9. Time series of 48-h averaged currents from Mila.

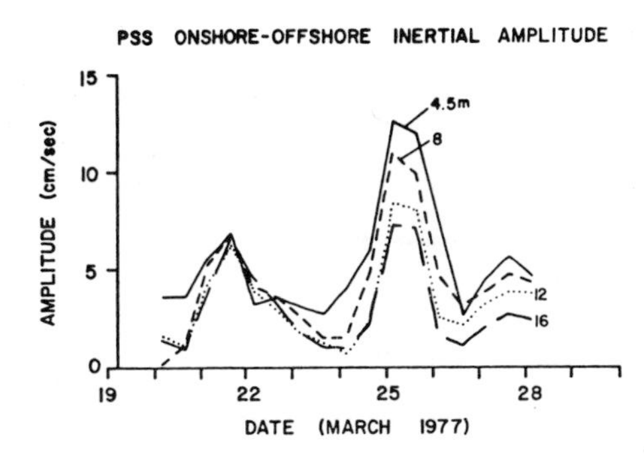

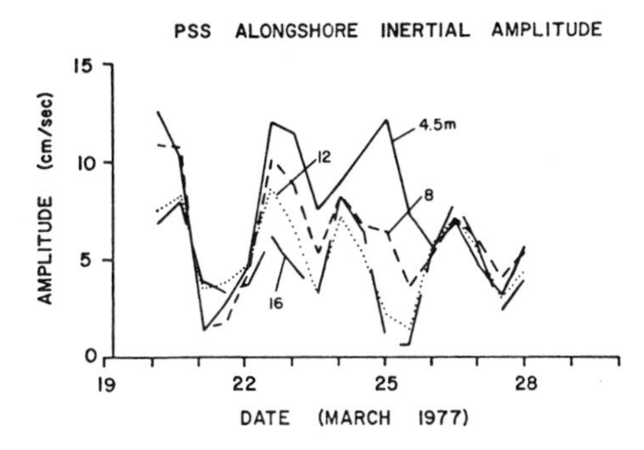

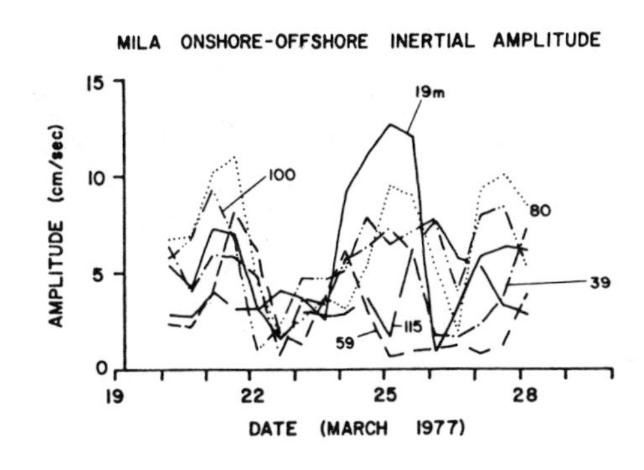

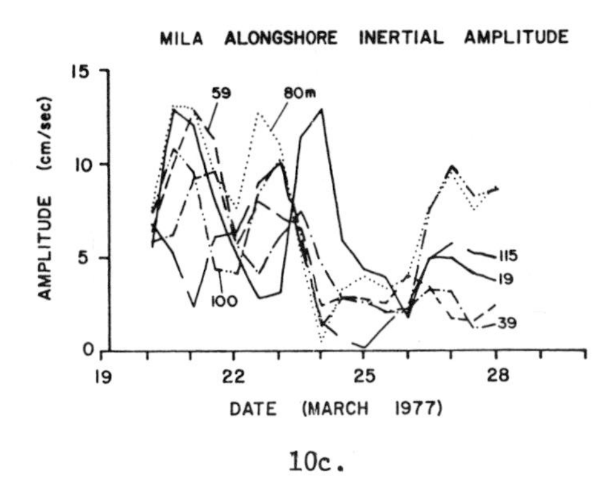

Fig. 10. Time series of inertial period complex demodulation amplitude from a) PSS, b) Agave, c) Mila, and d) Ironwood.

10c.

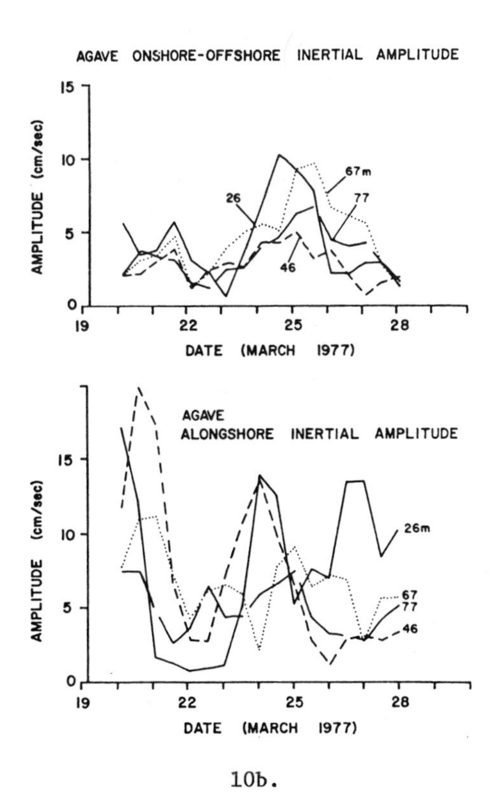

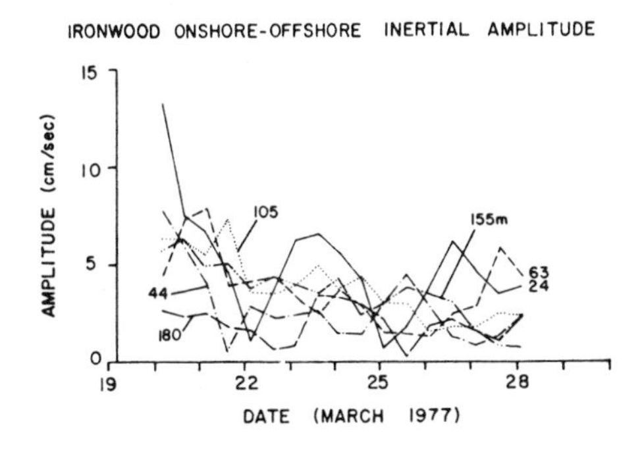

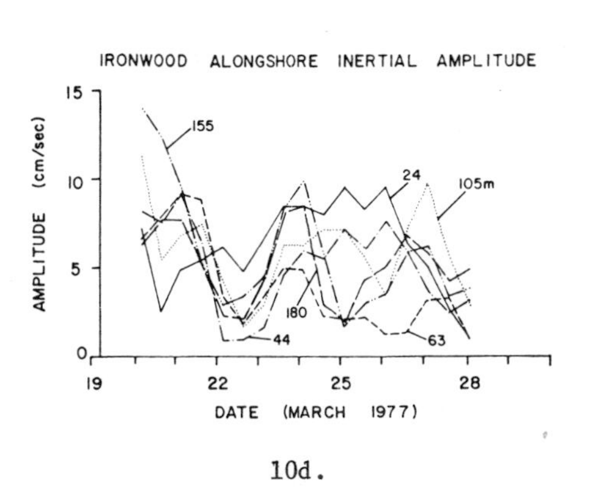

10b.

10d.

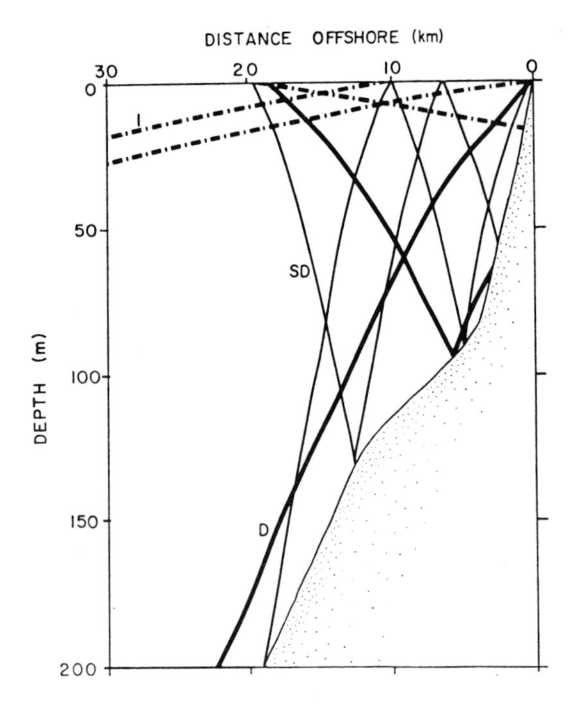

Fig. 11. Ray diagram for a) the case excluding the effects of horizontal density gradients and horizontal gradients of the alongshore flow and b) the case including the effects of horizontal density gradients and horizontal gradients of the alongshore flow.

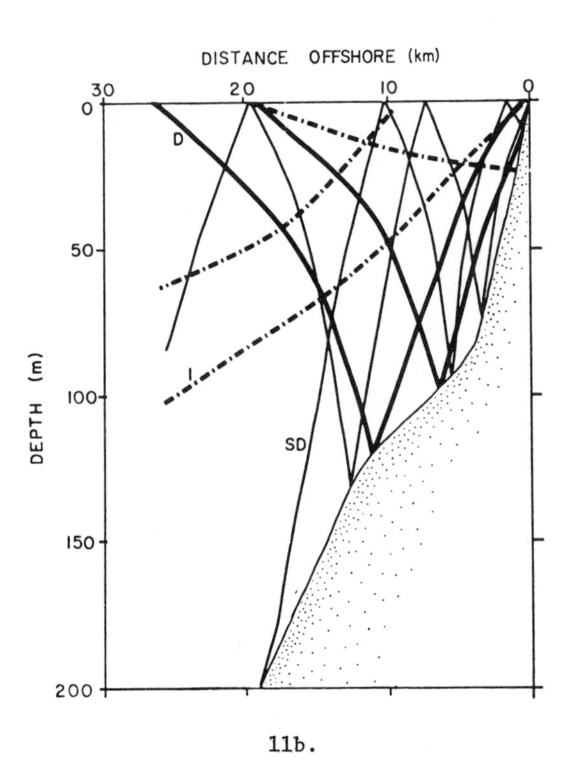

11b.

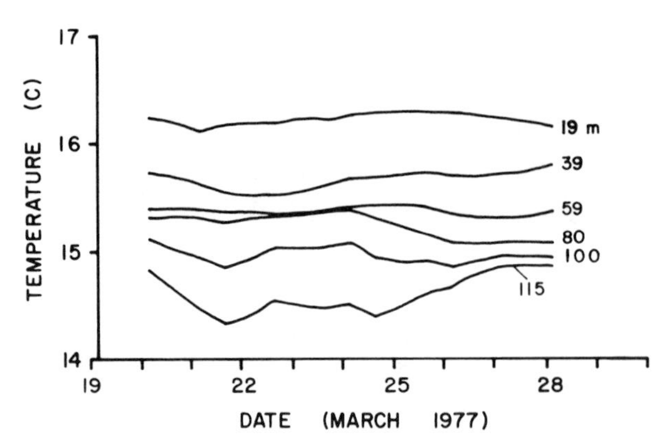

Fig. 12. Time series of temperature from Mila.

The response in the onshore-offshore component at Agave (Fig. 10b) had its largest amplitudes on 24 and 25 March at 26 and 67 m, with less amplitude at an intermediate depth of 46 m. The alongshore component response at Agave was maximum on 24 March and primarily at 26 and 46 m. A noticeable response was evident at Mila and Ironwood (Figs 10c and 10d). The response in the alongshore component leads that in the onshore-offshore component for most series. The results indicate that although the spectra from the current meters did not exhibit inertial peaks in many cases (Smith, 1978; Brink *et al.*, 1980), inertial motions were intermittently very energetic off Peru.

Ray Analyses

The method of characteristics was used to interpret some of the observed wave properties. The characteristics (rays) are lines of constant phase and are colinear with the group velocity and energy flux at any point (Mooers, 1975). Two sets of ray calculations were made, the first excluding the effects of horizontal density gradients and horizontal gradients of the alongshore flow (Fig. 11a), and the second including those effects (Fig. 11b) (Mooers, 1975). The inertial rays (dash-dotted lines) show much steeper slopes in Fig. 11b than in Fig. 11a, and indicate more penetration into the interior over the shelf, due to the influence of the frontal zone of the mean flow. The slopes of the inertial rays in Fig. 11b are roughly parallel to the isopleths of the inertial kinetic energy (Fig. 8), which suggests some consistency between observations and theory. The diurnal rays (heavy solid lines) also indicate steeper slopes (hence, a shortening of the horizontal wavelengths) due to the gradients (Fig. 11b). The slopes of the diurnal rays appear to be steeper than the isopleths of the diurnal kinetic energy (Fig. 7). Offshore propagating rays originating from the sea surface less than 2 km offshore would reflect from the bottom over the shelf. Thus, part of the energy from the diurnal motions forced by the sea breeze may be dissipa-

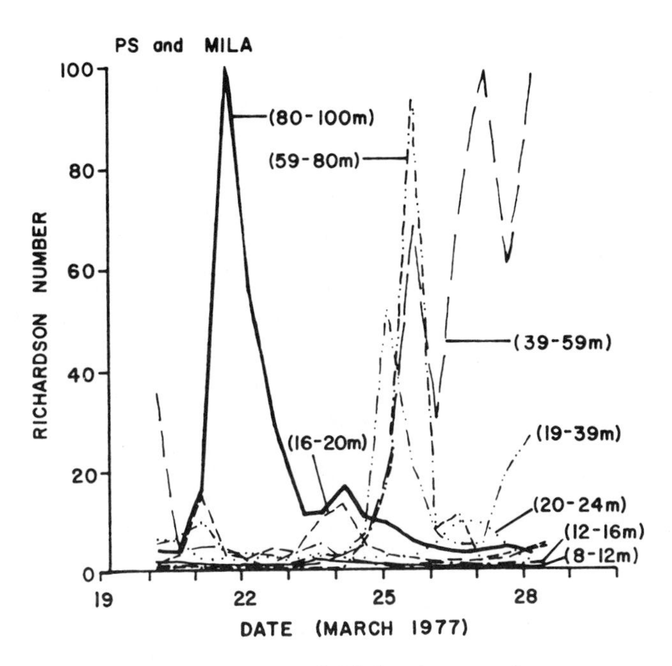

Fig. 13. Time series of Richardson numbers a) from PS and Mila. Calculations for PS were 8-12, 12-16, 16-20, and 20-24 m, and for Mila were 19-39, 39-59, 59-80, and 80-100 m; and b) from PSS and Agave. Calculations for PSS were 8-12 and 12-16 m and for Agave were 24-46, 46-67, and 67-77 m, and c) from Ironwood.

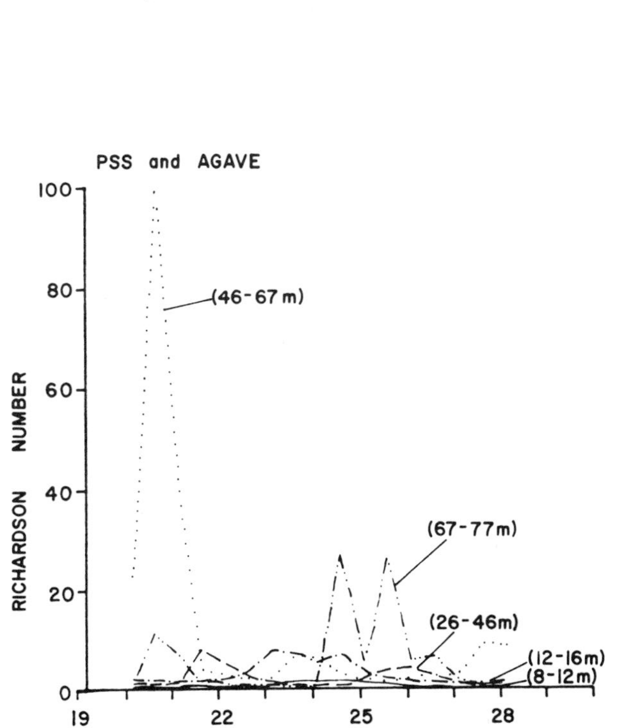

13b.

13c.

ted over the shelf. The semi-diurnal rays (thin solid lines) were included in Figs 11a and 11b only for completeness, because the alongshore semi-diurnal component indicated a predominately barotropic in-phase signal over the entire shelf, and they were not significantly affected by the mean flow gradients.

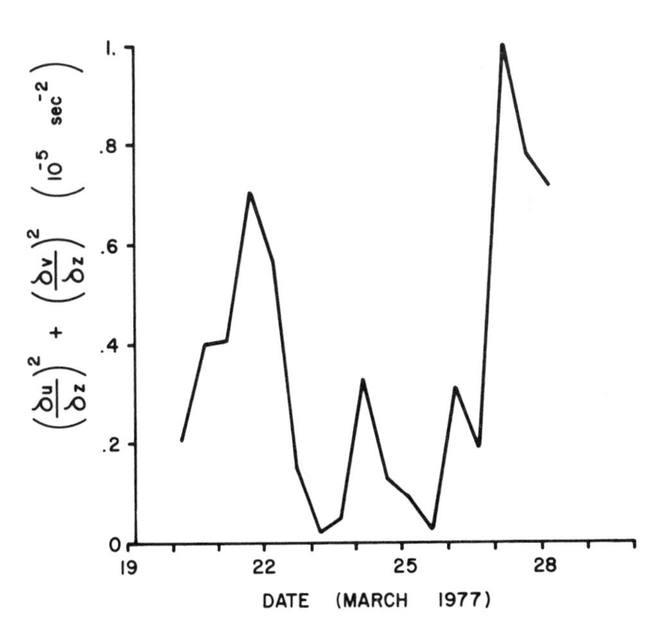

Fig. 14. Time series of current shear between 59 and 80 m from Mila.

Richardson Numbers

Gradient Richardson numbers were calculated for pairs of neighboring current meters, using the 48-h averaged temperature and velocity data. At mid-shelf (Mila-PS), a nearly isothermal layer (based on current meter temperature) from about 50 to 80 m formed on 22 March and remained until 24 March (Fig. 12), resulting in very small Richardson numbers over the period (Fig. 13a). (This vertically well-mixed layer was confirmed by CTD observations.) The inner shelf (Agave-PSS, Fig. 13b) Richardson numbers also incurred low values during the same period, 22 to 24 March, between 26 and 67 m. The mixing event occurred following a period of maximum shear of the alongshore flow on 21 and 22 March (Fig. 14). It also followed a period of high onshore-offshore inertial amplitude at Mila (Fig. 10c). The alongshore inertial amplitude was large at both Mila and Agave on 21 March and decreased on 22 March, except at 80 m at Mila, where high amplitudes continued. Probably both the shear of the slowly-varying alongshore flow and the energetic inertial motions contributed to the mixing event that followed. At the shelf break (Ironwood, Fig. 13c), Richardson numbers were somewhat lower between 105 and 155 m during 22 to 24 March, although their fluctuations were distinctly different than at mid-shelf. The near-surface Richardson number values at PS (Fig. 13a) and PSS (Fig. 13b) were low, especially following the wind event.

Conclusions

In summary, tidal and inertial motions were observed off Peru as part of JOINT II. The semidiurnal tide had larger amplitude in the alongshore component and had nearly uniform phase everywhere over the shelf. Diurnal motions were strongest near the surface near the coast, and were probably forced by the diurnal sea breeze. Inertial motions were strongest near the surface mid-shelf, as a response to a wind-forcing event. Johnson and Mooers (1980) have shown that inertial motions in upwelling regions can be dissipated in the interior over the shelf. From time series of the Richardson number, a mixing event occurred over the mid and inner shelf at the time of a wind event and the maximum equatorward flow resulting from a low frequency coastal trapped wave passage. The mixing event was related to the shear in the alongshore flow and a maximum in the alongshore inertial amplitude.

Acknowledgements. This research is a contribution to the Coastal Upwelling Ecosystems Analysis program, supported by the National Science Foundation, Office of the International Decade of Ocean Exploration, under grant OCE 77-28354 to the University of Delaware under the direction of C.N.K. Mooers, who also provided helpful editorial comments on a draft of the paper. I gratefully acknowledge the hospitality of R.L. Smith during a summer visit to Oregon State University. D. Halpern, R.L. Smith, and J.C. Van Leer generously provided the data used in this research.

References

Brink, K.H., J.S. Allen, and R.L. Smith, A study of low-frequency fluctuations near the Peru coast, *Journal of Physical Oceanography, 8,* 1025-1041, 1978.

Brink, K.H., D. Halpern, and R.L. Smith, Circulation in the Peruvian Upwelling System near 15° South, *Journal of Geophysical Research, 85,* 4036-4048, 1980.

Enfield, D.B., R.L. Smith, and A. Huyer, A compilation of observations from moored current meters, Volume 12, Wind, current, and temperature over the continental shelf and slope off Peru during JOINT II, *Data Report 70, Reference 78-4,* 347 pp., School of Oceanography, Oregon State University, Corvallis, 1978.

Johnson, W.R., and C.N.K. Mooers, A case study of inertial-internal waves in a coastal upwelling region, Submitted to *Journal of Physical Oceanography,* 1980.

Mooers, C.N.K., Several effects of a baroclinic current on the cross-stream propagation of inertial-internal waves, *Geophysical Fluid Dynamics, 6,* 245-275, 1975.

Smith, R.L., Poleward propagating perturbations in currents and sea levels along the Peru coast, *Journal of Geophysical Research, 83,* 6083-6092, 1978.

A SATELLITE SEQUENCE ON UPWELLING ALONG THE CALIFORNIA COAST

Laurence C. Breaker
Ronald P. Gilliland

National Environmental Satellite Service, Satellite Field Services Station, 660 Price Avenue, Redwood City, California 94063

Abstract. Infrared imagery from the Advanced Very High Resolution Radiometer aboard the TIROS-N and NOAA-6 polar-orbiting satellites is used to form a time-lapse sequence of upwelling along the California coast. The sequence was taken over an 8-day period in early September 1979, during which a brief wind event took place. The effects of the event are clearly evident in the sequence. The upwelling response to the event apparently took place in 9 h or less. Offshore surface velocities have been estimated from the displacement of selected frontal boundaries occurring between successive images. Sea-surface temperatures measured at several coastal locations within the area of satellite coverage were lower immediately following the wind event.

Introduction

Infrared (IR) remote sensing of the sea surface along the U.S. west coast during summer is usually made difficult by the presence of fog and stratus clouds associated with coastal upwelling. Along the California coast, it is quite uncommon during summer, particularly during July and August, for a week of even four or five days to go by without the accumulation of significant low clouds. Fog and stratus clouds form as moist marine air approaches the coast, becomes saturated, and condenses over the cooler upwelled waters. Occasionally the Pacific subtropical high-pressure cell is displaced to the northeast resulting in an offshore flow of dry continental air, which is highly favorable for remote sensing of the sea surface at IR wavelengths. Occasionally weak summer disturbances pass through the area and "flush out" the fog and stratus for brief periods.

During the first two weeks in September 1979, an exceptional period of clear skies occurred. Starting on 3 September, clear skies prevailed along the California coast for almost 10 days. The conditions came about not from the offshore flow of continental air but rather from weak westerly winds, which were not favorable to the maintenance of strong upwelling.

A time-lapse sequence of IR images was constructed from the Advanced Very High Resolution Radiometer (AVHRR) aboard the TIROS-N and NOAA-6 polar-orbiting satellites. Time-lapse techniques have been used with data from synchronous meteorological satellites for about 10 years in the field of operational meteorology (Plew and Thumin, 1975) and more recently in oceanography (Legeckis, 1975; Maul and Baig, 1975). The construction of a time-lapse sequence using data from polar-orbiting satellites brings to light problems that do not exist for data from the geosynchronous satellites.

The final loop contained 22 images over an 8-day period. Examination of the loop and associated wind observations revealed the occurrence of a brief period of increased winds that started on 8 September. Winds increased from 3 m/s or less to 10 m/s or greater in about 12 h at four coastal locations. The period of increased winds starting on September 8th subsequently will be referred to as the September 8th wind event although the event actually spanned several days and was experienced at slightly different times at different locations along the California coast.

Overall, it is our objective to demonstrate the utility of high-resolution IR satellite data in studying the upwelling process along the California coast.

The Setting

The area of interest along the California coast extends just beyond Cape Mendocino to the north and southward to Point Conception. Coverage off the coast varied but usually reached at least 400 km in the offshore direction.

During summer the Pacific subtropical high-pressure cell intensifies and moves north off the west coast. A thermal low also develops inland during the period and acts further to increase the coastal pressure gradient. The net result is a persistent coastal surface wind field that has a mean direction from the NNW.

The synoptic weather scene during the period 3 to 8 September was characterized by a low-pressure system in the Gulf of Alaska and the Pacific high displaced far to the northwest of its normal position. As a result, flow along the California coast was weaker and more westerly than normal because of the weaker pressure gradients there along with lower pressures to the north. A transition occurred on 8 and 9 September as the Pacific high intensified and moved eastward toward the Pacific Northwest. From 9 to 11 September, the Pacific high was characterized by a northeast-southwest trending ridge extending into the Pacific Northwest. For a brief period these conditions produced stronger-than-normal flow of northerly winds along the California coast.

The California Current flows south along the California coast during summer. Flow near the coast is generally southward during late summer and may resemble in some respects the coastal jet that occurs farther north off the Oregon coast. Farther off the coast is the main body of the California Current. A poleward countercurrent is frequently observed at depth over the continental shelf and slope (Reid, Roden and Wyllie, 1958; Wooster and Reid, 1963; Wickham, 1975). The flow is referred to as the California Undercurrent, a current that transports water of equatorial origin north along the California coast. The surface flow near the coast and that portion of the California Undercurrent flowing over the continental shelf may be related to coastal upwelling in the region if the process resembles that off the coast of Oregon where increased (decreased) flow in the alongshore direction is often required for continuity (Allen and Kundu, 1978).

Data Sources and Considerations

To examine the correspondence between surface winds and the surface thermal field, values of surface wind stress were superposed directly on the IR images used in constructing the loop. A boundary layer wind forecast produced by the National Weather Service (NWS) provided the values. The winds were derived from the Limited-Area Fine-Mesh model (LFM) of the NWS (NOAA NWS, 1976). LFM 12- and 24-h forecast mean boundary layer winds valid at 0000 GMT and 1200 GMT were used to compute surface wind stress. The values provided information on wind at three fixed locations over the coastal waters at regular intervals (Fig. 1). The forecast winds correspond to a level of 975 mb. Although frictional effects are often considered in extrapolating boundary layer winds to the surface, Riehl, Yeh, Malkus, and LaSeur (1951) and Kraus (1968) have shown that wind speed and direction over the eastern North Pacific are nearly constant with height in the lower layers during summer. Consequently no adjustments have been made to the boundary layer winds used here.

The following satellite and sensor specifications are appropriate to this study: The average altitudes of TIROS-N and NOAA-6 are 854 and 833 km respectively. Their orbits are near-polar with a nodal period of about 100 min. Considering both satellites, four overhead passes are received per day at the ground receiving station in Redwood City, California. The AVHRR is a multichannel scanning radiometer that receives radiation in four spectral bands. In this study, we have used the IR band from 10.5 to 11.5 microns. This spectral window is transparent to most constituents in the atmosphere except water vapor. Absolute measurements of sea-surface temperature, for example, require an atmospheric correction of the order of 2 to 4°C at mid-latitudes. The spatial resolution of the AVHRR is 1.1 km at the satellite subpoint but decreases away from nadir. The thermal resolution after processing is approximately 0.5°C. Digital data processing includes image enhancement*, rectification**, and enlargement. Processed images are shown in Fig. 1.

There are at least two reasons why time-lapse techniques are not usually applied to polar-orbiting satellite data. First, atmospheric motions are rapid compared to the sampling rates available from polar-orbiting satellites. Thus, film loops constructed from polar-orbiting satellite data appear to be better suited to resolving oceanic phenomena where motions take place more slowly. Second, geometric distortions in the data received from polar-orbiting satellites exist and make this kind of information difficult to work with. In more detail, several sources of geometric distortion exist in the data received from the AVHRR aboard the TIROS-N polar-orbiting satellites. The major source of geometric distortion is earth curvature, a non-linear distortion that has been removed by remapping (repeating) the picture elements along each scan line (Legeckis and Pritchard, 1976). A second-order systematic distortion arises from earth rotation and has not been removed. As the earth rotates beneath the path of a polar-orbiting satellite, a skewness (i.e., a cumulative displacement of successive scan lines) arises in the plane of the image that depends on the direction in which the spacecraft is moving. As the orbit is inclined, the skewness or distortion has both a horizontal and a vertical component. The vertical component results in a compression in the north-south direction. Overall, it is impossible to register ascending and descending passes exactly. The net effect after animation is a "jitter" that grows radially from the geographic point upon which the

*Image enhancement refers to the process of increasing contrast in the output display over a selected range of input values (radiances).

**Rectification refers to the removal of geometric distortion due to earth curvature (i.e., panoramic distortion).

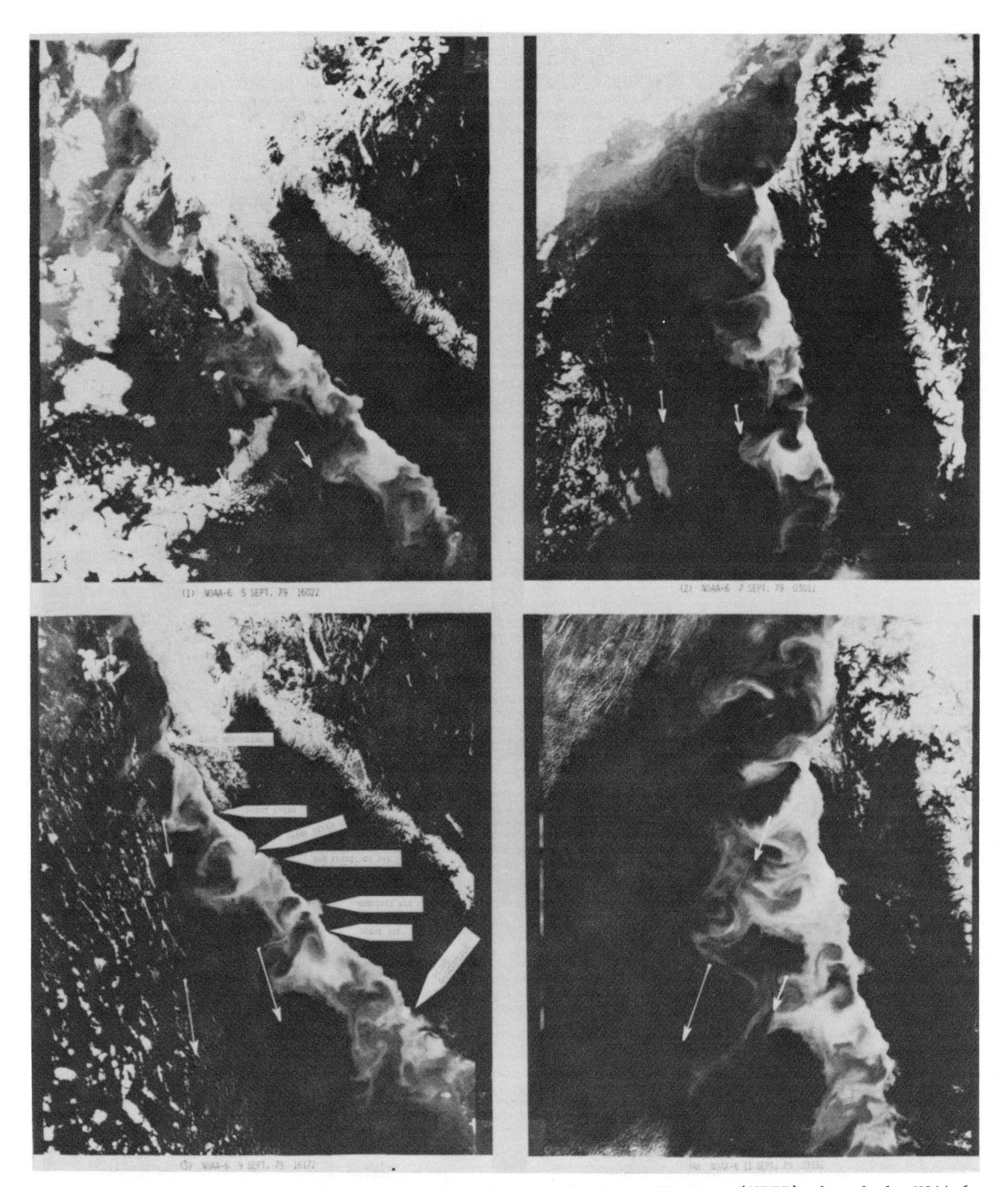

Fig. 1. Enhanced imagery from the Advanced Very High Resolution Radiometer (AVHRR) aboard the NOAA-6 satellite shows upwelling along the California coast before and after the September 8 wind event. Lighter shading near the coast represents lower sea-surface temperatures. Wind stresses overlain on the imagery are taken from the National Weather Service LFM forecast model.

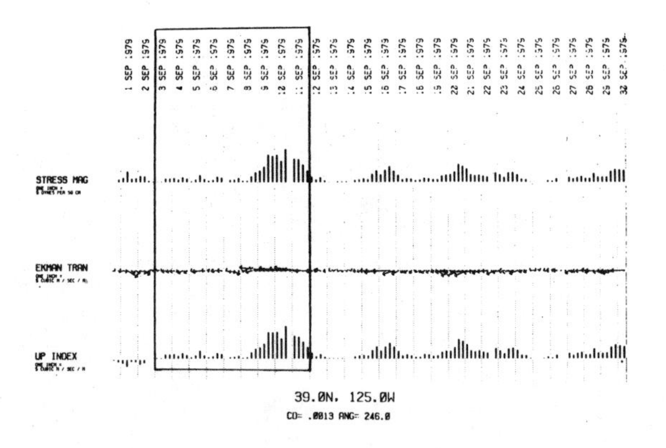

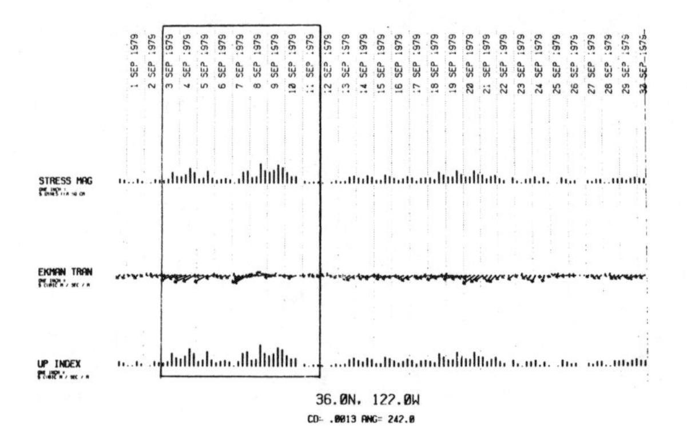

Fig. 2. Calculated offshore Ekman transport at Cape Mendocino (upper) and Point Sur (lower). The initiation of the September 8 wind event is more obvious at Cape Mendocino than at Point Sur (Bakun, 1979).

images are registered. Other sources of image distortion arise from random motions of the spacecraft, including roll, pitch, and yaw. Instantaneous records of the spacecraft attitude are required to apply corrections for these motions. Because the effects are usually small in comparison with the systematic errors already described, they have not been taken into account. Finally, a magnification error of less than 3% arises from the slight difference in altitude of the two spacecraft.

Results

The detailed patterns of upwelling (or upwelled waters) seen along the California coast in Figs 1 and 4 indicate the generally complex and spatially non-uniform character of the process. The areas covered by upwelled water and the convoluted patterns contained within the upwelled water are representative of upwelling conditions seen in previous satellite imagery for the area at this time of year. Further results are based mainly on observations taken from the film loop and as a result are not necessarily supported by the relatively small sample of still images contained in Figs 1 and 4.

Starting at Point Sur, intense upwelling activity appears to take the form of a developing anticyclonic circulation or gyre centered approximately 40 km off the coast. Previous satellite data in the area have shown upwelling patterns of similar scale but with the appearance of cyclonic circulation. The apparent anticyclonic circulation gradually increases in size mainly through a westward expansion during the period of study. However, the circulation within the feature appears to experience maximum intensification just prior to the September 8 wind event. The effect could be due to a brief increase in local winds about 24 h before the main wind event at Point Piños and Point Piedras Blancas (Fig. 3). Fig. 2 (lower half) shows the 6-hourly record of calculated offshore Ekman transport* for the Point Sur area. The September 8 wind event can be identified in the pattern of values of offshore Ekman transport shown in the figure starting at about 1500 local time on 8 September (Bakun, 1979).

Moving north to Monterey Bay, upwelled water appears to originate from within the bay itself. The upwelled waters appear to move farther offshore in response to the September 8 wind event. Near San Francisco, upwelled waters enter and leave San Francisco Bay, and, from the observed time-scales of motion, appear to reflect the strong tidal influence at this location (Conomos, 1979). The observation would probably not have been possible without examination of the animated sequence. Upwelling activity immediately following the September 8 wind event appears to be most intense between San Francisco and Cape Mendocino. The difference in area covered by upwelled water before and after the wind event is greater north of San Francisco than to the south. The 6-hourly pattern of offshore Ekman transport shown in Fig. 2 (upper half) clearly indicates the September 8 wind event. High values of offshore Ekman transport during the period appear to correlate well with the thermal response observed in the area. Finally, a small plume of upwelled water appears directly off Cape Mendocino (Figs 1 and 4). Previous satellite data for the area suggest that plumes of much greater extent often exist at this location.

Next, the ocean response time is considered in terms of changes in IR thermal patterns following the September 8 wind event. According to Yoshida (1967), "processes near coastal boundaries respond more, and faster, to time-variable winds than those in the central oceans". The observation that upwelling often takes place

*Offshore Ekman transport represents the wind-driven volume transport rate of upwelled water directed away from the coast per unit length of coastline. Ekman transport directed offshore is also referred to as the upwelling index (Bakun, 1975).

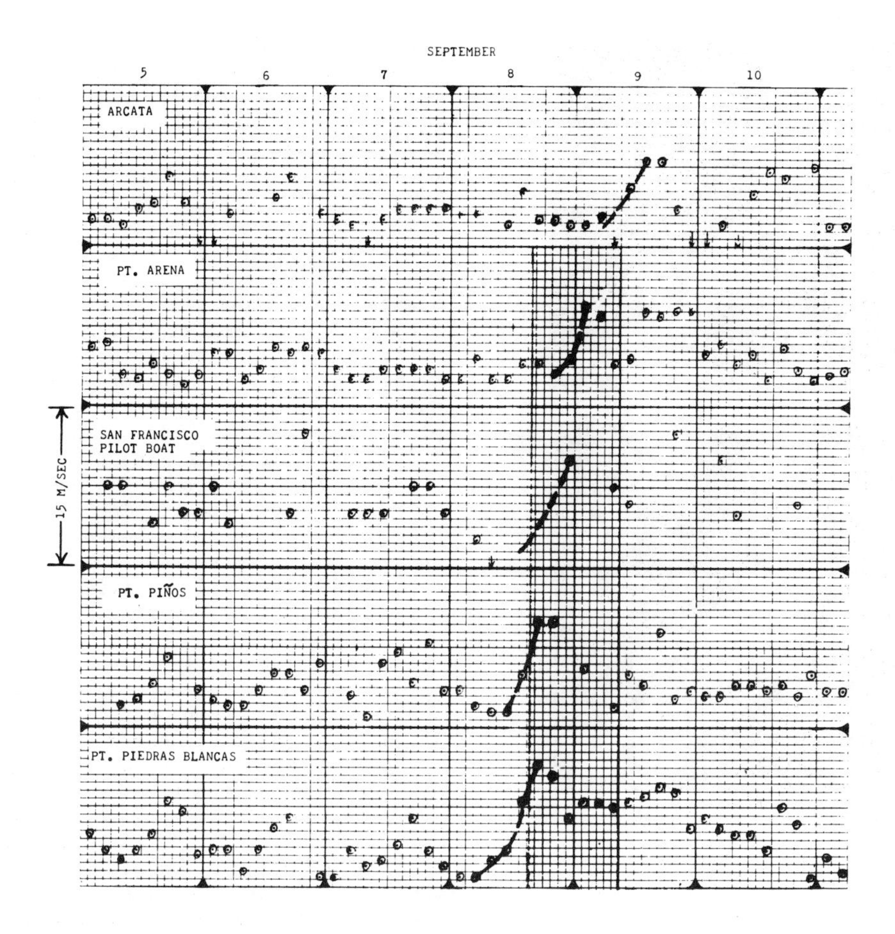

Fig. 3. Coastal wind speed observations at four locations. Shaded area represents a time window during which upwelling was observed to intensify based on satellite imagery. Wind observations were every 3 h, and the satellite images were received, on the average, every 7 to 8 h.

rapidly in response to intensified winds has been made on a number of occasions in the past. Smith, Pattulo, and Lane (1966) showed that upwelling developed rapidly over a 76-h period in early May 1963 off the coast of Oregon. In an area off southwest Africa, marked changes in SST patterns were observed within hours after a wind reversal (Bang, 1973). Smith (1974) indicated that changes in sea level and currents on the Oregon continental shelf associated with upwelling appear to be essentially simultaneous with changes in the wind. Fig. 3 shows wind speed at four coastal locations. The records show the September 8 wind event as it propagates northward along the California coast. At each location, wind speed increases from values of 3 m/s or less to 10 m/s or more in about 12 h. A shaded area or window, which brackets the wind event at each location, has been added. The trailing edge of the window corresponds to the last time just before or during the wind event when no apparent response was detectable in the surface thermal pattern. The leading edge of the window corresponds to the first time immediately following the wind event when a response in the SST field was apparent along the California coast as far north as Cape Mendocino. Because the images were not spaced closely enough in time, it was not possible to observe the local response as it most likely progressed up the coast. The window thus corresponds to an overall period of uncertainty of about 17 h. However, upon closer inspection the window can be used to estimate the upwelling response time associated with the wind event more closely. Point Arena represents the limiting case for estimating response time because it was affected by the wind impulse later than stations to the south. The time difference at Point Arena between the arrival of the wind impulse and the leading edge of the window is about 9 h (Fig. 3). Thus the upwelling response time appears to be about 9 h or less.

Next, an attempt is made to estimate offshore surface velocities associated with advective motion at two locations (Fig. 4). Measurements of this type may be useful because of the difficulty encountered in measuring large-scale surface circulation. Although ocean temperature is

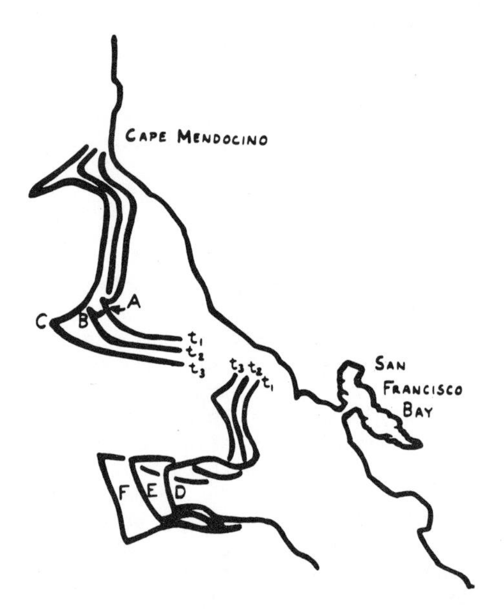

Fig. 4. The upper and lower images above the diagram correspond to times t_1 (9/8 @ 1637Z) and t_3 (9/10 @ 1555Z), respectively, and selected portions have been outlined in the diagram below. The image corresponding to time t_2 (9/9 @ 1617Z) is shown in the lower left-hand corner of Fig. 1. Advective motion has been estimated for the points shown in the diagram and the associated values are tabulated in Table 1.

nonconservative at the surface, thermal patterns associated with upwelling can often be recognized over periods of several days. The preservation of these patterns implies that advection is important at least locally. Because the displacement of individual points associated with selected patterns is estimated over known intervals of time, the technique is Lagrangian. Wind speeds are estimated in the same way at various levels in the atmosphere by tracking individual cloud elements (Anderson, 1974). Images were taken from the same satellite (NOAA-6) and for the same orbital geometries to minimize geometric distortions. Fig. 4 shows the areas and the points within those areas where the measurements of displacement were made. The first area lies about 150 km off Point Arena and the second about 300 km off Monterey Bay. The net speeds and directions are listed in Table 1. Because net displacements have been measured, the velocities so determined represent actual velocities only in the limiting case of rectilinear motion. We include an estimate of the absolute precision of the measurements (Table 1). We compare our values of offshore flow with those of Curtin (1979), taken by conventional means in the coastal upwelling zone off Oregon. Curtin estimated the movement of surface isotherms across the continental shelf to be between 5 and 15 cm/s during a period of active upwelling. Our values are generally significantly greater than those of Curtin. However Curtin's measurements indicate the mean cross-shelf speed of selected isotherms whereas our measurements represent the instantaneous speed of offshore flow associated with selected points along a front. Consequently, differences in offshore flow, above and beyond actual differences, should perhaps be expected in view of the different methods used in estimating these velocities.

A close inspection of the film loop gives the impression that the lower temperatures immediately following the September 8 wind event in certain offshore areas may be caused partly by local wind-driven divergences rather than by advection due to offshore Ekman transport. Offshore divergence and upwelling are to be expected for conditions of positive wind stress curl off the coast (Nelson, 1977). Positive wind stress curl occurs when the alongshore wind (northerly) increases in magnitude as one moves away from the coast. Positive values of wind stress curl generally occur along the California coast out to several hundred km from the coast during summer. Although the LFM-derived winds associated with the September 8 wind event do not indicate weaker winds near the coast, the data density is insufficient to establish whether or not such a wind distribution did exist. Therefore it is possible that the lower temperatures observed off the California coast following the September 8 wind event were partly due to divergence arising from positive wind stress curl.

In search of additional evidence for increased

TABLE 1. Estimates of Surface Velocity (± 4 cm/s)

Displacement	Speed (Apparent)	Direction* (Apparent)
A - B	18 cm/s	WSW/offshore
B - C	41 cm/s	W/offshore
D - E	28 cm/s	W/offshore
E - F	30 cm/s	W/offshore

*"Offshore" here refers to a direction approximately normal to the alongshore direction and directed away from the coast. The term "cross-shelf direction" is inappropriate because the flows to which we refer occur far beyond the continental shelf in the region.

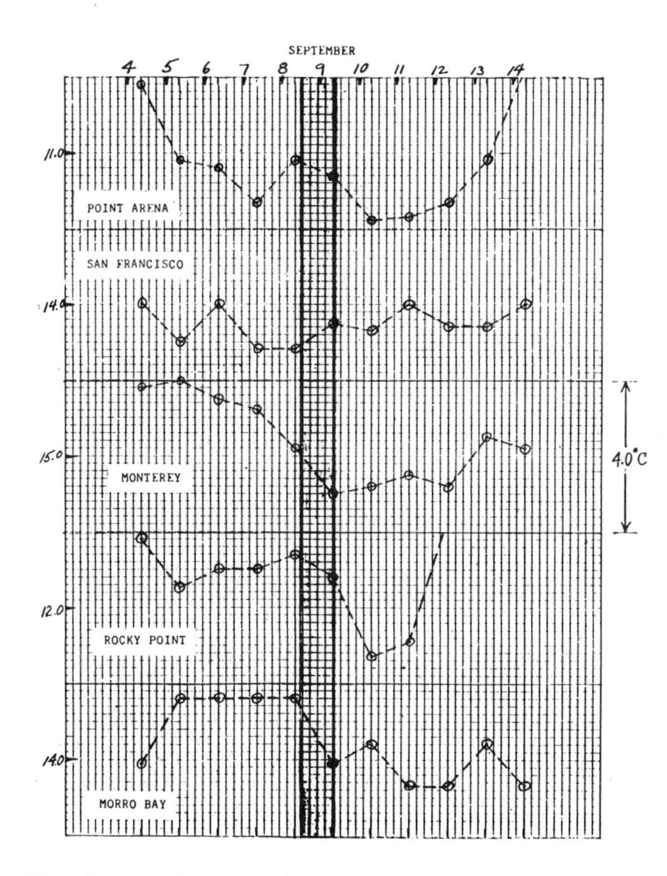

Fig. 5. Surface-truth sea temperatures at five coastal locations bracketing September 8 wind event. Same time window shown in Fig. 3 is shown.

upwelling along the California coast in response to the September 8 wind event, separate measurements of SST at five coastal locations have been examined. The daily temperature observations are shown in Fig. 5. The same window shown in Fig. 3 has been included in the figure. The temperature records show that the sea-surface temperatures fell by at least 0.4°C during the time window at four out of the five locations. Although a decrease in SST is not observed at San Francisco, "San Francisco" temperatures were actually observed at the Farallon Islands, about 45 km west of San Francisco. Hence, the observation point may have been just beyond the reach of recently upwelled waters at this particular time. At Monterey sea-surface temperatures had already been falling for several days prior to the wind event. Winds increased slightly (Fig. 3) prior to the wind event at Point Piños and Point Piedras Blancas and are characterized by three peaks each about 24 h apart with maximum values of about 8 m/s. These wind "pulses" could be responsible at least in part for the observed decrease in SST at Monterey before the main wind event. In summary, the measurements, which were not spatially or temporally coincident with the satellite data, still suggest the likely influence of the September 8th wind event on the temperature field along the California coast.

Acknowledgements. The authors thank J.T. Bailey, Manager of the NOAA NESS Satellite Field Services Station in Redwood City, California for making available the time and resources necessary to construct the film loop and write the paper. We thank A. Bakun of the Pacific Environmental Group of NOAA NMFS for making available his upwelling index calculations that were used in the text. We also thank the Data Collection and Processing Group at the Scripps Institute of Oceanography for making available the coastal measurements of sea-surface temperature that were used to support our satellite observations.

References

Allen, J.S., and P.K. Kundu, On the momentum, vorticity and mass balance on the Oregon Shelf, *Journal of Physical Oceanography, 8,* 13-27, 1978.

Anderson, R.K., Application of meteorological satellite data in analysis and forecasting, *ESSA Technical Report, NESC 51,* 1974.

Bakun, A., Daily and weekly upwelling indices, west coast of North America, 1967-73, *NOAA Technical Report, NMFS SSRF-693,* 114 pp., 1975.

Bakun, A., Upwelling index computations, NOAA NMFS, Monterey, CA., unpublished data, 1979.

Bang, N.D., Characteristics of an intense ocean frontal system in the upwelling regime west of Cape Town, *Tellus, 25,* 256-265, 1973.

Conomos, T.J., San Francisco Bay, the urbanized estuary, *Pacific Division of the American Assoc-*

iation for the Advancement of Science, 493 pp., 1979.

Curtin, T.B., Physical dynamics of the coastal upwelling frontal zone off Oregon, Ph.D. dissertation, University of Miami, Coral Gables, Fla., 317 pp., 1979.

Kraus, E.B., What do we not know about the sea-surface wind stress, *Bulletin of the American Meteorological Society, 49*, 247-253, 1968.

Legeckis, R., Application of synchronous meteorological satellite data to the study of time dependent sea-surface temperature changes along the boundary of the Gulf Stream, *Geophysical Research Letters, 2*, 435-438, 1975.

Legeckis, R., and J. Pritchard, Algorithm for correcting the VHRR imagery for geometric distortions due to the earth curvature, earth rotation, and spacecraft roll attitude errors, *NOAA Technical Memorandum, NESS 77*, 31 pp., 1976.

Maul, G.A., and S.R. Baig, A new technique for observing mid-latitude ocean currents from space, in Proceedings, 41st Annual Meeting, American Society of Photogrammetry, Washington, D.C., 713-716, 1975.

Nelson, C.S., Wind stress and wind stress curl over the California Current, *NOAA Technical Report, NMFS SSRF-714*, 87 pp., 1977.

NOAA NWS, National Weather Service forecasting handbook no. 1 facsimile products, Chapter 2, Numerical prediction models and techniques, 37 pp., 1976.

Plew, W., and A. Thumin, VISSR picture animation and movie loops, *NOAA Technical Memorandum, NESS 64*, 76-81, 1975.

Reid, J.L., G.I. Roden, and J.G. Wyllie, Studies of the California Current system, California Cooperative Oceanic Fisheries Investigation Progress Report, 1 July 1956 to 1 January 1958, 27-57, 1958.

Riehl, H., T.C. Yeh, J.S. Malkus, and N.E. LaSeur, The northeast trade of the Pacific Ocean, *Quarterly Journal of the Royal Meteorological Society, 77*, 598-626, 1951.

Smith, R.L., A description of current, wind, and sea level variations during coastal upwelling off the Oregon coast, July-August 1972, *Journal of Geophysical Research, 79*, 435-443, 1974.

Smith, R.L., J.G. Pattullo, and R.K. Lane, An investigation of the early stage of upwelling along the Oregon coast, *Journal of Geophysical Research, 79*, 1135-1140, 1966.

Wickham, J.B., Observations of the California Countercurrent, *Journal of Marine Research, 33*, 325-340, 1975.

Wooster, W.S., and J.L. Reid, Eastern boundary currents, in M.N. Hill (editor), *The Sea, 2*, 253-280, 1963.

Yoshida, K., Circulation in the eastern tropical oceans with special references to upwelling and undercurrents, *Japanese Journal of Geophysics, 4*, 1-75, 1967.

EQUATORIAL UPWELLING IN THE EASTERN ATLANTIC: PROBLEMS AND PARADOXES

B. Voituriez

Antenne ORSTOM, C.O.B. B.P. 337, 29273 Brest Cedex, France

Abstract. Recent oceanographic cruises have yielded new information that requires revision of ideas and hypotheses concerning equatorial upwelling in the Atlantic. Some of them concerned mainly with the eastern equatorial Atlantic are considered, including seasonal variations in equatorial upwelling, the physical processes accounting for seasonal sea-surface cooling and nutrient enrichment, and primary and secondary productions.

Seasonal variations in sea-surface temperature in the eastern equatorial Atlantic are comparable with those of the eastern equatorial Pacific. Variations of the zonal pressure gradient induced by variations of the wind stress along the equator explain the east-west asymetry of the sea-surface cooling during the upwelling season. In the eastern part of the Atlantic, where the thermocline is shallow in July and August, the southeasterly wind induces, south of the equator, an equatorial divergence causing marked cooling of the sea surface. During the same season, the thermocline in the western Atlantic is too deep to permit significant sea-surface cooling by the equatorial divergence. Unlike the eastern equatorial Pacific, nutrient enrichment of the surface layer is restricted to the period July to September, but this does not necessarily result in an increase in primary and secondary production during the period.

Introduction

The actual role of equatorial upwelling and the importance of variations in it have been underestimated for some time. The Equalant international cruises in 1963 and 1964 were the first aimed at a biological and physical description of the tropical Atlantic Ocean. Until recently the Equalant observations were considered as references of the Atlantic equatorial upwelling. However, it is now clear that 1963 was actually an exceptional year in that upwelling did not occur or was very weak, as suggested by Sedikh and Loutockhina (1971). This is clearly shown by comparison of the sea-surface temperatures in summer, 1963 (Fig. 1) and those in more recent atlases. Except in the far eastern part of the ocean, the SST was greater than 24°C everywhere in the equatorial zone in summer, 1963 whereas the atlases show that on the average temperatures are below 22.5 (Mazeika, 1968) or 22°C (Neumann, Beatty and Escowitz, 1975) in the Gulf of Guinea in summer. The anomaly is also shown by nutrient distributions in the surface layer. The maps of the distributions of phosphate in the surface layer (Fig. 2) show no equatorial enrichment either in February and March, 1963 (Equalant 1) or in summer 1963 (Equalant 2). However, the systematic observations made from the R.V. *Capricorne* in the eastern Atlantic along 4°W between 1971 and 1979 show nutrient enrichment of the surface layer to be the rule from July to September (Voituriez and Herbland, 1977).

Seasonal Variations of Equatorial Upwelling in the Gulf of Guinea

The observations of the R.V. *Capricorne* along 4°W in the Gulf of Guinea make it possible to define the cycle and amplitude of seasonal variations of the upwelling (Fig. 3). The sea-surface temperature falls from 29°C in March and April to 21°C in July and August, an 8°C amplitude. The anomaly of the summer of 1963 (Equalant 2) is clearly shown on Fig. 3. The figure shows the striking similarity of the seasonal variations in equatorial upwelling in both the eastern Atlantic and Pacific oceans. The seasonal cycle and the surface temperature signal are roughly the same in both oceans, a point to be emphasized. It will be shown that no such similarity exists in the variations of nutrient enrichment, which is much less in the Atlantic than in the Pacific. The difference between the temperature and the nutrient signals probably tended to mask the importance of upwelling in the equatorial Atlantic when the scientific community was more concerned with the consequences of upwelling on biological processes than on the climatic processes. Hisard

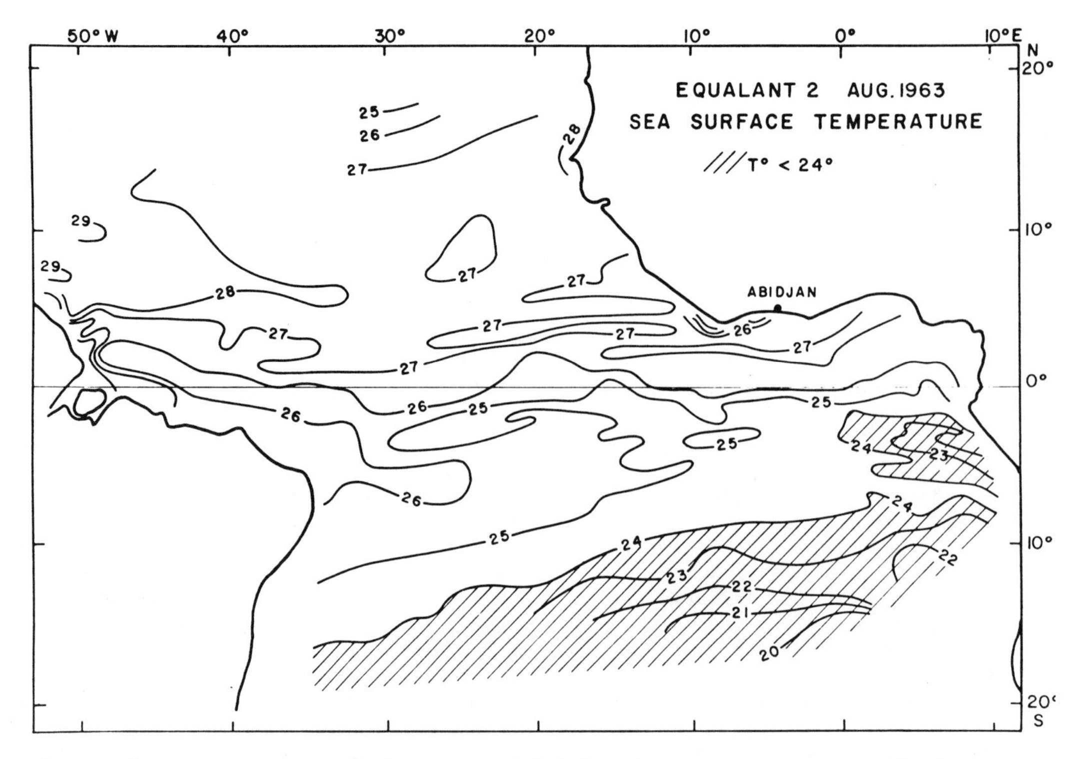

Fig. 1. Sea-surface temperature of the equatorial Atlantic Ocean in summer 1963 (Kolesnikov, 1973).

(1980) pointed out that the similarity between the eastern Atlantic and Pacific oceans probably included inter-annual temperature variations; he suggested that anomalies such as those observed in 1963 (Equalant 2) can be considered similar to the occurrence of *El Nino* in the Atlantic Ocean.

The Mechanisms of the Equatorial Upwelling in the Atlantic Ocean

The great interest in recent years in the equatorial upwelling of the eastern Atlantic stems from the idea that equatorial divergence could not account for the upwelling observed in the area, because the annual upwelling in the Gulf of Guinea was not associated with changes in the local winds (Hisard, Citeau, and Voituriez, 1977; Voituriez and Herbland, 1977; Moore *et al.*, 1978). The idea was supported by other studies showing that in the coastal upwelling of the Gulf of Guinea there is no relationship between variations of the sea-surface temperature and Ekman transport as deduced from the observed winds (Berrit, 1977; Bakun, 1978). Other hypotheses were proposed to explain the variations of the sea-surface temperature and summer cooling in the Gulf of Guinea such as vertical mixing in the vertical shear zone between the westward surface current and the equatorial undercurrent (Hisard *et al.*, 1977; Voituriez and Herbland, 1977) and an upwelling signal, generated by increased westward wind stress in the western Atlantic and travelling to the eastern Atlantic as an equatorially trapped Kelvin wave (Moore *et al.*, 1978; O'Brien, Adamec, and Moore, 1978). However, the hydrographic and meteorological observations along 4°W by R.V. *Capricorne* show that the new hypotheses are probably premature for two reasons: 1) There is good agreement between seasonal variations in sea-surface temperatures and the variations in the wind stress. This is clearly illustrated by Fig. 4; the sea-surface temperature is minimum when the wind stress is maximum. 2) The distribution of the sea-surface temperature agrees well with the qualitative model proposed by Cromwell (1953) linking the equatorial divergence to the trade winds.

This model is presented in Fig. 5. It shows that the equatorial divergence is right at the equator with easterly winds and is shifted southwards with southeasterly winds, exactly as observed during the CIPREA cruises in summer, 1978. Fig. 6 shows the divergence marked by the minimum sea-surface temperature to be south of the equator with southeasterly wind. The rotation of the wind and the increase of the zonal component of the wind stress between 4 and 9°W lead to an equatorward motion of the equatorial divergence in accordance with Cromwell's model. To conclude, in opposition to what was thought before, there is no apparent contradiction between equatorial upwelling events in the Gulf of Guinea and observed changes of the local wind and subsequent equatorial divergence.

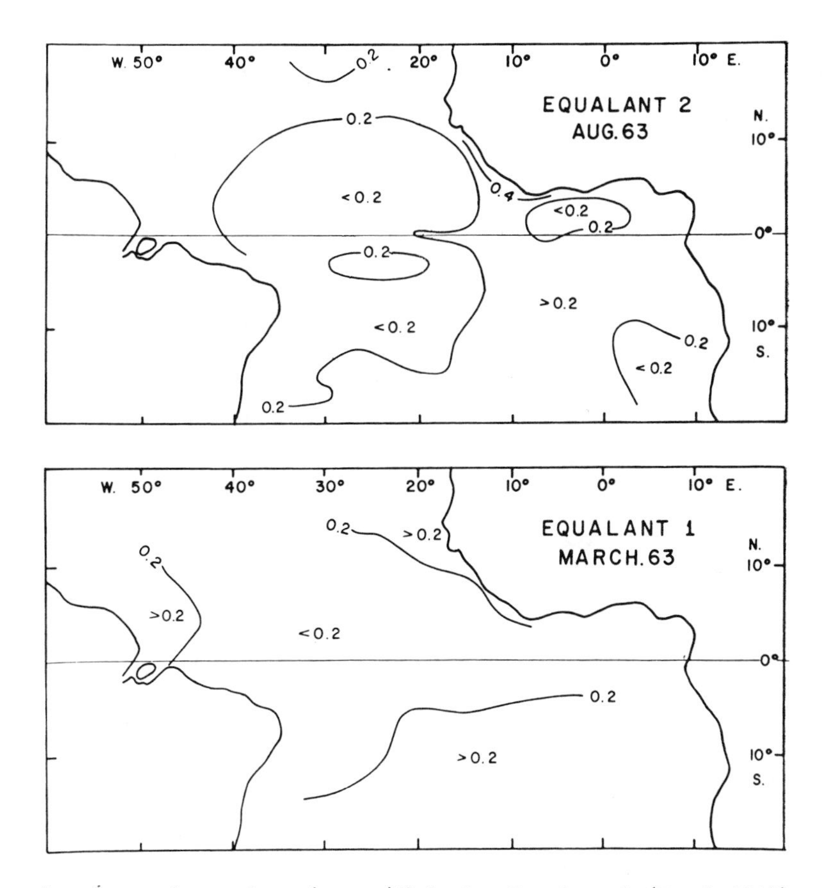

Fig. 2. Phosphate-phosphorus in surface (μg-at/l) during Equalant 1 (March 1963) and Equalant 2 (August 1963) (Kolesnikov, 1976).

A problem remains: the asymetry between the eastern and the western equatorial Atlantic during the upwelling season. Fig. 7 shows that the sea-surface temperature is lower in the eastern than in the western Atlantic whereas the wind stress along 2.5°S is greater in the west than in the east. Paradoxically, the cooling of the sea surface is the most important where the wind stress is the weakest. This seems to be in contradiction to the previous simple conclusions on the role of equatorial divergence in the upwelling events in the Gulf of Guinea, but the anomaly can be explained when the entire equatorial Atlantic is considered. Merle (1980) pointed out that the equatorial homogeneous surface layer is much thicker in the western than in the eastern Atlantic (Fig. 8) because of the accumulation against the American coasts of water carried westwards by the South Equatorial Current. Therefore the energy or the wind stress intensity needed to bring the cold, deeper waters to the surface is less in the eastern than in the western Atlantic. Thus the sea-surface temperature is not always a good indicator of the intensity of the equatorial divergence. In fact surface cooling in the equatorial divergence depends on two factors, the wind stress, which controls the inten-

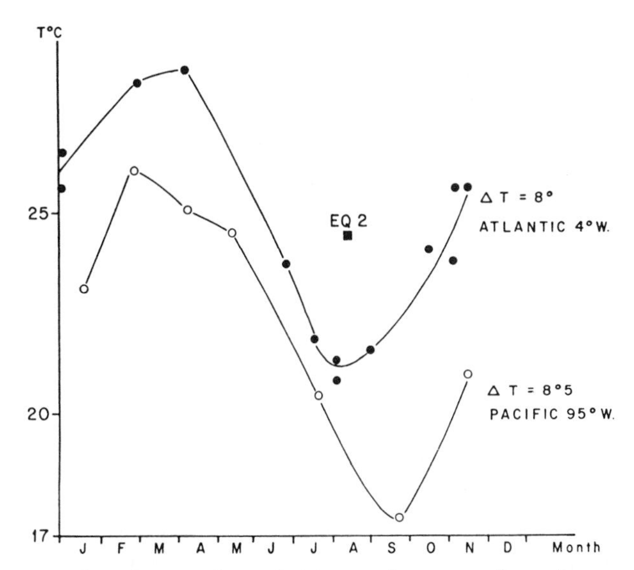

Fig. 3. Seasonal variations of sea-surface temperature in the eastern equatorial upwelling regions of the Atlantic and Pacific oceans: lowest temperature measured between 0° and 5°S along 4°W (1971-1979 R.V. *Capricorne*) and 95°W (1967-1968 Eastropac cruises).

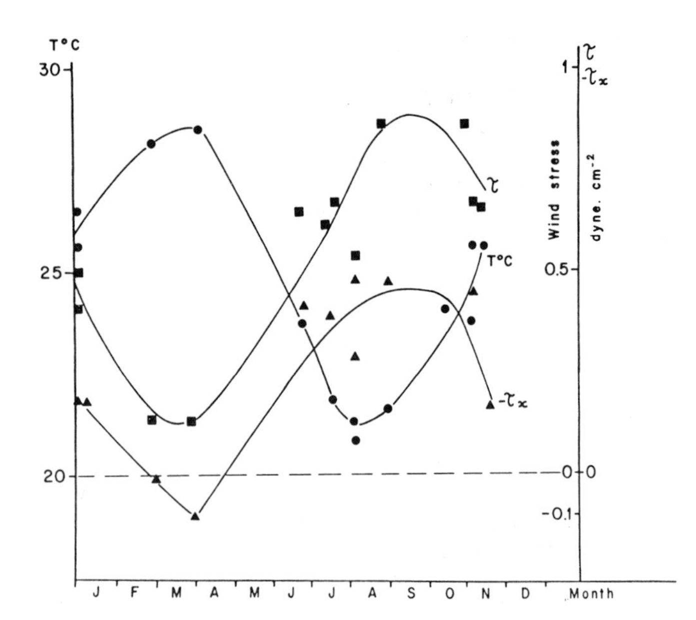

Fig. 4. Seasonal variations of the sea-surface temperature (lowest temperature between 0 and 5°S) and the wind stress (mean value between 0 and 5°S) in the equatorial upwelling region of the Atlantic Ocean along 4°W (Data collected by the R.V. *Capricorne* from 1971 to 1979.). τ: total wind stress, $τ_x$: zonal component of the wind stress.

sity of the divergence, and the depth of the thermocline. Consequently, in spite of a possibly more intense divergence in the western part of the Atlantic, surface cooling can be greater in the east than in the west.

The increase from east to west of the mixed layer depth accounts for the zonal pressure gradient along the equator as shown by Newmann *et al.* (1975) and Katz (1977) (Fig. 9). Katz (1977) has pointed out two facts: 1) There is a seasonal variation of the zonal pressure gradient. The maximum occurs in August and September and the minimum in March (Fig. 10). This means that the seasonal variations of the slope of the surface along the equator coincide with the variations of the sea-surface temperature in the eastern Atlantic as described before (Fig. 3). 2) The zonal pressure gradient is correlated with the wind stress; the stronger the wind stress, the greater the zonal pressure gradient. The summer increase of the wind stress has two consequences: 1) An increase of the intensity of the equatorial divergence over the ocean, and 2) An increase of the zonal pressure gradient, i.e., an increase of the depth of the thermocline in the west and a decrease of it in the east (Fig. 8). Accordingly the consequences of the equatorial divergence on the surface conditions (sea-surface temperature, nutrient concentrations) are amplified in the eastern Atlantic by the decrease of the thermocline depth and, on the contrary, can be neutralized in the western Atlantic by the increase of the thermocline depth. It can be concluded that the greater the zonal pressure gradient, the greater the cooling of the sea surface in the eastern equatorial Atlantic. Thus the unusually high sea-surface temperatures of August 1963 appear to correspond to an anomalously weak zonal pressure gradient (Fig. 10). Hisard (1980) suggested that such abnormal conditions in the Atlantic are similar to *El Niño* in the Pacific.

That rather simple explanation of the equatorial upwelling in the Atlantic does not exclude the other mechanisms, the trapped Kelvin wave and vertical mixing in the vertical shear zone between the surface current and the undercurrent; both can act to cool the eastern Atlantic. The Kelvin wave has not yet been detected in the equatorial area. It would act on the zonal pressure gradient and this would not change the previous conclusions on the equatorial divergence. The role of vertical mixing in summer in the western Atlantic (30°W) was pointed out by Kaiser and Postel (1979) and by Cornus and Meincke (1979). This is not in

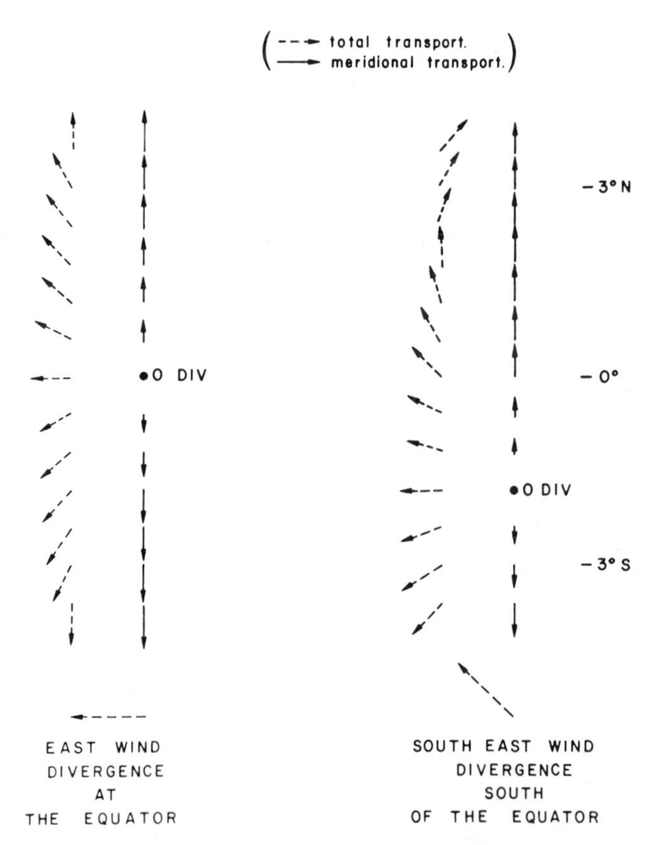

Fig. 5. The equatorial divergence model of Cromwell (1953). Intensity and directions of the wind and total transport are assumed constant. The angle of the wind with the total transport direction depends on the latitude so that the meridional transport pattern depends on the wind direction.

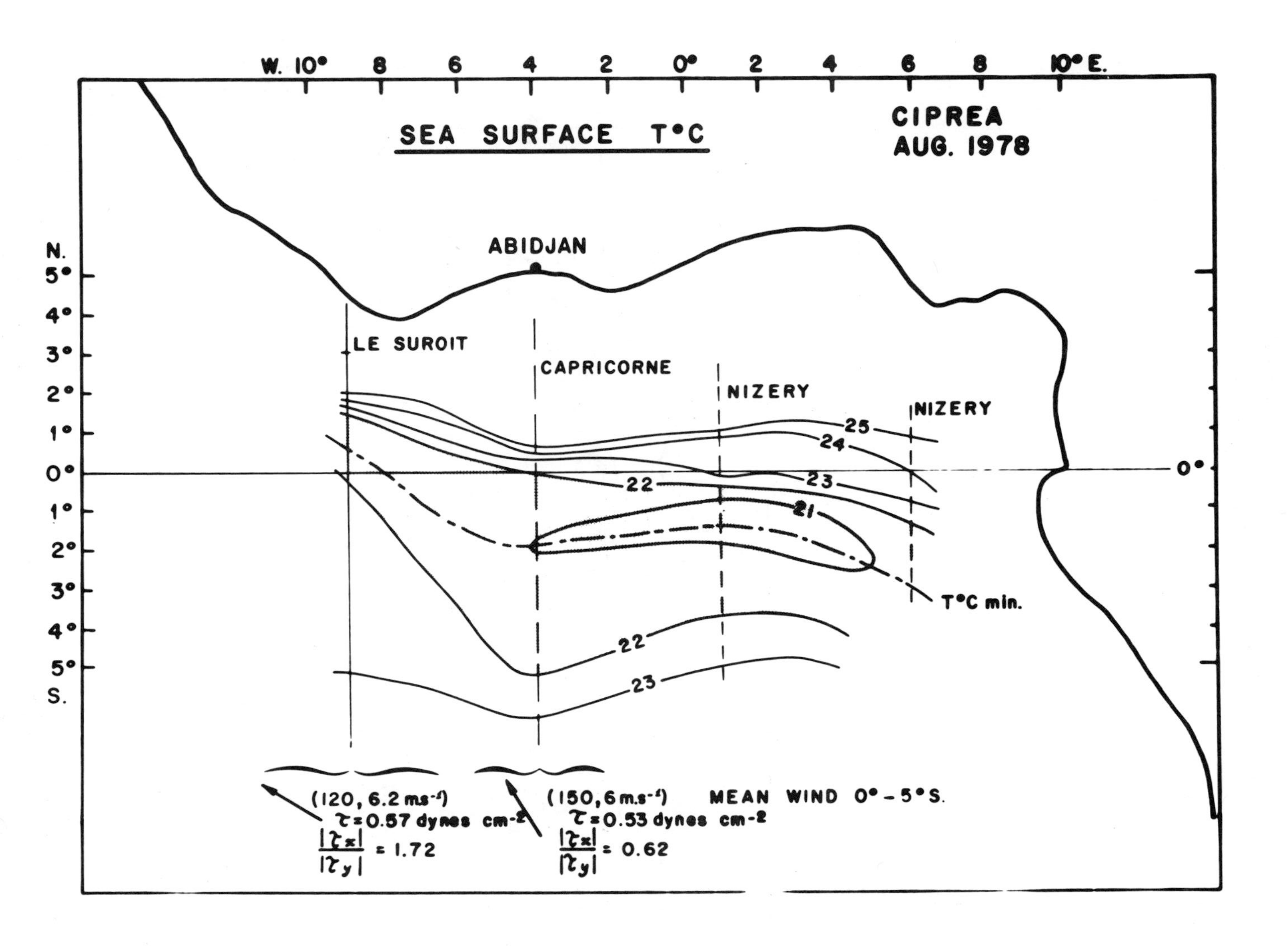

Fig. 6. Sea-surface temperature in the Gulf of Guinea in August, 1978 (CIPREA cruises of the research vessels *Suroit*, *Capricorne*, and *Nizery*). The mean wind (direction and speed) between 0 and 5°S is given for 9 and 4°W with the total wind stress and the ratios of the zonal to the meridional wind stress τ_x/τ_y.

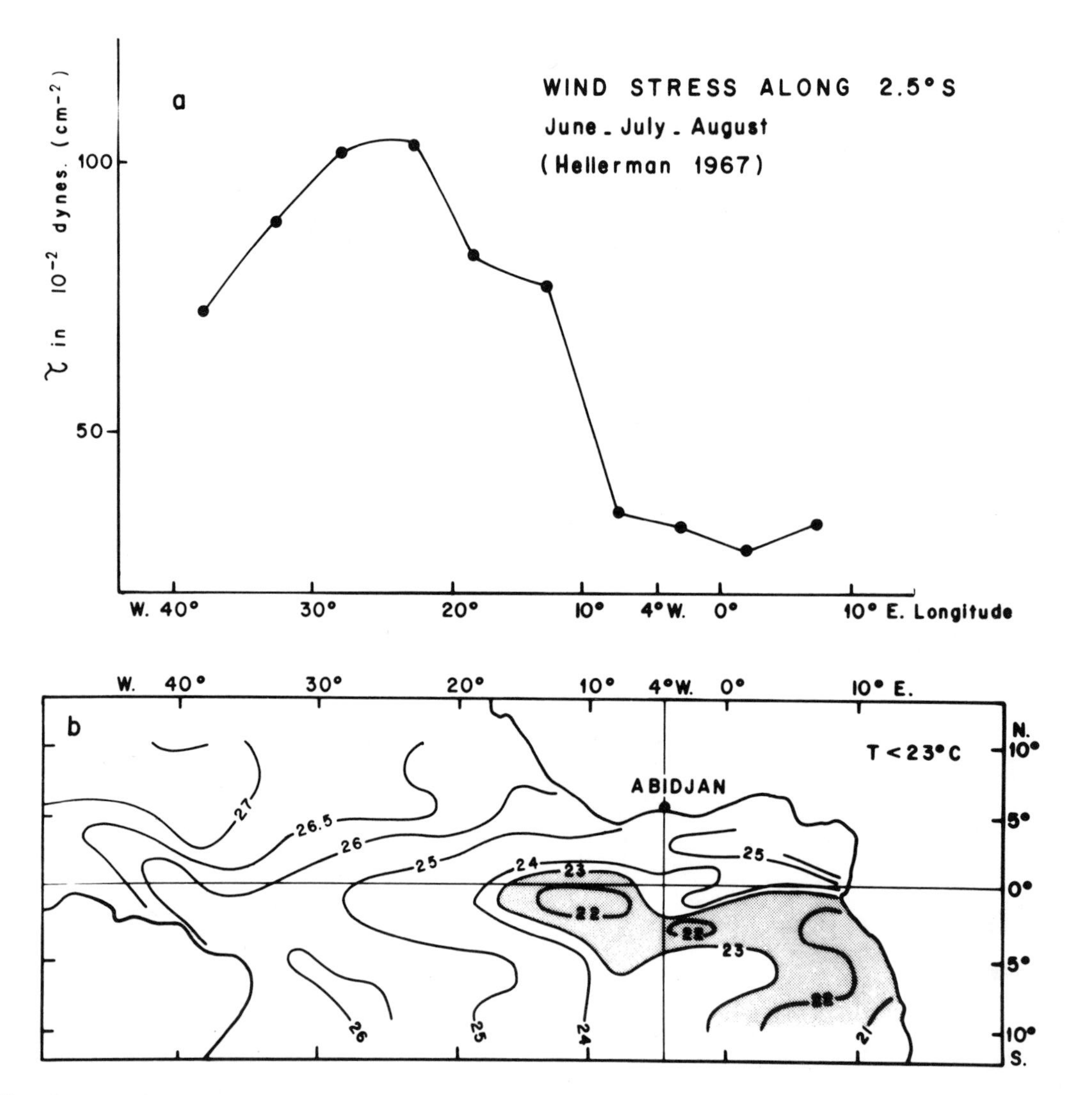

Fig. 7. Sea-surface temperature and total wind stress in the equatorial area of the Atlantic Ocean in summer. a: total wind stress along 2.5°S (data from Hellerman, 1967); b) sea-surface temperature in July from Neumann *et al.*, 1975).

contradiction with the observations in the Gulf of Guinea. In fact the relative importance of these processes on sea-surface conditions depends on the season and on the longitude. The equatorial divergence seems to be the most important factor in cooling the sea surface in the eastern Atlantic in summer when the thermocline is shallow. During other seasons in the same area and during all seasons in the western equatorial Atlantic, the thermocline is deeper and the influence of the equatorial divergence on the sea surface is probably negligible, so vertical mixing can become the primary cause of the small variations of the sea-surface temperature.

Nutrient Enrichment

Nitrate enrichment of surface waters along 4°W occurs only in July, August, and September (Fig. 11). The rest of the year, the mixed layer is nitrate impoverished. This contrasts with the eastern Pacific Ocean where surface nitrate concentrations higher than 6 μg-at/l were observed all the year during Eastropac observations at 95°W. However, there are definite seasonal variations in the eastern Atlantic when the total nitrate of the upper 100 m is considered. The variations are in good agreement with the variations in upwelling; enrichment is maximum in summer (August) and minimum in March and April. The differences between the Atlantic and the Pacific are interesting. While the effects of equatorial processes on the sea-surface temperature are similar, the effects on nutrient enrichment differ markedly. For the same sea-surface temperature the Pacific Ocean is richer than the Atlantic by 6 μg-at/l of nitrate-nitrogen, and the lowest

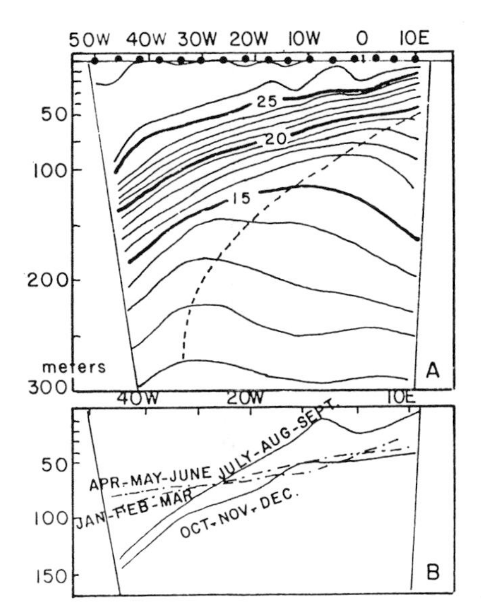

Fig. 8. a) Mean annual temperature in the 0 to 2°S band from the Brazilian Coast (50°W) to the African Coast (12°E). Data averaged by 4° longitude (from Merle, 1980); b) Depth variation of the 23°C isotherm for the four seasons of the year along the section shown in Fig. 8a (from Merle, 1980). The 23°C isotherm depth is taken for the thermocline depth.

temperature at which 6 µg-at/l of nitrate occurred was 20.5°C in the Atlantic and the highest (26°) was in the Pacific (Table 1).

Such differences may arise from the origin of the upwelled waters in the Atlantic, which is the equatorial undercurrent flowing in the thermocline layer. The undercurrent carries a salinity maximum from west to east (Fig. 12); its origin is subtropical, high-salinity, nutrient-depleted water; on a meridional section the equator shows a nitrate minimum. The upper 30 m of the thermocline can be nitrate-depleted (Voituriez and Herbland, 1977). For example (Fig. 13) at 6°E in 1971 between 20 May and 15 June the sea-surface temperature dropped 4°C and isotherms shoaled by 30 m. At the same time the nitrate isopleths showed a 30-m rise, but the nitrate content remained undetectable down to 20 m. This is important; Herbland and Voituriez (1979) used surface nitrate concentrations to define two types of situations in the tropical ocean: the typical tropical situation characterized by an absence of nitrate in the surface layer and the upwelling situation with significant surface nitrate concentrations. The eastern equatorial Pacific is usually in an upwelling situation whereas the eastern Atlantic is in a typical tropical situation nine months (October to June) out of twelve. With the same annual temperature signal nitrate does not limit primary production in the eastern Pacific, whereas it probably does most of the year in the eastern Atlantic.

Primary and Secondary Production

Seasonal variations in primary and secondary production in the equatorial Atlantic are much less than one might expect from the variations of the nutrient regime.

Primary production. The CIPREA cruises at 4°W in 1978 and 1979 show the usual seasonal variations of sea-surface temperature, nitrate at the surface, and integrated nitrate over 100 m. However, neither chlorophyll nor primary production followed the same pattern (Fig. 14). Variations in primary production were small, values being slightly more than 100 mg C $m^{-2}h^{-1}$. The chlorophyll content had the same value in April and August (Table 2). The conclusion is that in 1978 and 1979 primary production was rather high during all seasons in the equatorial area at 4°W and that the summer nutrient enrichment did not induce an increase of primary production and chlorophyll.

Secondary production. Similarly, zooplankton biomass does not always differ significantly between the cold season in July and the warm season in April (Fig. 15).

The equatorial area in the Gulf of Guinea is always rather productive, but the levels of production and biomass are not significantly increased by the appearance of nutrients in the surface layer. The absence of contrast of production between the two regimes has probably led to the underestimation of the importance of the interannual variations of the equatorial upwelling in the region. Indeed phenomenon similar to *El Niño* in the eastern Atlantic, as suggested by Hisard (1980) for 1963 (Equalant 2), has a marked effect on the sea-surface temperature but not on the biological production, which remains at a high level with or without upwelling.

Several authors such as Sorokin, Sukhanova, Konoyalova, and Pabelyeva (1975) have commented on the low level of production compared to nutrient availability in the equatorial upwelling regions of the Pacific Ocean. The anomaly was called "the paradox of the nutrients" by Walsh (1976); why are nutrients not taken up in greater quantity when light and nutrient conditions are not limiting? Several explanations have been proposed for the paradox. The lack of chelators was proposed by Barber and Ryther (1969), and intense vertical mixing in the vertical shear zone between the surface current and the equatorial undercurrent was proposed by Sorokin *et al.* (1975). Walsh (1976) suggested that the dominant frequencies of variability of the physical environment may define the importance of grazing stress as a constraint to nutrient utilization in the sea. For example the relatively high stability of the low-frequency regime of the equatorial upwelling could lead to an herbivore population exerting a quasi-continuous grazing pressure and consequently a low utilization of nutrients.

The discussion on this subject is probably not

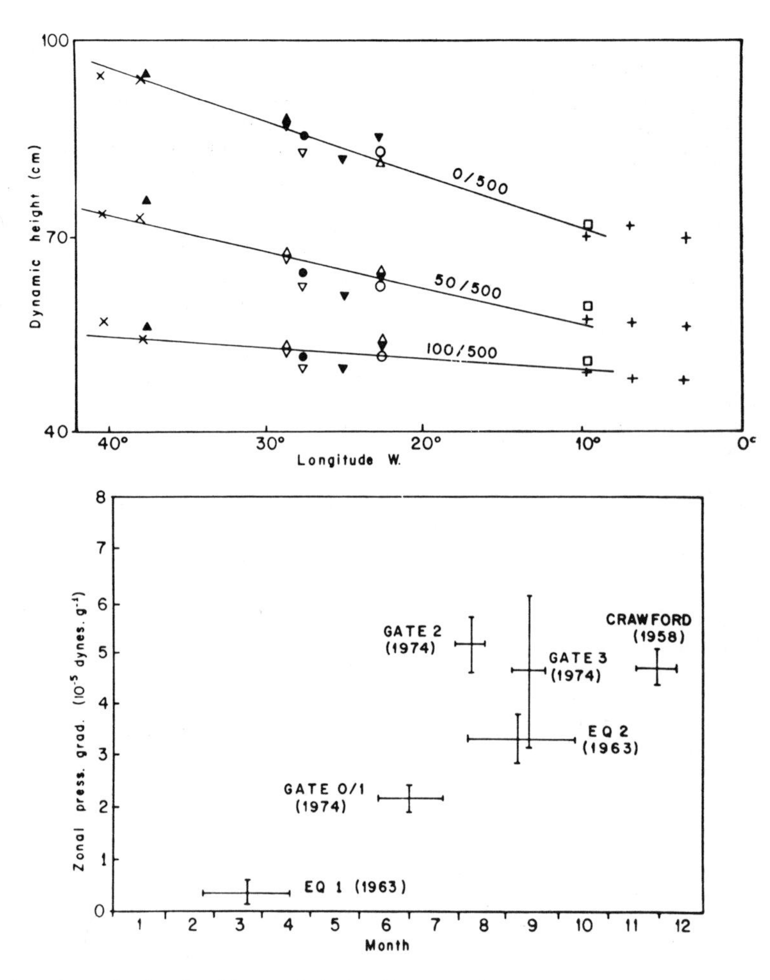

Fig. 9. Dynamic height of the 0, 50, and 100- dbar surfaces relative to 500 in the Atlantic during GATE-Phase 2 (27 July - 17 August 1974) (Katz, 1977).

over and the comparison of the two eastern oceans where the variations of the nutrient regime are quite different is a fascinating point which can help to solve the problem of the fertilization in the equatorial upwelling.

Conclusion

The several points discussed in this paper show that equatorial processes are complex. Many problems are still to be solved. What is the role of the different equatorial processes in nutrient enrichment and sea-surface cooling? What are the space and time variabilities of equatorial upwelling? Why do nutrient uptake and production seem relatively low in equatorial upwelling regions? To answer these questions we need more observations in the equatorial region, which must be considered as a whole because upwelling is not only a local phenomenon but it is the result of processes involving the whole equatorial basin. However some progress in our knowledge could be made now by making use of similarities and con-

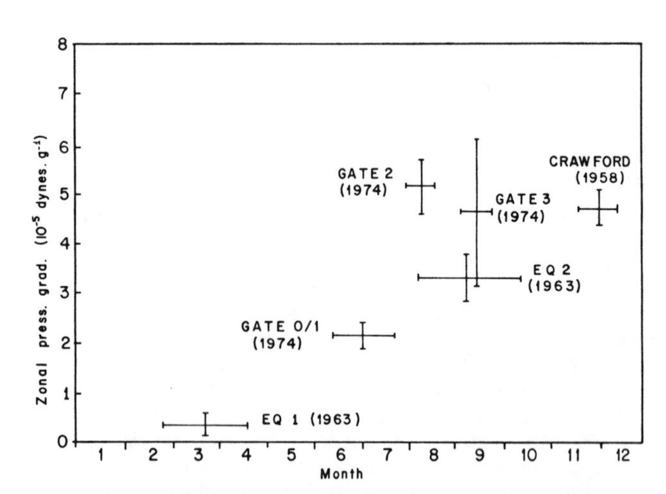

Fig. 10. The Atlantic equatorial zonal pressure gradient (50/500 dbar) west of 10°W as a function of the observational period and independent of year (Katz *et al.*, 1977).

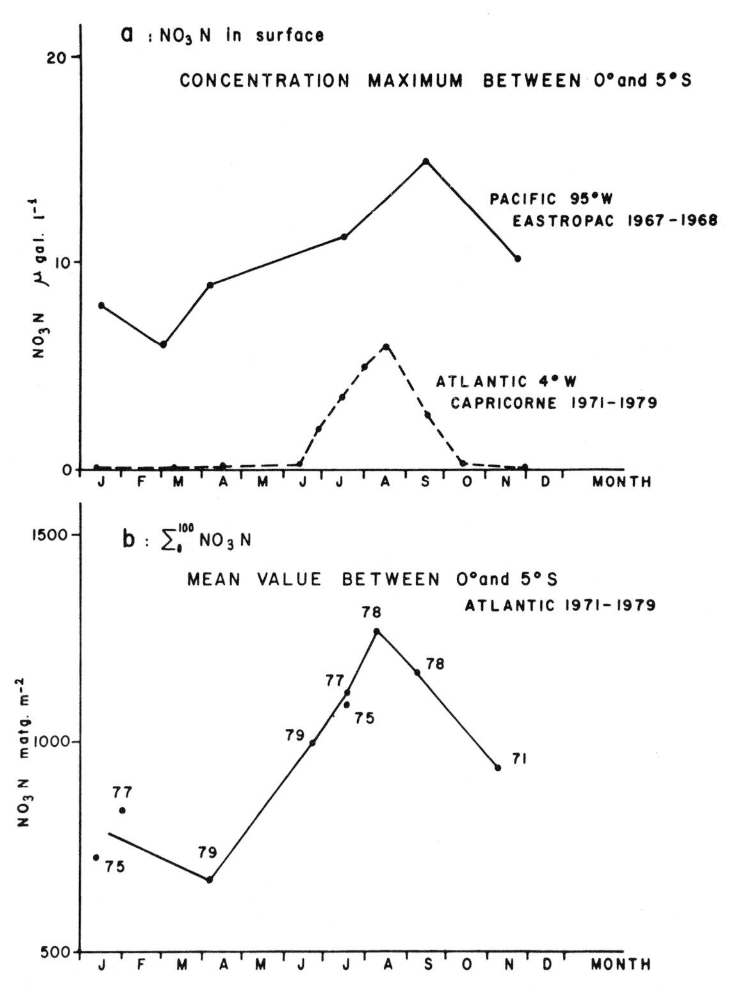

Fig. 11. Nitrate distribution in equatorial upwelling regions. a) Maximum concentration of nitrate in the surface between 0 and 5°S in the eastern equatorial Atlantic at 4°W (data collected by the R.V. *Capricorne* from 1971 to 1979) and in the Pacific at 95°W (Eastropac 1967-1968); b) Integrated value of nitrate over 100 m in the Atlantic Ocean along 4°W: mean value between 0 and 5°S (data collected by the R.V. *Capricorne* from 1971 to 1979).

TABLE 1. Nitrate concentrations at the surface in the equatorial upwelling regions of the eastern Pacific at 95°W (EASTROPAC 67-68) and the eastern Atlantic at 4°W (CIPREA 78-79)

Sea Surface (T^oC)	20.5	26
NO_3-N (surface) Pacific 95°W (μg-at/l)	12	6
NO_3-N (surface) Atlantic 4°W (μg-at/l)	6	0
ΔNO_3-N (μg-at/l)	6	6

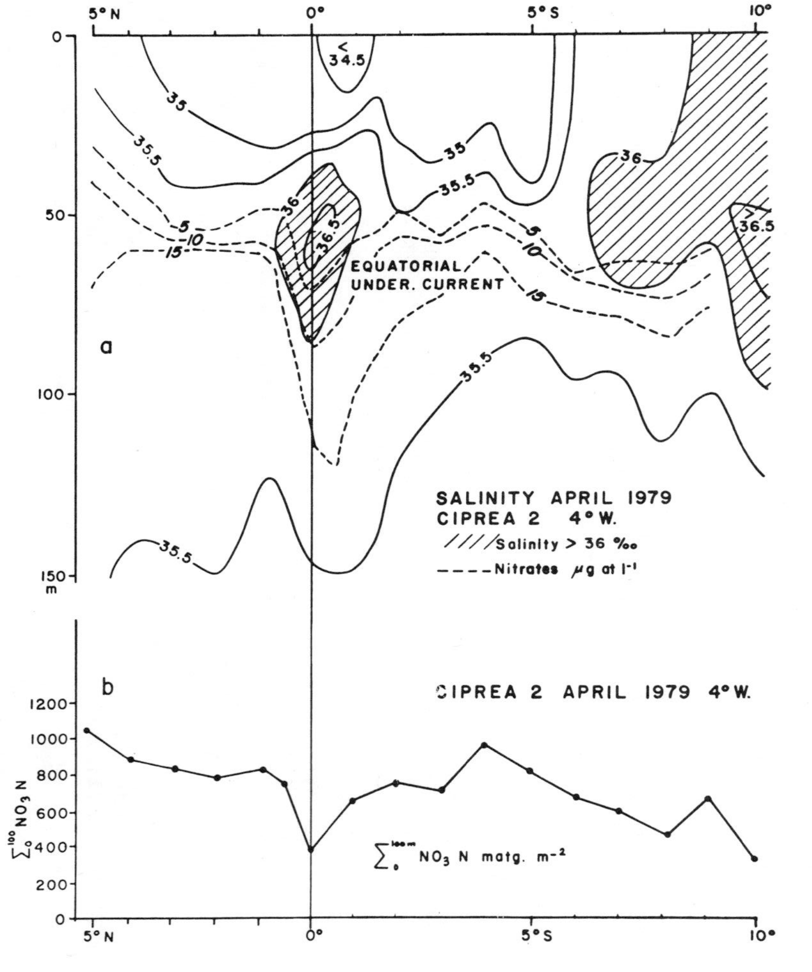

Fig. 12. Transect along $4^{O}W$ in April 1979, CIPREA 2. a) salinity and nitrate isopleths; b) integrated value of nitrate over 100 m.

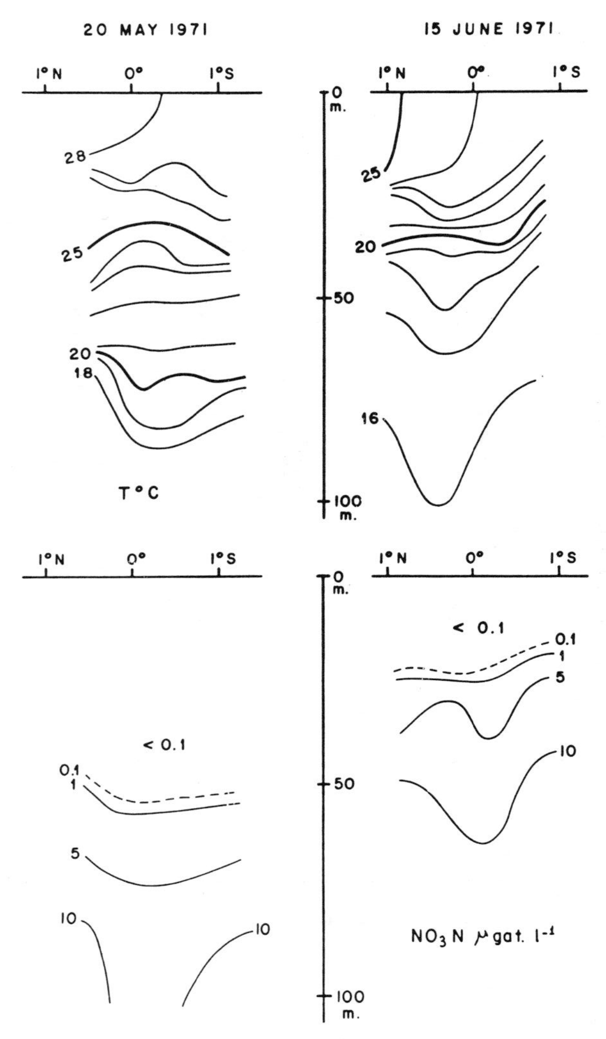

Fig. 13. Nitrate and temperature section along $6^{O}E$. Left: 20 May 1971. Right: 15 June 1971. Upper part: $T^{O}C$. Lower part: Nitrate in µg-at/l

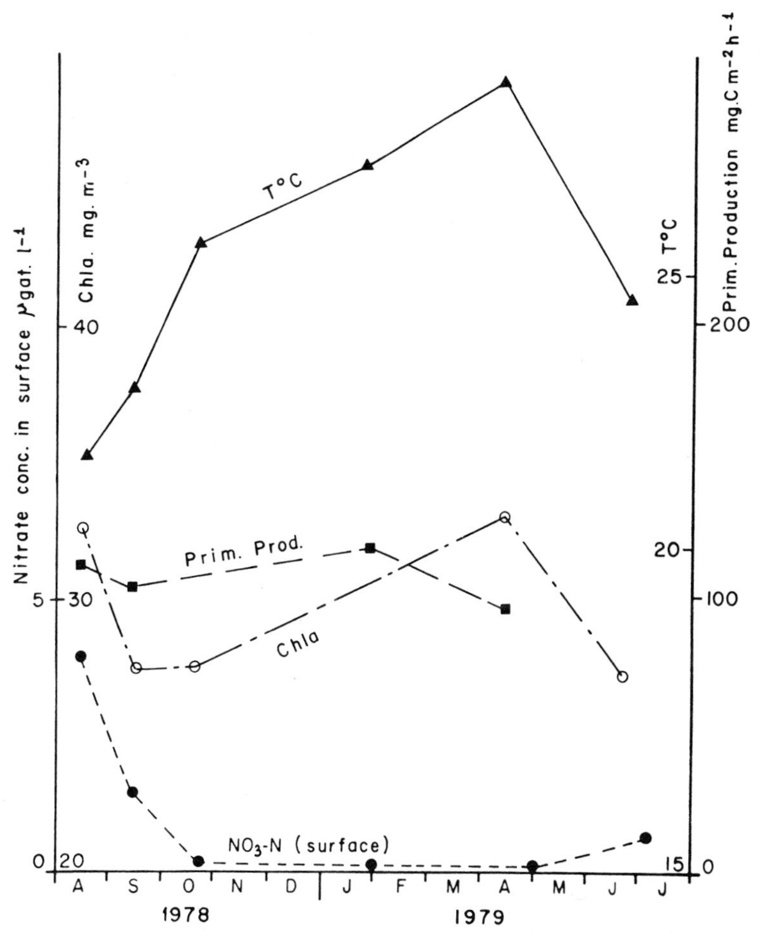

Fig. 14. Variations of temperature, primary production, chlorophyll, and surface nitrate concentration along 4°W in 1978-1979 (Mean values between 0 and 5°S of the CIPREA cruises).

TABLE 2. Mean values between 0° and 5°S in 1978-1979 at 4°W. CIPREA cruises of R.V. *Capricorne*.

Cruise	7802 Aug. 78	7802 Sept. 78	SOP 1 Jan. 79	7906 Apr. 79	7910 Jun. 79	7912 Oct. 79
T (Surface)	21.6	22.8	27	28.5	24.6	25.5
NO_3 (surface) µg-at/1	4	1.5	0	0	0.7	0.12
Integrated nitrate content, 0 to 100 m. (mg-at/m^2)	1275	1170		580	990	802
Integrated chlorophyll content, 0 to 100 m. (mg/m^2)	33	27.5		33.1	27.5	27.3
Primary productivity (mg C $m^{-2}h^{-1}$)	114	104	120	98		

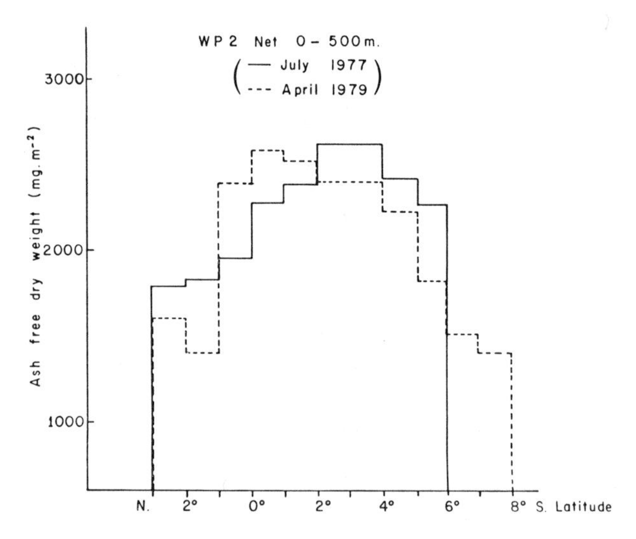

Fig. 15. Zooplankton biomass along 4°W in July 1977 (upwelling season) and April 1979 (warm season).

trasts observed between the equatorial areas of the three oceans on one hand and between the coastal and the equatorial upwelling on the other hand.

References

Bakun, A., Guinea current upwelling, *Nature, 271,* 147-150, 1978.

Barber, R.T. and J.H. Ryther, Organic chelators: factors affecting primary production in the Cromwell Current upwelling, *Journal of Experimental Marine Biology and Ecology, 3,* 191-199, 1969.

Berrit, G.R., Les eaux froides du Gabon à l'Angola sont-elles dues à un upwelling d'Ekman, *Cahiers O.R.S.T.O.M., Série Océanographie, 14,* 273-278, 1977.

Cornus, H.P. and J. Meincke, Observation of near-surface layer changes related to the Atlantic equatorial undercurrent, *Deep-Sea Research, 26A,* 1291-1299, 1979.

Cromwell, T., Circulation in a meridional plane in the central equatorial Pacific, *Journal of Marine Research 12,* 196-213, 1953.

Hellerman, S., An up-dated estimate of the wind stress on the world ocean, *Monthly Weather Review, 95,* 607-626, 1967.

Herbland, A. and B. Voituriez, Hydrological structure analysis for estimating the primary production in the tropical Atlantic Ocean, *Journal of Marine Research, 37,* 87-102, 1979.

Hisard, P., Observation de réponses de type "El Niño" dans l'Atlantique tropical oriental Golfe de Guinée, *Oceanologica Acta, 3,* 69-78, 1980.

Hisard, P., J. Citeau, and B. Voituriez, Equatorial undercurrent influence on enrichment processes of upper waters in the Atlantic Ocean, Report of the International Workshop on the GATE Equatorial Experiment, Miami, 28 February to 10 March, 1977.

Kaiser, W. and L. Postel, Importance of the vertical nutrient flux for biological production in the Equatorial Undercurrent region at 30°W, *Marine Biology, 55,* 23-27, 1979.

Katz, E.J. and collaborators, Zonal pressure gradient along the equatorial Atlantic, *Journal of Marine Research, 35,* 293-307, 1977.

Kolesnikov, A.G., Equalant I and II oceanographic atlas, Vol. 1, Physical Oceanography, UNESCO, Paris, 1973.

Kolesnikov, A.G., Equalant I and II oceanographic atlas, Vol. 2, Chemical and Biological Oceanography, UNESCO, Paris, 1976.

Mazeika, P.A., Mean Monthly Sea Surface Temperatures and Zonal Anomalies of the Tropical Atlantic, Serial Atlas of the Marine Environment, Folio 16, American Geographic Society, 1968.

Merle, J., Seasonal heat budget in the equatorial Atlantic Ocean, *Journal of Physical Oceanography, 10,* 464-469, 1980.

Moore, D., P. Hisard, J. McCreary, J. O'Brien, J. Picaut, J. Merle, J.M. Verstraete, and C. Wunsch, Equatorial adjustment in the eastern Atlantic, *Geophysical Research Letters, 5,* 637-640, 1978.

Neumann, G., W.H. Beatty III, and E.C. Escowitz, Seasonal changes of oceanographic and marine climatological conditions in the equatorial Atlantic, Department of Earth and Planetary Sciences of the City College of the City University of New York and CUNY Institute of Marine and Atmospheric Science, 211 pp., 1975.

O'Brien, J.J., D. Adamec, and D.W. Moore, A simple model of upwelling in the Gulf of Guinea, *Geophysical Research Letters, 5,* 641-644, 1978.

Sedikh, K.A. and B.N. Loutochkina, Aspects hydrologiques de la formation de la zone équatoriale productive du Golfe du Guinée, in *Les zones productives de l'Océan Atlantique équatorial et les conditions de leur formation,* Travaux Atlantniro, Kaliningrad, 37, 31-80, 1971 (Translated from Russian by H. Rotschi.).

Sorokin, Y.I., I.N. Sukhanoya, G.V. Konovalova, and E.V. Pabelyeva, Production primaire et phytoplancton de la divergence équatoriale du Pacifique oriental, *Travaux de l'Institut Océanologique Shirshov de l'Académie des Sciences de l'U.S.S.R., Tome 102,* 108-122, 1975 (Translated from Russian by H. Rotschi, 1976.).

Voituriez, B. and A. Herbland, Etude de la production pélagique de la zone équatoriale de l'Atlantique à 4°W, I - Relations entre la structure hydrologique et la production primaire, *Cahiers O.R.S.T.O.M., Série Océanographie, 15,* 313-331, 1977.

Walsh, J.J., Herbivory as a factor in patterns of nutrient utilization in the sea, *Limnology and Oceanography, 21,* 1-13, 1976.

Descriptions and Comparisons of Specific Upwelling Systems

A COMPARISON OF THE STRUCTURE AND VARIABILITY OF THE FLOW FIELD IN THREE COASTAL UPWELLING REGIONS: OREGON, NORTHWEST AFRICA, AND PERU

Robert L. Smith

School of Oceanography, Oregon State University, Corvallis, Oregon 97331

Abstract. The coastal upwelling regions off Oregon (near 45°N), northwest Africa (near 22°N), and Peru (near 15°S) are compared using nearly equivalent data sets from current meter moorings on the continental shelves. Wind measurements from anemometers mounted on buoys are used to determine the locally forced response in the currents. Intensive near-surface measurements allow good resolution in the surface Ekman layer. The alongshore flow throughout the water column is highly correlated with the wind stress except off Peru, where energetic coastal trapped waves are dominant. The deeper mean flow on the Oregon and Peru shelves is poleward, opposite to the wind. No evidence for a mean multi-cell cross-shelf circulation pattern, excepting the bottom Ekman layer, is observed in any of the three regions. The mean Ekman transport ($\overline{\tau_y f^{-1}}$), the mean offshore transport in the surface layer and, except off Oregon, the mean onshore transport below the surface layer, agree within a factor of two. The agreement between the variations in the surface layer offshore transport and in the wind stress is good: the correlations are all high and the regression coefficient between measured transport and Ekman transport is unity to within less than the estimated error. Simple Ekman dynamics seems to hold reasonably well for the surface layer in the mean and on the event time scale. The cross-shelf transport below the surface layer is also significantly correlated with the wind stress, but the variations in the lower layer cross-shelf flow are not significantly correlated with those in the surface layer. The findings are consistent with the observation that the onshore flow in the lower layer does not fully compensate (balance) the offshore surface layer flow on the event time scale. One must conclude that the upwelling process is essentially three dimensional.

Introduction

During the CUEA program three different coastal upwelling regions were intensively studied: Oregon in the CUE-2 experiment during July-August 1973, northwest Africa in the JOINT-1 experiment during March-April 1974, and Peru in the JOINT-2 experiment during March-May 1977. The regions differed in latitude, in stratification, and in the strength and variability of the wind driving the upwelling (Table 1), as well as in the bathymetry of the shelf and slope (Fig. 1). For several weeks during each experiment moored fixed-level current meters recorded the horizontal flow in the water column at a mid-shelf site near the seaward edge of the upwelling zone. Data from the current meters and the nearby meteorological buoys are used in this comparative study; the locations of the instruments are indicated in Fig. 1. Vector averaging (VACM) and Geodyne current meters suspended from a surface buoy were used to measure the flow in the upper 20 m; to measure the deeper flow, Aanderaa current meters were suspended on a second mooring beneath subsurface flotation at 15 to 20 m. Descriptions and analyses of the measurements have been published previously, e.g. Halpern (1976b); Halpern, Smith, and Mittelstaedt (1977), Brink, Halpern, and Smith (1980a). In this paper the structure, variability, and mass balance of the horizontal flow are compared using equivalent data sets and techniques. Details of the data processing and the statistical techniques and tests used are given in the Appendix. The emphasis is on the cross-shelf flow, because that "induces" the coastal upwelling. The recent reviews by Allen (1980) and Winant (1980) provide more general discussions of wind-driven currents on continental shelves.

The Mean State

Our first question is: Are the observed mean flow patterns consistent with the conceptual model of coastal upwelling? The classical conceptual model of coastal upwelling, which stems from McEwen's (1912) application of Ekman's (1905) theory and continues to be the basis of most contemporary mathematical models, states that water in the surface layer is driven offshore by the alongshore component of the wind stress (τ_y) and that water upwells near the coast to replace the

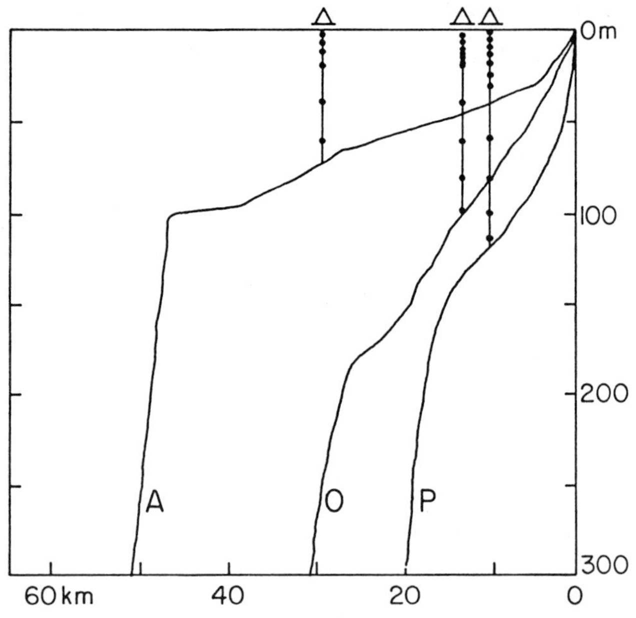

Fig. 1. Shelf and slope topography off northwest Africa (21°40'N), Oregon (45°N), and Peru (15°S) with location of moored instrument arrays shown.

offshore-directed Ekman transport. Theoretical considerations (Charney, 1955; Yoshida, 1955) suggest that the replacement or "upwelling" would occur within a coastal zone with an offshore scale given by the baroclinic radius of deformation: $R = H\overline{N}f^{-1}$, where H is the water depth, $\overline{N}$ is the vertically averaged Brunt-Väisälä frequency — a measure of the stratification, and f is the Coriolis parameter. Thus the offshore-directed Ekman transport ($\tau_y f^{-1}$) at the seaward edge of the coastal zone drives coastal upwelling by requiring water inshore to upwell into the surface layer to satisfy continuity. The horizontal flow beneath the surface layer must be convergent to satisfy the continuity equation. This can be done simply, but not necessarily, by cross-shelf flow beneath the surface layer equal and opposite to the Ekman transport in the surface layer.

The observations from the moored arrays (Fig. 1) can provide a test of the conceptual model. The mean alongshore component of the wind stress (Table 1) was favorable in the three regions studied (i.e., the wind blew equatorward along those eastern boundaries). The values for R estimated from the parameters given in Table 1, using $N^2 = g\Delta\rho\, H^{-1}\rho^{-1}$, and the hydrographic and biological evidence (Smith, 1974; Huntsman and Barber, 1977; Brink *et al.*, 1980b) support the assumption that the moored arrays in Fig. 1 were near the seaward edge of the upwelling zone.

The mean horizontal velocity profiles from the moorings are shown in Fig. 2. The alongshore-onshore directions are positive equatorward-eastward, i.e., toward 180° - 90°T for Oregon and northwest Africa and toward northwest (315°T)-northeast (45°T) for Peru, consistent with the general trends of the coastline (Plates I, II, and III). Near the surface the mean flow is strongly equatorward as is the wind. A poleward undercurrent is evident off Oregon and Peru but not on the northwest African shelf; a strong poleward undercurrent does exist over the slope off northwest Africa (Mittelstaedt, Pillsbury, and Smith, 1975). Poleward undercurrents are often associated with coastal upwelling regions (Wooster and Reid, 1963), but their dynamical relation to coastal upwelling is not well understood. Our principal interest in this paper is in the offshore-onshore flow. It is clear from Fig. 2 that the mean profile is consistent with our conceptual model of coastal upwelling: offshore flow in a surface layer of order of 20- to 40-m depth at mid-shelf causing an upwelling of water into the surface zone inshore — with the upwelled water being supplied by onshore flow in the deeper water. The question of whether there exists a quantitative agreement between the Ekman

TABLE 1. Comparison of Three Upwelling Regions (see Fig. 1 and Plates I, II, III). τ_y: Mean $\pm$ standard deviation of the alongshore (equatorward) component of the wind stress (dynes/cm^2) and maximum-minimum values during the experiments; $\Delta\rho$: Density difference (*sigma-t*) from surface to bottom at mid-shelf; R: estimate baroclinic radius of deformation (km). The $\Delta\rho$ values are estimated from data in Huyer (1976), Barton, Huyer, and Smith (1977), and Brink, Gilbert, and Huyer (1979).

	τ_y		$\Delta\rho$	R
Oregon (45°N)	0.50 ± 0.73;	3.48/-1.10	2.0	14
Northwest Africa (21°40'N)	1.50 ± 0.93;	3.20/-0.50	0.3	10
Peru (15°S)	0.55 ± 0.35	1.53/0.08	0.5	20

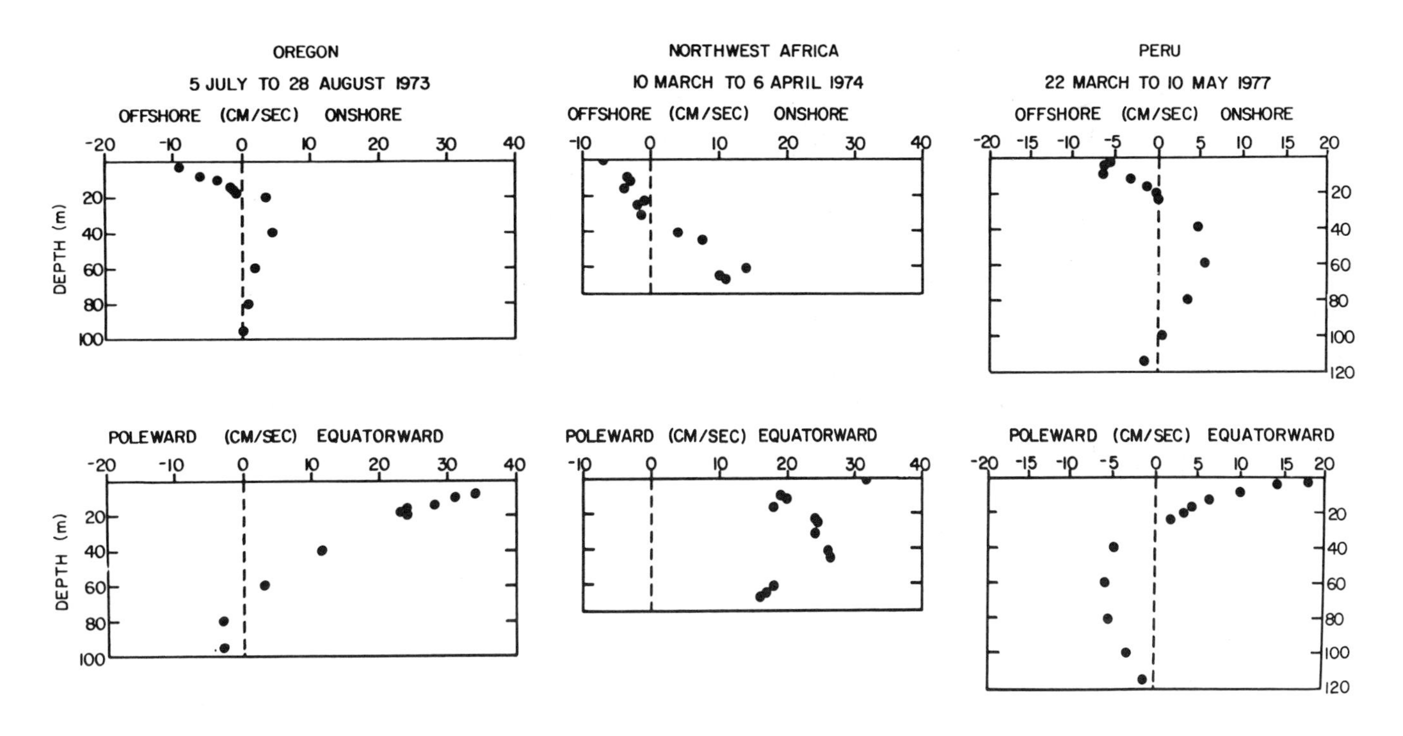

Fig. 2. Mean current profiles from the mid-shelf moored arrays (Fig. 1).

transport ($\tau_y f^{-1}$), the offshore transport in the surface layer, and the onshore transport beneath will be discussed in a later section.

The layer of mean offshore-directed flow is relatively thin: less than 20 m off Oregon, 35 m off northwest Africa, and about 30 m off Peru. The Ekman depth, a measure of the thickness of the Ekman layer, is $\delta = (2A_v f^{-1})^{\frac{1}{2}}$ from simple theory (Pedlosky, 1979, p. 177), where A_v is the vertical eddy viscosity. Halpern (1976b), Halpern (1977), and Brink *et al.* (1980a) estimated A_v by matching the wind stress to the near-surface stress due to vertical shear in the alongshore current, $\tau_y = A_v \partial v/\partial z$. Average values for A_v during periods of strong upwelling-favorable winds are 55 cm^2/s, 125 cm^2/s, and 70 cm^2/s for Oregon, northwest Africa, and Peru; the corresponding Ekman depths are 10, 22, and 19 m. An alternate estimate of the thickness of the surface layer comes from studies of turbulent boundary layers. Pollard, Rhines, and Thompson (1973) gave the thickness of the surface mixed layer as $\delta_s = 1.7\, u_* \,(fN)^{-\frac{1}{2}}$, where u_* is the friction velocity defined by $\tau = \rho u_*^2$. Using the parameters given in Table 1 and estimating mean N^2 as above, one obtains δ_s = 10, 35, and 26 m for Oregon, northwest Africa, and Peru, respectively. The values are in good agreement with the observed thickness of the layer of mean offshore flow.

The onshore flow below the surface layer, which could compensate for the upwelling inshore into the offshore-flowing surface layer, is clearly not in a bottom Ekman layer off Oregon and Peru. Off Oregon and Peru the deeper flow at mid-shelf is poleward and thus the bottom Ekman transport would be directed offshore. The onshore compensatory flow is shallow and presumably in geostrophic balance with the alongshore pressure gradient. A mean alongshore pressure gradient caused by an alongshore (equatorward) rise of sea level of 1 cm in 100 km would be sufficient to allow a mean geostrophic onshore flow to balance the mean offshore Ekman transport. Although such a mechanism is utilized in several theories and numerical models, the smallness of the pressure gradient has precluded experimental verification.

Off northwest Africa the onshore flow is adjacent to the bottom and is of a thickness above the bottom consistent with a bottom Ekman layer. On the basis of boundary layer theory the thickness of the bottom Ekman layer is often assumed to be $\delta_B = 0.3\, u_* f^{-1}$, where u_* is the friction velocity; u_* is estimated as 0.03 V_g, where V_g is the 'geostrophic speed' or speed outside the boundary layer (see Weatherly and Martin, 1978 for a discussion of the dynamics of the oceanic bottom boundary layers). For northwest Africa we take V_g to be the maximum vector mean speed beneath the surface Ekman layer, viz. V(45m) = 28 cm/s. This yields δ_B = 64 m and suggests that the bottom Ekman layer does play an important, perhaps dominant, role in the upwelling circulation. For comparison we compute δ_B for Oregon and Peru, using a more directly determined estimate u_* = 0.42 cm/s off Oregon (Caldwell and Chriss, 1979) and estimating u_* for Peru as

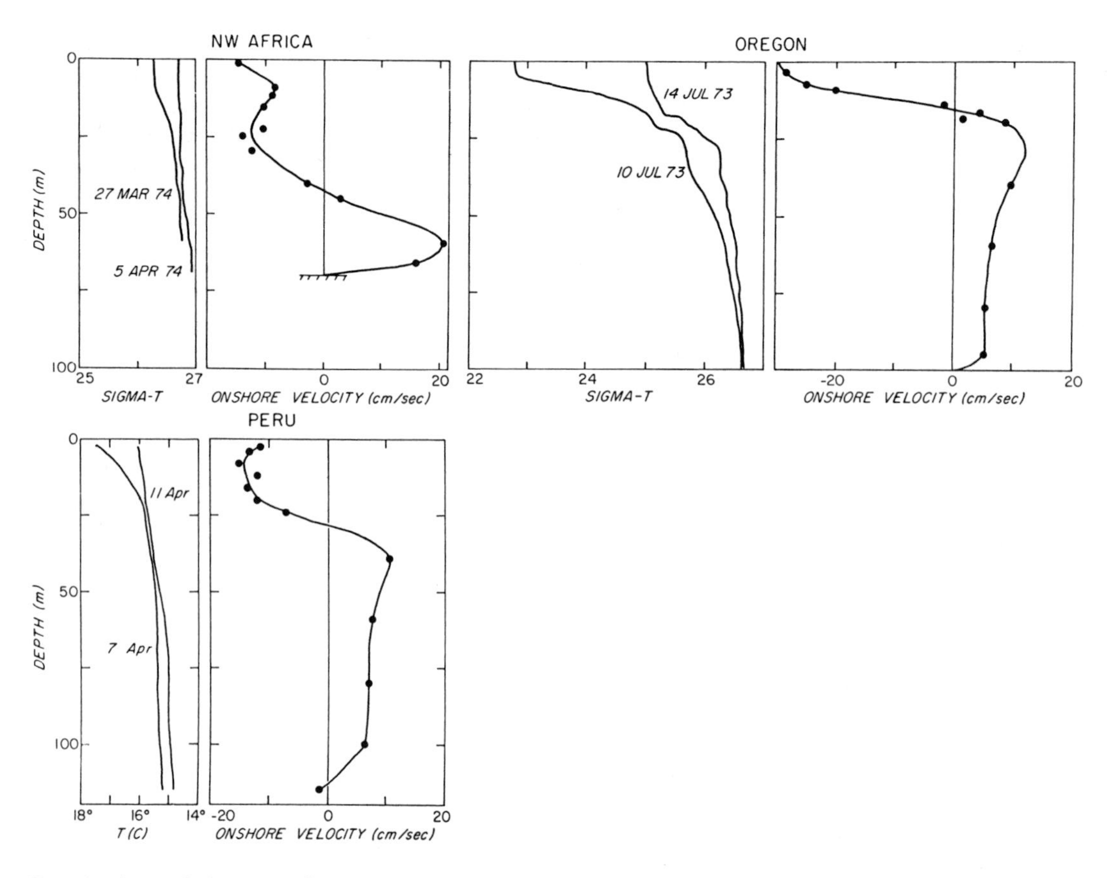

Fig. 3. Vertical profiles of *sigma-t* or density before and during an upwelling event and the cross-shelf flow at the height of an event (12 July 1973 off Oregon, 3 April 1974 off northwest Africa, and 9 April 1977 off Peru). The Oregon and northwest Africa figures are from Huyer (1976).

$0.03\ V_g$, using $V(59m) = V_g = 12$ cm/s; we obtain $\delta_B = 16$ m off Oregon and 38 m off Peru. These estimates ignore the effect of stratification, which decreases the thickness of the turbulent bottom boundary layer. An estimate of the thickness, taking into account stratification, can be made from the expression used above for the turbulent surface boundary layer, with the u_* values estimated for the bottom boundary layer and the mean N^2 values previously computed, except for Oregon where N^2 near the bottom is an order of magnitude less than the mean N^2. One now obtains $\delta_B = 10$ m off Oregon, 13 m off Peru, and 24 m off northwest Africa. The estimates of δ_B are consistent with the mean velocity profiles and suggest that an Ekman bottom layer provides the compensatory onshore flow off northwest Africa but plays a minor (and opposite, because of the mean poleward flow) role off Peru and Oregon.

The preceding discussion focused on the mean situation; the mean circulation is qualitatively consistent with the conceptual model under the mean upwelling-favorable winds experienced in the coastal upwelling regions off Oregon, northwest Africa, and Peru. The wind is, however, quite variable and "upwelling events", periods of strong upwelling associated with stronger than average upwelling-favorable alongshore winds, is now a common expression in upwelling literature. Do the mean onshore-offshore circulation patterns resemble those during upwelling events? To compare the event flow pattern to the mean we chose the wind event associated with the maximum wind stress in Table 1. The profiles of the cross-shelf flow (Fig. 3) show the low-pass filtered data at a time when the wind stress has been greater than the mean for more than one inertial period, the characteristic time scale for Ekman layer 'spin-up'. The σ_t or temperature profiles observed before the wind intensified and after the peak wind stress indicate that the water properties did change as one would expect during an upwelling event. The close similarity in the onshore-offshore circulation at the height of an upwelling event to the mean profile is clear. Even off Oregon, where the standard deviation of the wind stress is greater than the mean and the wind stress during the event is almost an order of magnitude greater than the mean, the event and mean profiles are very similar. We conclude that the conceptual model of upwelling holds both in the mean and during upwelling events.

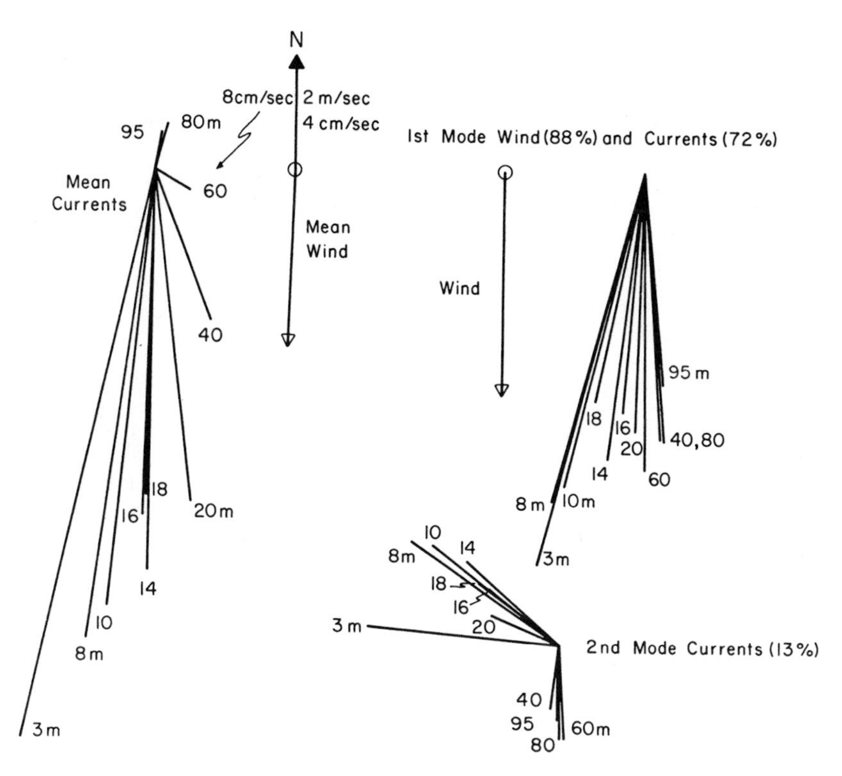

Fig. 4. Mean current and wind vectors and the major empirical orthogonal vector variance modes off Oregon. The percentage of the variance accounted for by a mode is stated.

The Patterns of Variability

To explore more thoroughly the vertical structure of the horizontal flow and the response of the flow to wind stress fluctuations we used principal component analysis, i.e., the vertical structure of the variability in the horizontal flow is described by means of empirical orthogonal modes. Specifically, we ask whether there are coherent patterns of flow variability that are similar in the three coastal upwelling regions, in spite of differences in the stratification, latitude, and wind strength and variability.

The vector vertical modes at each site were computed by treating the low-pass filtered northward and eastward components (with means removed) at each measurement depth as if they were scalar time series. The modes (for a site) are the eigenvectors of the matrix of correlation coefficients among the various northward and eastward current time series at the site; the variance associated with each mode is given by the corresponding eigenvalue. Each mode represents a part of the fluctuations (variance) that is coherent and in phase through the mid-shelf water column. The modes can be ranked in order of their contribution to the total variance. The advantage of this method is that a few (1 or 2) modes contain almost all of the variance contained in all the original time series.

The vector mean horizontal flow and the first two vertical modes of the vector variance are shown in Figs 4, 5, and 6. The orientation is such that upward (right) on the figure is northward alongshore (onshore). For presentation in Figs 4, 5, and 6 the eigenvector components of the vertical modes have been multiplied by the square root of the eigenvalue (i.e., scaled by the square root of the variance) and plotted as vectors in the same manner as the mean vectors. Thus the mean vector and the first two vector variance modes (which together make up more than 85% of the total variance) can be compared directly with the latter being interpreted as showing a standard deviation about the mean vector. A time series of a mode would show the magnitude of the individual vectors expanding and contracting in unison along their individual directions (which are constant) and going through zero (the origin), and expanding and contracting in the opposite direction. The time average of each vector in a mode would, of course, be zero. The mean wind vector and its first mode (i.e., the standard deviation along the major principal axis or direction of maximum variance) is shown. The directional sense of the wind and current modes are presented in the sense expected during an upwelling event.

The mean vectors show clearly, in the seagull's eye view of Figs 4, 5, and 6, the relative strength of the alongshore flow compared with cross-shelf flow; the latter, which presumably induces the coastal upwelling, is much weaker. The Peru vector mean shows as perfect an Ekman spiral in the surface layer as has been observed in an upwelling region; the bottom Ekman layer is also apparent in the offshore veering of the vectors below 59 m. The mean poleward undercurrent is clearly seen off Oregon and Peru, as is its absence off northwest Africa.

The first mode at each location contains nearly three quarters of the total variance and is directed alongshore with some slight evidence of Ekman veering in the surface and bottom boundary layers; the separation into onshore and offshore flow layers is most pronounced off northwest Africa. The first mode off Oregon is quasi-barotropic in spite of the large density gradient (n.b., the difference between the mean shear and the shear in the first mode in the region between the surface and bottom Ekman layers, roughly between 14 and 80 m). The barotropic response of the alongshore flow to wind fluctuations has been discussed by Smith (1974) and Kundu, Allen, and Smith (1975). The first modes off Oregon and northwest Africa are well correlated with the local wind: 0.56 with the current lagging the wind by 18 h off Oregon and 0.77 with a lag of 12 h off northwest Africa (the correlations are significantly different from zero at the 99% confidence level). Although the form of the first mode off Peru is similar to that off Oregon and accounts for nearly the same percentage of the total variance, it is uncorrelated with the local wind — the maximum correlation coefficient is 0.20 (to be significantly different from zero at 95% confidence level would require a CC = 0.325). The first mode however is correlated with the alongshore currents 400 km to the north and with sea level at San Juan (Plate III) at lags consistent with a free poleward propagating coastal trapped wave. The dominant variability in the alongshore flow at 15°S off Peru has been shown (Smith, 1978) to be associated with low-frequency (5- to 10-day period) internal Kelvin waves, which are uncorrelated with local winds; the first mode off Peru contains the wave variability and not a response to local winds.

The second mode in each region obviously represents a surface layer response to the wind: the significant contribution of the second mode to the cross-shelf flow is in a surface layer with a thickness consistent with our earlier estimates of δ. The correlation coefficients between the second mode and the wind fluctuations are maximum at zero lag and are 0.68 off Oregon, 0.57 off northwest Africa, and 0.60 off Peru (all significantly different from zero at the 99% confidence level) and in the sense that offshore flow near the surface is associated with an intensified equatorward wind.

Is the compensatory onshore flow below the surface layer associated with a particular mode? The third and higher modes (not shown) individually account for less than 5% of the vector variance, and in none of the regions do they clearly show a pattern that one could associate with compensatory onshore flow in response to an upwelling event. Except off Peru, the second mode shows no onshore flow at depth coupled to offshore flow in the surface layer. However, in all regions the first mode has an onshore contribution below the 'Ekman depth' that is at least 50% greater than the near-surface offshore flow in the mode. As this mode lags the wind by an inertial period, or less, off Oregon and northwest Africa we may assume it is part of the compensatory flow. Off Peru the first mode is not significantly correlated with the wind but some of the compensatory flow response may have leaked into the first mode; this is weakly suggested by the fact that the first mode has its maximum correlation with the second mode at a lag of about half an inertial period, although the correlation is weak and not significant at even the 90% level. The failure to isolate, in an empirical orthogonal mode, the onshore compensatory flow response to an upwelling event, in spite of the relatively obvious compensatory flow seen during events (e.g., Fig. 3 above, Fig. 6 in Huyer, 1976, and Fig. 5 in Brink *et al.*, 1980b) is probably due to the generally poor mutual correlation between the cross-shelf component of flow (Kundu and Allen, 1976; Brink *et al.*, 1980a). The compensatory flow below the surface layer is spread over a much greater depth range than the surface flow and thus has a weaker signal-to-noise ratio — resulting in lower mutual correlation functions that degrade the empirical orthogonal mode computations. Finally, the compensatory flow response to upwelling events may not result simply from variations in the cross-shelf component (u) but may result also from a convergence in the alongshore component (v) of the flow.

Volume Balances

Is the observed cross-shelf transport in the surface layer, $U_S = \int_{-\delta}^{0} u\,dz$, in quantitative agreement with the Ekman volume transport, $\tau_y(\ f)^{-1}$, computed from the wind observations? Is the offshore flow, U_S, balanced by an equal onshore flow below, $U_L = \int_{-b}^{-\delta} u\,dz$, or is the upwelling process more complex? The latter question can be restated: Is the convergence in the flow required to balance the Ekman layer divergence, caused by the wind stress along the coastal boundary, two-dimensional $(\frac{\partial u}{\partial x} + \frac{\partial w}{\partial z} = 0)$ or three-dimensional $(\frac{\partial u}{\partial x} + \frac{\partial v}{\partial y} + \frac{\partial w}{\partial z} = 0)$? As the wind is highly variable, the questions may have different answers with respect to the mean circulation than on the time scale of upwelling events.

The definition of the alongshore-onshore coordinate system is crucial in tests of volume flux balances because both the mean and variable alongshore flows are much greater than the cross-shelf flow. The net alongshore transports are nearly an order of magnitude greater than the offshore-onshore transports or the theoretical Ekman transports. Thus a slight rotation (only a couple

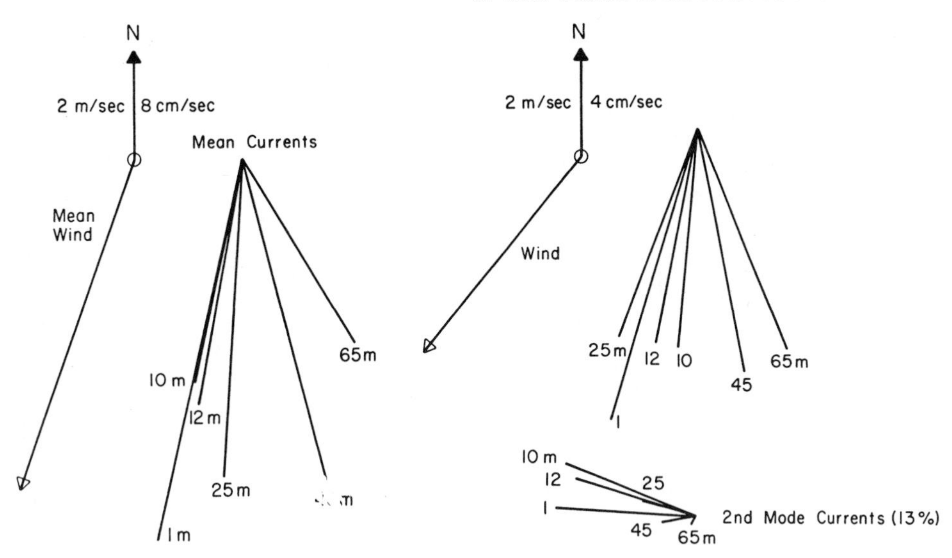

Fig. 5. Mean current and wind vectors and the major empirical orthogonal vector variance modes off northwest Africa.

degrees) in the nominal alongshore-onshore coordinate system can make an appreciable difference in the cross-shelf transport. The local isobath direction is the natural alongshore direction in coastal upwelling regions: The cross-isobath (u) velocity field forces coastal upwelling through continuity and, in time-dependent inviscid models, the variability in the alongshore flow (v) results from the $\frac{\partial v}{\partial t} + fu$ balance (Allen, 1980). In a stratified fluid the flow tends to organize in the along-isobath direction, and the CUEA results (e.g., Kundu and Allen, 1976) clearly show the tendency for the horizontal velocity fluctuations to align along isobaths. In general, bathy-

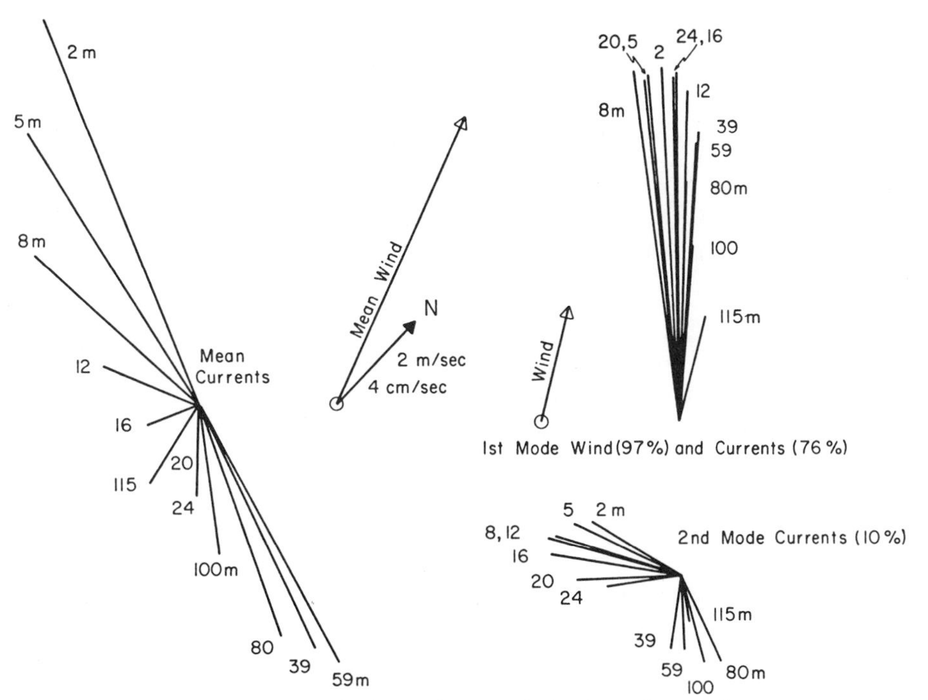

Fig. 6. Mean current and wind vectors and the major empirical orthogonal vector variance modes off Peru. Data interval: 1200 UT 7 March to 1200 UT 13 May 1977.

metric charts are not precise enough to allow estimates of isobath directions to within a few degrees and, furthermore, it is not apparent over what spatial scale the directions of the crooked and curving isobaths should be estimated. In this paper the major principal axis of the total volume transport vector (the velocity integrated from the bottom, $z = -b$, to the surface, $z = 0$: $U = \int_{-b}^{o} u dz$, $V = \int_{-b}^{o} v dz$) is chosen as an 'objective' definition of the alongshore direction, i.e., the direction of maximum variance of the depth-averaged velocity field is assumed to be the alongshore direction. The directions agree well with the isobath directions (see Appendix). A beneficial corrollary is that cross-shelf fluctuations are uncorrelated (at 0 lag) with the alongshore fluctuations in the principal axis system. Thus, cross-shelf transports are not contaminated with an alongshore transport contribution.

The volume transports were computed by numerical (trapezoidal) integration of the vertical array of velocity measurements used to compute the empirical orthogonal modes (Figs 4, 5, and 6). The contributions to the integration from current meters above and below a fixed depth D were assigned to the upper and lower layers respectively, i.e., a fixed D is used for δ in the computation of U_S and U_L. D was taken to be the depth mid-way between the deepest current meter showing offshore flow and the next deeper current meter. Except for Oregon, the mean profile, the mode patterns, and the event profiles gave a common "deepest" upper layer current meter, leading to the separation between surface and lower layers at $D_A = 35$ m for northwest Africa and $D_p = 32$ m for Peru. Off Oregon the distinction was made on the basis of the mean profile and $D_o = 19$ m was used, as was done by Bryden (1978). All the computations were also done with $D_o = 30$ m and $D_p = 26$ m as a check; the agreement between surface and Ekman transports, both mean and variable, was poorer with the "deeper" Oregon and the "shallower" Peru surface layers.

The mean Ekman transport computed from the wind stress and the measured mean transports in the upper and lower layers are given in the transport principal axis system in Table 2. (For comparison, the cross-shelf transport in the surface layer is also shown for the coordinate system defined by zero mean net cross-shelf transport). The large mean onshore transport, U_L, in the lower layer off Oregon is probably real and a result of a positive mean $\partial v/\partial y$; the mean southward flow along the 100-m isobath increases by about 2.5 cm/s from mooring C (used in this paper) to mooring P (about 50 km southward; see Plate II, this volume, and statistics in Kundu and Allen, 1976). An increased onshore flow in the region between C and P of nearly 1 cm/s would be required to supply the estimated increased transport over the 50% wider shelf inshore of 100 m at P. Thus about 100 units of the lower layer onshore transport off Oregon (Table 2) may be the result of a mean $\partial v/\partial y$ and not be part of the compensatory flow for the mean offshore Ekman transport. If this is the case, the Oregon lower layer transport would not appear anomalously large relative to the Ekman transport. A non-negligible mean $\partial v/\partial y$ may be ubiquitous in coastal upwelling regions as a result of bathymetric variations and thus preclude a mean two-dimensional mass balance except over space scales large compared to the bathymetric variations.

The observed mean offshore transport is in fair agreement with the mean Ekman transport (Table 2). The magnitudes of the respective standard deviations are in closer agreement. Does the Ekman relation hold for the variable flow? To test this, the correlations of the U_S series against the $\tau_y(\rho f)^{-1}$ series, with the means removed, were computed. The maximum correlations, all significant at >99.9% level of significance, are 0.66 with U_S lagging τ_y by 6 h off Oregon; 0.71 with 0 lag off northwest Africa, 0.54 with U_S lagging τ_y by 6 h off Peru. A quantitative answer to the question is provided by the linear regressions of U_S on $\tau_y(\rho f)^{-1}$, with the U_S series lagged to pro-

TABLE 2. Mean Ekman Transport Computed from the Alongshore Component of the Wind Stress, Mean Measured Cross-shelf Transport in the Surface and Lower Layers (Principal Axis System), and the Offshore-Onshore Transport in the Upper-Lower Layer in the Zero Net Cross-shelf Transport Coordinate System. Transport is given in 10^2 cm^3 cm^{-1}s^{-1} and is positive directed onshore.

	$\tau_y/\rho f$	U_S	U_L	$\lvert U_o \rvert$
Oregon	-48 ± 69	-65 ± 109	198 ± 177	143
Northwest Africa	-274 ± 170	-145 ± 184	248 ± 160	195
Peru	-141 ± 89	-82 ± 148	120 ± 260	90

vide the maximum correlation. The regression coefficient is unity to within the standard error in all cases (Table 3), i.e., the observed surface layer cross-shelf transport variations agree with those predicted from wind stress using simple Ekman theory.

The lower layer transports, U_L, were also significantly correlated with $\tau_y(\rho f)^{-1}$ (at the 99% significance level) but the regression coefficients were not unity to within the standard error except for Oregon. The results are also shown in Table 3. The cross correlations between U_S and U_L were considerably poorer than between τ_y and either U_S or U_L. Only for northwest Africa and Peru were the correlations significantly different from zero at the 90% level (northwest Africa: -0.39 with U_L lagging U_S by 36 h; Peru: -0.40 with U_L lagging U_S by 48 h). The negligible correlation between U_S and U_L off Oregon may be the result of energetic barotropic continental shelf waves, which are present in the Oregon data (Cutchin and Smith, 1973), obscuring the "baroclinic" cross-shelf circulation response to upwelling. In any case, the U_L, U_S relation cannot be characterized by a two-dimensional mass balance on the event time scale in any region. This is shown in Fig. 7 by plotting the cross-shelf direction (relative to east) that would be required for two-dimensional mass balance to hold (see Appendix). The direction is independent of the layer depths assumed; it is simply the normal to the direction of the net total transport. The Ekman transport is plotted to show that during events there is a tendency for the angles to be steady and near the cross-shelf direction defined by the 'principal' axis, but the variability is large. In nature, two-dimensional upwelling circulation seems to be ephemeral.

Conclusions

Measurements of the horizontal flow field in the coastal upwelling regions off Oregon, northwest Africa (Mauritania), and Peru show a flow pattern and variability consistent with the simple conceptual model of coastal upwelling: offshore flow in a surface layer on the order of 20 m thick; the offshore transport varies with the strength of the alongshore component of the wind stress; the onshore flow below the surface layer tends to compensate for the offshore flow. The mean Ekman transport ($\tau_y f^{-1}$), the mean offshore transport in the surface layer and, except off Oregon, the mean onshore transport below the surface layer, agree within a factor of two. Considering the uncertainties in estimating the wind stress from wind velocity measurements, the assumptions in computing the volume transports, and the vagaries involved in deciding on the 'offshore' direction, some may find the agreement of the various means surprisingly good. The agreement between the variations in the surface layer offshore transport and in the wind stress is good: the correlations are all high (significantly different from zero at greater than the 99.9% confidence level) and the regression coefficient of measured transport as a function of the Ekman transport is unity to within less than the estimated error. Simple Ekman dynamics seem to hold reasonably well for the surface layer in the mean and on the event time scale.

Although the onshore transport below the surface layer is also significantly correlated with the wind stress, the regression coefficient between observed onshore transport below the surface layer and the Ekman transport inferred for the surface layer is not unity to within the estimate of error. The variations in the onshore lower layer flow are not significantly correlated with the variations in the surface layer offshore flow at 95% confidence. These findings are consistent with the observation that the onshore flow in the lower layer does not fully compensate (balance) the offshore surface layer flow on the event time scale. One must conclude that the upwelling process is essentially three dimensional, i.e., the $\frac{\partial v}{\partial y}$ term in the continuity equation cannot be neglected.

TABLE 3. Coefficients from Linear Regression $\hat{U}_S = k{*}T$ and $\hat{U}_L = k{*}T$ where $T = \tau_y/\rho f$. (U positive onshore, τ_y positive equatorward alongshore).

	Dependent Variable (lags τ_y by n hours)	Correlation Coefficient	Regression Coefficient ± Standard Error
Oregon	$U_S(6)$	-0.66	-1.04 ± 0.33
	$U_L(0)$	0.49	1.25 ± 0.54
Northwest Africa	$U_S(0)$	-0.71	-0.77 ± 0.30
	$U_L(10)$	0.64	0.64 ± 0.25
Peru	$U_S(6)$	-0.51	-0.84 ± 0.33
	$U_L(0)$	0.63	1.85 ± 0.36

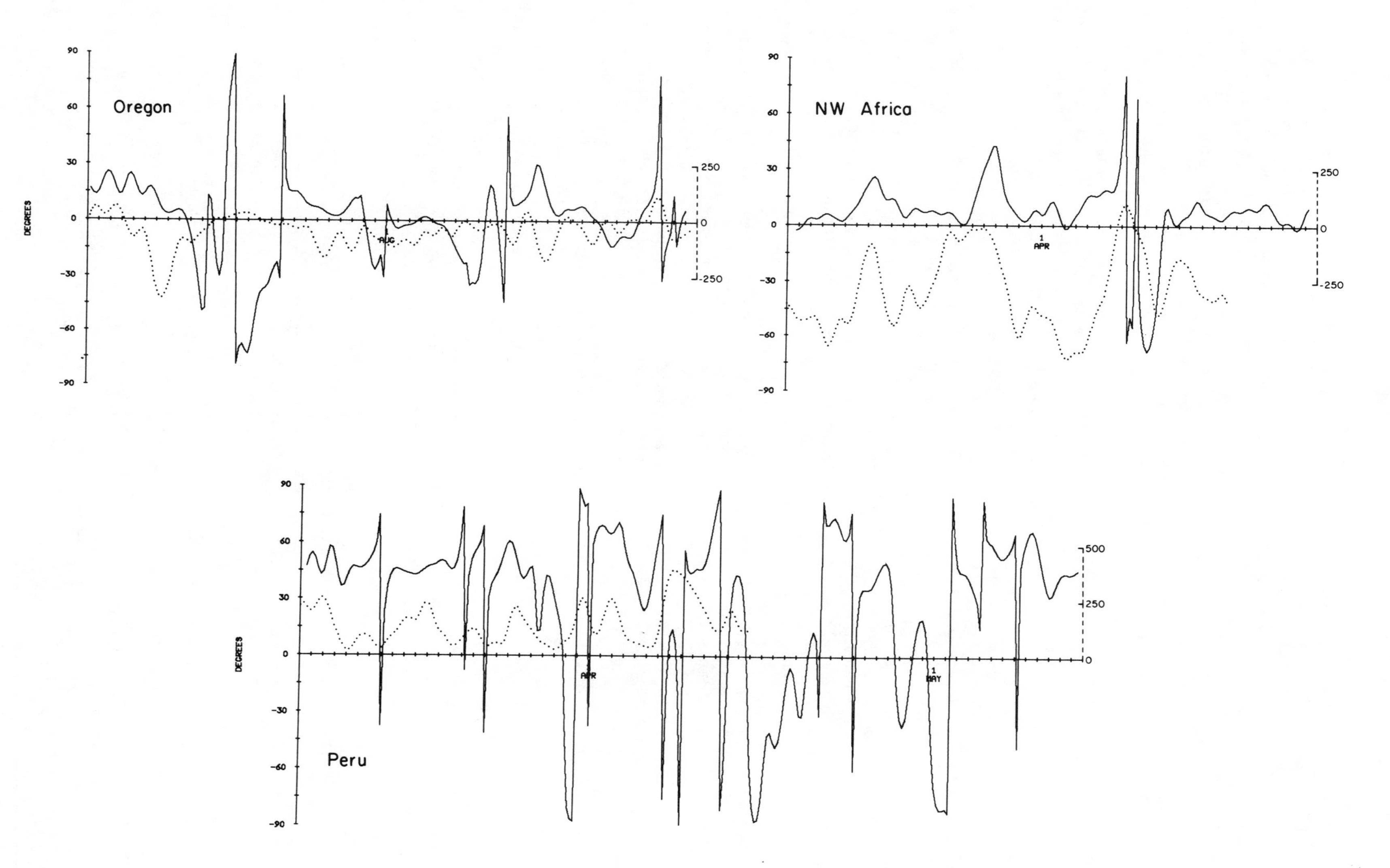

Fig. 7. Solid line: Onshore direction (relative to east) for which two-dimensional mass balance holds. Compare the solid curve with 'onshore' direction in Plates I, II, and III. Dashed line: northward alongshore component of wind stress (in units of Ekman transport: 10^2 cm^3 $cm^{-1}s^{-1}$.

Finally, the flow patterns (mean and variable) show no evidence of a two-cell circulation pattern such as envisioned by Mooers, Collins, and Smith (1976) and others. While this does not demonstrate that such circulation patterns do not occur, it does strongly suggest that they are, at most, ephemeral, migrant, and play no significant role.

Appendix

Data. The data used in this paper come principally from the Oregon B, C moorings (Plate II), the northwest Africa U, Li moorings (Plate I), and the Peru M, PS moorings (Plate III). The wind and current data were originally sampled at intervals of 20 min or less; the data were averaged to provide hourly data sets. Hourly wind stress was computed from the hourly wind velocity, extrapolated to the 10-m level assuming a logarithmic wind profile, using the aerodynamic square law with a constant drag coefficient of 1.3×10^{-3} (Halpern, 1976a). The hourly data sets were then filtered to eliminate tidal and higher frequencies by means of a symmetrical low-pass filter (half power at period of 40 h) spanning 121 h. The low-pass filtered series, decimated to 6-h values, were used as the basic data sets for this paper.

During an experiment not all instruments provided identical record lengths; the longest possible record lengths were used for computing a particular set of statistics or making a comparison. Except where stated the intervals are: Oregon (0600 UT 6 July to 0000 UT 27 August 1973), northwest Africa (1800 UT 10 March to 0600 UT 17 April 1974), Peru (1200 UT 7 March to 1800 UT 14 April 1977). Off Peru, the meteorological buoy (PS) directly over the current meters used in the comparison operated for a shorter period than one about 10 km inshore (PSS). Although the records are highly correlated, the wind stresses differ significantly (Brink *et al.*, 1980a), with the inshore wind stress being 20% higher due to orographic effects. Because the PS data were from directly over the mid-shelf site, the PS data with the shorter record lengths were used except for the modal analysis.

Cross Correlation Analysis and Regression Coefficients. Correlation coefficients at lags from 0 to 8 days at increments of 6 h were computed between the various time series. For each correlation the degrees of freedom, N_r, were found by using the method suggested by Davis (1976) and the significance level obtained from Pearson and Hartley (1970, p. 146). In general, the 95% significance level (i.e., the probability that a given correlation coefficient would be obtained from uncorrelated variables is less than 0.05) is used as the criterion of significance.

Following Allen and Kundu (1978, Appendix A) the degrees of freedom, N_r, were also used in the calculation of the standard error of the linear regression coefficients. The regression coefficient k between an observed transport, U, and Ekman transport (the independent variable), $T = \tau_y/\rho f$, in the regression equation $\hat{U} = kT$ is given by

$$k = \sum^{N} UT / \sum^{N} T^2,$$

where the time series U and T have N points and zero means. The standard error S_k of k is obtained from

$$S_k^2 = \sum^{N} (U-\hat{U})^2 / (N_r - 2) \sum^{N} T^2.$$

Coordinate Systems. The principal axis coordinate system is objective in that the orientation is determined by the characteristics of the data itself. In this paper, the total transport vector, or depth averaged velocity vector, is used to determine the principal axes. The procedure and consequences are discussed in Kundu and Allen (1976).

An alternate 'objective' procedure for determining the onshore-alongshore coordinate system is to assume *a priori* that two-dimensional cross-shelf mass balance holds. Two-dimensional cross-shelf mass balance holds in the coordinate system defined by $U' = 0 = U \cos\theta + V \sin\theta$ where U, V are the total transport components (cross-shelf, alongshore) in the original coordinate system and θ is the angle that the original coordinate system must be rotated to make the new net cross-shelf transport zero, $U' = 0$. It is instructive to compare the rotation angles from the original geographic (east-north) coordinate system to the onshore-alongshore systems defined by, 1) the principal axes, 2) mean two-dimensional mass balance, and 3) the 'visual' best fit to the isobaths on bathymetric charts. The respective angles, measured counterclockwise from east, are: Oregon (-1.1°, $+6.4^\circ$, $+2^\circ$); northwest Africa ($+3.4^\circ$, $+7.2^\circ$, $+3^\circ$); Peru ($+46.4^\circ$, $+48.8^\circ$, $+49^\circ$). The differences between the rotation angles are small, but the effect of the difference can be large because of the large alongshore transport. As can be seen in Fig. 7, the coordinate system that gives a mean two-dimensional mass balance does not hold on all time scales.

Acknowledgements. This paper represents an attempt to synthesize some data collected during the CUEA program. The work was supported by the National Science Foundation, most recently under grant OCE 78-03382. The author is grateful to David Halpern for permission to use the meteorological and surface layer current data collected by him during the CUEA experiments. An earlier version of the paper was prepared for a lecture at Institut für Meereskunde an der Universität Kiel during a visit sponsored by the Deutsche Forschungsgemeinschaft.

References

Allen, J.S., Models of wind-driven currents on the continental shelf, *Annual Review of Fluid Mechanics, 12,* 389-433, 1980.

Allen, J.S., and P.K. Kundu, On the momentum, vorticity, and mass balance off the Oregon coast, *Journal of Physical Oceanography, 8,* 13-27, 1978.

Barton, E.D., A. Huyer, and R.L. Smith, Temporal variation observed in the hydrographic regime near Cabo Corveiro in the northwest African upwelling region, February to April 1974, *Deep-Sea Research, 24,* 7-23, 1977.

Brink, K.H., W.E. Gilbert, and A. Huyer, Temperature sections along the C-line over the shelf off Cabo Nazca, Peru from moored current meters, 18 March - 10 May 1977 and CTD observations, 5 March - 15 May 1977, *CUEA Technical Report 49,* 78 pp., School of Oceanography, Oregon State University, Corvallis, OR, 1979.

Brink, K.H., D. Halpern, and R.L. Smith, Circulation in the Peruvian upwelling system near 15°S, *Journal of Geophysical Research, 85,* 4036-4048, 1980a.

Brink, K.H., B.H. Jones, J. Van Leer, C.N.K. Mooers, D. Stuart, M. Stevenson, R.C. Dugdale, and G. Heburn, Physical and biological structure and variability in an upwelling center off Peru near 15°S during March 1977. *This Volume*, 1980b.

Bryden, H.L., Mean upwelling velocities on the Oregon continental shelf during summer 1973, *Estuarine and Coastal Marine Science, 7,* 311-327, 1978.

Caldwell, D.R., and T.M. Chriss, The viscous sublayer at the sea floor, *Science, 206,* 1131-1132, 1979.

Charney, J.G., The generation of oceanic currents by wind, *Journal of Marine Research, 14,* 477-498, 1955.

Cutchin, D.L., and R.L. Smith, Continental shelf waves: low-frequency variations in sea level and currents over the Oregon continental shelf, *Journal of Physical Oceanography, 3,* 73-82, 1973.

Davis, R.E., Predictability of sea surface temperature and sea level pressure anomalies over the North Pacific Ocean, *Journal of Physical Oceanography, 6,* 249-268, 1976.

Ekman, V.W., On the influence of the earth's rotation on ocean currents, *Arkiv für Matomatik Astronomi och Fysik, 12,* 1-52, 1905.

Halpern, D., Measurements of near-surface wind stress over an upwelling region near the Oregon coast, *Journal of Physical Oceanography, 6,* 108-112, 1976a.

Halpern, D., Structure of a coastal upwelling event observed off Oregon during July 1973, *Deep-Sea Research, 23,* 495-508, 1976b.

Halpern, D., Description of wind and of upper ocean current and temperature variations on the continental shelf off northwest Africa during March and April 1974, *Journal of Physical Oceanography, 7,* 422-430, 1977.

Halpern, D., R.L. Smith, and E. Mittelstaedt, Cross-shelf circulation on the continental shelf off northwest Africa during upwelling, *Journal of Marine Research, 35,* 787-796, 1977.

Huntsman, S.A., and R.T. Barber, Primary productivity off northwest Africa: the relationship to wind and nutrient conditions, *Deep-Sea Research, 24,* 25-33, 1977.

Huyer, A., A comparison of upwelling events in two locations: Oregon and northwest Africa, *Journal of Marine Research, 34,* 531-546, 1976.

Kundu, P.K., J.S. Allen, and R.L. Smith, Model decomposition of the velocity field near the Oregon coast, *Journal of Physical Oceanography, 5,* 683-704, 1975.

Kundu, P.K., and J.S. Allen, Some three-dimensional characteristics of low frequency current fluctuations near the Oregon coast, *Journal of Physical Oceanography, 6,* 181-199, 1976.

McEwen, G.F., The distribution of ocean temperature along the west coast of North America deduced from Ekman's theory of the upwelling of cold water from adjacent ocean depths, *Internationale Revue der gesamten Hydrobiologie und Hydrographie, 5,* 243-286, 1912.

Mittelstaedt, E., R.D. Pillsbury, and R.L. Smith, Flow patterns in the northwest African upwelling area, *Deutsche Hydrographische Zeitschrift, 28,* 145-167, 1975.

Mooers, C.N.K., C.A. Collins, and R.L. Smith, The dynamic structure of the frontal zone in the coastal upwelling region off Oregon, *Journal of Physical Oceanography, 6,* 3-21, 1976.

Pearson, E.S., and H.O. Hartley (eds.), *Biometrika Tables for Statisticians,* 270 pp., Cambridge University Press, New York, 1970.

Pedlosky, J., *Geophysical Fluid Dynamics,* Springer-Verlag, New York Inc., 624 pp., 1979.

Pollard, R.T., P.B. Rhines, and R.O.R.Y. Thompson, The deepening of the wind-mixed layer, *Geophysical Fluid Dynamics, 3,* 381-404, 1973.

Smith, R.L., A description of current, wind, and sea level variations during coastal upwelling off the Oregon coast, *Journal of Geophysical Research, 79,* 435-443, 1974.

Smith, R.L., Poleward propagating perturbations in currents and sea level along the Peru coast, *Journal of Geophysical Research, 83,* 6083-6092, 1978.

Weatherly, G.L., and P.J. Martin, On the structure and dynamics of the oceanic bottom boundary layer, *Journal of Physical Oceanography, 8,* 557-570, 1978.

Winant, C.D., Coastal circulation and wind-induced currents, *Annual Review of Fluid Mechanics, 12,* 271-301, 1980.

Wooster, W.S., and J.L. Reid, Jr., Eastern boundary currents, *The Sea, Vol. 2,* 253-280, Wiley, 1963.

Yoshida, K., Coastal upwelling off the California coast, *Records of Oceanographic Works in Japan, 2,* 8-20, 1955.

LOCAL UPWELLING ASSOCIATED WITH VORTEX MOTION OFF OSHIMA ISLAND, JAPAN

Masayuki Takahashi

Institute of Biological Sciences, University of Tsukuba,
Sakura-mura, Ibaraki, Japan 305

Yoshibumi Yasuoka
Masataka Watanabe
Tadakuni Miyazaki

National Institute of Environmental Studies, Yatabe-machi,
Ibaraki, Japan 305

Shun-ei Ichimura

Institute of Biological Sciences, University of Tsukuba,
Sakura-mura, Ibaraki, Japan 305

Abstract. Upwelling around an island in a coastal current were observed by airplane remote sensing and ship observations. Vortices formed north and south of the island were detected downstream of the island. The vortices caused the upwelling of sub-surface water of low temperature, high salinity, high nutrients, and low chlorophyll. High chlorophyll concentrations in some of the upwelled plumes indicated that upwelled water was sometimes exposed to solar radiation for a prolonged time, allowing phytoplankton to respond.

Introduction

It is widely known that large-scale upwelling occurs along the western coasts of continents, but less attention has been paid to upwelling along the eastern continental margins of the world including that of the Asian Continent. However, as Uda (1959) pointed out, there may be local upwellings caused by current and topography even on the eastern side of continents.

Takahashi *et al.* (1980) observed nutrient-rich cold-water patches caused by upwelling off the eastern side of each of the Izu Islands, seven islands off the east coast of Japan; the Kuroshio flows near the islands. By analyzing detailed temperature profiles, Fukasawa and Nagata (1978) found extensive upwelling southeast of Kyushu Island where the Kuroshio flows over a seamount. Nakano (1977) also observed frequent upwelling created by the Kuroshio along the shelf edge in the East China Sea. Similar local upwelling was caused by the Gulf Stream flowing over a seamount and along the continental shelf edge, observed by Atkinson *et al.* (1980) off Georgia, southeastern United States. Pingree (1978) and Pingree, Holligan and Mardell (1979) pointed out the importance of cyclonic eddies in creating extensive upwelling and subsequent increases in chlorophyll.

The present paper will deal with possible upwelling mechanisms around an island near which the Kuroshio flows. Remote sensing from an airplane was used to obtain detailed temperature distributions at the surface. Chlorophyll, nitrate, salinity, and temperature were observed at hydrographic stations.

Methods

Sea-surface temperature was determined by a multi-spectro-scanner (MSS, Deadalus DS-1250) mounted on an airplane (Cesna 402B) at 3000 m above the sea on July 5, 1979 off Oshima Island. Observations were made along seven different flight paths within 2 h (11:16 to 13:08) (Fig. 1). An infrared spectrum (channel 11, 10.5 to 12.5 μm) was used for estimating temperature. Other channels were rejected because of the stormy glittering of the sun. The data were analyzed using an image processor (IPSEN, Image Processing Systems for the Environment, developed by the Japanese National Institute for Environmental Studies).

Surface water temperature, nitrate, and chlorophyll were also monitored continuously along a

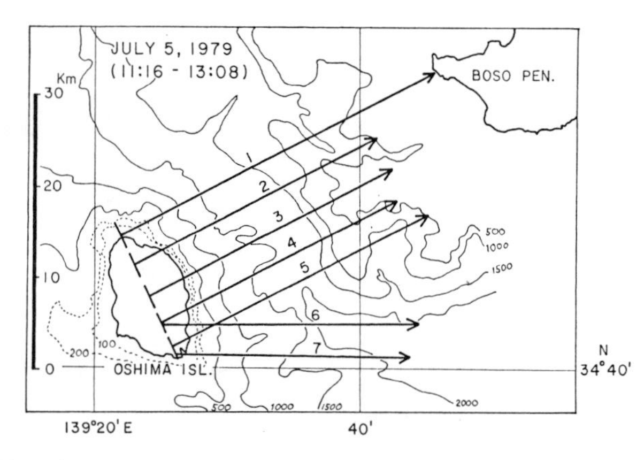

Fig. 1. Flight paths, shown by arrows with numbers from 1 to 7, for the MSS measurements off Oshima Island. Contours indicate depth in meters.

cruise track June 30 and July 1, 1979 by the research vessel *Tansei Maru* of the Ocean Research Institute of the University of Tokyo (KT-79-9). Water was pumped from 2 to 3 m below the surface. Temperature, nitrate, and chlorophyll were determined by a thermistor probe (Rigosha Co.), AutoAnalyzer®, and a fluorometer (Turner 111-003, with a 2-cm flow cell system), respectively. Details of the system are described by Takahashi *et al.* (1980).

Results

Fig. 2 shows the surface temperature distribution sensed by the MSS off Oshima Island. Seven strips of positive films obtained separately were joined by visual inspection, but due to possible differences in cruising speed, flight course, and time, precise matching was not always possible. However we observed several clear, cold patches each of Oshima Island (Fig. 2). One obvious patch (C1) was immediately off the northeast side of the island. It was almost circular and about 10 km in diameter (NE - SW direction). There were two other patches off the southeast side of the island, C2 and C3. C2 was also round and about 5 km in diameter (N - S direction); C3 was about 3 km across and near the island. Several other cold water patches such as C4, C5, and C6 were also observed.

The surface temperature distribution in some of the cold water patches, particularly C1, C2, and C3 suggested vortex structures. Vortex shedding of C1 indicated clockwise rotation, but C2 and C3 moved counterclockwise. Temperature distributions within the vortices were inhomogeneous in C1 and C3 (Fig. 3). One side of vortex C3 was fairly cold but the other side was warmer. However, C2 had uniform, lower temperatures than C1 and C3. There was a narrow thermal front extending eastward from C2, along which the temperature was fairly low.

Hydrographic observations of surface temperature, nitrate, and chlorophyll were made four days before the MSS observations. A cold-water patch was observed about 10 km along the N - S axis off the northeast side of the island (Fig. 4B). Surface temperature within the cold-water patch was below 20°C, and 17.5°C was measured at the center. During the 8-h observation, the ship came near the center of the cold-water patch (Stas 12 and 18) twice within 4 h. During the first visit (Sta. 12) the temperature varied between 17.5 and 18.8°C, but around Sta. 18 it was in the range of 18.2 to 20.4°C, suggesting that the cold-water patch was moving. Ship drift during bottle casts at a station also suggested easterly movement of the cold-water patch. Considering the possible eastward movement, the cold-water patch shown in Fig. 4 may have been elongated in the W - E direction and possibly distorted.

Surface nitrate concentrations in most of the study area were low, but they increased tremendously in the cold-water patch in a pattern similar to the temperature distribution (Fig. 4C), with low temperatures corresponding to high nitrate concentrations. Chlorophyll concentrations in the surface water were less than 1 μg/l in the study area, but they increased in the cold water patch to over 2 μg/l (Fig. 4D). Details of the pattern differed from those of nitrate and temperature, and high chlorophyll concentrations occurred in the 18.0 to 18.5°C temperature range. As there were still adequate nutrients left in the cold-water patch, chlorophyll could increase with time.

Discussion

The observations suggest that vortex motion may have formed the cold-water patches. It is known that islands in streams or currents can create vortices downstream (Uda, 1959). White (1971) showed closed eddies adjacent to a cylindrical island in an eastward current by using a model. Evidently there was a northeast current of ca. 5.4 km/h (3.0 knots) at the surface, possibly caused by the Kuroshio impinging on the western side of the island (Tokyo Metropolitan Fisheries Experimental Station, monthly data record, 1979) (Fig. 5). The current could have formed clockwise vortices northeast of the island and counterclockwise vortices off the southeast side. Two vortices at the southeast side suggested that a vortex street may have been generated.

Onishi and Nishimura (1979) made detailed analyses on micro-scale vortex formation of a few tens of meters in diameter around small islands in Naruto Strait. They found upward movement of water within the vortices and clearly demonstrated upwelling mechanisms associated with vortices in a model experiment. According to their results, upward motion prevailed in slow turning vortices in the entire vortex, but there was downward flow at the center of a fast turning vortex. Flierl (1979) simulated vortex-forced upwelling 40 km in diameter in a two-layer model.

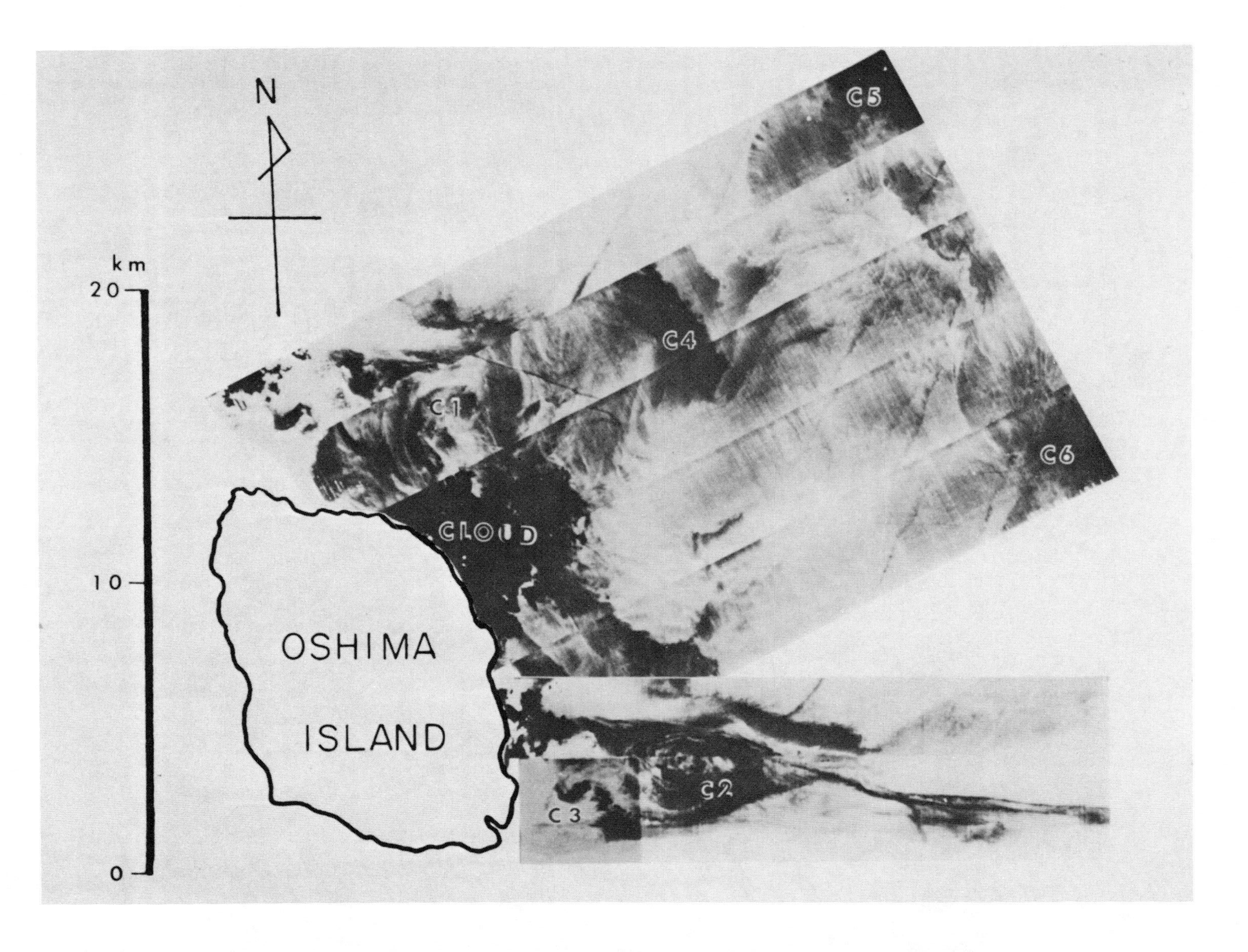

Fig. 2. Surface temperature distribution sensed by the MSS off Oshima Island. Darker color indicates colder water and light color is warmer water. C1, C2, C3, C4, C5, and C6 indicate cold-water patches.

Fig. 3. Surface temperature distribution in some cold-water patches, C1 (upper), C2 and C3 (lower). Data were treated by an image processor to enhance temperature differences, which were originally shown in color.

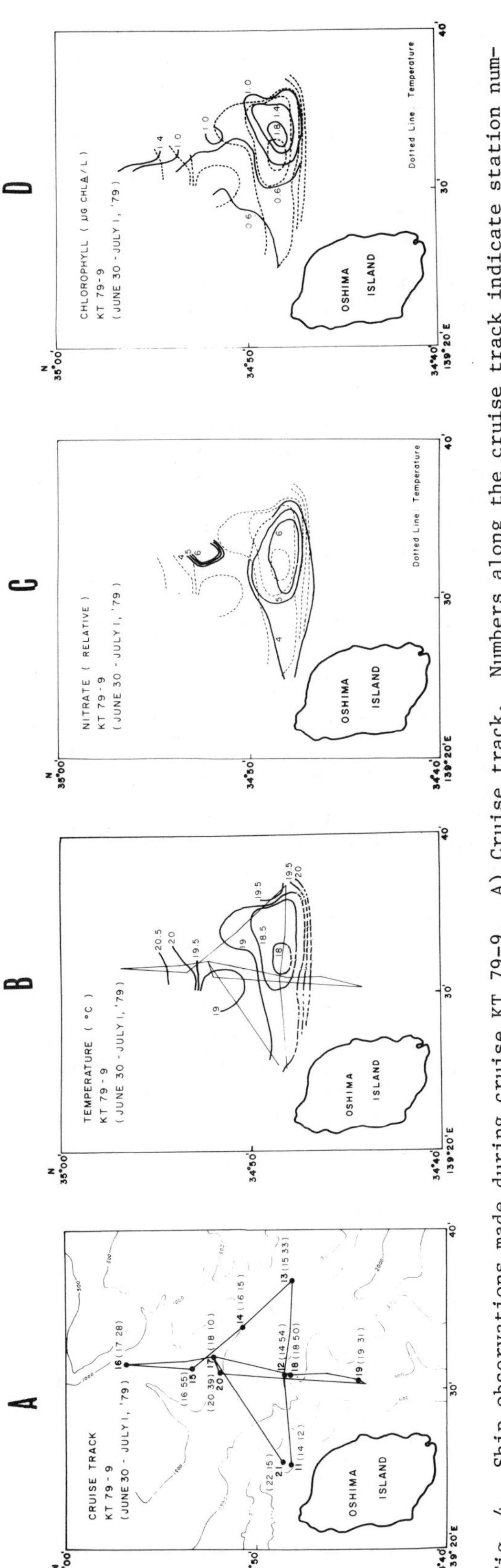

Fig. 4. Ship observations made during cruise KT 79-9. A) Cruise track. Numbers along the cruise track indicate station numbers for temperature profiles. Numbers in brackets indicate the time of arrival at the station. Depth contour interval, 500 m. B) Surface temperature distribution obtained by continuous monitoring along the cruise track. C) Nitrate distribution in the surface water. D) Chlorophyll distribution in the surface water.

It is highly probable that vortices cause upwelling of subsurface water. Along the transect between Stas 11 and 13 (Fig. 4a), an upward motion of cold water above 100 m was observed near Sta. 12 (unpublished observations).

We shall next estimate the depth from which water rises to the surface during upwelling from data obtained within the survey area shown in Fig. 4a but at different times. As vertical profiles of temperature, salinity, and nutrients inside and immediately outside cold-water patches showed no differences at 200 m, upwelling was believed to be from depths less than 200 m (Takahashi *et al.*, 1980). Profiles of nitrate and salinity showed a parallel change and did not diverge towards the surface, indicating that the upwelled water did not mix with surrounding water (Fig. 6). Similar vertical patterns of temperature and other nutrients were observed. Graphically shifting the inside profile towards greater depths until the profile matched the outside pattern, suggested that water below 60 m was brought up to the surface. Even though there was no direct evidence of vortices, similar surface current conditions (Tokyo Metropolitan Fisheries Experimental Station Monthly data record, 1978) and surface temperature distributions in the past could reflect similar upwelling (Takahashi *et al.*, 1980). The vertical temperature profile along the transects (Fig. 4a) also supports this view (Takahashi, unpublished).

Subsurface water contained high concentrations of various nutrients but low chlorophyll because of poor light penetration to the depths of high nutrients in the study area (c.f., Armstrong, Stearns, and Strickland, 1967). Upwelled water in the subsurface zone contained high concentrations of nutrients and low chlorophyll concentra-

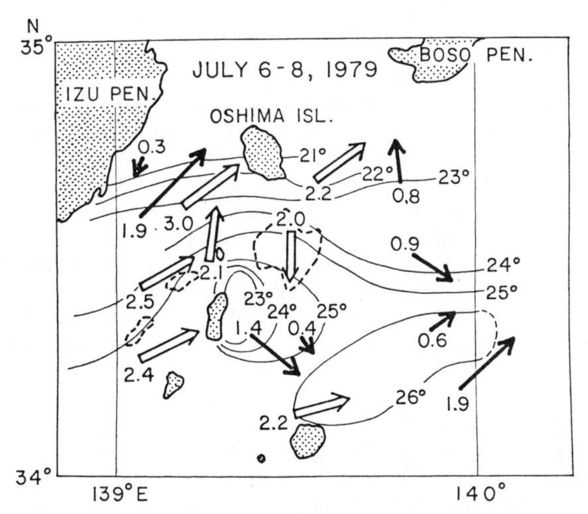

Fig. 5. Surface current direction (arrows) and speed (numbers by each arrow, in knots), and surface water temperature. Data were obtained from the Tokyo Metropolitan Fisheries Experimental Station, Oshima Branch (1979).

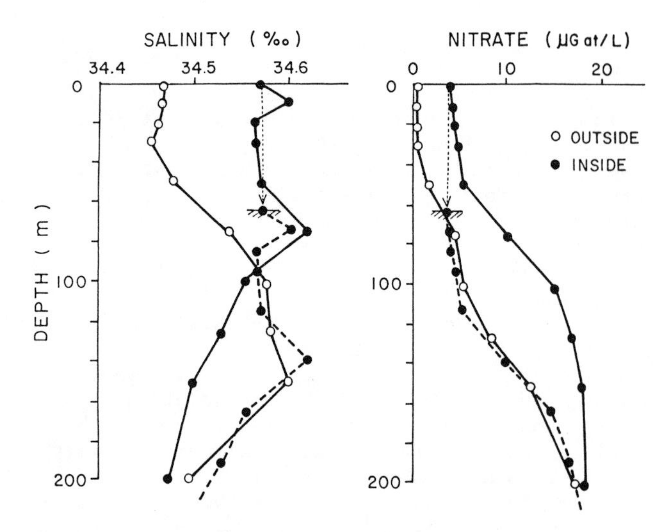

Fig. 6. Vertical profiles of salinity and nitrate inside and immediately outside a cold-water patch, November 5, 1978. The inside profile was shifted towards greater depth until the two profiles nearly matched, as shown by the dotted lines.

tions during the early stages after upwelling. Evidently several early-stage vortices, identified by their steep frontal structures, were characterized by high nutrient, low chlorophyll, and homogeneous water within the patches (Takahashi *et al.*, 1980). On the other hand, older cold-water patches, identified by more gradual frontal structures, contained more chlorophyll and smaller concentrations of nutrients, suggesting that cold-water patches were maintained for finite time periods, which allowed phytoplankton to respond. It has been observed that phytoplankton obtained from a newly upwelled patch started to grow with a lag of 1 day or less and, after 3 days at the surface, growth rates increased from $\mu = 1.0$ to 1.5/d, where $\mu = (\ln N_2 - \ln N_1)/(t_2 - t_1)$ and N_1 and N_2 are the algal biomass at time t_1 and t_2 (unpublished observations).

Local upwelling induced by currents, as observed off Oshima Island in the present study, has been observed around other of the Izu Islands, such as Kozu and Miyake islands (Takahashi *et al.*, 1980). Cold water with high nutrients and high chlorophyll was also observed for 100 km in the N-S direction east of the Izu Islands. The observations suggest that local upwelling, caused by current and topography, is common around the Izu Islands. Furthermore, as local upwelling was always observed during five different oceanographic cruises during spring and fall, upwelling was to be expected throughout that time period and possibly in winter. However, upwelling during winter might be difficult to observe because of active surface mixing due to surface cooling.

Acknowledgements. The research was supported by funds for environmental remote sensing from the Japanese Foundation for Shipping Advancement. Funds were also supplied by the Ocean Research Institute of the University of Tokyo, the University of Tsukuba, and the Ministry of Education (Grant-in-aid for Environmental Science #303015 and Grant-in-aid for Scientific Research #448008).

References

Atkinson, L.P., J.O. Blanton, R.B. Harrison, T.N. Lee, D.W. Menzel, G.A. Paffenhofer, L.J. Pietrafesa, L.R. Pomeroy, K.R. Tenor, and J.A. Yoder, Upwelling off the southeastern coast of the United States - An overview, IDOE International Symposium on Coastal Upwelling, Los Angeles, California, 4-8 February, 1980, Abstract, 1980.

Armstrong, F.A.J., C.R. Stearns, and J.D.H. Strickland, The measurement of upwelling and subsequent biological processes by means of the Technicon AutoAnalyzer ® and associated equipment, *Deep-Sea Research, 14,* 381-389, 1967.

Flierl, G.R., A simple model for the structure of warm and cold core rings, *Journal of Geophysical Research, 84,* 781-785, 1979.

Fukasawa, M., and Y. Nagata, Detailed oceanographic structure in the vicinity of the Shoal Kokosho-sone, *Journal of the Oceanographical Society of Japan, 34,* 41-49, 1978.

Nakano, T., Oceanic variability in relation to fisheries in the East China Sea and the Yellow Sea, *Journal of the Faculty of Marine Science and Technology, Tokai University,* Special Number, November, pp. 199-367, 1977.

Onishi, S., and T. Nishimura, Field observations of upwelling associated with vortex and considerations of generating mechanism of these phenomena, *Coastal Engineering in Japan, 22,* 61-76, 1979.

Pingree, R.D., Cyclonic eddies and cross-frontal mixing, *Journal of the Marine Biological Association of the United Kingdom, 58,* 955-963, 1978.

Pingree, R.D., P.M. Holligan, and G.T. Mardell, Phytoplankton growth and cyclonic eddies, *Nature, 278,* 245-247, 1979.

Takahashi, M., I. Koike, T. Ishimaru, T. Saino, K. Furuya, Y. Fujita, Y. Hattori, and S. Ichimura, Upwelling plumes in the Sagami Bay and sea areas around Izu Islands, Japan, *Journal of the Oceanographical Society of Japan, 36,* 209-216, 1980.

SHORT-TERM VARIATIONS OBSERVED IN THE CIRCULATION, HEAT CONTENT, AND SURFACE MIXED LAYER OF AN UPWELLING PLUME OFF CABO NAZCA, PERU

Merritt R. Stevenson

Instituto de Pesquieras Espaciais - INPE
Caixa Postal 515
Sao Jose dos Campos - SP
12200 Brazil

David W. Stuart

Department of Meteorology,
Florida State University,
Tallahassee, Florida 32306, U.S.A.

George W. Heburn

Department of Meteorology,
Florida State University,
Tallahassee, Florida 32306, U.S.A.

Abstract. Direct measurements of local horizontal current flow within a cool surface plume were made with drogue floats during the March 17 to 29, 1977 period of JOINT II. The trajectory data provided an opportunity to estimate horizontal divergence within the upper 4 m of the plume and on one occasion to 10-m depth. STD casts from a Lagrangian time series were used to estimate the thickness of the mixed layer and vertical stability within and adjacent to the plume. The determinations suggest that while limited to a thin surface layer, the upwelling velocity within the plume over 6-h intervals may be considerable, although 24-h values of $\underline{w}$ may average 2×10^{-2} cm/s. Because of the isotherm changes observed in the near-daily sea-surface temperature (SST) charts for the plume, calculations were made of the daily insolation and heat content in the upper water column within the plume and were compared to the horizontal and vertical motion in the plume. The apparent complex interrelations between the observed changes in SST and the horizontal circulation suggest that future studies of cool surface plumes or centers of upwelling be rigorously planned and completed in terms of sampling frequency, density, and duration.

Introduction

As part of the multidisciplinary CUEA program during JOINT II, drogue (Lagrangian drifter) experiments were conducted from the R.V. *Cayuse* in the vicinity of the C line off the southern coast of Peru (Plates III and IV). STD profiles were also made from the R.V. *Cayuse* before, during, and after tracking drogues to obtain information on the short-term three-dimensional distributions of temperature, salinity, and density. Because of the interactive mechanisms considered to be operating between the sea-surface temperature, local wind field, and surface water currents, a comparative study was made of the measurements together with measurements of solar radiation incident on the JOINT II area.

Studies by CUEA scientists and others have shown that coastal upwelling can exhibit two-dimensional circulation as off northwest Africa, or perhaps the three-dimensional circulation described for the Oregon coast and off the west coast of Baja California. One manifestation of three-dimensional coastal upwelling circulation is cool surface plume-like features, or centers of upwelled water at the sea surface. The purpose of this report is to describe variations in the circulation of such surface thermal features and how such motion relates to variations in the wind stress, the surface mixed layer depth, vertical stability, and heating by solar radiation.

Methodology

The first flight of the National Center for Atmospheric Research aircraft over the JOINT II area corresponded to the March 20-21, 1977 drogue

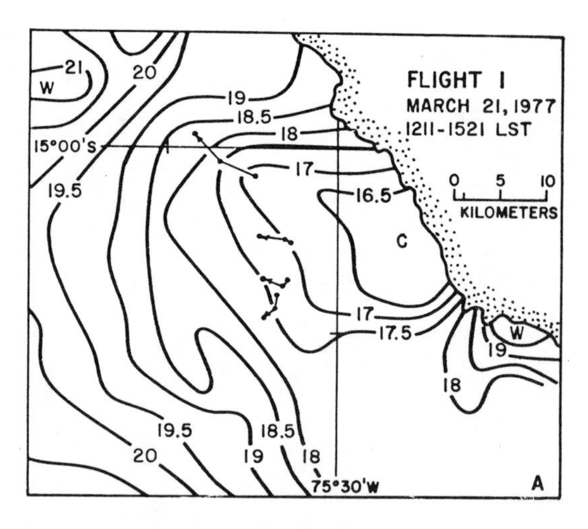

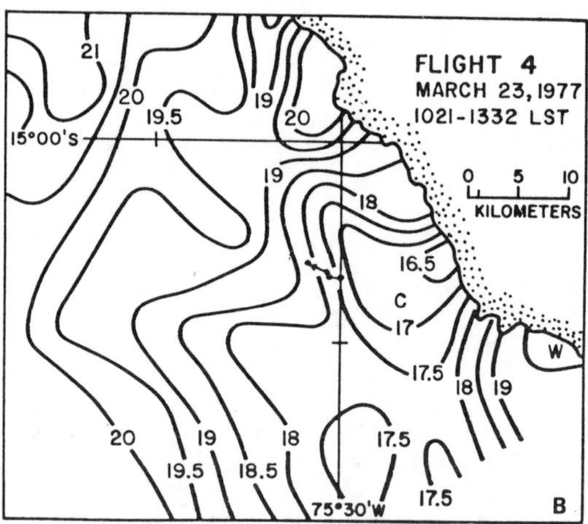

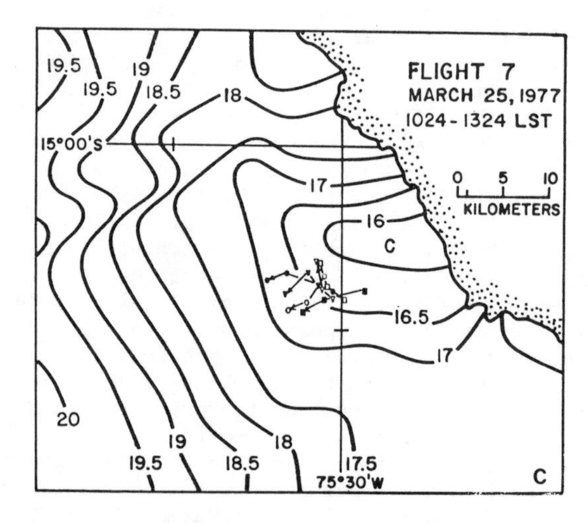

Fig. 1. Maps of SST as obtained via aircraft. A - Flight 1, March 21, 1977, B - Flight 4, March 23, 1977, C - Flight 7, March 25, 1977. Dots, squares, and triangles refer to 6-h displacements of drifters for a 12-h period centered around the flight time. Arrows indicate the sense of the drifter motion.

experiment. For this experiment four drogues were set for 4-m depth and launched along a line roughly normal to the dominant axis of the surface plume of cool water (Fig. 1A). The purpose of this experiment was to observe the effects of lateral shear on the surface plume structure. Six drogues were launched, one at a time, at 6-h intervals near Sta. C-3 for the March 23-25 experiment to observe variations in water motion down the axis of flow of the plume. The trajectories of the first three and the second three drogues are as shown in Fig. 1 B/C. Three drogues were set for 4-m depth and launched in a triangular configuration for the March 27-29 experiment. Unfortunately, no aircraft maps are available for this experiment due in part to fog over the region. At each of the vertices of the triangle another drogue was set for 10-m depth and similarly launched. One objective was to observe vertical shears in horizontal currents in the upper 10 m, another was to estimate divergence from changes in the areas encompassed by the drogues (Molinari and Kirwan, 1975) and from such measurements to calculate the vertical velocity in or around the plume. Additional details of the drogue experiments can be found in Stevenson and Wagner (1978).

Observations used in the analyses were made during the March 10 to 30 period. Data coverage, however, was not uniform in the time or space and a data subset with denser coverage for the intervals of March 23-25 and March 27-29 was used for the 24-h time series analyses. STD profiles were principally from Leg I data of the R.V. *Cayuse* (Stevenson and Wagner, 1978). STD profiles from 63 stations were made as a Lagrangian STD time series and were all within an area of about 2 by 2 km and extended over the period of March 23 to 29. Mixed-layer depth (MLD) was calculated from the temperature profiles of the Lagrangian STD series; MLD was defined as that thickness of the surface layer in which $\Delta T/\Delta Z \leq 0.2^{\circ}C/10$ m. Some additional STD and CTD data collected from the R.V. *Columbus Iselin* (Johnson, Koeb, and Mooers, 1979) and the R.V. *Melville* (Huyer, Gilbert, Schramm, and Barstow, 1978) were also used in the preparation of maps of sea temperature and mixed-layer depth.

Sea-surface temperature (SST) maps were based on infrared imagery of the sea surface made during aircarft flights centered over the C line and covering an area of 45 km offshore by 85 km alongshore. A Barnes Precision Radiation Thermometer (PRT-5) instrument aboard the aircraft produced a data record that was digitized at 1-min intervals. The data points were transcribed onto base charts prior to contouring of sea-surface temperatures. Further details on the collection and processing of the remotely sensed temperature data are in Stuart and Bates (1977).

Results

The cold surface plume varied in both intensity (evidenced by an expanding or contracting area and

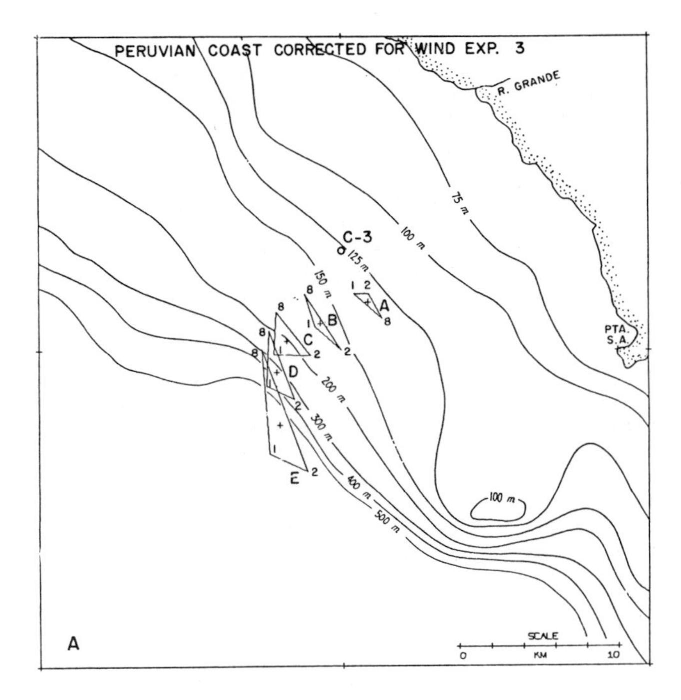

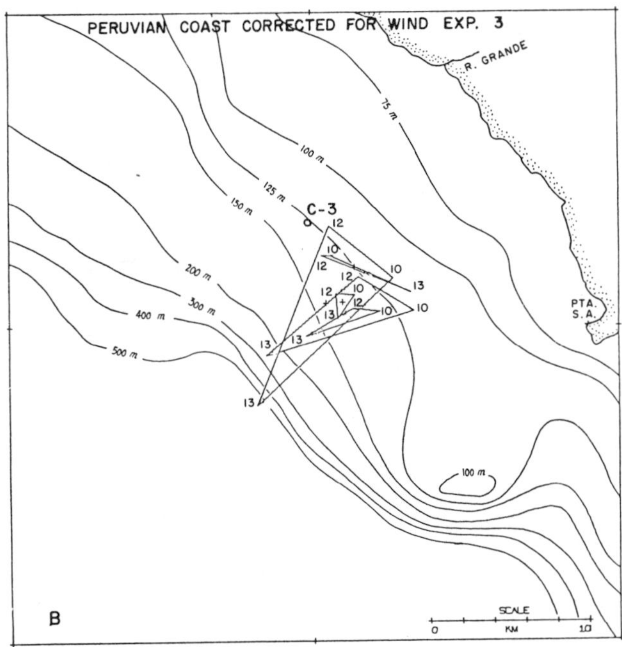

Fig. 2. Drogue Experiment 3, drogue depth 4-m. Drogue triangles plotted at 6-h intervals. A - March 23, 1977. Drogue numbers 1, 2, and 8. Progression of drifter ensemble shown after initial 36 h of experiment. The drogues launched at Sta. C-3 (marked on figure) drifted within the small triangle (A) for the first 36 h, and during the following 24 h moved successively through triangles B-E. B - March 24, 1977. Drogue numbers 10, 12, and 13. Progression of drifter ensemble after initial 24 h of experiment. Drogues in lower panel were launched in sequence at Sta. C-3 following the three drogues in the upper panel.

higher or lower core temperature) and to a lesser extent in location. During Flight 1, the drogue motions contained strong shears in the alongshore and offshore directions. The changes in motion between the drogues can best be seen in Fig. 1 A; the more offshore drogues had a strongly offshore component of motion. Because the mean direction of the inshore drogue was within about 8° of the geographic orientation of the coastline, the inshore drogue served as a local direction reference for the horizontal current shears tangent to the coast ($\partial v_t/\partial x_n$) and normal to the coast ($\partial v_n/\partial x_n$). Using the inshore and offshore drogue data we obtained 2.45×10^{-5}/s and 6.28×10^{-6}/s, respectively, for the shears tangent to and normal to the coast. In addition to the strong shears, however, horizontal divergence is evident when the movements of the drifters are plotted as polygons, accounting for the considerable increase in area of the polygons during the experiment. From March 23 to 25 the area of the plume increased and was accompanied by a lower core temperature (Fig. 1). Drogues nearer the coast during this period moved directly outward from the plume and alongshore; the drogues further from the coast continued to move offshore southward as they passed through the outer boundaries of the plume. During the March 23 to 25 experiment (Flights 4 and 7) the drogue polygons and local drogue motions (Fig. 2) suggested the presence of a convergence zone or front for up to 36 h before the front dissipated and the local parcel of water continued to diverge. Although no overflight was available for the March 27 to 29 experiment, the results are worth reporting. The drogue polygons (Fig. 3) indicate 1) the vertical current shear was strong ($\partial u/\partial z = -1.1 \times 10^{-2}$/s, $\partial v/\partial z = 1.3 \times 10^{-2}$/s) and indicated a current reversal between 4 and 10 m, and 2) changes in areas of the polygons at the two depths for the same time intervals were similar even though they did not expand consistently.

Fig. 4 (upper panel) is a time series of the vertical velocity estimates at a depth of 20 m and 2 km offshore along the C line, calculated using the variational method by O'Brien, Smith, and Heburn (1980). In the method measurements of horizontal currents are objectively adjusted and vertical velocities calculated while satisfying some physical constraint. In this case, the physical constraint is three-dimensional mass continuity. The calculations use the low-low passed onshore velocity components measured at Oregon State University's (OSU) AGAVE and MILA V moorings and Pacific Marine Environmental Laboratory's (PMEL) PS and PSS moorings. The alongshore divergence was estimated by using the alongshore components from OSU's PARODIA and MILA V moorings.

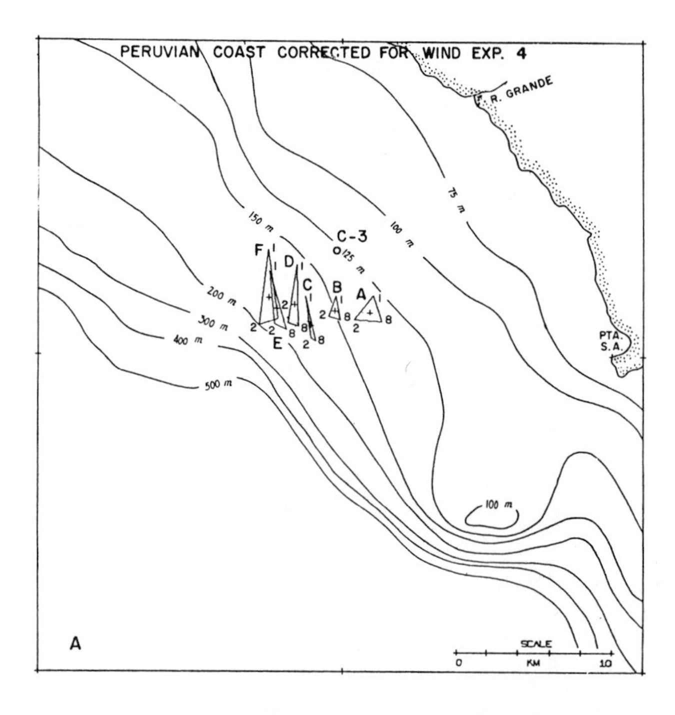

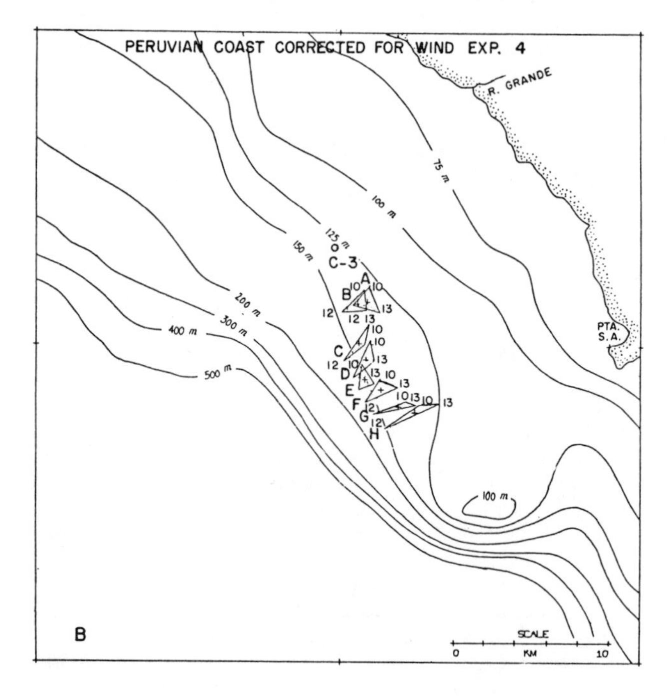

Fig. 3. Drogue Experiment 4, March 27, 1977. Drogue triangles plotted at 6-h intervals. A - Drogue numbers 1, 2, and 8. Drifter ensemble progressively moved across isobaths as shown by triangles A-F. Drogues set for 4-m depth. B - Drogue numbers 10, 12, and 13. Drifter ensemble moved along isobath and poleward (triangles A-H). Drogues set for 10-m depth.

Since the PARODIA data did not start until March 19, 1977, the alongshore divergence prior to March 19 is assumed to have been zero, i.e. essentially two-dimensional continuity, for lack of a better estimate. The lack of high-frequency fluctuations in the time series using the variational method is primarily due to the use of low-low passed current measurements where frequencies shorter than 24 h are suppressed.

The changes in polygon divergence that gave rise to variations in vertical velocity are also shown in Fig. 4 (lower panel). The areal changes of the polygons are evident in Figs 2 and 3. In addition to changes in area (signifying divergence or convergence) the drogues may undergo marked changes in position relative to nearby drogues, due to significant deforming and stretching shears and vertical vorticity. These shape and rotational changes, however, do not of themselves give rise to areal changes. Vertical velocity was estimated from calculations of divergence made from drogue displacements at 6-h intervals.

While a detailed discussion of measurement errors associated with drogue measurements is in the literature (e.g., Stevenson, 1974) estimates of divergence and vertical velocity errors are worth mentioning. For a typical 6-h interval, mean current speeds were typically 5 cm/s or more. Radar measurement errors for an equal period of time suggest an uncertainty of about $\pm$ 1.0 cm/s. Relating this value to divergence with a spatial scale of about 1 km would provide an uncertainty in divergence calculations of about 2×10^{-6}/s and an uncertainty in $\underline{w}$ of about 1×10^{-3} cm/s, adequate for our present study where upwelling velocities often appear to exceed 1×10^{-2} cm/s. The drogue data therefore appear sufficient to calculate divergence over 6-h intervals.

In addition to evaluating changes in local circulation, STD profiles were used to study diurnal changes in the upper 30 m of the water column. Data from the 63 stations were time-folded into a 24-h period and were used to construct the daily temperature curves for 0, 10, 20, and 30-m depths (Fig. 5). The curves show that diurnal changes in heating penetrate downward but occur primarily in the upper 20 m. A decrease in the amplitude of the temperature curve is also apparent with depth. In addition there may be a phase shift in the curves with time but the phase relation appears to change between the upper 10 m and 20 m and 30 m curves. The diurnal heating curve for the surface (Fig. 5) was used to correct or adjust surface temperature values obtained from ship measurements (Fig. 6). Besides data from the R.V. *Cayuse*, additional data from the R.V. *Columbus Iselin* and the R.V. *Melville* were used to increase the coverage on March 24 and 25. The cold water plumes between Cabo Nazca and Punta Santa Ana shown on the aircraft maps of SST are not so readily seen in the surface temperature map (Fig. 6) produced from STD measurements. A cool

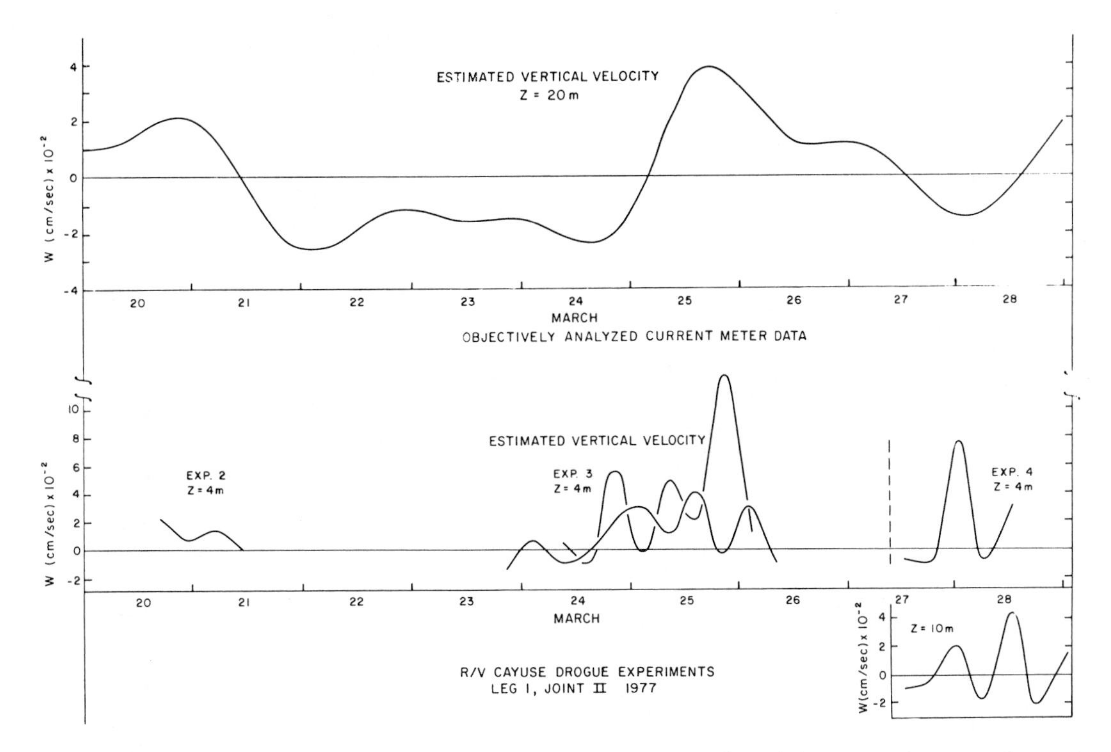

Fig. 4. Vertical velocity estimated from: Upper Panel - Numerical model using objectively analyzed (low-low passed) current meter data. Lower Panel - Changes in area (horizontal divergence) of drogue polygons over 6-h intervals. Curves in lower panel for March 23 to 26; each curve represents three drogues separated by 6 to 8 km and adjacent to the cool plume. Differences in the lower two curves for March 23 to 26 suggest that significant variations in w occur in the vicinity of the cool plume.

plume or core is still evident, however, near the coast (Fig. 6).

Wind data for the 3-day study period were resolved into tangential and normal wind components and have been converted into stress units. The tangential (alongshore) wind stress is shown in Fig. 7A. Even though the data were not smoothed the diurnal signal is quite apparent. The wind stress around local noon is only 1/5 that observed during the night (1900-2300 LST).

Because of the influence of solar heating on the water column, mixed-layer depth (MLD) data for the same 63 stations were used to construct a 24-h curve (Fig. 7B). The MLD decreased rapidly after 0700 LST and remained at or near zero meters from 1100 through 1300 (LST), after which time the MLD increased steadily until midnight, after which time it remained steady until 0600 LST. The MLD curve was then inverted and used to adjust the MLD data used to construct the MLD map for March 24-25 (Fig. 8). Larger MLD values are evident in the cool core and around the perimeter of the plume with intermediate values in between.

Static stability values (E) for 0 to 10, 10 to 20, and 20 to 30-m layers were also determined from the STD time series and the data time-folded into a 24-h period (Fig. 7C). Because of the large observed variations in stability, a simple 3-point smoothing was performed on the time series. While the mean stability for the 0 to 10 and 10 to 20-m layers was about the same, a strong diurnal signal was present in the surface layer. The stability for the 0 to 10-m layer was considerably less during the period of 2100-0700 LST than at other times of the day. The wind stress fluctuations appear to be inversely related to both stability and MLD curves. The reduced stability occurred in the absence of solar heating and during times of increased wind strength. The local effects of momentum transfer from the wind field and vertical mixing appear to be limited to the upper 20 m of water. While a slight diurnal signal is present in the 20 to 30-m layer, the changes in the water column due to diurnal effects are greatly diminished in going from the 10 to 20-m layer to the 20 to 30-m layer.

The level of daily insolation during the March 10 to 30 period (Fig. 9) was surprisingly constant with a mean value of 479.8 Ly/day and a standard deviation of only ± 30 Ly/day. The fluctuation represents an 8% change, 3 to 4% of which was due to the seasonal progression of the sun's altitude. A detailed examination of the solar traces for the period suggests that out of 20 traces only five showed significant levels of cloud contamination.

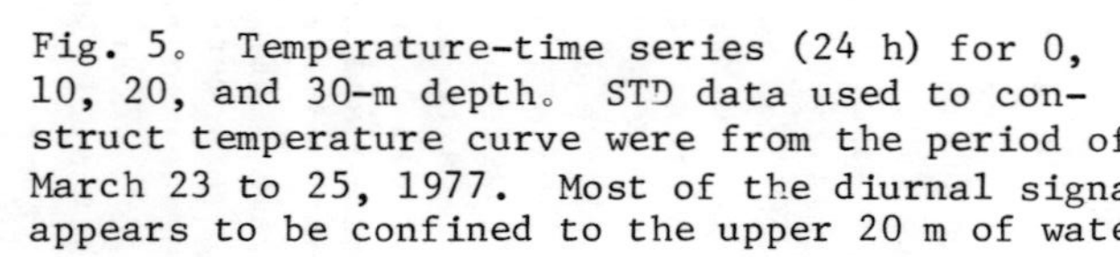

Fig. 5. Temperature-time series (24 h) for 0, 10, 20, and 30-m depth. STD data used to construct temperature curve were from the period of March 23 to 25, 1977. Most of the diurnal signal appears to be confined to the upper 20 m of water.

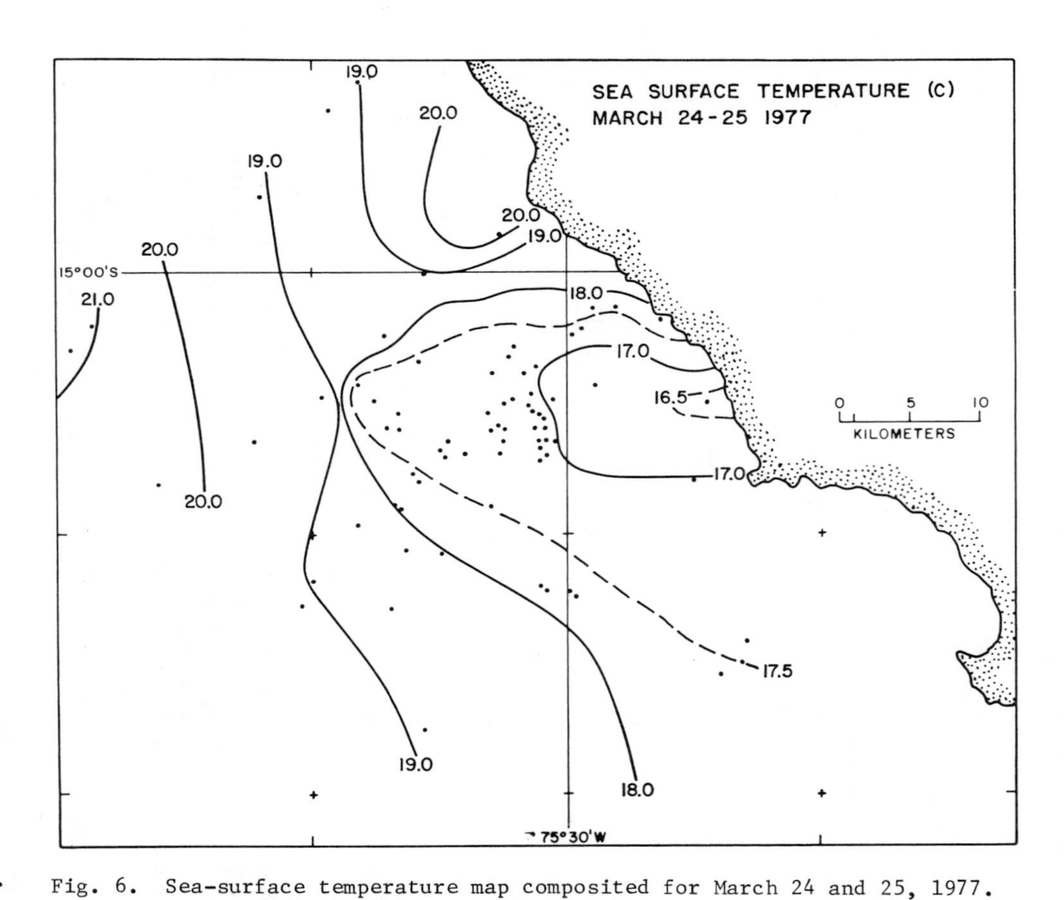

Fig. 6. Sea-surface temperature map composited for March 24 and 25, 1977. Data were corrected for diurnal changes in surface temperature using the surface temperature time series from Fig. 5.

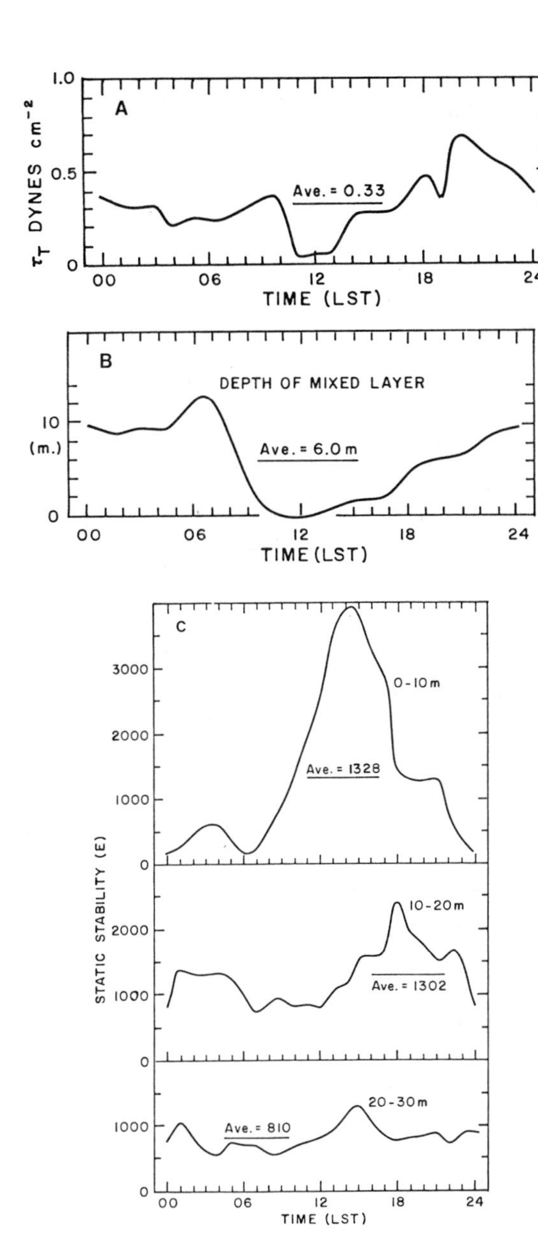

Fig. 7. Time series (24 h) for the period March 23 to 25, 1977. A - Tangential wind stress based on hourly wind observations from the R.V. *Cayuse*. B - Depth of mixed layer derived from 3-day STD time series. C - Static stability (E) obtained from the same STD time series for 0 to 10, 10 to 20, and 20 to 30-m layers.

Discussion and Summary

An evaluation of the SST maps based on data from aircraft suggests that they can show considerable detail, albeit of a thin layer of surface water. Other studies (Saunders, 1970) have shown that during relatively clear skies PRT data typically are within 0.5°C of the bulk surface temperatures. During periods of low winds combined with high levels of insolation, the surface skin of water may be considerably warmer than the upper meter of water; both bulk temperatures and PRT values may show diurnal signals on the order of 1 to 2°C (Stevenson and Miller, 1972). Although the NCAR overflights were made between 1100 and 1500 LST, a time of relatively weak winds, the radiation bias appears to have been minimal (e.g., 0.5°C) as evidenced from the calibration temperatures obtained during flights over the R.V. *Cayuse*.

A comparison of the drogue trajectories with aircraft SST charts suggests a strong association between near-surface drogue trajectories and water motions needed to develop the observed isotherm patterns in and around the cool surface plumes. During the March 21 to 30 period, vertical velocities varied and there was both upwelling and downwelling. Apparently the SST maps were sufficiently sensitive to indicate changes in the intensity of upwelling in the upper few meters of water on a time scale of 1 to 2 days, similar to the motions of near-surface drogues.

While experiments with a larger number of drogues would have been better, the experiments that were realized show the near-surface water (*i.e.*, 4-m depth) to be strongly coupled horizontally and vertically with the local wind field. Variations in the upper 30 m of the water column show a coupling with both the wind field and the insolation. Although short term (e.g., 6-h) time changes in vertical velocity result in unexpectedly large values of $\underline{w}$, such data averaged over 24-h periods produce values of $\underline{w}$ on the order of only 2 to 4 x 10^{-2} cm/s, which is of the same magnitude as the objective analysis model of upwelling based on current meter data (Fig. 4, upper panel). The vertical velocity time series of Fig. 4 suggests that both ensembles of suitably tracked drifters and data from arrays of current meters can provide realistic estimates of $\underline{w}$ in a coastal upwelling zone.

Temperatures at 10 and 20 m suggest that the coolest water lies along the coast around Cabo Nazca and extends southward toward Punta Santa Ana. This differs from the surface temperature maps, which show surface water off Cabo Nazca to be warmer than, and outside, the cool water plume north of Punta Santa Ana. If it is assumed that the coldest water represents the most recently upwelled water, then the most actively upwelled water begins its ascent near Cabo Nazca and flows poleward and alongshore. The water continues to rise as it moves into the vicinity of the surface plume, at which time it finally breaks into the surface mixed layer. There the momentum from locally applied wind stress advects the cool water equatorward and offshore. Ekman suction near the cool plume assures continuation of sub-surface water moving upward into the surface plume.

Near the cool surface plume the daily insolation amounts to about 465 Ly/day if a 3% reflec-

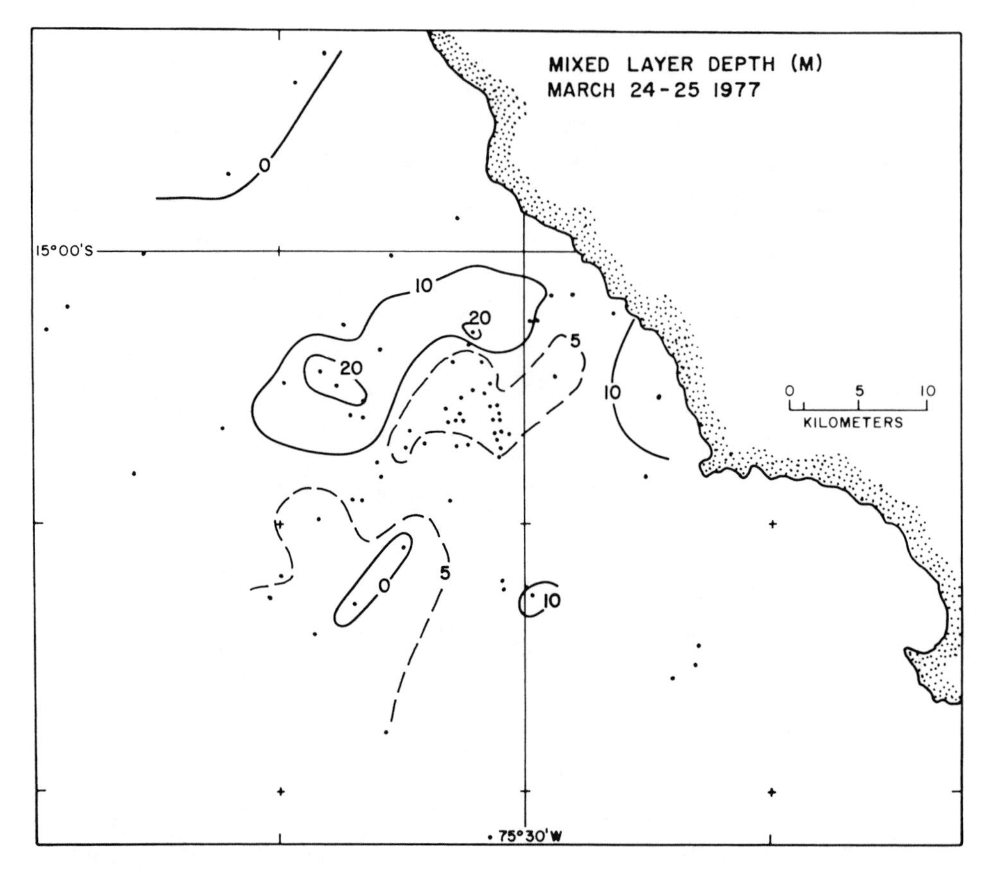

Fig. 8. Mixed layer depth (MLD) map composited for March 24 and 25, 1977. Data were corrected for diurnal changes in MLD using the MLD time series from Fig. 7B.

tion factor is assumed for the large solar altitude (68 to 75°). Using a surface mixed layer of 6 m for the maximum mixing depth of the incident solar radiation, the influx of solar energy is sufficient to raise the temperature of the MLD by 0.81°C/day. If we assume water at 4 m to flow out of the plume at 5 to 10 cm/s, then in one day the water will be advected 4.3 to 8.6 km from the immediate area of the upwelling and outward from the cold core toward the plume perimeter in 1 to 2 days. From the separation of isotherms on the SST maps together with the advection rate in the surface layer, the horizontal gradient of the isotherms approximately equals the heating rate of

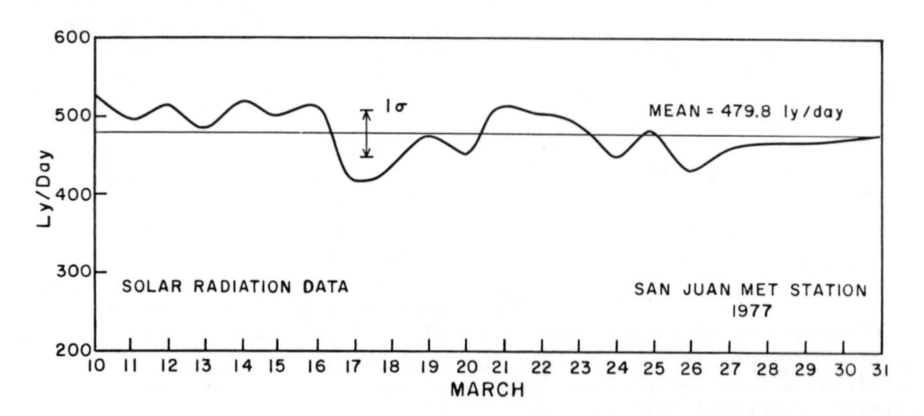

Fig. 9. Daily solar radiation for March 10 to 30, 1977, based on integration of daily radiation traces. Note the one standard deviation of ± 30 Ly/day about the mean radiation line.

the upper mixed layer and its advective rate out from the cool core area. Water ascending at 2 x 10^{-2} cm/s would bring water from 6 m to the surface in about 8 h. The effect of the cool ascending water is to reduce the net heat gain from solar insolation in the upper 6 m. Once a water parcel has moved away from the most intense upwelling area, *i.e.*, the cool core, upwelling becomes less and the vertical velocity may in fact indicate subsidence. By this time the water has been advected to the perimeter of the plume where it may remain for another day or two and absorb solar radiation. In summary, it appears that the warmer water immediately surrounding the cool core can be largely explained in terms of cumulative solar heating of the cool upwelled water as it moves away from the core of the plume.

Our analyses of local motions in the plume, mixed layer depth, heat content, and vertical stability in a coastal upwelling plume would have been more complete had the original field work been planned and conducted somewhat differently. While data from an array of current meters surrounding a plume can provide a sound basis for reference of local circulation, continuous measurements of a large number of some type of drifter are to be preferred. The Lagrangian drifter array can provide the necessary details of transient frontal features found within or adjacent to plumes and thereby provide improved insight into the dynamics of cool surface plumes or centers of upwelling.

Acknowledgements. The research and published results of this study were made possible through support of National Science Foundation (IDOE) Grants OCE 76-82834, OCE 76-82831, OCE 78-00611, and OCE 78-03042. The support of the Research Aircraft Facility of the National Center for Atmospheric Research (Boulder, Colorado) in providing the Queen Air aircraft during JOINT II 1977, is acknowledged with thanks. The authors also thank James Rigney and Robert Wagner for technical assistance. The authors extend their thanks to Drs. A. Huyer and R.L. Smith of Oregon State University, W. Johnson of the University of Delaware, and D. Halpern of NOAA/PMEL, Seattle, for use of their current meter and STD data.

References

Huyer, A., W.E. Gilbert, R. Schramm, and D. Barstow, CTD observations off the coast of Peru, R.V. *Melville* 4 March-27 May 1977, and R.V. *Columbus Iselin* 5 April-19 May 1977. Oregon State University, School of Oceanography, CUEA Data Report 55, Ref. 78-18, 409 pp. (Available at National Technical Information Services, Accession No. PB293377/AS), 1978.

Johnson, W.R., P.I. Koeb, and C.N.K. Mooers, JOINT II R.V. *Columbus Iselin* Leg I CTD measurements off the coast of Peru, March 1977, University of Delaware, College of Marine Studies, CUEA Data Report 56, 408 pp., 1979.

Molinari, R., and A.D. Kirwan, Jr., Calculations of differential kinematic properties from Lagrangian observations in the western Caribbean Sea, *Journal of Physical Oceanography, 5*, 483-491, 1975.

O'Brien, J.J., R. Smith, and G. Heburn, Determination of vertical velocity on the Continental Shelf, To be submitted to the *Journal of Physical Oceanography*, 1980.

Saunders, P.M., Corrections for airborne radiation thermometry, *Journal of Geophysical Research, 75*, 7596-7601, 1970.

Stevenson, M., Measurement of currents in a coastal upwelling zone with parachute drogues, *Exposure, 2*, 8-11, 1974.

Stevenson, M.R., and F.R. Miller, Application of high resolution infrared and visual data to investigate changes in and the relationship between surface temperatures and cloud patterns over the eastern Tropical Pacific, Inter-American Tropical Tuna Commission, c/o Scripps Institute of Oceanography, Final Report, Contract No. N62306-71-C-0120, 84 pp., 1972.

Stevenson, M., and R. Wagner, Physical/nutrient measurements, JOINT II March 5-30, 1977 Inter-American Tropical Tuna Commission, c/o Scripps Institute of Oceanography, CUEA Data Report 46, 196 pp. (Available at National Technical Information Services, Accession No. PB283333/AS), 1978.

Stuart, D.W., and J.J. Bates, Aircraft sea surface temperature data, JOINT II 1977, Florida State University, Department of Meteorology, CUEA Data Report 42, Ref. 77-1, 39 pp. (Available at National Technical Information Services, Accession No. PB278568/AS), 1977.

THE NORTHERN PERUVIAN UPWELLING SYSTEM DURING THE ESACAN EXPERIMENT

Eberhard Fahrbach

Institut für Meereskunde an der Universitat Kiel
Kiel, Federal Republic of Germany

Christoph Brockmann

Deutsches Hydrographisches Institut
Hamburg, Federal Republic of Germany

Nelson Lostaunau and Wilfredo Urquizo

Instituto del Mar del Peru
Lima, Peru

Abstract. From March to May, 1977 a joint German-Peruvian investigation was carried out in the northern Peruvian upwelling area between 4 and 6°S. Current meter and hydrographic measurements are used, together with data from land stations, to describe the main features of the area. The winds underlie an important seasonal trend with an increase to the southern hemisphere winter. The trend in the wind is reflected by a decrease in sea-surface temperature of more than 5°C within 10 days. A mean state over the 50-day period is given with a well developed shallow undercurrent reaching up to the shelf overlain by a current flowing towards the equator. Depending on the wind strength a one- or two-cell cross-shelf circulation is observed. Coupling between wind and current fluctuations is shown by spectral analysis. Internal seiches in the Bay of Paita play an important role in the frequency response of the ocean to the atmospheric forcing.

Introduction

One of the most important upwelling areas is off the Peruvian coast. Four regions are especially prominent, 4 to 6°S, 7 to 8°S, 11 to 12°S, and 14 to 15°S. The upwelling is most intense in the last area. In addition to the spatial change, there is a time variation with a pronounced annual period, with the maximum upwelling in the southern winter.

The northern upwelling area is of special interest because it may interact with features typical of the equatorial region, such as the Equatorial Front and the Equatorial Undercurrent. Special conditions may be generated by the Peru Current leaving the coast in this region. A dramatic change in the whole system occurs when *El Niño* takes place. The Equatorial Front advances far to the south and coastal upwelling is suppressed. From March to May 1977 the joint German-Peruvian investigation of the northern Peruvian upwelling area, ESACAN ("Estudio del Systema de Afloramiento Costero en el Area Norte") took place (Brockmann, Fahrbach, and Urquizo, 1978).

Observations and Data

Three current meter moorings were deployed on a section along 5°S in water depths of 104, 1360, and 3820 m for a period of 50 days. Sub-surface mooring techniques were used with Aanderaa current meters. Ten sections with hydrographic and nutrient casts and various current profiling measurements were carried out from the Peruvian research vessel *BAP Unanue* (Fig. 1). Salinity was determined with a Beckman salinometer, which failed after 8 sections. The samples from the 2 remaining sections were processed after the cruise. The large scatter of the corresponding T-S values lead us to reject these data. Wind speed and direction and atmospheric pressure were recorded at Talara and the sea surface temperature at Paita. The land station data are also available for before the observation period. The time series were low-pass filtered with a half-power point of 35 h (Fig. 2). As the smoothed isobaths run essentially north-south the expression onshore is used for currents to the east and positive alongshore to the north.

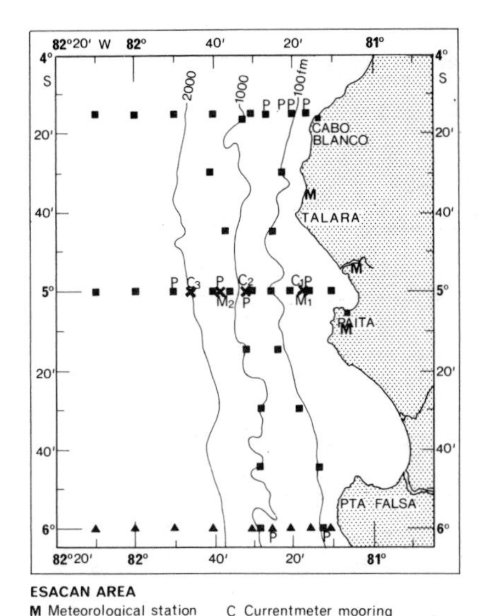

Fig. 1. The ESACAN area.

Discussion

The Seasonal Transition

Fluctuations of a large frequency range are superimposed on the system. The major fluctuation observed was the transition from the warm to the cold season. The SST record (Fig. 2) shows that the transition had begun before the ESACAN observations. The situation on March 30 and 31 was rather similar to the summer condition (e.g. Patzert, 1978). Upwelling was evident, but the upwelled water was covered by a 10- to 20-m layer of water warmer than 20°C (Fig. 3).

The lowering of the SST can be understood as the response of the sea to the changing meteorological conditions. At the beginning of the record (Fig. 2) the low-passed wind data show southerly winds of about 4 m/s with a slight onshore component. On March 24, a gradual increase of up to 8 m/s in the northern component took place. Also a change to a slight offshore component can be observed. Hidaka (1954) showed that the offshore Ekman transport depends on the wind speed and direction, and he calculated the most intense offshore transport for an angle of 21° between the wind direction and the coastline. During our observations the wind turned from 10° onshore to 14° offshore. This increased the offshore transport by 20%. In addition the increase of the wind speed by a factor of 2 increased the transport by a factor of 3.

To compare the summer and winter situation two sections occupied March 30 and 31 and May 9 and 10 (Fig. 3) are selected. The inshore temperature fell by 3.5°C but the offshore SST gradient over 74 km remained quite constant (from 2.11 to 1.92°C). Assuming an essentially two-dimensional heat balance with no increase in solar heating in winter, the SST observations and the increased Ekman transport lead to the conclusion that the mixed surface layer was deepening. This is reflected in the decrease of N^2 (Fig. 4) and leads to important consequences for the dynamics of the cross-shelf circulation. Whereas the horizontal near-surface temperature gradient remained constant, the horizontal density gradient decreased by a factor of 2 due to the increase of salinity in the offshore water from 34.80 to more than 35.0 (Fig. 3).

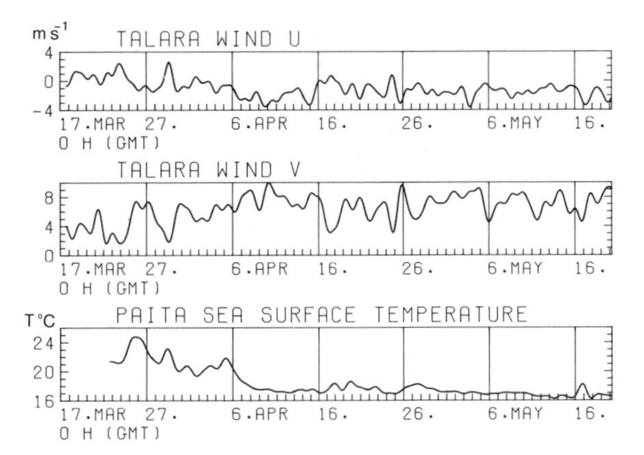

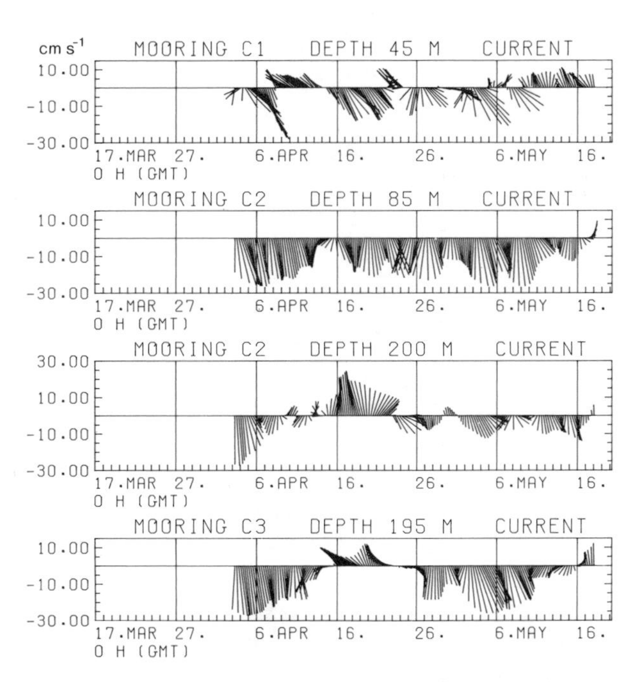

Fig. 2. Low passed time series of the onshore and offshore wind components at the airport station of Talara, the sea surface temperature in the Paita harbor, and moored current meters at the locations given in Fig. 1, at 45, 85, 200, and 195 m below the surface.

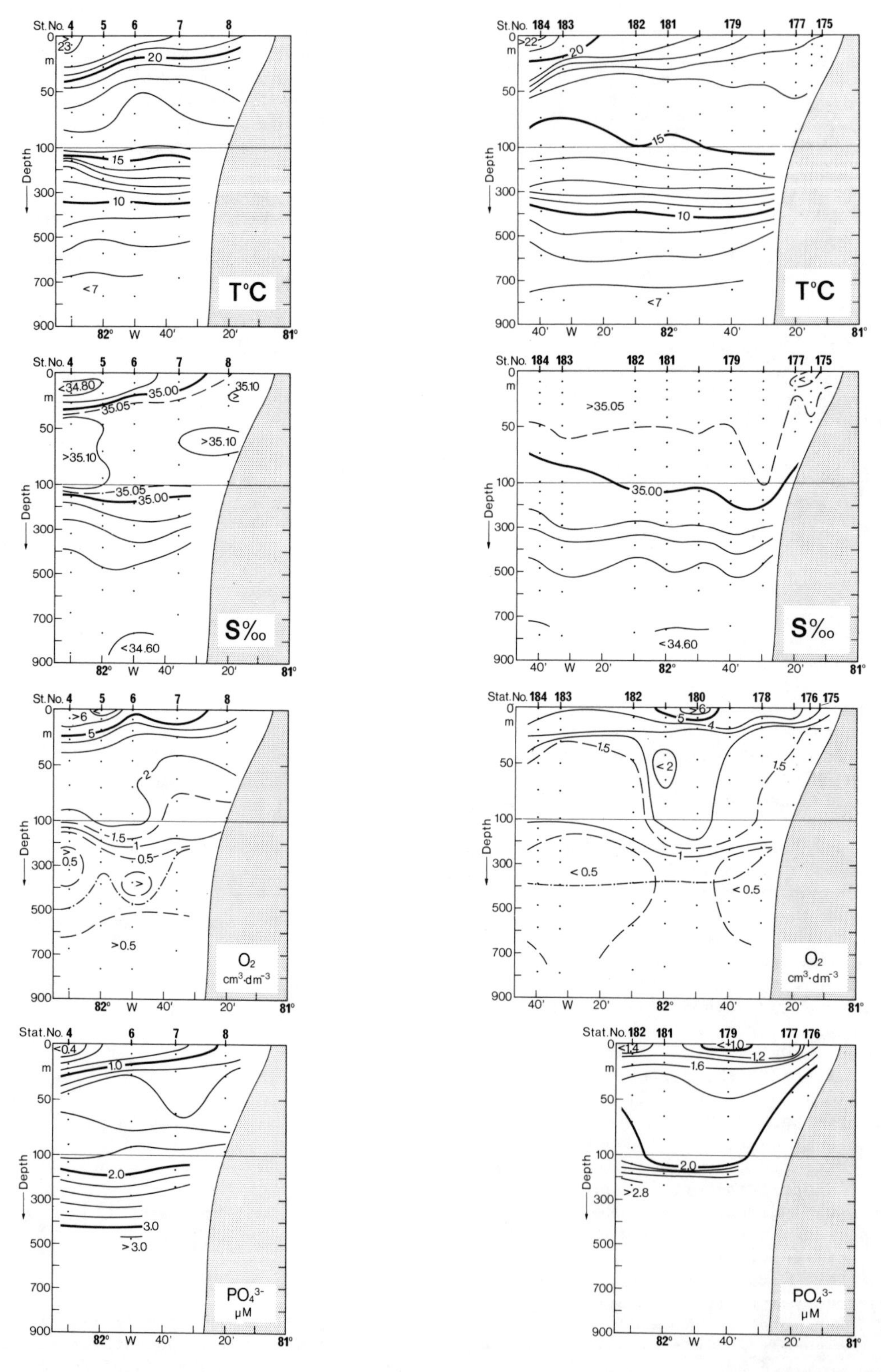

Fig. 3. Sections of temperature, salinity, oxygen, and phosphate along 5°S observed 30 and 31 March, 1977 (left) and 9 and 10 May, 1977 (right). The depth of the oxygen minimum is indicated by the dot-dash contour.

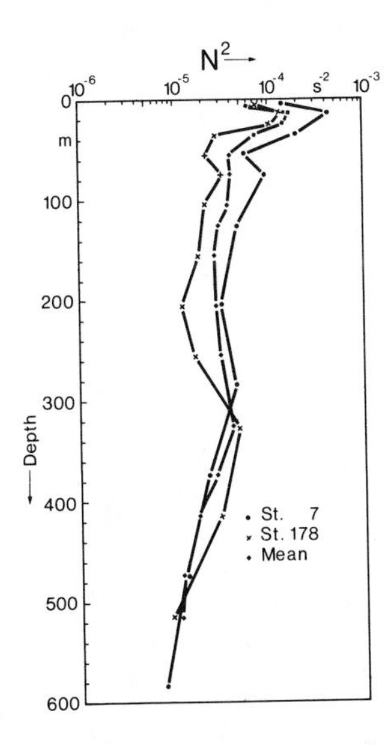

Fig. 4. Brunt-Väisälä frequency N^2 at Stas 7 and 178 on sections shown in Fig. 3. The mean N^2 is calculated for seven stations at 5°00'S, 81°31'W in the period between the two profiles. Sta. 178 is included.

Mean State

For technical reasons the two offshore current meter moorings were placed too deep, so that the uppermost current meters were below the real upwelling system. To obtain information about the currents above the current meters geostrophic calculations were carried out. Ten hydrographic sections were occupied, eight of them with reliable salinity determinations. Because of the small Coriolis parameter at 5°S ($f = 1.27 \times 10^{-5}$/s) even small internal waves or data errors can generate apparent high geostrophic current shear. Consequently, only strongly smoothed data give reasonable results. For this reason the mean of the eight complete sections was determined to calculate the geostrophic currents. Fig. 5 shows the corresponding sections of the means of temperature, salinity, dissolved oxygen, density, and geostrophic currents.

About 75 km offshore there was a northerly jet-like current associated with the horizontal density gradient. Below was a poleward undercurrent, which seems to have reached the surface between 35 and 55 km away from the coast. Below the undercurrent another equatorward flow lay close to the slope between 300 and 700 m. This is the level of the oxygen minimum observed by earlier authors (e.g. Wyrtki, 1966). In our observations the oxygen minimum clearly decreased some 55 km off the coast.

The transformation of calculated relative currents to absolute currents was made by using the measurements of the moored instruments. Each point in the mean current section of Fig. 5 represents an instrument. The calculated profiles were shifted to fit the measured mean profiles by a least squares method. This was possible with a maximum difference of 5 cm/s.

The observed mean currents, with the standard deviations, are given in Table 1. The mean cross-shelf circulation is much more difficult to detect because the means are very small compared to the standard deviation. A closer inspection of the time series (Fig. 2) shows that the means can be governed by short term events, e.g. C3 at 195 m, which were very intense (up to 15 cm/s) but short in time (8 days). This event caused an offshore mean current, whereas during most of the time the current was onshore.

To avoid the problems of showing a small mean current randomly overlain by large fluctuations, an indirect method to trace the mean onshore-offshore circulation was applied by using the $\Delta P/\Delta O$-atomic ratio (Fig. 6). The observed ratio of -1/295 is consistent with the number given by Redfield, Ketchum, and Richards (1963) of -1/276, which is accepted as a sign of the reliability of the nutrient data. Following the line from the source region with 2.2 μM PO_4^{3-} and 1.5 cm^3/dm^3 O_2, which was found between 150 m and 200 m (Fig. 3), a decrease of PO_4^{3-} and an increase of O_2 was found towards the sea surface near the coast and from there going offshore. This is an indication of upwelling near the shore and of offshore motion at the surface.

From the mean hydrographic casts on the 5°S-line a T-S diagram was drawn (Fig. 7). There are four water masses distinguishable in the terms used by Wyrtki (1966) and Enfield (1976). The Equatorial Surface Water (ESW), with a salinity less than 35 and temperatures higher than 20°C, was found on 82°00' and 82°10'W. This was observed only during the summer.

The Subtropical Surface Water (SSW), with salinities higher than 35.1 and temperatures higher than 19°C, was not observed in its original form, but it was evident as a source water that mixed with the Antarctic Intermediate Water to produce the Peruvian Coastal Water (PCW) and the Equatorial Pacific Central Water (EPCW), as given by Sverdrup, Johnson, and Fleming (1942).

The mean T-S diagram was used (Fig. 7) to find the source of the upwelled water. Source water can change its T-S properties by mixing with ambient water masses and by interaction across the sea surface. A mixing with SSW only can be excluded because the change of inclination of the T-S curve at 16°C requires ESW as an ingredient. This is possible only during the summer when ESW is present in the area. Solar heating can explain the change of inclination because water of a given salinity is heated close to the surface. Both processes modify the source water and act to give vertical changes in the T-S diagram. Therefore we accept the water mass at the change of the inclination as the unmodified source water with temperature of 16° ± 1°C and salinity of 35.05 ± 0.025.

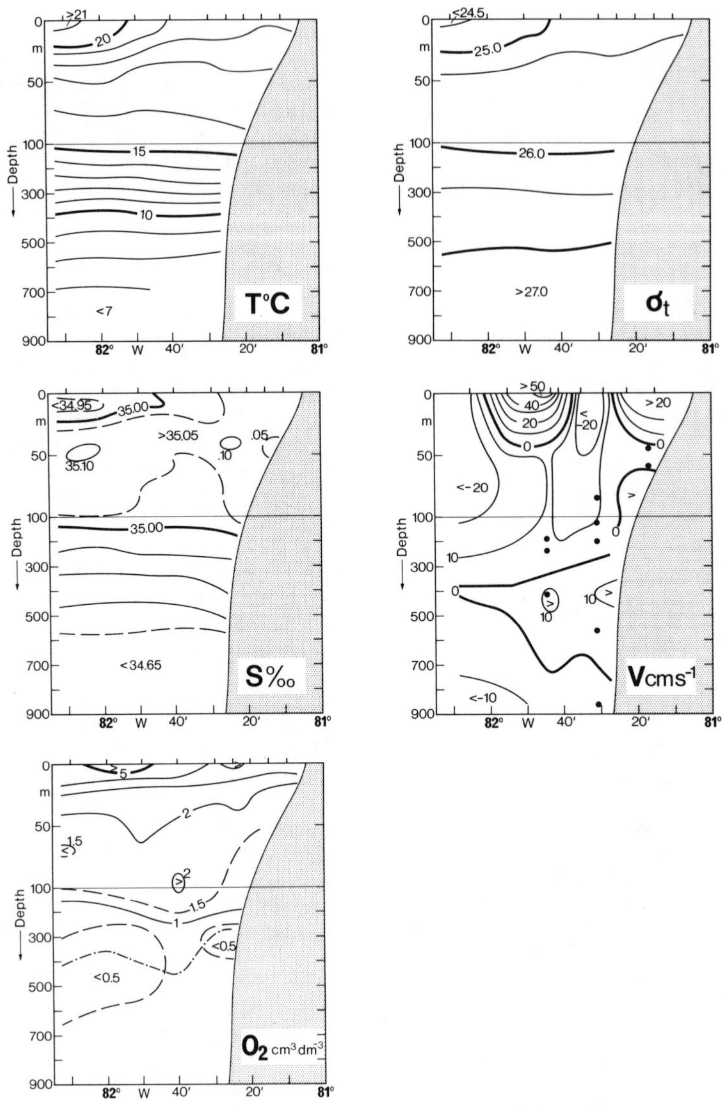

Fig. 5. Sections of mean temperature, salinity, oxygen, density, and geostrophic flow along 5°S calculated for 8 sections repeated during 40 days. The depth of the oxygen minimum is indicated by the dot-dash contour.

The Cross-Shore Circulation

At 110 km offshore the source water was at a depth of 80 m with a vertical extent of 65 m (Fig. 3). From the few sections reaching farther offshore it seems to have become even shallower with less vertical extent. Nearer the shore it kept its mean depth but spread vertically over 100 m (35 km offshore). This may be explained by an increase of mixing or by a cross-shelf circulation cell returning some of the warmed surface water to the source water level. Increased vertical mixing could be explained by the increase in current shear within the frontal jet (Fig. 5), but the fact that there is no significant horizontal change of the Richardson Number is an objection to this mechanism. The increasing shear was counteracted by increasing stratification:

$$Ri_{jet} = 2.5 \qquad Ri_{outside} = 3.1$$

Furthermore, the 16°C layer was well below the jet.

More information, at least for limited times, was given by direct measurement with a profiling current meter (Fig. 8). The turning of the currents to the left with depth suggests the presence of an Ekman layer. During the warm season the shallow thermocline (Fig. 3, left side) was intense enough to restrict the influence of the wind to a depth much less than the Ekman depth. The upper edge of the thermocline was at 17 m at

TABLE 1. Mean and Standard Deviations of Observed Time Series

	$\bar{u}$	σ_u	$\bar{v}$	σ_v	$\bar{T}$	σ_T
C 1						
45 m	2.2	8.9	2.8	10.8	16.61	0.63
60 m	-1.0	8.8	1.1	8.1	16.31	0.46
100 m	-0.7*	4.8*	0.4*	4.1*	15.59	0.47
C 2						
85 m	1.5	4.9	14.4	9.1	15.88	0.08
125 m	-0.0	5.0	11.0	9.9	15.42	0.34
200 m	1.4	5.0	2.2	12.3	13.83	0.52
305 m					12.08	0.73
560 m	0.1	4.1	-2.9	11.3	7.35	0.77
860 m	-0.0	3.7	-1.2	7.1	5.16	0.26
C 3						
195 m	-0.6	6.2	8.5	13.0	14.41	0.30
235 m	-0.8	5.4	4.4	13.2	13.59	0.40
410 m	-1.7	5.0	-3.4	10.1	9.80	0.81
650 m					6.91	0.25
Land stations						
Chiclayo	-0.0	0.7	-4.2	3.1		
Talara	-1.4	3.6	-6.9	2.3		
Paita					17.63	1.30

Temperature, $^{\circ}$C, wind, m/s, current, cm/s.
u positive to the east
v positive to the south
* current meter fouled after half the record length

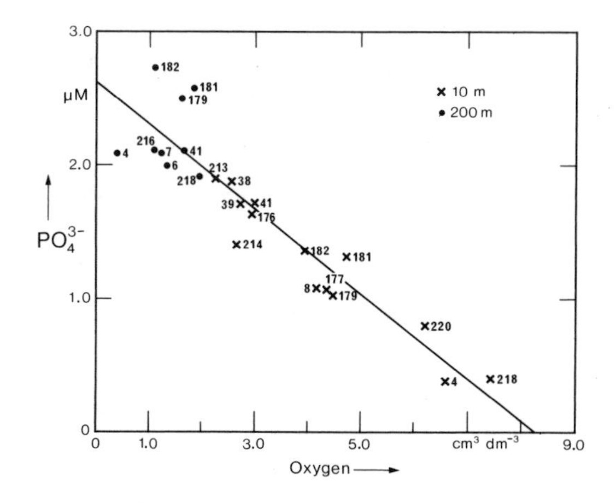

Fig. 6. PO_4^{3-}/O_2 diagram for 10- and 200-m samples from 4 sections. The numbers indicate the stations, crosses are values at 10 m, points at 200 m depth.

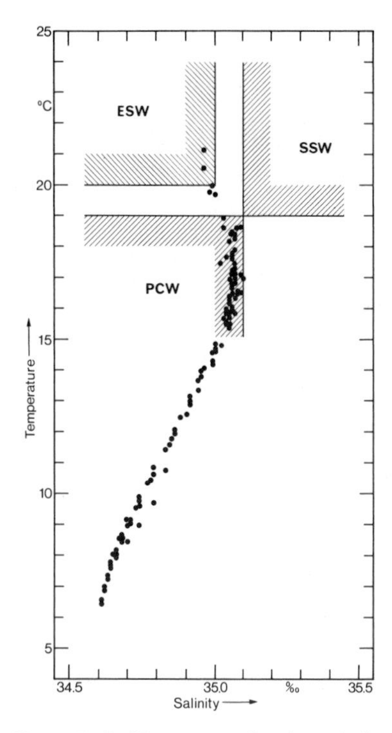

Fig. 7. Mean T-S diagram calculated from the mean temperature and salinity section given in Fig. 5. ESW = Equatorial Surface Water, SSW = Subtropical Surface Water, PCW = Peru Coastal Water.

Stas 10 and 11 and at 7 m at Sta. 13. The current vectors within this layer are shown in Fig. 9. The most shallow measurements were at 3.4 m. In comparison with the wind at Talara (Fig. 2, 7 m/s from 170°) the current vectors are essentially at 45° to the left. Therefore they are accepted as showing surface currents. At the top of the thermocline the currents turn through 90° to the left relative to the surface. According to Gonella (1971) this corresponds at Stas 10 and 11 to an Ekman layer in an unstratified ocean of 80 m. With this Ekman depth a surface current of about 40 cm/s and an Austausch coefficient of 40 cm^2/s is calculated. The calculated surface

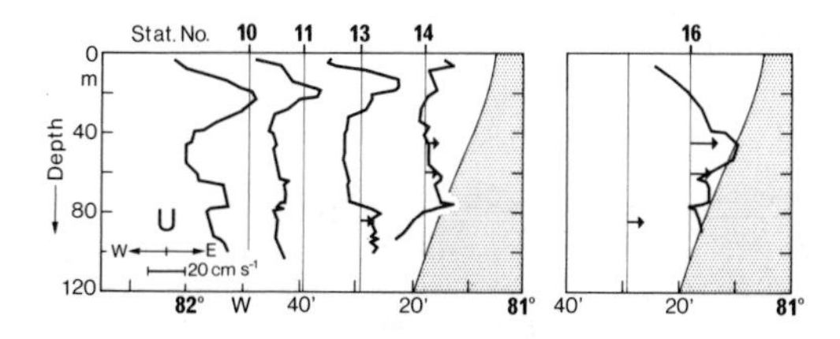

Fig. 8. Vertical profiles of onshore currents measured with a profiling current meter along 5°S. Left, 1 April, 1977; right, 4 April, 1977. The arrows indicate measurements of the moored current meters.

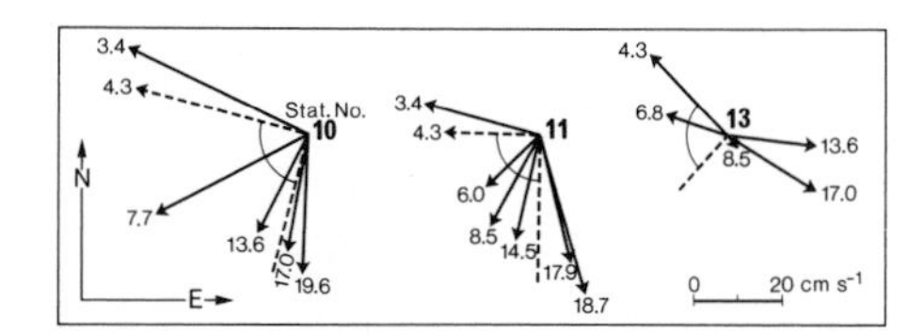

Fig. 9. The current vectors in the near surface layer at Stas 10, 11, and 13. The numbers at the arrows give the observation depths. The 4.3-m values at Stas 10 and 11 were interpolated to obtain the same reference depth. The dashed line gives a 90° angle to the direction of the current at 4.3 m.

current corresponds to the observed flow within the errors inherent in the use of measurements from a distant land station and a profiling current meter.

The underlying onshore flow must have been driven by the sea level inclination, while the inclination of the thermocline created the deep return flow. Even if the Coriolis force were taken into account in the surface layer with a well established Ekman flow, the geostrophic equilibrium cannot be attained in the deeper layer. This is evident by comparing the observed and calculated currents (Figs 10 and 11). As the wind increased from 4 to 8 m/s the Ekman transport, τ/f, became 7.4 10^4 cm^2/s, which is of the same order as the deep onshore transport of 6.0 10^4 cm^2/s (Fig. 8). In consequence the cold water rose, the shallow thermocline could not be maintained (Fig. 3), the stability decreased (Fig. 4), and the current shear increased resulting in an increase in the mixing. The near-shore oxygen content decreased and phosphate increased (Fig. 3).

A schematic representation of the cross-shelf flow for weak wind is given in Fig. 12. The

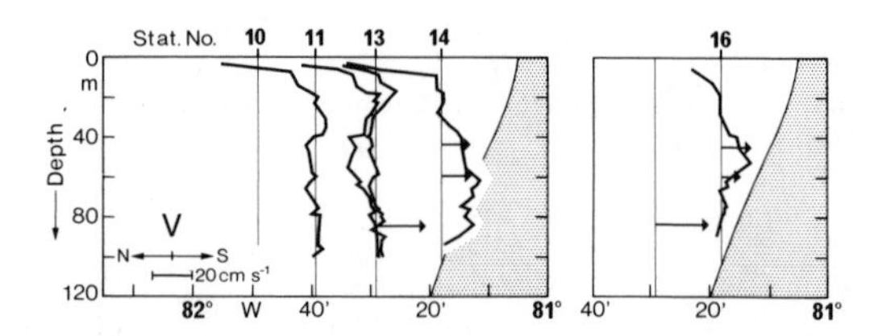

Fig. 10. Vertical profiles of alongshore currents measured with a profiling current meter along 5°S. Left, 1 April; right, 4 April, 1977. The arrows indicate measurements of the moored current meters. Generally both measurements correspond well, but an important deviation occurs on the v-component of Sta. 13. It cannot be concluded from our data whether this is due to small scale variability or to a measuring error. The moored current meter values fit the profiler measurements further offshore rather well.

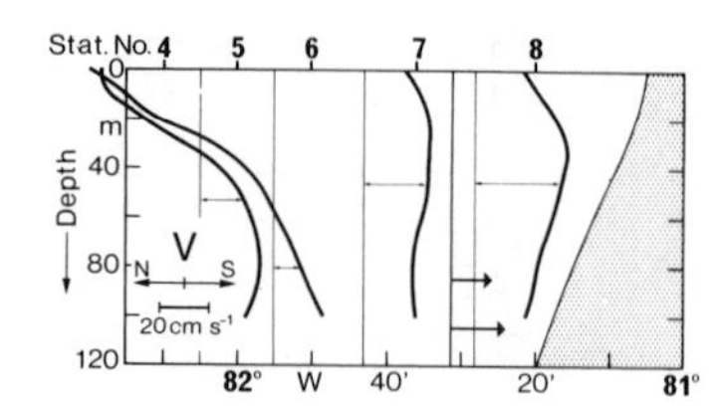

Fig. 11. Geostrophic alongshore currents calculated for the sections occupied 30 and 31 March, 1977. The arrows indicate the zero of the corresponding profile.

solid line gives the pattern from direct current meter measurements, the dashed line gives the deduced parts of the cells. With increasing wind the upper cell disappeared. These two-dimensional cells are not a permanent feature; there were times when onshore flow was observed in the whole water column, then a three-dimensional balance must be assumed. It is outside of the scope of the data to discuss situations which are the consequence of remote effects, such as internal Kelvin or continental shelf waves.

A similar circulation pattern has been deduced for the Oregon upwelling front by Mooers, Collins, and Smith (1976). They can explain an intermediate temperature maximum by the downwelling branch of the near-shore circulation cell. It cannot be concluded from our data that such a temperature maximum is not found off northern Peru because the spacing and accuracy of the temperature measurements are not good enough (Brockmann *et al.*, 1978).

The Interaction of Atmospheric and Current Fluctuations

The similarity between the seasonal trends in wind and SST gives an indication of the influence of the meteorological forcing on upwelling in the low-frequency range. Nonetheless the observation period was too short to treat the time scale statistically. The profiling current measurements indicate cases with well established Ekman flow. But it becomes evident that the directly wind-driven layer is shallower (30 to 40 m) than

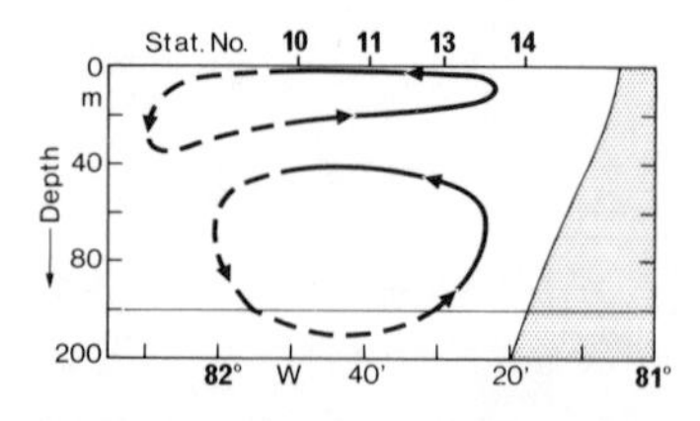

Fig. 12. Schematic cross-shore circulation for weak winds. The solid line indicates directly measured current, the dashed line is concluded from other observations.

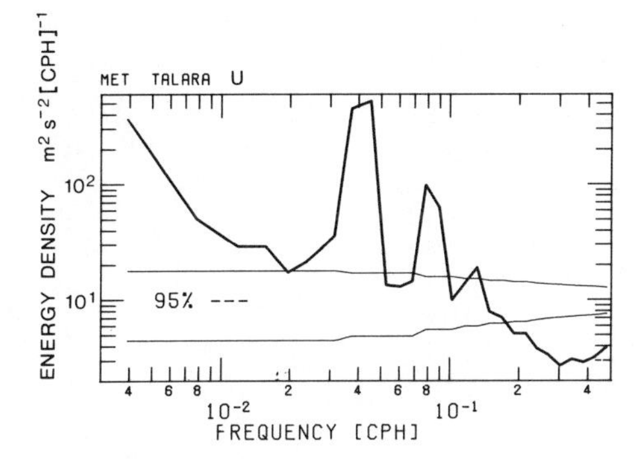

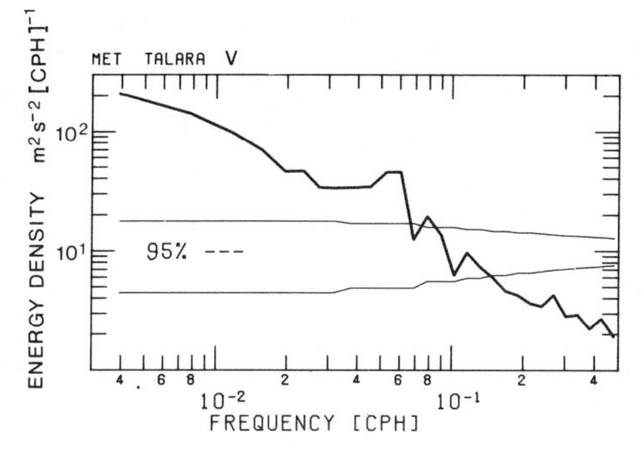

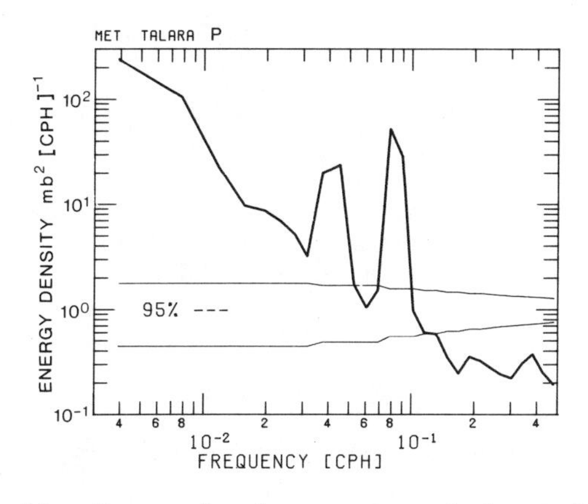

Fig. 13. Energy density spectra of the onshore, u, and alongshore, v, wind components and the atmospheric pressure, P, at Talara.

the depth of the shallowest current meter (45 m). The presence of a one- or two-cell circulation at different times complicates the relation between the wind and the water movement observed at a given depth. To investigate some aspects of the deeper reaching response of the currents to the atmospheric fluctuations, spectral analysis was applied to data from the land stations and the current meters. The study of the intertial waves was omitted even though they were seen to be present, because it was necessary to divide the time series into segments of 10 days to obtain a reasonable number of degrees of freedom. As the inertial period at 5°S is 5.7 days (0.7 x 10^{-2} cph), it occurs at the low-frequency edge of the spectra and is thus badly resolved, with the energy being spread over a wide frequency range.

The energy density spectra of the land station data from Talara (Fig. 13) are dominated by a land-sea-breeze peak with a period of one day, which is much stronger in the onshore (u) than longshore (v) component. Further, a 12-h peak and a shoulder-like structure at 50 h can be observed. The discussion of the meteorological spectra is not the aim of this paper; they are accepted as driving forces to which the sea reacts. It is worth mentioning that the spectra are not a local feature; they are rather similar at Chiclayo, 250 km south of Talara.

The oceanographic spectra reflect only some of the features. Over the shelf at 45 m below the surface (Fig. 14), the dominant peak is given by the semi-diurnal tide. The 50-h shoulder is at the same relative intensity as in the meteorological spectra. Proceeding offshore and deeper, the semi-diurnal tidal peak is maintained, but the intensity of the shoulder is reduced and it becomes clearer in the u-component (Fig. 15). The coherence between the meteorological time series and the currents over the shelf shows significant values only in the 50-h range for the onshore wind and air pressure to the longshore current

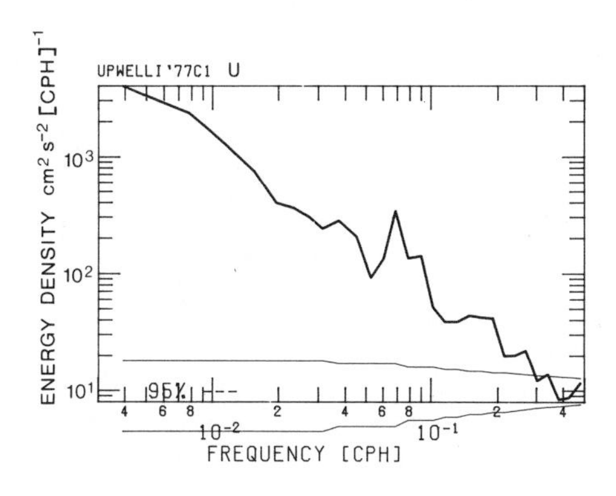

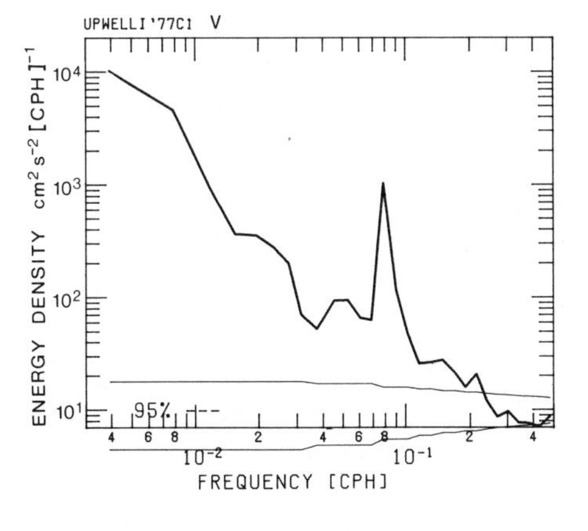

Fig. 14. Energy density spectra of the onshore, u, and alongshore, v, currents at the shelf mooring (C1) 45 m below the surface.

(Figs 16 and 17). Over the slope at 85-m depth the coherence between these components decreases below the 95%-significance level, but between various other components significant coherence is observed. As an example, the coherence between the u-components of wind and current is shown (Fig. 18).

Further offshore and deeper, coherence in this band becomes weaker. As an example, the energy spectra of the deepest current meter is shown (Fig. 19). It shows the 50-h shoulder as a small peak. Coherence in the range is significant between the atmospheric pressure at Talara and the longshore current (Fig. 20). There are various significant coherences in adjacent frequency ranges.

The fact that there is no coherence at the much more energetic 24-h peak, but that there is at the weak 50-h shoulder may be explained by the coastline configuration. The shelf mooring lay at the edge of Paita Bay (Fig. 1). During most sections a rather shallow thermocline was observed in the bay. If the wind or atmospheric pressure fluctuates over the bay, oscillations of the basin are enhanced. With a mean depth of H = 40 m and a length of L = 10 km the barotropic eigen-frequency is calculated for an open bay by Merians formula as

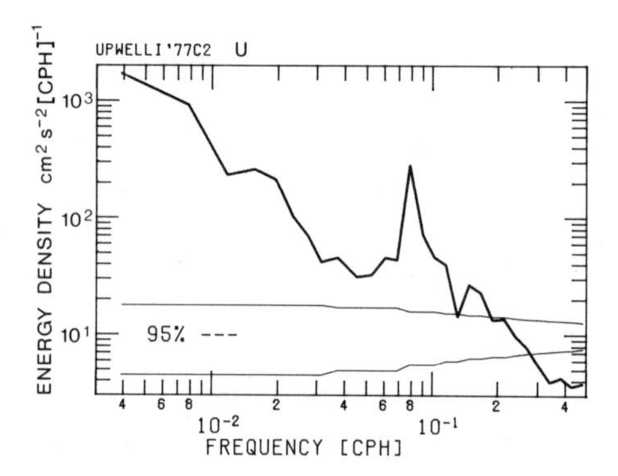

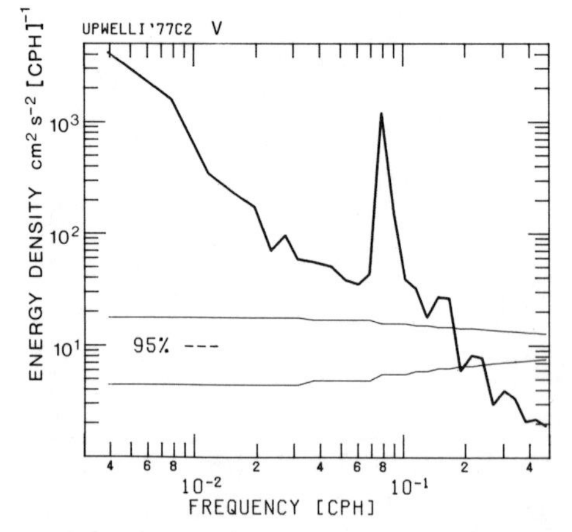

Fig. 15. Energy density spectra of the onshore, u, and alongshore, v, currents at the continental slope mooring (C2) 85 m below the surface.

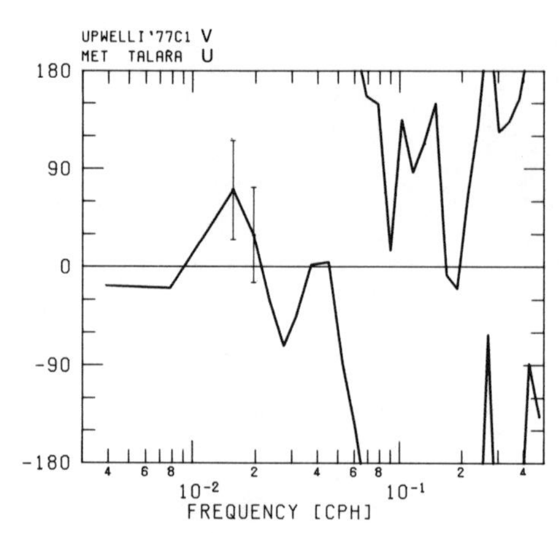

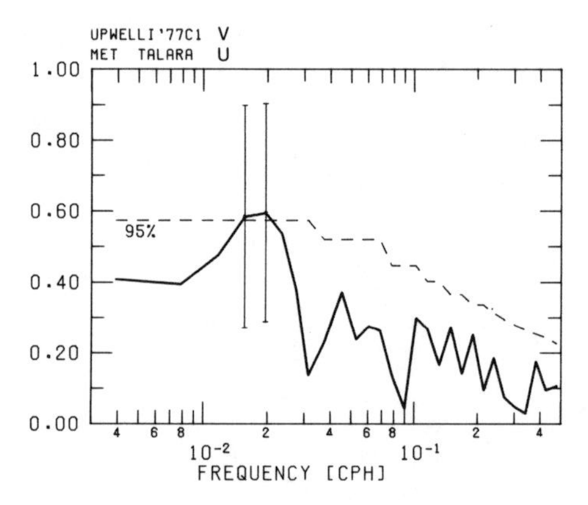

Fig. 16. Coherence and phase between the alongshore component at the shelf mooring and the onshore component of the wind at Talara.

$$1/T = \frac{\sqrt{gH}}{4L} = 1.8 \text{ cph}$$

(g = acceleration due to gravity),

which is too high to be observed in the data. For a two-layer model with a thermocline depth of h = 10 m, a density difference of $\Delta\rho = 0.5\ 10^{-3}$ and $\rho = 1.025$ the density of the lower layer, the frequency of the first horizontal mode results (e.g. Dietrich, Kalle, Krauss, and Siedler, 1975) in

$$1/T = \frac{\sqrt{g \frac{\Delta\rho}{\rho} \frac{(H-h)h}{H}}}{4L} = 0.017 \text{ cph}$$

i.e., T = 57 h.

A similar result is obtained using a method for continuous stratification given by Kraus (1966).

The oscillation of the bay incites the adjacent waters to co-oscillate because there is maximal horizontal movement at the node in the vertical

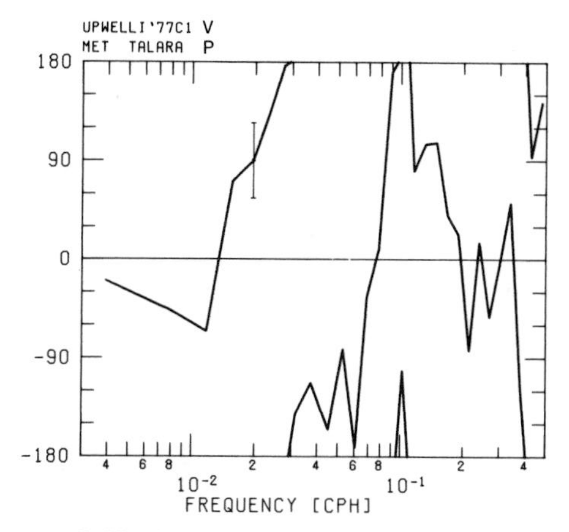

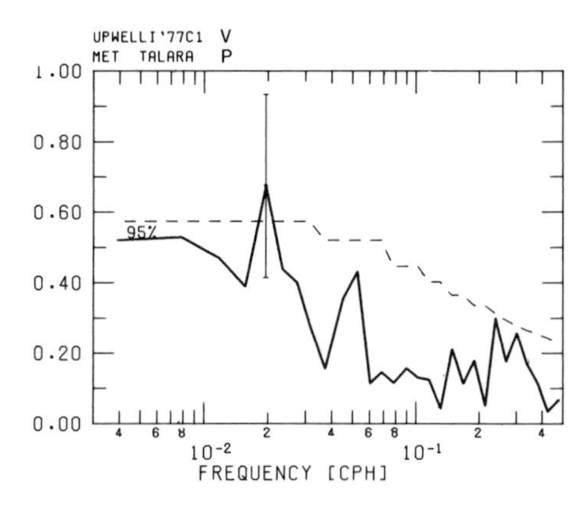

Fig. 17. Coherence and phase between the alongshore component at the shelf mooring and atmospheric pressure at Talara.

motion, which would be found at the outer edge of the bay. This is reflected in the more pronounced peak in the coherence between wind and offshore currents over the slope. If the fluctuations incite waves at the shelf edge, which propagate perpendicularly to the coast, they should not be easy to detect because the phase speed is similar to the offshore flow in the 40-m layer. As the onshore-offshore current changes with time and depth no clear coherence can be expected, but only spurious coherences over a wide frequency range.

The observed phases between the different components are due to the two-layer system. Onshore wind counteracts upwelling and deepens the upper layer. For this reason the current meters being in the upper part of the lower layer penetrate the upper layer, giving a transition from the southerly undercurrent to the northerly surface current.

Over the slope, onshore wind is in phase with offshore current (Fig. 18), which is consistent with the lower-layer movement.

The observation that the oscillations of a bay are transferred to the open ocean can be an interesting aspect, should the oscillations propagate along the coast.

Summary and Conclusions

The northern Peruvian upwelling region shows the features of a wind-driven upwelling circulation system. A clear relation between the windstress and the sea-surface temperature (SST) was observed. A complex interaction between offshore Ekman transport, onshore compensation flow, mixed layer depth, and heat gain lead, even with a changing SST, to a persistent offshore SST gradient. The cross-shelf flow pattern changed with increasing wind from a two-cell to a one-cell circulation, which reached as deep as 150 m (Fig. 12). In contrast to these two-dimensional cells, periods of evident lack of a two-dimensional balance were observed.

The mean longshore flow showed a shallow poleward undercurrent (Fig. 5). It reached to the

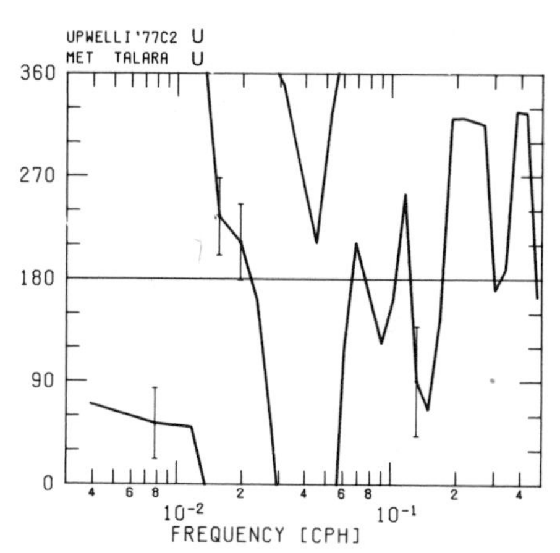

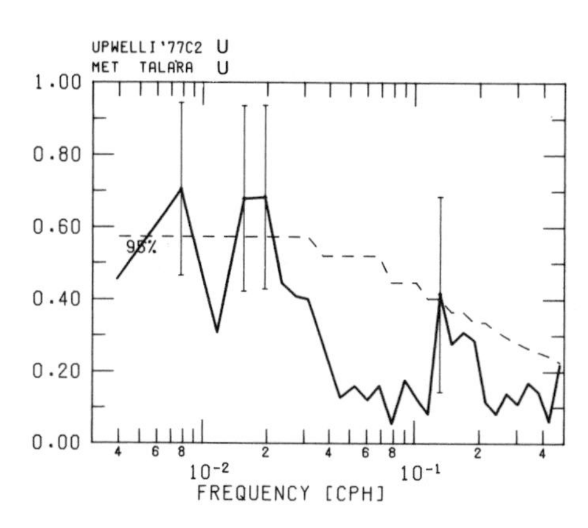

Fig. 18. Coherence and phase between the onshore component of the currents at the slope mooring 85 m below the surface and the onshore component of the wind at Talara.

shelf and extended beyond the range of our observations. At about 55 km offshore it reached the surface. Above and below there was flow towards the equator. The maximum values, up to 50 cm/s near the surface, should be regarded with care, because the strong velocity gradients disturb the assumed geostrophic equilibrium; further, an Ekman regime is to be expected.

Water mass analysis showed that the area of observation was influenced by equatorial water at the early part of the observation period. Further there is no indication that the Equatorial Undercurrent reached the area. The facts imply that the observations showed this upwelling area in a "normal" year, when equatorial waters leave the area during the spring transition, in contrast to a year when equatorial disturbances such as *El Niño* are observed far to the south.

The fluctuations of the system were dominated by various time scales. The most important was the seasonal transition. The change from the warm to the cold season fell in the period of observations and took place very quickly (within 10 days).

In general the coherence between current and wind fluctuations was low, but significant coherence between various current-meter time series and meteorological parameters was found in a frequency range of about 50 h. It is concluded that,

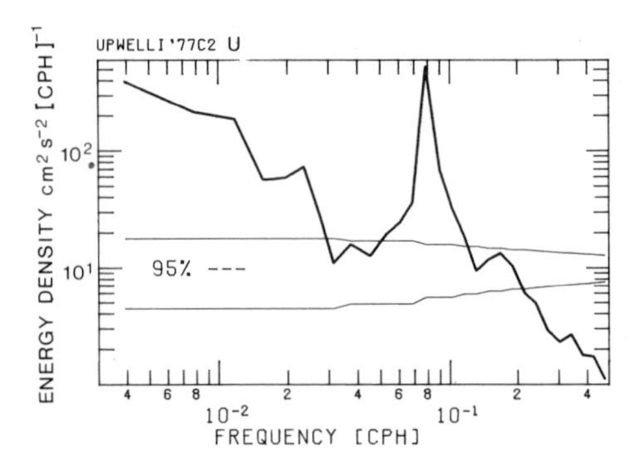

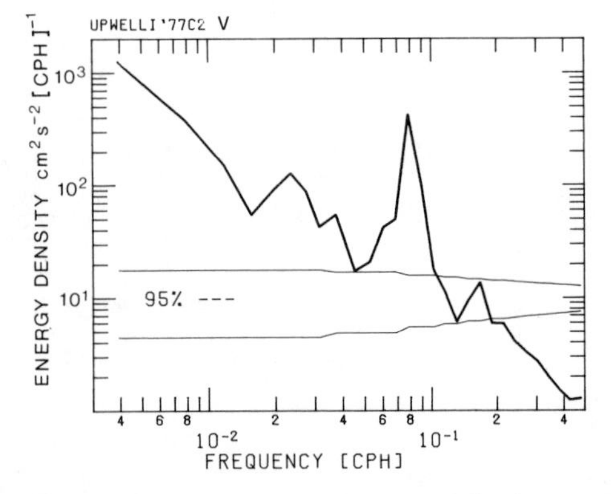

Fig. 19. Energy density spectra of the onshore, u, and alongshore, v, currents at the continental slope mooring (C2) 860 m below the surface.

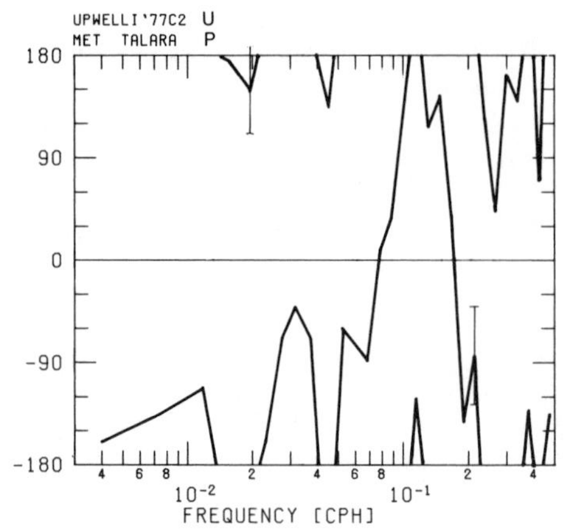

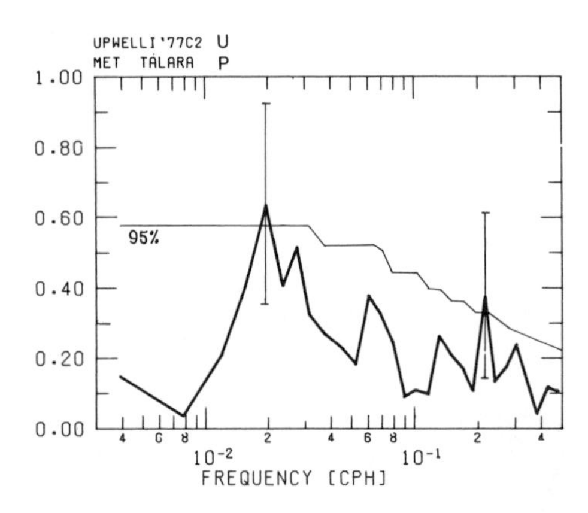

Fig. 20. Coherence and phase between the alongshore current, u, at the slope mooring 860 m below the surface and the atmospheric pressure, P, at Talara.

in this frequency range, energy transfer was more effective because of resonant interaction with the Bay of Paita, where internal seiches were generated. The oscillating bay enhanced the co-oscillation of the adjacent ocean. It appears that the oscillations could be the source of poleward-propagating disturbances.

There is some evidence that current fluctuations with a period of 5 days and longer are trapped coastal waves transporting energy to the south and imposing important fluctuations on the intensity of the upwelling. But a study of this frequency range is problematical because it includes the inertial frequency at this latitude. The limited record does not allow a reasonable resolution in the range. The area should be investigated with respect to longer time scales and three-dimensional aspects.

Acknowledgments. We gratefully acknowledge the help of the officers and crew of the *BAP Unanue*, as well as our colleagues from IMARPE, for the successful operation at sea. We thank the Corporation Peruana de Aerolineas Commerciales for making available the land station data. This work was supported by the Deutsche Forschungsgemeinschaft.

References

Brockmann, C., E. Fahrbach, and W. Urquizo, ESACAN-Data Report, *Berichte aus dem Institut für Meereskunde, Kiel, 51*, 54 pp., 1978.

Dietrich, G., K. Kalle, W. Krauss, and G. Siedler, *Allgemeine Meereskunde*, Gebrüder Borntraeger, 593 pp., 1975.

Enfield, D.B., Oceanografia de la región norte del frente ecuatorial: Aspectos fisicos, Actas de la reúnión de trabajo sobre el fenómeno conocido como "*El Niño*", Guayaquil, Ecuador, 4-12 de Diciembre de 1974, *FAO Informes de Pesca, 185*, 299-334, 1976.

Gonella, J., The drift current from observations made on the Bouee Laboratoire, *Cahiers Océanographiques, 13*(1), 19-33, 1971.

Mooers, C.N.K., C.A. Collins, and R.L. Smith, The dynamic structure of the frontal zone in the coastal upwelling region off Oregon, *Journal of Physical Oceanography, 6*, 3-21, 1976.

Hidaka, K., A contribution to the theory of upwelling and coastal currents, *Transactions of the American Geophysical Union, 35*, 431-444, 1954.

Krauss, W., *Methoden and Frgebnisse der theoretischen Ozeanographie II, Interne Wellen*, Gebrüder Borntraeger, 248 pp., 1966.

Patzert, W.C., *El Niño* Watch atlas, *Scripps Institution of Oceanography Reference Series 78-7*, 322 pp., 1978.

Redfield, A.C., B.H. Ketchum, and F.A. Richards, The influence of organisms on the composition of sea-water, in *The Sea 2*, Interscience, 26-77, 1963.

Sverdrup, H.U., M.W. Johnson, and R.H. Fleming, *The Oceans*, Prentice Hall, 1087 pp., 1942.

Wyrtki, K., Oceanography of the eastern equatorial Pacific Ocean, *Oceanography and Marine Biology, an Annual Review, 4*, 33-68, 1966.

HYDROLOGICAL AND METEOROLOGICAL ASPECTS OF UPWELLING IN THE SOUTHERN BENGUELA CURRENT

L.V. Shannon, G. Nelson and M.R. Jury

Sea Fisheries Institute, Private Bag,
Sea Point 8060, South Africa

Abstract. During the summer months, an upwelling plume forms west of the Cape Peninsula on the west coast of Africa (34°S), an area characterized by pronounced orographic, coastal, and bottom topographic features, and a longshore wind regime modulated by eastward moving cyclones driven by the circumpolar jet stream. During the upwelling season of 1978-1979 a synoptic study of the area was undertaken over a grid of 80 closely spaced stations using a CTD and profiling current meter. Additional information was provided by moored current meters, coastal weather stations, and by aircraft measurements of sea surface temperature, winds, and hydrothermal parameters. A well defined upwelling plume developed in response to the wind events. Superimposed on the secular growth of the plume during the summer were perturbations with a period of about two weeks caused by a slackening or reversal of the mesoscale wind field. The water entering the Benguela Current system from the south had identifiable T-S characteristics and was in geostrophic balance. As the water rose onto the southern edge of the shelf, the topography caused a convergence with the resultant development of an upwelling plume and associated oceanic front and frontal jet, while a thick bottom friction layer formed over the shelf area. The water moving onto the shelf region appeared to be controlled by horizontal pressure gradients established on a scale several times the length scale of the peninsula by both meteorological and oceanographic factors. An undersea canyon west of Cape Point is thought to play a significant role in the upwelling process.

Introduction

During the austral summer of 1978-1979 a synoptic study of upwelling was undertaken in the southern Benguela Current (Fig. 1). Known as CUEX-I, it was primarily a series of physical observations designed to provide information on boundary conditions, especially surface wind stress, to set up numerical models of upwelling in the region, but it was linked to a comprehensive program of ground truth measurements for the NIMBUS-7 Coastal Zone Color Scanner program.

The Benguela system, which is bounded by the Agulhas retroflection area southwest of Cape Point (34°S) and in the north, near Cape Frio (18°S), by the interaction area with the southward-flowing Angolan Current, is by no means uniform along its length. North of 32°S the coastline is fairly regular with a few small capes and embayments. The shelf is broad and the topography landwards rises a few hundred meters some 20 km inland with a coastal barrier of sand dunes so that the orographic influence on wind is not spatially variable. Perennial southerly winds blow with seasonal modulation and become diurnally pulsed by land-sea breezes. South of 32°S the situation is entirely different. Two pronounced capes, Cape Columbine and the Cape Peninsula, have a marked influence on the wind field and topographic enhancement of upwelling with the formation of plumes. In the area the shelf is narrow, two notable features being undersea canyons, namely the Cape Canyon near Cape Columbine and the Cape Point Canyon west of the Cape Peninsula. The region is meteorologically distinct from that to the north, coming under the influence of eastward-moving cyclones that periodically weaken the South Atlantic Anticyclone.

While the general features of the Benguela Current have been described by a number of investigators, *inter alia* Clowes (1950), Buys (1957), Hart and Currie (1960), Stander (1964), Shannon (1966), Schell (1968), Shannon and van Rijswijck (1969), Visser (1969), and De Decker (1970) the recognition of the Cape Peninsula upwelling plume as a separate entity was emphasized and described by Andrews and Cram (1969), Andrews (1974), Bang (1971, 1973), and Bang and Andrews (1974). In this region upwelling is affected strongly by topography and compression of the wind field. A pronounced upwelling plume of cold surface water is a feature of satellite imagery of the area as shown, for example, by Legeckis.

This paper reflects advances in knowledge of the physical processes in the region between

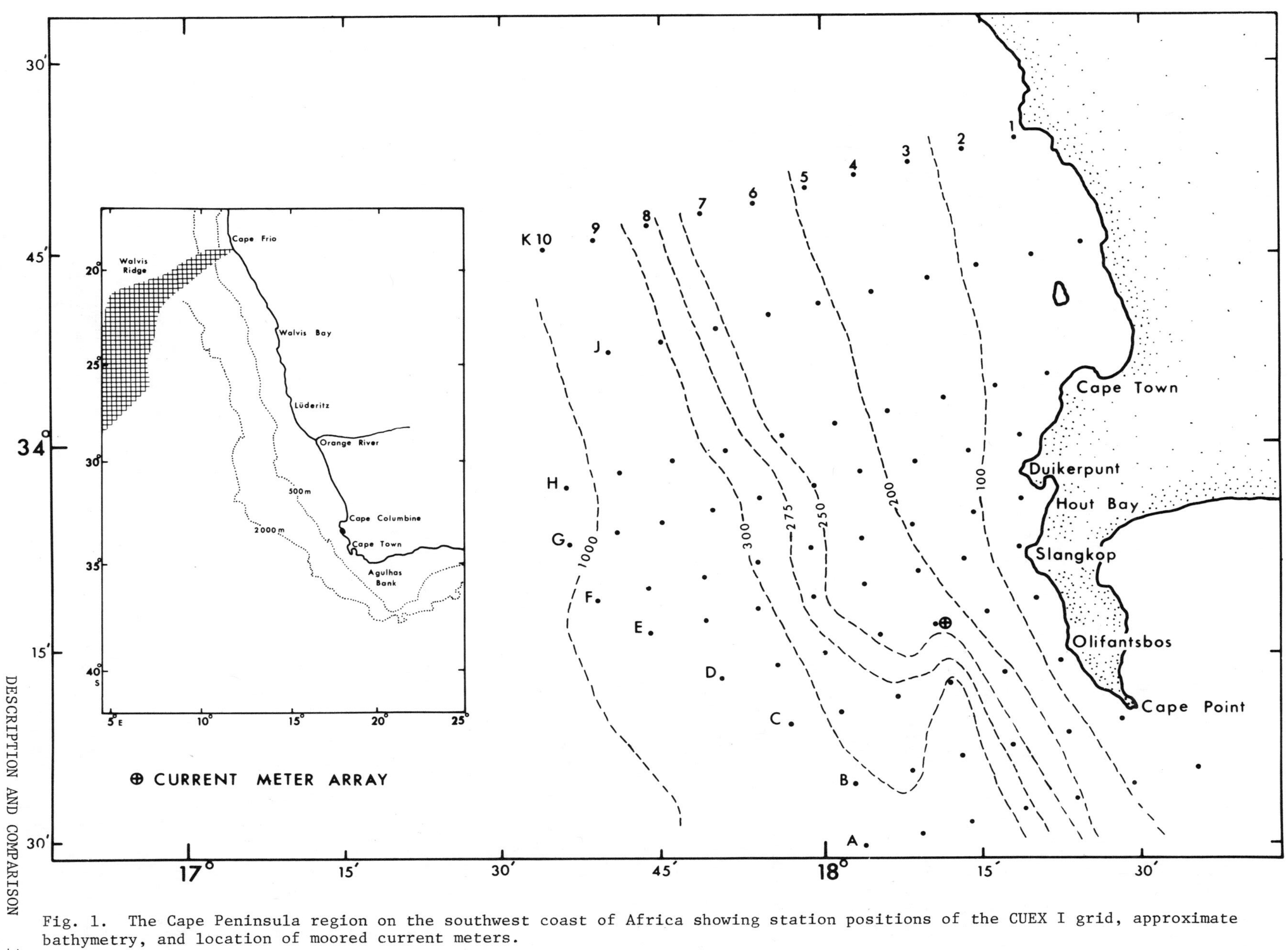

Fig. 1. The Cape Peninsula region on the southwest coast of Africa showing station positions of the CUEX I grid, approximate bathymetry, and location of moored current meters.

33°30'S and 34°30'S.

Plumes and Mesoscale Features of the Summer Wind Field

The mesoscale features of the summer wind field were investigated in 1978 and 1979 by means of aerial surveys conducted at an altitude of 150 m. Grid patterns were flown by a research aircraft fitted with a Barnes PRT 5 radiometer and a simple forced ventilation wet and dry bulb thermal probe using two Fluke 80T - 150 sensors and a digital voltmeter. Wind vectors at 150 m were calculated from drift sight, DECCA navigation, and ground track observations every 3 min of flight. Large spatial gradients and temporal fluctuations of the wind, sea-surface temperature, and air mass characteristics were observed to be related to local topographic and thermal effects and synoptic scale weather phenomena.

The mean surface pressure gradient during summer assumes a coastal alignment as a result of the South Atlantic Anticyclone (SAA) and an interior trough of lower pressure. The predominantly southeasterly winds are modulated by westerly waves in the circumpolar jet, which passes south of the continent. The pressure perturbations affect the mean pattern creating periodic wind events and reversals. A travelling anticyclone ridging south of the continent produces strong offshore pressure gradients along the southwest coast. Gale force winds occur for a few days, slackening with the approach of the next wave cyclone. Pressure gradients then collapse leading to onshore flow.

The upwelling plume develops as the travelling anticyclone pulls the SAA shorewards; an example was observed on December 11, 1978. Deep southeasterly flow became topographically accelerated over the peninsula in a classical cape effect (Fig. 2), initiating the formation of a plume west of Cape Town (Fig. 3).

With the anticyclone's influence receding, a coastal low tends to move southwards along the west coast. Zonally aligned pressure gradients under strong atmospheric inversion conditions produce shallow southeasterly flow. Orographic effects are magnified as the atmospheric "lid" promotes stratification of the air mass. A mountain range lying upwind of the peninsula shelters its northern half, creating a calm region and cyclonic wind vorticity. A flight on February 1, 1979 indicated these features (Fig. 4). The cold surface plume extending west of the peninsula (Fig. 5) curls cyclonically northwards in response to the wind field.

Following the passage of a coastal low east of Cape Town, westerly winds prevail and a shoreward relaxation of upwelled water compacts the oceanic thermal front over the shelf regions. Figs 6 and 7 show the features of this regime for January 16, 1979.

Superimposed on the wind event cycle are diurnal accelerations in the form of land-sea breezes. The pulsing of the longshore wind is amplified during subtropical inversion conditions.

Mesoscale Oceanographic Features

During November 1978 a grid of 80 stations on a 5-nautical mile spacing with lines bearing 255° was occupied by R.S. *Meiring Naudé* west of the Cape Peninsula (Fig. 1). A Neil Brown Mk3 CTD and rosette sampler were used on alternate lines. The rosette sampler was equipped with a Savonius rotor current meter for obtaining flow vectors at each sampling depth. Measurements were integrated over periods of approximately 1 min after excessive instability of the instrument package caused by lowering had died away. Ship drift was subtracted using a Motorola microwave range positioning system on stations within 20 nautical miles of the coast and using DECCA fixing on distant stations.

The grid was worked from south to north with lines being repeated about five days later on a second cruise, the object being to sample the area under different phases of the meteorological cycle. On both cruises the wind was blowing southeast at the start but swung to westerly further north so that the objective was not realized in full. Nevertheless, some useful comparisons were obtained.

Currents measured from R.S. Meiring Naudé

Even though optimum conditions of orthogonal range positioning may locate a ship to within a few meters, errors can be anticipated in spot current measurements from a drifting ship because of relative acceleration of the instrument package. The error increases with depth. However, measurements obtained from the two cruises of the *Meiring Naudé* in this way show a constancy with depth and for this reason confidence was placed in the data, with reservations as to absolute accuracy on stations where DECCA navigation was used to obtain drift.

Current measurements were made at 10-m intervals to 50 m, at 75 m, and then in steps of 50 m from 100 m. The field was remarkably baratropic away from stations nearest the coast and towards the southeast corner of the grid where Agulhas water rounded Cape Point from the east in the top 30 m. The observations cannot be treated as synoptic because unrealistic convergences and divergences would then be indicated. This is particularly true on Stas G1, 2, and 3 on Cruise 1 and on line G of Cruise 2 where two apparent strong convergences occurred, one close inshore, the other at the thermal front. The latter region, sampled over a period of about 4 h in a fairly homogeneous wind field, showed some baroclinicity. In particular, the current velocities at G8 were weak at depth and the orientation on G7 was variable.

On stations nearest the coast currents were

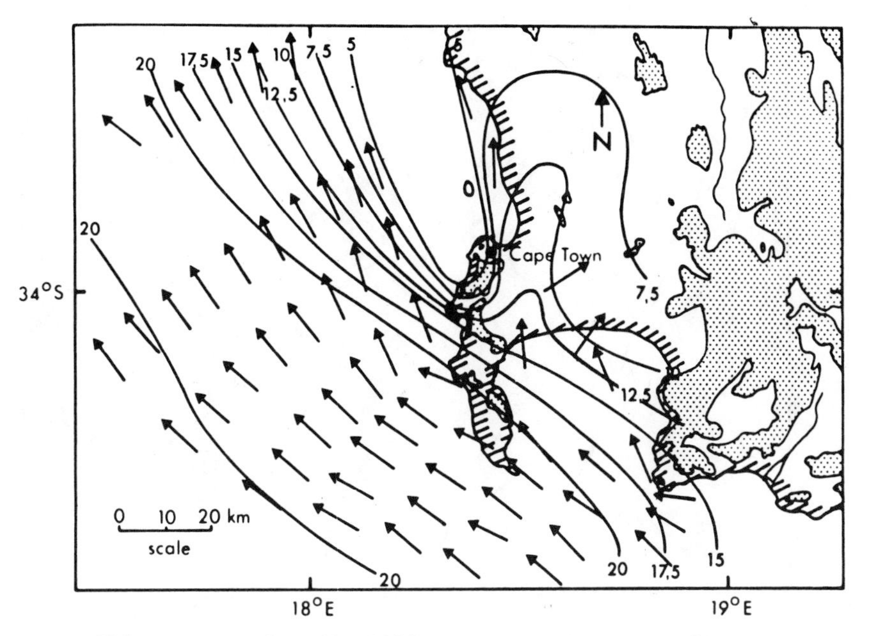

Fig. 2. Wind vectors at 150 m on December 11, 1978. Deep southeasterly flow was observed with maximum speeds (m/s) northwest of the peninsula.

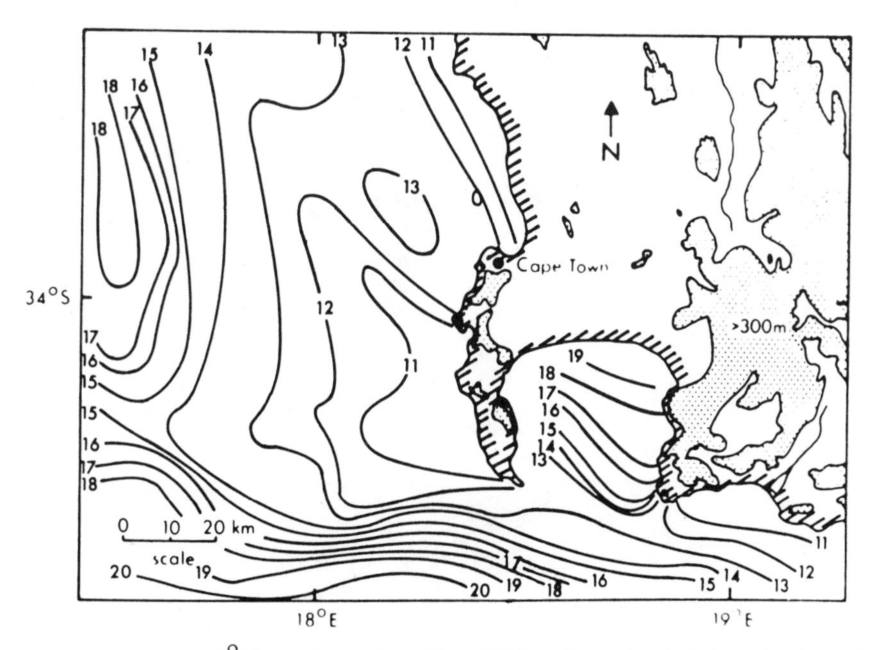

Fig. 3. Sea-surface temperatures (°C) on December 11, 1978. A newly developed plume is northwest of Cape Town with minimum temperature of 10°C. The flight grid is shown.

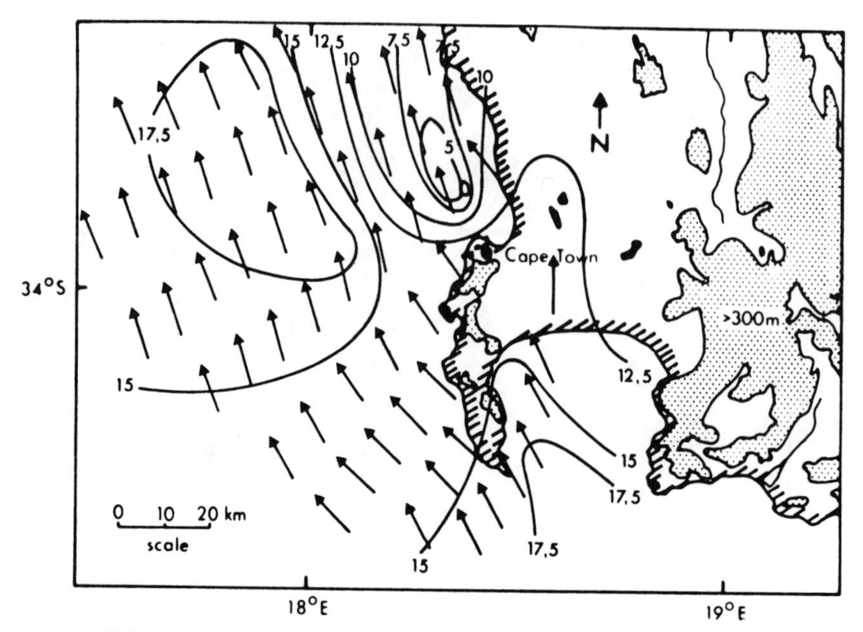

Fig. 4. Wind vectors at 150 m on February 1, 1979. Shallow southeasterly flow reached a peak speed (m/s) southwest of the peninsula with a large calm area downwind of the coastal mountain. High positive vorticity was observed.

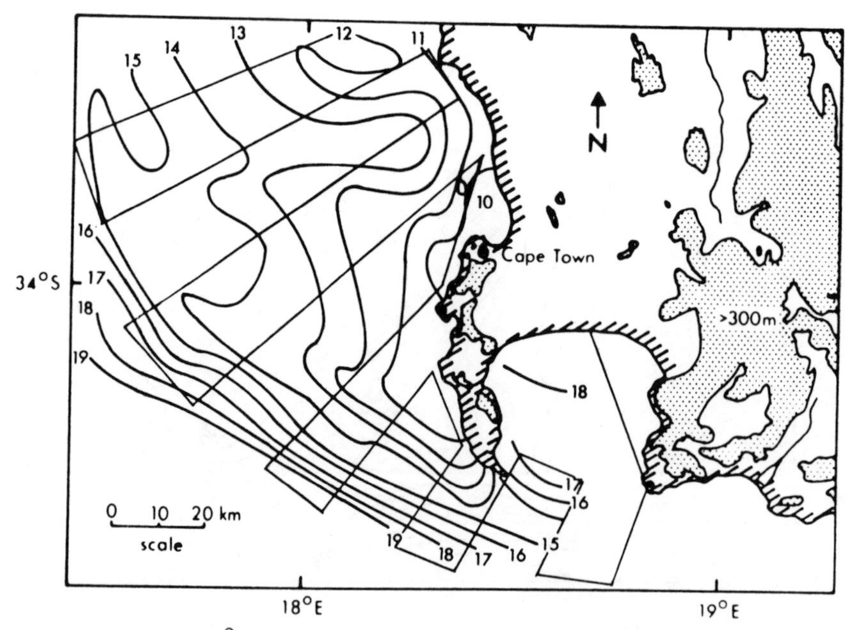

Fig. 5. Sea-surface temperatures (°C) on February 1, 1979. An upwelling plume extending west of the peninsula begins a cyclonic rotation with a warm pocket northwest of Cape Town.

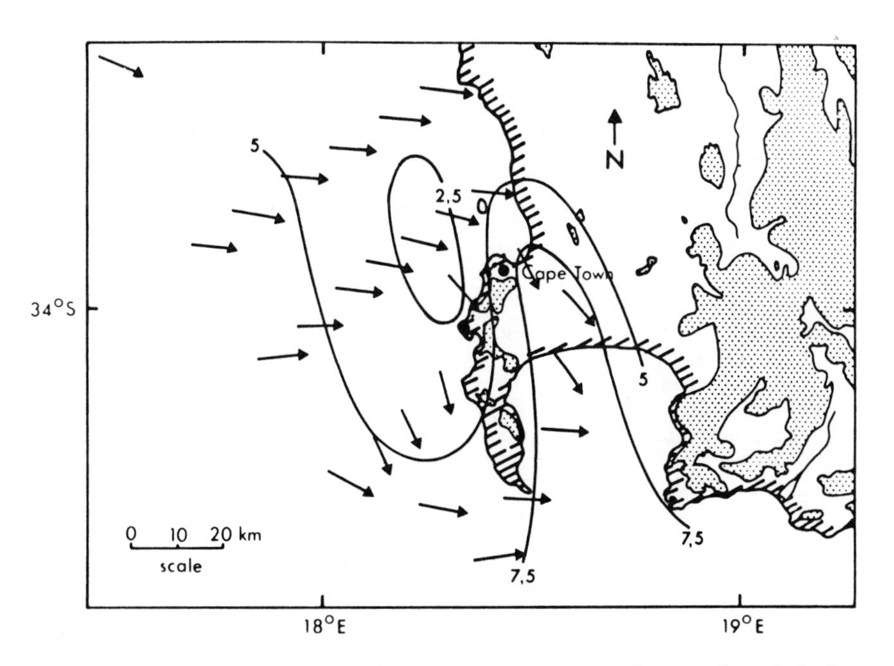

Fig. 6. Wind vector at 150 m on January 16, 1979. Light west-northwesterly winds became amplified southeast of the peninsula. Cloudy conditions prevailed.

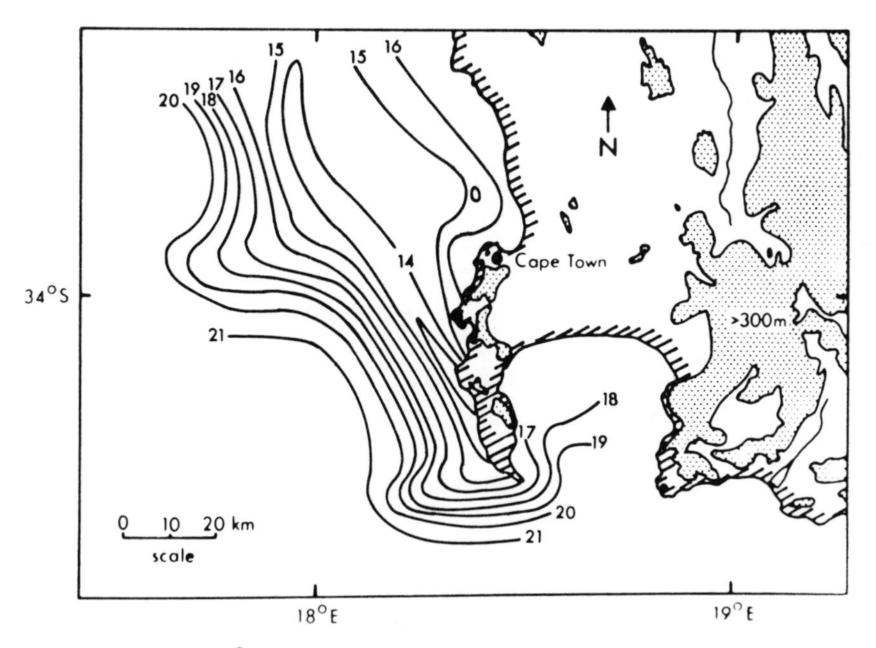

Fig. 7. Sea-surface temperatures (°C) on January 16, 1979. A shoreward relaxation of the upwelling area was noted with instabilities developing along the compressed thermal front.

weak but strongly baroclinic. The influence of tidal and possibly inertial currents in Hout Bay should be taken into consideration here. Near Sta. E2, two profiles observed at 17h29 and 21h59 on 18 November showed a complete change of current pattern within 4 h without significant change in the mesoscale wind. Currents at 14 m for Cruise 1 are shown in Fig. 8.

Towards the seaward front of the upwelling plume current speeds were high, suggesting the existence of a frontal jet. Bang and Andrews (1974) made current measurements from a drifting ship on a line bearing 255° from Cape Town, obtaining velocities of 1.2 m/s for the jet.

Fig. 9 shows the normal component of velocity on line C. Meteorological conditions on the line had been constant with southeasterly winds for three days and thus the data could be taken as synoptic. The thermal front was close inshore in November with water of 16° on Sta. C1. The pronounced deep jet on Sta. C3 was probably associated with the topography of the Cape Point Canyon, which shelves in the vicinity, rather than with the upwelling dynamics of the plume itself. As discussed below, corroboration of this was obtained from current meters moored nearby, which showed strong flow along the canyon. The strongest winds (12 m/s) were recorded on Stas C3 and C4, and this presumably accounts for the extension of the jet to the surface. The bulge west of the main jet below 75 m is most likely established by shear of the jet moving up the canyon and also extends to the surface under the action of wind stress.

Characteristics of salinity and temperature curves

As this is a region of confluence of distinct water masses, marked features may be anticipated on T-S plots. The ship-borne current meter showed a flow of water with a westward component in the upper 30 m on Stas A1, A2, and A3 during strong southeasterly winds (Fig. 8). Thus, water of Agulhas origin rounding Cape Point will mix with surface water from the Agulhas retroflection area and sub-tropical convergence zone.

The T-S curve of Sta. A6 (Fig. 10a) may be taken as representative of the body of water entering the upwelling area from the south at that time. A feature of the curve is that it approaches the T-S index for South Atlantic Central Water at 350 m, according to Defant and Wüst (1930). The characteristics of the water were not preserved within the upwelling area, with the most pronounced departures on nearshore stations on line D and E (Fig. 10b, c), but they remained coherent west of a line corresponding approximately to the thermal oceanic front. Along the line a frontal jet as described above can be expected, and evidence of such a jet was obtained on both November cruises from ship-board measurements when current speeds of 1 m/s were observed along the front compared to typical values of 0.5 m/s to the east and west.

Evidence of Agulhas water entering the area from the east round Cape Point was seen on Stas A1 to A5, B1 to B5, and C1 to C4. The T-S curves for the stations show a markedly lower salinity above 75 m and higher salinity below this depth referred to the curve for Sta. A6.

We thus conclude that during active wind-induced upwelling water from the south is accelerated along the western boundary of the plume while Agulhas water is entrained into the plume round Cape Point, mixing rapidly with upwelled South Atlantic Central water within the plume.

Moored current meters

Two Aanderaa RCM 4 current meters with temperature recorders were moored in water 253 m deep at the head of the Cape Point Canyon 13 nautical miles west of Olifantsbos (Fig. 1) during November 1978, at depths of 203 and 233 m. This was the first deep mooring on the west coast and thus knowledge of currents in the region was nonexistent at the time. The short time series was surprisingly informative.

The trends of temperature and flow structures shown by the two instruments were remarkably similar. The lower meter showed a slightly smoother flow of water about 0.5°C colder than that passing the upper meter moving with a slower component of westward velocity in the ratio 25:33. The lie of the canyon is 350° and presumably the flow near the lower meter set 20 m above the floor of the canyon was channeled more to the north.

The most striking feature was the sharp reversal of current from northward to southward on 10 and 18 November and from southward to northward on 14 and 21 November (Fig. 11). Synoptic weather charts show the formation of a southward-moving coastal low pressure cell at Luderitz (27°S) on 8 November. A marked weakening and separation of the South Atlantic anticyclone was observed on 9 November. Marine winds responding to changes in the pressure gradient would have swung from 170 to 120 and then to 350° as the coastal low passed east of Cape Town on 10 and 11 November. The high pressure cell strengthened again with strong southeasterly winds being re-established on 14 November. The process was repeated between 16 and 20 November but with a weaker pressure perturbation.

The abrupt changes in current direction correlate well with wind measured at the coastal weather station at Olifantsbos 13 nautical miles to the east (Fig. 1). The response time of a few hours shows that the flow of water onto the shelf at this point is controlled by horizontal pressure gradients established on a scale several times larger than the length scale of the plume by variation in sea level and the density gradient along the southern edge of the shelf. The

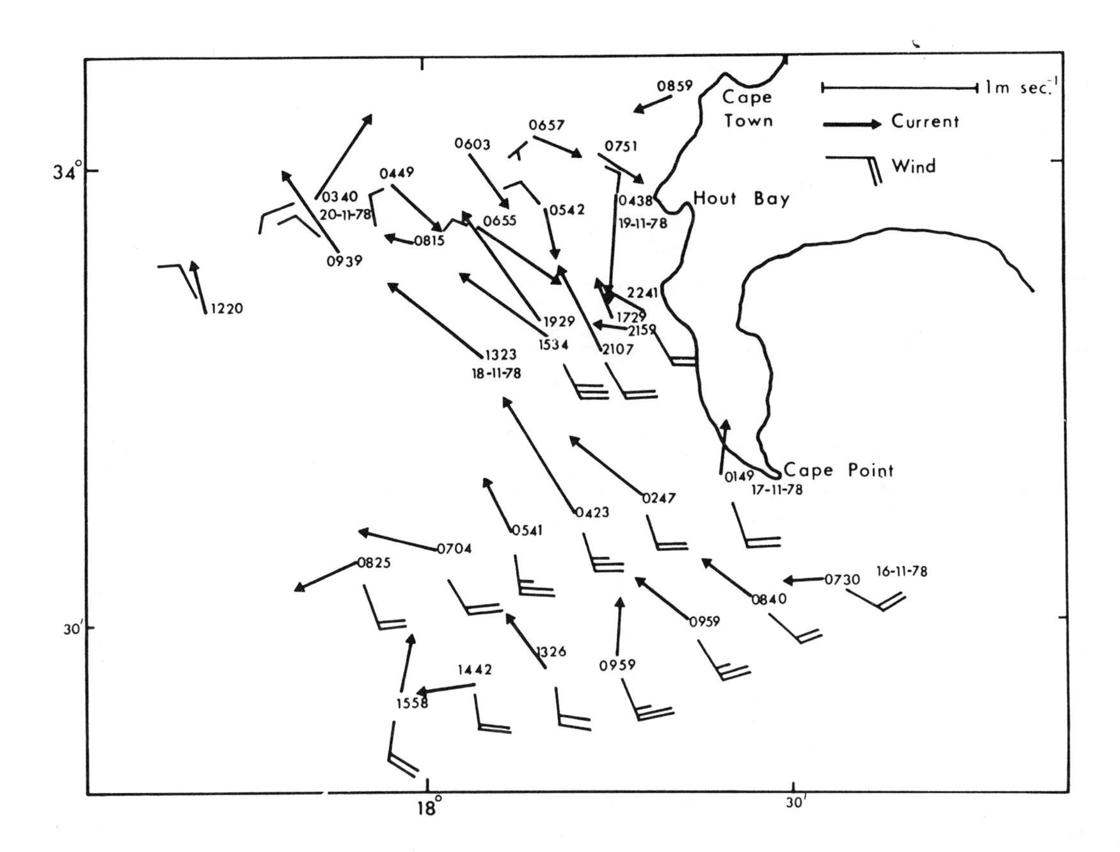

Fig. 8. Currents at 14 m from ship-borne current meter with drift subtracted and wind field. First cruise 16 - 20 November, 1978.

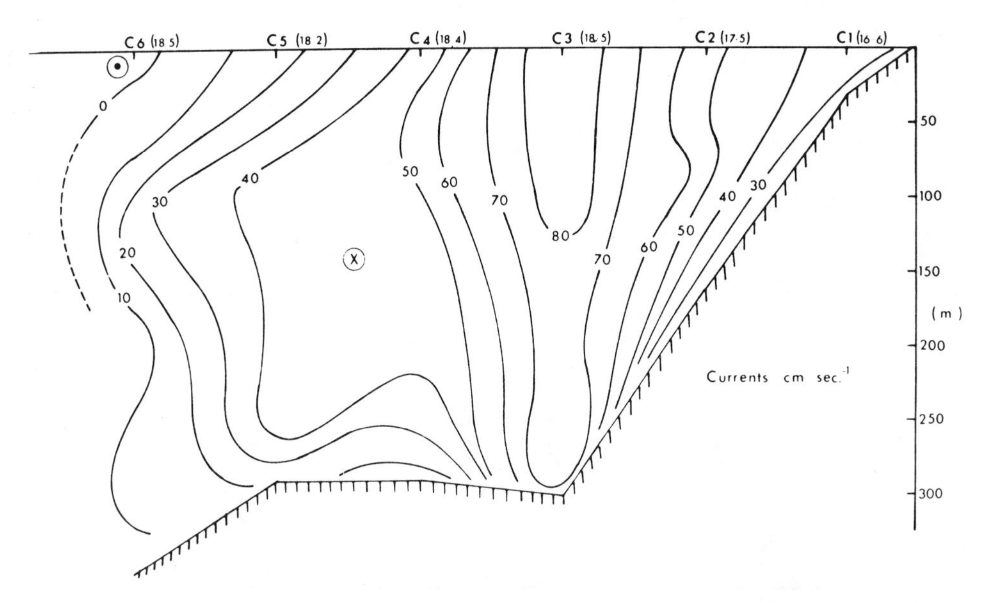

Fig. 9. Normal component of current from ship-borne current meter on line C and sea-surface temperatures.

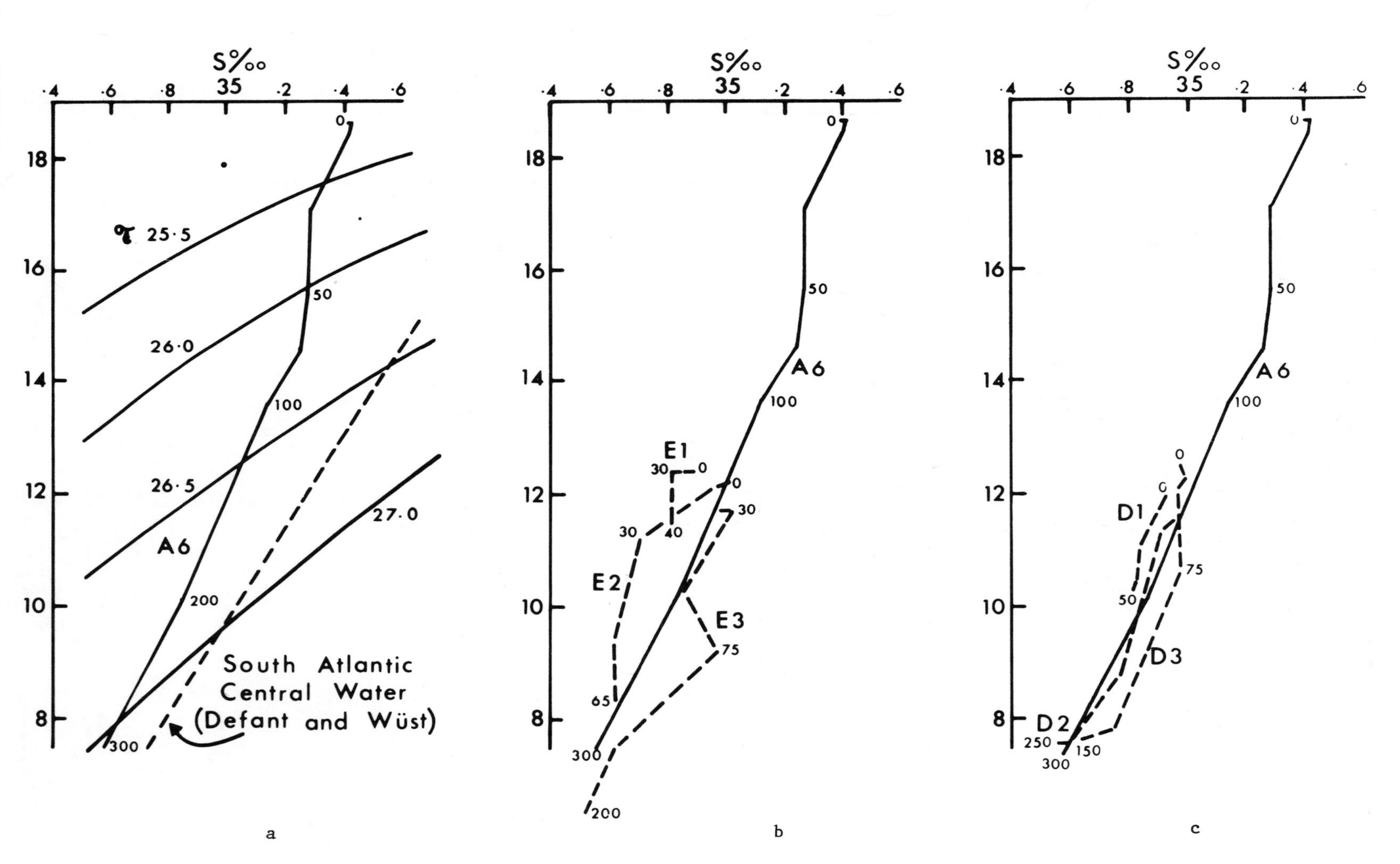

Fig. 10. a) T-S characteristic of water entering the Cape Peninsula upwelling zone from the south at Sta. A6 with envelope of zonally transformed South Atlantic Central water T-S characteristics of Defant and Wüst. b) Comparison of T-S characteristics of Stas E1, 2, and 3 and A6, and c) Comparison with D1, 2, and 3.

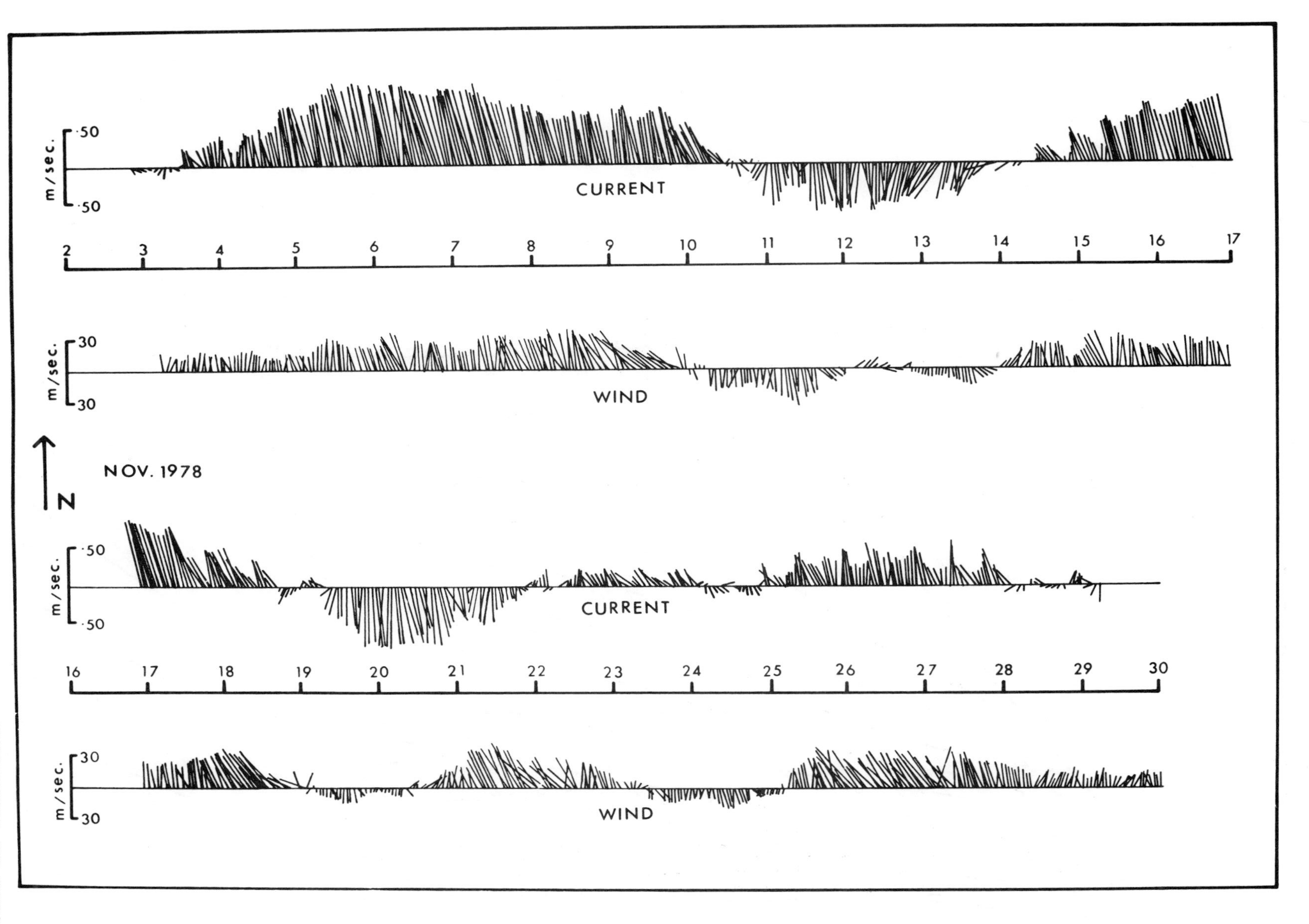

Fig. 11. Hourly filtered stick vectors of the current at 233 m from the moored Aanderaa array and winds at Olifantsbos November 1978. Winds are displayed according to direction of origin and currents according to destination.

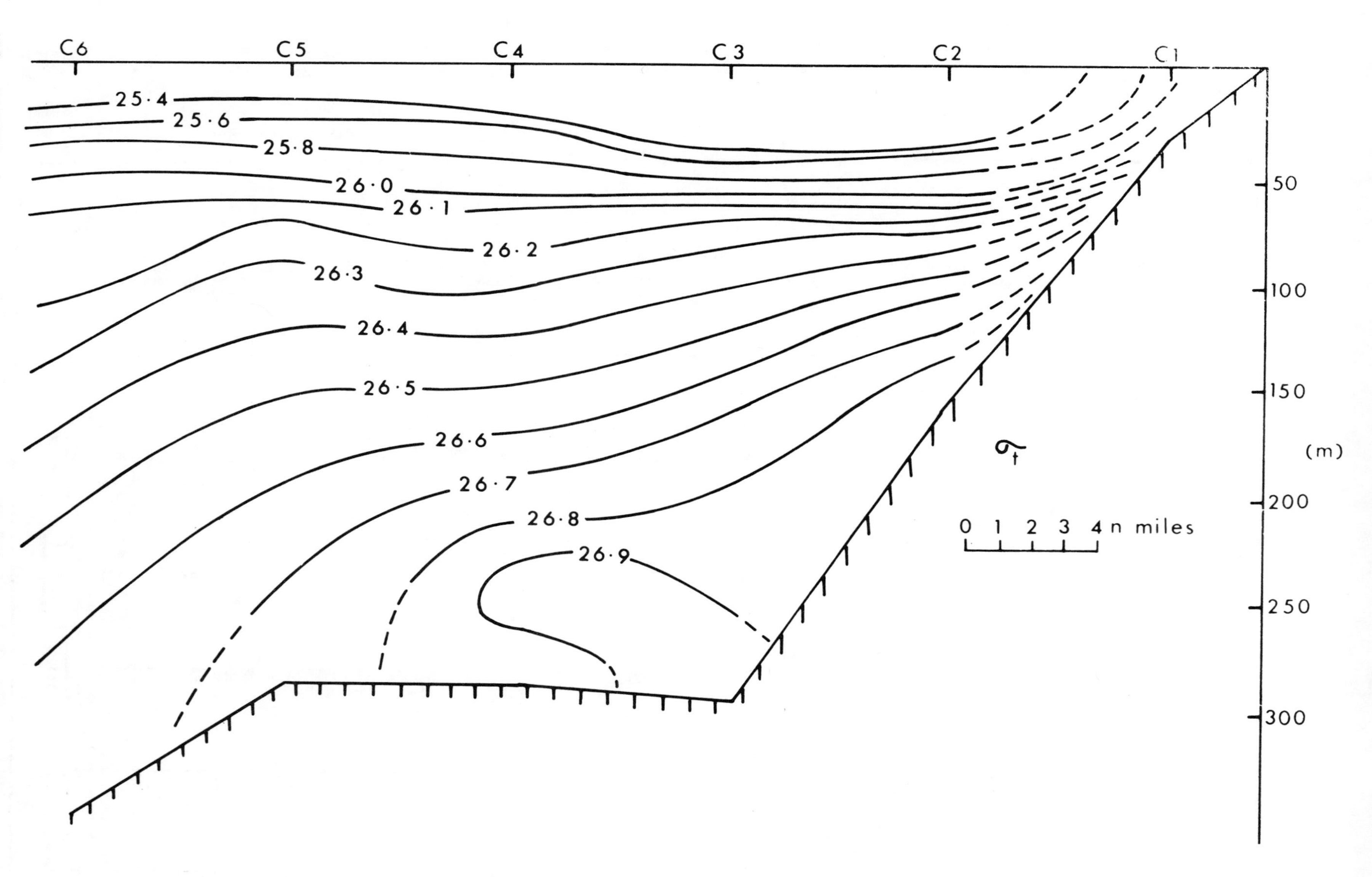

Fig. 12. Isopycnals on line C Cruise 1, November 1978.

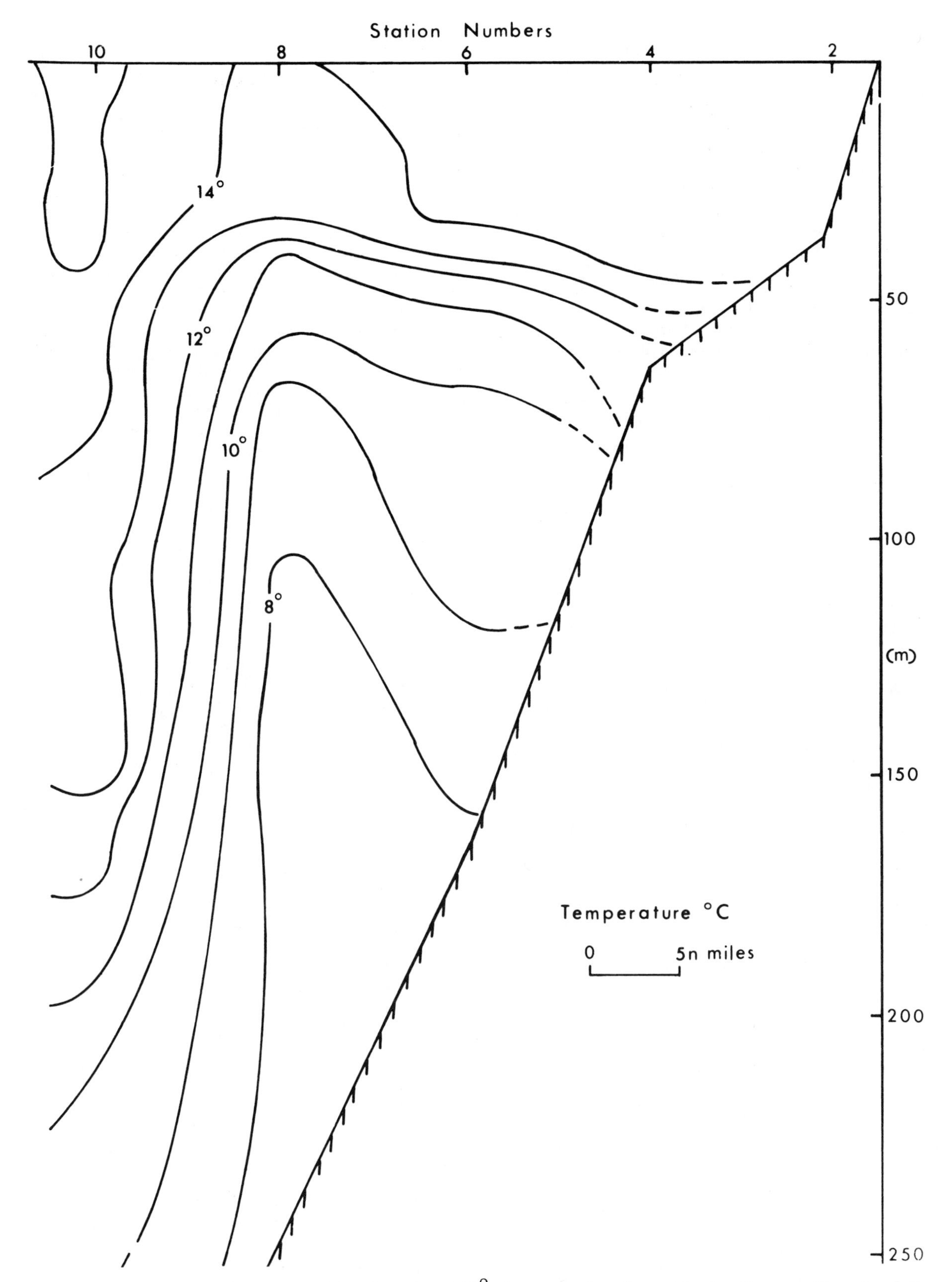

Fig. 13. Temperature profile on a line bearing 234° from a point 7 nautical miles northwest of Cape Town in August 1977 during surface downwelling conditions.

effect of the wind cannot reach a depth of 230 m on such a short time scale. The argument is strengthened by the wind reversal between 23 and 25 November associated with a weak low-pressure perturbation moving south of the continent. On the synoptic scale the pressure system remained high over the South Atlantic with southeasterly winds prescribed along most of the west coast. Although comparatively strong northeasterly winds were recorded near the current meters, only a weak reversal of water flow occurred.

Whether the reversal of flow occurs only near the canyon or along the whole shelf is unknown, but even if it were confined to the canyon a significant volume of water would have to be involved in a reversal to affect the deep current structure at the southern end of the plume.

Thermohaline Features

The temperature and salinity characteristics off the Cape Peninsula are such that density is strongly temperature dependent. Thus, isotherms and isopycnals show a similarity in both the vertical and horizontal planes. Salinity sections show localized patches of anomalously high or low values indicative of the mixing of South Atlantic Central water of maximum recorded salinity and temperature, 34.40 at 6.5°C, and surface Agulhas water of maximum values 35.47 at 18.5°C. This was noted particularly north of Slangkop when the wind relaxed and surface upwelling ceased.

Isopycnals show an upward tilt towards the coast characteristic of upwelling zones. Typical is line C (Fig. 12). Below 50 m the feature is perennial, as exemplified on a line west of Cape Town in August 1977 (Fig. 13), and is consistent with the north-northwestward geostrophic flow. Downwelling during northwesterly winds appears to be confined to the upper 50 m. During both cruises of the *Meiring Naudé* in November 1978 there was evidence of a well-mixed tongue of water at 7°C and 34.5 salinity on the shelf originating in the Cape Point Canyon at line A and extending to line F at successively shallowing depths. Horizontal sections show water of 9°C moving over the shallow depression at the head of the canyon and spreading out in a broad patch between Slangkop and Duikerpunt.

Thus it appears that dynamically induced upwelling is an important feature of the area south of Duikerpunt. The propulsion of water from a region of geostrophic flow south of Cape Point onto the wedge-shaped southern end of the shelf would result in a tilting and compression of isotherms. Accompanying this must be a westward displacement of water and acceleration along the shelf edge. Clearly, the ingredients of the plume are present in the sub-surface layers without any wind inducement. If, in fact, the tongue of cold water reaching the top of the shelf is mainly funneled up the canyon, then this topographic feature will have a significant influence on the whole Cape Peninsula system.

North of Slangkop the wind is the dominant feature in the formation of the near-surface plume. This is the region of maximum wind stress. Vertical sections show a well-mixed layer extending to 50 m indicating that cold water is brought to the surface directly by vertical mixing. Southeasterly winds are predominantly offshore along the coast north of Duikerpunt and upwelling from 30 m is likely to occur as a consequence of a strong surface divergence. The time and spatial scales in the area are too small for an Ekman-Sverdrup interior to contribute extensively to the upwelling.

Capes along the coastline, especially at Duikerpunt, appeared to enhance upwelling according to a process discussed by Arthur (1965) and Yoshida (1967). The change in relative vorticity along a streamline with appreciable curvature near a cape will add to the planetary vorticity equatorward, thereby assisting wind-induced upwelling. These features were apparent on horizontal temperature sections.

References

Arthur, R.S., On the circulation of vertical motion in the eastern boundary currents from determination of horizontal motion, *Journal of Geophysical, Research, 70*, 2799-2803, 1965.

Andrews, W.R.H., Selected aspects of upwelling research in the southern Benguela Current, *Tethys, 6*, 327-340, 1974.

Andrews, W.R.H., and D.L. Cram, Combined aerial and shipboard upwelling survey in the Benguela Current, *Nature, 244*, 902-904, 1969.

Bang, N.D., Characteristics of an intense ocean frontal system in the upwell regime west of Cape Town, *Tellus, 25*, 256-265, 1973.

Bang, N.D., The Southern Benguela Current region in February 1966. Part II, Bathythermograph and air-sea interactions, *Deep-Sea Research, 18*, 209-225, 1971.

Bang, N.D., and W.R.H. Andrews, Direct current measurements of shelf edge frontal jet in the southern Benguela system, *Journal of Marine Research, 32*, 405-417, 1974.

Buys, M.E.L., Temperature variations in the upper 50 m in the St Helena Bay area September 1950 to August 1954, *Investigational Report of the Division of Sea Fisheries, South Africa, 27*, 114 pp., 1957.

Clowes, A.J., An introduction to the hydrology of South African waters, *Investigational Report, Marine Biological Survey, Division of Sea Fisheries, South Africa, 12*, 42 pp., 1950.

De Decker, A.H.B., Notes on an oxygen-depleted subsurface current off the west coast of South Africa, *Investigational Report, Division of Sea Fisheries, South Africa, 84*, 24 pp., 1970.

Defant, A., and G. Wüst, Die Mischung von Wasserkorpen im System S=f(t), *Rapports et Procès-verbaux des Réunions, Conseil permanent international pour l'exploration de la Mer, 67*, 40-48, 1930.

Hart, T.J., and R.L. Currie, The Benguela Current, *Discovery Reports, 31,* 123-289, 1960.

Legeckis, R., A survey of Worldwide Sea Surface Temperature fronts detected by environmental satellites, *Journal of Geophysical Research, 83,* C9, 4501-4522, 1978.

Schell, I.L., On the relation between the winds off South West Africa and the Benguela Current and the Agulhas penetration in the south Atlantic, *Deutsche hydrographische Zeitschrift, 21,* 109-117, 1968.

Shannon, L.V., Hydrology of the south and west coasts of South Africa, *Investigational Report, Division of Sea Fisheries, South Africa, 58,* 22 pp., 1966.

Shannon, L.V., and M. Van Rijswijck, Physical oceanography of the Walvis Ridge region, *Investigational Report, Division of Sea Fisheries, South Africa, 70,* 19 pp., 1969.

Stander, G.H., General hydrography of the waters off Walvis Bay, South West Africa, *Investigational Report, Marine Research Laboratories, South West Africa, 12,* 43 pp., 1964.

Visser, G.A., Analysis of Atlantic waters off the coast of Southern Africa, *Investigational Report, Division of Sea Fisheries, South Africa, 75,* 26 pp., 1969.

Yoshida, K., Circulation in eastern tropical oceans with special references to upwelling and undercurrents, *Japanese Journal of Geophysics, 4,* 1-75, 1967.

UPWELLING IN THE GULF OF LIONS

Claude Millot

Laboratoire d'Océanographie Physique du Muséum
National d'Histoire Naturelle, BP 2, 83501 LaSeyne, France

Lucien Wald

Centre de Télédétection et d'Analyse des Milieux Naturels,
Ecole Nationale Superieure des Mines de Paris,
Sophia-Antipolis, 06560 Valbonne, France

Abstract. The hydrological and meteorological characteristics of the Gulf of Lions are such that upwelling occurs with no bias due to tides or strong longshore circulation. The sky is generally cloud-free, an uncommon feature in an upwelling area that allows extensive use of satellite infrared data. The observations are adequate to compute mean maps of the sea-surface temperature during upwelling events. Undoubtedly upwelling is much more intense along straight coastal segments 10 to 20 km long than near capes and small bays; upwelling locations are mainly related to the coastline shape. The imagery also suggests that the surface circulation varies markedly in space and time; this has been verified by *in situ* measurements, as has the existence of wind-induced eddies in the surface layer. Satellite images from the largest upwelling regions (northwest Africa, Oregon, Peru) show similar spatial variability of the sea-surface temperature, but because of the rectilinear coastline, and except for some specific areas characterized by highly variable wind stress or bottom topography, plumes and edies move slightly alongshore and are not characteristic of a mooring point. The geomorphology of the Gulf of Lions is such that eddies have a fixed location and mean currents are significant. Upwelling studies in such areas might be simplified by the reduction of the perturbations associated with the general circulation and geostrophic turbulence.

Introduction

Arrays of *in situ* recorders have recently been set along sections perpendicular to the coast in some of the largest upwelling areas (northwest Africa, Oregon, Peru); many of the papers presented during the 1980 International Symposium on Coastal Upwelling dealt mainly with the circulation in a vertical plane, although it is now generally accepted that upwelling is a three-dimension phenomenon. Some of the papers discussed observations from minor upwelling areas (northeast America, southwest Africa, etc.) where there are generally significant relationships between upwelling location and bathymetry. A fine-scale study of upwelling in the Gulf of Lions (Millot, 1979) shows that its occurrence is not related to the very smooth bathymetry (Fig. 1) but only to the coastline configuration. The purpose of this paper is to demonstrate that, at least in the Gulf of Lions, the trend of the coastline is the main determinant of the sea-surface temperature and the surface-current distribution in the coastal zone.

The first concepts of the structure of upwelling in the Gulf of Lions were suggested by infrared satellite imagery collected from 1975 to 1978. A four-mooring experiment was conducted near an upwelling center for a 3-month period during the 1978 summer. The sub-surface moorings supported two Aanderaa current meters at depths of 10 m beneath the surface and 5 m above the bottom and a thermistor chain between them; the four moorings were in depths of 43 (A), 98 (B), and 74 m (C and E) (Fig. 1). The mooring at Sta. C was destroyed by a trawler two days after its setting, and bottom current meters at Stas A and E failed after 2 months and 2 weeks, respectively.

An image from the NOAA 5 radiometer is formed by a combination of the satellite's motion and the scanning movement of the radiometer from horizon to horizon in a direction perpendicular to the track. The spatial resolution at the nadir is ca. 1 km^2 when the radiometric resolution is 0.5^oC (noise is ca. 1.5^oC). Data provided by the Centre de Météorologie Spatiale in Lannion are processed by the Centre de Télédétection et d'Analyse des Milieux Naturels in Sophia Antipolis (both are French centers). The

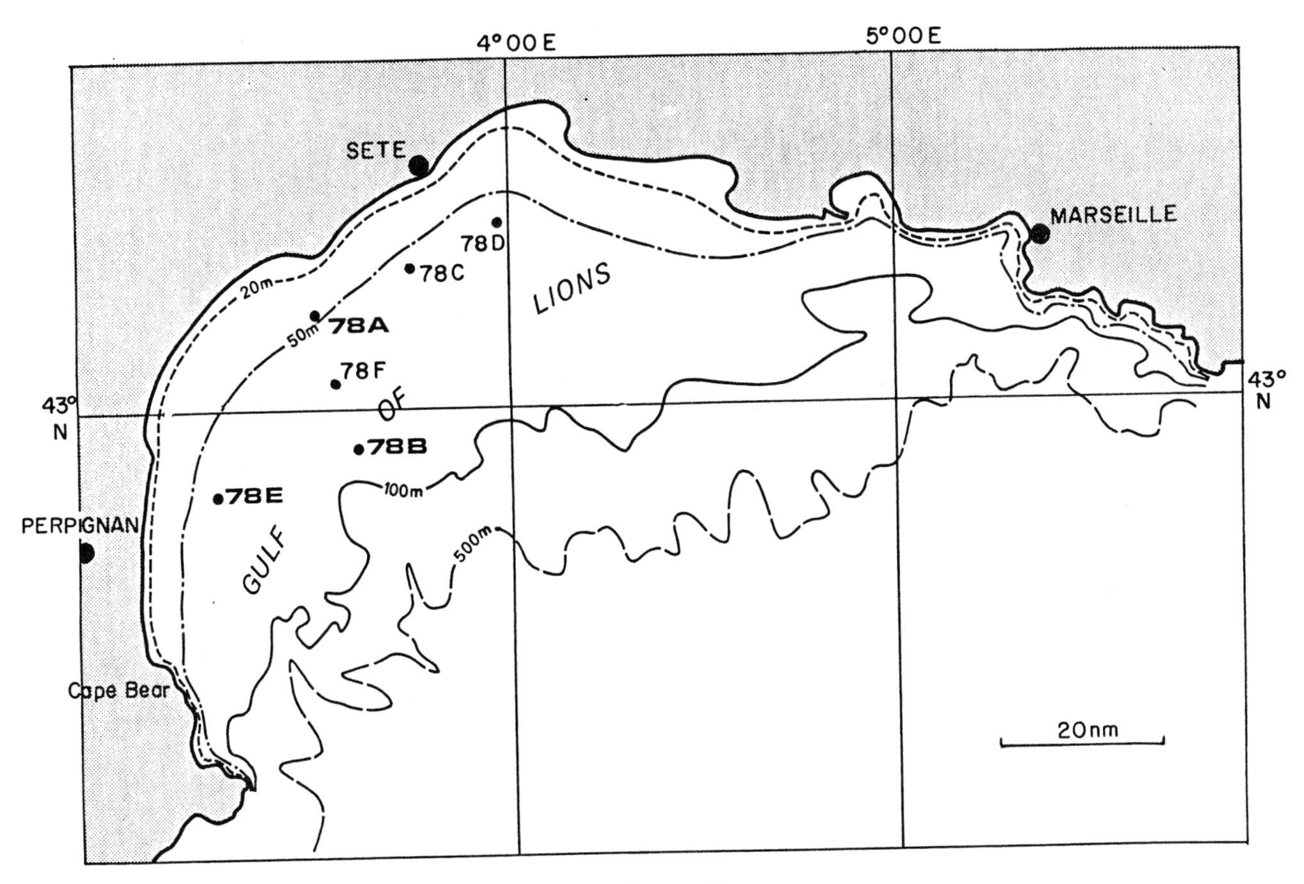

Fig. 1. The study area.

data processing includes the computation of temperature, the smoothing of images, the drawing of thermographic maps, and the geometric correction due to the earth's spin and the curvature of the surface. Noisy data are smoothed with a bi-dimensional filter of the moving average type, which is conditioned by the importance of local gradients with respect to noise (Albuisson, 1976).

The relative sea-surface temperatures are determined with an accuracy of 0.5°C for a 3-km resolution (Albuisson, Pontier, and Wald, 1979) so the data are reliable in the study of horizontal thermal gradients and their spatial distribution. To make the images superimposable ($\pm$ 1 km^2) on each other, they are geometrically corrected from landmarks (Albuisson, 1978). Images are two-dimensional arrays, and it is possible to compute a time-averaged value for each geographic point. Problems such as the elimination of small clouds, the calibration of the radiometer, and the atmospheric absorption have been solved (Wald, 1980), and it is shown that mean thermal gradients are defined with satisfactory accuracy.

The Observations

Details of the meteorological and hydrological regimes in the Gulf of Lions, the observations, and specific features of the upwelling events are given in Millot (1979); only the main characteristics will be discussed briefly.

The Gulf of Lions (Fig. 1) is in the northwestern Mediterranean Sea. It has a roughly semi-circular shape with a radius of about 100 km; the bathymetry of the continental shelf is quite smooth, as is the topography of the coast. The gulf is the most windy region of the entire Mediterranean basin; associated with cyclogenesis over the northern part of Italy, strong (daily speeds range from 10 to 20 m/s) and transient northwesterlies frequently blow for periods of 1 to 10 days all year long. The main geographic feature is a set of three mountain masses separated by two valleys, which make the wind stable in direction. During summer, the thermocline in the coastal zone lies at about mid-depth and separates a bottom layer (minimum temperature about 13.5°C) from an upper layer with a mean temperature of about 20°C. Surface temperatures reach maxima around 25°C.

Satellite data. The temperatures (from infrared sensing) in Fig. 2 follow by about 1 day the onset of a northwest pulse. On a larger scale, it is clear that upwelling is occurring along coastlines where an observer with his back to the wind has the sea mainly on his right; winds per-

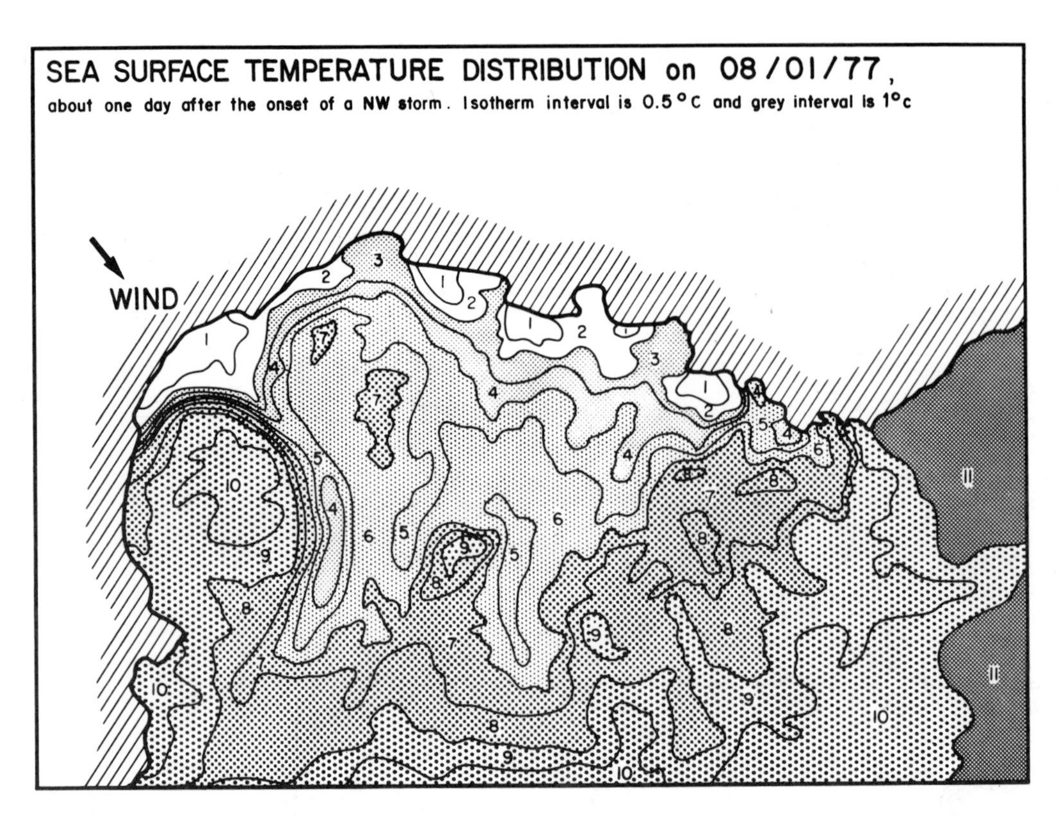

Fig. 2. The infrared satellite view taken on the 1st of August 1977 at about 09:00 UT. Note the discontinuity of upwelling and the surface circulation suggested by the structure of the isotherms.

pendicular to the coast also create upwelling, as in the vicinity of Sta. A. As a rule, no sea-surface temperature gradient appears in the southwestern part of the gulf (downwelling). On a smaller scale, upwelling is highly discontinuous, and actual source-points of cool water are clearly distinguishable. If it is considered that the coastline forms a series of curves (segments 10 to 20 km long are separated by capes and small bays), it appears that upwelled water spreads out along straight segments more than near smaller features; this statement implys that the upwelling centers are on the leeward side of some capes and on the windward side of others. The largest sea-surface temperature gradients induced by the wind in the gulf occur near shore in the vicinity of capes and bays; they are of the order of 1°C/km and are directed alongshore. Plumes of cool water drift seaward from some of the upwelling centers. Among the 100 views of the area we have, 15 are cloud-free and show all the centers; the time-lag between the onset of the wind and the passage of the satellite ranges from 1 to 3 days, but this does not seem to be important with respect to the sea-surface temperature structure. To produce a statistical picture of upwelling in the gulf, we have used the summation program presented in the introduction. The results of the summation are shown in Fig. 3.

The strongest temperature gradients in the various views are not in exactly the same place, and they are smoothed by the summation. Nevertheless, the main features described above appear clearly in Fig. 3: distinct upwelling centers (i.e., low temperature centers) are along straight coastal segments and large temperature gradients are observed in the vicinity of capes and bays; plumes of cool water that have drifted seaward from the largest upwelling centers also appear to have a roughly fixed location.

In situ *data*. Figs 1 and 3 show that mooring points were located near an upwelling center (A), roughly in the axis of a plume of cool water that drifted seaward (B), and on the edge of the upwelling zone (C and E). The array was intended to define with a sufficient accuracy features suggested, for instance, by Fig. 2, such as an anticyclonic eddy in the southwestern part of the gulf, and complex wind-induced motions on the edges of the upwelling center. Wind measurements were obtained in Sète, Perpignan, and at Cape Bear (Fig. 1).

Fig. 4 shows daily winds, currents, and some temperature records during a pulse from September 5 to 13. We choose to describe this event because it shows both classical and uncommon features. The onset of the wind was between 00:00 and 06:00 h on September 5. Its direction was roughly the same at the two southern stations, but the wind blew at Sète from NW to N (a classical feature). The wind stopped on September 14

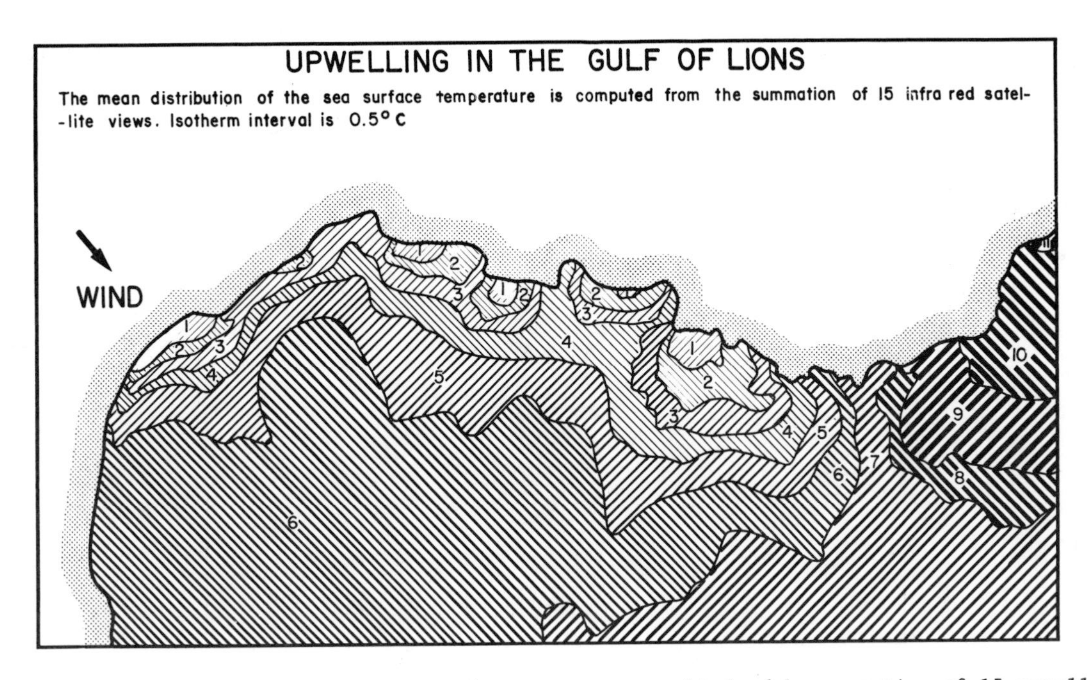

Fig. 3. The mean distribution of the sea-surface temperature obtained by summation of 15 satellite views. The location of upwelling along straight coastal segments is actual and offshore jets of cool water also have a roughly stable location.

and then blew again for 1 day. Wind speeds over the studied area during the pulse ranged from 10 to 15 m/s.

The temperature measured 10 m beneath the surface at point A was decreasing after 1 day of wind, and the whole column was only slightly stratified after about 3 days. The 30-m deep record is disturbed only by inertial internal waves (Millot and Crepon, 1981). At B, the deeper record reveals that the temperature was increasing quickly on September 7, when a record from a shallower depth shows a smooth decrease up to the 8th; both records then increase and nearly reach the values measured at the 10-m depth before the pulse. Such features show that temperature variations due to advection and oscillations of the thermocline are larger than those due to mixing. Downwelling is evident at E and the shallower temperature is roughly constant: this supports the idea that the thermocline at Sta. E was deeper than at Sta. B. The temperature records and wind data are consistent with current measurements.

All the current speeds were small before the onset of the wind and on the 5th when the wind was moderate. As the northwest wind increases, the current pattern becomes consistent with the classical scheme (see below), i.e., surface currents are to the southeast at points A and B when the surface current at E is large and northward (opposed to the theoretical Ekman transport); currents in the whole bottom layer are northeastward. So upwelling was increasing near A and the cool water upwelled to the surface drifted roughly from A to B in the direction of the wind. Warm water moving northward at E turns to the right and joins the offshore current of cool water. From the 8th, the wind blew from the north-northeast quadrant in a region north of the study area, and the surface current at A rotated to the southwest from the 10th. The current was then opposed to the current at E, which deviated offshore. Near B the current was not very different, but the water characteristics had changed. The offshore jet of cool water passed across B in the afternoon of the 8th, and then was located between B and E. Waters moving seaward at B and E from the 10th were warm, i.e., waters of the surface layer. Speeds were significant and corresponded to displacements larger than 30 km a day at B, about 40 km offshore. So, during the second part of the wind event, a cyclonic wind-driven circulation occurred in the northern part of the study area. But the most surprising feature described above is the occurrence of an upwind surface current at E when the wind field was roughly homogeneous, that is when the wind was blowing from the northwest over this part of the gulf.

To demonstrate that the above wind-induced feature is common we have computed mean currents during two sets of typical meteorological conditions (northwesterlies either blow or not, whatever the wind stress may be). Preliminary results were obtained by considering only the strongest winds (28 days out of 85). Upon considering all days and parts of days when strong or moderate northwesterlies were blowing, the statistics for

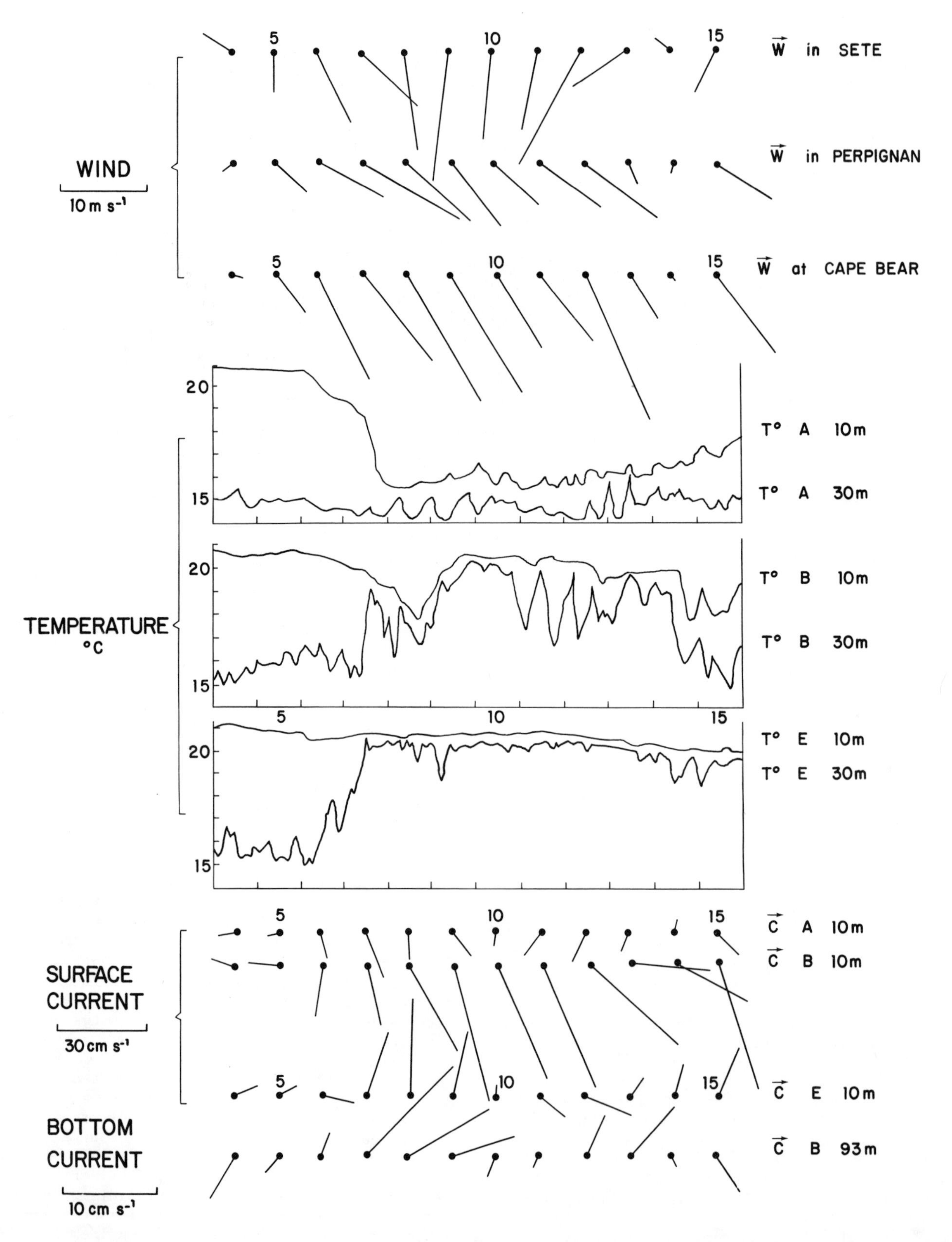

Fig. 4. The daily winds and currents and some temperature records from September 4 to 15. Vectors are plotted in such a way that the vertical axis is north-south.

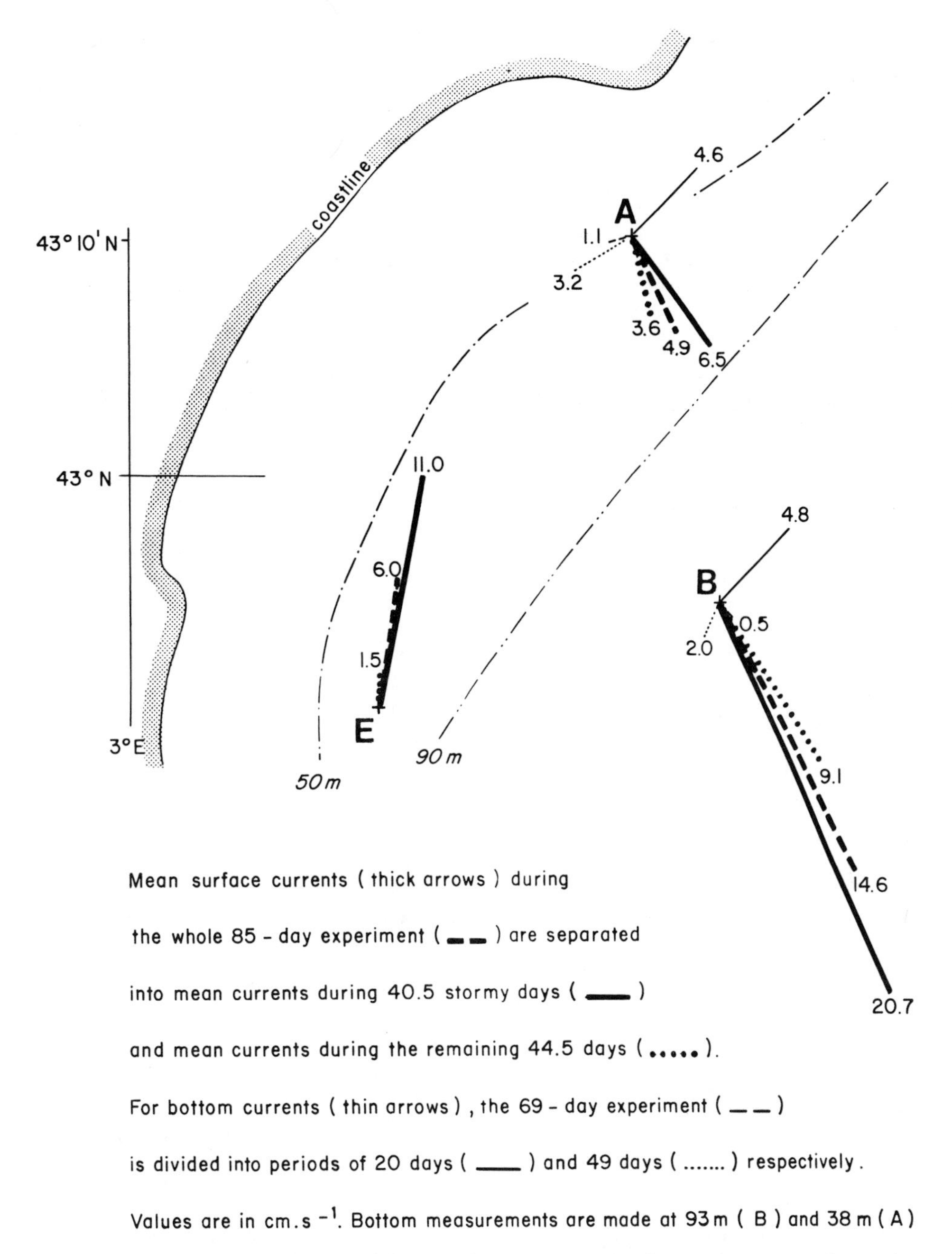

Fig. 5. Dependence of mean surface and bottom currents upon the occurrence of northwest storms.

40.5 days out of 85 are more convincing than the previous ones (Fig. 5) mainly at E. It should be emphasized that the mean circulation observed out of the stormy periods is also influenced by the wind-induced circulation (mainly at A and B), because such eddies do not disappear as soon as the wind drops.

Measurements were obtained at A and B in the bottom layer (Fig. 5) (at depths of 38 and 69 m, respectively) for a 69-day period. As expected, the layer reacts only to the strongest winds and with about 1-day delay, so we have considered 20 days of wind. The wind-induced mean currents in the bottom layer appear to be strong and opposed to the currents observed during quiet periods, which are in agreement with the general circulation in the region (Millot, 1979). The mean temperatures at 10 m during all the experiments were 17.44°C at A, 19.28°C at B, and 19.48° at E. Thus, upwelling was an influence at A, but water moving northward at E and southeastward at B had roughly the same temperature as the undisturbed surface layer.

Discussion

A first question is why does upwelling, at least in the Gulf of Lions, spread out along straight coastal segments more than in the vicinity of smaller scale features? A second question is how can the wind-induced current field be so heterogeneous in such an area?

Numerical models (Hua and Thomasset, 1980) have been used to test the relationship between upwelling location and the coastline shape; even if the geographic condition is schematically approximated, results are satisfyingly coherent with observations. However, the models do not show the northward flow at Sta. E. The hypothesis that might be advanced is that when an upwelling-favorable wind blows in the vicinity of a straight coast, all the particles of the surface layer are subjected to the same forcing; without perturbations due to geostrophic turbulence, the result is that the upwelling must spread out. In a small bay, however, the wind may be upwelling-favorable in a small area and downwelling-favorable not far away. The current field is much less homogeneous than in the vicinity of a straight coast; complex circulations probably occur and reduce both up- and downwellings both in small bays and near capes. Let us discuss, for instance, the features occurring in a non-rotating fluid in which a semi-infinite barrier is supposed to schematize a cape. If the wind is blowing in a direction perpendicular to the barrier orientation, it is clear that upwelling (downwelling) will spread out along the leeward (windward) coast. But around the origin of the barrier, neither upwelling nor downwelling is expected, and strong currents will characterize this singular area.

Considering from our data set the structure of the current field induced by the wind in the surface layer, we have to start from the upwelling location; that this location is related to the coastline, to the wind field, or to the bathymetry is not to be taken into account. The fact is that upwelling is there because the offshore current in the area is stronger than in the neighboring areas. In the Gulf of Lions, an offshore drift current is regularly observed at B, generally at A, and infrequently at E. The temperatures in Fig. 2 show an anticyclonic eddy, which is consistent with the northward wind-induced current at E. How such an eddy is created by the wind is not easily explained, and we must consider parameters other than the wind stress. First, the special structure of the gulf: warm waters accumulate in the southwest part and perhaps tend to flow toward areas where the sea level is lower as, for instance, upwelling areas. Second, the density gradients between the upwelling center and the other areas: the offshore drift current is compensated by the pumping of heavy bottom water, or by the advection of surface water, the density of which is equal to or lower than the density in the seaward flow. What is the largest compensating phenomenon? In our opinion, it is probable that the pumping of bottom water occurs first; counter-circulations and eddies may then arise in the surface layer to compensate the offshore drift. The satellite imagery of all the upwelling areas shows that the sea-surface temperature distribution and the associated circulation are highly discontinuous. Along the straight coastlines of the largest upwelling regions, the structures may move alongshore and make difficult a detailed analysis of current records, explaining why only mean currents can readily be discussed.

Acknowledgements. The authors are obliged to Miss E. Nantois for her help in computing the map of the mean distribution of the sea-surface temperature during upwelling. The Centre de Météorologie Spatiale in Lannion generously made available the satellite data. This work was supported by the Centre National pour l'Exploitation des Océans and the Centre National de la Recherche Scientifique.

References

Albuisson, M., Traitement des thermographies des satellites NOAA. Rapport CNEXO 75-1516, 1976.

Albuisson, M., Traitement sur mini-ordinateur des images VHRR pour la thermographie. Rapport CNEXO 77-1671, 1978.

Albuisson, M., L. Pontier, and L. Wald, A comparison between sea-surface temperature measurements from satellite NOAA4 and from airborne radio-meter Aries, *Oceanologica Acta, 2,* 1-4, 1979.

Hua, B.L., and F. Thomasset, Modelling of upwelling in the Gulf of Lions by non-conforming elements, *IDOE International Symposium on Coastal Upwelling, Los Angeles,* Feb. 4-8, Abstract, 1980.

Millot, C., Wind-induced upwellings in the Gulf of Lions, *Oceanologica Acta, 2,* 261-274, 1979.

Millot, C., and M. Crepon, Inertial oscillations in the upwelling area of the Gulf of Lions. Observations and theory. Submitted to *Journal of Physical Oceanography,* in press, 1980.

Wald, L., Utilisation du satellite NOAA 5 à la connaissance de la thermique océanique. Etude de ses variations saisonnières en Mer Ligure et de ses variations spatiales en Méditerranee, *Thèse 3ème Cycle, Université Pierre et Marie Curie,* Paris, 1980.

INTERACTION BETWEEN THE CANARY CURRENT AND THE BOTTOM TOPOGRAPHY

Antonio Cruzado

and

Jordi Salat

Instituto de Investigaciones Pesqueras de Barcelona (C.S.I.C.),
Paseo Nacional s/n, Barcelona, Spain

Abstract. Temperature and salinity distributions show important longshore structures that help define the circulation pattern in the widening shelf area south of Cabo Bojador, off Western Sahara. Two mechanisms, one a two-cell cross-shelf circulation and another based on the bending of the flow lines over the bottom contours, interact to fertilize the region. The two-cell circulation acts in the northern sections under the pressure of a frontal zone created by warm surface water that forces sinking of upwelled water. The circulation changes to a one-cell type in the middle sections and is less affected by the front. In the southern sections, the onshore bottom flow practically vanishes, giving rise to a longshore flow pattern that follows the bottom contours at all depths. In the north a tongue of more saline water protrudes from the frontal zone; the tongue is surrounded by less saline deep water that upwells into the coastal zone and offshore at the surface. Both waters are finally mixed in the southern sections.

Introduction

Coastal upwelling off northwest Africa and especially its biological consequences have been considered to be less intense than that in other eastern boundary current systems such as those off Peru and Baja California. The low nutrient concentrations in the North Atlantic Central Water (NACW) (Margalef, 1978), the lack of a rugged topography (Codispoti and Friederich, 1978), and the high turbulence (Dugdale and MacIsaac, 1980) relative to the other regions have been suggested as direct causes of the weak dynamics and lower standing crops of phytoplankton. However, the northwest African upwelling system, extending over more than 15 degrees of latitude, is far from homogeneous and thus should not be considered as a single entity.

Although it is true that the rather smooth morphology differentiates the shelf off northwest Africa from those off North and South America and off southwest Africa, parts of the northwest African shelf have major morphological structures that may produce important interactions with the Canary Current, such as the Banc d'Arguin Canyon, which strongly affects the upwelling intensity off Mauritania (Mittelstaedt, 1974). Other such features may play a similar role in the area of the JOINT I studies between Cap Blanc and Cabo Barbas, but they have not yet been sufficiently documented. The effect on the intensity of upwelling of coastal morphology, both the shape of the shorelines and the bottom topography, can hardly be estimated because the upwelling occurs over relatively large distances, but the scale of upwelling off northwest Africa, with its relatively smooth bottom topography, should be larger than off Peru.

Enhancement of the biological productivity of the waters over the shelf off Western Sahara appeared to be associated with the alongshore rise of the bottom between Cabo Bojador and Punta Durnford as a single synoptic-scale upwelling entity (Cruzado, 1974a, 1974b; Velasquez and Cruzado, 1974). Alongshore bottom slope was considered as a cause of deep water uplifting in this area, where the B.O. *Cornide de Saavedra* conducted various cruises at several times of the year between 1971 and 1975. The slope is similar to that often observed in the lee of capes and sills or islands. Whenever there are large displacements of water, the relative importance of this 'sill' mechanism in causing fertilization is comparable to that of the wind stress (Cruzado, 1975). Evidence that this 'sill' mechanism was the main cause of fertilization along the Western Sahara shores, at least as far south as Cabo Leven, was obtained more recently when water with salinity below 36×10^{-3} was observed to rise near the coast in spring. As the water flowed equatorwards south of Cabo Bojador, the tongue of water was totally unrelated to the main body of deep water that lay off the shelf at depths of

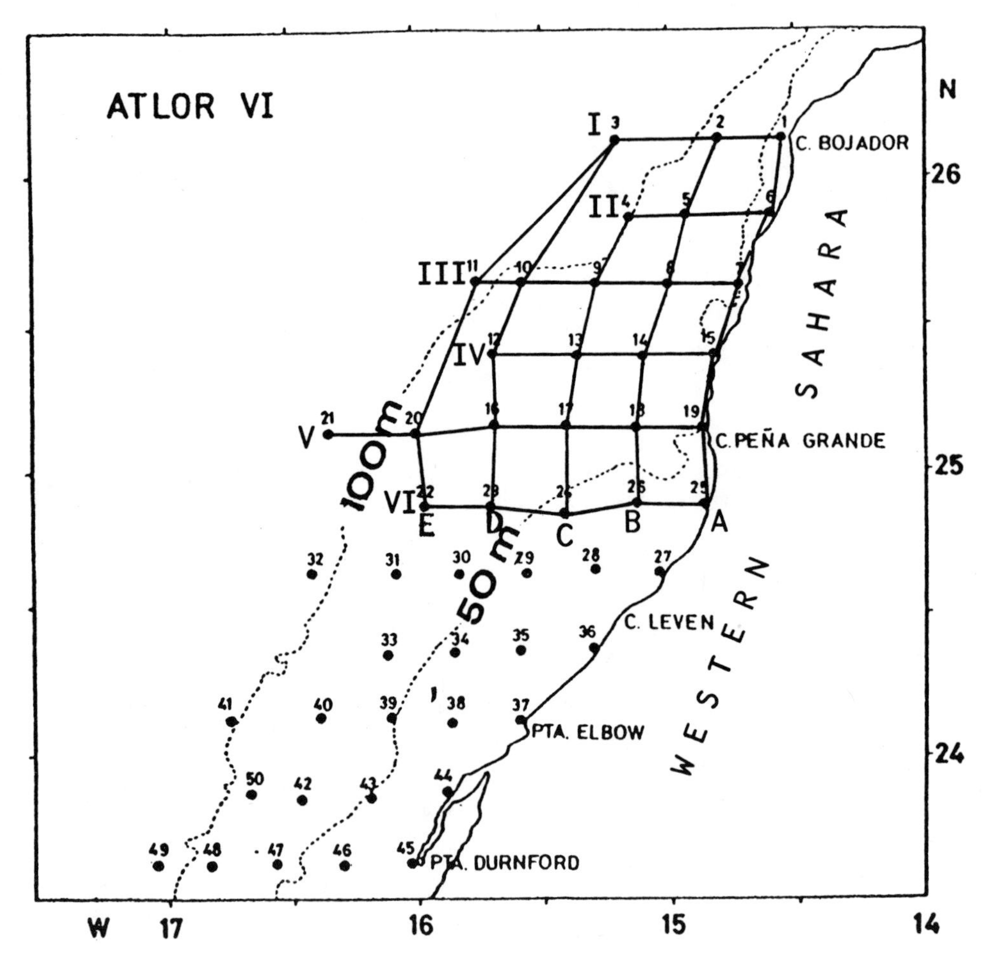

Fig. 1. Location of hydrographic stations during cruise ATLOR VI off Western Sahara. Dashed lines indicate bottom contours. Solid lines indicate longshore sections (A to E) and cross-shelf sections (I to VI).

about 300 m, except north of Cabo Bojador where the shelf is very narrow (Cruzado, 1979). The conclusion is further supported by evidence presented in this paper of the existence of longshore structures in the water. Interaction between the Canary Current and the bottom topography, at the synoptic-scale level, is important for the northern part of Western Sahara, but similar mechanisms may be operative in other regions where simple surface Ekman dynamics do not fully explain the distribution of the various chemical and biological characteristics. Coastal upwelling is naturally a three-dimensional phenomenon and cannot be treated in two dimensions.

Methods and Materials

This paper is based on temperature and salinity data at standard depths at stations occupied during the cruise *ATLOR VI* (Fig. 1) carried out on board the Spanish B.O. *Cornide de Saavedra* off Western Sahara in October of 1975 (Manriquez and Rucabado, 1976). Nearly all the casts at shallow stations were to within 10 m of the bottom. Sigma-t values are not included in the present analysis because they showed a pattern similar to that of the temperature field and, as no current measurements were made, water flow is inferred either from the hydrographic observations or from previous observations in the region (Johnson, Barton, Hughes, and Mooers, 1975). Maps of surface distributions of temperature and salinity (Fig. 2) and vertical distributions of these properties in the longshore direction (sections A to E, Fig. 3) and across the shelf (sections I to VI, Fig. 4) were drawn with the aid of an 1130 IBM computer.

Results

The cruise, carried out just after the summer, revealed meteorological and water-mass conditions somewhat different from those observed during spring cruises (Cruzado, 1974b, 1975, 1979).

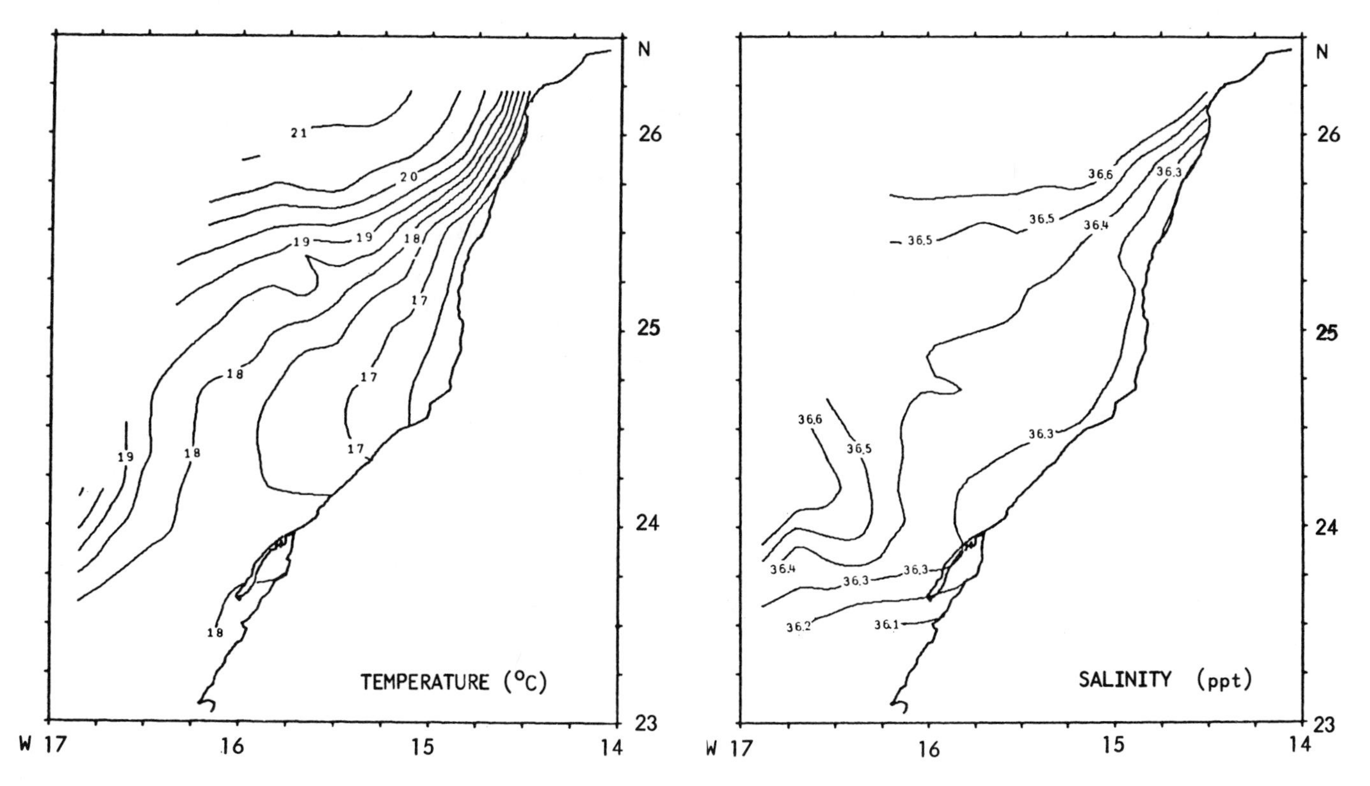

Fig. 2. Surface distribution of temperature (left) and salinity during cruise ATLOR VI off Western Sahara.

Winds changed, after the first week, from the less favorable northwest quadrant to the more favorable northeast quadrant, maintaining speeds of 10 to 25 m/s. The offshore surface layer above 75 to 100 m was occupied by water warmer in relation to its salinity than had been the case in spring, and at the northernmost stations the water was more saline than expected.

The distributions of surface temperature and salinity (Fig. 2) illustrate the structure of the coastal upwelling in the area. Surface isotherms and isohalines define a horizontal front close to the shoreline off Cabo Bojador diverging from the coastline along the bottom contours and bending to a more east-west direction at about 25°30'N. The warm, high-salinity water found close to the shore in the north is located offshore in the central part of the region, although it appears more shoreward again as a wedge at about 24°30'N. The 18°C and the 36.4×10^{-3} isopleths practically coincide, but the 36.3×10^{-3} isopleth intersects the 17.5 and 16.5°C isotherms. This may be because the upwelled water comes from two different sources on either side of the thermohaline front to the northwest.

The alongshore sections (Fig. 3) show a general rise of the isotherms from north to south, following the rise of the bottom. When the bottom becomes shoaler than about 25 or 30 m, the cold water (indicated in the figure by the shaded area) completely fills the water column. The salinity distribution shows a somewhat different pattern, complicated by the presence of the high-salinity water (above 36.4×10^{-3} at all depths at the stations north of 25°30'N. This water extends southwards along section B, as an intermediate tongue, underlain by low-salinity water that extends shorewards over the bottom at Sta. 8 and probably also at Stas 5 and 2, where the deepest bottles were too far above the bottom to sample the bottom layer. The high-salinity tongue is overlain by low-salinity water extending up to the surface and enveloping the high-salinity water at some positions (Stas 8, 14, and 17) and mixing with it.

A similar pattern is shown in all the cross-shelf sections (Fig. 4). The isotherms, except in the surface mixed layer, generally tilt roughly parallel to the bottom. On the other hand, the isohalines frequently intersect the bottom, and the shelf water of salinity less than 36.4×10^{-3} seldom maintains continuity with offshore water of the same salinity. Interleaving of various water parcels with different origin and history renders the picture somewhat more complicated than was generally the case in spring. The clear connection between the deep offshore and shallow inshore waters in section I might still be evident in section II if Sta. 5 had gone closer to the bottom. However, it is completely disrupted in sections IV and V where, at Stas 13 and 16, slightly more saline water was observed.

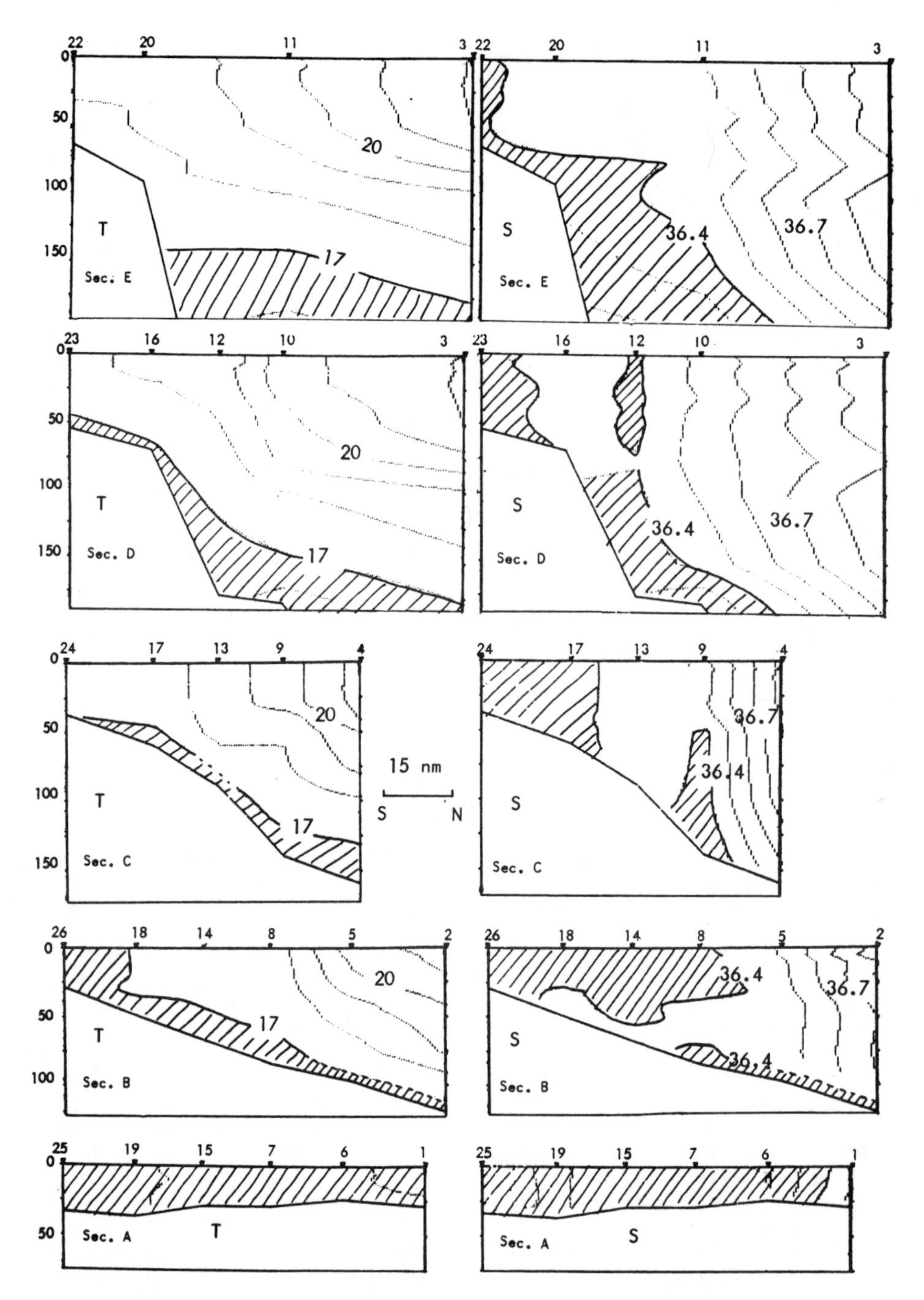

Fig. 3. Vertical distribution of temperature (left) and of salinity (right) in longshore sections A (below) to E (above). Isotherm interval is $1^{o}C$, shaded areas represent temperatures below $17^{o}C$. Isohaline interval is $0.1x10^{-3}$, shaded areas represent salinity lower than $36.4x10^{-3}$.

To explain the large volume of water with salinity greater than $36.4x10^{-3}$ at these stations, one may assume that it is produced by the shorewards flow of upwelled water that is thoroughly mixed with water from the high-salinity tongue of section B. On the other hand, part of the low-salinity 'spike' at Sta. 9 (sections C-III) and the low-salinity balloon at Sta. 12 (sections D-IV) are parcels of previously upwelled water, judging from the values of other variables such as nitrate and chlorophyll *a*. The water has sunk as it moved seawards, remaining beneath the more oceanic surface water of Stas 10, 11, 20, and 21.

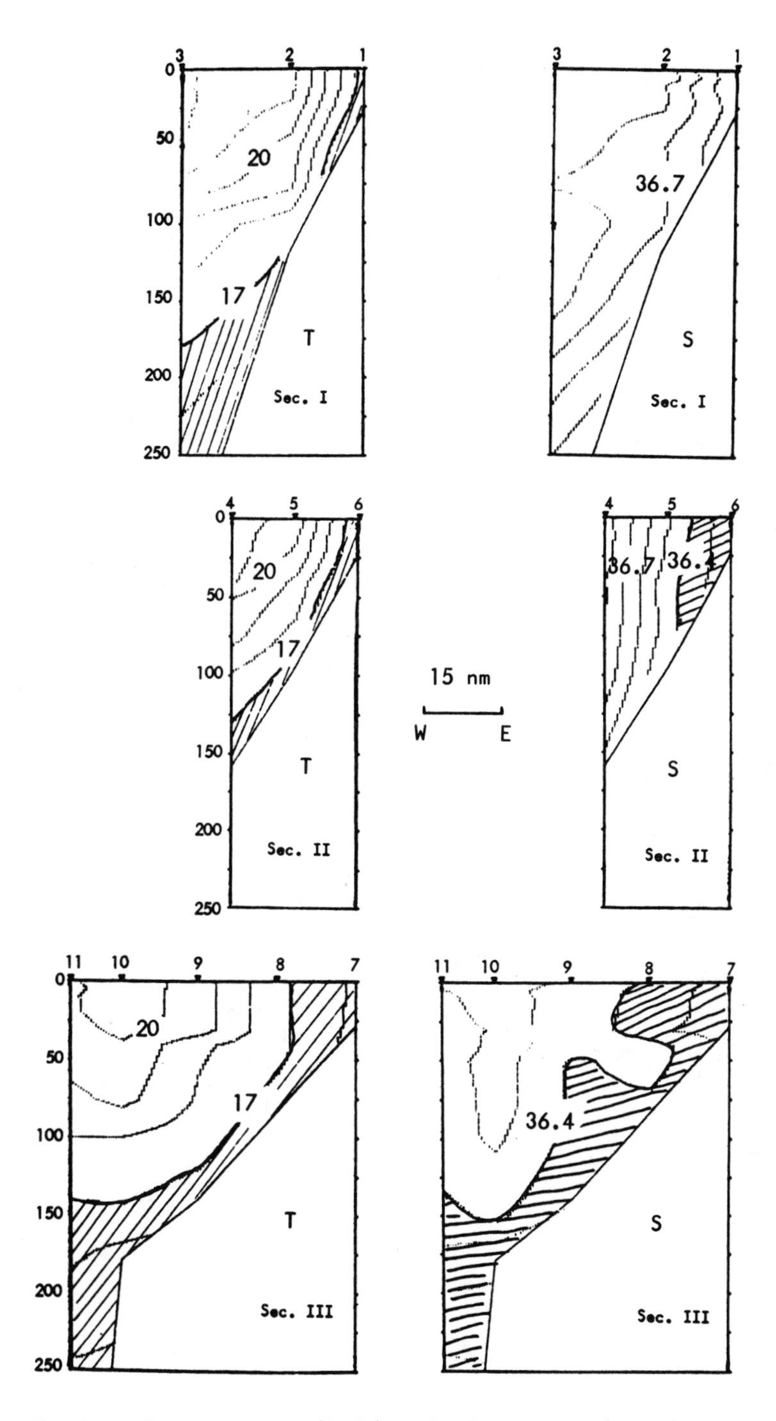

Fig. 4. Vertical distribution of temperature (left) and of salinity (right) in cross-shelf sections I to III (a) and IV to VI (b). Units as in Fig. 3.

Discussion

The present analysis agrees with the analysis of the spring data (Cruzado, 1979). Water from between 150 and 300 m upwells off the northern half of Western Sahara and spreads over a large area along the coast in the wide shelf region south of Cabo Bojador. The process is controlled, as it would be in areas of similar geomorphology, by two mechanisms:

1. The water is pumped as it flows shorewards over the bottom as a compensation flow of a one- or two-cell type of cross-shelf circulation.
2. The water rises as it flows equatorwards

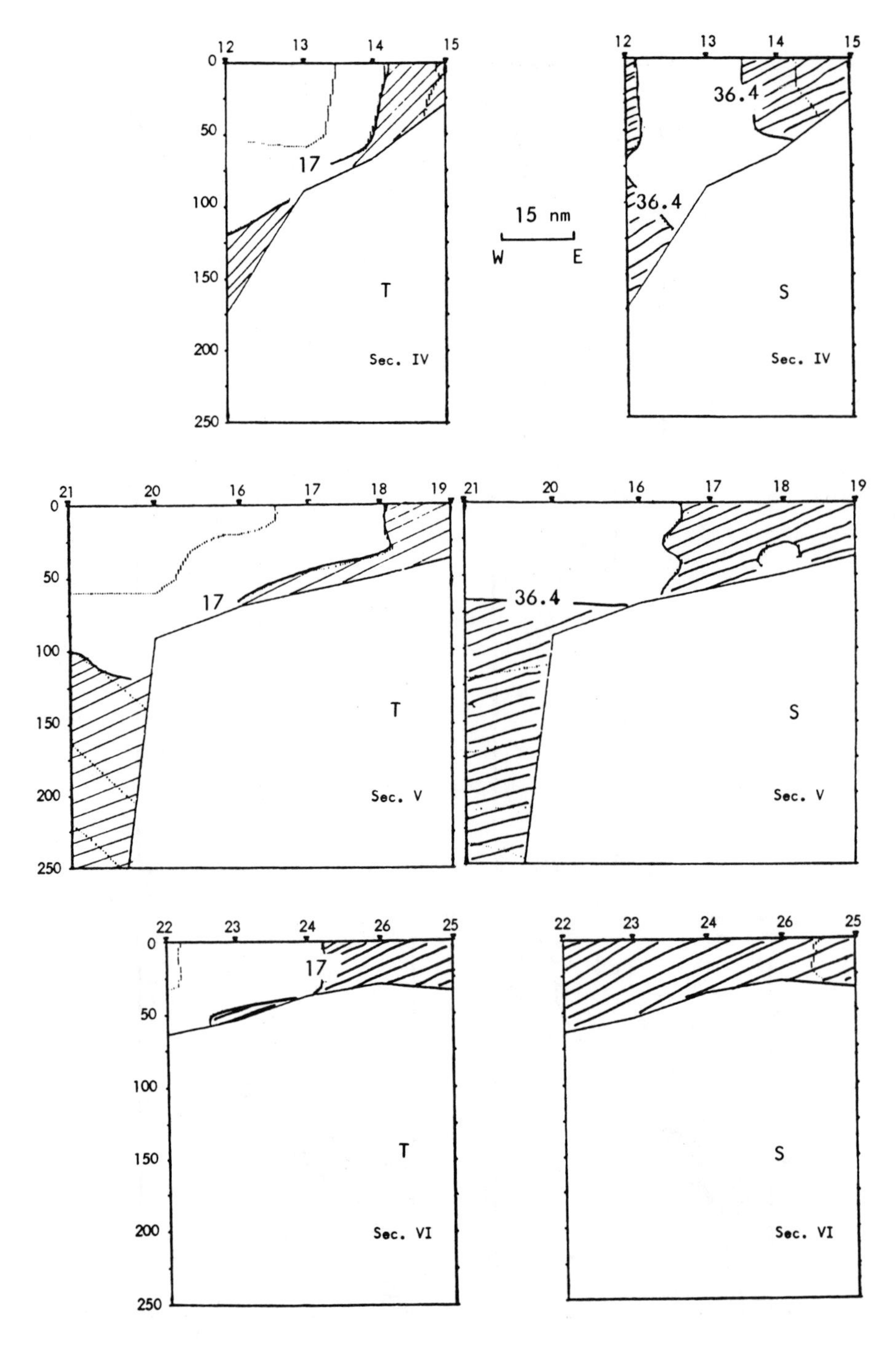

Fig. 4b

over the bottom with little or no shorewards component, because shoaling of the bottom forces the water to turn seawards, following the bottom contours.

In summer, however, the pattern is more complex (Fig. 5). High-salnity frontal zone water (*fzw*) (shown white in the figure) is present in the north from the surface to about 300-m depth offshore but is restricted to a surface layer at mid-shelf and is practically absent near shore. The shaded areas, the low-salinity water, show the deep water (*dw*) in a near-bottom layer that rises shorewards and equatorwards, and the upwelled water (*uw*) in a surface layer that flows seawards, sinking in the north beneath the *fzw* and upwelling again further offshore. In the

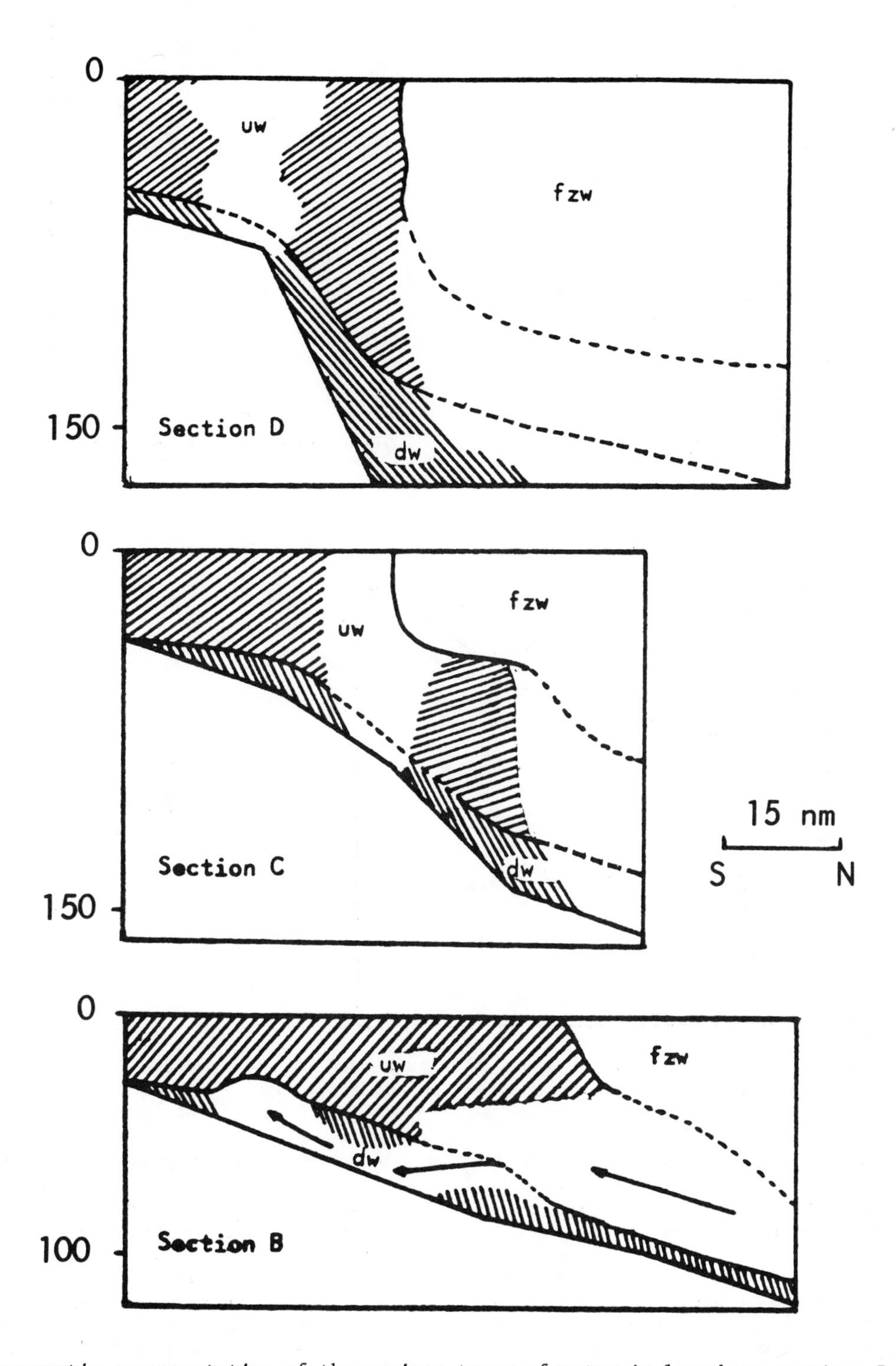

Fig. 5. Diagrammatic representation of the various types of water in longshore sections B to D. *Fzw* = frontal zone water; *dw* = deep water flowing onshore; *uw* = upwelled water flowing offshore; shaded areas correspond to salinity lower than 36.4 $^{0}/oo$.

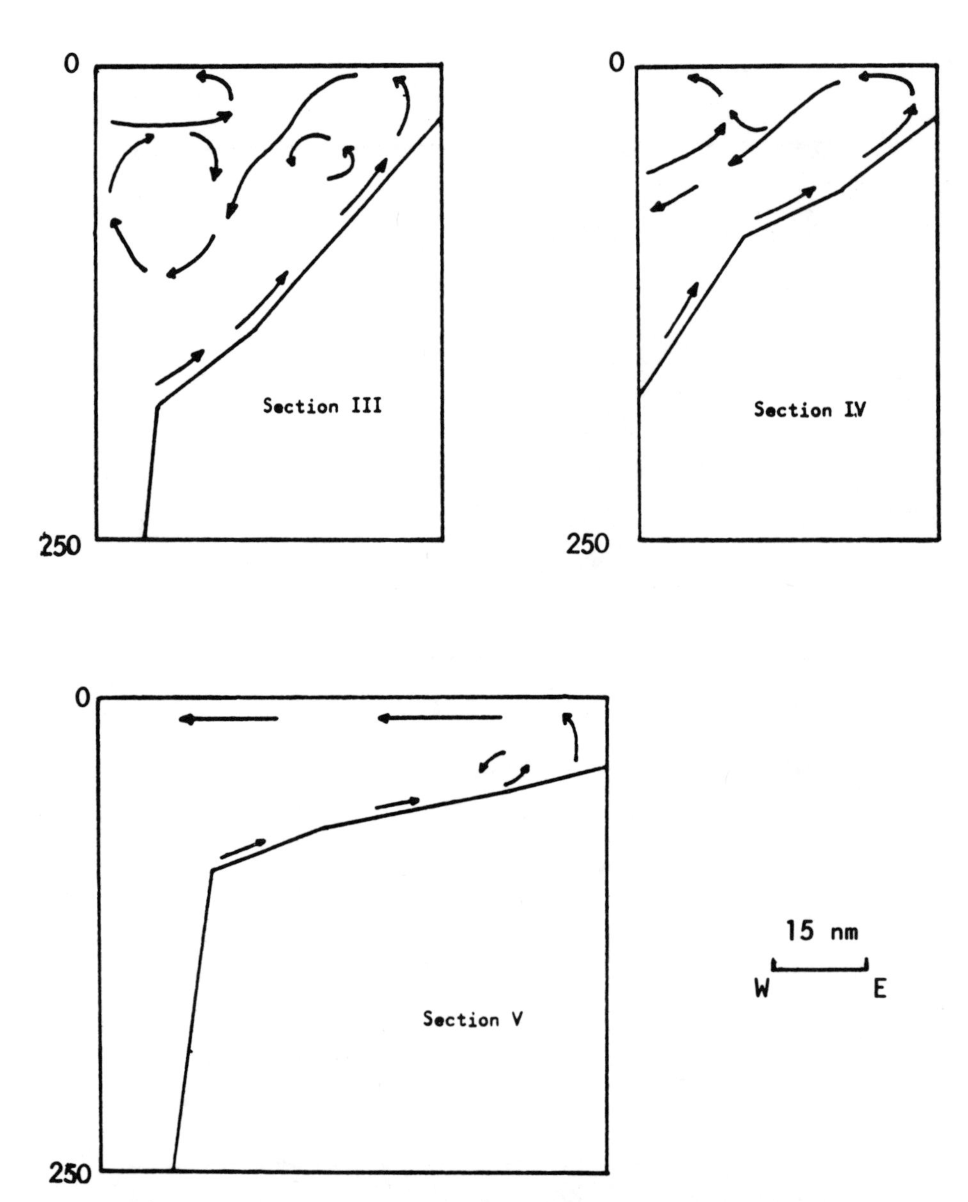

Fig. 6. Diagrammatic representation of the water flow patterns in cross-shelf sections III to V. This and Figs 3 to 5 are drawn to the same scales in both the horizontal and vertical directions.

near-shore section B, the *dw* includes some of the high-salinity water flowing equatorwards. In section B, the *uw* in the south corresponds to the low-salinity *dw* observed in the north. In the mid-shelf section C, only part of the *uw* corresponds to low-salinity *dw*, the remaining being strongly influenced by high-salinity *fzw*. In section C, some of the *uw* from section B has sunk below the *fzw*. In section D, the entire system has moved equatorwards (to the left of the figure) and the low-salinity water, trapped under the *fzw* in section C, upwells again.

According to this scheme, the water moves along a bending, right-handed helicoidal pattern made up of a longshore component and a roughly circular onshore-offshore component centered at about 50-m depth and 15 nm offshore (Fig. 6). Cross-shelf circulation in sections III to V, drawn taking into account the distribution of properties (they are superimposable with the salinity and temperature sections of Fig. 4), shows a primary cell, with bottom shorewards flow and surface seawards flow, enveloping the tongue of *fzw* that flows southwards along section B and constitutes the axis of the helix, and sinks at mid-shelf beneath the surface *fzw* associated with a shallow secondary cell of the same sign. This two-cell circulation pattern seems to break up over the wide shelf section V where most of the water flowing seawards within the Ekman layer is provided by the longshore component. The upwelled water, sinking beneath the *fzw* at mid-shelf in section III, would be associated with a reverse cell that weakens as the surface front vanishes to the south.

Conclusion

It can be concluded that the first mechanism is restricted to the narrow shelf area in the north off Cabo Bojador with a one-cell cross-shelf circulation in spring (Cruzado, 1979) and a two-cell circulation in summer when there is a strong surface front close to the shore. The second mechanism is responsible for most of the upwelling on the wide shelf south of Cabo Bojador. Smaller features such as the reverse cell beneath the frontal zone and transient phenomena such as internal gravity waves, which seldom appear over the shelf but are often detected at the shelf break (Salat and Font, 1977), may contribute to the uplifting of water parcels, especially in narrow shelf areas.

References

Codispoti, L.A., and G.E. Friederich, Local and mesoscale influences on nutrient variability in the northwest African upwelling region near Cabo Corbeiro, *Deep-Sea Research, 25,* 751-770, 1978.

Cruzado, A., Resultados del análisis continuo en Africa del NO entre 23°N y 28°N, *Resultados Expediciones Cientificas B/O Cornide, 3,* 117-128, 1974a.

Cruzado, A., Coastal upwelling between Cabo Bojador and Punta Durnford (Spanish Sahara), *Tethys, 6,* 133-142, 1974b.

Cruzado, A., Is wind stress the main driving force of coastal upwelling, *Third International Symposium on Upwelling Ecosystems,* Kiel (F.R.G.), 25-28 August, 1975.

Cruzado, A., Coastal upwelling off Western Sahara, *Investigaciones Pesqueras, 43,* 149-160, 1979.

Dugdale, R.C., and J.J. MacIsaac, Variability and control of nitrate and ammonia uptake, *IDOE International Symposium on Coastal Upwelling,* Los Angeles (U.S.A.), 4-8 February, Abstract, 1980.

Johnson, D.R., E.D. Barton, P. Hughes, and C.N.K. Mooers, Circulation in the Canary Current upwelling region off Cabo Bojador in August 1972, *Deep-Sea Research, 22,* 547-558, 1975.

Margalef, R., Phytoplankton communities in upwelling areas. The example of NW Africa, *Oecologia aquatica, 3,* 97-132, 1978.

Manriquez, M., and J. Rucabado, Area de afloramiento del NO de Africa, 23°30'N - 26°10'N, Octubre de 1975 (Campaña *ATLOR VI*), *Datos Informativos I.I.P., 1,* 184 pp., 1976.

Mittelstaedt, E., Some aspects of the circulation in the NW African upwelling area off Cap Blanc, *Tethys, 6,* 89-92, 1974.

Salat, J., and J. Font, Internal waves off northwest Africa, in: *Bottom Turbulence,* J.C.J. Nihoul (ed.), Elsevier Oceanographic Series, Elsevier, Amsterdam, 269-274, 1977.

Velasquez, Z.R., and A. Cruzado, Distribución de biomasa fitoplanctónica y asimilación de carbono en el NO de Africa, *Resultados Expediciones Cientificas B/O Cornide, 3,* 147-168, 1974.

UPWELLING OFF THE GALICIAN COAST, NORTHWEST SPAIN

F. Fraga

Instituto de Investigaciones Pesqueras de Vigo,
Muelle de Bouzas, Vigo, Spain

Abstract. Along the Galician coast, northwestern Spain, North Atlantic Central Water (NACW) upwells from 150 m in summer with a maximum intensity northwest of Galicia. There is a distinct difference between sub-surface waters of the western and northern coasts. The sub-surface water along the western coast is NACW proper, while along the northern coast the NACW is highly modified. The upwelling along the northern coast is displaced when invaded by eastern surface water.

High concentrations of nitrate arising from the upwelling of coastal waters inside the *rías* occur when the upper NACW reaches the depth of the bottom of the *rías*. On the other hand, the estuarine circulation in the *rías* gives rise to an increase in the nutrient concentrations of the bottom water, which also affects the exterior coastal water.

Introduction

Deep water upwells along the Galician coast and influences the water inside the *rías*. The cold, upwelled water represents marked contrast because summer is the period of maximum insolation and air temperature. Despite this, the first paper presenting vertical sections of temperature showing this upwelling appeared only recently (Molina, 1972), although the upwelling along the Portuguese coast was previously known (Madelain, 1967a). The upwelling along the Galician coast is not a local phenomenon but can be considered part of the general upwelling produced by the North Atlantic anticyclonic gyre that extends from Galicia to Cape Vert (Wooster, Bakun, and McLain, 1976). Seasonal variations in the intensity of the Galician upwelling coincide with the annual latitudinal displacement of the whole system (Wooster *et al.*, 1976).

An intensification appears at the northern and southern boundaries of the large-scale North Atlantic coastal upwelling. At both boundaries the topography is similar as is the coincidence with a discontinuity in the sub-surface water. At the southern boundary the intensification is off Cape Blanc, 21°N lat., coinciding with the frontal meeting of two different water masses, the North Atlantic Central Water (NACW) and the South Atlantic Central Water (SACW). At the northern boundary, northwest of Galicia, the upwelling intensifies between Cape Finisterre, 42°52'N lat. and Punta Roncudo, 43°17'N lat., where it coincides with the boundary of two sub-surface water masses, although the differences between them are noticeably less than the difference between the NACW and SACW off Cape Blanc. South of Cape Finisterre the sub-surface water is typically NACW, while off Cape Ortegal the upper part of the NACW, just below the surface layer of about 100 m, disappears and the 100- to 400-m layer is occupied by water with a density (σ_t) range of 27.17 to 27.23. The σ_t range normally occurs in NACW from 400 to 500 m so that off Cape Ortegal the whole water column from 100 to 500 m is relatively homogeneous. The T-S diagram of two extreme structures can be seen in Fig. 1. At the south, (Sta. C, 42°10'N lat.; 9°59'W long.) the NACW is represented by the fragment of the straight line II, 75 to 500 m, while at the northeast (Sta. B, 44°16'N lat.; 6°56'W long.), the portion of NACW between 75 and 500 m is rather homogeneous, with salinity varying little from 35.56×10^{-3}.

The upwelling

Along the Galician coast the wind is variable due to the change of position of the meteorological perturbation centers. However in winter, from December to February, the prevailing winds come from the south, while in summer, from April to August, northerly winds prevail, favoring upwelling. Thus in summer upwelling is intense but disappears during winter. The meteorological situation in summer of the region around Galicia was studied by Chateau-Thierry (1974), who computed the velocity of upwelling in a large oceanic zone.

The seasonal development of coastal upwelling was followed off the Ria of Vigo by making casts at a single hydrographic station fortnightly

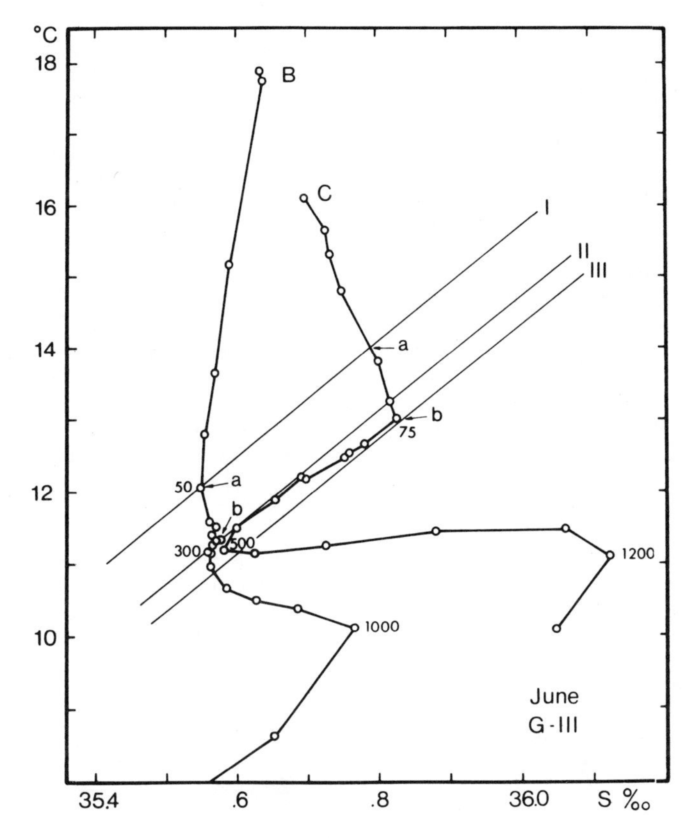

Fig. 1. T-S diagrams from a northern station, B, at 44°16'N lat., 6°56'W long. and a southern station, C, 42°10'N lat., 9°59'W long. I, the line along which the points belonging to NACW are distributed along the Galician coast. I and III, limits of the NACW according to Sverdrup, Johnson, and Fleming (1942).

during 1962 and 1977. To establish the evolution of the system, the depth of the boundary between the surface mixed water and sub-surface water (consisting of the upper part of NACW) was taken as a reference point; it generally appeared somewhat below the thermocline. (This reference was selected because the upwelling of NACW supplies nutrient-rich water to the upper layers.) The boundary appears near the coast below 100 m from November to January but by March it is shoaler than 50 m, reaching the surface in August when the NACW is evident on the T-S diagram. We have no data on events in the northern region of Galicia except from cruises during June to October.

On the western coast marked by the *rías*, identification of the boundary between the NACW and surface water is not clear cut because of mixing inside the *rías*, because energy interchanges with the atmosphere cause *in situ* temperature changes, and because salinity is altered by evaporation and the addition of fresh water. Neither can total inorganic and organic nutrient contents be used because there is a transport of particulate matter from one layer to another.

The surface temperature distribution for June is represented in Fig. 2. The low temperatures show where the upwelling is more intense. From the temperatures of the deep water and of typical surface water that mix, the fraction of upwelled water in the surface mixture can be estimated. The temperature of the upwelling water, the upper part of NACW in this case (point B, Fig. 1), differs from place to place, changing from 11.3°C at Sta. B in the northeast (Fig. 2) to 13.0°C at Sta. C in the southwest. At Punta Roncudo (Fig. 2) it is 12°C. In this zone the temperature of the typical surface water is difficult to establish, but it is assumed to be about 16°C at the time. Thus the 13°C isotherm corresponds to 63% NACW and the 14°C isotherm corresponds to 50% NACW.

The upwelling structure can be seen in the vertical density section off Punta Roncudo (Fig. 3). The upwelling includes a narrow coastal fringe that forms a small circulation cell with a sinking border 18 km from the coast. Off Cape Finisterre the sub-surface structure is similar to that off Punta Roncudo, although the surface temperature distribution is rather different.

In August the warm surface water along the northern coast of the Iberian Peninsula, the Cantabrian Sea, moves towards the west, preventing the deep water from reaching the surface along

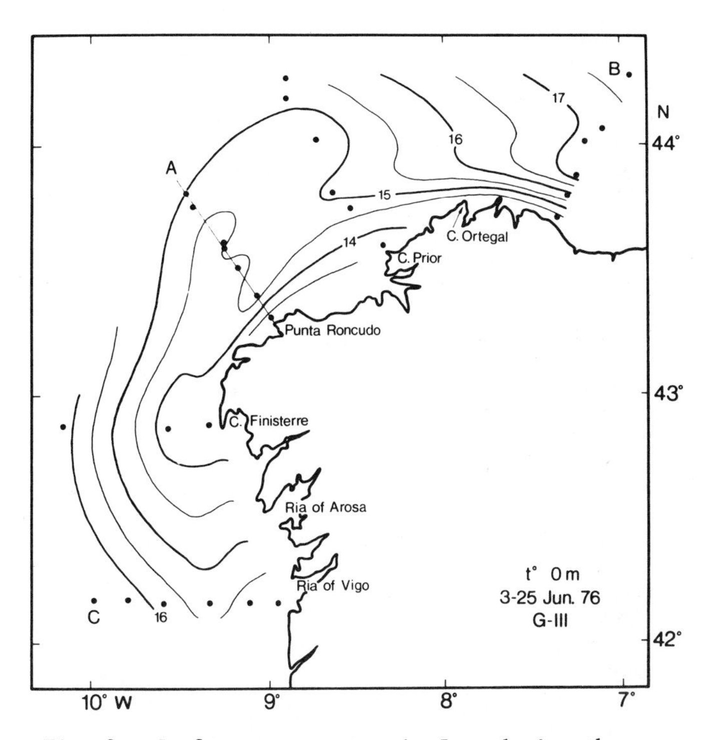

Fig. 2. Surface temperature in June during the "Galicia III" cruise. The vertical distribution of density along section A is shown in Fig. 3.

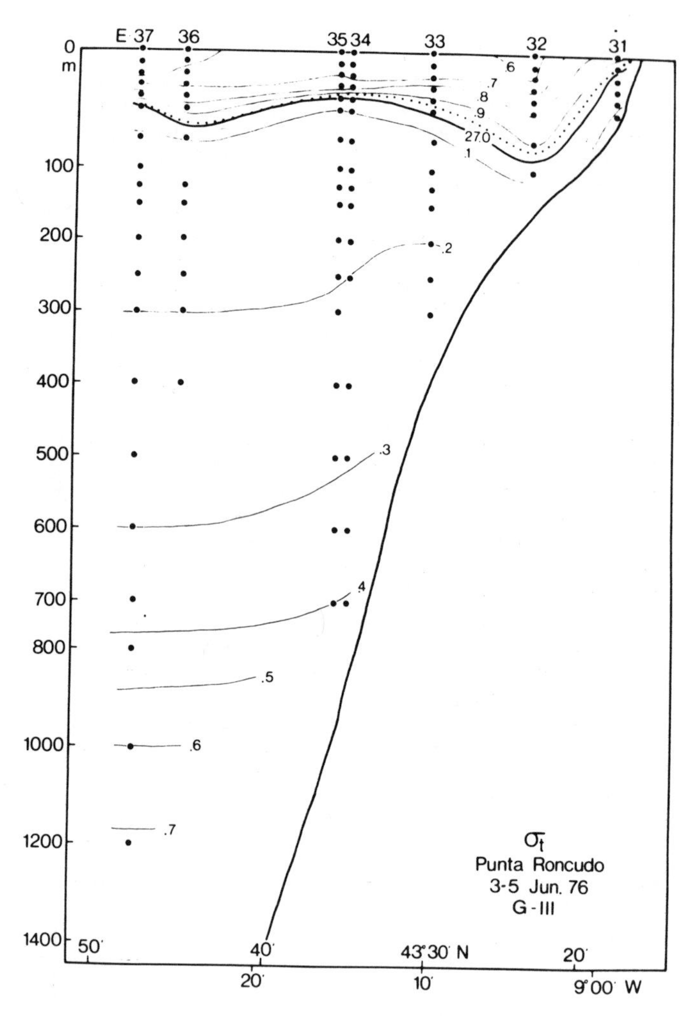

Fig. 3. Vertical distribution of density off Punta Roncudo, section A, Fig. 2, "Galicia III" cruise. The dotted line close to σ_t 26.95 indicates the upper boundary of NACW.

the northern coast of Galicia. The isotherms (Fig. 4) are compressed because the upwelling is even more intense than in June, with surface temperatures decreasing to 13°C. The vertical section in the zone (Fig. 5) clearly shows the slope of the isopycnals and coastal upwelling from 150 m of water that is totally NACW with its upper boundary (dotted line) at about σ_t 26.9. The upper boundary of NACW (Fig. 6) was determined for each station by finding the intersection of the T-S diagram with the straight line I. $T = 7.8 + 8(S\ ^{o}/oo - 35)$ at point a (Fig. 1), defining the separation between the warmer surface water and the NACW, and interpolating the depth between the two points closest to the intersection. Above this level the water is a mixture of NACW and surface water.

Line II, Fig. 1, $T = 6.8 + 8(S\ ^{o}/oo - 35)$, corresponds to NACW along the Galician coast, and the upper part of the NACW corresponds to point b, but in some T-S diagrams the straight segment appears above line II. So that the T-S diagrams always cut the straight line and to simplify the calculation, line I was chosen because it is one of the limits given by Sverdrup, Johnson, and Fleming (1942). Also, on choosing line I the point a comes closer to the level of the nitratocline and therefore is more interesting from a biological point of view.

To understand better the situation during August, a map of dynamic topography was drawn for the period. As the depths in the continental slope are less than 2000 m, the reference level was chosen, following Madelain (1967b), as the intermediate level between the Mediterranean water (MW) flowing northward and NACW, which was supposed to flow southwards, except that we chose the level where the fractions of NACW and of the core of MW are the same. The mean salinity of the two values was calculated at every station using the minimum salinity from the NACW (at about 400 m) and the maximum salinity from the core of the MW vein (at about 1100 m). The depth corresponding to the mean salinity was interpolated. The average for 39 stations was 729 m, so 730 m was used as the reference level. The dynamic topography (Fig. 7) shows a surface current along the coast at a velocity of 8 to 25 cm/s, but measurements by Swallow, Gould, and

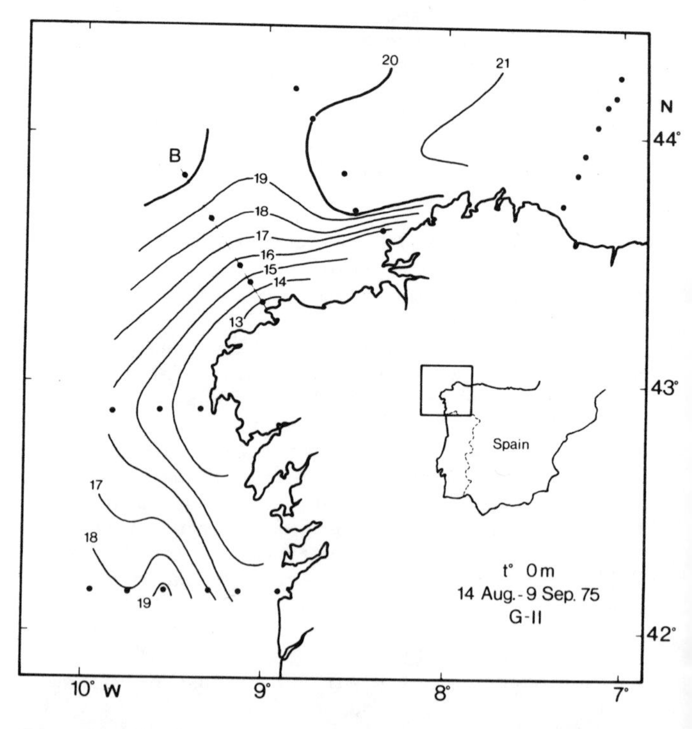

Fig. 4. Surface temperature in August and September during the "Galicia II" cruise. The vertical distribution of density along section B is shown in Fig. 5.

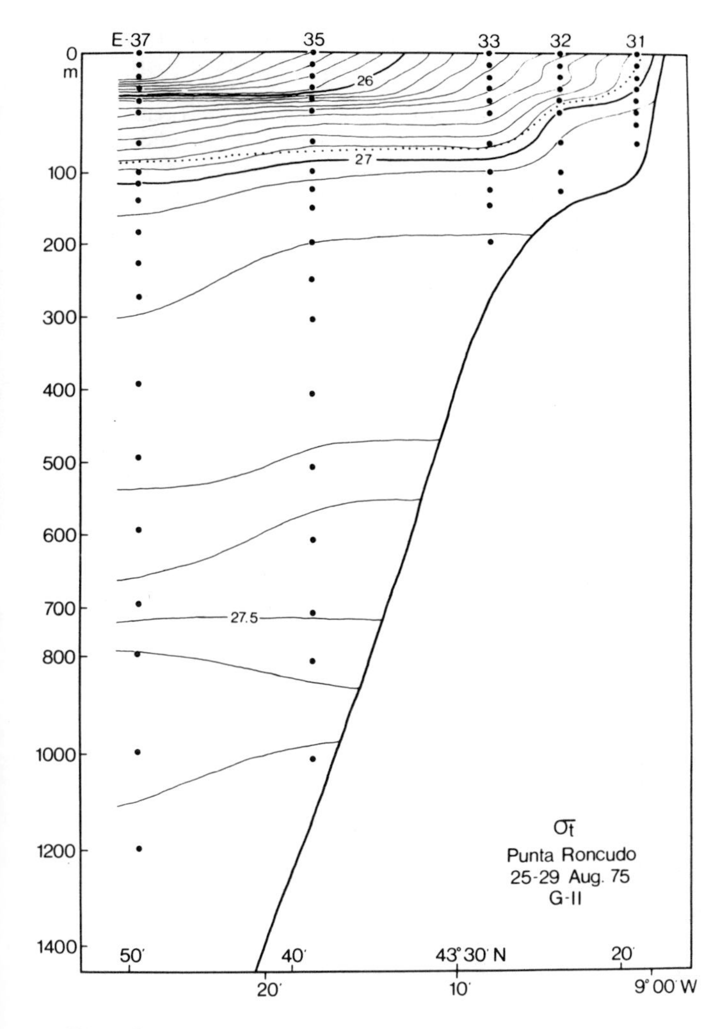

Fig. 5. Vertical distribution of density in August along section B, Fig. 4.

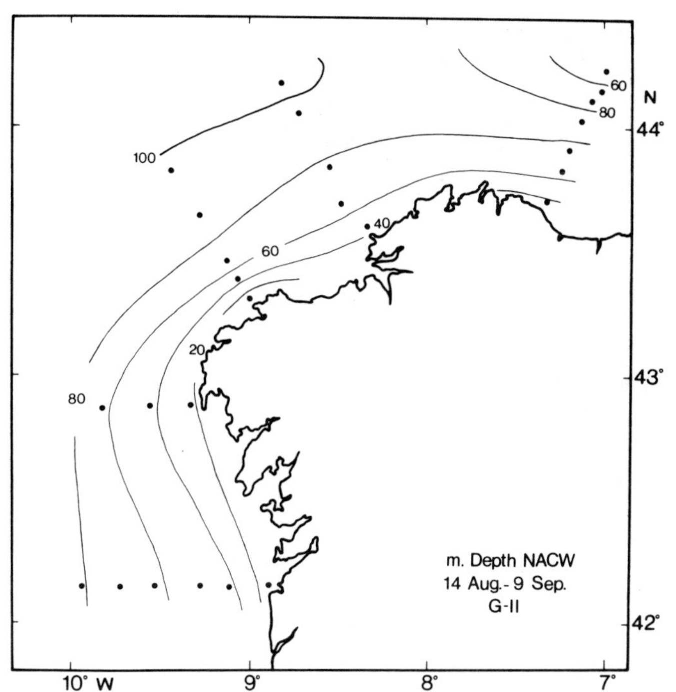

Fig. 6. Depth of the surface separating the NACW and the surface water during August and September. Below the level the water is 100% NACW and above it NACW is mixed with surface water.

Saunders (1977) show a poleward flow, not confined to the core of the MW, extending through most of the water column, including the reference level chosen as above. The absolute velocity estimated by Swallow *et al.* in January 45 miles off Vigo was 5 cm/s at 730 m, therefore to estimate absolute velocities that value must be subtracted from the southern component of the relative surface velocities. Despite lacking a reference level of zero velocity, the surface currents are faster than those of the level chosen, and the dynamic topography of Fig. 7 agrees quite well with the location of the upwelling and the distribution of surface temperature.

The surface circulation derived from Fig. 7 is similar to that given by Fruchaud-Laparra, Le Floch, Le Tareau, and Tanguy (1976) for the Phygas 34 cruise during a month later in a different year. The principal difference is that the dynamic topography for cruise G-II is more compressed against the coast in the southeast

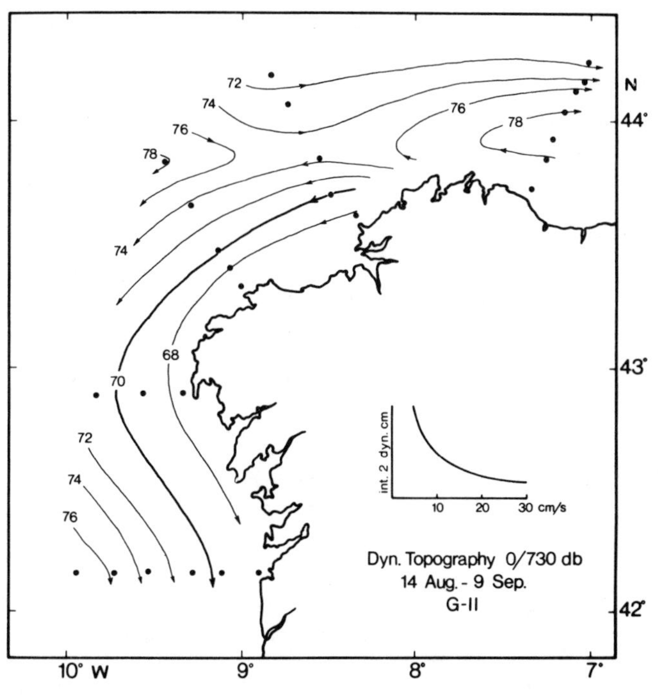

Fig. 7. Dynamic topography of the sea surface referred to the level of 730 m during August and September. Heights are given in anomalies of dynamic centimeters.

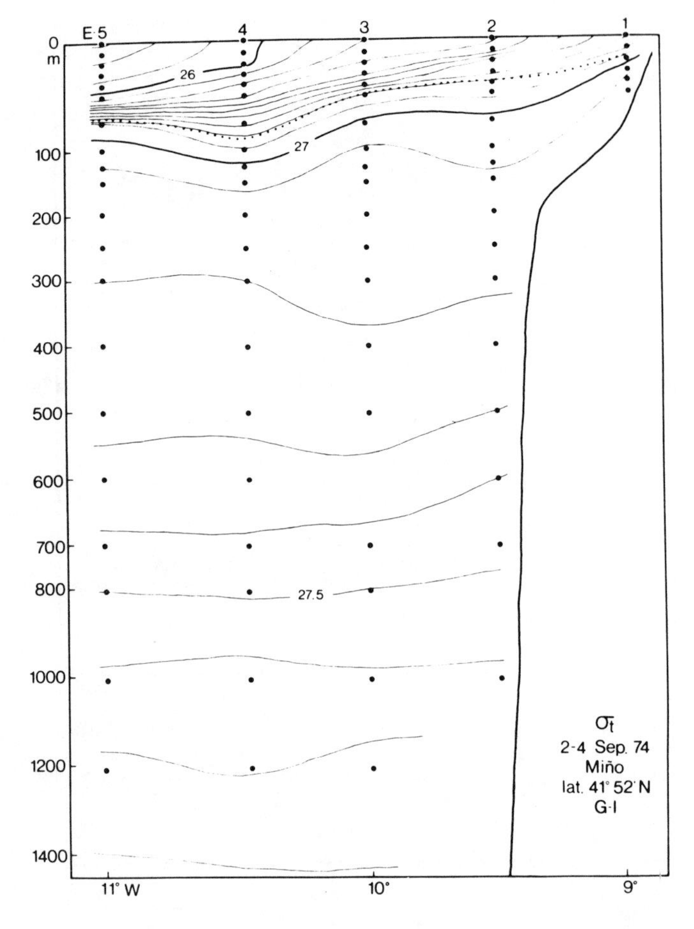

Fig. 8. Vertical distribution of density off the River Miño, 41°52'N lat., in September. "Galicia I" cruise.

direction.

In September, continuous measurements of surface temperature off Cape Prior, along a zig-zag course between 43°30' and 44°N lat. and 8 to 9°W long. gave uniform values averaging 17.5°C; even 3 miles off the coast the temperature was not lower than 17°C, indicating the absence of upwelling. However in September of a different year, Molina (1972) observed upwelling off Cape Ortegal, suggesting year-to-year variability.

To the south, the upwelling was less pronounced but persisted as is evident in the vertical sections occupied 15 days apart at 41°52'N lat. (Figs 8 and 9). Central water did not reach the surface but remained at a depth of 20 m near the coast. In October, some traces of upwelling remained between the Ría of Arosa and Cape Finisterre about 15 km from the coast (Fig. 10). Sta. 17 in the section was inside the Ría of Arosa, therefore apart from the north-south coastal circulation, so upwelling was not apparent because of the addition of fresh water at the beginning of the rainy season.

In describing changes in upwelling along the Galician coast from June to October, data from four different years have been used, and some of the interpretations may be influenced by interannual variations. However, judging from the seasonal variations observed for several years in the mouth of the Ría of Vigo the variations follow a general pattern.

Hydrographic data from along the Galician coast from November to May are scarce and the hydrographic structure during the period is poorly known.

Interaction of coastal water and the inner part of the rías

The western coast of Galicia is formed by the Rías Bajas. They are about 18 to 33 km long with an average depth of 40 to 60 m at the mouth. The inner circulation is similar to a positive estuary; coastal water flows in along the bottom, rises by mixing, and flows out along the surface. The intensity of the circulation depends on various factors (Fraga and Margalef, 1979), one of which is the wind (Chase, 1975). The average

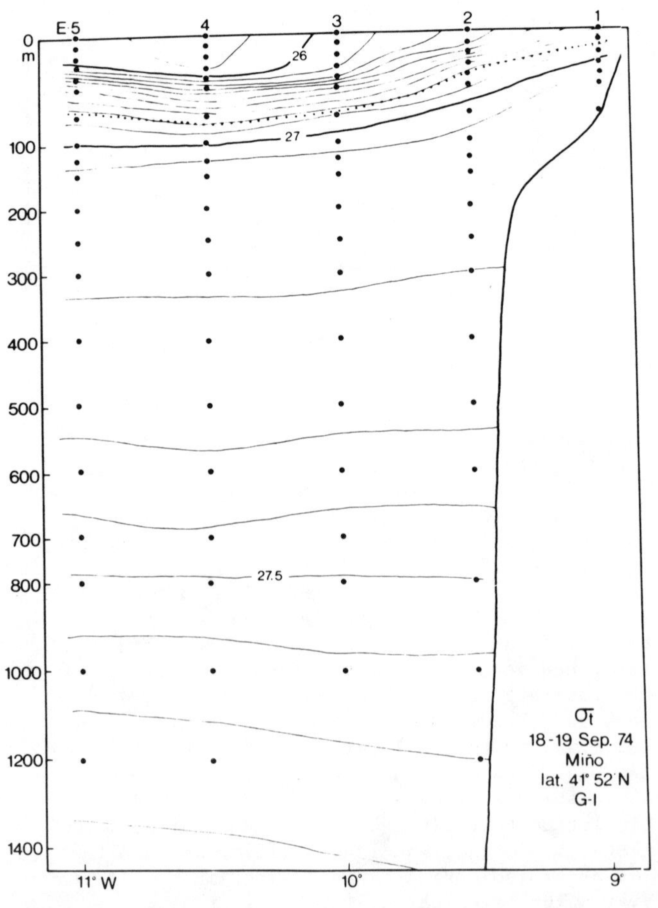

Fig. 9. Vertical distribution of density along the same section shown in Fig. 8, 15 days later.

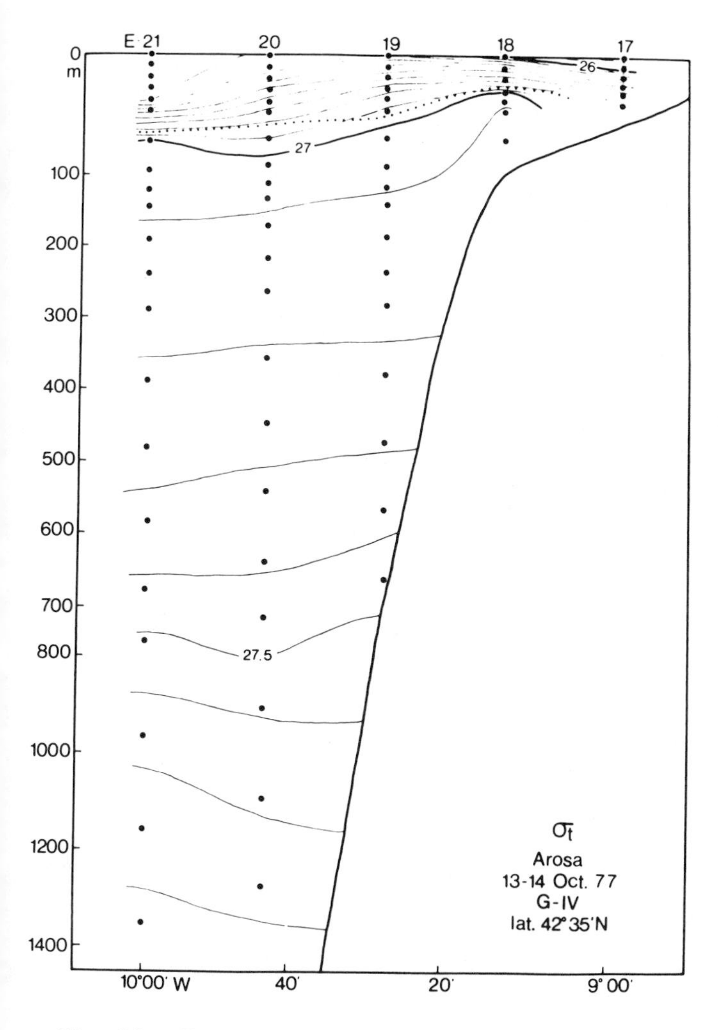

Fig. 10. Vertical distribution of density off Ría of Arosa at latitude 42°33'N, in October. "Galicia IV" cruise.

velocity of upwelling calculated by Margalef and Andreu (1958) is 3.1 m/day in summer in the Ría of Vigo and somewhat faster (4.7 m/day) in winter, due to river runoff.

The high primary production of the *rías* depends on the contribution of nutrients which, in turn, depends on the velocity of upwelling and especially on the coastal water that flows in along the bottom. During winter, when the coastal water surface layer is more than 100 m thick, there is an exchange of water between the *rías* and the nutrient-poor coastal water. Combined with the small amount of insolation, the winter is a season of low primary productivity.

In spring the upper levels of the NACW rise above the level of the bottom of the *rías*, and nutrient-rich water reaches the photosynthetic layer inside the *rías* leading to high primary production. Part of the biomass produced is transported in the surface layers towards the mouth of the *ría*, sinks to the bottom, and decomposes. This enriches the deep water, which flows back into the *ría*. Nutrient regeneration takes place not only inside the *ría* but also in the innermost zone of the continental shelf, giving rise to nutrient concentrations that could lead to incorrect estimates of the original depth of the upwelled water.

Off the Ría of Arosa (Fig. 11) the water at 70 m at Sta. 18 appears to have a deeper origin and the density distribution (Fig. 10) indicates that it is from around 200 m. The concentration of nitrate, 16.7 µg-at N/l, at 70 m at Sta. 18 is not found in the nearby open ocean at depths less than 1400 m, indicating biological enrichment by a mechanism such as that suggested above. Deep water with 16.7 µg-at N/l has a concentration of 4.4 ml O_2/l while at Sta. 18 the concentration is only 2.9 ml/l. The difference, 1.5 ml/l, has been used to mineralize the particulate organic matter from the upper layer. If a molar relation of $O_2:NO_3^-$ = 144:15 is assumed for the mineraliza-

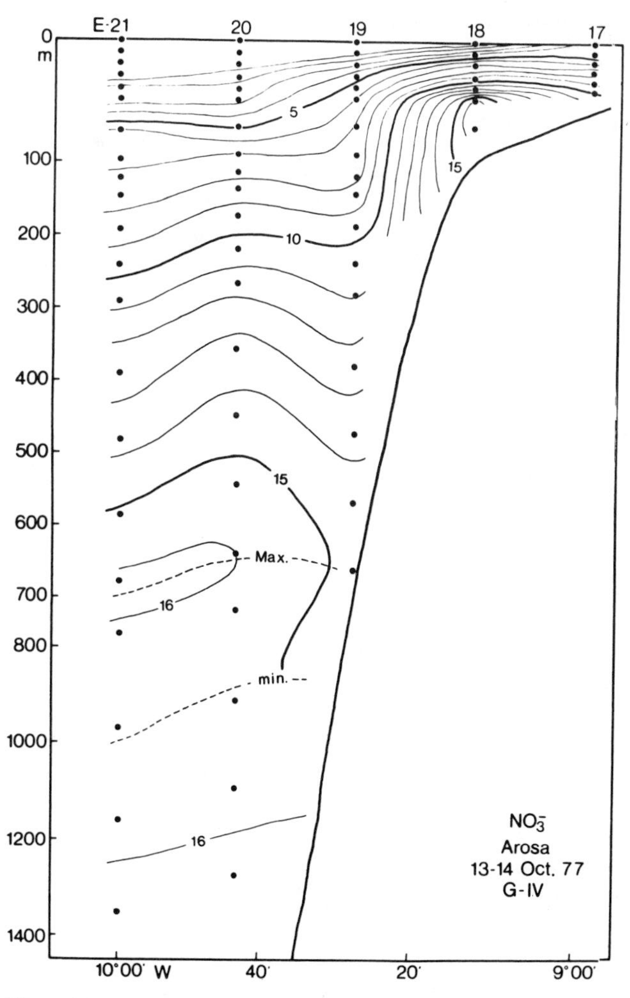

Fig. 11. Vertical distribution of nitrate along the same section shown in Fig. 10.

tion of organic matter, the amount of nitrate produced during the consumption of 1.5 ml O_2 would be 7.0 μg-at/l. At Sta. 21 the water at the depth where σ_t = 27.11, the density at 70 m at Sta. 18, has 8.1 μg-at/l of nitrate-nitrogen, which, when added to 7.0 μg-at N/l of organic nitrogen, gives rise to 15.1 μg-at, nearly the same as at Sta. 18 at 70 m. Similar conditions are present off the other *rías* south of Cape Finisterre.

Acknowledgements. I thank Miss Aida Fernandez and C. Mourino for the calculation of the anomalies of specific volume, drawing the figures, and translation of the text into English and to Dr. Dolors Blasco for correcting the text.

References

Chase, J., Wind-driven circulation in a Spanish estuary, *Estuarine and Coastal Marine Science, 3,* 303-310, 1975.

Chateau-Thierry, H. de, Mise en évidence de mouvements verticaux induits par certaines situations météorologiques dans le proche océan, *Tethys, 6,* 341-348, 1974.

Fraga, F., and R. Margalef, Las *rías* gallegas. In *Estudio y explotación del mar en Galicia*, 487 pp., Ed. Univ. Santiago de Compostela, 1979.

Fruchaud-Laparra, B., J. Le Floch, J.Y. Le Tareau, and A. Tanguy, Etude hydrologique et variations saisonnières dans le proche Atlantique en 1973, *Rapports scientifiques et techniques CNEXO, 26,* 111 pp., 1976.

Madelain, F., Etude hydrologique au large de la peninsula Iberica, *Cahiers Océanographiques, 19,* 125-136, 1967a.

Madelain, F., Calculus dynamiques au large de la peninsula Iberique, *Cahiers Océanographiques, 19,* 181-193, 1967b.

Margalef, R., and B. Andreu, Componente vertical de los movimientos del agua de la Ría de Vigo y su posible relación con la entrada de sarina, *Investigación Pesquera, 11,* 105-126, 1958.

Molina, R., Contribución al estudio del "upwelling" frente a la costa noroccidental de la península Ibérica, *Boletín del Instituto Español de Oceanografía, 152,* 1-39, 1972.

Sverdrup, H.V., M.W. Johnson, and R.H. Fleming, *The Oceans, their Physics, Chemistry and General Biology,* Prentice Hall, 1087 pp., 1942.

Swallow, J.C., W.J. Gould, and P.M. Saunders, Evidence for a poleward eastern boundary current in the North Atlantic Ocean, *International Council for the Exploration of the Sea,* C.M./C, 32 Hydrography Committee, 1977.

Wooster, W.S., A. Bakun, and D.R. McLain, The seasonal upwelling cycle along the eastern boundary of the North Atlantic, *Journal of Marine Research, 34,* 131-141, 1976.

WEAKLY STRATIFIED SHELF CURRENTS

John A. Johnson

School of Mathematics and Physics, University of East Anglia,
Norwich NR4 7TJ, England

Abstract. An analytical model describes the shelf currents produced by an upwelling-favorable wind stress in a weakly stratified ocean. The barotropic flow driven by the wind stress causes an adjustment in the density field that generates a baroclinic component in the current. Single-cell upwelling is produced by a uniform wind. Double-cell cross-shelf circulation is generated by more complex spatial distributions of wind stress.

Introduction

Coastal currents over continental shelves are often associated with coastal upwelling, as in the observations by Barton, Huyer and Smith (1977) in the Atlantic ocean off northwest Africa. Three important features present in the observations are topography, stratification, and variability. An ideal model of coastal currents ought to incorporate all these physical characteristics.

Although many theoretical models have been constructed of upwelling regions it is unusual for realistic attempts to be made to cover all three features. A number of unsteady stratified models use normal modes to represent the stratification, but the topography is unrealistic as it has a vertical coastal boundary to the ocean. These models, which follow Gill and Clarke (1974), are based on the method introduced by Lighthill (1969). Other models take account of topography with a sloping shelf but omit stratification, as in Johnson and Manja (1980). Numerical models, for example Peffley and O'Brien (1976), usually have a vertical coast and often discrete layers, which produce difficulties after large times have elapsed if interfaces intersect the surface.

In the subsequent sections the effects of stratification are introduced into a sloping shelf model by considering a weakly stratified ocean as a perturbation to a basic barotropic model. The stratification interacts with the barotropic flow and generates a baroclinic component in the velocity field.

The Model

This slightly stratified shelf model is a modification of the barotropic solution discussed by Johnson (1976), where precise justifications are given for the equations used here. The dimensionless coordinates have "x" eastward, "y" northward, and "z" upward from the surface at z=0. Over the shelf the bottom is at $z=\alpha(x-1)$ with the eastern boundary of the ocean model at x=1, and the actual shelf slope is $\alpha D/L$ where D is the depth and L is the width of the ocean. Ekman layers at the surface and on the sloping shelf are separated by a shelf layer. The only properties required of the Ekman layers are that they are well mixed and carry Ekman transport. Other details are unimportant.

In the shelf region observations show that the flow V alongshore is much larger than the flow U onshore and that variations across the shelf are much larger than changes along the shelf. Therefore, let

$$V \gg U \text{ and } \partial/\partial x \gg \partial/\partial y.$$

Then the equations of momentum, continuity, and heat have the following dominant terms in the shelf region:

$$fV = P_x, \qquad fU = -P_y, \tag{1}$$

$$P_z = S\theta, \tag{2}$$

$$U_x + V_y + W_z = 0, \tag{3}$$

$$U\theta_x + V\theta_y + W\theta_z = 0, \tag{4}$$

where subscripts denote derivatives, P is pressure, W is the vertical velocity, θ is temperature, $f = 1 + \beta y$ is the Coriolis parameter, and the β-plane and Boussinesq approximations are used. In the shelf layer advection of heat by the current dominates mixing giving the heat equation (4). S is a stratification parameter that represents the ratio of thermal driving to wind driving and is defined by

$$S = Dg\bar{\alpha}(\Delta\theta)/(f_0 L\bar{U})$$

where g is the acceleration due to gravity, $\bar{\alpha}$ is the coefficient of thermal expansion, and $\Delta\theta$, f_0, $\bar{U}$ are scales for the temperature field, the Coriolis parameter, and the wind-driven horizontal velocity respectively. If S=0 the equations (1) to (3) reduce to the barotropic equations in Johnson (1976).

The solutions of (1) to (4) have to satisfy certain conditions. The matching with the Ekman layers is achieved by using the Ekman suction conditions

$$W = \tau^y_x/f \text{ at } z = 0 \qquad (5)$$

and

$$\frac{W - \alpha U}{(1 + \alpha^2)^{\frac{1}{2}}} = \frac{V_x}{(2f)^{\frac{1}{2}}} \text{ at } z = \alpha(x-1), \qquad (6)$$

where τ^y is the longshore component of the surface wind stress. The latter condition relates the normal suction into the Ekman layer to the gradient of the alongshore velocity just above the Ekman layer and includes the effect of bottom friction.

The Ekman layers join at the coast and continuity of transport requires that

$$\frac{\tau^y}{f} = \{\frac{1 + \alpha^2}{2f}\}^{\frac{1}{2}} V \text{ at } x = 1 \qquad (7)$$

The left hand side is the onshore transport in the surface Ekman layer and on the right is the transport in the lower Ekman layer, proportional to the northward velocity V just above the layer. Further details of (5) to (7) were given by Johnson (1976).

The Slightly Stratified Solution

(i) The first order barotropic solution. For slightly stratified flow (S < 1) expand the pressure and velocity as a power series in S by setting

$$\begin{aligned} P &= P_1 + SP_2 + \ldots, \\ \underset{\sim}{V} &= \underset{\sim}{V}_1 + S\underset{\sim}{V}_2 + \ldots \end{aligned} \qquad (8)$$

Substitution in (1) to (3) leads to lowest order in S to

$$P_{1z} = 0, \; fV_1 = P_{1x}, \; fU_1 = -P_{1y}, \qquad (9)$$

$$U_{1x} + V_{1y} + W_{1z} = 0$$

for steady flow. The equations imply that

$$W_{1z} = (\beta/f^2)P_{1x}. \qquad (10)$$

From (9), (10), and the suction conditions (5) and (6) it can be shown that the pressure P_1 satisfies

$$\{\frac{1 + \alpha^2}{2f}\}^{\frac{1}{2}} P_{1xx} - \frac{\alpha\beta(x - 1)}{f}P_{1x} - \alpha P_{1y} = \tau^y_x. \qquad (11)$$

A similar equation was solved by Johnson (1976) using Laplace transforms for a homogeneous shelf model. Thus the lowest order velocity field in this slightly stratified model is the basic barotropic flow.

Substituting (8) to (10) into (4) gives

$$\theta_z\{\tau^y_x + (\beta z/f)P_{1x}\} + \theta_y P_{1x} - \theta_x P_{1y} = 0. \qquad (12)$$

In situations where the wind stress does not vary rapidly across the continental shelf, then $\tau^y_x = 0$ and (12) can be rewritten

$$\frac{\partial(\theta,\, P_1,\, z/f)}{\partial(x,\, y,\, z,)} = 0 \qquad (13)$$

which determines the effect on the temperature distribution θ produced by the first order barotropic flow. Some properties of the solution of (13) can be deduced without a formal solution. It follows from (13) that

$$\theta = \theta(P_1,\, z/f). \qquad (14)$$

At the surface, z=0, θ is a function of P_1 only, and the isotherms follow the isobars. This emphasizes the importance of advection of heat in the model, as indicated in the heat equation (4). Furthermore at the bottom, $z = -h(x)$, (14) requires that θ is a function of only P_1 on contours of h/f, showing the attempt by the flow to conserve potential vorticity but being forced away from the contours by the driving wind stress.

The general method of solution of (12) for θ is by numerical integration along the characteristics of (12) for a given temperature distribution in the deep water offshore of the shelf. A simpler solution that can be used to illustrate the adjustments to the barotropic flow due to the temperature field is obtained by choosing

$$\theta = \theta_s(x,\, y)\exp(cz/f), \qquad (15)$$

where θ_s is the surface temperature distribution and c is a positive constant. Some solutions for θ_s and the velocity field are presented below for particular wind fields.

(ii). The Second-order Baroclinic Solution. Substitution of (8) into (2) shows that the second-order correction to the pressure distribution is

$$\begin{aligned} P_2 &= P^*_2(x,\, y) + \int\theta dz \\ &= P^*_2 + (f/c)\theta_s \exp(cz/f) \end{aligned} \qquad (16)$$

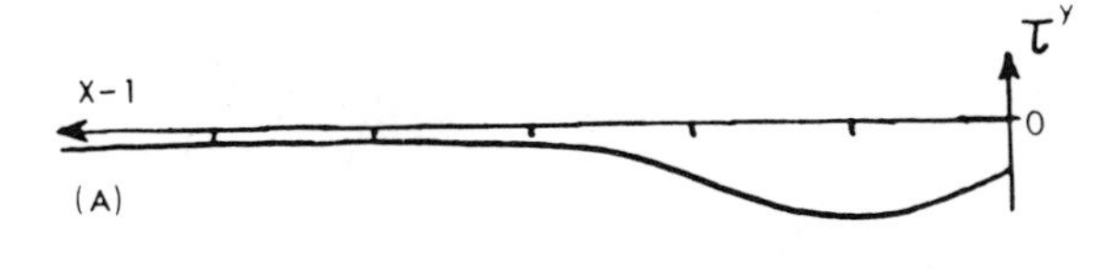

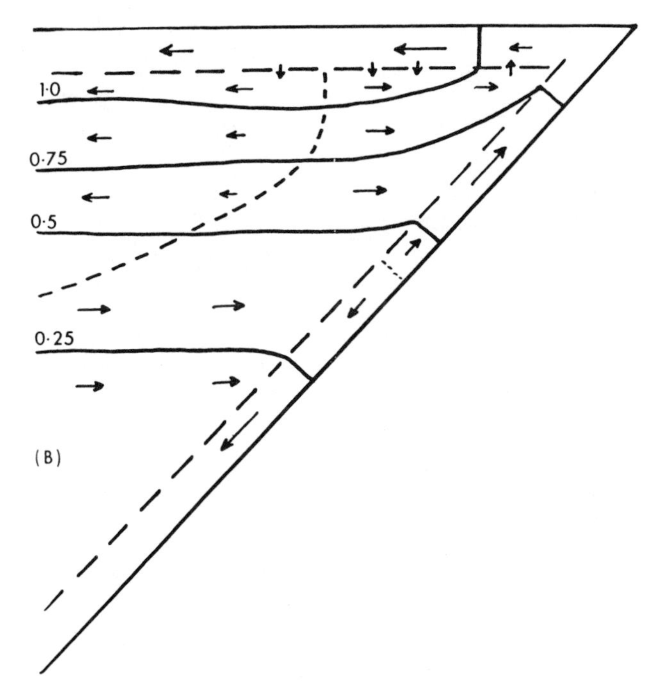

Fig. 1. (a) The wind stress τ^y over the shelf. (b) A cross section through the shelf region at y = 0. The short broken lines indicate the boundary between the onshore and offshore flow. The solid lines represent the isotherms.

using the special form (15) for θ and where P_2^* is an arbitrary function that is the barotropic correction to P. Substituting (16) into (1) gives

$$\left.\begin{aligned} V_2 &= P^*_{2x}/f + (\theta_{sx}/c)\exp(cz/f), \\ U_2 &= -\frac{P^*_{2y}}{f} - \left\{\frac{\beta\theta_s}{fc} + \frac{\theta_{sy}}{c} - \frac{\beta z\theta_s}{f^2}\right\} x \\ &\quad \exp(cz/f). \end{aligned}\right\} \quad (17)$$

Thus the baroclinic adjustment to the velocity field depends on the horizontal gradients of the temperature field. The variation with depth of the onshore flow produces a two-cell circulation over the shelf for some wind fields.

(iii). The Barotropic Second-order Solution. The remaining unknown in (16) and (17) is the barotropic correction to the flow and it can be shown that P_2^* satisfies

$$\left\{\frac{(1 + \alpha^2)^{1/2}}{2f}\right\} P^*_{2xx} - \frac{\alpha\beta(x - 1)}{f} P^*_{2x} - \alpha P^*_{2y} \quad (18)$$
$$= (\text{function of } \theta_s, \theta_{sx}, \theta_{sy}).$$

Comparison of this equation with (11) demonstrates that the second order correction is driven by variations in the temperature field whereas the first order basic flow is driven by the wind stress. To complete the solution for P_2^* requires use of the transport condition (7) which, to order S, reduces to $V_2 = 0$ at the coast. Therefore, as z = 0 at the coast, (17) implies that

$$P^*_{2x} = -(f/c)\theta_{sx} \quad \text{at } x = 1 \quad (19)$$

Through (18) and (19) a feedback is provided from the temperature field to the basic barotropic flow.

Examples to Illustrate the Solution

The procedure described in the above section has been followed to determine the shelf currents produced by various upwelling-favorable wind stress distributions. The numerical solution is started with zero coastal current at latitude y = -0.5

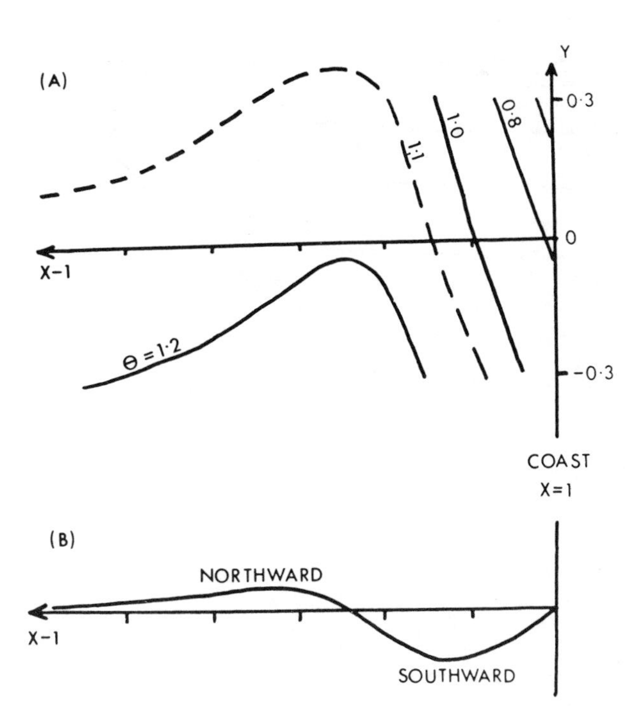

Fig. 2. (a) The barotropic transport in the shelf current at y = 0 and (b) surface isotherms for the wind stress shown in Fig. 1(a).

and then integrated for increasing y to determine the current and temperature distributions induced at latitude y = 0. In the two examples that follow, a value of S = 0.3 is used for the stratification parameter.

(i) In this case the wind stress at each latitude is as shown in Fig. 1 and has a maximum southward component just offshore of the coast. Distributions of wind that are non-uniform across the shelf region have been reported by Halpern (1976) for the Oregon upwelling region. A cross-section through the shelf region at y = 0 showing some isotherms and indicating diagrammatically the cross-shelf flow is presented in Fig. 1. The Ekman suction caused by the variations in the wind stress cause the near surface isotherms to bow downwards. The onshore and vertical motions form a two-cell circulation similar to the observations reported by Halpern, Smith and Mittelstaedt (1977).

Fig. 2 shows the longshore transport at y = 0. The effect of the southward coastal current and the offshore countercurrent on the surface isotherms can be noted. The colder water at the coast is caused by a combination of coastal upwelling and of advection of cold water along the coast. Offshore a tongue of warmer water is formed by the northward advection in the countercurrent.

(ii) In the second case the wind stress is uniform. The cross-shelf circulation is illustrated in Fig. 3 and the longshore transport and isotherms in Fig. 4. The uniform southward wind distribution produces a southward coastal current without a countercurrent and a single-cell pattern develops for the onshore flow. Colder water, produced by coastal upwelling and southward advection, occurs at the coast.

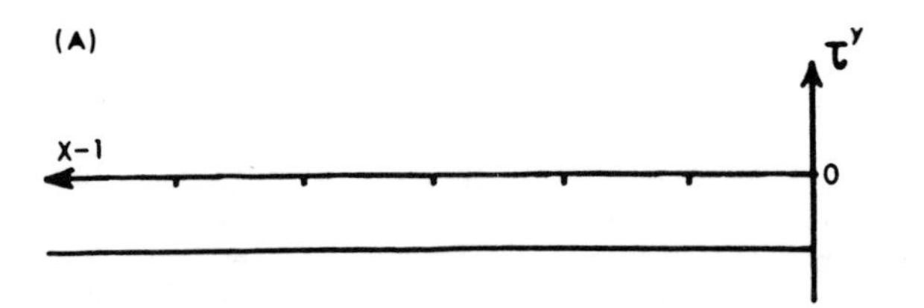

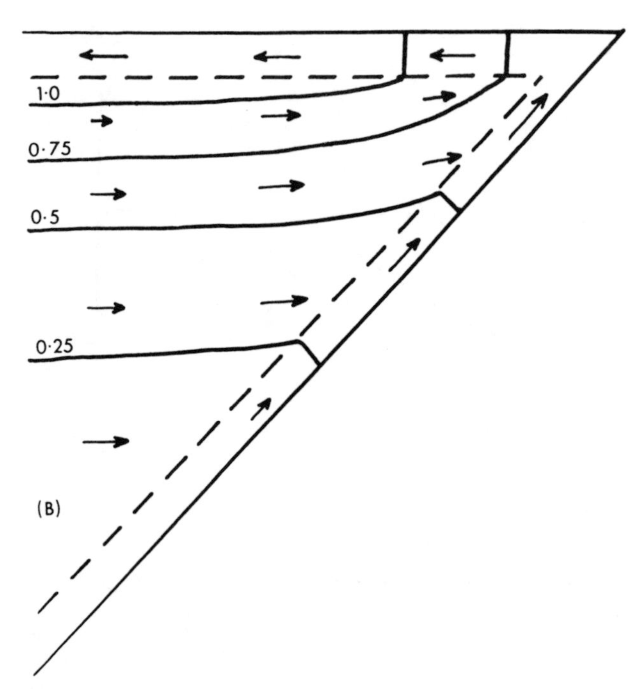

Fig. 3. (a) The wind stress τ^y over the shelf. (b) A cross section through the shelf region at y = 0. The solid lines represent the isotherms.

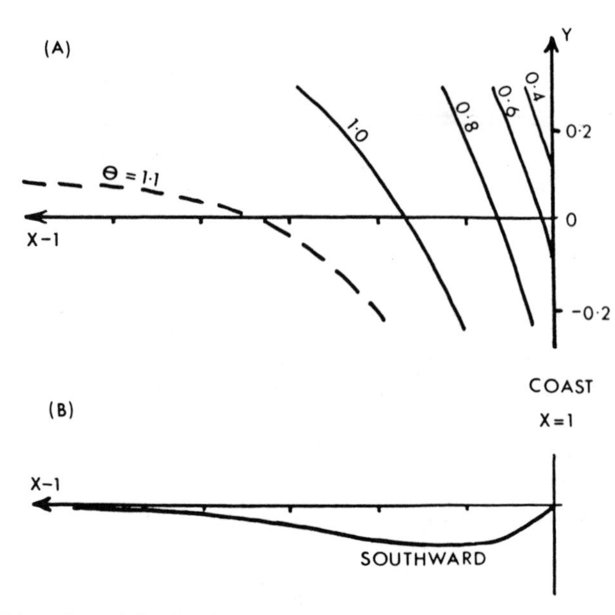

Fig. 4. (a) The barotropic transport in the shelf current at y = 0 and (b) surface isotherms for the wind stress in Fig. 3(a).

Discussion

The formation of countercurrents in models of shelf currents over sloping shelves was noted by Johnson (1976) when the wind stress varies offshore with x and by Johnson and Manja (1980) for winds that change sign with latitude y or with time t. The change in direction of the longshore velocity causes a reversal of the Ekman transport in the bottom boundary layer. The reversal is associated with the two-cell circulation and is illustrated in Figs 1 and 3 for the two examples presented here. A set of solutions calculated for different wind fields suggests that two-cell circulations are most likely to occur in time and space near a change in direction of the longshore component of the wind.

This article has concentrated on the steady shelf currents in a slightly stratified ocean. A complementary study of unsteady shelf currents in a homogeneous ocean is presented by Johnson and Manja (1980). Their numerical solutions contain evidence of Kelvin and shelf wave propagation along the coast and of Rossby wave propagation offshore. The next development in this series of models is to include a time-dependent wind stress for the slightly stratified ocean. This should follow in a straightforward manner as a perturbation of the unsteady barotropic flow given by Johnson and Manja (1980) using methods similar to those pursued in this article.

Acknowledgements. I am grateful to the Royal Society and the American Geophysical Union for travel grants that enabled me to attend the IDOE International Symposium on Coastal Upwelling at Los Angeles in February 1980.

References

Barton, E.D., A. Huyer, and R.L. Smith, Temporal variations observed in the hydrographic regime near Cabo Corbeiro in the northwest African upwelling region in February-March 1974, *Deep-Sea Research, 24*, 7-23, 1977.

Gill, A.E., and A.J. Clarke, Wind-induced upwelling, coastal currents and sea-level changes, *Deep-Sea Research, 21*, 325-345, 1974.

Halpern, D., Measurements of near-surface wind stress over an upwelling region near the Oregon coast, *Journal of Physical Oceanography, 6*, 108-112, 1976.

Halpern, D., R.L. Smith, and E. Mittelstaedt, Cross-shelf circulation on the continental shelf off northwest Africa during upwelling, *Journal of Marine Research, 35*, 787-796, 1977.

Johnson, J.A., Modifications of coastal currents and upwelling by offshore variations in wind stress, *Tellus, 28*, 254-260, 1976.

Johnson, J.A., and B.A. Manja, Unsteady currents over a continental shelf, *Tellus, 32*, to appear, 1980.

Lighthill, M.J., Dynamic response of the Indian Ocean to onset of the Southwest Monsoon, *Philosophical Transactions of the Royal Society, A265*, 45-92, 1969.

Peffley, M.B., and J.J. O'Brien, A three-dimensional simulation of coastal upwelling off Oregon, *Journal of Physical Oceanography, 6*, 164-180, 1976.

INTERNAL CROSS-SHELF FLOW REVERSALS DURING COASTAL UPWELLING

Donald R. Johnson[1]

Rosenstiel School of Marine and Atmospheric Science
University of Miami, 4600 Rickenbacker Causeway, Miami, Florida 33149

and

Christopher N.K. Mooers

Department of Oceanography, Naval Postgraduate School, Monterey, California 93940

[1]Present affiliation: Old Dominion University,
Norfolk, Virginia 23508

Abstract. A theoretical balance for the cross-shelf flow in a coastal upwelling frontal zone is developed and elucidated. The Coriolis force due to the cross-shelf flow is balanced by the vertical friction associated with the geostrophic along-shelf flow. Using the "thermal wind" relation, the cross-shelf flow for constant vertical eddy viscosity is proportional to the horizontal variation of the static stability. Hence, there is a critical streamline (cross-shelf flow reversal) in the center of an inclined pycnocline (frontal zone). For a vertical eddy viscosity inversely dependent on the Richardson number, the critical streamline generally occurs in the lower half of the frontal zone.

Introduction

One of the most controversial and as yet uncertain aspects of coastal upwelling circulation concerns reversals with depth in the cross-shelf flow. The reversals considered are "internal" to the water column, i.e., located away from the direct influence of either the surface or the bottom turbulent boundary layers. As defined for this study, the reversals consist of a shoreward flow above a seaward flow in the vicinity of the uplifted pycnocline. A model that concentrates on the internal flow reversals and is consistent with some of the observations reported in the literature will be presented. An example of a cross-shelf internal flow reversal is presented in Fig. 1, taken from Cabo Bojador, northwest Africa.

Several common characteristics are noted in the coastal upwelling regions where such reversals have been observed (Johnson, Barton, Hughes, and Mooers, 1975; Johnson and Johnson, 1979; Johnson, Fonseca, and Sievers, 1980): 1) They occur most prominently near mid-shelf where a near-surface pycnocline forms a frontal zone that frequently outcrops at the sea surface forming a surface front; 2) They are weak or non-existent near the shelf break-slope; and 3) They occur just below the level of the maximum density gradient in a layer of large vertical shear of the alongshore flow, in conjunction with low dynamic stability due to mean and transient flows.

The similarities in the current profiles and in their conditions of occurrence are strong evidence that the flow reversals are characteristic features. The state of the cross-shelf flow at any instant, including the existence of the flow reversals, depends on the synoptic scale wind systems, as well as other factors, which produce transient coastal upwelling. Thus, considerable spatial and temporal variability can be expected in the cross-shelf flow field, and some delicacy is required in observation and analysis to perceive the essence of this flow component.

Dynamics

With a two-layered model in which a geostrophic, baroclinic alongshore jet is in equilibrium with an upwelled density interface, Csanady (1972) found that frictional stress across the interface produces a secondary circulation (due to a double Ekman layer along the interface) resembling the observed internal flow reversals. It was implied that upwelling-favorable wind maintains the circulation and density front. In a two-layered numerical simulation model with similar interface conditions of matching velocity and stress, para-

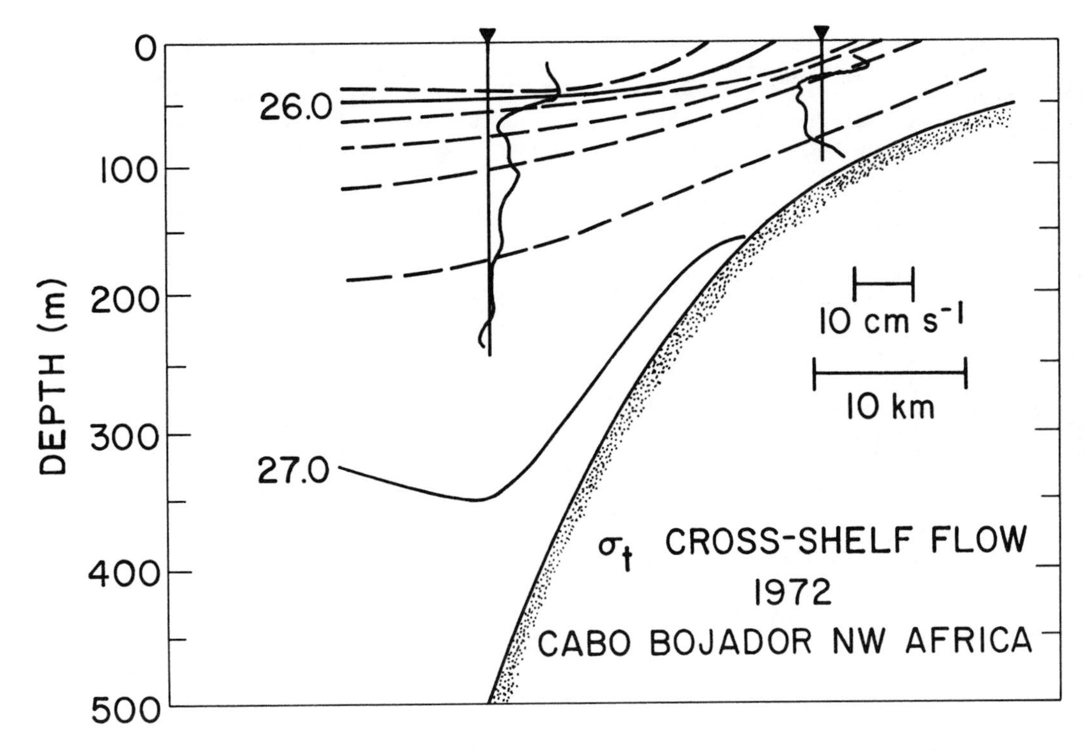

Fig. 1. Cross-shelf current profiles superimposed on σ_t section. Adapted from Johnson, Barton, Hughes, and Mooers (1975). The shallow station profile has been averaged over a 2-day period and the deeper station profile over a continuous 3-day period, each with casts at 6-h intervals.

meterization of turbulent entrainment, and upwelling-favorable winds, Thompson (1974) found that flow reversals occurred near the steeply inclined interface (frontal zone) over the shelf and that they did not extend to the shelf break.

The two-layered models do not fully describe the frontal zone in coastal upwelling situations because the no-slip condition at the interface is not fully realistic. This objection can be avoided with a local model of an inclined pycnocline for which it can be assumed that the following conditions apply: (1) steadiness, (2) alongshore uniformity, (3) small Rossby number, (4) negligible horizontal friction, and (5) small Ekman number E_v; $E_v = A/H^2f$, where f is the Coriolis parameter, A is the vertical eddy viscosity coefficient, and H is the thickness of the shear layer (the thickness of the pycnocline in the model provides a rough estimate of H). Then the first-order alongshore flow is in geostrophic ("thermal wind") balance, and the cross-shelf flow results from relatively slight frictional veering of the velocity vector:

$$-fv_z = \alpha_o g \rho_x \qquad (1)$$

$$fu = (Av_z)_z \qquad (2)$$

$$N^2 \equiv -\alpha_o g \rho_z \qquad (3)$$

where x is positive eastward (shoreward), y is positive northward, z is positive upward, ρ is the density, and α_o is the reference specific volume. Subscripted variables represent partial derivatives. Assuming constant A, the following equation for u is obtained from equations (1) to (3):

$$u = \frac{A}{f^2} (N^2)_x \qquad (4)$$

In a typical upwelling situation, $(N^2)_x$ is positive near the top of the inclined pycnocline and u is expected to be positive (shoreward), $(N^2)_x$ is negative near the bottom of the pycnocline, and u is expected to be negative (seaward). Also, there is a stronger tendency toward internal secondary circulation at lower latitudes and with stronger vertical mixing.

Although the above model is simplified for a clear statement of the pertinent physics, Endoh (1977), Suginohara (1977), and Matsuno, Takeuchi, and Suginohara (1978) achieved similar results in more complete, time-dependent, numerical models*.

* Hamilton and Rattray (1978) also applied eddy viscosity models, but with little indication of internal flow reversals. The difference may be due to the small vertical variation in N^2, hence weak second derivative in v.

Endoh used a density contrast of $\Delta\rho \sim 2 \times 10^{-3}$ g/cm^3 across a pycnocline approximately 20 m thick, a wind stress of 1.0 dyne/cm^2 equatorward and parallel to the coastline, $f = 10^{-4}$/s, and A = 1,12.5, and 100 cm^2/s for three representative cases. The internal circulation pattern did not develop with A = 1 cm^2/s ($E_v \sim 0.0025$), although cellular patterns developed above the pycnocline. However, the internal flow reversals developed in the other cases, namely A = 12.5 cm^2/s ($E_v \sim 0.03$) and A = 100 cm^2/s ($E_v \sim 0.25$).

In effect, the model is something like the double Ekman layer model, but it does not require the no-slip interface: the vertical shear in alongshore flow generated by the no-slip condition of the Csanady and Thompson models is replaced by geostrophic shear near the inclined pycnocline. Vertical turbulence acting on the alongshore shear leads to veering of the velocity vector producing the internal flow reversal. However, the model can be criticized on the basis that the internal flow reversal has a critical streamline near the center of the pycnocline while the observed reversals seem to occur with the critical streamline below the maximum density gradient.

Although equation (4) provides an understanding of the expected internal cross-shelf flow based upon the several assumptions, the cross-shelf component is provided directly in equation (2). Proceeding diagnostically, u profiles were calculated from equation (2) using observed v profiles and a constant A. Only limited success was achieved in verification against observed u profiles. A region of negative v_{zz} (offshore u) occurred below the pycnocline in the observations; however, a region of strong positive v_{zz} (shoreward u) within (or above) the pycnocline was missing, probably due to overlapping with the near-surface turbulent layer.

Observations have shown that reversals occur in conjunction with low Richardson number (dynamic stability), R_i, where $R_i = N^2/S^2$ and $S^2 = (du/dz)^2 + (dv/dz)^2$. If the vertical eddy viscosity is inversely dependent on R_i (Munk and Anderson, 1948), then the use of a constant A in equation (2) is not generally appropriate. Although the inverse relationship between A and R_i is not fully established, it is not unreasonable to assume that such a relationship must exist under conditions of turbulence generated by local shear. Furthermore, although mixing will only occur when R_i reaches a critically low value, the probability of overturn is surely enhanced in the presence of low $|O(1)|R_i$ based upon average conditions. Hence, lacking information on the vertical distribution of internal wave shears, the R_i - dependent A, calculated from average N^2 and geostrophic shears, can be considered an "expected" value.

Munk and Anderson (1948) derived the relationship

$$A = \frac{A_o}{(1 + 10R_i)^{1/2}}$$

where A_o is the vertical eddy viscosity in the limit of neutral static stability. With the realistic assumption that $R_i \geq 1$ in the interior, and that the vertical shears are principally given by v_z, the alongshore shear,

$$A \approx \frac{A_o}{3R_i^{1/2}} \approx \frac{A_o|v_z|}{3|N|}$$

with $\rho_x \approx -\gamma\rho_z$ and γ the local slope of an isopycnal (positive in the interior during upwelling), then

$$v_z \approx -\frac{\gamma N^2}{f} \text{ and } A \approx \frac{A_o\gamma N}{3f} .$$

Then, from equation (2),

$$u \approx -\frac{A_o\gamma N^2}{3f^3}(3\gamma N_z + 2N\gamma_z) \qquad (5)$$

The first term in equation (5) is negative in the region (interior) below maximum N^2. In coastal upwelling fronts, isopycnals tend to converge as they upwarp toward shore, i.e., the slope of an isopycnal is greater below the pycnocline than above it. There γ_z is negative and the second term is positive. Hence, the internal flow reversal occurs where these two terms balance, which will be at some level below the center of the pycnocline but above its base.

From equation (5), cross-shelf flow reversals are expected to be (1) restricted to frontal regions, and (2) highly sensitive to latitude and turbulent intensity, being stronger at lower latitudes and where vertical mixing is strongest if everything else is equal. The theory depends implicitly on the existence of intermittent packets of internal wave energy that enhance the interior mixing (Johnson, Van Leer, and Mooers, 1976; Wang and Mooers, 1976), which was accounted for by the implication that internal wave shears may reduce already low R_i values to critical values.

Although to elucidate the pertinent physics the model has been simplified by assuming $E_v^2 << 1$, the qualitative results would have been similar with a larger Ekman number scaling. In that case, the term $(Au_z)_{zz}$ would have to be added to equation (1) and the shear in cross-shelf flow would have to be included in calculations of R_i. The remaining four assumptions, as previously listed, are not so easily dismissed. Consider the following, expanded set of equations for horizontal motion:

$$v_t + uv_x + fu = -\alpha_o P_y + Av_{zz} + A_h v_{xx} \qquad (6)$$

$$u_t + uu_x - fv = -\alpha_o P_x + Au_{zz} + A_h u_{xx} \qquad (7)$$

TABLE 1. Scale estimates

Scale Values	Scaled Terms in Equation (8)
$E_v \sim 0.2$	Local acceleration ~ 1
$E_h \sim 0.1$	Nonlinear acceleration ~ 0.2
$u_o/v_o \sim 0.2$	Coriolis ~ 1
$n \sim 5$	Pressure gradient ~ 0.5
$R_o \sim 0.2$	Vertical friction ~ 1
$L_x/L_y \sim 0.1$	Horizontal friction ~ 0.5

Equations (6) and (7) are scaled in the following manner:

$$u = u_o u' \qquad v = v_o v' \qquad t = \frac{n}{f} t' \qquad x = L_x x' \qquad y = L_y y' \qquad z = Hz'$$

$$E_v \equiv \frac{A}{H^2 f} \text{ (vertical Ekman no.)}$$

$$E_h \equiv \frac{A_h}{L_x^2 f} \text{ (horizontal Ekman no.)}$$

$$R_o \equiv \frac{V_o}{f L_x} \text{ (Rossby no.)}$$

$$\alpha_o \frac{\partial P}{\partial x, \partial y} = \frac{\Delta P}{L_x, L_y} \cdot \alpha_o' \frac{\partial P'}{\partial_x', \partial_y'}$$

The acceleration time scale nf^{-1} is convenient in that energetic wind fluctuation periods, which are assumed to be the dominant time scales, are expressed as n inertial periods in length. Using the scale values given in Table 1, it is easy to verify that the x-component equation (7) is dominated by the geostrophic relation. Then by eliminating the pressure term with the help of the geostrophic balance of equation (7), dividing through by the vertical friction term, and dropping the primes, the scaled y-component equation becomes

$$(nE_v)^{-1} v_t + \frac{(R_o u_o)}{(E_v v_o)} u v_x + \frac{(u_o)}{(v_o E_v)} u = - \frac{(L_x)}{(L_y E_v)} \alpha_o P_y + v_{zz} + (E_h E_v^{-1}) v_{xx} \qquad (8)$$

It should be emphasized that the scales given in Table 1 are estimated for a location near the frontal zone and, in particular, for that area at the base of the pycnocline where mixing is relatively intense. It is also noted that alongshore variations in the velocity field have been ignored. This may not be valid in certain cases. Applying these estimates to equation (8), it is clear that none of the terms (with the possible exception of the non-linear acceleration term) can be safely excluded. The point of the exercise is easily taken: any of the terms can be important and probably are important at different locations and at different times.

Although the internal flow reversals have been observed on at least four different occasions and in three widely separated upwelling areas, the relatively short observational periods, i.e., a few days, make it unclear if the reversals are an average or a transient feature. However, considering the consistent observational characteristics of the reversals, the relatively straightforward physical explanation, and the fact that reasonable scales do not exclude them, it is realistic to expect that they are part of the upwelling sequence.

Summary

Internal cross-shelf flow reversals of similar characteristics have been observed in widely diverse coastal upwelling regions. A local model has been developed that is consistent with some of the observed features. At the core of the model is a vertical eddy viscosity parameterization that depends inversely on the gradient Richardson number R_i, which in turn depends upon the vertical shear of the alongshore geostrophic velocity. Shoreward convergence of upwarped isopycnals causes a minimum in R_i just below the level of the maximum vertical density gradient, which results in enhanced turbulent mixing toward the base of the pycnocline and veering of the velocity vector such as to produce the cross-shelf flow reversals.

Theoretical considerations together with observations support the inference that cross-shelf flow reversals are restricted to relatively strong frontal regions with their concomitant strong vertical shears in geostrophic velocity. According to the model presented, the strength of the cross-shelf flow is greater at lower latitudes and where turbulent mixing is greatest.

Acknowledgements. This manuscript is a contribution to the Coastal Upwelling Ecosystems Analysis Program, supported by the National Science Foundation, Office of the International Decade of Ocean Exploration, under grant OCE 77-28354 to the University of Delaware where C.N.K. Mooers was the principal investigator. Partial support has been provided to D.R. Johnson by the National Science Foundation under grant ATM 76-82520.

References

Csanady, G.T., Frictional secondary circulation near an upwelled thermocline, Paper presented at the Canadian Committee of Oceanography Symposium, Burlington, Ontario, May 4, 1972.

Endoh, M., Double-celled circulation in coastal upwelling, *Journal of the Oceanographical Society of Japan, 33,* 30-37, 1977

Hamilton, G.P., and M. Rattray, A numerical model of the depth-dependent, wind driven circulation on a continental shelf, *Journal of Physical Oceanography, 8,* 437-457, 1978.

Johnson, D.R., E.D. Barton, P. Hughes, and C.N.K. Mooers, Circulation in the Canary Current upwelling region off Cabo Bojador in August, 1972, *Deep-Sea Research, 22,* 547-558, 1975.

Johnson, D.R., T. Fonseca, and H. Sievers, Upwelling in the Humboldt Coastal Current near Valparaiso, Chile, *Journal of Marine Research, 38,* 1-16, 1980.

Johnson, D.R., and W.R. Johnson, Vertical and cross-shelf flow in the coastal upwelling off Oregon, *Deep-Sea Research, 26,* 399-408, 1979.

Johnson, D.R., J.C. Van Leer, and C.N.K. Mooers, A cyclesonde view of coastal upwelling, *Journal of Physical Oceanography, 6,* 566-574, 1976.

Matsuno, K., K. Takeuchi, and N. Suginohara, Upwelling fronts and two-cell circulation in a two-dimensional model, *Journal of the Oceanographical Society of Japan, 34,* 217-221, 1978.

Munk, W.H., and E.R. Anderson, Notes on a theory of the thermocline, *Journal of Marine Research, 7,* 276-295, 1948.

Suginohara, N., Upwelling front and two-celled circulation, *Journal of the Oceanographical Society of Japan, 33,* 115-130, 1977.

Thompson, J.D., The coastal upwelling cycle on a beta-plane: Hydrodynamics and thermodynamics, Mesoscale Air-Sea Interaction Group Technical Report, Florida State University, Tallahassee, 141 pp., 1974.

Wang, D.-P., and C.N.K. Mooers, Evidence for interior dissipation and mixing during a coastal upwelling event off Oregon, *Journal of Marine Research, 35,* 697-713, 1976.

A TWO-DIMENSIONAL DIABATIC ISOPYCNAL MODEL OF A COASTAL UPWELLING FRONT

E-Chien Foo

Meteorology Annex, The Florida State University
Tallahassee, Florida 32306

Claes Rooth

Rosenstiel School of Marine and Atmospheric Science,
University of Miami, 4600 Rickenbacker Causeway,
Miami, Florida 33149

Rainer Bleck

Rosenstiel School of Marine and Atmospheric Science,
University of Miami, 4600 Rickenbacker Causeway,
Miami, Florida 33149

Abstract. A two-dimensional high resolution isopycnal model is designed to simulate coastal upwelling frontogensis under favorable longshore wind stress. Diabatic effects are generated by vertical mixing of density. The vertical eddy coefficients are determined explicitly from the mean field gradient by either a Richardson number related formula or a second order moment closure turbulence model. Munk and Anderson's empirical formula is chosen for the first method. For the second method, a simplified diagnostic version of a turbulence closure model derived from Worthem and Mellor's model (1980) is used.

Using the moment closure turbulence model, the eddy coefficients have strong gradients across the front. The turbulent mixing works to intensify an existing front and then serves as a dissipating mechanism.

This study suggests that there is a strong interaction between the turbulent mixing and the frontal structure. Certain phenomena on the frontal scale emerge from all the case studies. Two phenomena are particularly important: First, the velocity and density structure of the front confine the turbulence events to localized regions; second, the localized turbulent events largely affect the frontal structure and secondary circulation on the frontal scale. Instead of forming a double Ekman layer in the interior, the density gradient of the surrounding area is affected. Strong upwelling occurs at the position of strong downward mixing. The secondary circulation pattern induces concentrated upwelling flow away from the coast. The flow may be important in increasing local nutrient concentrations and inducing patches of biologically highly productive water.

1. Introduction

The formation of a surface front during an upwelling period is not always supported by observations. On the contrary, a strong surface convergence zone associated with a sharp front was observed by Stevenson, Garvine, and Wyatt (1974) during the relaxation period. However, the convergence was confined to a very shallow layer. Curtin (1979) found that during the upwelling period the surface front, though moderate, had longshore uniformity, while during the relaxation the front sharpened, it developed a three-dimensional irregularity.

In this study, we try to simulate a two-dimensional surface front during an upwelling period. Frontogensis in a two-dimensional field can be greatly enhanced by applying a strain rate to the system (Blumen, 1978). A deformation field with a scale much larger than the local frontal scale can induce non-linear advection and cause ultimate frontogenesis. Hoskins and Bretherton (1972) used a convergent barotropic flow while Orlanski and Ross (1977) used a baroclinic cross-front flow to force frontogenesis in atmospheric studies.

Pedlosky (1978) used the interior return flow generated by surface Ekman transport to force frontogensis in a coastal upwelling study. A sur-

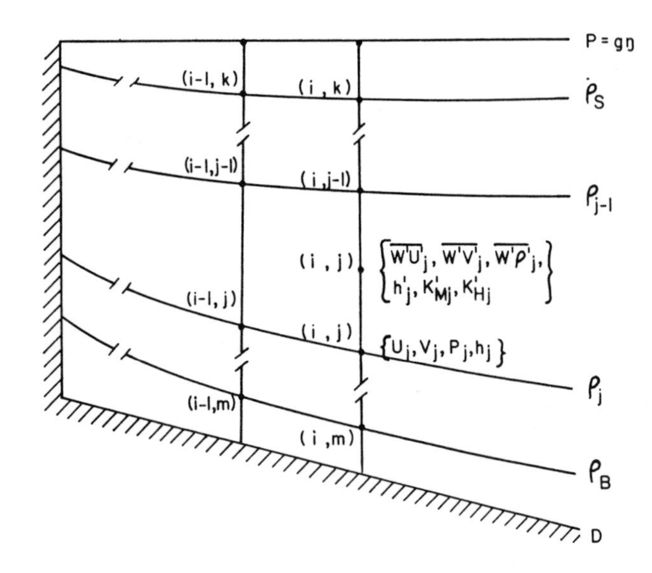

Fig. 1. The grid layout of the model. Only two interior grid levels and the surface and bottom boundary layers are shown. Indices in the parenthesis indicate the grid location. The parameters are explained in the text.

face density discontinuity is formed along the coast within a few days after the onset of a wind event. The vertical stretching of the density layer decreases the local stratification, which in turn decreases the horizontal scale of the upwelling zone. Thus, a vertical stretching is further intensified. Over a finite depth along the coast, the non-linear feedback mechanism forces the vertical velocity to escalate without restraint. The density field in the region is lifted to the surface and a discontinuity is formed. To generate a front with realistic structure and location, Pedlosky (1978) suggested that a higher order mechanism such as turbulence should be important dynamically.

The objective of this study is to simulate frontal development during an upwelling event. Emphasis has been put on evaluating non-linear and turbulent effects, especially within a front, so that frontal dynamics can be simulated adequately.

Turbulent diffusion is basically induced by vertical mixing. From one-dimensional boundary layer studies, it is well known that, although constant eddy coefficient formulations work as diffusive mechanisms, vertically varying eddy coefficients can sometimes sharpen an existing density gradient. To obtain some spatial distributions of turbulence intensity, the eddy coefficient will be determined explicitly from the mean field structure.

In this study, we simulate the evolution of a coastal upwelling front with a pre-existing strong pycnocline. The inhomogeneity helps to initiate a surface front as it surfaces. How does a surface front affect the turbulence structure? Is a turbulence event feedback to the frontal dynamics? Does positive feedback exist between the frontal dynamics and the turbulent events? What is the final circulation pattern?

2. Model

A diabatic isopycnal model is designed with the following characteristics:

(1) In an adiabatic system, the isopycnal coordinates are material. A material coordinate improves the vertical grid resolution in frontal regions in a natural manner. The diabatic effect will generate vertical velocity in this coordinate system. The velocity is physical and exists in every coordinate system. Furthermore, the grid levels run parallel to the frontal structure so that false horizontal diffusion across a sloping front is avoided.

(2) The diabatic effect is estimated from vertical mixing of the density field. A sophisticated moment closure scheme is used to derive vertical eddy coefficients so that sharp boundaries exist between the fully turbulent regions and the surrounding quiescent regions. We use a simplified diagnostic version derived from Worthem and Mellor's (1980) model.

To demonstrate the effect of localized turbulent mixing, a comparison has been made with Munk and Anderson's (1948) empirical formula, which gives a much diffused eddy coefficient distribution.

(3) Due to the frequent turbulence at boundary regions, two homogeneous regions are developed at the surface and at the bottom of the ocean. The non-isopycnal boundary layers use average velocity and density while they are assumed to vary linearly between two adjacent isopycnal grid levels.

2.1 *Structure of the Model*

Fig. 1 shows the grid layout for the variables, the surface boundary layer, two adjacent interior grid levels, and the bottom boundary layer. The index i = 1, ..., N, indicates the horizontal grid location running from left to right, while j = 1, ..., M, indicates the vertical grid location running from top to bottom. The variables, u_j, v_j, h_j, ρ_j, and P_j are the horizontal cross-shelf velocity, longshore velocity, depth, density, and pressure at level j, respectively, while $\overline{wu}_j$, $\overline{wv}_j$, $\overline{w\rho}_j$, K_{Mj}, and K_{Hj} are the vertical fluxes and eddy coefficients at the midpoint of level j and j-1. The depth of the midpoint is h'_j; D is the total depth.

Details of the model structure are given by Foo (1980). The equations are given here without furthere explanation.

2.2 Governing Equations

2.2.1 Interior

$$\frac{\partial u_j}{\partial t} + u_j \frac{\partial u_j}{\partial x} + \dot{\rho} \frac{\partial u_j}{\partial \rho} = A_M \frac{\partial^2 u_j}{\partial x^2} + \frac{\partial \rho}{\partial h} \frac{\partial}{\partial \rho} K'_{Mj} \frac{\partial u_j}{\partial \rho} - \frac{\partial M_j}{\partial x} - f v_j \tag{2-1}$$

$$\frac{\partial v_j}{\partial t} + u_j \frac{\partial v_j}{\partial x} + \dot{\rho}_j \frac{\partial v_j}{\partial \rho} = A_M \frac{\partial^2 v_j}{\partial x^2} + \frac{\partial \rho}{\partial h} \frac{\partial}{\partial \rho} K'_{Mj} \frac{\partial v_j}{\partial \rho} - \frac{\partial M_j}{\partial y} + f u_j \tag{2-2}$$

$$M_j = \frac{P_j}{P_j} + g h_j \tag{2-3}$$

$$\begin{bmatrix} K'_{Mj} \\ K'_{Hj} \end{bmatrix} = \begin{bmatrix} K_{Mj} \\ K_{Hj} \end{bmatrix} \frac{\partial \rho_j}{\partial h} \tag{2-4}$$

$$\dot{\rho}_j = \frac{d\rho}{dt} = \frac{\partial}{\partial h} K_{Hj} \frac{\partial \rho}{\partial h} = \frac{\partial \rho}{\partial h} \frac{\partial}{\partial \rho} K'_{Hj} \tag{2-5}$$

$$\frac{\partial P_j}{\partial t} = \sum_{i=2}^{j} \frac{\partial}{\partial x} \left[P_i - P_{i-1}\right] \left[u_i + u_{i-1}\right] /2 + \dot{\rho}_j \frac{\partial P_j}{\partial \rho} + A_H \frac{\partial^2 P_j}{\partial x^2} \tag{2-6}$$

where A_M and A_H are horizontal eddy coefficients.

2.2.2 Bottom Boundary Layer:

$$\frac{\partial u_m}{\partial t} + u_m \frac{\partial u_m}{\partial x} = A_M \frac{\partial^2 u_m}{\partial x^2} - \frac{C_d u_m^2}{(D-h'_m)} + \frac{\overline{w'u'}_m}{(D-h'_m)} - \frac{\partial M_m}{\partial x} + \frac{P_m}{\rho_B^2} \frac{\partial \rho_B}{\rho x} + f V_m \tag{2-7}$$

$$\frac{\partial v_m}{\partial t} + u_m \frac{\partial v_m}{\partial x} = A_M \frac{\partial^2 v_m}{\partial x^2} - \frac{C_d v_m^2}{(D-h'_m)} + \frac{\overline{w'v'}_m}{(D-h'_m)} - \frac{\partial M_m}{\partial y} + \frac{P_m}{\rho_B^2} \frac{\partial \rho_B}{\partial y} - f u_m \tag{2-8}$$

$$\frac{\partial \rho_B}{\partial t} = - \frac{\overline{w'\rho'}_m}{h'_m} - u_m \frac{\partial \rho_B}{\partial x} \tag{2-9}$$

$$\frac{\partial P_m}{\partial t} = \frac{\partial \rho_B}{\partial t} \frac{\partial P}{\partial \rho} \Big|_{P_m} = \frac{\partial \rho_B}{\partial t} \frac{P_m - P_{m-1}}{\rho_B - \rho_{m-1}} \tag{2-10}$$

$$h_m = P_m / \rho_B \tag{2-11}$$

Subscript m is the vertical index of the bottom boundary layer; Cd is the bottom drag coefficient. The Reynolds fluxes are defined from the region adjacent to the boundary layer.

$$\overline{w'u'}_m = K_{Mm} \frac{u_m - u_{m-1}}{h_m - h_{m-1}} = K'_{Mm} \frac{u_m - u_{m-1}}{\rho_B - \rho_{m-1}}$$

$$\overline{w'v'}_m = K_{Mm} \frac{v_m - v_{m-1}}{h_m - h_{m-1}} = K'_{Mm} \frac{v_m - v_{m-1}}{\rho_B - \rho_{m-1}}$$

$$\overline{w'p'}_m = K_{Hm} \frac{\rho_m - \rho_{m-1}}{h_m - h_{m-1}} = K'_{Hm} \tag{2-12}$$

2.2.3 Surface Boundary Layer

$$\frac{\partial u_k}{\partial t} + u_k \frac{\partial u_k}{\partial x} = A_M \frac{\partial^2 u_k}{\partial x^2} - \frac{\overline{w'u'}_{k+1}}{h'_{k+1}} - \frac{\partial M_k}{\partial x} + \frac{P_k}{\rho_S^2} \frac{\partial \rho_S}{\partial x} + f v_k \tag{2-13}$$

$$\frac{\partial v_k}{\partial t} + u_k \frac{\partial v_k}{\partial x} = A_M \frac{\partial^2 v_k}{\partial x^2} + \frac{\tau_y}{h'_{k+1}} - \frac{\overline{w'v'}_{k+1}}{h'_{k+1}} - \frac{\partial M_k}{\partial y} - \frac{P_k}{\rho_S^2} \frac{\partial \rho_S}{\partial y} - f u_k \tag{2-14}$$

$$\frac{\partial \rho_S}{\partial t} = \frac{Q + \overline{w'p'}_{k+1}}{h'_{k+1}} - u_k \frac{\partial \rho_S}{\partial x} \tag{2-15}$$

$$\frac{\partial P_k}{\partial t} = \frac{\partial \rho_S}{\partial t} \frac{\partial P}{\partial \rho} \Big|_{P_k} = \frac{\partial \rho_S}{\partial t} \frac{P_{k+1} - P_k}{\rho_{k+1} - \rho_S} \tag{2-16}$$

$$h_k = P_k / \rho_S, \tag{2-17}$$

where k is the vertical grid number for the surface boundary layer, τ_y is the longshore wind stress, and Q is the surface heat input.

2.3 Vertical Eddy Coefficients

One major part of this study is to model the distribution of turbulent events within a frontal zone and how it affects the cross-front circulation. Two different methods are used:

2.3.1 Richardson number dependent formula

Munk and Anderson (1948) used experimental data to derive the following empirical eddy coefficient formula:

$$K_M = 5 + 50\,(1 + 10R_i)^{-1/2} \tag{2-18}$$

$$K_H = 50\,(1 + 3.3R_i)^{-2/3}, \tag{2-19}$$

where Ri is the Richardson number; the constants are adopted from Hamilton and Rattray (1978).

2.3.2 Second order moment closure scheme

A simplified diagnostic approach is derived from Worthem and Mellor (1980) by dropping the tendency term of the turbulent kinetic energy equation:

$$\frac{\partial}{\partial h}\left(K_M \frac{\partial}{\partial h} q^2\right) = -K_M \left(\frac{\partial u}{\partial h}\right)^2 - K_M \left(\frac{\partial v}{\partial h}\right)^2 - K_H \frac{g}{\rho}\frac{\partial \rho}{\partial h} + 2\frac{q^3}{B_1 \ell}, \tag{2-20}$$

where q^2 is the turbulent kinetic energy, B is an empirical constant, and ℓ is the mixing length scale that has to be determined by an empirical equation. No time derivative term is involved. The eddy coefficients are determined by turbulent energy, mean field gradient, and turbulent length scale:

$$K_M = \ell q S_M \tag{2-21}$$

$$K_H = \ell q S_H \tag{2-22}$$

$$S_M = f\left[\ell,\ q,\ \frac{\partial \rho}{\partial h},\ \frac{\partial u}{\partial h},\ \frac{\partial v}{\partial h}\right] \tag{2-23}$$

$$S_H = g\left[\ell,\ q,\ \frac{\partial \rho}{\partial h},\ \frac{\partial u}{\partial h},\ \frac{\partial v}{\partial h}\right] \tag{2-24}$$

The functions f and g are given by Worthem and Mellor (1980). An empirical formula is used for the length scale, ℓ,

$$\ell = \left(\ell_S^{-1} + \ell_B^{-1} + \ell_V^{-1} + \ell_D^{-1}\right)^{-1}, \tag{2-25}$$

where

$$\ell_S = \kappa h \tag{2-26}$$

$$\ell_B = \kappa (D-h) \tag{2-27}$$

$$\ell_V = \frac{\partial u}{\partial h} \Big/ \frac{\partial^2 u}{\partial h^2} \tag{2-28}$$

$$\ell_D = \left(R_{ic}\, q^2/N^2\right)^{1/2},\quad R_{ic} = 5.43 \tag{2-29}$$

ℓ_S and ℓ_B are controlled by physical boundaries. ℓ_V is a local length scale defined by Von Karman (Tennekes and Lumley, 1972), and ℓ_D was determined by Dickey (1977) in an unforced laboratory experiment. The parameter is the critical maximum length scale for a given turbulence to be sustained.

The model is combined with the isopycnal model to update the eddy coefficients. Coordinate transformation is involved and boundary conditions are changed. In Foo (1980) a one-dimensional comparison test has been run with the original model and shows similar results, which include deepening rate, intensity of turbulent energy and eddy coefficients, velocity, and density structure.

2.4 Integration Scheme

To resolve the fast response of a wind event, a free surface is used to include the fast barotropic gravity wave mode. The horizontal velocity is divided into barotropic and baroclinic components. By formally separating the momentum equation, the surface elevation appears only in the barotropic part of the equation. Therefore, fast surface waves exist only in the barotropic mode. A short time step is used to integrate the barotropic component while a long time step is used to integrate the baroclinic component:

$$\frac{\partial U}{\partial t} + U\frac{\partial U}{\partial x} + (A) - fV = A_M \frac{\partial^2 U}{\partial x^2} + g\frac{\partial \eta}{\partial x} \tag{2-30}$$

$$\frac{\partial V}{\partial t} + U\frac{\partial V}{\partial x} + (B) + fU = A_M \frac{\partial^2 V}{\partial x^2} + g\frac{\partial \eta}{\partial y} \tag{2-31}$$

$$\frac{\partial u}{\partial t} + u\frac{\partial u}{\partial x} + \frac{\partial uU}{\partial x} + \dot{\rho}\frac{\partial u}{\partial \rho} - (A) - fV = -\frac{\partial M}{\partial x} + A_M \frac{\partial^2 u}{\partial x^2} + \frac{\partial \rho}{\partial h}\frac{\partial}{\partial \rho}\left(K_M \frac{\partial u}{\partial \rho}\right) \tag{2-32}$$

$$\frac{\partial v}{t} + u\frac{\partial v}{\partial x} + U\frac{\partial v}{\partial x} + u\frac{\partial V}{\partial x} + \dot{\rho}\frac{\partial v}{\partial \rho} - (B) + fu = -\frac{\partial M}{\partial y} + A_M \frac{\partial^2 v}{\partial x^2} + \frac{\partial \rho}{\partial h}\frac{\partial}{\partial \rho}\left(K_M \frac{\partial v}{\partial \rho}\right) \tag{2-33}$$

$$(A) = \Sigma\, u\frac{\partial P}{\partial t} + \Sigma\, \Delta P \left[U\frac{\partial u}{\partial x} + u\frac{\partial u}{\partial x} - A_M \frac{\partial^2 u}{\partial x^2} + \frac{\partial M}{\partial x}\right] \Big/ \rho g D + \frac{\tau_S^x - \tau_B^x}{\rho D} \tag{2-34}$$

$$(B) = \Sigma v \frac{\partial P}{\partial t} + \Sigma \Delta P \left[U \frac{\partial v}{\partial x} + u \frac{\partial v}{\partial x} - A_M \frac{\partial^2 v}{\partial x^2} + \frac{\partial M}{\partial y} \right] / \rho g D + \frac{\tau_S^y - \tau_B^y}{\rho D}, \quad (2\text{-}35)$$

where U and V are the barotropic components, u and v are the baroclinic components of velocity, and η is the surface elevation. The summations indicate the sum over the vertical layers.

After each long-time step, the barotropic velocity generated in this time step is subtracted from u and v and added to U and V uniformly at each short-time step.

3. Results

Three separate cases are presented to illustrate the frontal development and the importance of turbulent mixing. The first uses a flat bottom, the other two use a sloping bottom.

Certain conditions have been imposed on the model:

1. The model domain is 30 km wide and 150 m deep with a vertical coastal wall at the left hand side. For a sloping bottom, the depth changes linearly from 100 to 150 m.
2. A non-slip boundary is assumed at the coastal wall.
3. The model starts from a state of rest with some horizontal density distribution. A pycnocline centered around 52 m is assumed initially to provide structure for later frontogenesis.
4. A positive, horizontally uniform longshore wind stress is applied, which increases from 0 to 1 dyn/cm^2 within 34 h and then is constant.
5. Two homogeneous boundary layers 10 m thick are assumed.
6. A balance between the surface wind stress and longshore pressure gradient is applied at the ocean boundary so that the surface Ekman transport is balanced by a uniform interior return flow. No bottom boundary current develops in this region.

3.1 The Formation of An Upwelling Front

To illustrate the evolution of an upwelling front, four consecutive figures are shown to give the flow patterns at different stages of an upwelling simulation in which different physical processes are important. Two plots are shown in each figure. The tick marks around the frame represent 1 km in the horizontal and 3 m in the vertical. The horizontal grid spacing is 2 km. The dashed lines are density surfaces; every other line is a grid level. The uneven dashed line indicates the mixed layer depth. In plot (a), the solid lines contour longshore velocity field with 10-cm/s interval. Plot (b) shows the cross front streamfunction with solid lines of 0.1/s interval. Positive value indicated clockwise circulation.

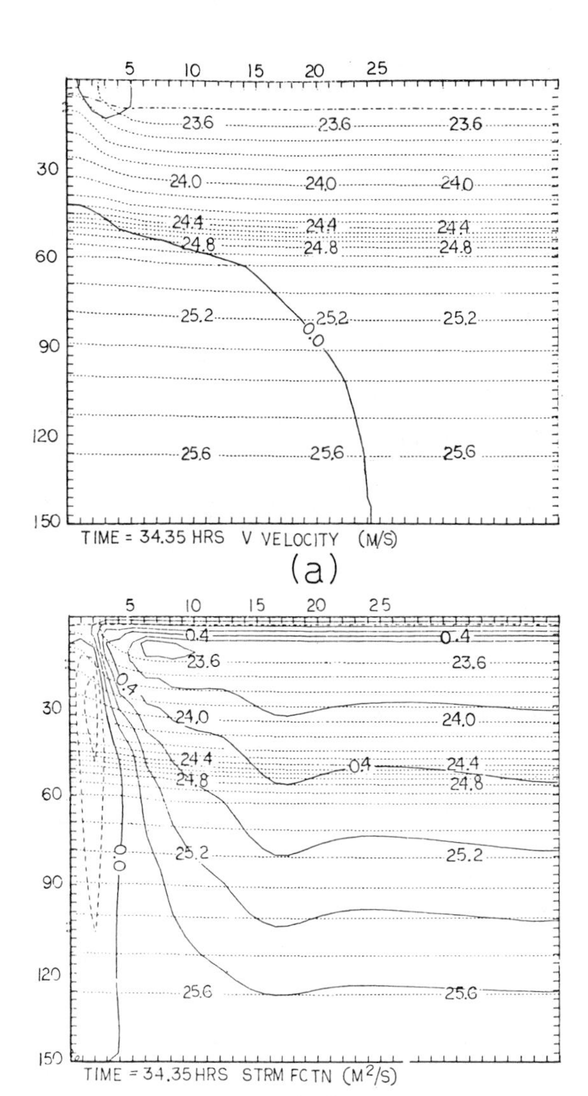

Fig. 2. The upwelling pattern for a flat bottom case at 34.3 h, (a) the longshore velocity with contour interval of 10 cm/s, positive value indicates flow into the plane of the drawing, (b) the cross-front stream function with contour interval of 0.1/s, positive value indicates clockwise circulation.

Fig. 2 shows the flow patterns after 34 h of integration. The coastal jet is weak with maximum velocity of 14 cm/s. The cross-front circulation is quite linear and consists of a uniform surface Ekman outflow and a uniform interior inflow. The two transports are connected through upwelling near the coast. The upwelling region, defined as the region where interior water penetrates into the surface layer, is concentrated in a narrow area.

Fig. 3 shows the flow pattern at 102 h. The longshore current has already developed to full

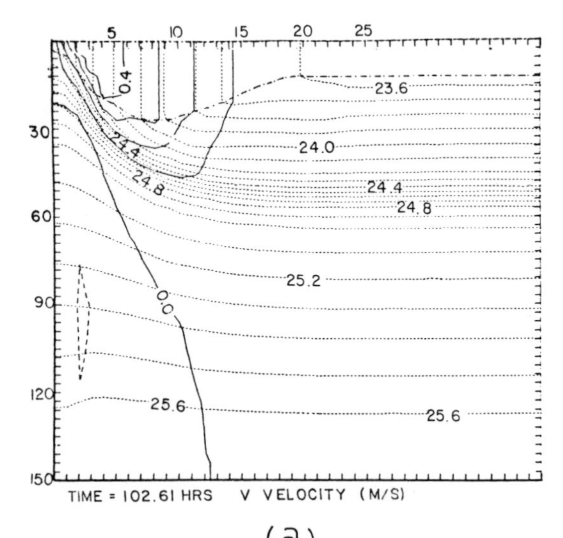

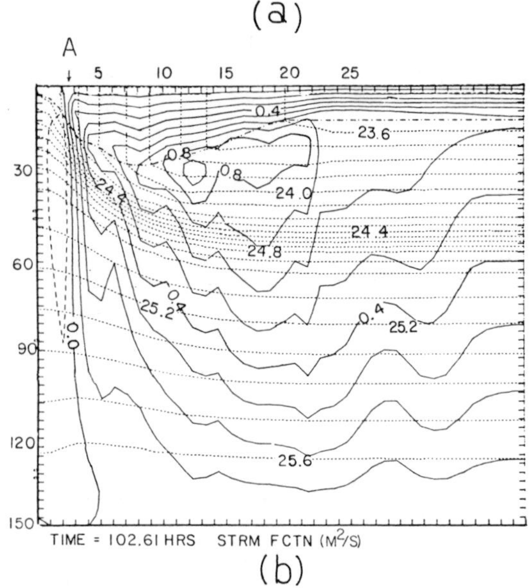

Fig. 3. The same as Fig. 2, but for 102.6 h.

strength. The core of the current has accelerated to 42 cm/s but stays near the coast. The interior density front associated with the original pycnocline just reaches the surface.

The non-linear advection of the longshore momentum by the cross-shelf circulation can no longer be neglected. Additional cross-shelf velocity develops due to the imbalance of the longshore momentum. At the anticyclonic side of the jet, advection increases the longshore velocity in the surface mixed layer. Part of the extra momentum will turn offshore, increasing the cross-shore velocity. At the cyclonic side of the jet, the longshore and cross-shore velocities decrease in the mixed layer. Therefore, a streamfunction maximum is developed 13 km offshore beneath the mixed layer because of the extra offshore transport. A closed circulation develops around the maximum. The upwelling region expands to include the whole area inside the streamfunction maximum. A weak downwelling exists further offshore.

After another 34 h, the pycnocline surfaces and forms a surface front (see Fig. 4). The core of the coastal jet advects away from the coast with this front. Turbulent mixing takes place under the front and at the inshore side of it as indicated by the downward diffusion of the density surfaces. The closed cross-front circulation cell shown in Fig. 3b begins to break into two parts (Fig. 4b).

Turbulent mixing becomes the dominant mechanism at 170 h (Fig. 5). At column B, intensive vertical mixing persists. The mixed layer depth increases 20 m within 34 h. The density surfaces

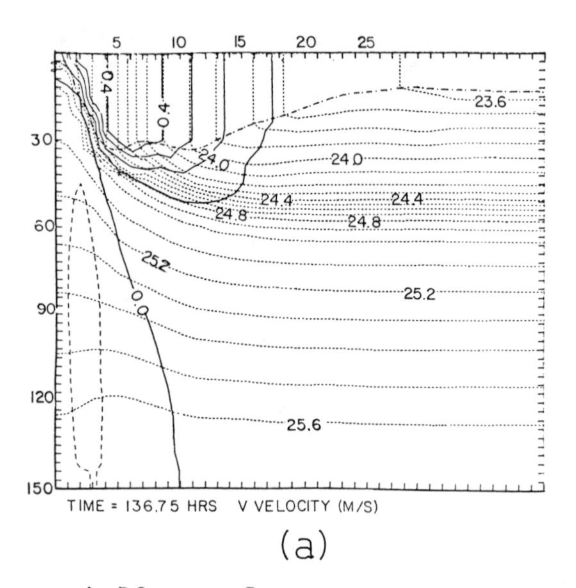

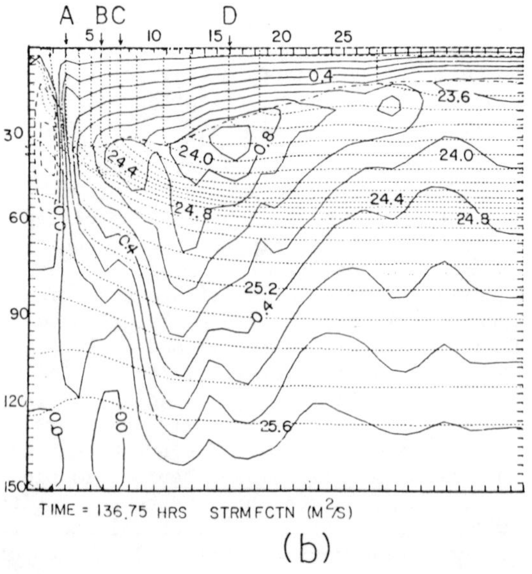

Fig. 4. The same as Fig. 2, but for 136.8 h.

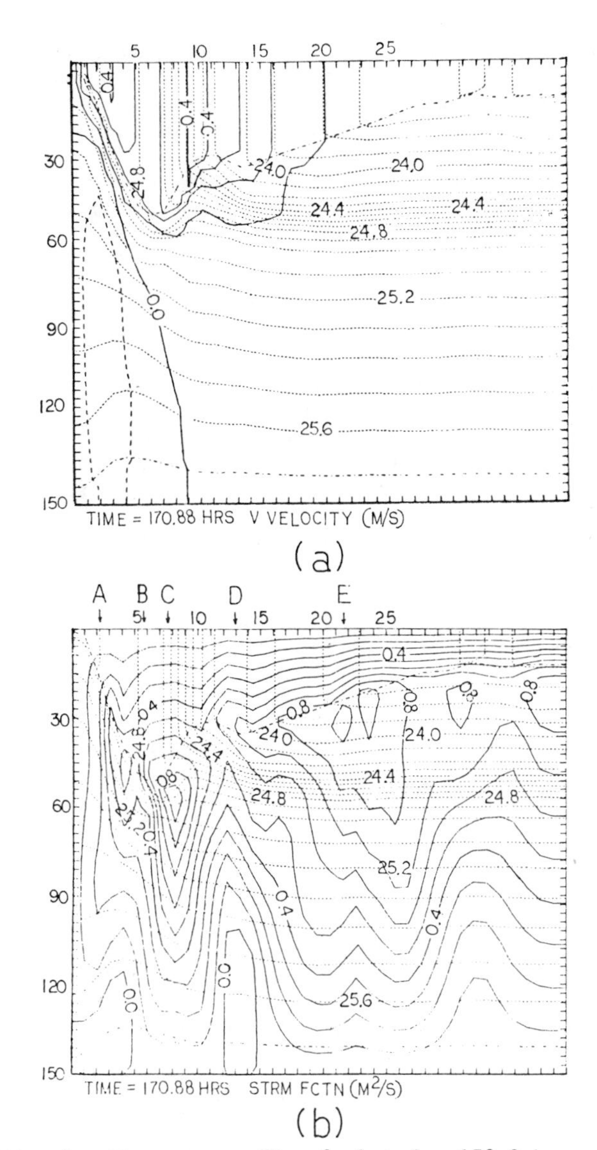

Fig. 5. The same as Fig. 2, but for 170.9 h.

beneath it form a local dip. The surface density gradient greatly decreases at the inshore side and increases at the offshore side. Therefore, the surface density front is intensified by the vertical mixing.

The localized turbulent mixing not only diffuses heat and momentum downward but also changes the horizontal density gradient of the surrounding area. Geostrophic adjustment takes place in these areas. The effect of vertical mixing on a thermal wind balanced field has been suggested to develop an interior double Ekman layer and has been used to account for the double-cell circulation. It has been simulated numerically in a two-layer model (Thompson, 1974). However, such a process cannot explain the strong upwelling at the location of strong downward mixing and the decrease of longshore velocity locally. An explanation will be given in the next section.

3.2 *Interaction Between Turbulent Mixing and a Front*

A comparison test is carried out to illustrate the importance of turbulence distribution on frontal structure and the circulation pattern. Two simulations were run under exactly the same conditions, except that vertical eddy coefficients were determined differently. Case I uses the turbulence closure scheme similar to the previous case. Case II uses the Munk and Anderson empirical for-

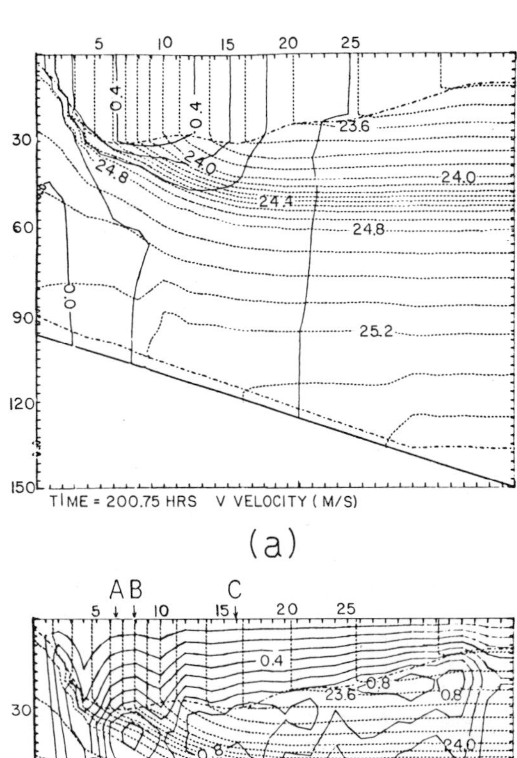

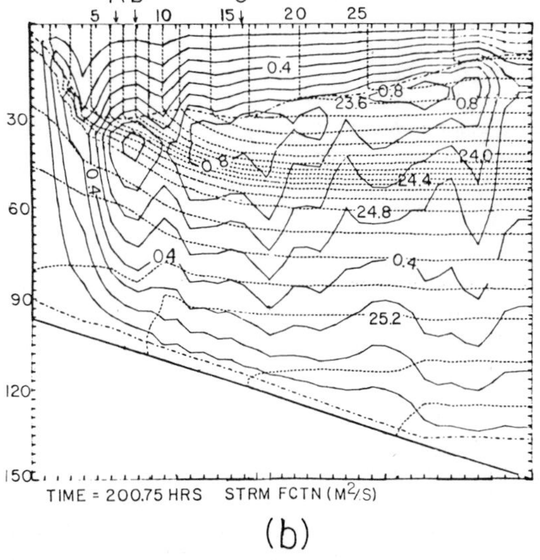

Fig. 6. The flow pattern for Case I at 200.8 h. A linear sloping bottom is used, and vertical eddy coefficients are determined by the moment closure scheme.

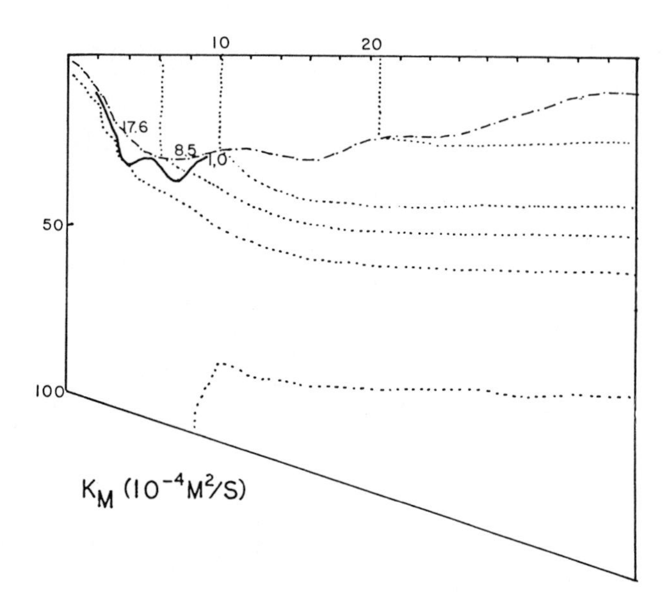

Fig. 7. Distribution of the eddy coefficient, K_M, at the time of Fig. 6. The dashed lines represent density surfaces; the uneven dashed line represents the base of the surface mixed layer; the solid line indicates regions where K_M is larger than 1 cm^2/s. The numbers indicate the extreme values of K_M in cm^2/s.

mula. Fig. 6 shows the flow patterns after 200 h for Case I. Due to minor changes of the model, the turbulent mixing does not develop so strongly as in the previous case. However, mixing is still confined to the region under and inside the surface front (Fig. 7).

The basic flow pattern is the same in Figs 5 and 6. The similarities in the two figures can be summarized as follows: (1) A surface front is formed. (2) The mixed layer depth has a maximum at column A at the inshore side of the surface front. (3) The turbulent mixing occurs primarily between column A and the coast. (4) Two closed clockwise circulations centered at column B and C are located roughly at the same position. (5) Upwelling is also split into three localized regions at the coast and the inshore and offshore sides of the surface front.

Using the Richardson number related formula, the circulation pattern and frontal structure are quite different at 200 h (Fig. 8). The major reason is the stronger mixing over a larger area than in Case I (Fig. 9).

Strong vertical mixing is indicated by the following phenomena:

(1) The mixed layer is about 15 m deeper than in case VI over most of the model domain.

(2) The surface density outside the upwelling zone is increased not only by horizontal advection but also by vertical mixing.

(3) The interior density surfaces are diffused downward and piled on top of the interior stable layer.

(4) The counterbalance of the downward diffusion of heat and upwelling of cold water keeps the interior density surface horizontally oriented. Interior density surfaces shows a frontal structure only within 8 km of the shore.

(5) The interior density surfaces are entrained into the surface layer by convection and mixing rather than by upwelling. This generates a widely spread density gradient without a surface front.

A much weaker, more diffused longshore velocity develops. The cross-front circulation shows a diffused pattern, too.

The strong frontal scale circulation near the surface front in the first two cases is generated by localized turbulent events. Particularly strong upwelling develops at the location where downward mixing is the strongest. A conceptual model is made to explain the process:

Assuming a region with the density structure shown by the dotted lines in Fig. 10a, the associated geostrophic current has a constant vertical shear as shown by the thin lines in Fig. 10b. If local turbulence takes place at column B, the density structure changes from a straight thin line to a thick curve as shown in Fig. 10c. The density increases near the surface and decreases in a lower region. For the geostrophic adjustment to develop, this localized density change can influence a wider area (Fig. 10a). The horizontal density gradients at two chosen depths change from two straight lines to the dashed curves.

At column C, the geostrophic shear increases near the surface and decreases further below as shown by the dashed curve in Fig. 10b. The total geostrophic shear is controlled by the total mass between column B and C and is not changed by the mixing. To conserve the total momentum at column C, barotropic adjustment takes place. By developing a surface pressure gradient, the whole curve is shifted to the right. The actual velocity becomes super-geostrophic at the upper layer and turns away from B; it becomes sub-geostrophic at the lower layer and turns toward B. At column A, the process is reversed. As a result, the upper layer diverges and the lower layer converges. A strong upward motion develops right at the center of downward mixing. This localized upwelling away from the coast will increase the longshore current offshore and decrease it inshore. In the more extreme case of the previous section, the coastal jet is split by the process.

4. Conclusion

From this study, we find that both non-linear and turbulent effects are important in the coastal upwelling front. During a wind-driven upwelling event, the non-linear effect becomes important first, then the turbulent mixing becomes dominant. Dramatic changes in the circulation pattern occur, especially after the onset of shear instability and turbulent mixing.

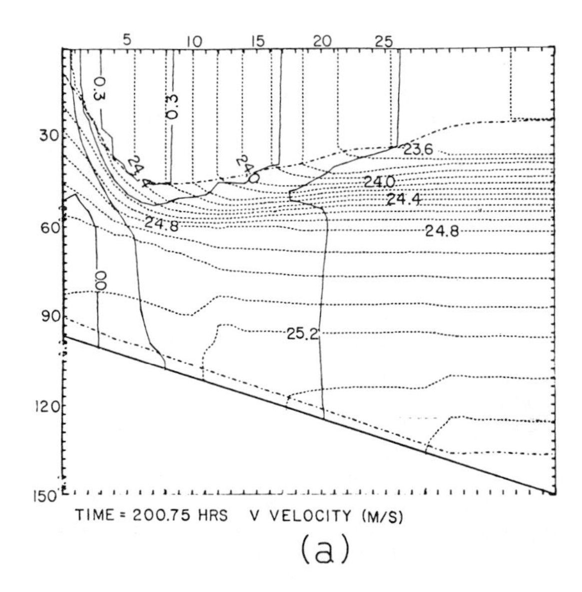

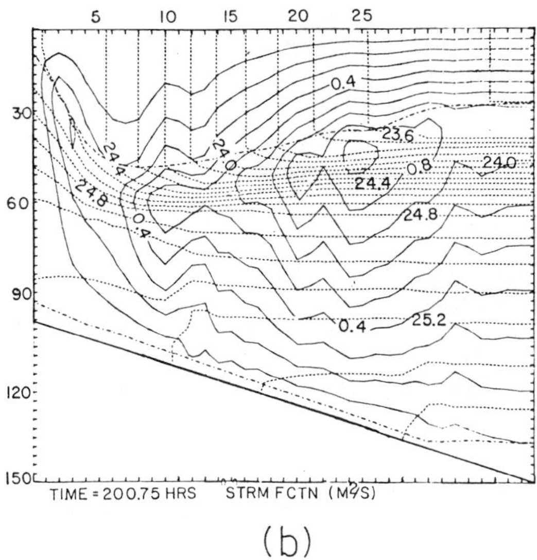

Fig. 8. The flow pattern for Case II at 200.8 h. Vertical eddy coefficients are determined by Munk and Anderson's formulae.

Due to the spatial variation of density and momentum in a frontal region, turbulent mixing is confined to regions at the inshore side of the surface front. The local vertical mixing changes both the vertical and horizontal density structure, which induces a dynamical interaction between the geostrophic shear and the turbulent mixing. Strong upwelling develops at the location of the strongest downward mixing. A frontal-scale closed circulation is thus generated.

The inclusion of a surface layer induces an interaction with the interior that does not exist in analytical models. The density field is lifted to the surface, advects offshore, and is mixed with the interior through free convection. Therefore, both the density and the momentum field in the near-surface region are changed. Furthermore, offshore advection of longshore momentum in the surface layer causes imbalance of the longshore momentum. An additional divergent field develops in the cross-front circulation to change the upwelling pattern into a complicated structure over a wide area. The point sink assumption used in the analytical model is no longer valid.

We have also shown that it is the distribution, not the intensity of turbulent mixing, which is responsible for the intensification of a front. Using the Munk and Anderson formula, the eddy coefficient is a much smoother function of the Richardson number than that of the turbulence closure model. The surface front is destroyed by the vertical mixing over the whole mixed layer. On the other hand, by using the turbulence model, a surface front initiated by the surfacing of the pycnocline is actually intensified by the localized turbulent mixing.

The model is designed to elucidate two-dimensional upwelling phenomena, especially the circulation pattern in the frontal scale. The phenomena associated with three-dimensional forcing cannot be generated by this model. For instance, the locally trapped wave generated by the irregularity of coastline and bottom topography, the propagating coastal wave generated by distant forcing, and the instability of the coastal jet are all beyond the limits of this model. Not only coastal waves and longshore variations but also the boundary conditions in general are excluded. In Foo (1980) several cases have been

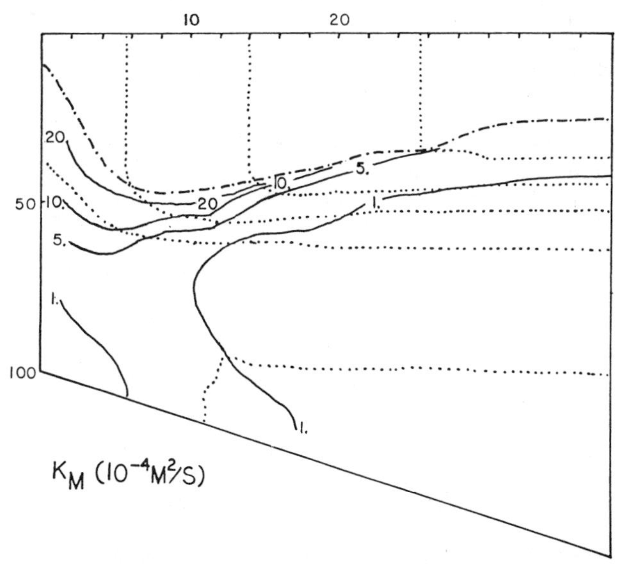

Fig. 9. Same as Fig. 7, except that the distribution of the eddy coefficient is shown for Case II at the time when Fig. 8 is drawn. Solid curves indicate areas where K_M is larger than 1, 5, 10, and 20 cm^2/s.

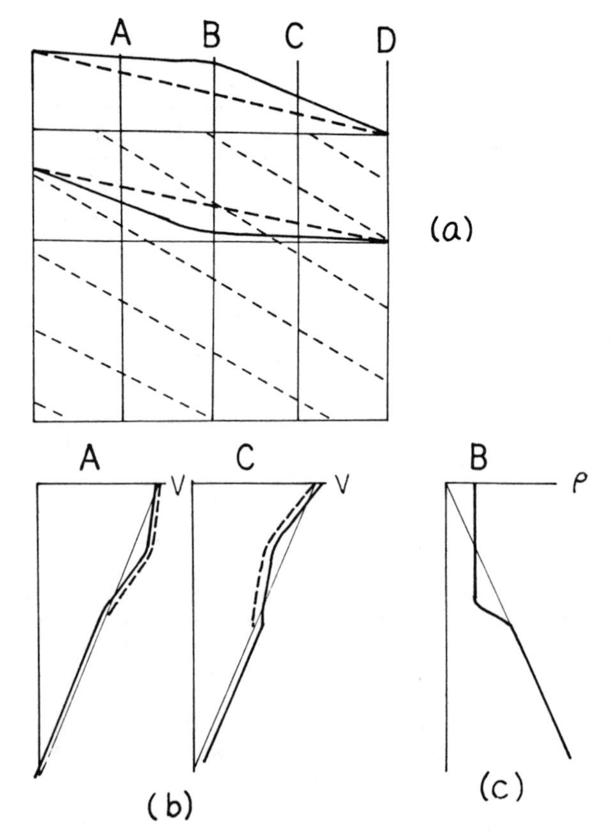

Fig. 10. The geostrophic adjustment around a localized vertical mixing maximum. a) The density structure of the area of interest. The thin dashed lines represent the density surfaces. The thick dashed lines indicate the original horizontal density gradient at two chosen depths. The solid thick curves indicate the horizontal density gradient after the turbulent mixing at column B. b) The velocity gradient at columns A and C; the solid thin lines are the actual velocity. The dashed curves are the geostrophic shear. The solid thick curves indicate the geostrophic shear after adjustment of total momentum. c) The density structure at column B. The thin curve indicates the initial structure. The thick curve indicates the structure after turbulent mixing.

run with different boundary conditions, such as f-plane assumption and slipping coastal wall. The results suggest that although the overall upwelling pattern changes with different boundary conditions, the circulation within the frontal region is not largely affected by them.

Acknowledgements. The results of this paper are part of the Ph.D. project of E-Chien Foo. The research was sponsored by the Office of Naval Research under contract number N000-14-75-C-0173 and by the National Science Foundation under grants OCE 77-26801 and ATM 78-25396. The work could not have been accomplished without the computer facility of the National Center for Atmospheric Research under the project number 35191009. The final preparation was at the Florida State University with the support of the Office of Naval Research under contract number N000-14-80-C-0201.

References

Blumen, W., Uniform potential vorticity flow: Part I, Theory of wave interactions and two-dimensional turbulence, *Journal of Atmospheric Science, 35,* 774-783, 1978.

Curtin, T., Physical dynamics of the coastal upwelling front zone off Oregon, Ph.D. Dissertation, University of Miami, Florida, 1979.

Dickey, T., An experimental study of decaying and diffusing turbulence in neutral and stratified fluid, Ph.D. Dissertation, Princeton University, 108 pp., Princeton, New Jersey, 1977.

Foo, E., A two-dimensional diabatic isopycnal model of a coastal upwelling front, Ph.D. Dissertation, University of Miami, 113 pp., Miami, Florida, 1980.

Hamilton, P., and M. Rattray, Jr., A numerical model of the depth-dependent, wind-driven upwelling circulation on a continental shelf, *Journal of Physical Oceanography, 8,* 437-457, 1978.

Hoskins, B.J., and F.P. Bretherton, Atmospheric frontogenesis models, mathematical formulation, and solution, *Journal of Atmospheric Science, 29,* 11-37, 1972.

Munk, W.H., and E.R. Anderson, A note on the theory of the thermocline, *Journal of Marine Research, 7,* 276-295, 1948.

Orlanski, I., and B.B. Ross, The circulation associated with a cold front, Part I, Dry case, *Journal of Atmospheric Science, 34,* 1619-1633, 1977.

Pedlosky, J., A nonlinear model of the onset of upwelling, *Journal of Physical Oceanography, 8,* 178-187, 1978.

Stevenson, M., R.W. Garvine, and B. Wyatt, Lagrangian measurement in a coastal upwelling zone off Oregon, *Journal of Physical Oceanography, 4,* 321-336, 1974.

Tennekes, H., and J.L. Lumley, *A First Class in Turbulence,* The M.I.T. Press, 300 pp., 1972.

Thompson, J.D., The coastal upwelling cycle on a beta-plane: hydrodynamics and thermodynamics, Ph.D. Dissertation, Florida State University, Tallahassee, Florida, 141 pp., 1974.

Worthem, S., and G. Mellor, Turbulence closure model applied to upper tropical ocean, in *GATE-1, Oceanography and Surface Meteorology in the B/C-Scale,* G. Siedler and J.D. Woods (eds.), Pergamon Press, Oxford and New York, pp. 237-272, 1980.

A COASTAL UPWELLING CIRCULATION MODEL WITH EDDY VISCOSITY DEPENDING ON RICHARDSON NUMBER

Masahiro Endoh*

NASA Wallops Flight Center, Wallops Island, Virginia 23337

Christopher N.K. Mooers

Department of Oceanography, Naval Postgraduate School, Monterey, California 93940

Walter R. Johnson

College of Marine Studies, University of Delaware, Lewes, Delaware 19958

* Present Address: Ocean Research Institute
University of Tokyo
Minamidai 1-15-1
Nakano, Tokyo, 164 Japan

Abstract. A two-dimensional numerical model (Endoh, 1977) with constant eddy viscosity is extended to study the possible influence of variable eddy viscosity on the dynamics and cross-shore circulation in an upwelling regime.

Variable vertical eddy viscosity and diffusivity depending upon local Richardson number are introduced using Munk and Anderson's (1948) formulation with fine grid interval in the vertical direction (2.5 m). Initial value problems are solved numerically. The alongshore wind stress, uniform in time and space, is imposed on a two-layered body of water at rest in a uniformly rotating straight channel with flat bottom topography.

Due to the local low Richardson number in the thermocline, i.e., due to the maximum of the vertical eddy viscosity there, double cell type circulation, as in the constant eddy viscosity case, is still possible. Study of momentum and vorticity balance in the stage after the initial spin-up shows that, in the thermocline, the thermal wind relationship for vertical shear of the alongshore velocity, and alongshore momentum balance between the Coriolis force and the vertical eddy viscosity term, are satisfied. Effects of an imposed alongshore barotropic pressure gradient are also described.

Introduction

Double cell circulation associated with coastal upwelling regimes (Mooers, Collins, and Smith, 1976) has been related either to the frictional layer in the tilted thermocline or to the local dynamics of the upwelling frontal region. The former approach, with an internal Ekman layer concept, was made by Thompson (1974), using a homogeneous two-layered model with the assumed frictional layer. Numerical level models with more general dynamical conditions were developed by Endoh (1977), hereafter referred to as I.

Endoh's model in I (also called model I) with constant eddy viscosity shows that the momentum balance of the Coriolis force with the eddy viscosity term, due to the large eddy coefficient, makes a closed double cell circulation possible prior to the formation of the surface front.

As the eddy viscosity coefficient is not constant, especially in the vertical direction, and the constant eddy viscosity may distort the dynamics involved in nature, we extend model I to include a variable eddy viscosity formulation that is dependent on the Richardson number. First, we briefly introduce model I as a prototype and then describe our present experimental procedures.

Description of the Model

The model is a level model with equations of motion, (1) to (3), an equation of continuity, (4), and a density equation, (5). For simplicity, buoyancy forcing is neglected. The major assumptions are two-dimensionality (vertical and onshore-offshore directions), the Boussinesq approximation for an imcompressible fluid, the rigid-lid approximation, which suppresses the external mode of motion, and the hydrostatic assumption

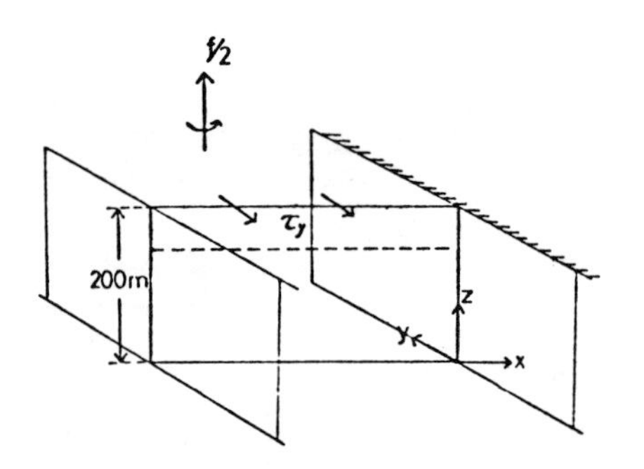

Fig. 1. Diagram of the rotating straight channel.

for the vertical motion. Then, the equations are

$$\frac{\partial u}{\partial t} + \frac{\partial}{\partial x}(u \cdot u) + \frac{\partial}{\partial z}(u \cdot w) - fv = -\frac{1}{\rho_o}\frac{\partial p}{\partial x} + A_H \frac{\partial^2 u}{\partial x^2} + \frac{\partial}{\partial z}(A_V \frac{\partial u}{\partial z}) \quad (1)$$

$$\frac{\partial v}{\partial t} + \frac{\partial}{\partial x}(v \cdot u) + \frac{\partial}{\partial z}(v \cdot w) + fu = -\frac{1}{\rho_o}\frac{\partial p}{\partial y} + A_H \frac{\partial^2 v}{\partial x^2} + \frac{\partial}{\partial z}(A_V \frac{\partial v}{\partial z}) \quad (2)$$

$$0 = -\frac{1}{\rho}\frac{\partial p}{\partial z} - g \quad (3)$$

$$\frac{\partial u}{\partial x} + \frac{\partial w}{\partial z} = 0 \quad (4)$$

$$\frac{\partial \rho}{\partial t} + \frac{\partial}{\partial x}(\rho u) + \frac{\partial}{\partial z}(\rho w) = K_H \frac{\partial^2 \rho}{\partial x^2} + \frac{\partial}{\partial z}(K_V \frac{\partial \rho}{\partial z}), \quad (5)$$

where the x direction is shoreward, the y direction is alongshore, and the z direction is upward (Fig. 1). The other variables are conventional. A_H and K_H are the horizontal eddy viscosity and diffusivity coefficients, respectively. A_V and K_V are the vertical eddy viscosity and diffusivity coefficients, respectively. Convective overturning is introduced by vertical mixing if the water column becomes statically unstable.

Fig. 1 shows a schematic view of a straight channel rotating uniformly at the rate of 1.0×10^{-4}/s. The depth of the channel is uniform, 200 m. The width is dependent on the particular case, either 50 or 100 km. The alongshore wind stress (-1.0 dyne/cm^2), constant in time and uniform in space, is applied to the surface of the stratified water at rest.

Model I is based on the constant eddy viscosity coefficients. When the vertical eddy coefficient is larger than ca. 10 cm^2/s with the horizontal eddy coefficient of the order of 10^5 to 10^6 cm^2/s, the two circulation cells are formed with a separation streamline in the thermocline. Fig. 2 shows a typical case for the double cell circulation in the almost steady state with A_V = 100 cm^2/s. The viscosity term, $A_V(\partial^2 v/\partial z^2)$, in the alongshore momentum equation (2) is balanced by the Coriolis term, fu, in the thermocline as in the surface and bottom Ekman layers. In the alongshore vorticity equation, the vertical shear of the alongshore velocity, $f(\partial v/\partial z)$, is not large enough to balance the horizontal density gradient, $g(\partial\rho/\partial x)$, but the difference is compensated for by the vertical diffusion of alongshore vorticity, $A_V\, \partial^2/\partial z^2 (-\partial u/\partial z)$.

In this paper, we show the results calculated in the model with an extended domain and finer resolution. The width of the channel is increased from 50 to 100 km. The horizontal and vertical resolutions are 2 km and 2.5 m, respectively. The major change from model I is the introduction of the variable vertical eddy viscosity and diffusivity, which depend on the Richardson number based on the local field variables.

The formulations for eddy viscosity A_V and diffusivity K_V are based on Munk and Anderson (1948):

$$A_V = \frac{A_o}{(1 + 10\,Ri)^{1/2}} \quad (6)$$

and

$$K_V = \frac{K_o}{(1 + 3.3\,Ri)^{3/2}} \quad (7)$$

$$\text{where } Ri = -\frac{g}{\rho_o} \cdot \frac{\partial \rho}{\partial z} \Big/ \left| (\frac{\partial u}{\partial z})^2 + (\frac{\partial v}{\partial z})^2 \right| .$$

These are the trial parameterizations that show the most reasonable functional dependence of the mixing coefficients on the dynamic stability. The coefficients for the neutral static stability, A_o and K_o, are 50 cm^2/s. The horizontal eddy viscosity and diffusivity are 1.0×10^6 cm^2/s.

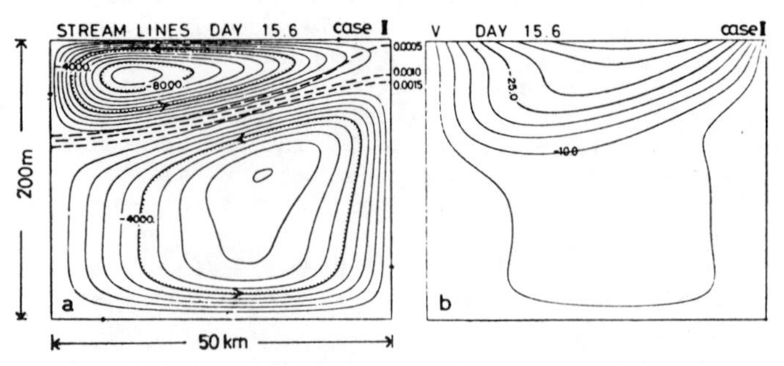

Fig. 2. a) Streamlines with density field (dashed lines) and b) alongshore velocity field. Model I with constant A_V = 100 cm^2/s (Endoh, 1977).

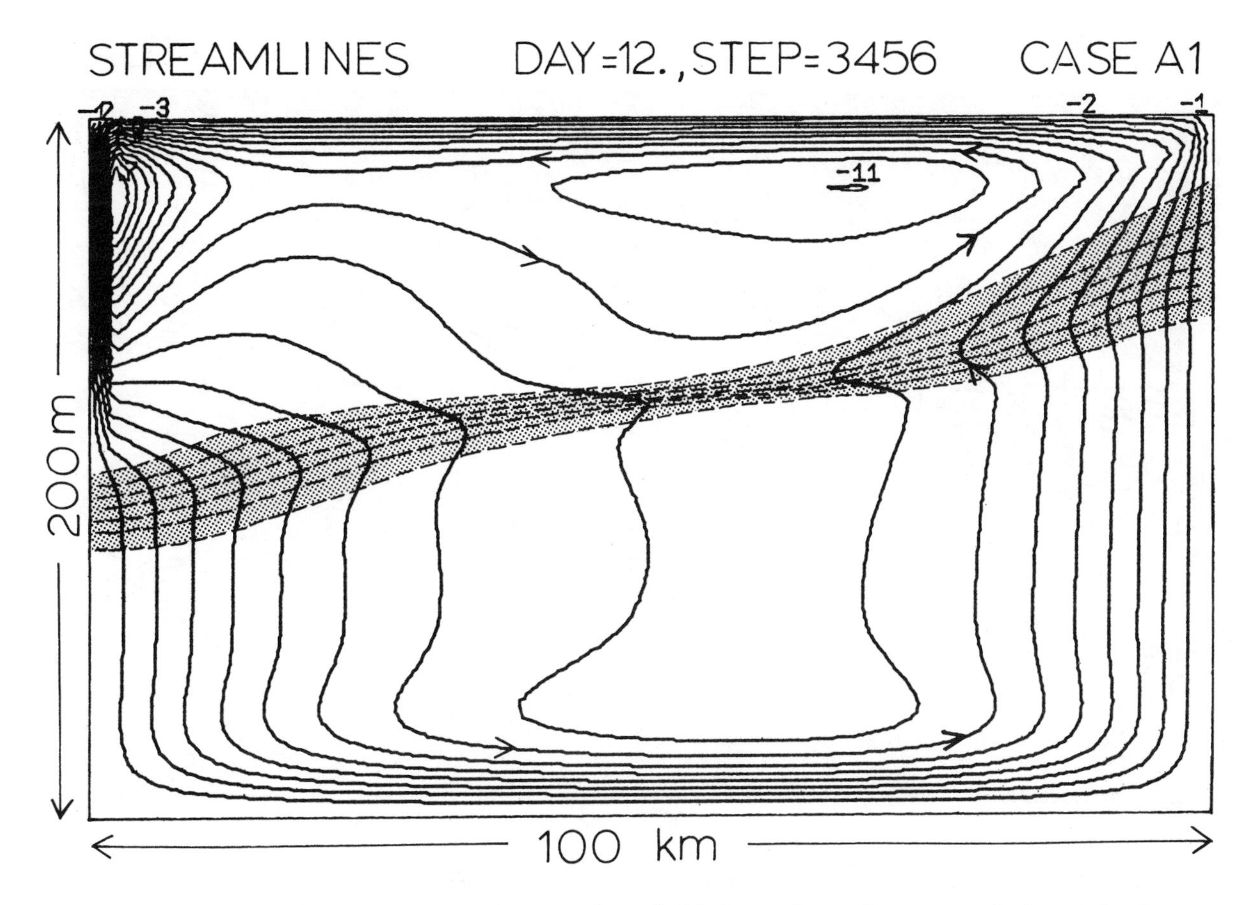

Fig. 3. Streamlines with density field (dashed lines) 12 days after the onset of the wind. Contour interval for streamlines is 1.0×10^3 cm^2/s and for density $5 \times 10^{-4}\rho_o$.

From the vertical profiles of the different variables at 20 km offshore from the upwelling coast (Fig. 6), the eddy viscosity maximum is associated with the minimum of the Richardson number through the vertical shear of the alongshore velocity.

Momentum and Vorticity Balance

The momentum and vorticity ($-\partial u/\partial z$) balances are briefly discussed in order to clarify roles of the eddy term in the dynamical balances.

The vertical profile of each term in the cross-shore momentum equation (1) is shown in Fig. 7. In addition to the Ekman balance in the surface and bottom layers, the geostrophic balance in the thermocline is obvious. Vertical differentiation of (1) indicates that the thermal wind relationship $-f\ \partial v/\partial z = g\ \partial\rho/\partial x$ holds very well, with only a small contribution from the vertical frictional term.

The vertical profile of each term in the alongshore momentum equation (2) is shown in Fig. 8. In the thermocline, the gross balance between the Coriolis term, $-fu$, and the vertical frictional term, $\partial/\partial z\ (A_V\ \partial v/\partial z)$, is apparent. Although the non-linear, temporal change and horizontal friction terms are not negligible, correlation between the Coriolis and vertical eddy terms is evident. Moreover, $-fu$ is balanced by $A_V\ \partial^2 v/\partial z^2$, thus the change of the sign of u is associated

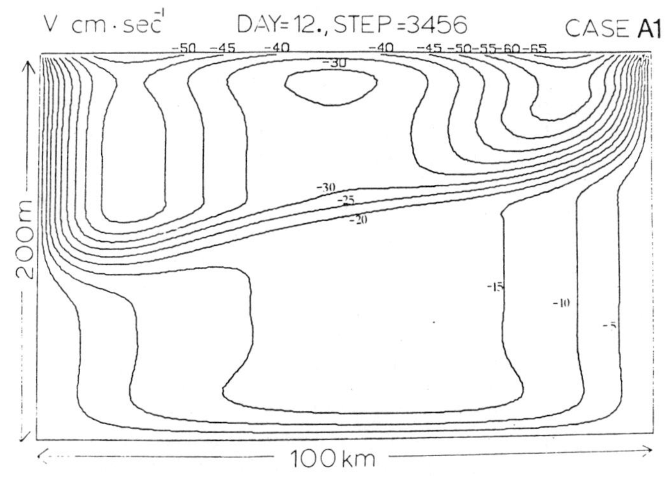

Fig. 4. Alongshore velocity field at the 12th day. Contour interval is 5 cm/s.

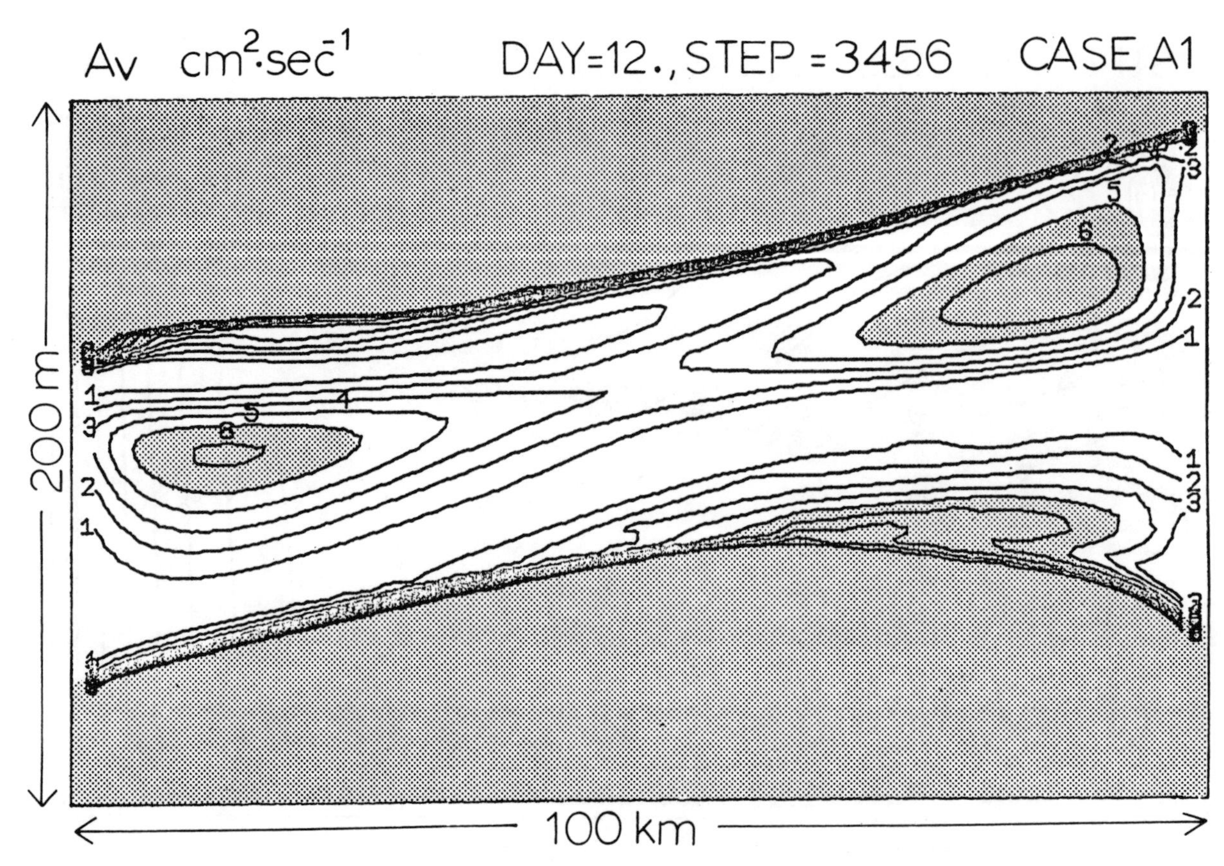

Fig. 5. Eddy viscosity field (A_V) at the 12th day. Contour interval is 1 cm^2/s. Contour lines for A_V larger than 10 cm^2/s are omitted. A_V for stippled areas is larger than 5 cm^2/s.

with the vertical structure of the alongshore velocity. These balances are found in the coastal upwelling region off Oregon (Curtin, 1979).

Comments

1. The present results are not general enough to suggest that we always have such two cell circulations, because they may depend on the particular formulation chosen for parameterization of eddy mixing. However, if the formulation reflects a possibility of dynamical instability, the eddy viscosity can be expected to have a maximum in the thermocline. A more sensitive formulation of the eddy coefficient dependence on the Richardson number can lead to a greater maximum value A_V (of the order of 10 cm^2/s) which would result in a more enhanced (closed) double cell circulation. As an example of such formulation, Johnson and Mooers (1980) have suggested another local dynamic model using the dependence $A_V = A_0 Ri^{-1}$. The initial relative density difference between the two layers is 4.0×10^{-3}, which gives 18 km for the internal radius of deformation $(g\ \Delta\rho/\rho_0\ h)^{\frac{1}{2}}/f$ with the upper layer depth h = 80 m and g = 980 cm/s^2. With A_V = 50 cm^2/s the vertical scale depth $(A_V/f)^{\frac{1}{2}}$ for the Ekman layers is 7 m, which is consistent with observed mixed layer thickness in the upwelling regions.

Results

In this section, we describe transient motions in the upwelling regime, where the thermocline is being continuously uplifted and geostrophic balance for the alongshore velocity has been achieved since 3 days after the onset of the wind. At this stage, the cross-shore circulation is almost steady except for a small internal gravity-inertial seiche with the period of 16.2 h, where the pure inertial period is 17.5 h. The alongshore velocity is still gradually increasing.

Fig. 3 shows streamlines and the density field 12 days after the onset of the wind. Although closed double cell circulation is not seen, the cross-shore velocity, however, changes sign in the thermocline. Thus, there is an open double cell circulation.

Comparison of the alongshore velocity field (Fig. 4) with the isopleths of the eddy viscosity (Fig. 5) indicates that the eddy viscosity maximum in the thermocline is due to the large vertical shear of the alongshore velocity.

2. An experiment (not shown here) with baro-

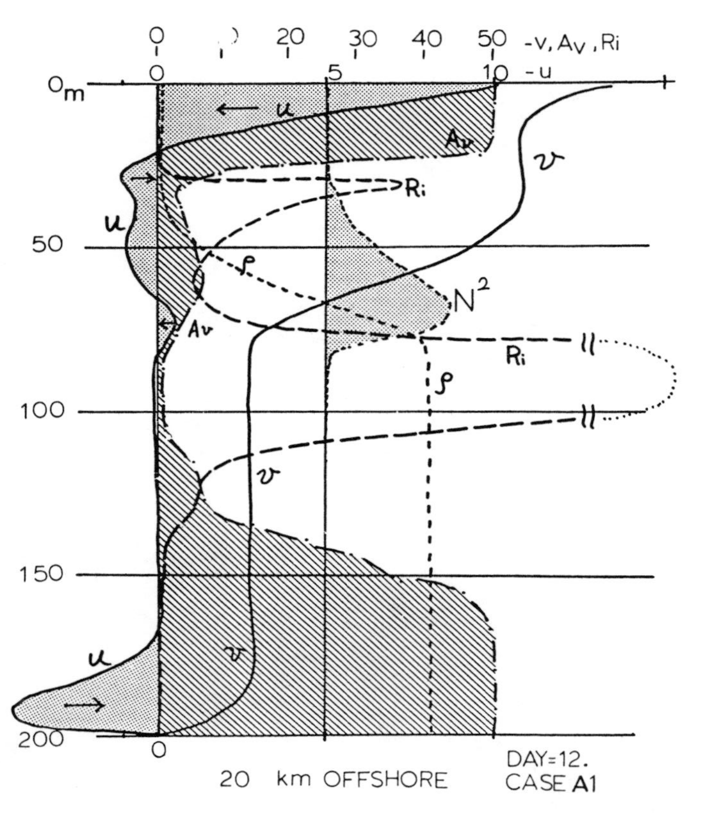

Fig. 6. Vertical profiles of u and v in cm/s (solid lines), A_v in cm^2/s (hatched area), density ρ and vertical density gradient N^2 in arbitrary units, and Richardson number Ri at 20 km offshore from the upwelling coast for the 12th day.

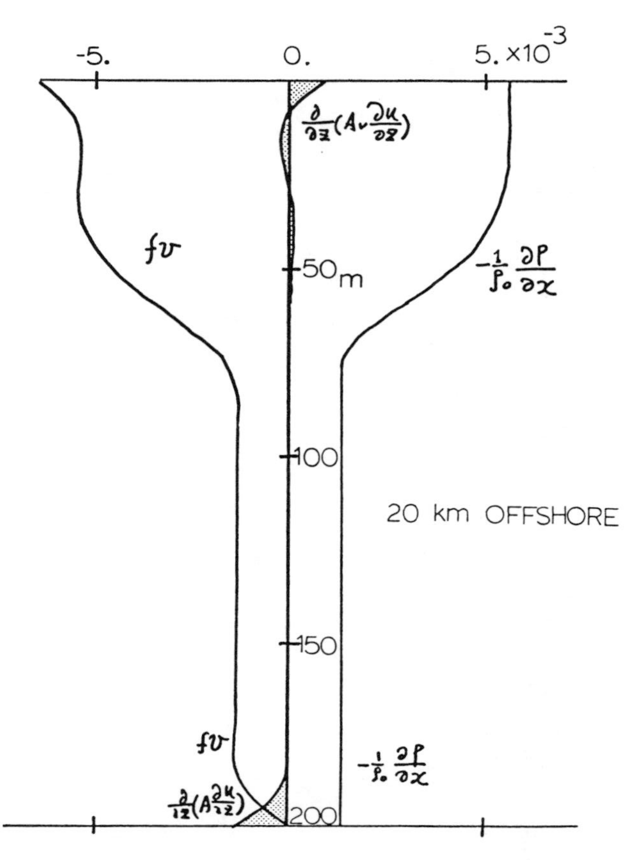

Fig. 7. Vertical profiles of terms in the cross-shore momentum equation (1) at 20 km offshore for the 12th day. Unit is cm/s^2.

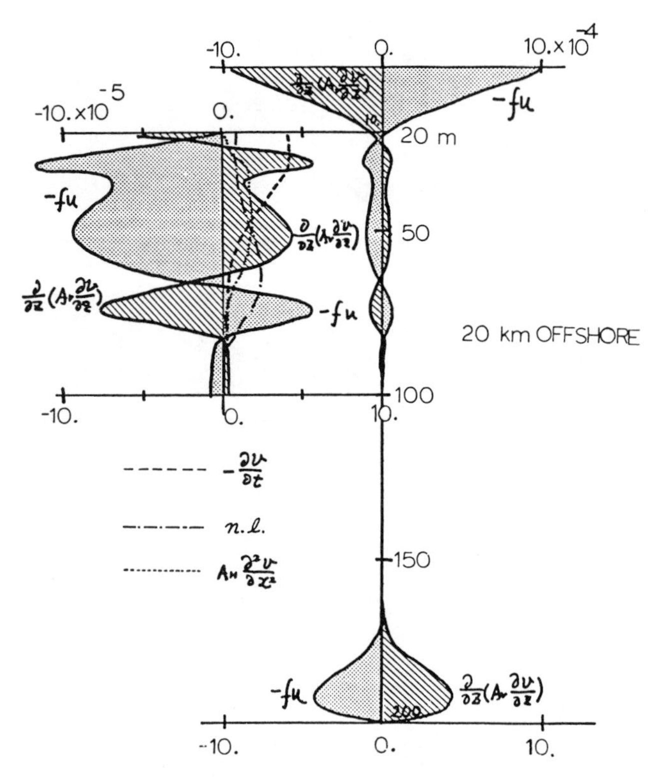

Fig. 8. Vertical profiles of terms in the alongshore momentum equation (2) at 20 km offshore for the 12th day. Profiles for the depth 20 to 100 m are enlarged at the left. Unit is cm/s^2.

tropic alongshore (uniform in the vertical direction) pressure gradient, equivalent to a 4-cm increase of the sea surface level in the equatorward direction over 1000 km, results in a weaker (equatorward) alongshore current with poleward undercurrent beneath the surface jet. The cross-shore circulation and the alongshore velocity structure are not changed qualitatively.

3. The change of sign of the cross-shore velocity seems to occur in the middle of the thermocline. In the present model, it does not occur at the base of the thermocline as it does in some cases of the upwelling experiments (Johnson and Mooers, 1980).

Acknowledgements. This work was done while the first author was visiting NASA Wallops Flight Center as a NRC Research Associate in 1979 and 1980. He is most indebted to Dr. Norden E. Huang for his sincere support of the present study. The research is a contribution to the Coastal Upwelling Ecosystems Analysis Program, supported by the National Science Foundation, Office of the International Decade of Ocean Exploration, under grant OCE 77-28354 to the University of Delaware.

References

Curtin, T.B., Physical dynamics of the coastal upwelling frontal zone off Oregon, *Coastal Upwelling Ecosystems Analysis Technical Report 59*, 317 pp., September, 1979.

Endoh, M., Double-celled circulation in coastal upwelling, *Journal of the Oceanographical Society of Japan, 33*, 30-37, 1977.

Johnson, D.R. and C.N.K. Mooers, Internal cross-shelf flow reversals during coastal upwelling, *This Volume*, 1980.

Mooers, C.N.K., C.A. Collins, and R.L. Smith, The dynamic structure of the frontal zone in the coastal upwelling region off Oregon, *Journal of Physical Oceanography, 6*, 3-21, 1976.

Munk, W.H. and E.R. Anderson, Notes on a theory of the thermocline, *Journal of Marine Research, 7*, 276-295, 1948.

Thompson, J.D., The coastal upwelling cycle on a beta-plane: hydrodynamics and thermodynamics, *Mesoscale Air-Sea Interaction Group Technical Report*, Florida State University, 141 pp., 1974.

Some Chemical Consequences of Upwelling

TEMPORAL NUTRIENT VARIABILITY IN THREE DIFFERENT UPWELLING REGIONS

L.A. Codispoti

Bigelow Laboratory for Ocean Sciences
West Boothbay Harbor, Maine 04575

Abstract. The nutrient regimes in upwelling systems off northwest Africa, Peru, and Somalia suggest differences in the dominant types of temporal nutrient variability. For example, wind variations over several-day time scales can explain much of the variability observed at ∿21°N off northwest Africa during the JOINT-I experiment whereas seasonal differences are dominant off Somalia.

Observations from the upwelling zone at ∿15°S off Peru suggest significant variability over a wide range of time scales. Recent observations suggest major changes in the main secondary nitrite maximum and the occurrence of massive red tides on a frequency even lower than El Niños which last for ∿12-15 months and occur on a several-year scale. There is a significant seasonal alternation of the upwelling source water composition near 15°S, several-day perturbations may arise from coastal trapped waves, and some responses can be related to daily changes in the local winds. Episodes of complete denitrification and hydrogen sulphide production in the open ocean waters off Peru also occur, but their frequency is not known.

Introduction

It is generally recognized that the marine community arising from a given input of light and nutrients will not depend solely on the average rate at which these essentials are supplied. Among the additional factors that may be important is the nature of the variability of the environment (e.g., Walsh, 1976). For this reason, knowledge of the important temporal scales of nutrient variability in different upwelling systems may be useful. In this paper, nutrient variability is reviewed for three coastal upwelling regions located within the tropics: northwest Africa at ∿22°N, Peru at ∿15°S, and the regions of strong upwelling that occur off Somalia between ∿5 and 10°N. In addition to the apparent differences in dominant types of temporal nutrient variability, the zones differ significantly in other important ways outside the scope of this paper. Some of the additional differences between northwest Africa and Peru have been previously discussed by Codispoti, Dugdale, and Minas (in press).

While there is evidence suggesting significant differences among the three regions, the following caveats should be kept in mind:

1. The wider range of significant time scales observed off Peru (see below) could arise partially from the more detailed data base available from the region.
2. The data from northwest Africa and Peru are primarily from mesoscale observations at ∿22°N and ∿15°S, respectively (Figs 1 and 2). The variability may be different at the other latitudes. In particular, it is known that off northwest Africa, the seasonal variation is minimal near 22° (e.g. Wooster, Bakun, and MacLain, 1976).
3. Variability on the less than several-day scale has not been considered.
4. While the populations in a region may be determined to a considerable extent by the nature of the nutrient supply, they exert their own influence on nutrient concentrations. The data bases for estimating biological rates are often insufficient for determining the biological influence on nutrient variability, and such effects are given only slight attention here.

Despite the above difficulties, this review will summarize much of what is known about the temporal variability of the nutrient fields in the three regions.

Data and Methods

The data for the analysis come mostly from the JOINT-I (March-May 1974), JOINT-II (March 1976-May 1977) and INDEX (March-August 1979) programs. The JOINT-I observations were concentrated along a line (at ∿21°40'N) off northwest Africa (Fig. 1). Although JOINT-II included some large-scale surveys, the data are also concentrated along a mesoscale line (near 15°S) off Peru (Fig. 2). The INDEX observations were designed

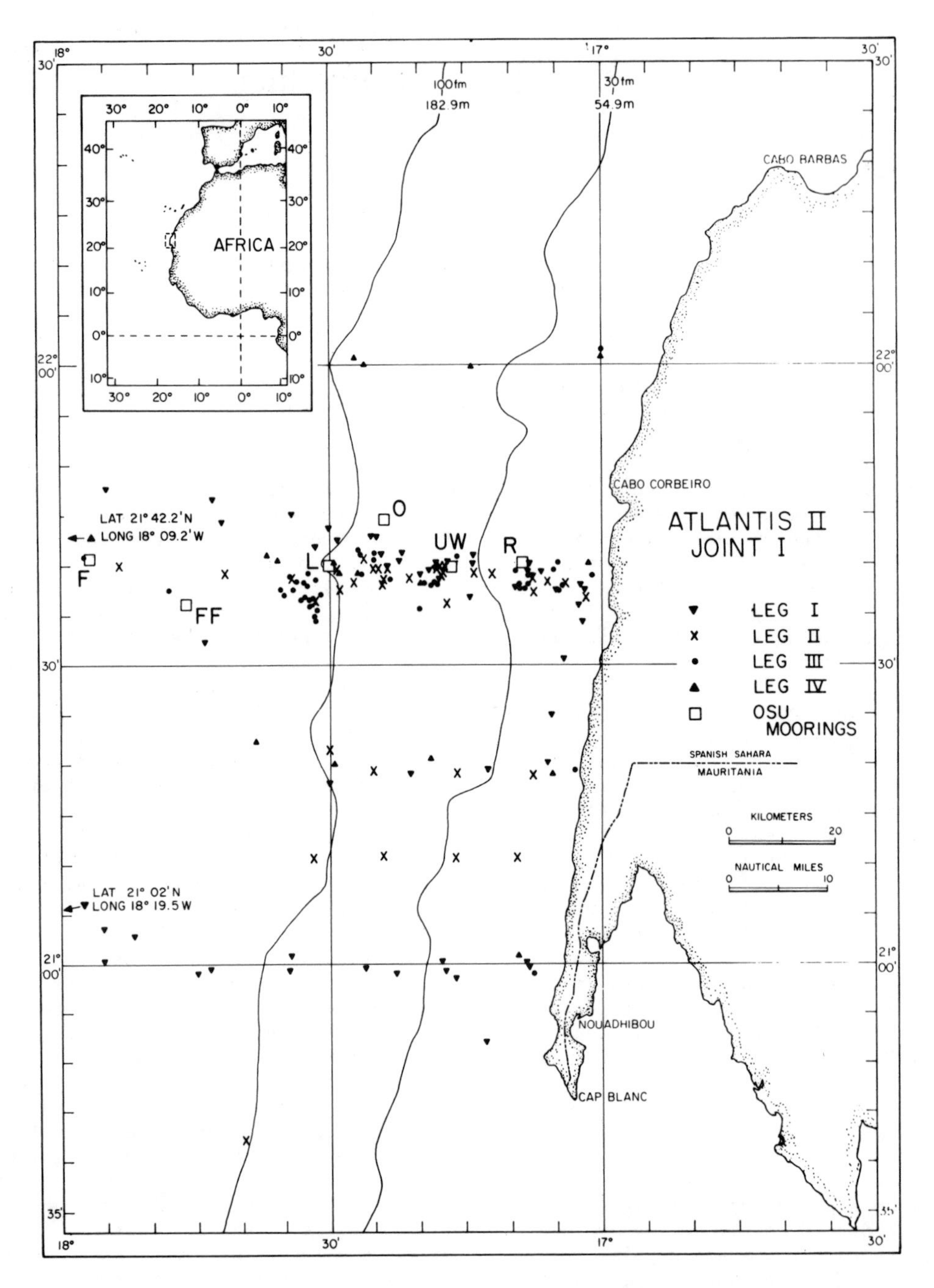

Fig. 1. Station locations for R.V. *Atlantis-II* cruise 82 (JOINT-I), March-May 1974, and the Oregon State University (OSU) current meter array locations.

to investigate conditions off Somalia before, during the onset, and at full development of the Southwest Monsoon (the upwelling season). Fig. 3 shows the region sampled by the R.V. *Iselin* during the last phase of this program.

The papers and reports listing the pertinent data and methods from the above experiments were summarized by Codispoti and Friederich (1978), Friederich and Codispoti (1979), Codispoti *et al.* (in press), Codispoti and Packard (1980), and Smith and Codispoti (1980).

Results and Discussion

Nutrient Variability on the Several-Year Scale

Northwest Africa. The available information (e.g. Boje and Tomczak, 1978; Minas, Codispoti

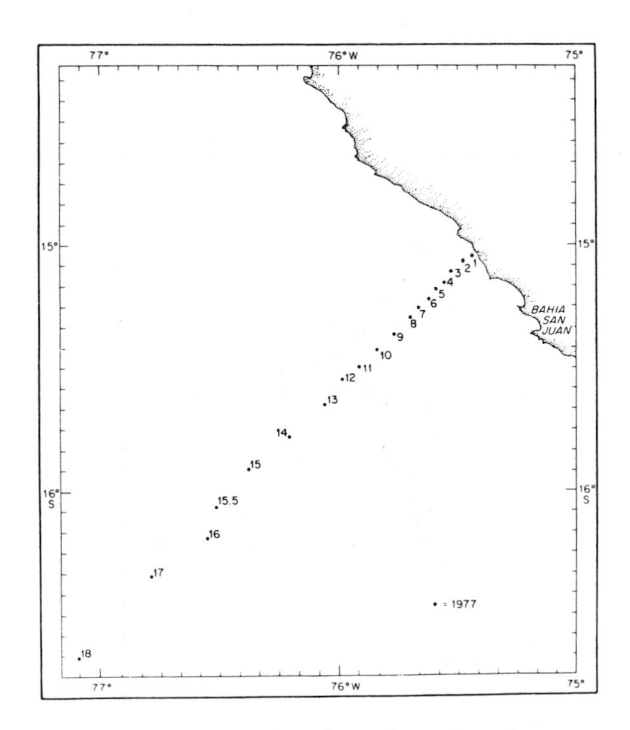

Fig. 2. Average station locations for the main hydrographic line (the "C" line) during the March-May 1977 parts of the JOINT-II observations.

and Dugdale, in press) provides little evidence for variability in the nutrient regime on the inter-annual or greater scale off northwest Africa, but it is possible that some observable differences would be revealed by extensive long-term observations. For example, Schulz, Schemainda and Nehring (in press) discuss inter-annual variations in upwelling winds and sea surface temperatures in the southern reaches of the northwest African upwelling region (∿15°N), and it is likely that the differences could be correlated with variations in the nutrient regime.

Peru. Three types of long-term variability have been identified off Peru. Of these, the El Niño condition is the best known. The phenomenon is a response to large-scale meteorologic forces (Quinn, 1974), and during El Niños the surface waters off Peru are altered by an influx of abnormally warm and nutrient-poor water from the equator. Subsurface (>100 m) changes in the nutrient regime may also be caused by this phenomenon, but there is no firm evidence for this (e.g., Codispoti *et al.*, in press). Major El Niños last for ∿10 to 16 months (Wooster and Guillen, 1974; Wyrtki, 1975, 1978), and since 1950 they have begun in 1957, 1965, 1972, and 1976. Long-term temperature records show considerable variation in average sea surface temperatures even among non-El Niño years (D.B. Enfield, personal communication). Maximum nutrient concentrations at the surface are probably higher during the colder non-El Niño years, but aside from this, little can be said about the response of the nutrient regime to these less dramatic changes.

Most El Niños are accompanied by an increased incidence of red tides (Wooster, 1960). However, one that occurred during the 1976 Niño was exceptionally large, and probably should be classified as an additional type of long-term variability. The feature extended from the equator to at least 15°S. Dugdale, *et al.* (1977) suggested that such events occur less frequently than El Niños. As dinoflagellates do not require dissolved silicon and function differently than diatoms in other ways, one must expect anomalies in the surface nutrient field during a major red tide. Nutrient uptake rates were lower than normal within the dinoflagellate bloom (Codispoti, *et al.*, in press), and nitrate-dissolved silica ratios were sometimes quite low (Codispoti, Bishop, Friebertshauser and Friederich, 1976). However, it is not clear whether the low ratios arose from the presence of so many dinoflagellates or from the anomalous occurrence of complete denitrification in some of the upwelling source waters.

Normally, denitrification occurs within the oxygen-deficient strata off Peru (Codispoti, *et al.* 1980) but it seldom goes to completion. The production of hydrogen sulphide off Peru is often mentioned, particularly within the enclosed waters near Callao (e.g., Sverdrup, Johnson, and Fleming, 1942), but only two cases of complete denitrification leading to the onset of sulphate reduction have been documented for the open waters off Peru. In 1974, a non-El Niño year, a bottom layer containing hydrogen sulphide was encountered over the shelf near 7°30'S (Sorokin, 1978), and hydrogen sulphide production was observed near 15°S for a brief period during the 1976 El Nino (Fig. 4). Because of the rarity of such observations it is tempting to conclude that episodes of complete denitrification with subsequent hydrogen sulphide production occur infrequently (less than once a year) in the open ocean off Peru. However, modest changes in the organic matter and dissolved oxygen supplies can have large effects on the denitrification regime (Codispoti and Packard, 1980). Therefore, there exists the other possibility that hydrogen sulphide production occurs frequently, but on temporal and spatial scales too small to be resolved by most sampling programs.

The third type of long-term variability that has been observed off Peru is the recent intensification of the nearshore part of the nitrite maximum associated with the main denitrification zone between 10 and 25°S. A comparison of pre- and post-1972 nitrite concentrations (Fig. 5) suggests that maximum nitrite concentrations were considerably higher in the post-1972 period. Codispoti and Packard (1980) speculate that the increased concentrations indicate higher denitrification rates arising from an increase in the

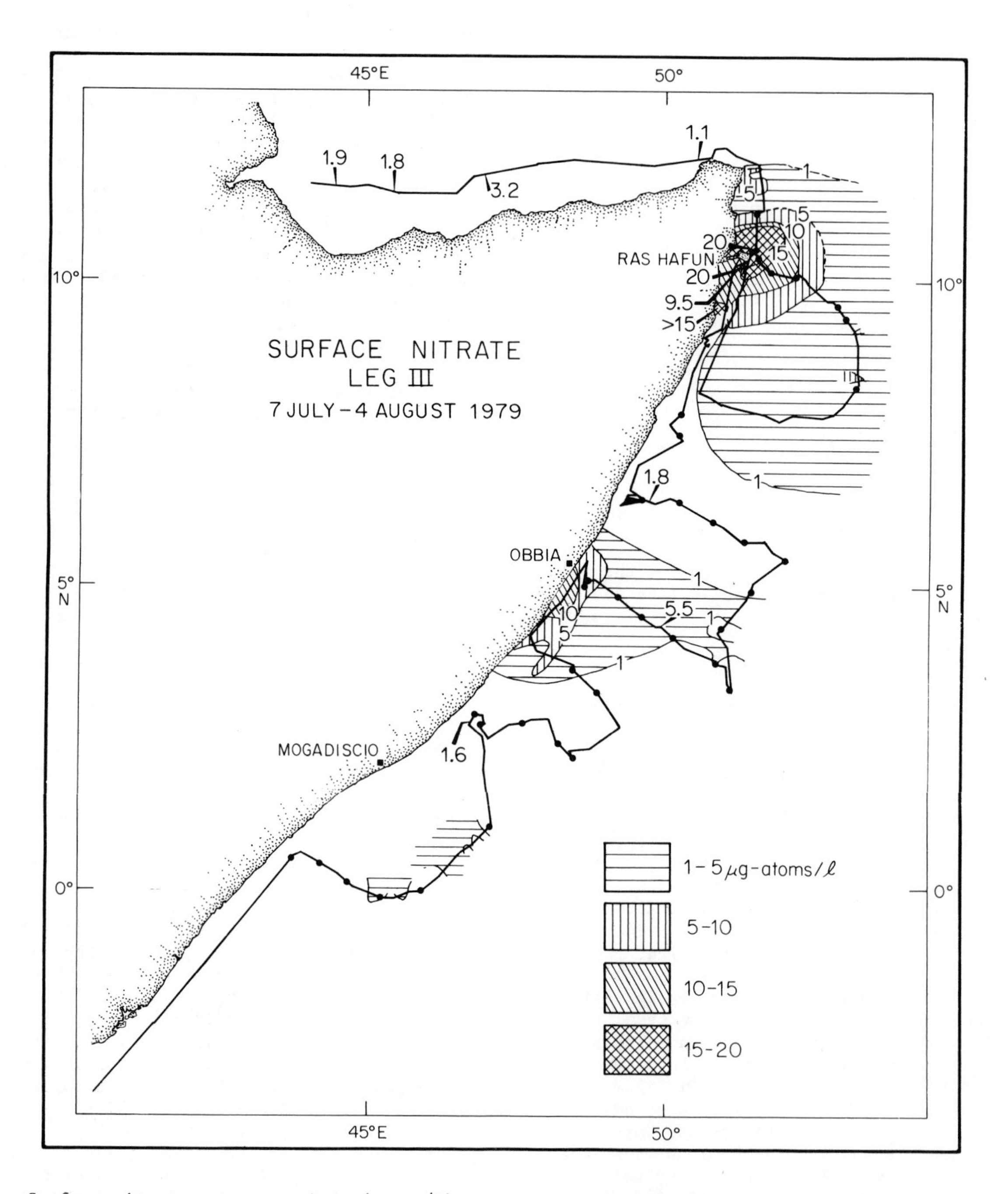

Fig. 3. Surface nitrate concentrations (μg-at/l) observed off Somalia during R.V. *Iselin* Leg III, 7 July to 4 August 1979. Black dots indicate station positions. Surface samples were also taken along the indicated track about every half hour (from Smith and Codispoti, 1980).

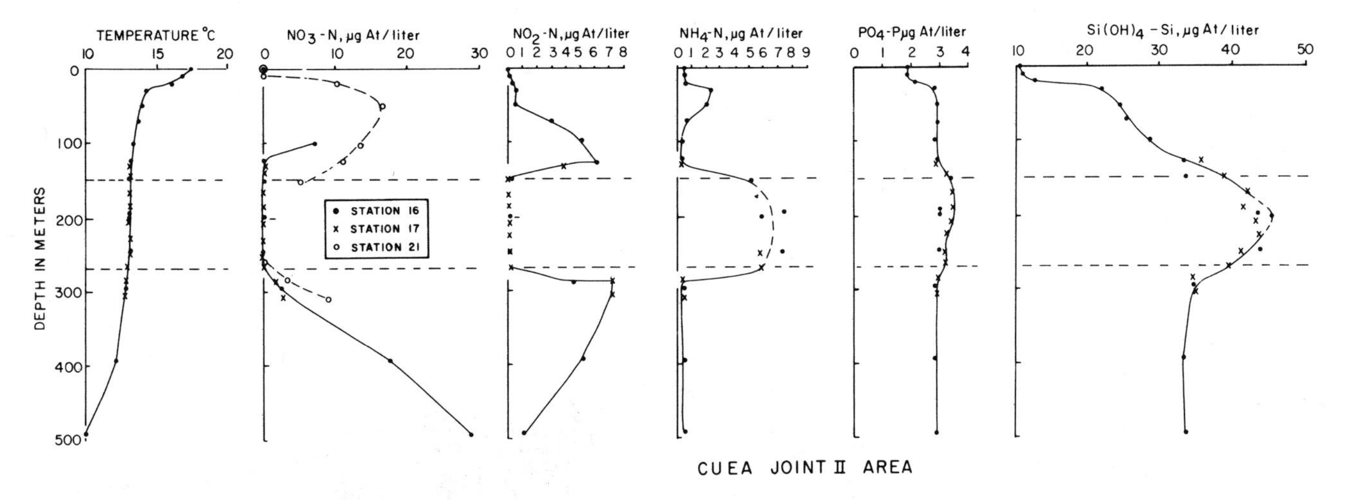

Fig. 4. Profiles of temperature and nutrients observed at Sta. C-5 (see Fig. 2) from the R.V. *Alpha Helix* during 1 to 3 April 1976 (from Dugdale *et al.*, 1977). The dotted line indicates the depth range within which sulphide was detected.

fraction of primary production reaching the denitrification zone. The massive decline of the anchoveta population since the El Niño of 1972 indicates that a significant change in the distribution of organic matter is possible. The absence of a large before and after 1972 difference in salinities within the nitrite maximum (Fig. 6), suggests that the change is not related to changes in large-scale advection, but a transient increase in current velocities over the shelf may have resuspended large amounts of organic material and allowed some of it to enter the nitrite maximum (A. Huyer, G.T. Rowe and J.C. Van Leer, personal communication: Codispoti and Packard, 1980). The change in nitrite concentrations may be a low frequency type of variability since the first nitrite observations off Peru were made in 1960 (Wooster, Chow and Barrett, 1965).

Somalia. Major inter-annual differences in the nutrient regime may occur off Somalia. Usually one site of intense coastal upwelling (near 10°N) is found during the upwelling season (Fig. 3), but sea-surface temperature data have revealed the presence of two large upwelling plumes in some years (Brown, Bruce, and Evans, 1980). Whether or not this apparent absence of the southern plume is a real inter-annual difference or the result of insufficient data is not yet certain.

In any event, until the 1979 INDEX experiment, nutrient data from such a two plume situation were not available, but two plumes were encountered during this period (Fig. 3). The INDEX observations suggest that maximum nutrient concentrations in the southern plume were lower than in the north, probably due to a combination of weaker upwelling and lower nutrient concentrations (at a given σ_t) in the southern upwelling

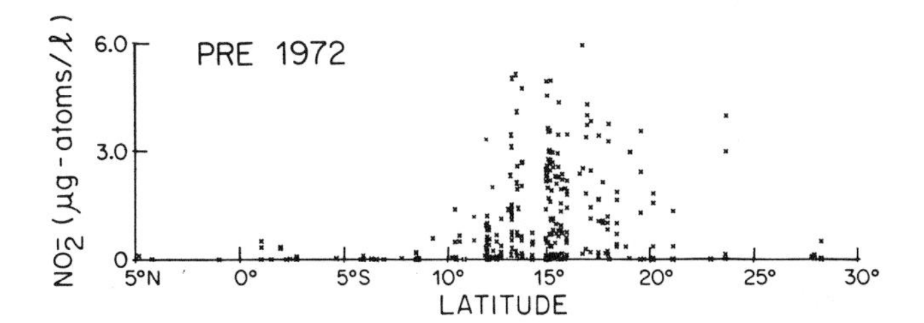

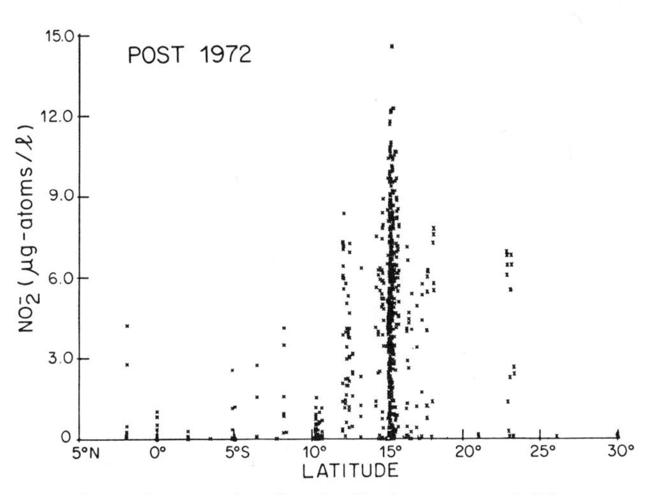

Fig. 5. Nitrate vs. latitude in water within ∿200 km of the western coast of South America pre-1972 vs. post-1972, $O_2 \leq 0.25$ ml/l, σ_t = 26.2 to 26.8. Data in the two 1° squares closest to the coast were used so the offshore boundary varies somewhat with latitude (from Codispoti and Packard, 1980).

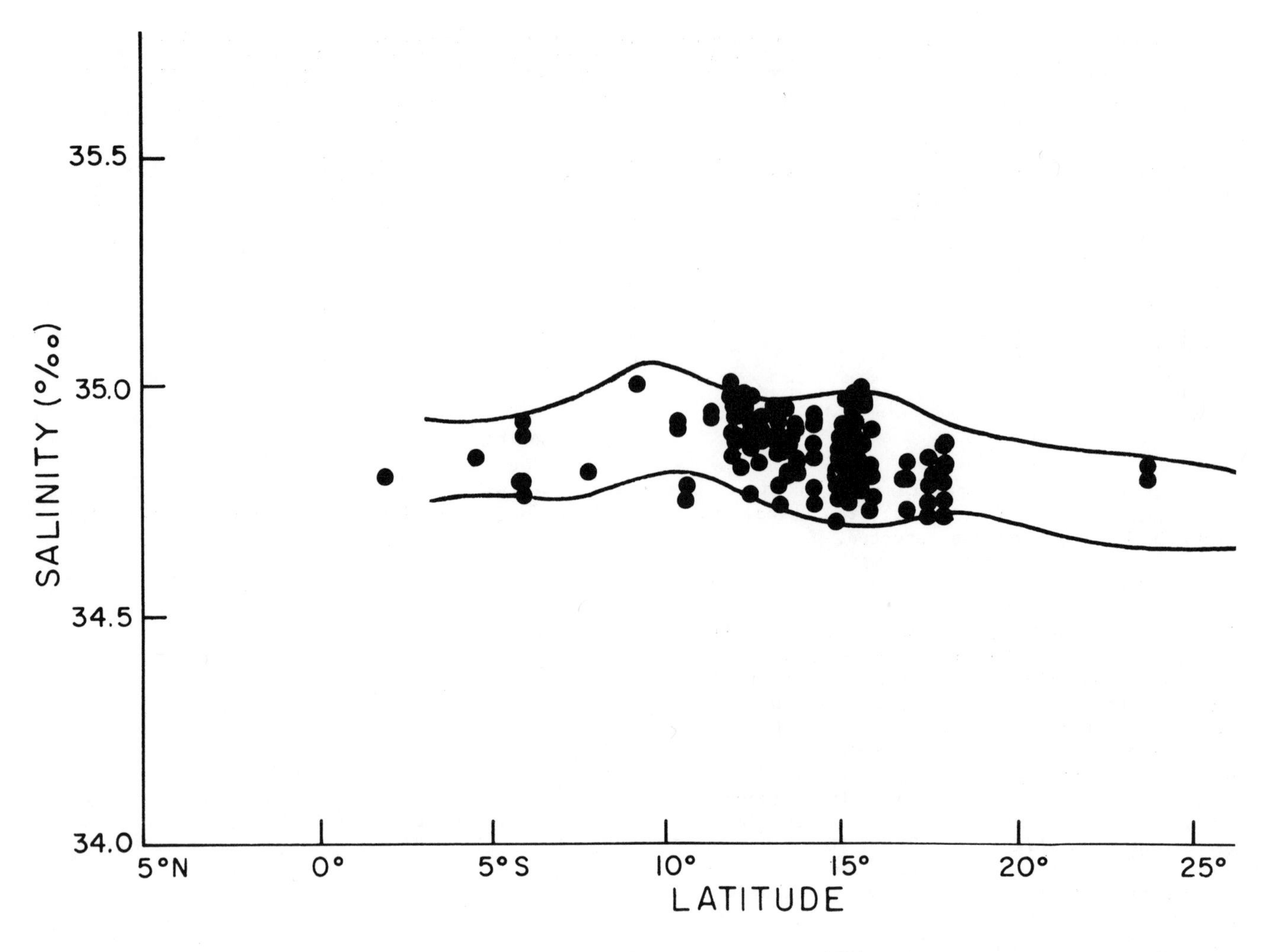

Fig. 6. Salinities vs. latitude in the 26.2 to 26.8 σ_t range. Pre-1972 values are plotted and the envelope is based on post-1972 values.

TABLE 1. Somalia: Southwest Monsoon of 1979. Selected Values from the Upwelling Regions near 5 and 10°N *

Number of Observations	T (^{o}C)	S (^{o}/oo)	σ_t	NO_3^- (µg-at/l)	Si (µg-at/l)	$PO_4^{=}$ (µg-at/l)
6[a]	18.6	35.3	25.4	13.1	10.8	0.92
4[b]	17.2	35.4	25.8	16.1	14.6	1.15
7[c]	18.7	35.6	25.6	19.9	15.7	1.53

* From Smith and Codispoti (1980).
a Surface water with nitrate >12.9 µg-at/l near 5°N
b Surface water with nitrate >15 µg-at/l south of 10°12'N
c Surface water with nitrate >19 µg-at/l north of 10°12'N

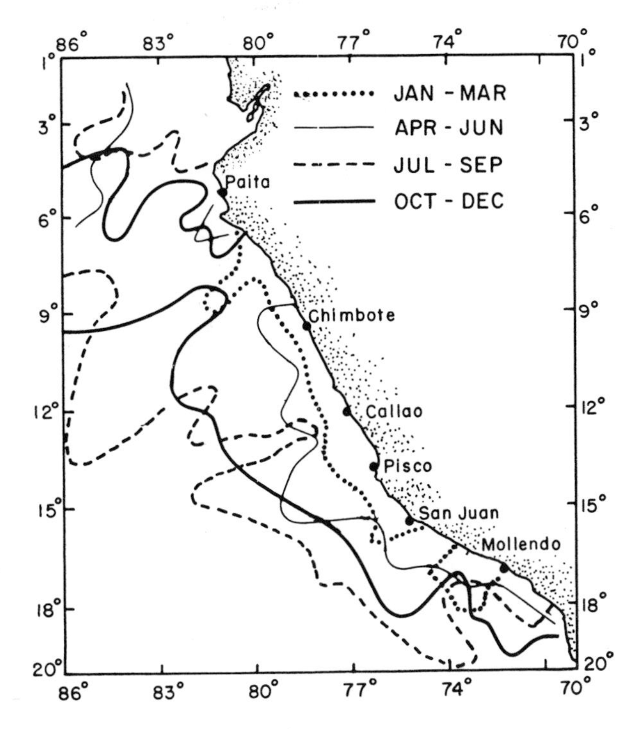

Fig. 7. The 1.0 μg-at/l reactive phosphorus isopleths at the sea surface off Peru for January-February-March, April-May-June, July-August-September, and October-November-December. The concentrations in the lower left-hand corner of the figure are always less than 1.0 μg-at/l. The figure is based on data published by Zuta and Guillen (1970).

waters (Table 1, and Smith and Codispoti, 1980). The maximum offshore depth of the upwelling water in the southern plume also seemed to be less than in the north (∿100 vs. 200 m).

Nutrient Variability on the Several-Month Scale

Northwest Africa. The seasonal variability in coastal upwelling off northwest Africa was summarized by Wooster, *et al*. (1976) and Schulz *et al*. (in press). The JOINT-I region (Fig. 1) is characterized by strong upwelling throughout the year. As one proceeds south of 20°N the length of the upwelling season becomes progressively shorter with the strongest upwelling occurring in the winter-spring period. Upwelling may occur during all seasons north of 25°N, but it is strongest from June to November.

Peru. Although upwelling occurs throughout the year off Peru there are seasonal variations in its intensity. Some aspects of the seasonal variability were summarized by Wooster and Sievers (1970) and Zuta and Guillen (1970). Surface temperature and reactive phosphorus distributions suggest that, in contrast to northwest Africa, the seasonal variability is similar along the entire Peruvian coast with maximum upwelling in southern winter and spring and weakest upwelling in summer and autumn (Fig. 7).

There are also significant seasonal variations in upwelling source water composition. The details of the response are gone into in another paper (Friederich and Codispoti, this volume), but they can be summarized as follows:

1. During all seasons the upwelling waters along most of the coast are dominated by relatively high salinity and nutrient rich waters carried southward in a poleward undercurrent. However, near 15°S there is a front created by the meeting of the equatorial waters and subantarctic waters carried equatorward by the Peru coastal current.
2. During the season of strong upwelling an equatorward flow that is almost always present offshore of the undercurrent moves closer to the coast.
3. The equatorward current contains subantarctic waters that are fresher and have lower reactive phosphorus and dissolved silicon concentrations than the northern waters, and mixing between the northern and southern water reduces nutrient levels (at a given value of σ_t) in the upwelling source waters.
4. Because the northern waters are more affected by denitrification, the differences in nitrate concentrations between the northern and southern waters are not so great as the differences in dissolved silicon and reactive phosphorus. An additional fact that may contribute to this situation is the differences in silicon/nitrate ratios in the source waters. The equatorial source waters have higher ratios than the southern waters even before the onset of denitrification.

Somalia. As Düing (1970) pointed out, the circulation of the northern part (north of 20°S) of the Indian Ocean is remarkable for its semiannual variability. Off Somalia's western shores, the situation is reflected in a change from equatorward surface flow during the Northeast Monsoon to the vigorous (max. speeds >350 cm/s) anticyclonic flows (Düing, Molinari and Swallow, 1980) observed during the Southwest Monsoon period (April-November).

The marked semiannual change also effects the character of the upwelling off Somalia's western coast. Strong upwelling occurs during the Southwest Monsoon, but little or none occurs during the Northeast Monsoon. Fig. 3 shows conditions during the Southwest Monsoon of 1979. Although the situation was unusual because of the upwelling near 5°N (see above), the upwelling zone observed near 10°N was fairly typical. In con-

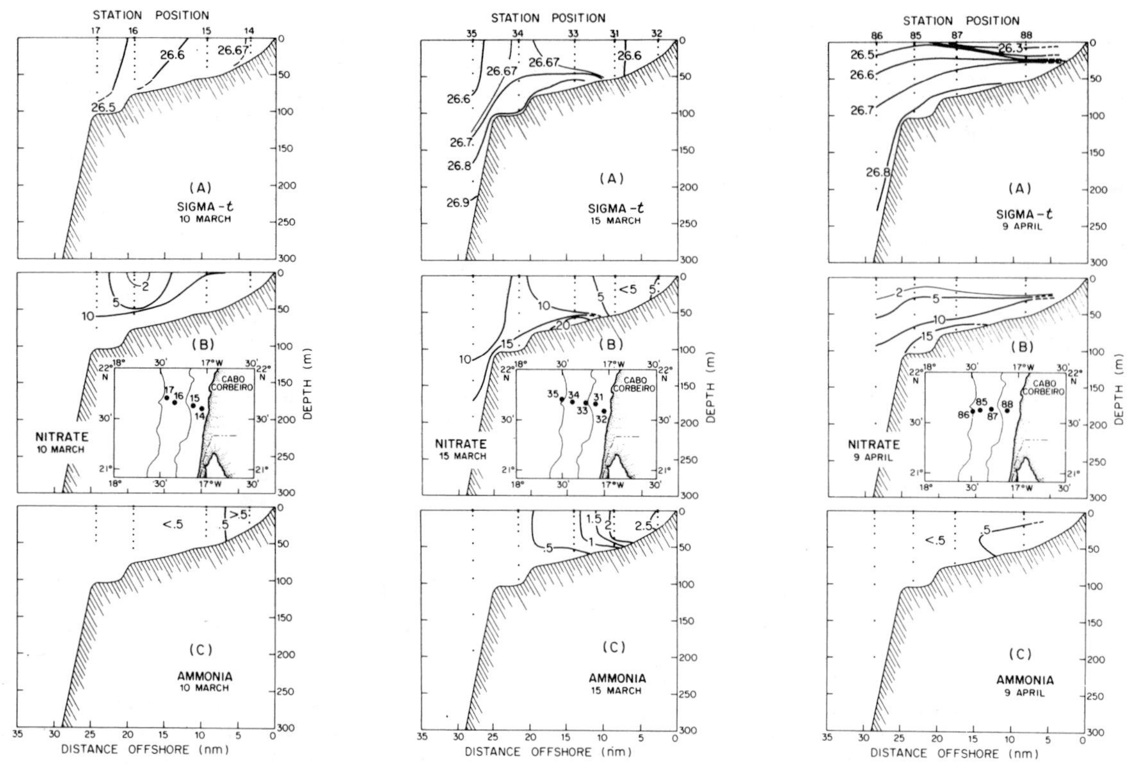
Fig. 8. Sigma-t (A), nitrate (B), and ammonia (C) off northwest Africa during 10 March, 15 March, and 9 April 1974. Nitrate and ammonia concentrations are in µg-at/l.

trast, the surface nitrate concentrations observed during the Northeast Monsoon of 1979 seldom exceeded 1 µg-at/l (Smith and Codispoti, 1980).

Nutrient Variability on the Several-Day Scale

Northwest Africa. The JOINT-I observations (Fig. 1) demonstrate significant changes in the nutrient regime on the several-day scale at ~21°40'N off northwest Africa and show that much of the variability can be correlated with variations in the local winds (Barton, Huyer and Smith, 1977; Codispoti and Friederich, 1978). Five periods of enhanced upwelling ("events") were identified during February to April 1974; the two largest had similar life cycles. In brief, the nutrient response would lag the onset of upwelling-favorable winds by about one day and nutrient-rich water would first reach the surface over the inner shelf. With continuing favorable winds the center of most intense upwelling would migrate to the vicinity of the shelf break where it would remain until the winds relaxed. Fig. 8 shows conditions at the beginning and at the height of one of the two major "events". Conditions during one of the periods with winds unfavorable for upwelling are also shown.

Another way in which wind variations may contribute to the several-day nutrient variability in the JOINT-I region is by their effect on a

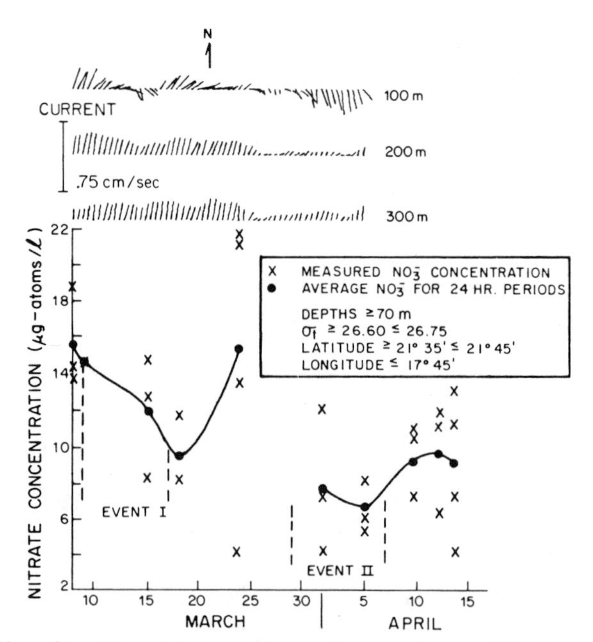

Fig. 9. Time history for nitrate (µg-at/l) in the upwelling source waters near 21°40'N off northwest Africa and currents over the slope (array L., Fig. 1). The time axis is common for both currents and concentrations. Observations on 0059 24 March have been included with the 23 March data.

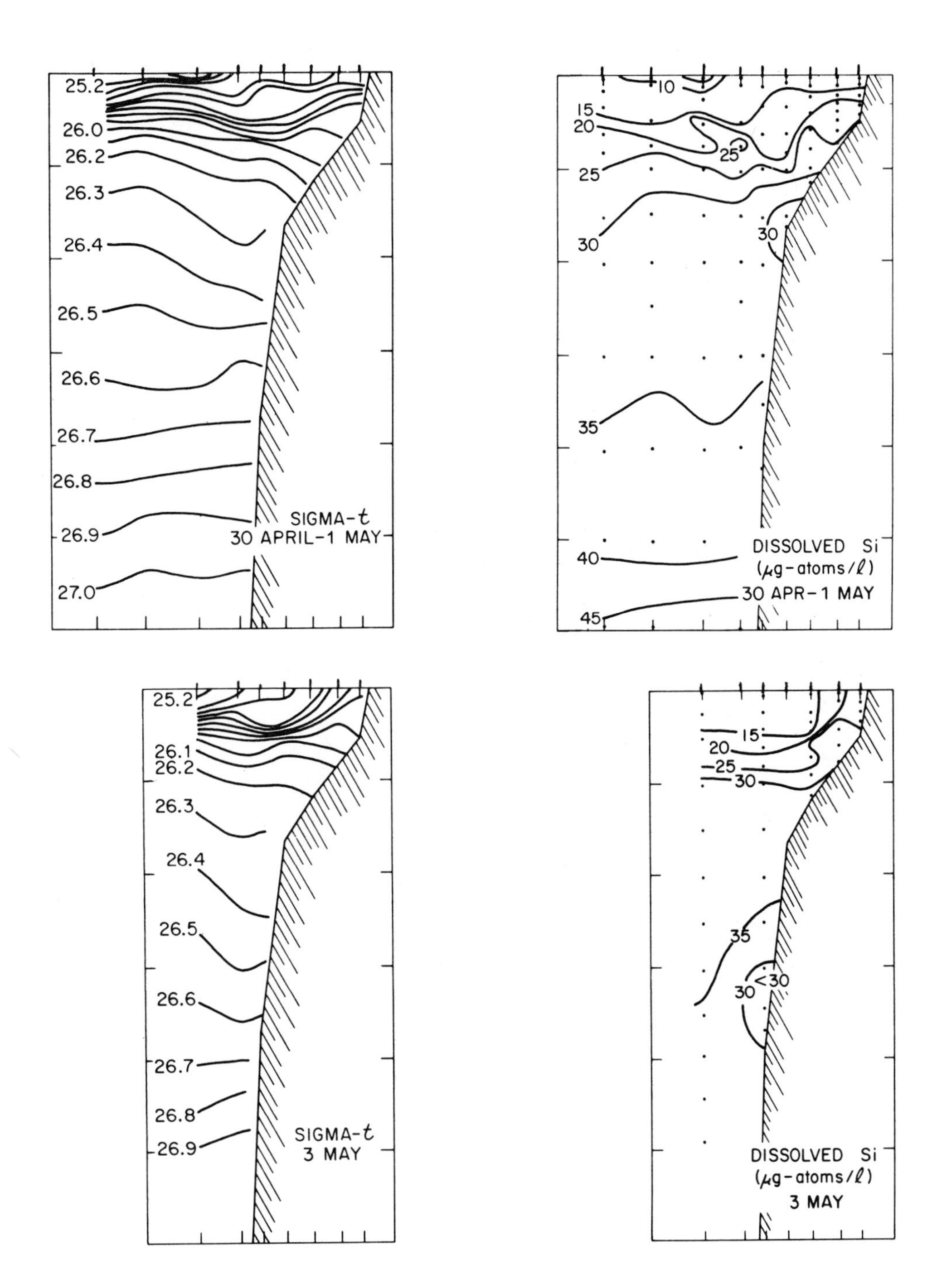

Fig. 10. The "C"-line sections of 30 April to 1 May and 3 May 1976. Dissolved silica and ammonia are in μg-at/l. The "tick" at the top of the figures denotes the average "C"-line positions shown in Fig. 2.

poleward undercurrent that carries nutrient-rich South Atlantic Central Water into the region. The undercurrent may be relatively weak when upwelling winds are strongest (Mittelstaedt, Pillsbury and Smith, 1975), and nutrient concentrations in the upwelling source water are relatively low during these periods (Fig. 9 and Codispoti and Friederich, 1978).

Peru. Winds at 15°S off Peru may be steadier than those encountered off northwest Africa during JOINT-I (Codispoti, *et al.*, in press; Barton,

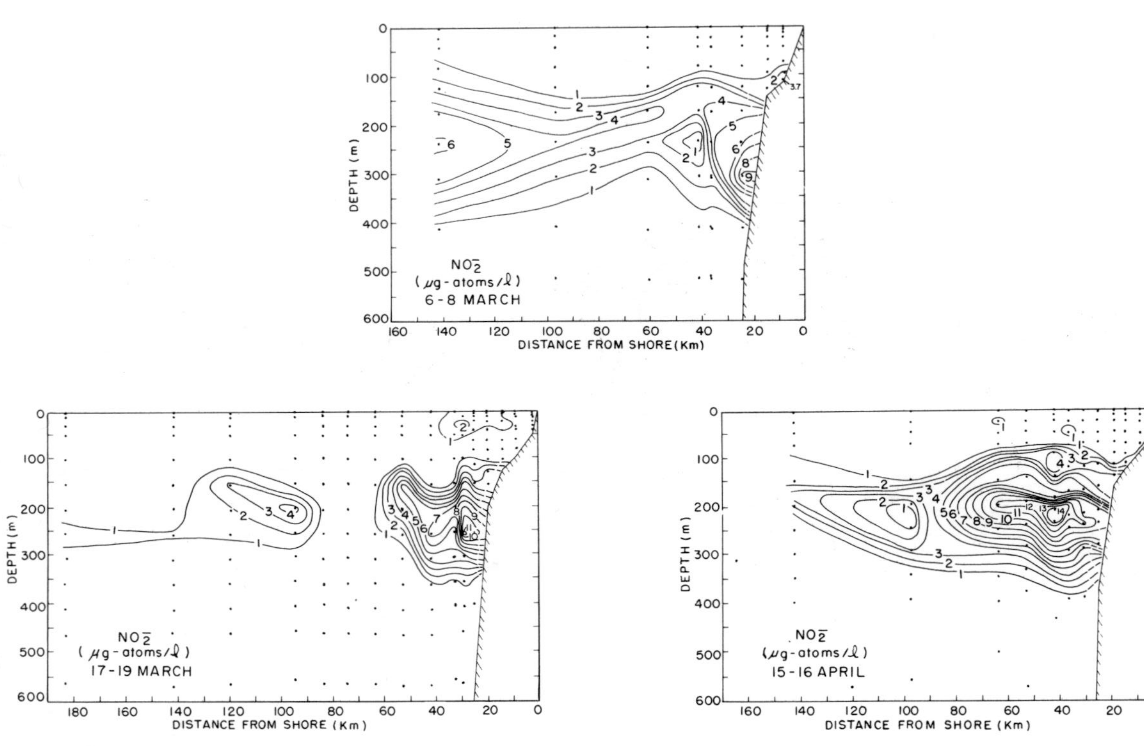

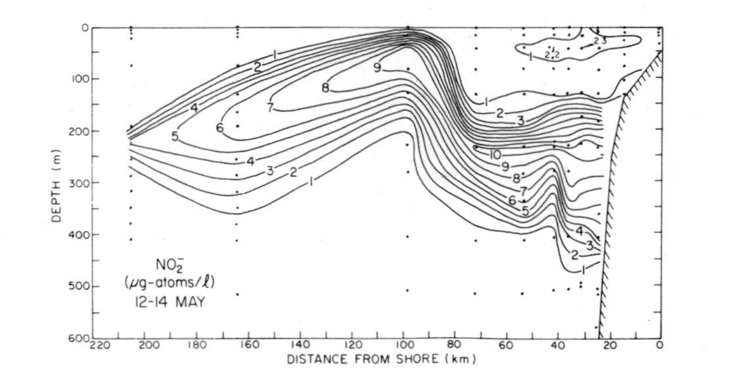

Fig. 11. Anomalous nitrite distributions observed during the passage of the most pronounced coastal trapped waves observed off Peru during March to May 1977. The waves influenced the nearshore region at ~15°S off Peru in mid-March and mid-May. More normal distributions observed on 6 to 8 March and 15 to 16 April are shown for comparison. Nitrite concentrations are in µg-at/l. The "C"-line sections are from Hafferty, Lowman and Codispoti (1979).

Pillsbury and Smith, 1975; Brink, Smith and Halpern, 1978; Enfield, Smith and Huyer, 1978), and correlations between local winds and nutrient fields do not appear to be so dramatic off Peru (at $\sim 15^{\circ}S$) as they are off northwest Africa (at $\sim 21^{\circ}40'N$). Nevertheless, some correlations do exist. For example, the average daily longshore wind (upwelling favorable) increased from about 2 m/s on 30 April to about 7 m/s on 3 May, and the nutrient fields appeared to respond to the increase (Fig. 10).

Another reason why correlations between wind and nutrient fields off Peru may be weaker than off northwest Africa is that longshore internal waves with sub-inertial periods may also influence the several-day nutrient variability off Peru. Some of the effects of such a wave are discussed by Brink, *et al.* (this volume). Although no causal relationship has been established, it is interesting to note that conditions in the nitrite maximum $\sim 15^{\circ}S$ off Peru were anomalous during the passage of the two largest coastal trapped waves observed during 1977 (Fig. 11).

A several-day biological effect on the nutrient regime near $15^{\circ}S$ was evident as the system recovered from the episode of hydrogen sulphide production mentioned above. Either as a result of the lesser degree of denitrification in the upwelling source waters or a change in phytoplankton species (from predominantly dinoflagellates to mostly diatoms), nitrate/dissolved silicon ratios increased after the hydrogen sulphide episode (Codispoti, *et al.*, 1976).

Somalia. The results of physical observations (e.g., Warren, Stommel and Swallow, 1966; Brown, *et al.*, 1980) suggest that there should be considerable several-day variability in the nutrient fields during upwelling off Somalia, but there are no nutrient data to test this hypothesis.

Conclusion

Although the above analysis suffers from insufficient data, particularly from the Somali upwelling zone, it does suggest some differences in the types of temporal variability that are important in the three zones. Briefly, much of the nutrient variability at $\sim 21^{\circ}40'N$ off northwest Africa occurs on the several-day scale; the upwelling zone off Peru is affected by processes with time scales ranging from days to many years; and the Somali upwelling zone is dominated by the several-month life span of the Southwest Monsoon, but is probably also significantly affected by processes acting on other scales.

Acknowledgements. I thank A.J. Hafferty for his assistance in the preparation of this paper. Financial support was provided by the National Science Foundation under grants OCE-7727128 and OCE-7825456 and by the Office of Naval Research under contract N-00014-76-C-0271. This is contribution 80024 from the Bigelow Laboratory for Ocean Sciences.

References

Barton, E.D., A. Huyer, and R.L. Smith, Temporal variation observed in the hydrographic regime near Cabo Corbeiro in the Northwest African Upwelling Region, February to April, 1974, *Deep-Sea Research, 24,* 7-24, 1977.

Barton, E.D., R.D. Pillsbury, and R.L. Smith, A compendium of physical observations from JOINT-I, School of Oceanography, Oregon State University, Reference 75-17, 60 pp., 1975.

Boje, J., and M. Tomczak, *Upwelling Ecosystems,* Springer-Verlag, 303 pp., 1978.

Brink, K.H., B.H. Jones, Jr., J.C. Van Leer, C.N.K. Mooers, D. Stuart, M. Stevenson, R.C. Dugdale, and G. Heburn, Physical and biological structure of the upwelling plume off Peru at $15^{\circ}S$, *This Volume,* 1980.

Brink, K.H., R.L. Smith, and D. Halpern, A compendium of time series measurements from moored instrumentation during the MAM'77 Phase of JOINT-II, School of Oceanography, Oregon State University, Reference 78-17, 72 pp., 1978.

Brown, O.B., J.G. Bruce, and R.H. Evans, Sea surface temperature evolution in the Somali Basin during the southwest monsoon of 1979, *Science, 209,* 595-597, 1980.

Codispoti, L.A., R.C. Dugdale, and H.J. Minas, A comparison of the nutrient regimes off northwest Africa, Peru, and Baja California, *Rapports et Proces-Verbaux des Reunions, International Council for the Exploration of the Sea,* in press.

Codispoti, L.A., and G.E. Friederich, Local and mesoscale influences on nutrient variability in the northwest Africa upwelling region near Cabo Corbeiro, *Deep-Sea Research, 25,* 751-770, 1978.

Codispoti, L.A. and T.T. Packard, Denitrification rates in the eastern tropicl South Pacific, *Journal of Marine Research, 38,* 453-477, 1980.

Codispoti, L.A., D.D. Bishop, M.A. Friebertshauser, and G.E. Friederich, JOINT-I - R/V *Thomas G. Thompson* cruise 108 bottle data April-June, 1976, *IDOE-Coastal Upwelling Ecosystems Analysis Data Report 35,* 370 pp., 1976.

Dugdale, R.C., J.J. Goering, R.T. Barber, R.L. Smith, and T.T. Packard, Denitrification and hydrogen sulfide in the Peru upwelling region during 1976, *Deep-Sea Research, 24,* 601-608, 1977.

Düing, W.O., *The Monsoon Regime of the Currents in the Indian Ocean,* East-West Center Press, University of Hawaii, Honolulu, 68 pp., 1970.

Düing, W.O., R.L. Molinari, and J.C. Swallow, Somali Current: Evolution of surface current, *Science, 209,* 588-590, 1980.

Enfield, D.B., R.L. Smith, and A. Huyer, A compilation of observations from moored current meters, Vol. XII, Wind, currents and temperature over the continental shelf and slope off Peru during JOINT-II, School of Oceanography, Oregon State University, Reference 78-4, 343 pp., 1978.

Friederich, G.E., and L.A. Codispoti, On some factors influencing dissolved silicon distribution over the northwest African shelf, *Journal of Marine Research, 37,* 337-353, 1979.

Friederich, G.E., and L.A. Codispoti, The effect of mixing and regeneration on the nutrient content of the upwelling waters at 15°S off Peru, *This Volume*, 1980.

Hafferty, A.J., D. Lowman, and L.A. Codispoti, JOINT-II, *Melville* and *Iselin* bottle data sections, March-May 1977, *Coastal Upwelling Ecosystems Analysis Technical Report, 38,* 130 pp., 1979.

Minas, H.J., L.A. Codispoti, and R.C. Dugdale, Nutrients and primary production in the northwest African upwelling region, *Rapports et Proces-Verbaux des Reunions, International Council for the Exploration of the Sea,* in press.

Mittelstaedt, E., D. Pillsbury, and R.L. Smith, Flow patterns in the northwest African upwelling area, *Deutsche hydrographische Zeitschrift, 28,* 145-167, 1975.

Quinn, W.H., Monitoring and predicting El Niño invasions, *Journal of Applied Meteorology, 13,* 825-830, 1974.

Schulz, S., R. Schemainda, and D. Nehring, Seasonal variations in the physical, chemical and biological features in the CINECA regions, *Rapports et Proces-Verbaux des Reunions, International Council for the Exploration of the Sea,* in press.

Smith, S.L., and L.A. Codispoti, Southwest monsoon of 1979: Chemical and biological response of Somali coastal waters, *Science, 209,* 597-599, 1980.

Sorokin, Y.I., Description of primary production and of heterotrophic microplankton in the Peruvian upwelling region, *Oceanology, 18,* 62-71, 1978.

Sverdrup, H.V., M.W. Johnson, and R.H. Fleming, *The Oceans*, Prentice-Hall, 1087 pp., 1942.

Walsh, J.J., Herbivory as a factor in patterns of nutrient utilization in the sea, *Limnology and Oceanography, 21,* 1-13, 1976.

Warren, B., H. Stommel, and J.C. Swallow, Water masses and patterns of flow in the Somali Basin during the Southwest Monsoon of 1964, *Deep-Sea Research, 13,* 825-860, 1966.

Wooster, W.S., El Niño, *California Cooperative Oceanic Fisheries Investigations Reports, 7,* 43-45, 1960.

Wooster, W.S., A. Bakun, and D.R. McLain, The seasonal upwelling cycle along the eastern boundary of the North Atlantic, *Journal of Marine Research, 34,* 131-141, 1976.

Wooster, W.S., T.J. Chow, and I. Barrett, Nitrite distribution in Peru Current waters, *Journal of Marine Research, 23,* 210-221, 1965.

Wooster, W.S., and O. Guillen, Characteristics of El Niño in 1972, *Journal of Marine Research, 32,* 387-403, 1974.

Wooster, W.S., and H. Sievers, Seasonal variations of temperature, drift and heat exchange in surface waters off the west coast of South America, *Limnology and Oceanography, 15,* 595-605, 1970.

Wyrtki, K., El Niño - The dynamic response of the equatorial Pacific Ocean to atmospheric forcing, *Journal of Physical Oceanography, 5,* 572-584, 1975.

Wyrtki, K., Advection in the Peru Current as observed by satellite, *Journal of Geophysical Research, 82,* 3939-3944, 1978.

Zuta, S., and O. Guillen, Oceanography of the coastal waters off Peru, *Instituto del Mar del Peru Boletin, 2,* 157-324, 1970.

THE EFFECTS OF MIXING AND REGENERATION ON THE NUTRIENT CONTENT OF UPWELLING WATERS OFF PERU

G.E. Friederich

L.A. Codispoti

Bigelow Laboratory for Ocean Sciences
West Boothbay Harbor, ME. 04575, U.S.A.

Abstract. During July to October, 1976 and March to May, 1977 the nutrient content of the upwelling source waters near 15°S off Peru exhibited variability not solely related to density. On a given σ_t surface, mixing between subtropical and subantarctic waters affected the nutrient concentration, and significant changes also arose from nutrient regeneration.

High-salinity, nutrient-rich water carried southward in a poleward undercurrent was the source for most of the upwelled water, and between 5 and 15°S, nitrate and dissolved silica concentrations within the flow increased by as much as 8 µg-at/l and 16 µg-at/l, respectively, by regenerative processes. The increases were opposed by the effects of mixing with low-salinity, nutrient-poor subantarctic water advected northward some distance off the coast. The dilution was especially evident during July to November, 1976 when low-salinity, nutrient-poor water appeared to influence the composition of the upwelling source water as far north as 10°S. During March to May, 1977, the undercurrent appeared to transport nutrient-rich subtropical water to at least 15°S before substantial mixing with subantarctic water occurred.

Nitrate increases by regeneration were counteracted to some extent by denitrification.

Introduction

The supply of inorganic nutrients to the photic zone is one of the major factors facilitating the high primary productivity of upwelling regions. It is also generally acknowledged that inorganic nutrients are often the limiting factor for phytoplankton growth in tropical upwelling systems. We can therefore assume that the supply of nutrients to the photic zone is intimately related to the productivity of an upwelling region. While the vertical velocity field may be the primary factor controlling the nutrient supply to the near-surface layers, significant variability is added by varying water mass composition and recycling (Codispoti and Friederich, 1978; Friederich and Codispoti, 1979).

Using some of the JOINT-II observations (Fig. 1) collected during July to October, 1976 and March to May, 1977 (Kogelschatz, Paul, Codispoti, Whitledge, and Huyer, in press; Hafferty, Codispoti, and Huyer, 1978), we have investigated factors controlling nutrient concentrations in the upwelling source waters off Peru. The rising

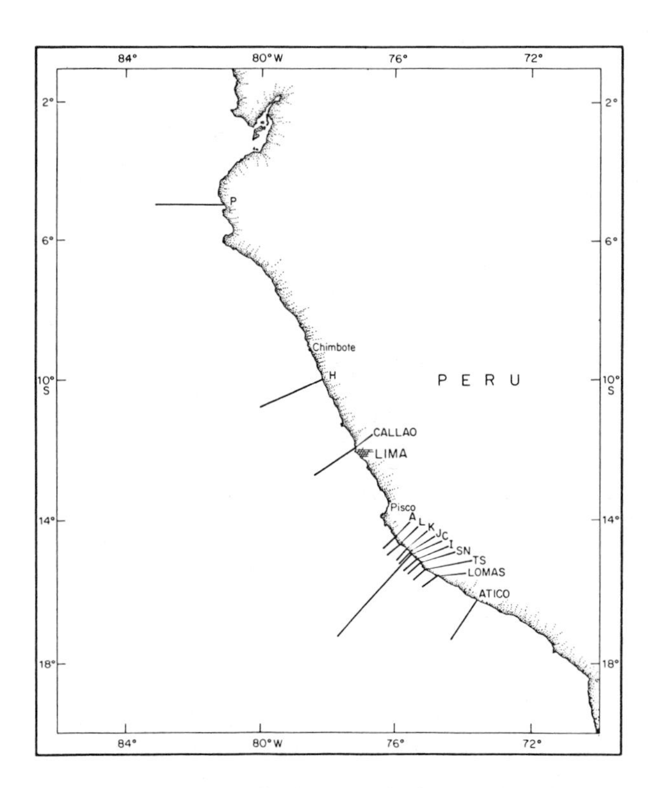

Fig. 1. Location of the main hydrographic lines during the JOINT-II experiment, April, 1976, May, 1977.

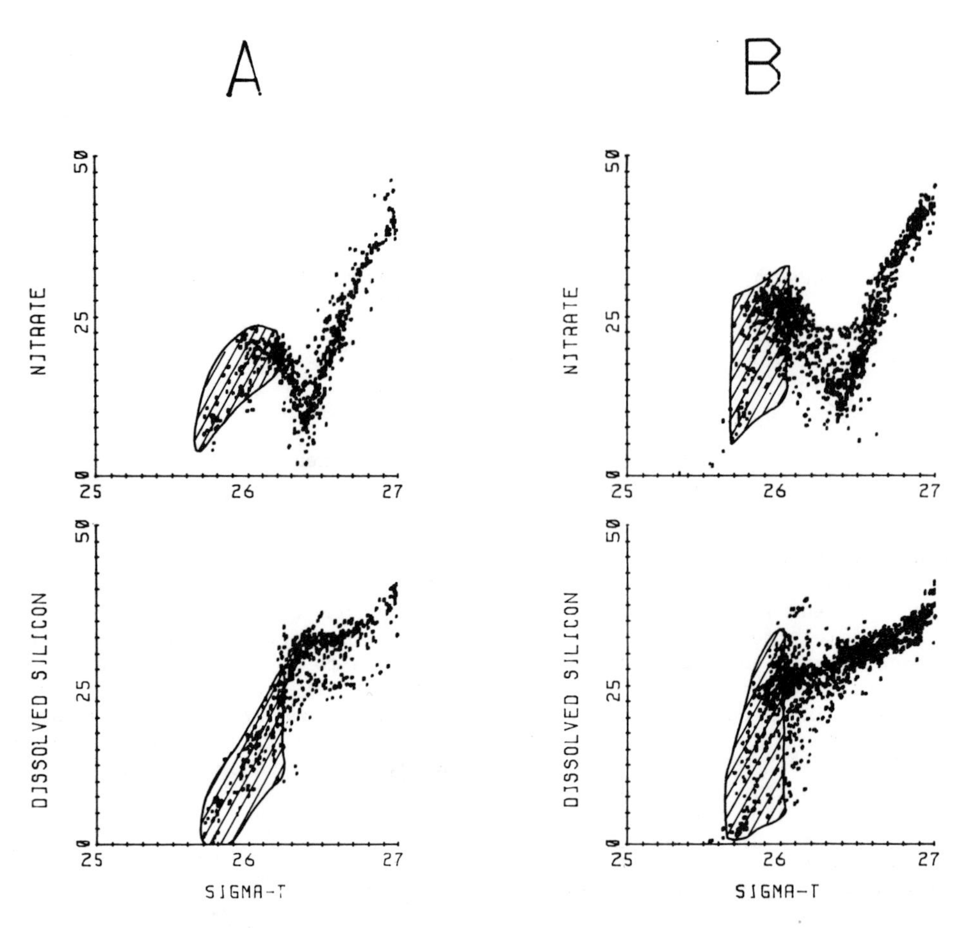

Fig. 2. The relationship between density and nutrient concentrations in the subsurface waters (depth $\geq$ 40 m) near 15°S (latitude $\geq$ 14°S and $\leq$ 17°S) during two phases of JOINT-II. (A) Observations from July to October, 1976 and (B) March to May, 1977. Nitrate and dissolved silica concentrations are in μg-at/l. Shaded areas indicate potential upwelling source waters.

waters appear to originate predominantly in the equatorial region and are carried southward in a poleward undercurrent over the shelf and slope (Enfield, Smith, and Huyer, 1978). There are two potential "equatorial" sources for the rising waters, the Equatorial Undercurrent and the South Equatorial Subsurface Countercurrent that flows eastward between 4 to 8°S (Tsuchiya, 1975). Although the equatorial input dominates the upwelling source water characteristics along most of the coast, there is a significant input of subantarctic water, an influence most pronounced south of 15°S.

Data and Results

Although there is a strong relationship between *sigma-t* and nutrient content within the *sigma-t* range (26.0) of the rising waters, the effects of origin and *in situ* processes also cause significant variability. For example, during July to October, 1976, nitrate ranged from 11 to 22 μg-at/l on the 26.0 σ_t surface below 40 m and the range of dissolved silica on the same surface was 7 to 18 μg-at/l (Fig. 2). Even greater variability was observed on the same isopycnal during March to May, 1977, 16 to 32 μg-at/l for nitrate and 6 to 36 μg-at/l for dissolved silica. While the data in Fig. 2 cover periods of several months, some changes may occur over shorter periods. For example, a time series station at midshelf occupied during May 1976 (Codispoti, Bishop, and Friebertshauser, 1976) showed fluctuations of 9 μg-at/l in nitrate and 8 μg-at/l in dissolved silica on the 26.0 σ_t surface during a 24-h period.

In general, the equatorial source waters have higher salinities than the subantarctic waters (Wooster and Gilmartin, 1961) that rise to the surface, and plots of salinity vs. latitude show a salinity front near 15°S (Fig. 3). Consequently, the waters north of 15°S were only slightly influenced by subantarctic water during March to May, 1977 (Fig. 3) and moderately influenced during July to October, 1976 (Fig. 4). The data in Fig. 4 were observed when upwelling-

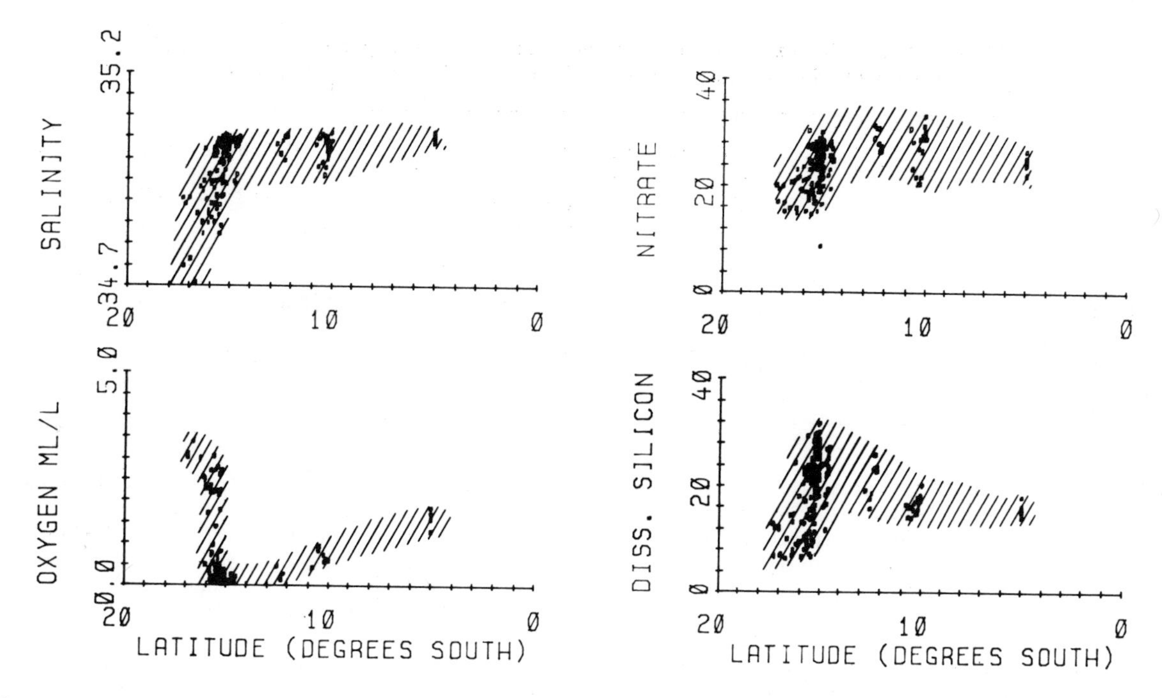

Fig. 3. Trends of salinity, dissolved oxygen, nitrate, and dissolved silica on the 26.0 σ_t surface along the coast of Peru during March to May, 1977. Depth $\geq$ 40 m. Nitrate and dissolved silica concentrations are in μg-at/l.

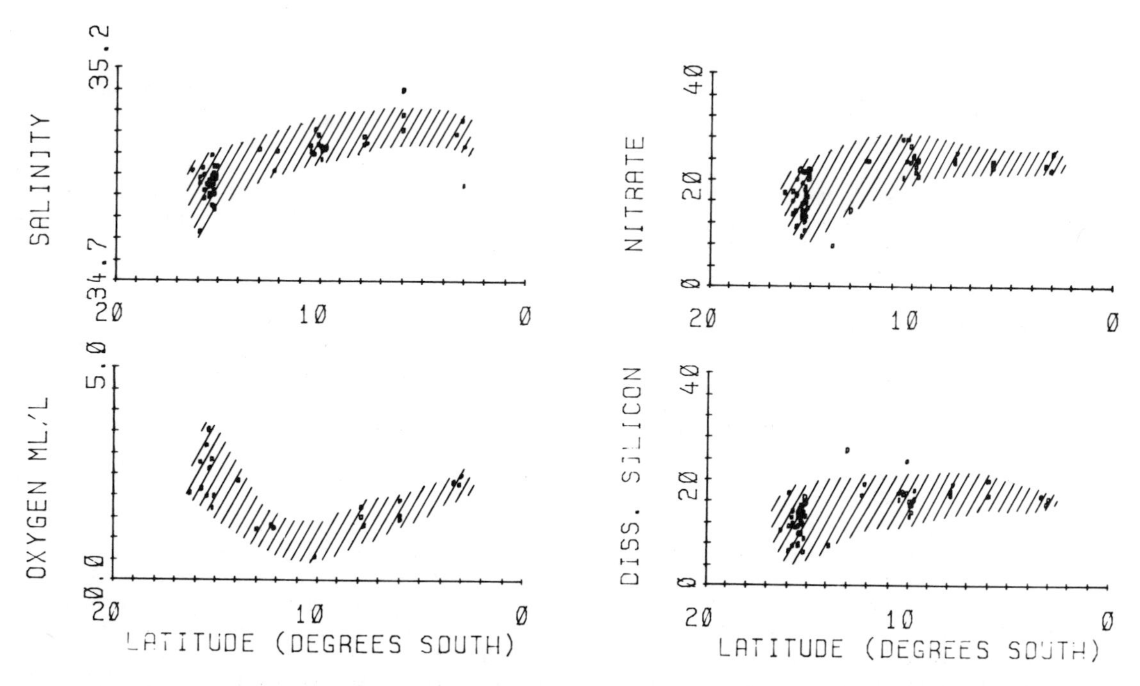

Fig. 4. Trends of salinity, dissolved oxygen, nitrate, and dissolved silica on the 26.0 σ_t surface along the coast of Peru during July to October, 1976. Depth $\geq$ 40 m. Nitrate and dissolved silica concentrations are in μg-at/l.

TABLE 1. Source Water Characteristics Derived from Oxygen, Nitrate, and Dissolved Silica vs. Salinity Plots for the 26.0 σ_t Surface

	July-October 1976		March-May 1977	
	Equatorial Water	Subantarctic Water	Equatorial Water	Subantarctic Water
Salinity	35.055	34.865	35.050	34.825
Oxygen, ml/l	1.8	2.8	1.6	3.0
Nitrate, μg-at/l	23	14	25	16
Dissolved silica, μg-at/l	17	6	15	7

favorable winds are normally at their peak, while the March to May, 1977 (Fig. 3) cruise was during a period between the months of strongest (June to October) and weakest (January, February) upwelling (Wooster and Sievers, 1970; Zuta, Rivera, and Bustamante, 1975).

The seasonal variability was evident in the dissolved oxygen and silica distributions (Figs 3, 4). During the season of moderate upwelling, oxygen concentrations decreased and dissolved silica increased more or less continuously within the southward flow between 5 and 15°S. During the season of stronger upwelling, the continuous dissolved silica increase was small relative to the weak upwelling period because of the penetration of low-salinity, nutrient-poor and dissolved oxygen-rich subantarctic waters north of 15°S. The meridional increase of nitrate between 10 and 15°S was relatively small during both periods, probably because of the large denitrification zone between ∿100 and 400 m south of ∿10°S (Codispoti and Packard, 1980). Denitrification removes nitrate and appears to prevent a rise in nitrate concentrations similar to the dissolved silica increase observed during March to May, 1977. The range of nitrate concentrations (16 to 32 μg-at/l) in the frontal zone on the 26.0 σ_t surface was less than that for dissolved silica (6 to 36 μg-at/l) because of denitrification, and because the lowest nitrate values (16 μg-at/l) in the subantarctic water were higher than the lowest dissolved silica values (6 μg-at/l).

In an attempt to quantify the extent of *in situ* processes (between 5 and 17°S), the upwelling waters were treated as a simple two-component mixture of equatorial and subantarctic waters. The extreme northern equatorial and southern subantarctic water mass characteristics observed (Table 1) were selected as end-points for a mixing diagram; salinity was considered to be a conservative property, and expected values of nutrients and dissolved oxygen were calculated for the 26.0 σ_t surface. We then compared the values with the observations; the differences, assumed to be due to *in situ* processes, are shown in Fig. 5. Dissolved silica and oxygen were probably not significantly altered by vertical mixing because the concentrations of both change monotonically with density. Sample calculations of mixtures that could result from vertical mixing confirmed the assumption. For example, when we examined Stas 380, 451, and 459 in Hafferty *et al.* (1978) and 192 and 258 in Kogelshcatz *et al.* (in press) and mixed waters over a depth interval of 50 to 100 m, we produced a new 26.0 σ_t water with dissolved silica concentrations only slightly different (1 to 7%) from the origianl 26.0 σ_t water; dissolved oxygen concentrations differed by 3 to 7%. South of 10°S nitrate concentrations would generally be lowered by vertical mixing because waters both of higher and lower density contain less nitrate (Fig. 2). Once dissolved oxygen concentrations fall below ∿0.1 ml/l denitrification can also reduce nitrate concentrations even if there is no vertical mixing.

Despite the above complications, most of the nitrate deviations for July to October, 1976 could be related to changes in oxygen (for ΔNO_3^- vs. ΔO_2, m = -4.1 μg-at NO_3^-/ml O_2, and r = -0.83) in much the same ratio as the usual relationship of Redfield, Ketchum, and Richards (1963) (m = -3.9 μg-at NO_3^-/ml O_2 for ΔNO_3^- vs. ΔO_2). For example, it was possible to estimate nitrate concentrations on the 26.0 σ_t surface with a standard deviation of ± 1.6 μg-at/l. Since the above relationship appears to represent the data rather well, even though both vertical mixing and denitrification could cause some deviations, we assume that lateral mixing and regeneration were the major processes controlling nitrate concentrations on the 26.0 σ_t surface during July to October, 1976. The distribution of dissolved silica, although following similar trends, was not so closely related to changes in dissolved oxygen (as one might expect) and a meaningful relationship could not be established.

During March to May, 1977 some of the high-salinity water of equatorial origin penetrated almost to 16°S without significant interaction with subantarctic water (Fig. 3), while 15°S appeared to be the northern boundary of the lower-salinity subantarctic water on the 26.0 σ_t surface during this season. As relatively pure equatorial water (σ_t = 26.0, salinity ≥ 35.03)

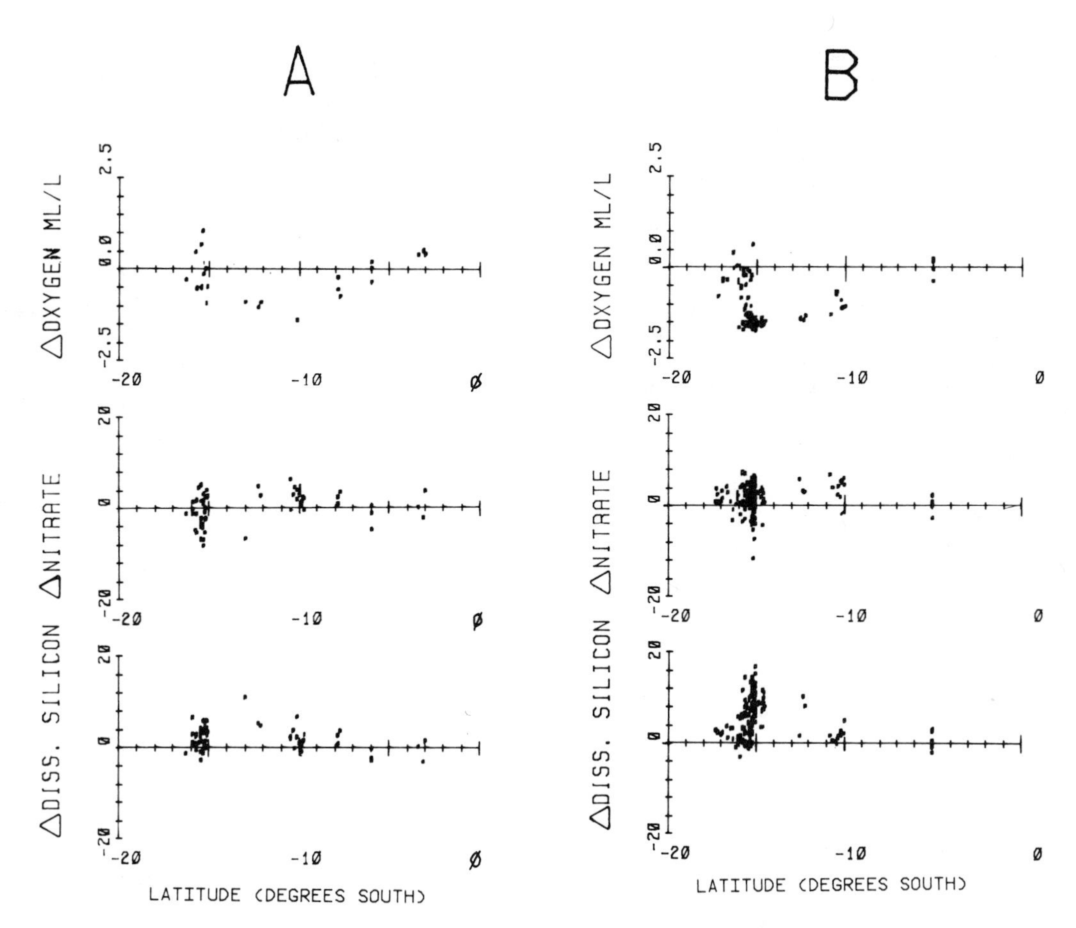

Fig. 5. Deviations from two-component mixing of equatorial and subantarctic waters on the 26.0 σ_t surface below 40 m. (A) Observations from July to October, 1976 and (B) March to May, 1977. Δ nitrate and Δ dissolved silica are in μg-at/l. Δ dissolved oxygen is in ml/l.

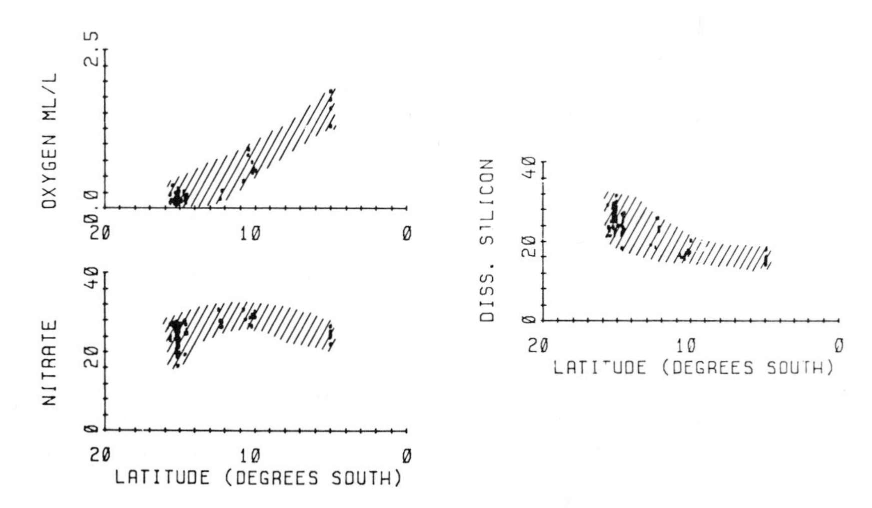

Fig. 6. Changes in dissolved oxygen, nitrate, and dissolved silica content with latitude in waters of equatorial origin on the 26.0 σ_t surface along the Peru coast during March to May, 1977. Depth $\geq$ 40 m, salinity $\geq$ 35.03. Nitrate and dissolved silica concentrations are in μg-at/l.

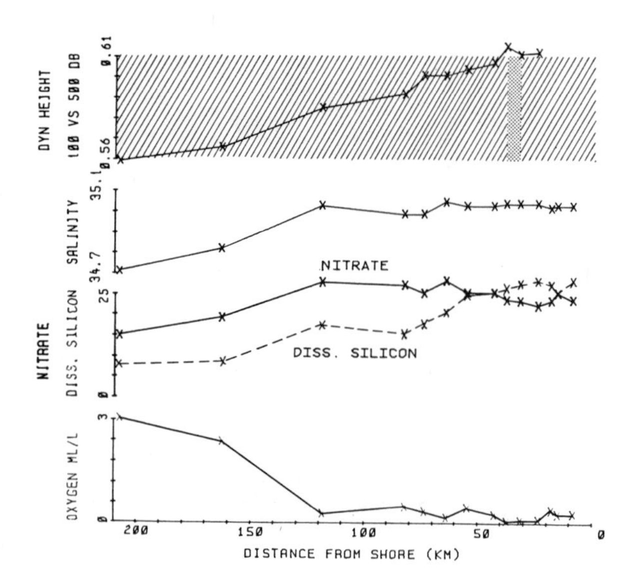

Fig. 7. Data from the section of 5 to 8 May, 1977 near 15°S. Cross-hatched areas indicate southward flow, dotted areas indicate northward flow at a depth of ∿100 m. Salinity, nitrate, dissolved silica, and oxygen content are for the 26.0 σ_t surface. Nitrate (solid line) and dissolved silica (broken line) concentrations are in µg-at/l.

progressed southward, the dissolved oxygen content decreased continuously from a value of ∿1.6 ml/l near 5°S until it was exhausted near 15°S (Fig. 6). According to Redfield *et al.* (1963) ∿8.2 µg-at/l of nitrate should be generated between 5 and 15°S, which agrees with the maximum ΔNO_3^- values of ∿7.6 µg-at/l obtained between 10 and 15°S (Fig. 5B). In general, however, ΔNO_3^- could not be related to ΔO_2 during this season. Apparently denitrification had become significant in altering the nitrate distribution on the 26.0 σ_t surface (Codispoti and Packard, 1980). Both the most positive and the most negative ΔNO_3 value at the time were associated with near zero oxygen values.

Although a large fraction of the variability of dissolved silica (∿8 µg-at/l, σ_t = 26.0) during March to May, 1977 could be attributed to the mixing of high- and low-salinity water, the effects of *in situ* processes were even more significant. The equatorial water (σ_t = 26.0, salinity ≥ 35.03) that reached 15°S in the poleward undercurrent had gained an average of 10 µg-at/l and an estimated maximum of 16 µg-at/l near the coast (Fig. 6).

Between 5 and 10°S the gradients of dissolved oxygen and nutrients may be influenced by an eastward extension of the South Equatorial Subsurface Countercurrent. Although it and the Equatorial Undercurrent appear to have indistinguishable water-mass characteristics at some distance offshore (300 to 400 km) it is not known to what extent *in situ* processes alter dissolved oxygen and nutrient concentrations near the coast. Therefore, it is possible that our estimated *in situ* changes do not occur solely within the meridional flows between 5 and 15°S; some could arise within nearshore portions of the two zonal currents mentioned above.

It is implicit in the above discussion that variations in the character of the poleward undercurrent can have a significant effect on the characteristics of the upwelling water. For example, a strong, broad undercurrent would increase the proportions of equatorial water while a weakly developed undercurrent might permit the effects of the subantarctic water to extend further towards the equator. Because of this we have examined the correlation between geostrophic flow and water mass characteristics at the approximate depth of the 26.0 σ_t surface near 15°S (Section C, Fig. 1).

While direct current measurements on the shelf indicated that the monthly average flow throughout JOINT-II (Enfield *et al.*, 1978) was southward, geostrophic current calculations suggest that the flow further offshore changed considerably. Fig. 7 shows the situation in early May 1977 when the poleward undercurrent appeared to have had almost complete dominance over the composition of the upwelling source waters. At that time, the undercurrent extended ∿200 km offshore, and salinity data suggest that the inner 70 km were composed of equatorial waters that had undergone little, if any, mixing with subantarctic water. Nutrient concentrations were closely correlated with the salinities between 70 and 200 km offshore. Inside of 70 km, regeneration became predominant and dissolved silica concentrations continued to increase towards the west while nitrate began to decrease because oxygen supplies were nearly exhausted.

In the winter months (June to September) the undercurrent near 15°S at the approximate depth of the upwelling source-waters (σ_t = 26.0) appears to have shrunk to a width of less than 30 km. The section of August 12, 1976 (Fig. 8) showed a strong northward flow of subantarctic water between the shelf and 160 km offshore. On the shelf, however, a strong southward flow was detected at a mid-shelf current meter installation (Enfield *et al.*, 1978). Low-salinity nutrient-poor water was observed near the shelf break and the mixing zone between this water and higher salinity water of the undercurrent appeared to extend across the entire shelf. The situation led to a continual lowering of salinity and nutrients in the upwelling waters in a southerly direction during the winter months between 10 and 15°S. Intermediate situations between the extremes portrayed in Figs 7 and 8 suggest that the poleward undercurrent has to extend ∿70 km offshore for the upwelling equatorial waters near 15°S not to be significantly modified by the subantarctic waters.

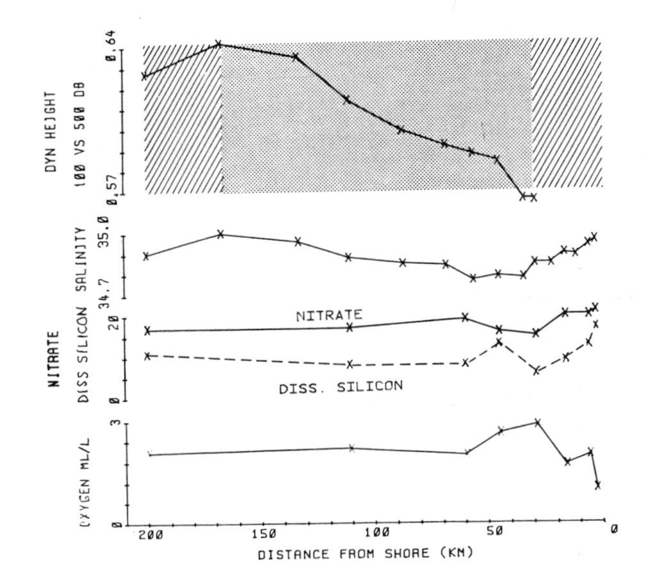

Fig. 8. Data from the section of 12 August near 15°S. Cross-hatched areas indicate southward flow, dotted areas indicate northward flow at a depth of ∿100 m. Salinity, nitrate, dissolved silica, and oxygen content are for the 26.0 σ_t surface. Nitrate (solid line) and dissolved silica (broken line) concentrations are in µg-at/1.

Acknowledgements. The study depended heavily on results from the JOINT-II program, which was organized and conducted by the Coastal Upwelling Ecosystems Analysis (CUEA) program with substantial support from the Instituto del Mar del Peru.

We wish to thank R.T. Barber and R.C. Dugdale for their consistent support and encouragement, and we are grateful for the technical assistance provided by A.J. Hafferty, D. Lowman, and D. Wisegarver. Financial support was provided by the National Science Foundation's International Decade of Ocean Exploration Program (IDOE) under grant OCE-7727138 and the Office of Naval Research, Contract No. N00014-76-C-0271.

Contribution No. 80025 from the Bigelow Laboratory for Ocean Sciences.

References

Codispoti, L.A., D.D. Bishop, M.A. Friebertshauser, and G.E. Friederich, JOINT-II R.V. *Thomas G. Thompson* cruise 108, Bottle data, April-June, 1976, *Coastal Upwelling Ecosystems Analysis Data Report 35*, 370 pp., 1976.

Codispoti, L.A., and G.E. Friederich, Local and mesoscale influences on nutrient variability in the northwest African upwelling region near Cabo Corbeiro, *Deep-Sea Research, 25*, 751-770, 1978.

Codispoti, L.A. and T.T. Packard, Denitrification rates in the eastern tropical South Pacific, *Journal of Marine Research, 38*, 453-477, 1980.

Enfield, D.B., R.L. Smith, and A. Huyer, A compilation of observations from moored current meters: wind, currents and temperature over the continental shelf and slope off Peru during JOINT-II: March 1976 - May 1977, *Coastal Upwelling Ecosystems Analysis Data Report 52*, 343 pp., 1978.

Friederich, G.E., and L.A. Codispoti, On some factors influencing dissolved silicon distribution over the northwest African Shelf, *Journal of Marine Research 37*, 337-353, 1979.

Hafferty, A.J., L.A. Codispoti, and A. Huyer, JOINT-II R.V. *Melville* Legs I, II, and IV and R.V. *Iselin* Leg II, bottle data, March - May 1977, *Coastal Upwelling Ecosystems Analysis Data Report 45*, 779 pp., 1978.

Kogelschatz, J.E., J.C. Paul, L.A. Codispoti, T.E. Whitledge, and A. Huyer, JOINT-II JASON 1976 Hydro Data R.V. *Eastward* cruises E-5F-76 through E-51-76, *Coastal Upwelling Ecosystems Analysis Data Report 38*, in press.

Redfield, A.C., B.H. Ketchum, and F.A. Richards, The influence of organisms on the composition of seawater, in *The Sea, Vol. 2*, N.N. Hill, ed., Interscience, New York, 554 pp., 1963.

Tsuchiya, M., Subsurface countercurrents in the eastern equatorial Pacific Ocean, *Journal of Marine Research, 33*, 145-175, 1975.

Wooster, W.S., and M. Gilmartin, The Peru-Chile Undercurrent, *Journal of Marine Research, 19*, 97-120, 1961.

Wooster, W.S., and H. Sievers, Seasonal variations of temperature, drift and heat exchange in surface waters off the west coast of South America, *Limnology and Oceanography, 15*, 595-605, 1970.

Zuta, S., T. Rivera, and A. Bustamante, Hydrological aspects of the main upwelling areas off Peru, in *Upwelling Ecosystems*, R. Boje and M. Tomczak, eds., Springer-Verlag, Berlin, 303 pp., 1975.

SATELLITE OBSERVATIONS OF A CYCLONIC UPWELLING SYSTEM AND GIANT PLUME IN THE CALIFORNIA CURRENT

E.D. Traganza and J.C. Conrad

Department of Oceanography, Naval Postgraduate School
Monterey, California 93940

L.C. Breaker

NOAA/NESS
Redwood City, California 94063

Abstract. Studies off Pt. Sur, California employing satellite thermal imagery and *in situ* underway sampling methods show that coastal upwelling waters may become cyclonic or occur as elongated plumes extending more than 250 km across the California Current with sharp thermal and chemical fronts. The occurrence of microplanktonic blooms adjacent to the sharp gradients suggests a relationship between the horizontal nutrient flux and primary production and between chemical mesoscale and biological patchiness. Because of the recurrence of upwelling systems in favored locations, an oceanic climatology could be developed for predicting upwelling phenomena from satellites.

Introduction

Chemical Mesoscale and Sea-Surface Thermal Structure

Satellite infrared (IR) images of the eastern North Pacific reveal surface features that appear to be thermal fronts and eddies associated with the California Current and coastal upwelling (Bernstein, Breaker, and Whritner, 1972; Fett *et al.*, 1979; Traganza, Nestor, and McDonald, 1980). Studies using satellite remote sensing of the temperature structure of the sea surface and *in situ* underway sampling methods (Traganza *et al.*, 1980) show that some of the features result from upwelling pulses, which produce nutrient structure of the same scale. Distinctively structured upwelling systems with sharp thermal and chemical fronts form recurrently off Pt. Sur, California. At times the upwelling waters appear to have cyclonic or anticyclonic motion or to advect as elongated plumes extending more than 250 km across the California Current. The objective of this paper is to examine the relationships of sea-surface thermal patterns (from satellite thermal imagery) to nutrients such as nitrate and phosphate and to show how they can help to explain the relationships between chemical mesoscale and pelagic ecosystems in regions of upwelling.

Upwelling Systems

While the correlation of satellite thermal gradients with *in situ* thermal gradients may hold in all cases, correlations with chemical gradients is by inference and may not always hold. The following is a comparison of two strikingly different upwelling features: the first, a cyclonic upwelling system, is in an early stage of development and has an excellent correlation between nutrients and temperature, the second, a giant plume, is in an advanced stage of development in which the correlation is poor.

Cyclonic Upwelling System

The first upwelling feature was observed in satellite infrared (IR) imagery on 17 April, 1979 as a seaward extension of coastal upwelling water off Pt. Sur, California. In two days it had developed the appearance of cyclonic motion (Fig. 1). Ten days later the feature was investigated in some two-dimensional (i.e., three transects; Fig. 2) and three-dimensional aspects (i.e., expendable bathythermographs were dropped at ca. 5-km intervals along the cruise track). Nutrients and temperatures were measured *in situ* every 0.6 km at a depth of 2.5 m. Adenosine triphosphate (ATP) was used as an estimate of living microplanktonic biomass (bacteria, algae, and microzooplankton). Samples (50 ml) were collected ca. every 10 min or every 3 km over most of the cruise track, filtered through a 200-μm (approximate pore size) nylon screen and a 0.5-μm (approximate pore size) glass filter. ATP

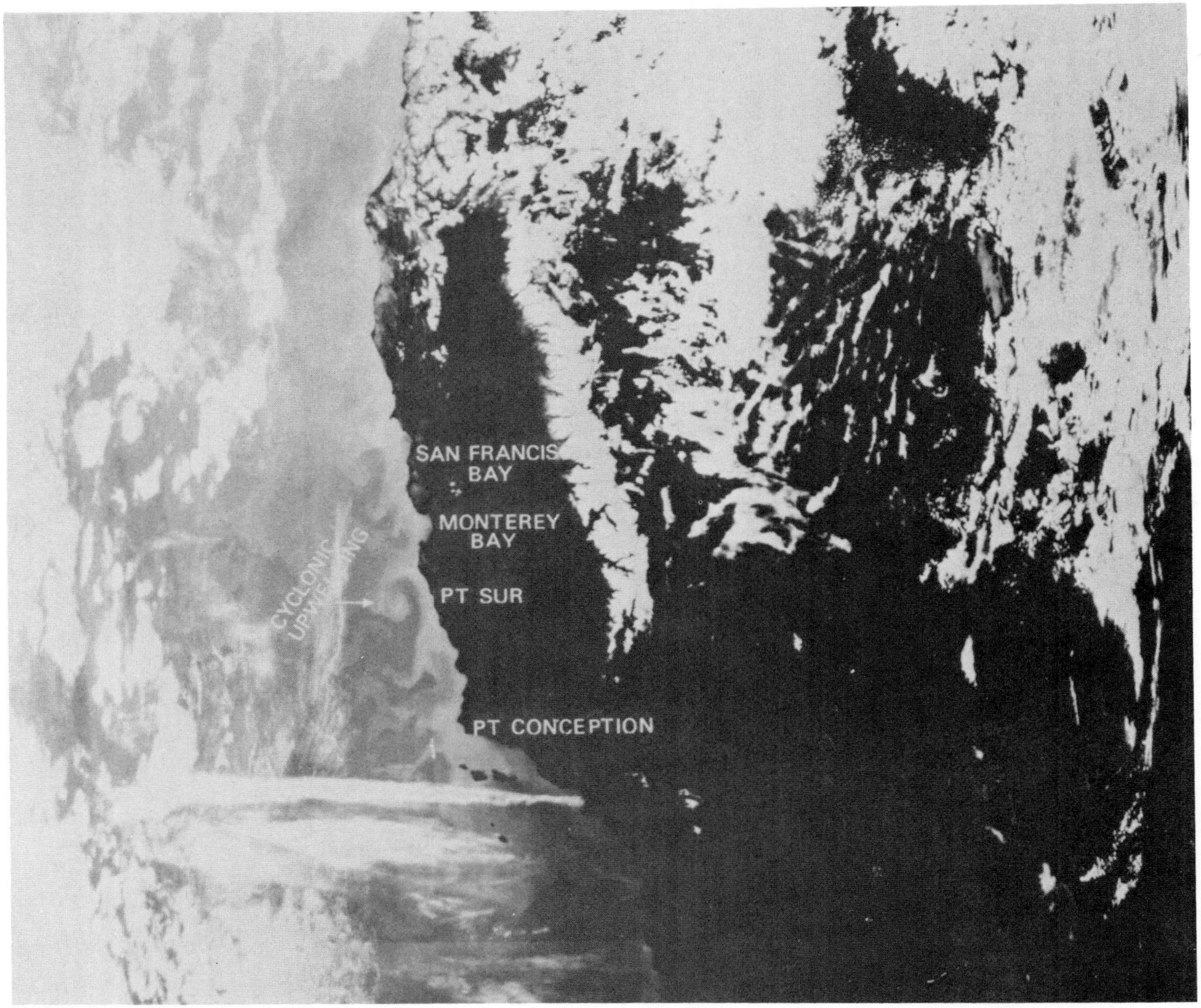

Fig. 1. Eastern North Pacific, 19 April 1979, TIROS N Satellite, VHRR infrared image enhanced for water temperature (white $\sim 10^{o}C$; black $\sim 20^{o}C$.

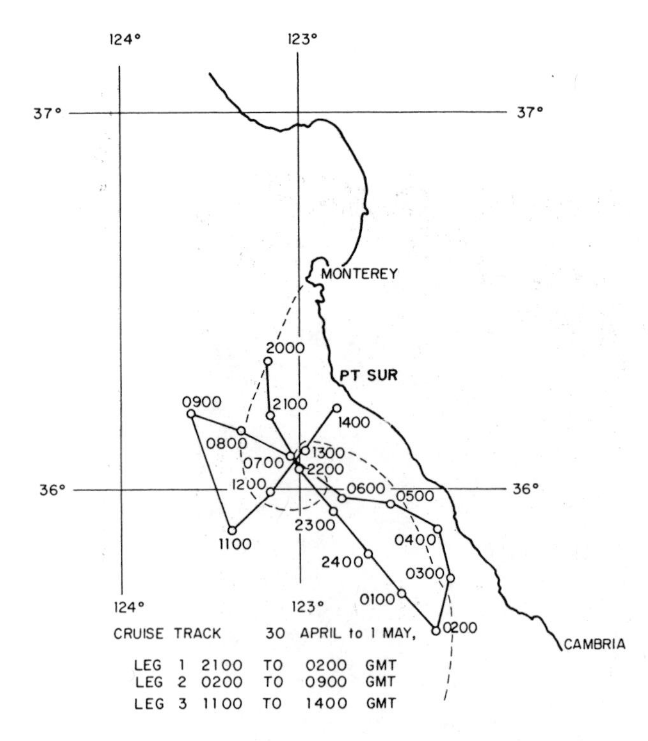

Fig. 2. Track of the 30 April, 1979 cruise (solid line) and outline of cyclonic upwelling of cold water (dashed line) as interpreted from a satellite image and *in situ* data.

was extracted from the biomass retained on the glass filter, analyzed (Holm-Hansen and Karl, 1976), and converted to mg C/l using the average conversion factor (250) proposed by Holm-Hansen (1969). The living phytoplanktonic biomass was estimated from continuous fluorescence measurement as described by Lorenzen (1966). Discrete samples (250 ml) were taken every 0.5 h to calibrate the fluorescence record in terms of plant pigment (principally chlorophyll *a*), which was determined by the method of Strickland and Parsons (1968). Chlorophyll *a* concentrations were converted to mg C/l using the average conversion factor (100) proposed by Holm-Hansen (1969). The combination of frequent underway *in situ* sampling and satellite thermal imagery made it possible to locate the center of the feature on the first attempt and transect it from three different directions, providing a two-dimensional impression of its structure.

Chemical Fronts, Thermal Fronts, and Biomass

In situ data from Leg 1 (Fig. 3) revealed sharp chemical gradients inversely correlated with strong thermal gradients. The inverse correlation was excellent in both large and small scale detail. No microplankton were found in the feature. In comparing Figs 3 and 4 there is the impression that microplankton are concentrated on the inside edge of the cyclonically curled feature in the thermal gradient from upwelled cool coastal water to warmer oceanic water and adjacent to the sharpest nutrient gradient. The pattern is repeated on Legs 2 and 3, supporting this impression.

Conceptual Model

The vertical thermal structure (Fig. 5), the satellite image, and the *in situ* thermal, chemical, and biological transect data, suggest a conceptual model of the cyclonic upwelling system (Fig. 6). The system was centered ca. 60 km southwest of Pt. Sur, near the Sur Canyon where similar systems have been observed frequently. Satellite images have shown the region to be an active area of upwelling, corroborating the suggestions of earlier investigators (Reid, Roden, and Wyllie, 1958) that upwelling occurs in geographically preferred locations south of capes and points along the eastern boundary of the North Pacific Ocean (a west coast with equatorward flow) and near the axis of a mesoscale canyon. The data suggest that the system developed into a large (∿100 km diameter) feature from a ridge of upwelled water that moved seaward, was fed by coastal upwelling water, and interacted with the California Current (possibly an eddy) to form a cyclonic upwelling system. The term cyclonic upwelling system was chosen because of the cyclonic appearance of the sea surface temperature pattern; the system appears to be divergent because the isotherms rise sharply to the surface beneath the feature, which is coldest at its center, and it maintains a horizontal integrity with respect to the adjacent surface layer because of sharp gradients in the horizontal direction. As a result sharp thermal fronts and chemical fronts are characteristic components of the system.

Horizontal Flux and Primary Production

Whether or not there is a flow of properties across the fronts because of the sharp gradients in the horizontal direction is an important question. The occurrence of biomass along the fronts may be a result of horizontal dispersion, literally an intertwining of an existing coastal band of microplanktonic blooms within the equatorward edge of the cyclonic curl of upwelled water, and dispersion on the poleward edge. (There is a hint of downwelling on the poleward edge from the dip of the isotherms at elapsed distance 242 and 306 km in Fig. 5). However, the strong chemical gradient is a potential source of nutrients and there may be a quantitative relationship between the nutrient flux and the rate of primary production in the adjacent transition zone between upwelling cool, nutrient-rich coastal water and warmer nutrient-poor oceanic water. The explanation may be that where there

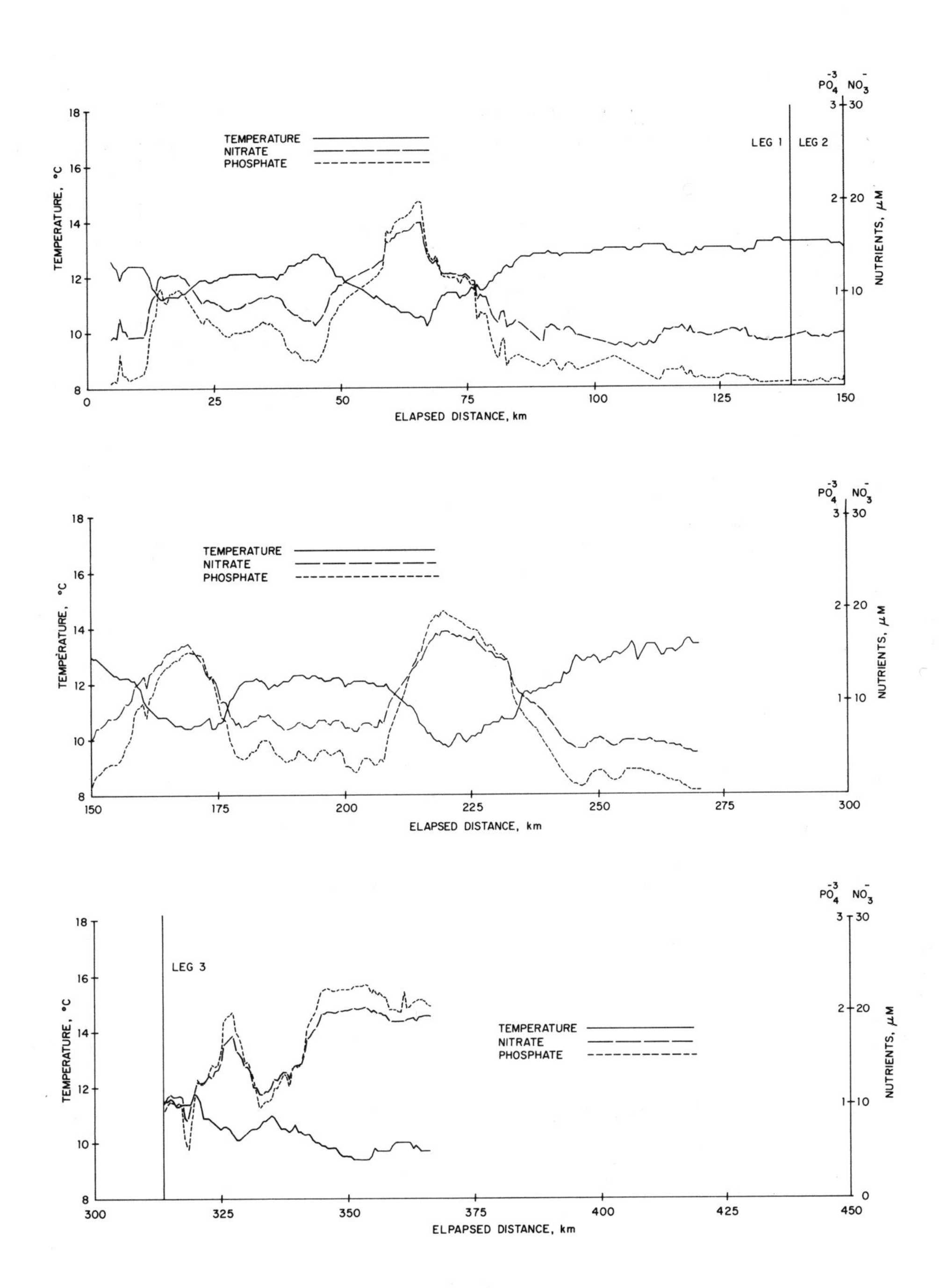

Fig. 3. Nitrate, phosphate, and temperature at 2.5 m along the 30 April, 1979 cruise track (Fig. 2).

is a horizontal chemical flux it will be a result of a horizontal physical flux that maintains the gradient (a natural chemostat) and simultaneously provides a mechanism for biological seeding of newly upwelled organisms. Satellite images suggest that the systems form in a matter of days and persist for days to weeks.

Chemical Inferences from Satellite Thermal Imagery

Except in rare cases when the distribution of atmospheric moisture is not uniform in the area of interest, satellite-derived sea-surface temperatures can be corrected to with $\pm 1^{\circ}C$ when *in*

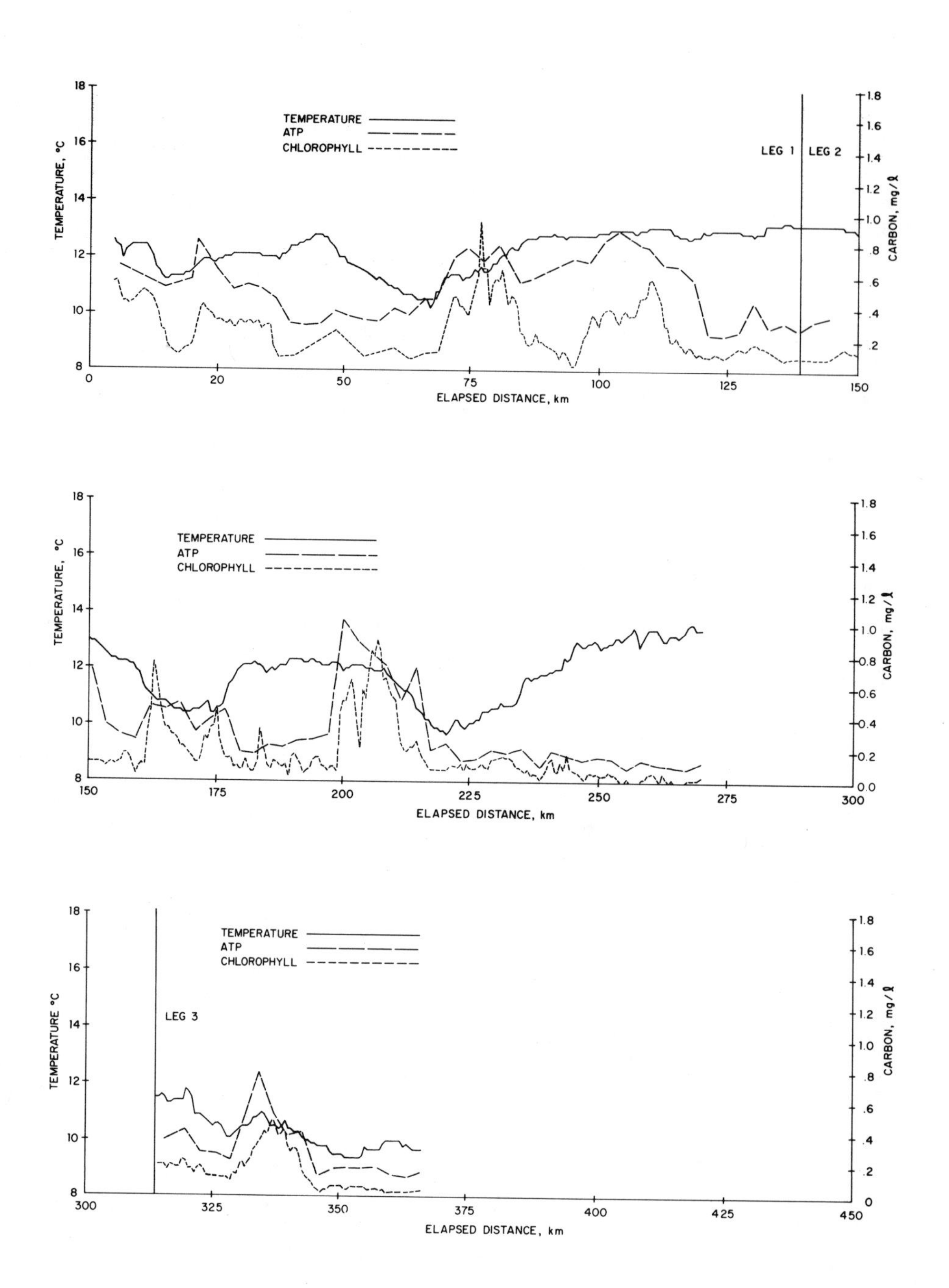

Fig. 4. Adenosine triphosphate-biomass, chlorophyll-biomass and temperature along the 30 April, 1979 cruise track (Fig. 2).

situ temperatures are available for calibration. As a satellite image is an almost instantaneous view of the thermal structure of the sea-surface with good spatial (∿1.0 km) and thermal (∿0.5°C) resolution, upwelling systems can be detected and observed from space. Changes in the thermal structure of the sea surface will be a function of the dynamic processes (namely mixing, advection, and heat transfer), the boundary conditions (atmospheric forcing and bathymetry), and the initial conditions (temperature and advection of the source water). No direct satellite measurements of nitrate and phosphate are made, so one must consider whether quantitative inferences can be drawn from satellites by the correlation of the nutrients with temperature. In the case of

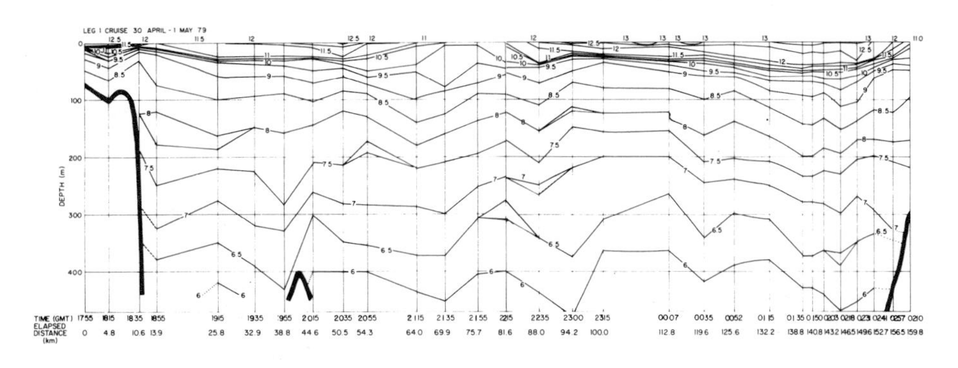

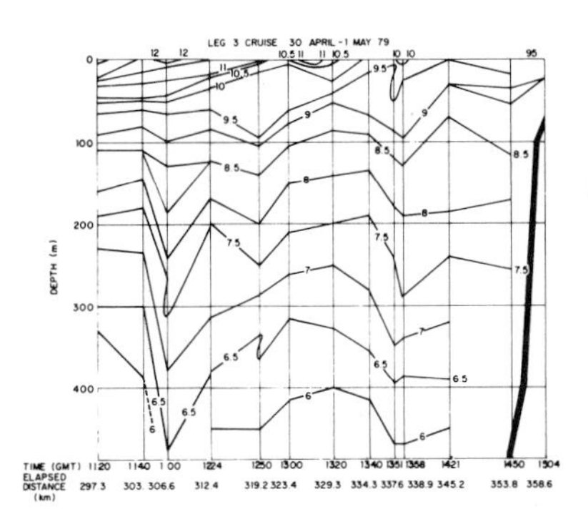

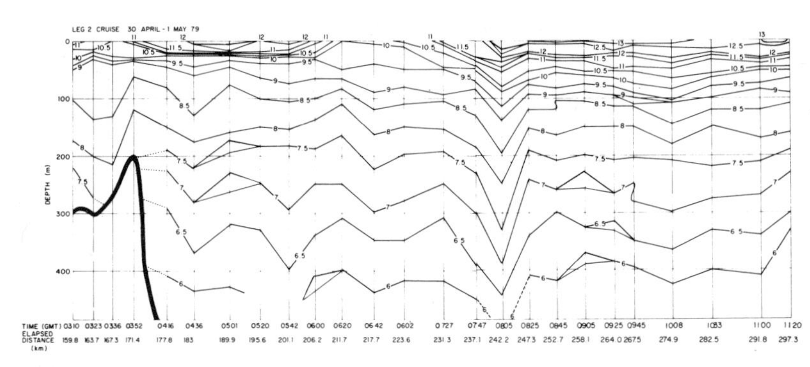

Fig. 5. Vertical temperature structure along the 30 April, 1979 cruise track (Fig. 2). Vertical lines represent XBT drops. Thick lines represent the bottom.

the cyclonic upwelling system, which, judging from the high N/P ratios and correlation coefficient, r = 0.98 (Figs 7 and 8a), is recently upwelled and "biochemically new" (Traganza *et al.*, 1980), there are excellent correlations of nitrate with temperature, r = -0.96 (Fig. 8b), and of phosphate with temperature, r = -0.94 (Fig. 8c). The regression lines are steep and a small error in temperature could make a large difference for the specified nitrate or phosphate value. However, if a significant number of *in situ* temperature measurements were available, average nutrient concentrations could be predicted. While this suggests the possibility of calibrating satellite thermal data in terms of nutrient concentrations, nitrogen and phosphorous are nonconservative constituents. The initial relationship between temperature and nutrients must change as the system changes dynamically and from differential biological utilization and regeneration of nutrients.

Giant Plume Across the California Current

The first observations of a system in a more advanced stage of development came in a surprising way. On 30 July, 1979, a giant plume was discovered extending ca. 150 km southwest from Pt. Sur. By 5 August (Fig. 9) the giant plume had extended to ca. 250 km. With satellite guidance a cruise track was planned with the first leg within the plume on the outbound transit (Fig. 10), Legs 2, 3, 4, and 5 to pass back and forth through the southerly or equatorward edge, and Leg 6 to stay within the plume on the inbound leg. Sharp thermal and nutrient gradients persisted far offshore (Fig. 11a) across the California Current with evidence (ATP-biomass) of intermittent concentrations of microplankton along the equatorward front (Fig. 11c). The dominance of low N/P ratios, their variability, r = 0.42 (Figs 11b and 8d), and the poor correlation of nitrate with temperature, r = -0.11 (Fig. 8e) and phosphate with temperature, r = -0.39 (Fig. 8f), indicated a system in an advanced stage of development. That the processes of advection, mixing, heat and salt transfer, and nutrient utilization and regeneration had been at work is supported by the following.

Chemical and Thermal Variability

The plume contained recently upwelled water at 50 km (elapsed distance) where water temperatures were below 13°C and nutrient concentrations were high (phosphate > 2 μM). The area has a vertical thermal structure (Fig. 12) similar to that in the cyclonic system (i.e., beneath the feature the isotherms rose to the surface and the sharp surface-layer thermocline was absent) and may have represented the source of upwelling water from which the plume developed. However, on the outbound transit there was a sharp thermocline beneath and outside the feature, indicating that

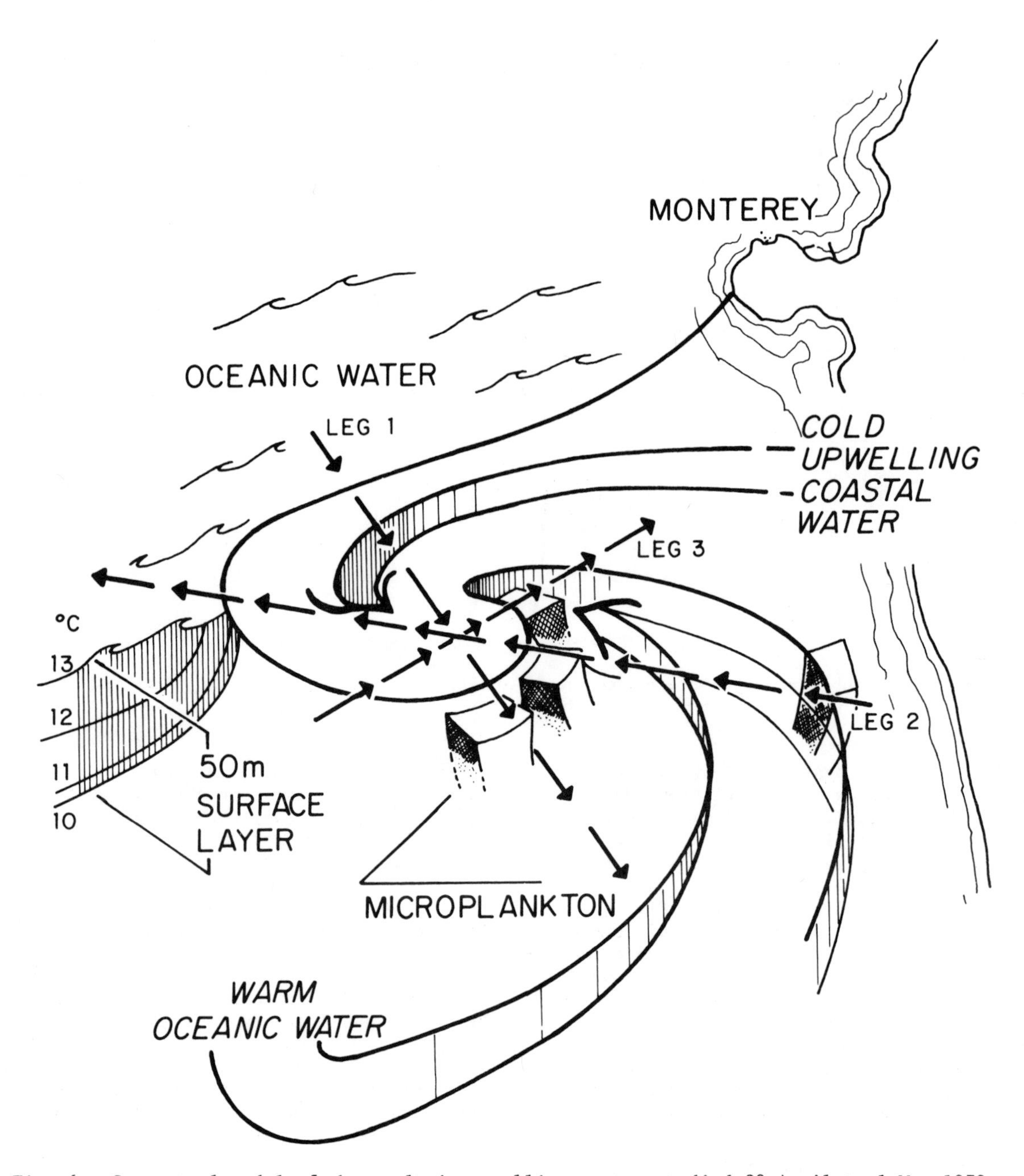

Fig. 6. Conceptual model of the cyclonic upwelling system studied 30 April to 1 May 1979.

a distinct surface layer was present everywhere except at the apparent source. Nutrients decreased with increasing temperature but with considerable variability. A sharp thermal front was encountered at the equatorward edge of the feature on Leg 2 (Fig. 11a, distance 255 km). The nutrients mirrored the thermal signature. In some areas at the western extent of the track, nitrate concentrations were below the level of detection and presumably insufficient to sustain photosynthesis. Nevertheless, an albacore fisherman (personal communication), who was using the same satellite information, found his catch just outside the equatorward edge of the plume (in the general vicinity of the ATP-biomass peaks), but the fish would stop biting each time he crossed the front into the feature.

Variable Wind Stress

Although it is impossible to separate the dynamical from the biological processes that modi-

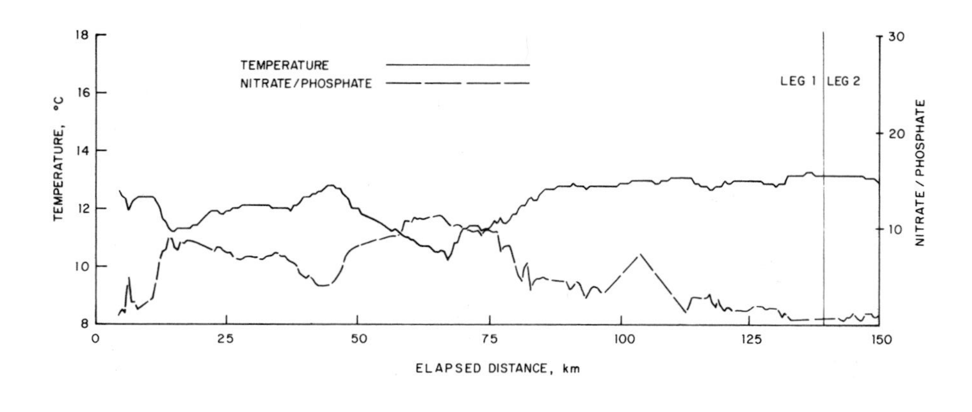

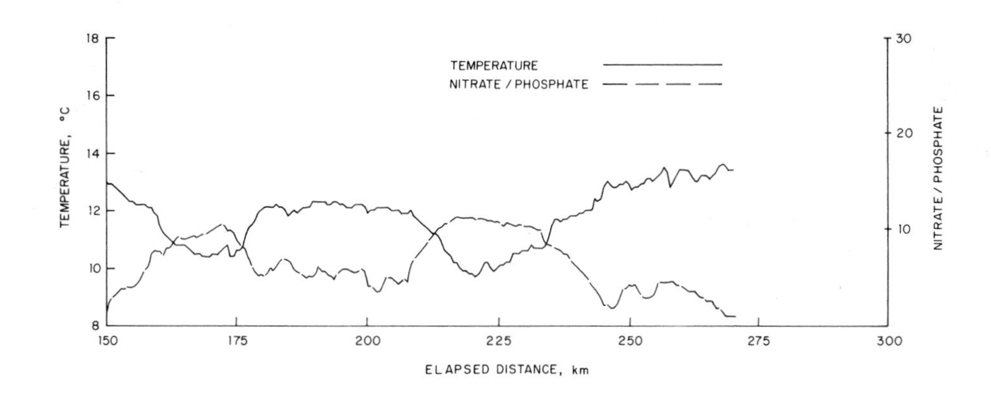

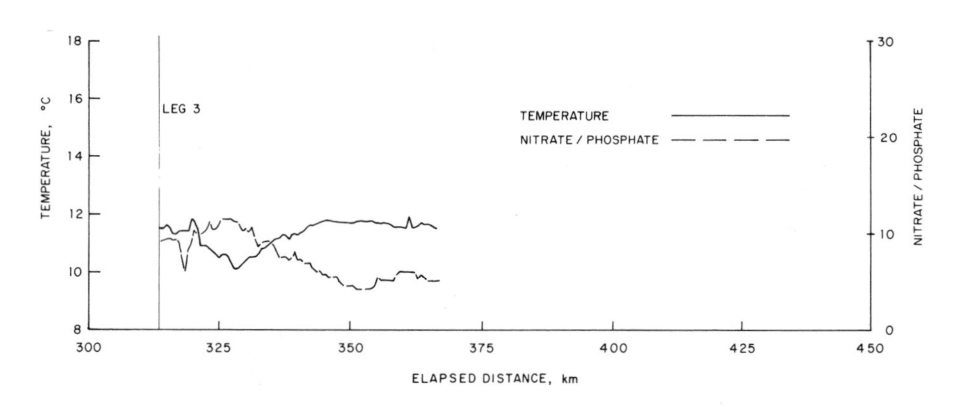

Fig. 7. Nitrate:phosphate ratio and temperature along the 30 April, 1979 cruise track (Fig. 2).

fied the chemical characteristics of the plume, there is an indication of the effects of atmospheric forcing. Wind stress analyses (A. Bakun, unpublished data, 1980) from early April through August imply that a cyclical pattern of upwelling events probably occurred in the area of the source water. Upwelling periods were of the order of 1 to 3 weeks and relaxation periods were of the order of 2 to 6 days. Ekman transport varied from west-southwest to southwest. With the long intermittent pattern of offshore transport it is possible that the plume consisted of upwelled water of varying ages. Other research (Barton, Huyer, and Smith, 1977) indicates that during an upwelling event the coldest water will migrate to the edge of the continental shelf and remain there until the system relaxes. Upon return of favorable winds upwelling will recur along the inner shelf. In this case, nutrient-rich, cold water was observed at the surface, at 50 km (elapsed distance) just inside the shelf break. The center of an upwelling zone may stop

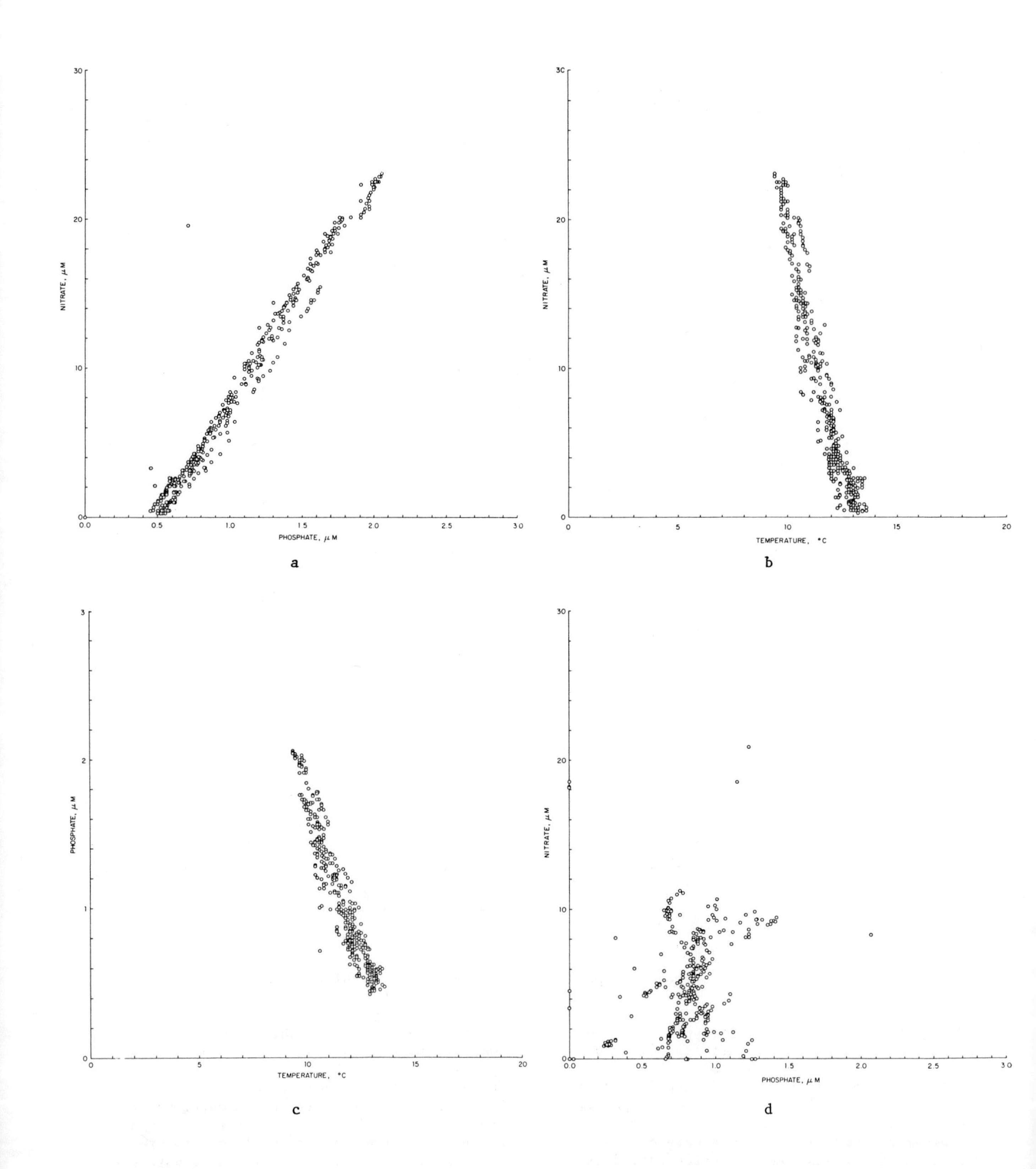

Fig. 8. a) Nitrate vs. phosphate, b) nitrate vs. temperature, and c) phosphate vs. temperature for the 30 April, 1979 cruise; d) nitrate vs. phosphase, e) nitrate vs. temperature, and f) phosphate vs. temperature for the 7 to 9 August, 1979 cruise.

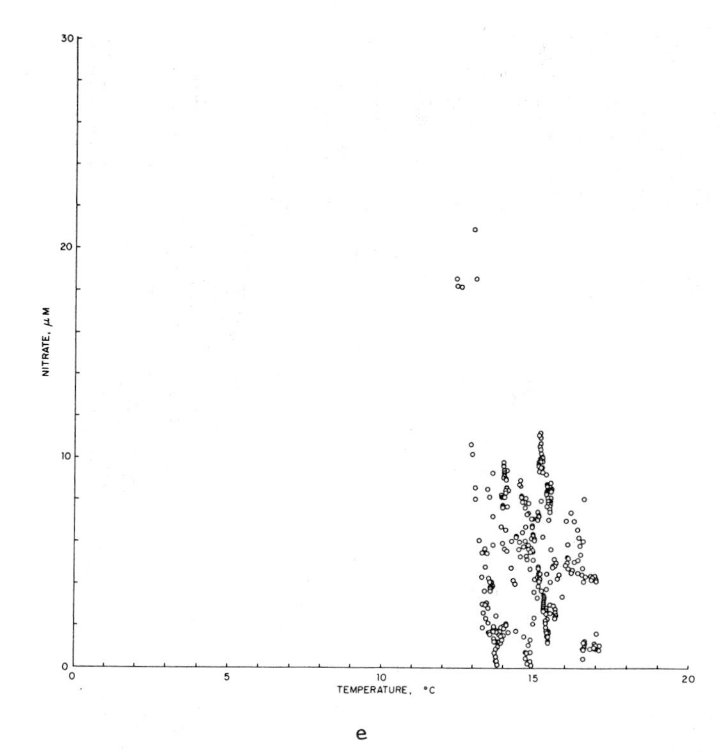

e

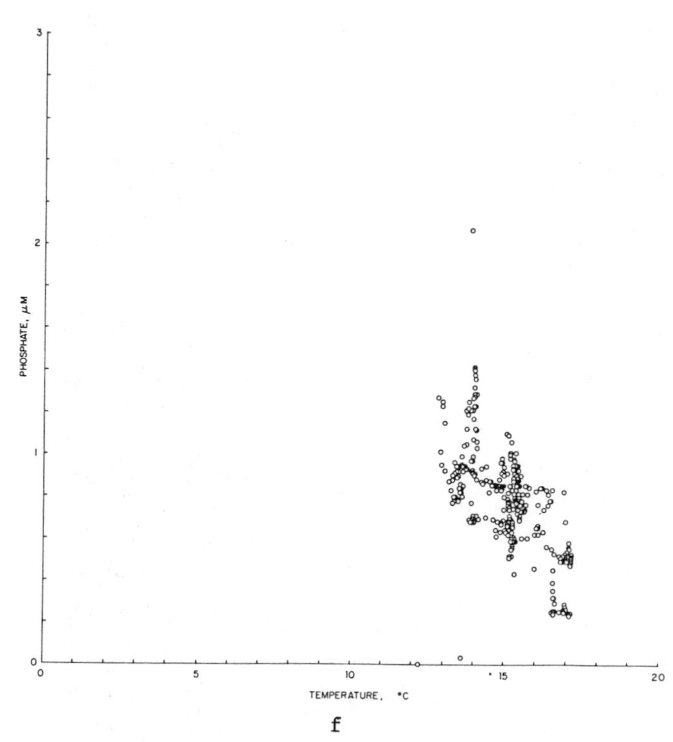

f

at the shelf edge. With reintensification of favorable winds, upwelled water will continue to be transported offshore in a surface layer (Huyer, Smith, and Pillsbury, 1974). Patchiness or banding can result from this start-stop action between close upwelling events (Barton *et al.*, 1977). The satellite images and *in situ* data show some patchiness. From the two-image sequence, the feature moved ca. 90 km southwest in agreement with wind data, which show Ekman transport to be west-southwestward during the period. The above may partially explain the variability in the correlation between nutrients and temperature.

Conclusions

Chemical, Biological, and Physical Relationships

We conclude that satellite remote sensing of the thermal structure of the sea surface will be a valuable tool in upwelling studies, especially because upwelled systems have been found to be distinctively structured and at considerable distances from the coast. Upwelling features ranging in their seaward extent from tens to hundreds of kilometers are evident in the images. The occurrence of microplanktonic blooms adjacent to sharp thermal and chemical gradients suggests a relationship between the horizontal nutrient flux and primary production, between chemical mesoscale and biological processes, and between physical and biological processes that control chemical processes in the upper layer.

Oceanic Climatological Model

Clearly, while inferences of the mean nutrient flux in the surface layer can be drawn from satellite imagery for recently upwelled water, calibration of satellite data for application to all phases in the formation and evolution of an upwelling system is a more difficult problem. Even two systems in the same relative phase of evolution will not have the same characteristics if the initial conditions of source water masses differ. This was illustrated in the analysis of a well defined cyclonic upwelling system that appeared in September, 1979, off Pt. Sur (unpublished data). Both nutrients and biomass were low. In short, though a strong upwelling regime can be well defined by satellite imagery and *in situ* temperatures, it may not be nutrient-rich nor associated with increased biological activity.

However, with the recurrence of characteristic frontal systems in favored locations (a function of bathymetry and other boundary conditions), given a knowledge of source water characteristics (from *in situ* monitoring and historical data), stage of development (inferred from satellite images and *in situ* monitoring), and wind stress, an oceanic climatological model could be developed for prediction of upwelling phenomena from satellite data.

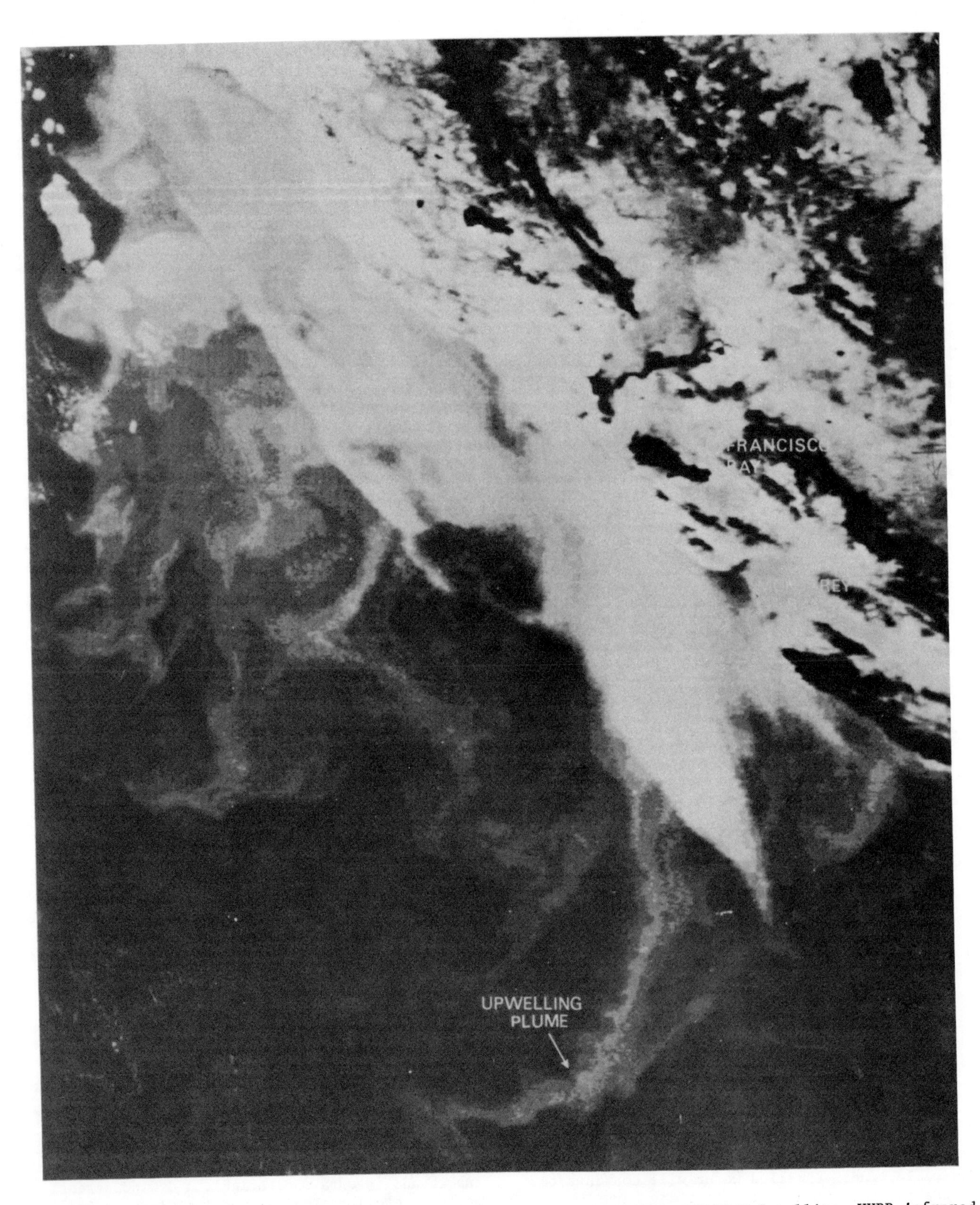

Fig. 9. Eastern North Pacific off Central California, 5 August 1979, TIROS N Satellite, VHRR infrared image enhanced for water temperature (white ∿10°C; black ∿20°C).

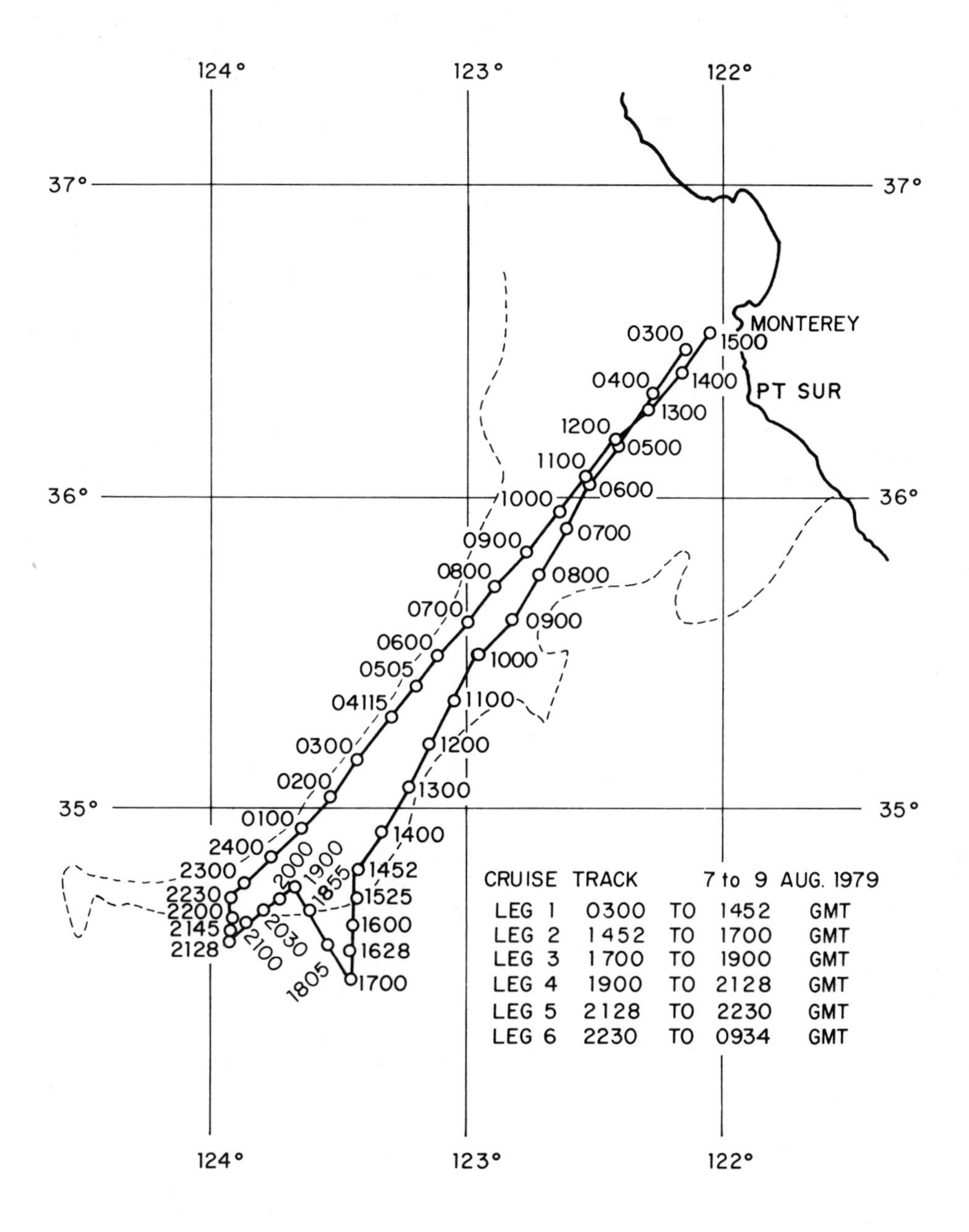

Fig. 10. Track of the 7 to 9 August cruise (solid line) and outline of the upwelled plume of cold water (dashed line) as interpreted from a satellite image and *in situ* data.

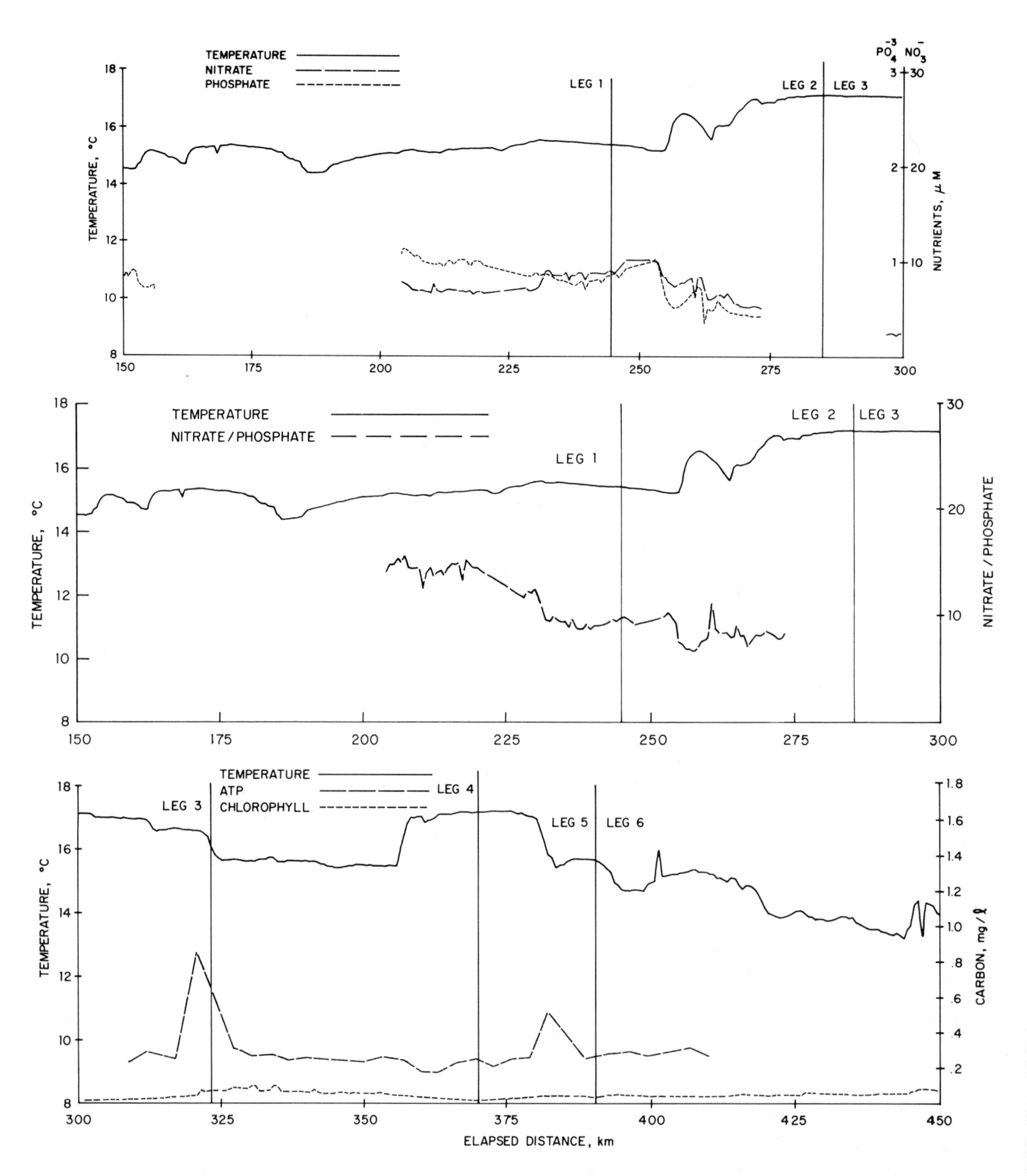

Fig. 11. a) Nitrate, phosphate, and temperature, b) nitrate:phosphate ratio and temperature at 2.5 m between 150 and 300 km along the 7 to 9 August cruise track (Fig. 10), and c) adenosine triphosphate-biomass, chlorophyll-biomass, and temperature between 300 and 450 km along the 7 to 9 August cruise track (Fig. 10).

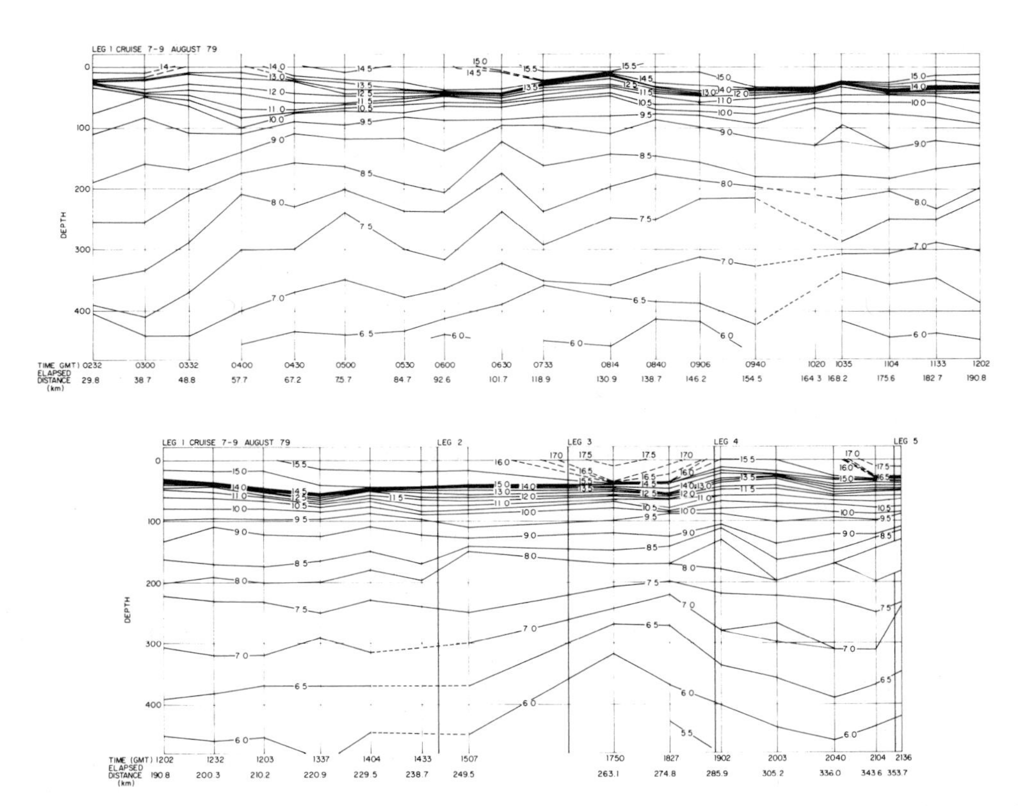

Fig. 12. Vertical temperature structure along the 7 to 9 August, 1979 cruise track (Fig. 10). Vertical lines represent XBT drops.

Acknowledgements. This was supported by the Office of Naval Research, CODE 482, NSTL, Bay St. Louis, MS., NPS contract 57751-5765-146314. The authors thank all individuals whose assistance made the study possible, particularly Captain W.W. Reynolds and the crew of the R.V. *Acania*, C.N.K. Mooers for his helpful critique of the manuscript, and E. Johnson and R. Phoebus who participated in the project.

References

Barton, E.D., A. Huyer, and R.L. Smith, Temporal variations observed in the hydrographic regime near Cabo Corbeiro in the northwest African upwelling region, *Deep-Sea Research, 24,* 7-25, 1977.

Bernstein, R.L., L.C. Breaker, and R. Whritner, California Current eddy formation: ship, air and satellite results, *Science, 195,* 353-359, 1972.

Fett, R.W., P.E. La Violette, M. Nestor, J.W. Nickerson, and K. Rabe, *Navy tactical applications guide, volume 2,* Environmental phenomena and effects, Defense Meteorological Satellite Program (DMSP), Naval Environmental Prediction Research Facility, Monterey, California, 93940, NEPRF Tech. Rept. 77-04, 1979.

Holm-Hansen, O., Determination of microbial biomass in ocean profiles, *Limnology and Oceanography, 14,* 740-747, 1969.

Holm-Hansen, O., and D.M. Karl, Effects of luciferin concentration on the quantitative assay of ATP using crude luciferase preparations, *Analytical Biochemistry, 75,* 100-112, 1976.

Huyer, A., R.L. Smith, and R.D. Pillsbury, Observations in a coastal upwelling region during a period of variable winds (Oregon coast, July 1972), *Tethys, 6,* 1-2, 1974.

Lorenzen, C.J., A method for the continuous measurement of *in vivo* chlorophyll concentration, *Deep-Sea Research 13,* 223-227, 1966.

Reid, J.L. Jr., G.I. Roden, and J.G. Wyllie, Studies of the California Current system, *California Cooperative Fisheries Investigation Progress Rept.,* 1 July 1956 to 1 January 1958, 27-56, 1958.

Strickland, J.D.H., and J.R. Parsons, *A Practical Handbook of Seawater Analysis,* Fisheries Research Board of Canada, Queen's Printer and Comptroller of Stationery, Ottawa, Canada, 311 pp., 1968.

Traganza, E.D., D.A. Nestor, and A.K. McDonald, Satellite observations of a nutrient upwelling off the coast of California, *Journal of Geophysical Research, 85,* 4101-4106, 1980.

CONSUMPTION AND REGENERATION OF SILICIC ACID IN THREE COASTAL UPWELLING SYSTEMS

D.M. Nelson

School of Oceanography, Oregon State University, Corvallis, Oregon 97331

J.J. Goering and D.W. Boisseau

Institute of Marine Science, University of Alaska, Fairbanks, Alaska 99701

Abstract. Tracer studies of the dynamics of silicon in the Peru, northwest Africa, and Baja California upwelling systems were conducted between 1969 and 1978 using the stable isotope ^{30}Si to measure uptake and regeneration rates directly. Uptake by phytoplankton can proceed at substantial rates in the dark in all three systems, extends to approximately twice the depth to which photosynthesis and nitrate uptake penetrate, and continues at significant rates throughout the 24-h day. The uptake rate within the euphotic zone is nearly uniform vertically when the average daily winds exceed 5 m/s off Peru, and under all conditions off northwest Africa and Baja California. Under lighter winds in the Peru system a strong near-surface maximum in uptake develops, and the overall rate of uptake within the euphotic zone increases significantly. Silicic acid is regenerated at high rates in Baja California and northwest African surface waters, and the result is the prevention of silicon limitation in those systems. Near-surface regeneration appears to be much slower off Peru, resulting in depletion of silicic acid to limiting concentrations as the upwelled water is transported offshore. The large differences observed among the three systems with respect to near-surface recycling of silicon may be related to differences in the characteristics of the predominant herbivores.

Introduction

In an upwelling area, the cold water transported into the euphotic zone is rich in nitrate, phosphate, and silicic acid. These nutrients normally support a production cycle of high amplitude, often dominated by diatoms (Blasco, 1971; Estrada and Blasco, 1979). Diatoms require silicon for growth (Richter, 1906; Lewin, 1955), and their uptake processes occasionally remove all analytically detectable silicic acid from surface water as it is transported offshore in upwelling regions (Dugdale and Goering, 1970). Thus silicon may at times play a dominant role in controlling the production cycle and phytoplankton species succession in upwelling systems.

The dynamics of the silicon cycle in the upwelling regions off Baja California, northwest Africa, and Peru were investigated during cruises of the IDOE Coastal Upwelling Ecosystems Analysis (CUEA) Program. The main objectives of the studies were to delineate the processes effecting the utilization and regeneration of silicic acid in each region and to examine the role silicon plays in the production cycle in coastal upwelling areas.

Goering, Nelson, and Carter (1973), Nelson and Goering (1977a, b, 1978), and Nelson and Conway (1979) reported studies of silicic acid uptake by natural phytoplankton populations in the Peru, Baja California, and northwest Africa upwelling regions. The studies indicated that silicon behaves differently from other primary nutrients in coastal upwelling systems and that each region studied is unique with respect to silicon dynamics. In this paper we summarize the earlier studies, present more recent silicic acid uptake observations from 1976 and 1977 cruises in the Peru upwelling region, and present an intersystem comparison of silicon dynamics.

Methods

Niskin bottles (30-ℓ) were used to collect seawater samples at depths of 100, 50, 30, 15, 5, 1, 0.1, and 0.01 % of surface incident light penetration. To determine the vertical profile of silicic acid uptake and dissolution, samples were drawn from the appropriate Niskin bottle into Plexiglas bottles shaded with neutral density screens (Perforated Products Inc.) approximating the light intensity at each sampling depth. Each sample was inoculated with a concentration of $H_4{}^{30}SiO_4$ high enough to prevent substrate limitation of uptake in diatoms (20 to 40 μM) and incubated under natural light in running seawater incubators. Incubation periods varied with the experimental design. Upon termination of the incubation, each sample was filtered through a 0.8-μm polycarbonate membrane filter (Nuclepore, Inc.) and the particulate fraction retained for silicon isotope analysis and particulate silicon determination. The dissolved silicon fraction was extracted from the aqueous phase into n-butanol and retained for isotopic analysis. Silicic acid uptake rates were determined from the increase in the ^{30}Si content of the particulate silicon and regeneration rates from the increase in the ^{28}Si content of the dissolved silicic acid during the incubation periods. Details of the experimental designs, isotopic sample preparation, mass spectrometry, and calculations to convert isotopic data into uptake and regeneration rates were presented by Nelson and Goering (1977a, b, 1978).

Experiments to determine the light dependence of silicic acid uptake were conducted on samples from the 50% light depth. The samples were drawn and divided into a 1-ℓ Plexiglas bottle fitted with a neutral density screen approximating 50% light transmission and into an identical 1-ℓ opaqued Plexiglas bottle. The bottles were inoculated and incubated, and samples were collected and analyzed as described above.

The substrate concentration dependence of silicic acid uptake was determined by collecting large (30 ℓ) water samples from the 50% light depth, splitting them into six identically screened 1-ℓ bottles and incubating with $H_4{}^{30}SiO_4$ at a series of concentrations ranging from 0.1 to 20 μM above the ambient level. The experiments were conducted only at those stations where near-surface concentrations of silicic acid were the most severely depleted.

Vertical Distribution of Silicic Acid Uptake

Peru

During the March, April, and May, 1977 Peru cruises silicic acid uptake rates and particulate silicon concentrations were determined at standard light penetration depths at 25 stations. The vertical profiles of uptake and of particulate silicon from these stations fall into two distinct classes. In one, both the specific uptake rate (V) and the particulate silicon concentration (PSi) are low and nearly constant with depth. In the other, both V and PSi show pronounced near-surface maxima and decrease sharply between the surface and the 1% light penetration depth.

Records of surface and 30-m water temperatures, the on and offshore flow at 2.5, 4.6, 8.1, 12, 16, 20, 24, and 39 m, and the average surface winds prevailing at the San Juan CUEA weather station (Brink, Smith, and Halpern, 1978) during the day of sampling were examined statistically to determine whether any of these environmental features differed significantly between days when nearly uniform vertical profiles of V and PSi were obtained and days showing strong near-surface maxima. No correlation could be established relating the shape of the V or PSi profile to surface temperature, thermal stratification in the upper 30 m, or onshore-offshore flow at any depth. However, the daily average alongshore wind velocity appears to play a major role in determining the shape of the uptake rate and particulate silicon profiles: uniform vertical profiles were observed predominantly at stations occupied on days when the average alongshore winds were >5 m/s (here designated as strong wind stations) while pronounced surface maxima occured only at stations where the average daily winds were <5 m/s (weak wind stations). The profiles of mean specific uptake rates of silicic acid (${}^{V}H_4SiO_{4av}$), mean particulate silicon concentrations (PSi_{av}), and mean absolute silicic acid uptake rates (${}^{\rho}H_4SiO_{4av}$) obtained at strong and weak wind stations during the period are presented in Figs 1, 2, and 3. Averaged data, rather than individual station profiles, are presented to show the vertical distributions of silicic acid uptake and particulate silicon that prevailed generally under strong and weak wind conditions, respectively.

Silicic acid uptake at both strong and weak wind stations in the Peru system extended well below the 1% light penetration depth, as has been observed in both the Baja California and northwest Africa upwelling regions. (Depth profiles from the Baja California and northwest Africa CUEA studies described by Nelson and Goering, 1978 are reproduced in Figs 4 and 5.) At strong wind stations, ${}^{V}H_4SiO_{4av}$ and ${}^{\rho}H_4SiO_{4av}$ at sampling depths above the 1% light penetration depth were not only nearly constant vertically, but significantly lower than at weak-wind stations. At weak-wind stations uptake and particulate silicon were, on the average, maximum at the upper light depths and decreased to values similar to those at strong-wind stations at or near the 1% light penetration depth. ${}^{V}H_4SiO_{4av}$ values in the 1 to 0.1% light penetration region were similar at strong- and weak-wind stations and decreased with depth at stations of both types.

The absolute rate of silicic acid uptake (${}^{\rho}H_4SiO_4$), integrated from the surface to the 1% light depth, was significantly greater at weak wind than at strong wind stations (t-test significant at the 98% confidence level; see Table 1). Thus, significantly more biogenic silica was being produced in the Peru system on days when the average alongshore wind velocity was <5 m/s.

Baja California and Northwest Africa

The vertical extent of silicic acid uptake in the Baja California and northwest Africa upwelling systems has been discussed by Nelson and Goering (1978). They described the following features of silicic acid uptake in the surface

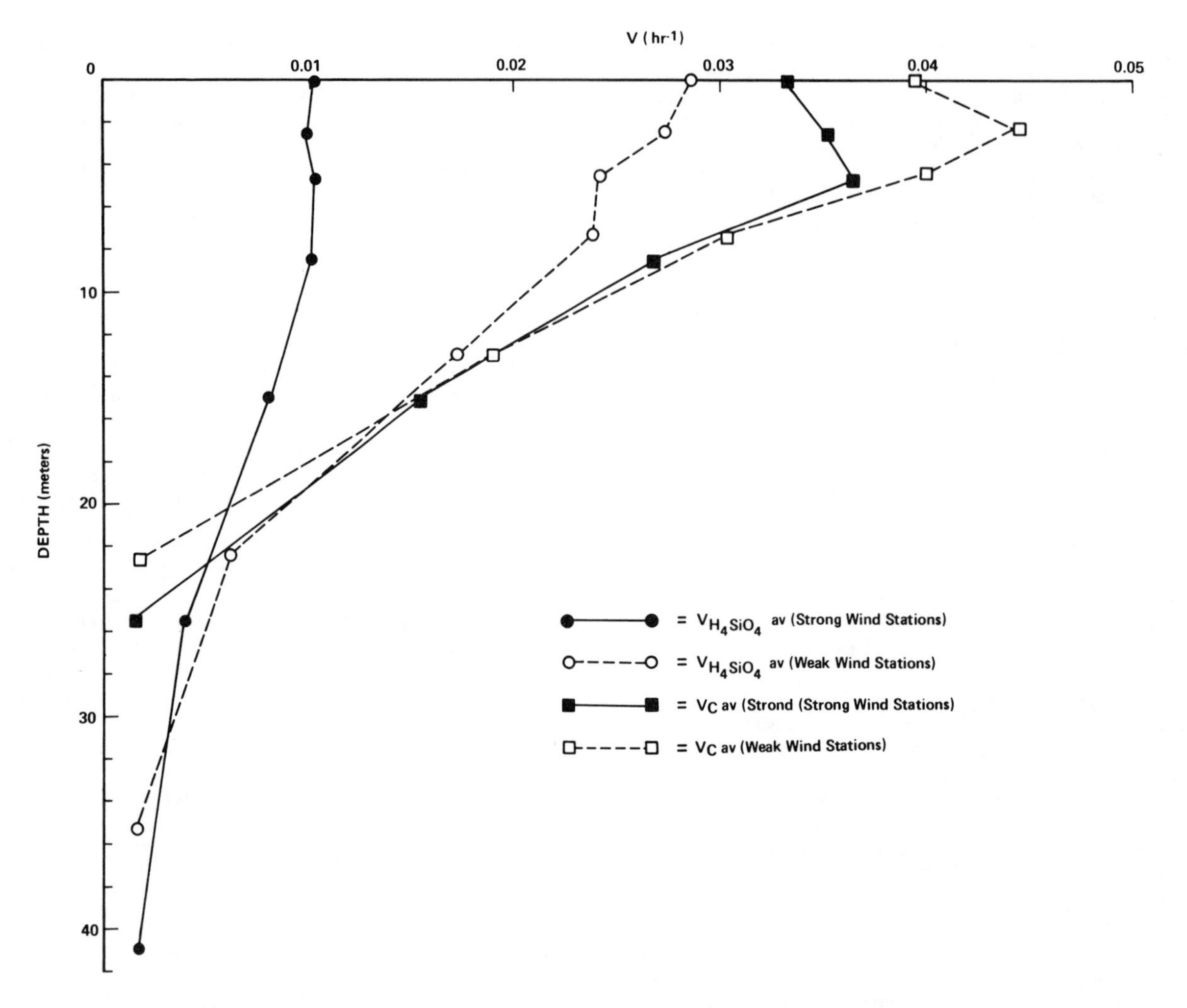

Fig. 1. Averaged vertical profiles of the specific uptake rate of silicic acid and carbon at the mean depths of 100, 50, 30, 15, 5, 1, and 0.1% surface light penetration under strong-wind (solid lines) and weak-wind (dashed lines) conditions off Peru during March and April, 1977. Carbon data from Barber *et al.* (1978) and Kogelschatz *et al.* (in press).

water of these two systems: Silicic acid uptake in both systems was much more uniform, both with depth (Figs 4 and 5) and through time than were carbon or nitrogen assimilation. Silicon uptake also extended to about twice the depth to which carbon and nitrogen uptake were significant and continued at substantial rates through the day and night.

Intersystem Comparison

Figs 1 to 5 present the average vertical profiles of silicic acid uptake determined on cruises to the Baja California, northwest Africa, and Peru upwelling systems. In a previous communication, Nelson and Goering (1978) discussed the similarities and differences between the Baja California and northwest Africa systems. In this discussion we shall compare all three systems.

In all three upwelling systems the rate of silicic acid uptake is generally much more uniform with depth and time than is either photosynthesis or nitrogen uptake. (Nelson and Goering, 1978; Barber *et al.*, 1978; Kogelschatz, MacIsaac, and Breitner, in press; H.L. Conway, unpublished data, 1978). The maintenance of a nearly constant elemental composition of the phytoplankton requires that specific uptake rates of all elements, integrated over depth and time, approximately balance. The fact that specific uptake rates of carbon and nitrogen are, on average, greater than that of silicon during 6-h daylight incubations (ibid.) reflects the temporally more discontinuous uptake of carbon and nitrogen. Uptake rates of silicic acid fluctuate much less from day to night than do rates of carbon and nitrogen assimilation (Goering *et al.*, 1973; Nelson and Goering, 1978; Nelson and Conway,

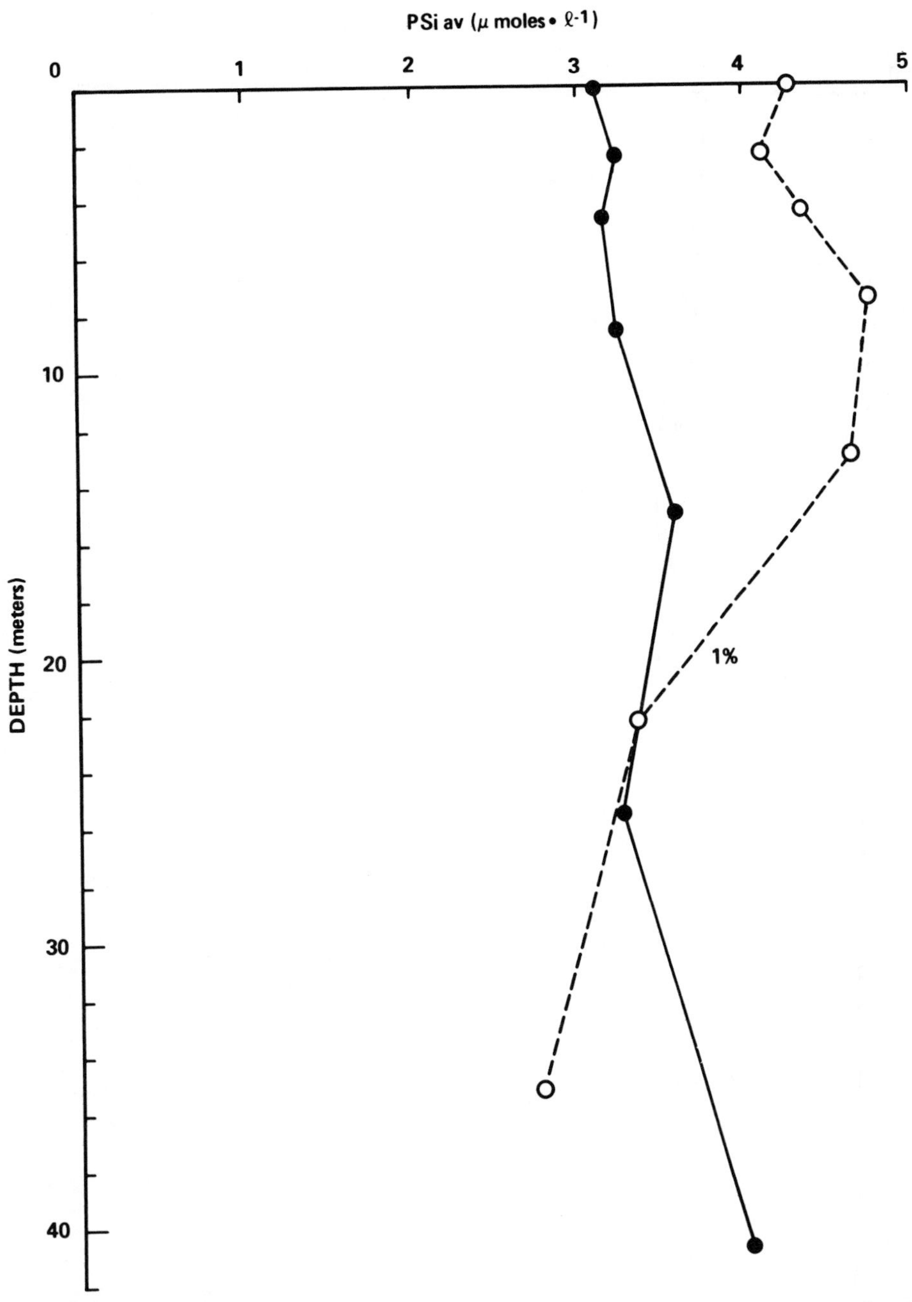

Fig. 2. Averaged vertical profiles of particulate silicon concentration in Peru surface water during March and April, 1977; the solid line represents strong wind conditions, dashed line weak wind.

1979). Phytoplankton populations in the three upwelling regions thus maintain the appropriate silicon composition through a much less pronounced diel variation in silicic acid uptake than in carbon or nitrogen uptake.

A general pattern of substantial silicic acid uptake at and below the 1% light penetration depth was observed in all three upwelling systems. In general, nitrogen assimilation was much reduced at these depths, and carbon assimilation was negligible (Barber and Huntsman, 1975; Barber, Huntsman, and Kogelschatz, 1976; Barber *et al.*, 1978; Kogelschatz *et al.*, in press; H.L. Conway, unpublished data, 1978). Silicic acid uptake thus exhibits a much less pronounced vertical variation than does that of nitrogen or carbon. This fact, together with the much more uniform diel cycle of silicic acid uptake, reflects the ability of energetically healthy natural populations of upwelling phytoplankton to

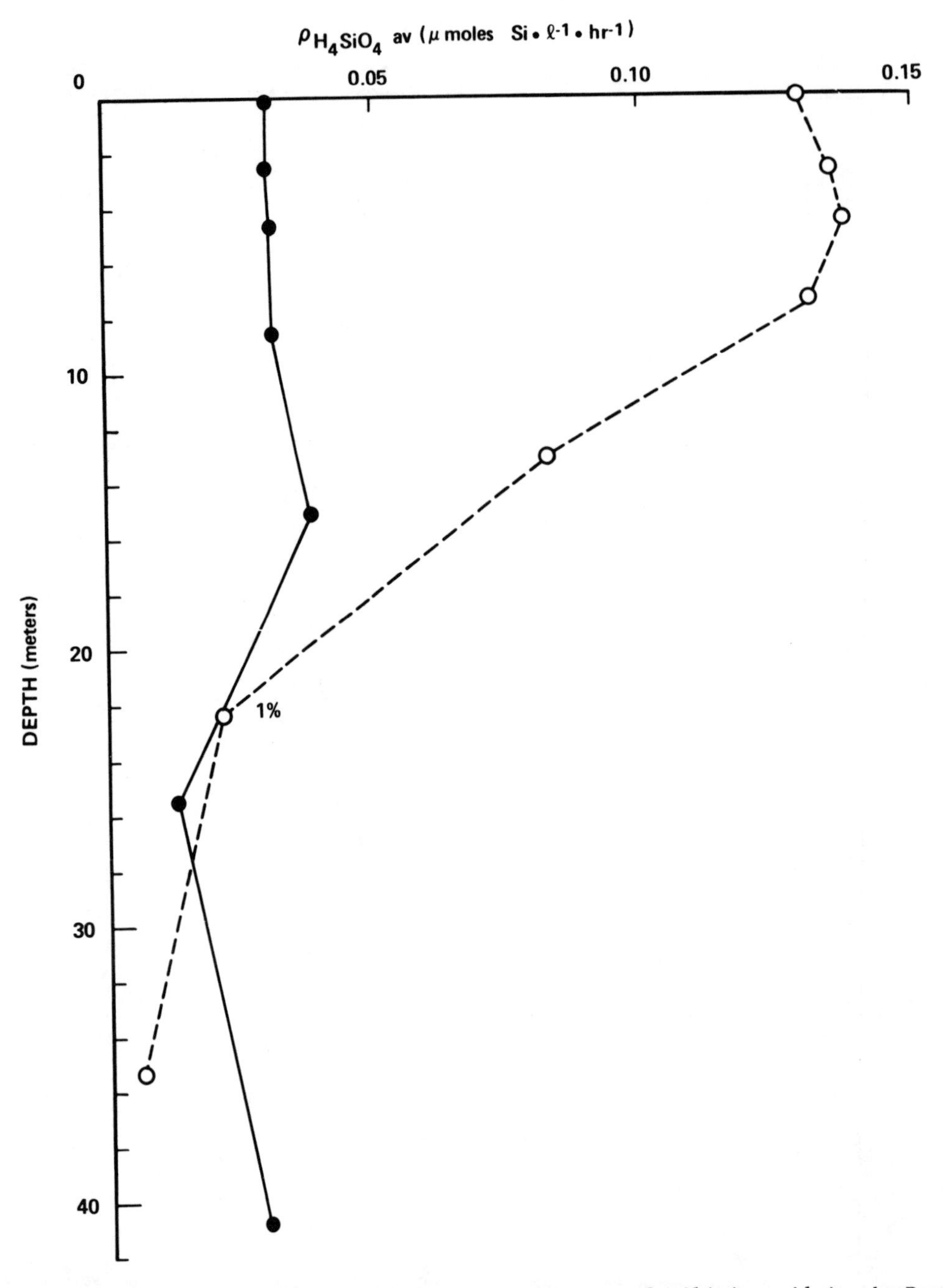

Fig. 3. Averaged vertical profiles of the absolute uptake rate of silicic acid in the Peru upwelling system during March and April, 1977; the solid line represents strong wind conditions, dashed line weak wind.

take up silicic acid in the dark or when exposed to low light intensities (Nelson and Conway, 1979).

The average vertical profiles of silicic uptake off Baja California and northwest Africa are similar in shape to those observed in the Peru system during periods of strong winds (>5 m/s) (Figs 1 to 5). Periods of weak winds off Peru represent the only definable condition under which the vertical profile of silicic acid uptake shows any persistent structure within the surface layer in any of the three systems.

The observed difference between strong and weak wind periods off Peru with respect to silicic acid uptake profiles may result from the exposure of the phytoplankton to different average light

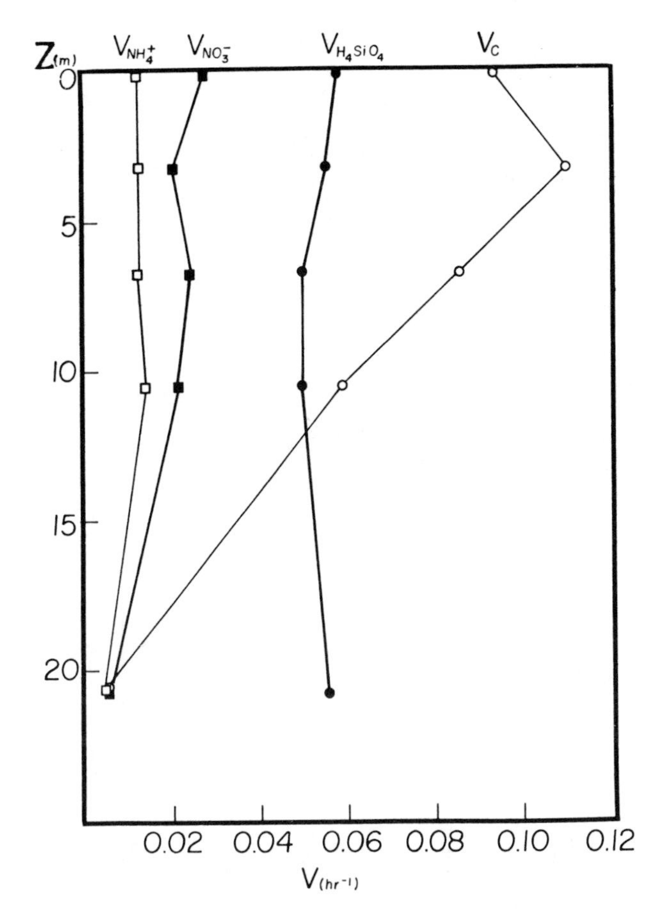

Fig. 4. Averaged vertical profiles of the specific uptake rates of silicic acid, nitrate, ammonium, and carbon at the mean depths of 100, 50, 25, 10, and 1% surface light penetration off Baja California during a 10-day period of continuous strong upwelling in 1973 (from Nelson and Goering, 1978).

regimes. During weak wind periods the depth of the wind mixed layer did not extend far below the 1% light penetration depth (20 to 30 m). Also, mixing within the surface layer was presumably less vigorous, allowing cells in the upper part of the mixed layer to remain there for longer periods of time. Such cells would thus have been exposed, on average, to longer periods of photosynthetically saturating light intensities during weak wind periods than any cells would be during strong wind periods, when there is more rapid mixing throughout the surface layer and all cells would thus spend considerable time at low light levels. The decreased uptake of both silicic acid and carbon in surface waters during strong wind periods (Table 1) suggests that such periods provide an energetically poor environment for phytoplankton growth because of increased vertical mixing. Thus as long as winds, and wind-driven upwelling, are strong enough to prevent nutrient limitation, the optimum conditions for phytoplankton growth are found when the winds are relatively quiescent and turbulence within the surface mixed layer is at a minimum.

Light Effects on Silicic Acid Uptake

Silicic acid uptake by phytoplankton is ultimately dependent upon photosynthesis as an energy source, and the light regime may therefore play an important role in controlling uptake. However, previous studies suggest that silicic acid uptake is usually not so closely coupled to photosynthesis as is nitrate uptake (MacIsaac and Dugdale, 1972; Nelson, Goering, Kilham, and Guillard, 1976; Nelson and Conway, 1979) and there have been numerous reported cases of substantial dark uptake of silicic acid by natural phytoplankton (Goering *et al.*, 1973; Azam and Chisholm, 1976; Nelson and Goering, 1978).

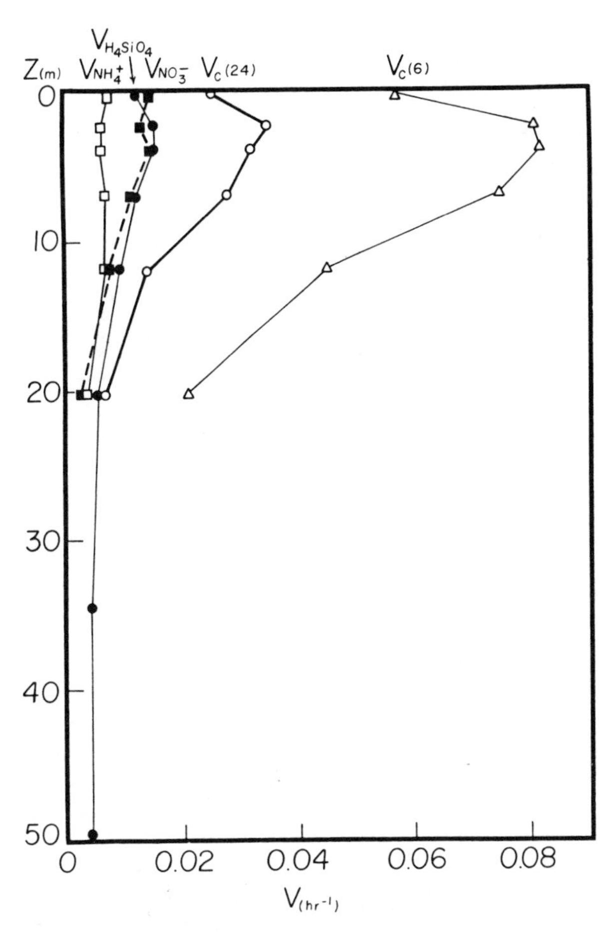

Fig. 5. Averaged vertical profiles of the specific uptake rates of silicic acid, nitrate, ammonium, and carbon (6 and 24 h incubation) at the mean depths of 100, 50, 30, 15, 5, 1, 0.1 and 0.01% surface light penetration off northwest Africa, March-May, 1974 (from Nelson and Goering, 1978).

TABLE 1. Statistical Summary of Peru Upwelling Particulate Silicon and Silicic Acid and Carbon Uptake Data Collected at Stations which had Average Winds During Day of Sampling in Excess of 5 m/s (strong wind stations) and those with average winds Less than 5 m/s (weak wind stations).

	Number of Stations	Units	Highest	Lowest	Mean	Standard Deviation
Strong Wind Stations	14					
$V_{H_4SiO_4}$ (50% light depth)		h^{-1}	0.026	0.001	0.010	0.008
V_C*(50% light depth)		h^{-1}	0.042	0.029	0.035	0.010
$\rho_{H_4SiO_4}$ (50% light depth)		μ moles l^{-1} h^{-1}	0.084	0.003	0.030	0.026
$\rho_{H_4SiO_4}$ (integrated over 100-1.0% light depth interval)		m moles $m^{-2}h^{-1}$	2.106	0.109	0.694	0.575
PSi (integrated over 100-0.01% light depth interval)		m moles m^{-2}	205.0	40.0	115.4	39.7
Weak Wind Stations	11					
$V_{H_4SiO_4}$ (50% light depth)		h^{-1}	0.058	0.007	0.027	0.016
V_c*(50% light depth)		h^{-1}	0.058	0.026	0.044	0.010
$\rho_{H_4SiO_4}$ (50% light depth)		μ moles $l^{-1}h^{-1}$	0.550	0.015	0.136	0.152
$\rho_{H_4SiO_4}$ (integrated over 100-1.0% light depth interval)		m moles $m^{-2}h^{-1}$	4.047	0.140	1.679	1.153
PSi (integrated over 100-0.01% light depth interval)		m moles m^{-2}	157.0	66.0	115.3	29.8

Significant t-tests

	t	Degrees of Freedom	Confidence Level (%)
- $V_{H_4SiO_4}$ (50% light depth) greater at weak wind stations than at strong wind stations	3.45	23	99
- V_C (50% light depth) greater at weak wind stations than at strong wind stations	2.30	23	95
- $\rho_{H_4SiO_4}$ (50% light depth) greater at weak wind stations than at strong wind stations	2.96	23	99
- $\rho_{H_4SiO_4}$ (integrated over 100-1.0% light depth interval) greater at weak wind stations than at strong wind stations	2.65	23	98

* Carbon data from Barber *et al.* (1978) and Kogelschatz *et al.* (in press).

At 25 stations in the Baja California upwelling system, 20 in the northwest Africa system, and 27 in the Peru system, the uptake of silicic acid was measured at simulated 50% of surface light intensity (V_{Lt}) and in the dark (V_{Dk}). A statistical summary of the V_{Dk}/V_{Lt} ratios at these stations is given in Table 2. The analytical precision of ^{30}Si uptake measurements is *ca.* ± 10% (Nelson and Goering, 1977a). Thus an analytically detectable effect of light is present in those comparisons where V_{Dk}/V_{Lt} differs from unity by more than 0.10. The incubation period required for $H_4{}^{30}SiO_4$ tracer experiments is long relative to the response time of planktonic algae to changes in light conditions (Eppley and Coatsworth, 1968; Sournia, 1974; Beardall and Morris, 1976). In addition, culture experiments have indicated that uptake of silicic acid is not directly coupled to illumination and suggest that uptake may proceed at undiminished rates initially, then decay through the course of the dark incubation period (Nelson *et al.*, 1976). Because the ^{30}Si tracer experiments summarized in Table 2 represent the amount of silicic acid taken up during a 6-h incubation, the V_{Dk}/V_{Lt} values represent a relative integrated potential for dark uptake over a 6-h period. V_{Dk}/V_{Lt} can be interpreted as having been constant over the 6-h interval only when it approaches either 0 or 1. When $V_{Dk}/V_{Lt} = 0$ it shows that the dark uptake rate was low throughout the incubation, and when $V_{Dk}/V_{Lt} = 1$, uptake in the dark was virtually the same as that in the light at all times. For intermediate values, V_{Dk}/V_{Lt} may represent the fraction of the incubation period during which silicic acid uptake proceeded in the dark at undiminshed rates and may thus be a measure of the cells' stored energetic reserves available for H_4SiO_4 uptake (Nelson and Conway, 1979).

The ratio of dark to light uptake of silicic acid varied from 0 to >1 in the Baja California and northwest Africa upwelling systems and from 0 to 0.73 in the Peru system (Table 2). Complete light independence of silicic acid uptake ($V_{Dk}/V_{Lt} = 1$) was never observed in the Peru system while at some stations in the Baja California and northwest Africa systems it was. Inhibition of silicic acid uptake by light ($V_{Dk}/V_{Lt} > 1.10$) was never observed in any of the systems.

The V_{Dk}/V_{Lt} data for the Baja California, northwest Africa, and Peru upwelling regions were examinated statistically for significant differences within and between data sets. At the 95% confidence level (as determined by t-test), the mean V_{Dk}/V_{Lt} for silicic acid was lower off both Peru and northwest Africa than off Baja California and, within the Baja California system, lower during thermally stratified periods than during upwelling periods. The higher overall mean V_{Dk}/V_{Lt} for Baja California is due entirely to the high V_{Dk}/V_{Lt} ratios observed during upwelling periods, as no significant difference exists between the mean value for Baja California during stratified periods and the overall mean value for either Peru or northwest Africa. The high ratios observed off Baja California during upwelling periods reflect a genuinely enhanced capacity for uptake in the dark. A similar effect would have been obtained if V_{Lt} were depressed during upwelling periods, but this was not the case; both V_{Dk} and V_{Lt} tended to be greater during upwelling periods than during stratified periods (Nelson and Conway, 1979). In the Peru values there were no significant differences among the overall values and those for the weak-wind and strong-wind stations.

Nelson and Conway (1979), considering only Baja California and northwest Africa data, could offer no comprehensive explanation for the systematic differences observed in the capacity of natural phytoplankton assemblages to take up silicic acid in the dark. The data become even less compatible with any simple hypothesis when the Baja California, northwest Africa, and Peru data sets are considered together. Nelson and Conway (1979) suggested that the capacity for uptake of silicic acid by natural phytoplankton is affected by both the instantaneous light intensity and the cells' stored energy reserves, which reflect their previous light-exposure history. If this were true, then the optimum environment for silicic acid uptake would be a strongly stratified water column and a light-saturated condition. This view is supported by the fact that the overall production rates of both biogenic silica and organic carbon were significantly higher off Peru when the average daily wind was <5 m/s. When only the very near-surface (50% light penetration depth) uptake rates are considered the difference between strong-wind and weak-wind stations is even more pronounced (see Table 1). However, comparison of V_{Dk}/V_{Lt} ratios shows no significant differences in the light dependence of uptake between strong-wind and weak-wind stations off Peru. Also there were no significant differences in mean V_{Dk}/V_{Lt} ratios among data sets from Baja California during stratified periods, northwest Africa, and Peru, even though the first of these represents a strongly structured water column, the second a vigorously mixed water column, and the third an intermediate condition. The idea that V_{Dk}/V_{Lt} reflects the stored energetic reserves of the diatom assemblage may still be correct. However, such an interpretation would mean that different areas with different degrees of vertical mixing develop diatom floras of different species composition but with a tendency toward similar energetic reserves. It is not clear why this should be, nor is it clear why the phytoplankton had an apparently much greater capacity for dark uptake of silicic acid during upwelling periods off Baja California than under any other definable hydrographic condition in any of the regions.

Silicic Acid Regeneration

Early studies of the nutrient dynamics of the Peru upwelling system showed that silicic acid

TABLE 2. Statistical Summary of Silicic Acid V_{Dk}/V_{Lt} in Seawater Collected from the 50% Light Penetration Depth at Stations in the Baja California[1], Northwest Africa[1], and Peru Upwelling Regions.

	Number of Stations	High	Low	Mean	Standard Deviation
Baja California					
overall	25	1.10	0.00	0.627	0.380
upwelling	16	1.10	0.10	0.800	0.305
stratified	9	1.00	0.00	0.319	0.338
Northwest Africa	20	1.00	0.00	0.298	0.303
Peru					
overall	27	0.726	0.00	0.340	0.239
1976	11	0.726	0.00	0.368	0.212
1977	16	0.708	0.00	0.321	0.260

Significant t-tests[2]

	t	Degrees of Freedom	Confidence Level (%)
Mean V_{Dk}/V_{Lt} for Baja California upwelling greater during upwelling periods than during stratified periods	3.65	23	99
Mean V_{Dk}/V_{Lt} greater for Baja California overall than for Peru overall	3.16	45	99
Mean V_{Dk}/V_{Lt} greater for Baja California during upwelling periods than for Peru overall	5.27	41	99
Mean V_{Dk}/V_{Lt} greater for Baja California overall than for northwest Africa	3.105	43	99

1 Baja California and northwest Africa data from Nelson and Conway (1979).
2 All other mean V_{Dk}/V_{Lt} differences within each region and between regions were tested and found not to be significant at the 95% confidence level.

was the first nutrient to be depleted to potentially limiting concentrations as surface water was transported offshore from the region of strongest upwelling (Dugdale and Goering, 1970). It was hypothesized at that time that the depletion of silicic acid took place primarily because it was regenerated more slowly in the surface water than nitrogen or phosphorus (Dugdale, 1972). Furthermore, direct measurements of nutrient excretion by the dominant herbivore in the Peru system, the anchoveta, *Engraulis ringens*, have shown it to excrete reduced nitrogen compounds at a rate some 20 times higher than silicic acid (Whitledge and Packard, 1971). Reduced nitrogen, chiefly as ammonium (Conway, 1974) and urea (McCarthy, Taylor and Taft, 1977) is taken up by phytoplankton in preference to nitrate. Thus, regeneration of ammonium and urea at much higher rates than silicic acid, combined with their preferential use, may explain the fact that nitrate remains at *ca.* 4 µM at 70 to 80 km downstream from the source of upwelling, where silicic acid is depleted to nearly undetectable levels (Dugdale and Goering, 1970) and its uptake is clearly limited by the substrate concentration (Goering *et al.*, 1973).

From 18 through 21 March, 1977, a surface (4 m) drogue was followed off Peru in an effort to monitor the biological events in a parcel of recently upwelled water. Silicic acid uptake rates and particulate and dissolved silicon concentrations were determined at the approximately 50% light depth at four stations along the drogue path

(Table 3). In 72 h the ambient silicic acid concentration decreased from 12.1 to 1.1 μM, a net decrease of 11.0 μM. Gross production of particulate silica, calculated from ^{30}Si tracer measurements of the absolute uptake velocities of silicic acid obtained at the stations, was 11.8 μM, suggesting that the rate of return of silicic acid to the surface water by the combined effects of advection and dissolution was less than 10% of the uptake rate. This approximate budget tends to support the idea that the rate of dissolution of particulate silicon in the Peru system is insignificant relative to the silicic acid uptake.

This general picture of nutrient dynamics in the Peru system is clearly inapplicable to either northwest Africa or Baja California. Direct measurements have shown *in situ* dissolution of particulate silica over the continental shelf off northwest Africa to proceed at a rate high enough to satisfy the entire metabolic demand of the phytoplankton for silicic acid (Nelson and Goering, 1977b). No direct measurements of silicic acid regeneration rates have been made in the Baja California system. However, consideration of the relative advective input rates of silicon and nitrogen via upwelling (Walsh, 1976), silicic acid uptake rates, particulate silicon concentrations (Nelson and Goering, 1978), nitrate and ammonium uptake rates, and particulate nitrogen concentrations (H.L. Conway, unpublished data, 1978) shows that severe depletion of silicic acid from the surface waters should take place in the Baja California system if silicic acid regeneration proceeds at a rate less than or equal to that of nitrogen remineralization (Nelson and Goering, 1978). Exactly the opposite actually takes place, with total combined nitrogen frequently depleted to very low levels (<0.5 μM) while silicic acid concentrations are almost never <3 μM (Whitledge and Bishop, 1973). Thus, if proposed nutrient budgets for the Baja California system (Walsh, 1976; Walsh, Whitledge, Kelley, Huntsman, and Pillsbury, 1977) are even approximately correct, *in situ* regeneration of silicic acid must proceed at a substantially higher rate than nitrogen remineralization, with the net result that nitrogen rather than silicon is the first nutrient depleted to potentially limiting concentrations off Baja California.

The reasons for the dramatic differences observed in overall nutrient dynamics among the Peru, northwest Africa, and Baja California upwelling systems are poorly understood at present, but differences in the characteristics of the predominant herbivores offer a plausible explanation. In the Peru upwelling system the anchoveta, *Engraulis ringens*, excretes reduced nitrogenous compounds and small amounts of silicic acid directly into the water (Whitledge and Packard, 1971), while almost all ingested diatom silica is released in the form of large and rapidly sinking fecal pellets. In contrast, the two upwelling regions where rapid *in situ* regeneration of silicic acid takes place are not grazed heavily by

TABLE 3. *Wecoma* Drogue Experiment 1

Date	Sta.	$V_{H_4SiO_4}$ (h^{-1})	$\rho_{H_4SiO_4}$ (μ moles $l^{-1}h^{-1}$)	Calculated Daily Particulate Si Production (= $\rho_{H_4SiO_4} \times 24$) (μ moles $l^{-1}d^{-1}$)	PSi (μ moles l^{-1})	Ambient (H_4SiO_4) (μM)
18 March	15	0.016	0.084	2.0	5.4	12.1
19 March	16	0.026	0.173	4.2	6.6	8.8
20 March	17	0.033	0.233	5.6	7.0	6.6
21 March	19				9.4	1.1

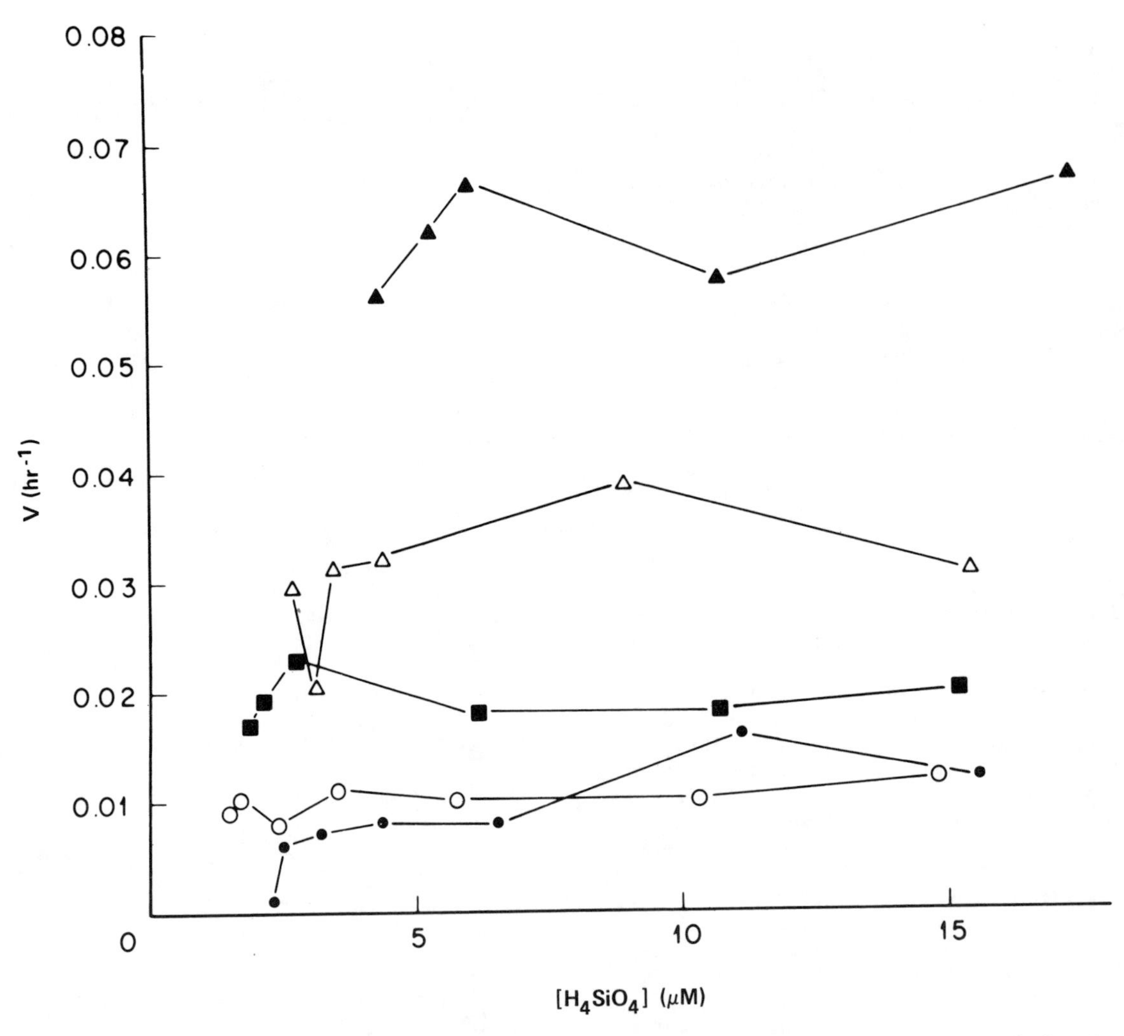

Fig. 6. Concentration dependence of silicic acid uptake at five stations in the Baja California upwelling system in 1973 (Δ, ▲) and the northwest Africa upwelling system in 1974 (O, ●, ■). Stations were selected for the experiments on the basis of surface silicic acid concentrations, which approached the lowest encountered in each of the two systems.

herbivorous fishes, and it is likely that the fecal material produced by zooplankton remains in suspension much longer than the large fecal pellets of *E. ringens*. Thus silicic acid regenerated *in situ* from fecal pellet material can be a much greater fraction of the silicon reserve where zooplankton provide the major grazing stress.

Concentration Dependence of Silicic Acid Uptake

Kinetic experiments in 1973 and 1974 indicate no detectable concentration dependence of silicic acid uptake off Baja California and only one case of weak concentration dependence off northwest Africa (Nelson and Goering, 1978)(Fig. 6). Similar experiments off Peru during the 1969 *Anton Bruun* cruise (Goering *et al.*, 1973), and again in 1976 and 1977, have shown clear limitation of uptake by silicic acid concentrations, but only offshore from the region of strongest upwelling and following depletion of near-surface silicic acid concentrations to <2.5 μM. In all three systems kinetic experiments were performed, not at randomly selected or predetermined stations, but at those where near-surface silicic acid concentrations were among the lowest encountered. Thus, our kinetic data provide strong evidence that substrate limitation of silicic acid uptake is confined to offshore areas of nutrient depletion off Peru, is extremely uncommon off northwest Africa, and probably does not occur off Baja California.

The ambient silicic acid concentrations at the stations selected for kinetic experiments ranged from 1.6 to 5.5 μM, with limitation of uptake detectable only at stations where the surface

TABLE 4. Silicic Acid Uptake Data from 22 Kinetic Experiments Performed on Unialgal Diatom Cultures and Natural Phytoplankton Populations. V_1 and V_2 = Uptake Rate at Highest and Second Highest Silicic Acid Concentrations Used in Each Experiment, respectively. S_1 and S_2 = Highest and Second Highest Silicic Acid Concentrations at which Uptake Rates were Determined.

Experiment	V_1	(units)	S_1 (μM)	V_2	(units)	S_2 (μM)	V_1/V_2	Reported K_s (μM)
Goering *et al.* (1973								
TT 36-59 (Peru)	0.034	h^{-1}	23	0.030	h^{-1}	17	1.133	-
TT 36-51 (Peru)	0.058	h^{-1}	13	0.056	h^{-1}	7.5	1.036	2.93
Paasche (1973)								
Skeletonema costatum	0.085	$\frac{pgSi}{cell \cdot h}$	12	0.092	$\frac{pgSi}{cell \cdot h}$	9	0.924	0.84
Thalassiosira pseudonana	0.037	"	12	0.036	"	10	1.028	1.40
Licomorpha sp.	2.0	"	15	2.1	"	10	0.952	2.06
Thalassiosira decipiens	3.1	"	11	2.9	"	8	1.069	3.69
Ditylum brightwelli	21.0	"	13	19.5	"	9	1.077	2.35
Nelson *et al.* (1976)								
Thalassiosira pseudonana:								
clone 3H, Si-replete	0.095	h^{-1}	15	0.076	h^{-1}	10	1.250	2.3
clone 3H, Si-deplete	0.059	h^{-1}	15	0.061	h^{-1}	10	0.967	0.8
clone 13-1, Si-replete	0.028	h^{-1}	15	0.029	h^{-1}	10	0.966	1.4
clone 13-1, Si-deplete	0.026	h^{-1}	15	0.025	h^{-1}	10	1.040	1.5
Azam and Chisholm (1976)								
Gulf of California	0.72	$\frac{\mu moles^*}{l \cdot h}$	23	0.075	$\frac{\mu moles^*}{l \cdot h}$	14	0.960	2.53
Baja California coast	1.02	"	24	0.082	"	14.5	1.243	1.54
Kilham *et al.* (1977)								
Diatoma elongatum	1.59	$\frac{\mu moles}{cell \cdot h}$	31	1.85	$\frac{\mu moles}{cell \cdot h}$	16	0.859	6.13
Diatoma elongatum	1.62	"	27	1.51		13.5	1.019	-
Nelson and Goering (1978)								
TT 78-25 (Baja Calif)	0.031	h^{-1}	16	0.038	h^{-1}	9	0.816	-
TT 78-67 (Baja Calif)	0.065	h^{-1}	18	0.058	h^{-1}	11	1.138	-
AII 82-31 (NW Africa)	0.011	h^{-1}	16	0.016	h^{-1}	11	0.688	-
AII 82-37 (NW Africa)	0.020	h^{-1}	15.5	0.018	h^{-1}	11	1.111	-
AII 82-38 (NW Africa)	0.011	h^{-1}	15	0.010	h^{-1}	10.5	1.100	-
Boisseau (unpublished data)								
WE2-124 (Peru)	0.089	h^{-1}	17.6	0.092	h^{-1}	13.6	0.967	3.7
WE2-42 (Peru)	0.024	h^{-1}	23.7	0.026	h^{-1}	13.6	0.923	-
Mean			17.43			11.51	1.012	2.38
Standard Deviation			5.26			2.99	0.131	1.41
95% Confidence Limit (2x st error or mean)			2.24			1.27	0.056	0.755

* All uptake rates other than those of Azam and Chisholm (1976) are normalized to cell number or particulate silicon concentrations, and are thus specific, or biomass independent, rates. Azam and Chisholm's data were not reported as specific rates, but were observed at constant biomass (i.e., on subsamples from a large seawater sample), so the ratio V_1/V_2 accurately reflects the ratio of the specific rates.

silicic acid concentration was <2.5 μM. Numerous previous studies of kinetics of silicic acid uptake by unialgal diatom cultures and natural phytoplankton populations have indicated that the uptake rate (V) can be related to the cells' maximum or nutrient-saturated uptake rate (V_{max}) to a good approximation by the Michaelis-Menten equation:

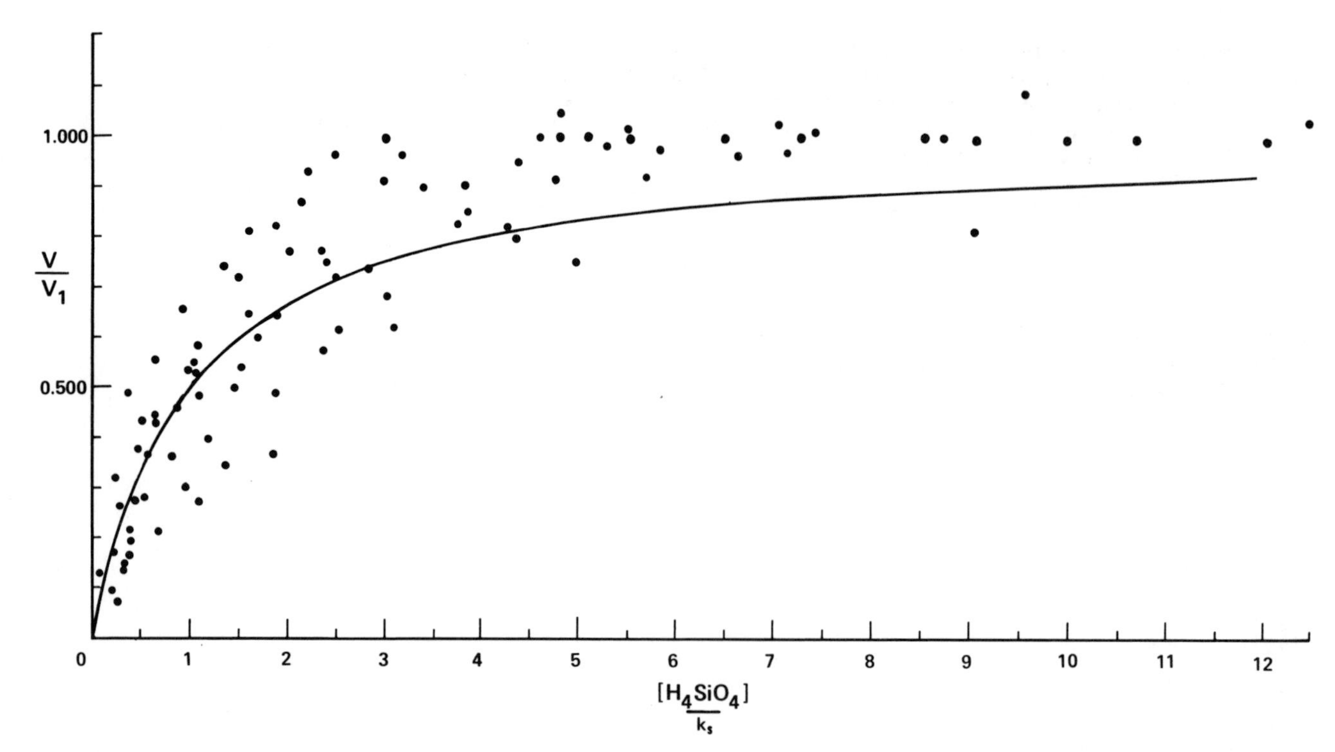

Fig. 7. Concentration dependence of silicic acid uptake. Figure includes data from 14 experiments by various investigators on unialgal diatom cultures and natural phytoplankton populations. See text and Table 4 for details. The hyperbola generated by the Michaelis-Menten equation (eq. 1) is plotted with, but not fitted to, the data.

$$V = V_{max} \frac{[H_4SiO_4]}{K_s + [H_4SiO_4]} \quad (1)$$

with values of the half-saturation constant (K_s) ranging from 0.8 to 6.1 μM (Paasche, 1973; Goering *et al.*, 1973; Azam, 1974; Nelson *et al.*, 1976; Azam and Chisholm, 1976; Kilham, Kott, and Tilman, 1977; Nelson and Goering, 1978). Therefore, the ability of natural phytoplankton in the Baja California, northwest Africa, and Peru upwelling systems to take up silicic acid at or near their maximum rates when the silicic acid concentration was in the 2.5 to 5.5 μM range indicates that one of two things must be true: either the diatoms in upwelling systems have substantially lower K_s values for silicic acid uptake than do most of those for which kinetic data have been reported, or the Michaelis-Menten hyperbola is not a rigorously accurate description of the uptake kinetics at relatively high silicic acid concentrations.

We have re-examined the available literature on silicic acid uptake by diatoms, attempting to consider the high concentration range in a way that is independent of curve fitting. Table 4 lists data from 22 experiments on short-term uptake by unialgal diatom cultures or natural phytoplankton assemblages in batch culture. Only the data from Azam's (1974) study of *Nitzschia alba* have been excluded from this analysis. In Azam's study the highest silicic acid concentrations used were higher (60 to 80 μM) than those used in other experiments. *N. alba* is a benthic, heterotrophic, unpigmented diatom that appears to have silicic acid uptake kinetics substantially different from those of planktonic, photosynthetic diatoms. For the remaining 22 experiments, we have determined the ratio of the uptake rate at the highest silicic acid concentration used in each experiment (V_1) to that at the second highest (V_2), as a measure of concentration dependence of uptake when the silicic acid concentrations significantly exceed K_s. No statistically detectable departure of V_1/V_2 from unity emerges in this treatment of the available data, indicating that the maximum rate of uptake is actually achieved at substrate concentrations of (on average) <10 to 12 μM, rather than being approached asymptotically as the Michaelis-Menten equation (1) predicts.

In 14 of the 22 experiments the Michaelis-Menten equation was fitted to the data by the original investigators and K_s values were reported (Table 4). For the 14 experiments, the ratio V/V_1 is plotted versus the ratio $[H_4SiO_4]/K_s$ in Fig. 7. By assuming that $V_1 = V_{max}$ (which appears to be the case; see Table 4) we can represent the Michaelis-Menten hyperbola on the same axes. The curve generated in this way is plotted with, but not fitted to, the data points in Fig. 7. The figure indicates that Michaelis-Menten kinetics provides a reasonably good fit to silicic acid uptake data when silicic acid concentration, $[H_4SiO_4]$ is less than 2 K_s, but that uptake is substantially less concentration-dependent than the Michaelis-Menten model predicts at higher substrate concentrations. Regression analysis indicates no detectable increase in V with increasing substrate concentration after $[H_4SiO_4]$ exceeds 4 K_s: a concentration range over which equation (1) predicts a 25% increase in V.

On the basis of this analysis we conclude that uptake of silicic acid at the cells' maximum rate is genuinely achieved, rather than approached as a limit, by planktonic diatoms and that uptake at less than this maximum is extremely rare in the natural phytoplankton assemblages of upwelling regions.

Summary

Knowledge of the dynamics of silicon in the near-surface of the ocean is still in a primitive stage compared with current understanding of the cycling of carbon, nitrogen, and phosphorus. However, consideration of data on uptake and regeneration of silicic acid in the surface waters of the Peru, northwest Africa, and Baja California upwelling systems between 1969 and 1977 allows the following generalizations and conclusions:

1) Silicic acid uptake by phytoplankton in upwelling regions extends at least to the depth of 0.01% surface light penetration, approximately twice the depth to which photosynthesis and nitrogen uptake are observed.
2) In the Baja California and northwest Africa systems the uptake rate tends to be quite uniform vertically between the surface and the 1% light penetration depth. The same is true in the Peru system when the average daily alongshore winds exceed 5 m/s, but a strong near-surface maximum in uptake is generally evident when winds are <5 m/s.
3) The integrated production rate of biogenic silica in the euphotic zone off Peru is significantly higher on days when winds are <5 m/s, probably due to a decreased rate of vertical wind mixing between the surface and poorly lit depths. Near-surface (50% light depth) photosynthetic rates are also significantly higher off Peru under the low-wind conditions.
4) The rate at which silicic acid uptake can proceed in the dark is highly variable, and at times substantial, in all three systems. There is systematic variability in the capacity of phytoplankton in the three systems to take up silicic acid in the dark, but the causes of the variability are not clear.
5) Silicic acid is regenerated rapidly in the surface layer via dissolution of particulate silica in the northwest Africa and Baja California systems, and the result is that silicon limitation does not occur. In the Peru system, near-surface regeneration appears minimal, and silicic acid is the first nutrient depleted to potentially limiting concentrations as the upwelled surface water is transported offshore.
6) Limitation of the silicic acid uptake rate by substrate availability is confined to offshore areas of silicic acid depletion off Peru, is extremely uncommon off northwest Africa, and probably does not occur off Baja California. It appears that uptake at the cells' maximum rate is genuinely achieved by phytoplankton in upwelling regions, rather than being approached as a limit as most current kinetic models of uptake would predict.

Acknowledgements. We are grateful to Susan Banahan and Frank Flynn of the Institute of Marine Science, University of Alaska, for their invaluable advice and assistance both in the field collection and in the laboratory preparation and analysis of samples. This work was conducted under the auspices of the IDOE Coastal Upwelling Ecosystems Analysis (CUEA) program and supported by the National Science Foundation Grants OCE 78-02019, OCE 76-00593, and GA-37963 to the University of Alaska and GX-33502 to the University of Washington. Part of the work was supported by a Woods Hole Oceanographic Institution Postdoctoral Fellowship to DMN.

References

Azam, F., Silicic-acid uptake in diatoms studied with (^{68}Ge) germanic acid as a tracer, *Planta (Berlin)*, *121*, 205-212, 1974.

Azam, F. and S.W. Chisholm, Silicic acid uptake and incorporation by natural marine phytoplankton populations, *Limnology and Oceanography*, *21*, 427-435, 1976.

Barber, R.T. and S.A. Huntsman, JOINT-I carbon, chlorophyll and light extinction - R/V *Atlantis II* cruise 82, IDOE Coastal Upwelling Ecosystems Analysis Data Report 14, University of Washington, Seattle, 165 pp., 1975.

Barber, R.T., S.A. Huntsman, and J.E. Kogelschatz, Reduced productivity data: Outfall I, April 1972: MESCAL-II, April, 1973; Outfall II, May, 1978; F.S. *Meteor* Riese 36, February, 1975; *El Niño* Watch Cruise, February-May, 1975, IDOE Coastal Upwelling Ecosystems Analysis Data Report 32, Duke University Marine Laboratory, Beaufort, North Carolina, 173 pp., 1976.

Barber, R.T., S.A. Huntsman, J.E. Kogelschatz, W.O. Smith, B.H. Jones, and J.C. Paul, Carbon, chlorophyll and light extinction from JOINT-II 1976 and 1977, IDOE Coastal Upwelling Ecosystems Analysis Data Report 49, Duke University Marine Laboratory, Beaufort, North Carolina, 476 pp., 1978.

Beardall, J. and I. Morris, The concept of light intensity adaptation in marine phytoplankton, some experiments with *Phaeodactylum tricornutum*, *Marine Biology*, *37*, 377-387, 1976.

Blasco, D., Composición y distribución del fitoplancton en la región del afloramiento de las costas peruanas, *Investigacion Pesquera, 35,* 61-112, 1971.

Brink, K.H., R.L. Smith, and D. Halpern, A compendium of time series measurements from moored instrumentation during MAM '77 phase of JOINT-II, IDOE Coastal Upwelling Ecosystems Analysis Technical Report 45, Oregon State University Reference 78-17, Corvallis, Oregon, 1978.

Conway, H.L., The uptake and assimilation of inorganic nitrogen by *Skeletonema costatum* (Grev.) Cleve, 125 pp., Ph.D. Dissertation, University of Washington, Seattle, 1974.

Dugdale, R.C., Chemical oceanography and primary productivity in upwelling regions, *Geoforum, 11,* 47-61, 1972.

Dugdale, R.C. and J.J. Goering, Nutrient limitation and the path of nitrogen in Peru current production, *Anton Bruun* Report No. 4, Texas A&M Press, pp. 5.3-5.8, 1970.

Eppley, R.W. and J.L. Coatsworth, Uptake of nitrate and nitrite by *Ditylum brightwelli* - kinetics and mechanisms, *Journal of Phycology, 4,* 151-156, 1968.

Estrada, M. and D. Blasco, Two phases of phytoplankton community in the Baja California upwelling, *Limnology and Oceanography, 24,* 1065-1080, 1979.

Goering, J.J., D.M. Nelson, and J.A. Carter, Silicic acid uptake by natural populations of marine phytoplankton, *Deep-Sea Research, 20,* 777-789, 1973.

Kilham, S.S., C.L. Kott, and D. Tilman, Phosphate and silicate kinetics for the Lake Michigan diaton *Diatoma elongatum, Journal of Great Lakes Research, 3,* 93-99, 1977.

Kogelschatz, J.E., J.J. MacIsaac, and N.F. Breitner, JOINT-II R/V *Wecoma*: March-May 1977 General Productivity Report, IDOE Coastal Upwelling Ecosystems Analysis Data Report 50, Duke University Marine Laboratory, Beaufort, North Carolina, 170 pp., in press.

Lewin, J.C., Silicon metabolism in diatoms, III. Respiration and silicon uptake in *Navicula pelliculosa, Journal of General Physiology, 39,* 1-10, 1955.

MacIsaac, J.J. and R.C. Dugdale, Interactions of light and inorganic nitrogen in controlling nitrogen uptake in the sea, *Deep-Sea Research, 19,* 209-232, 1972.

McCarthy, J.J., W.R. Taylor, and J.L. Taft, Nitrogenous nutrition of the plankton in the Chesapeake Bay, I. Nutrient availability and phytoplankton preferences, *Limnology and Oceanography, 22,* 996-1011, 1977.

Nelson, D.M. and H.L. Conway, Effects of the light regime on nutrient assimilation by phytoplankton in the Baja California and northwest Africa upwelling systems, *Journal of Marine Research, 37,* 301-319, 1979.

Nelson, D.M. and J.J. Goering, A stable isotope tracer method to measure silicic acid uptake by marine phytoplankton, *Analytical Biochemistry, 78,* 139-147, 1977a.

Nelson, D.M. and J.J. Goering, Near surface silica dissolution in the upwelling region off northwest Africa, *Deep-Sea Research, 24,* 65-73, 1977b.

Nelson, D.M. and J.J. Goering, Assimilation of silicic acid by phytoplankton in the Baja California and northwest Africa upwelling systems, *Limnology and Oceanography, 23,* 508-517, 1978.

Nelson, D.M., J.J. Goering, S.S. Kilham, and R.R.L. Guillard, Kinetics of silicic acid uptake and rates of silica dissolution in the marine diatom *Thalassiosira pseudonana, Journal of Phycology, 12,* 246-252, 1976.

Paasche, E., Silicon and the ecology of marine plankton diatoms, II. Silicate-uptake kinetics in five diatom species, *Marine Biology, 19,* 262-269, 1973.

Richter, O., Zur Physiologie der Diatomeen, I. Mitteilung, *Akademie der Wissenschaften Mathematisch-Natururssenschafliche Klasse, 115,* 27-119, 1906.

Sournia, A., Circadian periodicities in natural populations of marine phytoplankton, *Advances in Marine Biology 12,* 325-389, 1974.

Walsh, J.J., Models of the Sea, in: *Ecology of the Sea,* D.H. Cushing and J.J. Walsh (eds.), pp. 388-407, Blackwell, 1976.

Walsh, J.J., T.E. Whitledge, J.C. Kelley, S.A. Huntsman, and R.D. Pillsbury, Further transition states of the Baja California upwelling ecosystem, *Limnology and Oceanography, 22,* 264-280, 1977.

Whitledge, T.E., and T.T. Packard, Nutrient excretion by anchovies and zooplankton in Pacific upwelling regions, *Investigacion Pesquera, 35,* 243-250, 1971.

Whitledge, T.E., and D. Bishop, MESCAL II hydrography, STD and underway maps -- R/V *Thomas G. Thompson* cruise 78, IDOE Coastal Upwelling Ecosystems Analysis Data Report 11(1), University of Washington, Seattle, 560 pp., 1973.

NITROGEN RECYCLING AND BIOLOGICAL POPULATIONS IN UPWELLING ECOSYSTEMS

Terry E. Whitledge

Oceanographic Sciences Division
Brookhaven National Laboratory
Upton, New York 11973

Abstract. Nutrient recycling has been studied in the upwelling areas of Baja California, northwest Africa, and Peru. Biological regeneration contributes significant quantities of recycled nitrogen, which are utilized in productivity processes. Each area has a different combination of organisms, which leads to differences in the relative contributions of zooplankton, nekton, or benthos to the nutrient cycles.

Comparisons of ammonium regeneration rates of zooplankton and nekton-micronekton populations in the three upwelling areas show that zooplankton recycle less nitrogen in the Baja California and Peru systems than nekton do. In the northwest African upwelling region, however, zooplankton, fish, and benthic inputs are all substantial. In recent years the Peruvian upwelling system has been altered with the decline of the anchoveta population and an increase in the importance of zooplankton in nutrient recycling. The distribution of recycled nitrogen (ammonium and urea) in transects across the shelf at 10 and 15°S suggests that regeneration is important in both areas, but the distributions of ammonium and urea are not entirely coincident, indicating differences in their production and utilization.

Introduction

Many upwelling areas in the world's oceans are off the west coasts of continents and are typified by cold, nutrient-rich waters that maintain large phytoplankton populations, especially in the low latitudes where the quantity of light does not undergo large seasonal fluctuations. Upwelling transports nutrient-rich subsurface water into the euphotic zone where phytoplankton utilize the nutrients during growth. The primary uptake of nutrients derived from upwelling has been called "new production" to show that biological processes have not utilized the nutrient since it was introduced into the euphotic zone (Dugdale and Goering, 1967). The uptake of nutrients derived from herbivore excretion or bacterial regeneration has been termed "regenerated production" to show that the production was based on the uptake of nutrients that previously had been used in a biological process in the euphotic zone.

With vertical advection bringing nutrient-rich water into the euphotic zone of upwelling areas, large concentrations of nutrients are available to stimulate primary productivity; however, recycled nutrients are also important from the standpoint of increasing the production efficiency of the ecosystem. The topic has only recently been examined in upwelling ecosystems and is the central focus of this paper.

Northwest Africa

The JOINT-I program was concerned with the upwelling ecosystem off northwest Africa in the area of Cape Blanc off Spanish Sahara (Barber, 1977). The region was of interest because numerous countries utilize the fishery resources of the area.

Zooplankton. The distribution of zooplankton at 21 stations across the shelf off Cape Blanc was determined using vertical tows of bongo nets equipped with 102-μm mesh netting. The samples were corrected for phytoplankton contamination using chlorophyll as an index (Smith and Whitledge, 1977). The cross-shelf transect of four size fractions (>1000 μm, 505 to 1000 μm, 223 to 505 μm, and 102 to 223 μm) showed maxima inshore and immediately offshore of the shelf break (Fig. 1). The inshore maximum was observed on a station at the 45-m isobath while the offshore maximum was in water about 200 m deep. The total biomass was largest (13.19 g dry wt/m^2) offshore; it was composed primarily of euphausiids, copepods, and salps (Blackburn, 1979). The inshore maximum of zooplankton biomass (3.78 mg dry wt/m^2) was dominated by the 223 to 505-μm size class and was composed mainly of copepods. Excretion rate measurements ranged from 1.68 to 32.26 μg N (mg dry wt)$^{-1}$d^{-1} for ammonium excretion and 0.67 to 5.38 μg N (mg dry wt)$^{-1}$d^{-1} for urea release. When the excretion rates were combined with biomass values at each of the locations, the regen-

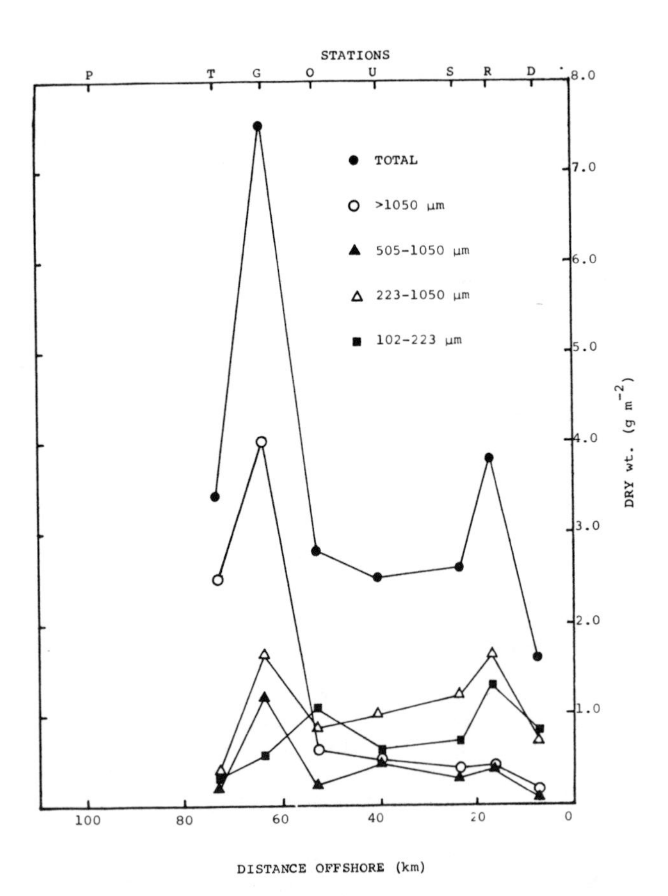

Fig. 1. Biomass of zooplankton (g dry wt/m^2) collected on a transect near 21°40'N off northwest Africa during JOINT I, March to May, 1974 (after Smith and Whitledge, in press).

eration rates of zooplankton across the shelf (Table 1) were largest at the areas of high zooplankton biomass both inshore and offshore. The inshore rates were higher than the offshore rates as a consequence of the smaller organisms there and the lesser depth. An independent estimate of zooplankton ammonium excretion, using respiratory electron transport activity, resulted in a mean value equivalent to 4.75 mg-at $m^{-2}d^{-1}$ (Packard, 1979), a value that is only slightly smaller than the mean of 5.4 mg-at $m^{-2}d^{-1}$ calculated for the four inshore shelf locations.

Nekton. Nekton biomass was determined in the Cape Blanc area by acoustic mapping surveys and bottom trawls. Results of the acoustic surveys indicated that the mean pelagic biomass over the shelf was 10 to 15 g dry wt/m^2 (Thorne *et al.*, 1977). Analysis of net samples further indicated that the relative abundance of sardine and anchovy eggs and larvae in the study area was about 4:1 (Blackburn and Nellen, 1976). Demersal fish stocks, sampled by bottom trawls, were estimated to be equivalent to 0.55 g dry wt/m^2, cephalopods were about 0.17 g dry wt/m^2, and shrimp were about 0.22 g dry wt/m^2 to give a total of 0.94 g dry wt/m^2 for demersal biomass (Haedrich, Blackburn, and Brulhet, 1976). This is about an order of magnitude smaller than the pelagic biomass estimates.

Nekton excretion rates were measured in experiments using several kinds of demersal and pelagic fish, sharks, and mollusks (Table 2). Samples for nitrogen compounds (ammonium and urea) were collected every 10 min. The concentrations in the experimental tank showed a nearly linear increase with time with ammonium accounting for more than 50% of the excreted nitrogen. The rate measurements ranged from 0.42 to 4.62 μg NH_4-N (mg dry wt)$^{-1}d^{-1}$. Excretion rates were combined with biomass values to produce nutrient regeneration estimates over the shelf and beyond the shelf break (Table 3). The regeneration by nekton was largest near the outer part of the shelf by a factor of 2. The ammonium and urea nitrogen regeneration rates by nekton are equal in contrast to zooplankton, which regenerate mostly ammonium.

Sediments, Benthos, and Bacterioplankton. Nutrient release into the water column from sediments was investigated by placing bell jars on the bottom, collecting box cores, and obtaining pore water nutrient samples (Rowe, Clifford, and Smith, 1977). In the bell jar experiments 5.64 mg-at NH_4-N $m^{-2}d^{-1}$ was released into the water column, a quantity nearly equivalent to the zooplankton regeneration rate at the nearest inshore station. There were only two stations so the variability could be quite large. However, the southern shelf station, which had a larger ammonium flux rate than the northern shelf station, also had a larger ammonium gradient between the water column and the sediments. The concentrations in pore water were about 150 μg-at N/1 throughout the top 20 cm. This suggests that the benthic release of ammonium is probably very constant over time, neglecting any seasonal temperature effects.

The impact of bacteria in the northwest Africa upwelling ecosystem was investigated by Watson (1978) using a new bacterial biomass technique. It was found that phytoplankton biomass in the water column was much larger than bacteria in the shallow shelf stations while bacteria had the greater biomass when water depths were larger than 350 m; therefore, bacterial rates in the water column were neglected because other samples used in the analysis were from depths of 100 m or less. A larger biomass of bacterial organic carbon was found in the sediments on the shallow shelf station (3580 mg C m^{-2}) than in samples from the slope (3030 mg C m^{-2}). It was estimated that the mean bacterial biomass in the water column and sediment of the shelf and slope regions (6.68 g C m^{-2}) was about 10% of the yearly amount of carbon incorporated into bacterial cells. This

TABLE 1. Ammonium (mg-at N $m^{-2}d^{-1}$) Regenerated by Size Fractions of Zooplankton (Smith and Whitledge, 1977). See Fig. 1 for Station Locations.

Location	Size Fraction (μm)				
	102-223	223-505	505-1000	1000	Total
Offshore	1.82	2.26	0.14	6.58	10.80
G	1.44	4.26	0.15	2.56	8.41
O	2.72	2.21	0.03	0.39	5.35
U	1.49	2.52	0.06	0.28	4.35
R	3.29	4.21	0.05	0.27	7.82
D	1.93	1.93	0.01	0.08	3.95

could result in a yearly consumption of 133 g C m^{-2}, 44% of the yearly phytoplankton production. On a daily basis this would recycle about 0.06 μg-at N $l^{-1}d^{-1}$ assuming a C:N ratio of 5 by atoms. The sediment bacterial regeneration value is already included in the sediment release rate estimate because the bell jar technique measures the net change over the sampling period.

Nutrient Regeneration Budget. The major ecosystem components related to regeneration are summarized in Table 4 for two areas of the northwest African upwelling system. The inshore area represented by Sta. R is in a water depth of about 40 m. The zooplankton are dominated in biomass by copepods in the 223 to 505-μm size range so their recycling rate is relatively large compared to the mid shelf region. Nekton biomass is lower than the shelf-break area where most of the pelagic biomass was located. The release rate of the sediments was measured directly in bell jars and was similar to the nekton recycling rate. Bacterial recycling estimates were not used because the major fraction of bacterial biomass is in the sediments, rather than the water column, so the bacterial rate is included in the sediment release rate. The total recycling rate of 16.9 mg-at N $m^{-2}d^{-1}$ is large compared to the measured phytoplankton ammonium uptake rate of 7.5 mg-at $m^{-2}d^{-1}$. Thus the total recycling rate of all of the trophic levels represents 225% of the phytoplankton ammonium requirement and produces large concentrations of ammonium nearshore (Fig. 2). This oversupply of available ammonium

TABLE 2. Nitrogen and Phosphorus Release by Selected Species of Fish Collected over the Shelf off Northwest Africa. The number of organisms in each experiment is shown in parentheses (Smith and Whitledge, in press)

Species	Condition	Ammonium-N	Urea-N
		(μg mg dry wt^{-1} d^{-1})	
Diplodus senegalensis (12)	Fresh	1.44	0.76
Glyphis glaucus (2)	Fresh	0.44	0.55
Octopus vulgaris (4)	Fresh	0.78	0.11
Sardinella (5)	Fresh	4.61	4.78
Pomadasys incisus (1) and *Diplodus senegalensis* (9)	Fresh	1.22	0.33
Pagellus couperi (2) and *Cantharus cantharus* (1)	Fresh	0.91	-
Pagellus couperi			

TABLE 3. Nutrient Regeneration by Zooplankton and Nekton and Nutrient Uptake by Phytoplankton in the Upwelling System off Northwest Africa. Units are mg-at $m^{-2}d^{-1}$. See Fig. 1 for station locations (after Smith and Whitledge, in press)

Location	Ammonium-N			Urea-N	
	Uptake*	Zooplankton	Nekton	Zooplankton	Nekton
P	-	7.46	-	0.41	-
T	-	7.17	-	0.56	-
OFF	15.5	10.80	-	2.20	-
G	5.6	8.41	7.11	2.95	7.22
O	16.2	5.35	6.45	1.35	6.54
U	11.9	4.35	6.78	1.53	6.88
S	-	1.30	-	0.95	-
R	7.5	7.82	3.58	2.29	3.48
D	8.3	3.95	3.15	0.97	3.14

* unpublished data from J.J. MacIsaac and R.C. Dugdale.

was observed in the nearshore area off Cape Corveiro by Coste and Slawyk (1974) and off Cape Blanc, where ammonium concentrations >2 μg-at/l were observed as far as 15 km offshore (Codispoti and Friederich, 1978). The relatively low phytoplankton uptake rate due to poor *in situ* light conditions was probably the reason for the large observed ammonium concentrations (Huntsman and Barber, 1977).

The outer shelf region, near Sta. O, had one of the lowest zooplankton biomasses of the transect and contained a mixture of large and small organisms. The zooplankton regeneration rate was about 70% of the inshore value. Because of the large pelagic nekton biomass in the region, the nekton rate was nearly twice that observed inshore. The release of nitrogen from the offshore sediments was only 33% of the inshore rate, and was based largely on the lower sediment bacterial biomass estimates offshore (Watson, 1978). Pore

TABLE 4. Ammonium Budget for Northwest Africa Upwelling Ecosystem. Units are mg-at $m^{-2}d^{-1}$

	Ammonium Regeneration	Phytoplankton Uptake
Shelf (40-m water depth)		
Zooplankton	7.82	
Nekton	3.48	
Sediments	5.64	
Total	16.94	7.5*
Outer Shelf (150-m water depth)		
Zooplankton	5.35	
Nekton	6.45	
Sediments	1.88	
Total	13.68	16.2*

* unpublished nitrogen uptake data of J.J. MacIsaac and R.C. Dugdale.

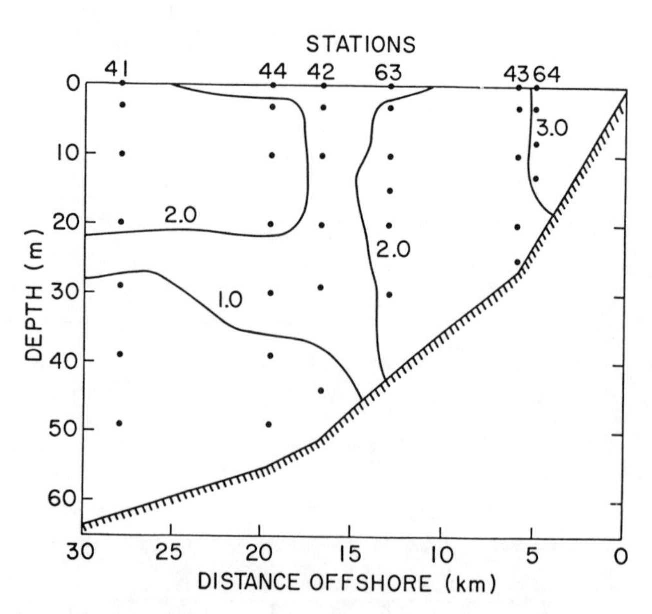

Fig. 2. Concentration of ammonium (μg-at/l) on a cross shelf transect near 21°40'N off northwest Africa.

TABLE 5. Zooplankton Dry Weight and Regeneration Rates by Size Fraction in the Peru Upwelling System at 10 and 15°S in April 1977. See Fig. 3 for station locations

		Size Fractionated Zooplankton Dry Weights (g/m²)					Zooplankton Nitrogen Regeneration Rates (mg-at m⁻² d⁻¹)				
Sta. No.	Depth (m)	102-223 (μm)	223-505 (μm)	505-1050 (μm)	>1050 (μm)	Total (μm)	102-223 (μm)	223-505 (μm)	505-1050 (μm)	>1050 (μm)	Total (μm)
10°S											
182	60	0.408	3.047	0.752	5.008*	9.215*	0.91	7.01	0.53	-	8.45
183	90	0.383	1.308	1.058	2.594*	5.343*	0.85	3.01	0.74	-	4.60
184	100	0.171	0.558	0.419	1.770	2.918	0.38	1.28	0.29	0.67	2.62
185	100	0.123	0.586	0.660	1.114	2.483	0.27	1.35	0.46	0.42	2.50
186	100	0.092	0.676	0.431	0.895	2.094	0.21	1.55	0.30	0.34	2.40
188	100	0.120	0.545	0.376	0.645	1.686	0.27	1.25	0.26	0.25	2.03
189	100	0.103	0.655	0.598	2.619	3.975	0.23	1.51	0.42	1.00	3.16
190	100	0.105	0.458	0.432	1.363	2.358	0.23	1.05	0.30	0.52	2.10
191	100	0.072	0.405	0.174	2.692	3.343	0.16	0.93	0.12	1.02	2.73
15°S											
192	35	0.175	0.443	1.020	0.719	2.357	0.39	1.02	0.71	0.27	2.39
193	60	0.780	1.750	1.517	2.291	6.338	1.74	4.03	1.06	0.87	7.70
194	100	0.175	0.465	0.135	1.814	2.589	0.39	1.07	0.09	0.69	2.24
195	100	0.227	1.726	0.506	15.793	18.252	0.51	3.97	0.35	6.00	10.83
196	100	0.448	1.256	0.280	7.845	9.829	1.00	2.89	0.20	2.98	7.07
197	100	0.611	1.748	0.338	1.174	3.871	1.36	4.02	0.24	0.45	6.07
198	100	0.104	0.875	1.357	6.189	8.525	0.23	2.01	0.95	2.35	5.54
236	85	0.241	1.838	1.326	3.765	7.170	0.54	4.23	0.93	1.43	7.13

* phytoplankton contamination in sample.

water nutrient concentrations also decreased offshore. The total recycling rate in this deep shelf region was 13.68 mg-at $m^{-2}d^{-1}$, 84% of the 16.2 mg-at $m^{-2}d^{-1}$ phytoplankton uptake rate estimates.

Peru

Several sets of data have been collected in the Peru upwelling ecosystem during the years 1966, 1969, 1976, and 1977. Most of the descriptions have been for the years 1976 and 1977 when zooplankton rate measurements were collected, but in 1976 and 1977 the anchoveta biomass had decreased to about 10% of its previous level. The decrease in anchoveta biomass was probably accompanied by changes in other nekton stocks such as an increase in the biomass of sardine and hake. In addition, the decline in anchoveta biomass probably resulted in an increase in zooplankton biomass (Walsh *et al.*, 1980).

Zooplankton. The zooplankton stocks were determined for the Peru upwelling ecosystem at 10 and 15°S using vertical bongo net tows (Dagg *et al.*, 1980). The zooplankton dry weight biomass at 10°S, where the shelf is relatively wide, was larger than 8 g/m² inshore. (There was an overestimate on the two innermost stations because of phytoplankton contamination, Table 5). The counts were dominated by *Paracalanus* spp., *Oncaea* spp., *Oithona* spp., and Appendicularia (Judkins, 1980; Geynrikh, 1973). Just beyond the shelf break, about 100 km offshore, the 1050-μm size class had more *Calanus* spp. and reflects the same general biomass distribution pattern observed in March, 1978 (Flint and Timonin, 1980). Zooplankton dry-weight biomass was larger on the 15°S transect where the shelf is only about 25 km wide. Larger zooplankton (e.g., *Eucalanus inermis*, *Centropages brachiatus*, and *Calanus chilensis*) were found at most stations at 15°S. The three species comprised 31% of the biomass on the 15°S transect and 9.6% on the 10°S transect. Zooplankton dry weights over the shelf were similar at 10 and 15°S. The 100-km wide shelf at 10°S was characterized with a mean (± standard deviation) dry weight biomass of 2.30 ± 0.53 g/m² (Table 5). At 15°S the 25-km wide shelf had a biomass of 3.76 ± 2.23 g/m², an increase of about 60%. In the offshore area at 10°S there was more biomass (3.23 ± 0.81 g/m²) than over the shelf, however, at 15°S the mean dry weight biomass offshore was 9.53 ± 5.36 g/m², 295% of the shelf value.

Zooplankton excretion rates were measured on

TABLE 6. Nitrogen Release (μg N (mg dry wt)$^{-1}$d^{-1}) by Zooplankton off Peru. Value are means ± standard deviation, with number of observations in parenthesis, -: no data (after Smith, 1978)

Size Fraction (μm)	Mean Size (mg dry wt per individual)	Ammonium Nitrogen Released	Urea Nitrogen Released
102 - 223	0.002	31.25 ± 23.86 (27)	38.98 ± 19.15 (5)
223 - 505	0.006	32.26 ± 17.14 (29)	22.18 ± 19.15 (5)
505 - 1050	0.042	9.47 ± 9.41 (25)	-
>1050 *(Eucalanus inermis)*	0.128	5.38 ± 6.38 (3)	-
>1050 *(Euphausia* sp.)	4.955	0.07 (1)	-
<505	-	31.92 ± 1.34 (56)	27.22 ± 16.13 (10)
>505	-	8.40 ± 7.39 (27)	-

large organisms that were also being used in ingestion experiments. The rates ranged from 4.4 to 29.9 μg N (mg dry wt)$^{-1}$d^{-1} for ammonium and 0.63 to 21.9 μg N (mg dry wt)$^{-1}$d^{-1} for urea. The ranges are similar to values in the northwest Africa upwelling experiments. Size class excretion rates were also determined in 1976 at the same location (Table 6). The rates for Peru were nearly identical to those for northwest Africa with the exception of the smallest size class, 102 to 223 μm, which is about 50% larger than the same size range in northwest Africa.

Nekton. With the introduction of acoustic surveys of the pelagic fish stocks, the Peru upwelling ecosystem has been the focus of attention since the decline of the anchoveta in 1972. Speculation exists on the quantity of anchoveta remaining and the possible increase of hake and sardine (Walsh *et al.*, 1980). Acoustic methods were used to estimate the total pelagic nekton biomass in the study area at 15°S. Zig-zag mapping tracks were followed both day and night. In the area within 13 km of the coast (to about the 150-m isobath) there was an estimated mean pelagic biomass of 127 g wet wt/m^2, which was probably composed of anchovy and sardine (Lee, Mathisen, and Thorne, 1979). Between 13 and 26 km offshore (within the 150 to 700-m isobaths) there was about 83.3 g wet wt/m^2 of pelagic nekton, which were thought to be jack mackerel, mackerel, and saury. The estimates are subject to daily variations, but nevertheless represent a composite of several runs.

The excretion rate measurements of some of the pelagic nekton were made on the anchoveta (*Engraulis ringens*), sardine (*Sardinops sagax*), and silverside (*Austromenidia regia regia*) (Table 7) using the methods described by Whitledge (1978). An additional measurement for jack mackerel (*Trachurus symmetricus*) was made off California (McCarthy and Whitledge, 1972). The largest ammonium excretion value measured for the silverside (5.2 μg N (mg dry wt)$^{-1}$d^{-1}) was comparable to the rate measured for *Sardinella* spp. off northwest Africa. The sardine and anchoveta were characterized by lower rates, 2.2 and 1.7 μg N (mg dry wt)$^{-1}$d^{-1}.

Nutrient Regeneration Budget. A nitrogen budget was constructed for the shelf and offshore Peruvian upwelling areas at 15°S using the appropriate nekton ammonium excretion rates combined with mean nekton biomass estimates (Table 8). The calculation assumed that 33% of the inshore nekton biomass was anchoveta and 67% was sardine (Walsh *et al.*, 1980). The shelf nitrogen budget, which used the three inshore stations, estimated the zooplankton regeneration rates to be 4.11 mg-at m^{-2}d^{-1} (Table 5). Based on ^{35}S-sulfate reduction experiments, the quantity of carbon utilized by the bottom community was estimated to be ∿91 g C m^{-2}y^{-1} (G.T. Rowe, personal communication). Converting to nitrogen using a C:N ratio of 5, about 4.16 mg-at NH_4-N m^{-2}d^{-1} were released from the sediments. The sediments were relatively rich in carbon, i.e., the C:N ratio was greater than 5, so this is probably a conservative estimate for the release rate. The ammonium-N regeneration rate of 12.9 mg-at m^{-2}d^{-1} by the three components (Table 8) compares favorably with the measured ammonium uptake rate of about 15 mg-at m^{-2}d^{-1} as estimated by ^{15}N uptake measurements on

TABLE 7. Ammonium Excretion Rates for Nekton Inhabiting the Peru Upwelling Ecosystem.

	Mean Dry Wt. (g)	NH_4 Excretion Rate $\pm$ std dev (μg N mg dry wt^{-1}d^{-1})	
Engraulis ringens (7)	1.70	1.74 $\pm$ 1.48	ref 1
Sardinops sagax (4)	1.52	2.19 $\pm$ 0.24	
Trachurus symmetricus (1)	47.5 *	1.26	ref 2
Austromenidia regia regia (14)	3.26	5.17 $\pm$ 1.78	

* Calculated from 190 g wet wt.
ref 1. Whitledge and Dugdale, 1972.
ref 2. McCarthy and Whitledge, 1972.

nine stations occupied by the R.V. *Wecoma* in April 1977 (unpublished results of J.J. MacIsaac and R.C. Dugdale). The difference between these two estimates (∿2.1 mg-at $m^{-2}d^{-1}$) is about 14%, a value near the expected experimental variability.

The off-shelf nitrogen budget includes the estimated offshore pelagic nekton biomass and jack mackerel excretion rates. The demersal nekton regeneration rate, calculated from an estimated standing stock for hake (Gulland, 1970) and excretion rates for starry flounder (Wood, 1958), was very low (0.07 mg-at $m^{-2}d^{-1}$). It is possible that the population was grossly underestimated. The zooplankton ammonium regeneration rate, calculated using the five offshore stations (Table 5), was 7.33 mg-at $m^{-2}d^{-1}$. The rate was larger than the shelf value (∿178%) mainly because of the large zooplankton biomass just offshore of the shelf break.

The benthic regeneration rate for the shelf sediments was used for the offshore budget because there were no strong cross-shelf gradients in the ^{35}S data used to estimate the sediment release rates. The total of 13.28 mg-at NH_4-N $m^{-2}d^{-1}$ regenerated in the offshore region is 74% of the measured phytoplankton ammonia uptake rate of 17.88 mg-at $m^{-2}d^{-1}$ at three stations occupied by the R.V. *Wecoma* in April, 1977 (J.J. MacIsaac and R.C. Dugdale, unpublished results).

Distribution of Recycled Nutrients. As a result of the large quantity of regenerated ammonium recycled by the biological population in the Peruvian upwelling ecosystem the ambient concentrations of ammonium and other substances are often quite large. The cross-shelf distributions of ammonium and urea at 15 and 10°S (Figs 3a-3d) each exhibited a maximum concentration about 50 to 100 km offshore. At 15°S the maximum concentrations of ammonium and urea were in the surface waters beyond the shelf break about 25 km offshore. The high concentrations of ammonium and urea occurred over the shelf at 10°S (shelf break at ∿100 km) near the bottom of the euphotic zone. The urea distributions at 15 and 10°S show that the maxima were inshore of the respective ammonium maxima. This could be the result of different utilization patterns for ammonium and urea. Previous measurements have shown that urea may be utilized in primary productivity almost as rapidly as ammonium and nitrate (McCarthy, 1972). An alternate explanation could be related to the relative rates of release of ammonium and urea by nekton and zooplankton populations. The fraction of urea in excreted nitrogen was larger for nekton than for zooplankton (Table 3). The zooplankton ammonium excretion over the shelf is nearly the same as for nekton (4.11/3.86 = 106%), while offshore the zooplankton ammonium regeneration was comparatively much larger (7.33/1.72 = 426%).

TABLE 8. Ammonium Budget for the Peru Upwelling Ecosystem. Units are mg-at $m^{-2}d^{-1}$

	Ammonium Regeneration	Phytoplankton Uptake
Shelf		
Zooplankton	4.11	
Pelagic Nekton	4.61	
Sediments	4.16*	
Total	12.88	15.0**
Offshore		
Zooplankton	7.33	
Pelagic Nekton	1.72	
Demersal Nekton	0.07***	
Sediments	4.16*	
Total	13.28	17.88**

* From ^{35}S estimate (G. Rowe, pers. commun.).
** Unpublished ^{15}N data of J.J. MacIsaac and R.C. Dugdale.
*** Based on 1970 stock estimate (Gulland, 1970) and flounder excretion rates (Wood, 1958).

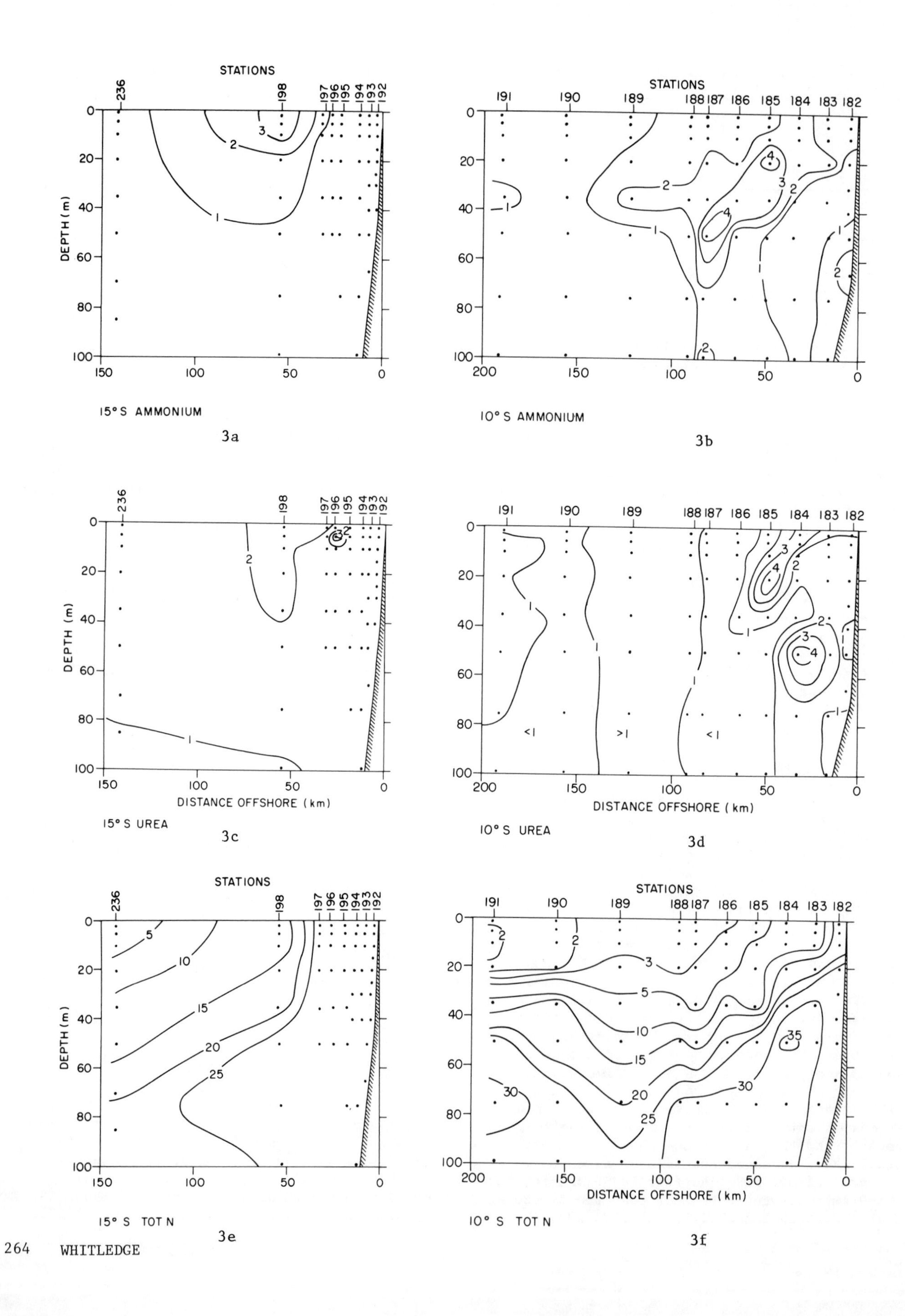
STATIONS
236
198
197
196
195
194
193
192
DEPTH (m)
15° S AMMONIUM
3a
10° S AMMONIUM
3b
15° S UREA
3c
10° S UREA
3d
DISTANCE OFFSHORE (km)
15° S TOT N
3e
10° S TOT N
3f

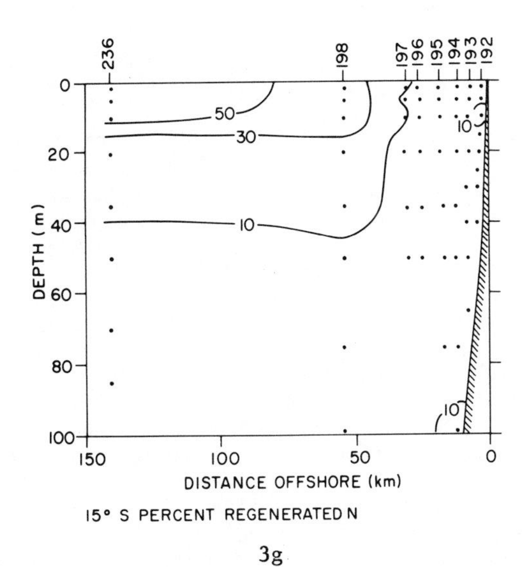

3g

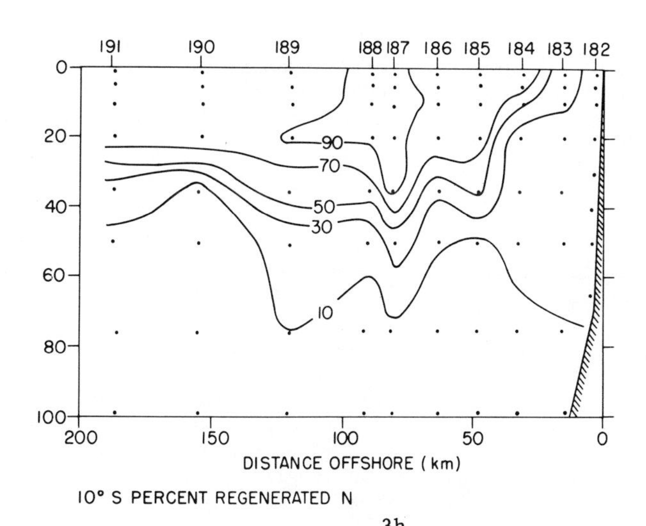

3h

Fig. 3. Concentrations of dissolved nitrogen nutrients (μg-at/l) in cross-shelf transects off Peru during April, 1977 observed during JOINT II, *Melville* leg 3. a) ammonium at 15°S; b) ammonium at 10°S; c) urea at 15°S; d) urea at 10°S; e) Total nitrogen (NO_3 + NO_2 + NH_4 + UREA) at 15°S; f) Total nitrogen (NO_3 + NO_2 + NH_4 + UREA) at 10°S; g) Percent regenerated nitrogen (NH_4 +UREA/Total nitrogen) x 100 at 15°S; h) Percent regenerated nitrogen (NH_4 + UREA/Total nitrogen) x 100 at 10°S.

So the inshore urea maximum might be the result of regeneration by the anchovy and sardine populations while the more offshore ammonium maximum could be created by zooplankton.

The distributions of total nitrogen (the sum of nitrate, nitrite, ammonium, and urea) across the shelf at 15 and 10°S (Fig. 3e, f) show the dominance of the upwelling of nitrate-laden water near the coast and decreasing concentrations in the surface layers as the water mass moves offshore. The surface water nearshore at 15°S was characterized by a total nitrogen concentration of more than 25 μg-at/l with a sharp gradient about 125 km offshore, where the concentration decreased to less than 5 μg-at/l. The nearshore area at 10°S appeared not to be upwelling so rapidly as at 15°S, possibly because of the much wider shelf at 10°S, which would affect the residence time of the water over the shelf. Consequently the nitrogen could go through more uptake-regeneration cycles while over the shelf. This suggests that the upwelling area off 10°S might be more dependent on regenerated nitrogen than the region off 15°S.

The relative amounts of regeneration are demonstrated well in the fractions of regenerated nitrogen in the water column (ammonium plus urea/ total nitrogen). The total nitrogen concentration at 15°S consisted of less than 10% regenerated nitrogen in the nearshore area where nitrate concentrations were high (Fig. 3g), but the fraction increased to greater than 50% beyond the shelf break. Similarly at 10°S the inshore water had a small fraction of regenerated nitrogen, but it increased to values greater than 90% in the surface layer at the shelf break (Fig. 3h).

Dissolved organic nitrogen (DON), as measured by an ultraviolet irradiation technique (Armstrong, Williams, and Strickland, 1966) was in uniformly high concentrations nearshore as if its source were in deep water; concentrations were lower near the surface offshore where up-

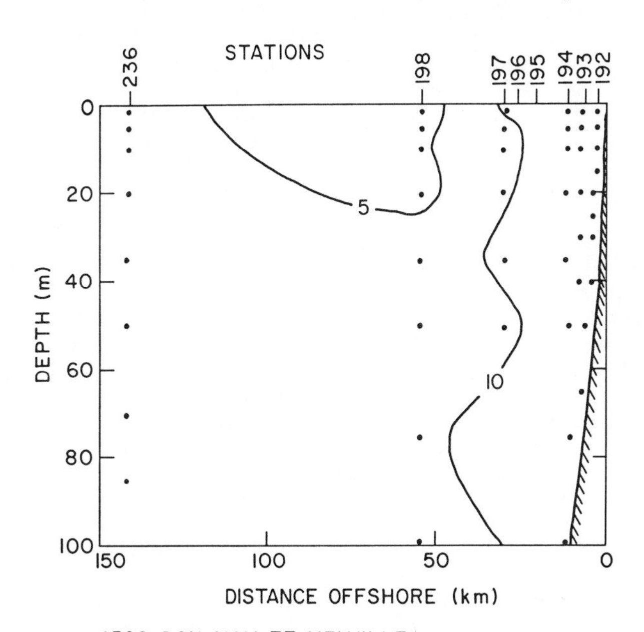

Fig. 4. Concentration of dissolved organic nitrogen (μg-at/l) on a cross-shelf transect near 15°S off Peru during April, 1977 observed during JOINT II, *Melville* leg 3.

take may have occurred (Fig. 4). The distribution of DON differed distinctly from ammonium or urea so the source of DON must not be pelagic organisms. The bottom sediments of the shelf could be a source, but there are no known measurements from the area.

Baja California

The upwelling off Baja California is seasonal with strong upwelling starting in March or April (Walsh *et al.*, 1974). The populations appear to be adapted to an area where upwelling occurs for only part of the year and is structured to withstand fluctuations in available food. In the upwelling area off Baja California a pelagic red crab, *Pleuroncodes planipes*, was probably the dominant herbivore during the upwelling season (Walsh *et al.*, 1977). During the non-upwelling time of the year the red crabs can remain on the bottom and switch to detritus as an alternate food source (Boyd, 1962).

Zooplankton. The change in zooplankton biomass in March, 1972 was from 0.2 to 1.07 g wet wt/m^3 over 14 days and was composed principally of stage V nauplii of *Calanus helgolandicus* (Walsh *et al.*, 1974). So it appears that in the 50-m water column the zooplankton biomass in the Baja California upwelling system increased from about 1.15 to 6.38 g dry wt/m^2 during the onset of upwelling. In April, 1973 the zooplankton community consisted of 72% copepodites, 22% adult *Acartia* spp., and 6% *Calanus* spp., with a mean estimated biomass equivalent to 6.61 g dry wt/m^2. The biomass similarity in the two sampling periods is being treated as a confirmation that the values are representative of the upwelling ecosystem. Ammonium regeneration rates in both time periods for the zooplankton were estimated using an excretion factor of 14.3 μg N (mg dry wt)$^{-1}$d^{-1} (Walsh *et al.*, 1977; Martin, 1968).

Micronekton. Using acoustic gear, the red crab, *P. planipes*, was estimated to have a mean pelagic dry weight biomass of 0.35 g/m^3 in the entire study area although consistently higher biomass values of 0.85 g dry wt/m^3 were observed off Punta San Hipolito (Blackburn and Thorne, 1974), where biological rate measurements were concentrated in March. At the same locations mean biomass values of 1.2 g dry wt/m^3 were estimated from net tows (Walsh *et al.*, 1977). So biomass values of 1.2 and 0.85 g dry wt/m^3 were used in computing the March 1972 and April 1973 regeneration rates by using an excretion rate of 2.9 μg N (mg dry wt)$^{-1}$d^{-1}.

The red crabs also gather in large concentrations on the bottom, especially those three years old or older (Boyd, 1962; Smith, Harbison, Rowe, and Clifford, 1975). Bottom photographs were used to estimate a density of about 8.2 red crabs/m^2; the mean weight of the crabs was 5.58 g (Walsh *et al.*, 1977). The distribution of the crabs on the bottom was patchy, but the biomass could be as high as 8.06 g dry wt/m^2.

Nekton. The net hauls used to calibrate the acoustic assessment surveys of nekton and micronekton in the Baja California upwelling region showed that *P. planipes* and the northern anchovy, *Engraulis mordax*, inhabited the area in a ratio of 9:1 by weight (Blackburn and Thorne, 1974). Thus the sardine had an estimated biomass of about 2.5 g dry wt/m^2.

Nutrient Regeneration Budget. A nitrogen budget for the shelf area of the Baja California upwelling system was constructed for March, 1972 and April, 1973 (Table 9). The budget for March has a relatively small zooplankton ammonium regeneration rate of 1.2 mg-at $m^{-2}d^{-1}$. The pelagic red crabs recycled 12.7 mg-at $m^{-2}d^{-1}$ and the benthic red crabs recycled 1.68 mg-at $m^{-2}d^{-1}$ for a total of 14.38 mg-at $m^{-2}d^{-1}$. The regeneration from the small biomass of anchovies was about 0.30 mg-at $m^{-2}d^{-1}$. The carbon utilized by benthic respiration was estimated to be 200 mg m^{-2} d^{-1}, so the nitrogen release rate from the sediments was 2.86 mg-at N $m^{-2}d^{-1}$ if a C:N conversion factor of 5 is used (G.T. Rowe, personal communication). The minimum ammonium uptake rate estimated from ^{15}N tracer experients at six stations was equivalent to 22.3 mg-at $m^{-2}d^{-1}$ (Walsh *et al.*, 1974). Ammonium regeneration by the combined sources produces 18.71 mg-at $m^{-2}d^{-1}$, 84% of the conservatively estimated phytoplankton ammonium uptake rate.

TABLE 9. Ammonium Budget for the Baja California Upwelling Ecosystem. Units are mg-at $m^{-2}d^{-1}$.

	Ammonium Regeneration	Phytoplankton Uptake
March 1972		
Zooplankton	1.17	
Nekton	0.30	
Pleuroncodes-Pelagic	12.70	
Pleuroncodes-Benthic	1.68	
Sediments	2.86	
Total	18.71	22.3*
April 1973		
Zooplankton	6.28	
Nekton	0.30	
Pleuroncodes-Pelagic	9.10	
Pleuroncodes-Benthic	1.68	
Sediments	2.86	
Total	20.22	17.5**

* From Walsh *et al.* (1974).
** From Whitledge and Conway (1977).

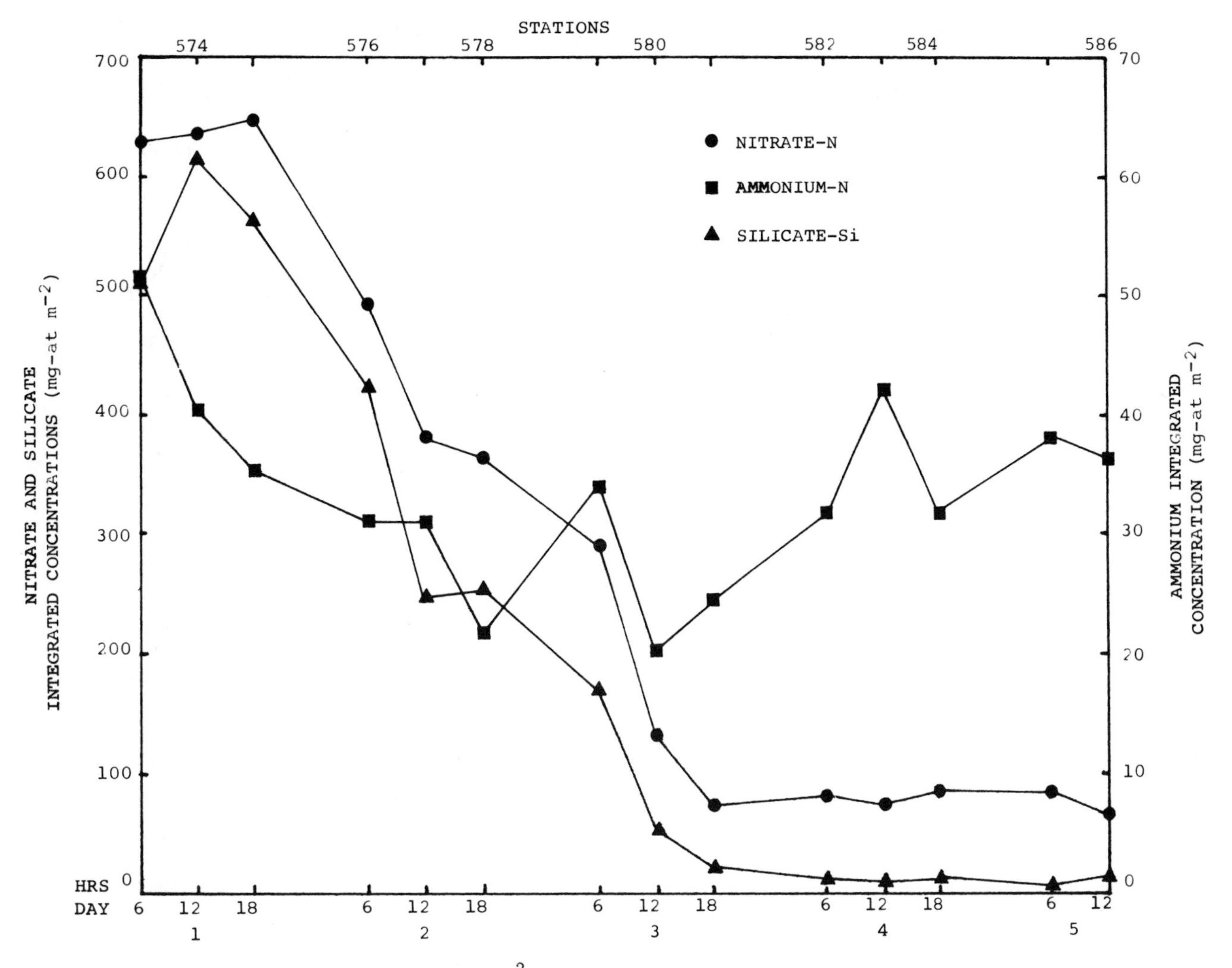

Fig. 5. Integrated concentrations (mg-at/m^2) of nitrate, silicate, and ammonium in the euphotic zone along drogue track of *Anton Bruun* 15 near 15°S off Peru in April, 1966.

The April budget had a larger zooplankton regeneration rate (6.28 mg-at N $m^{-2}d^{-1}$) than March because the biomass had increased almost fivefold. The pelagic red crab regeneration decreased to 9.1 and the benthic estimate remained constant at 1.68 for a total of 10.78 mg-at NH_4-N released $m^{-2}d^{-1}$ by all of the *P. planipes*. The decrease in the red crab regeneration rate in April was due to the loss of red crab biomass, possibly to foraging tuna. The anchovy biomass and nitrogen release from the sediments remained the same as in March. A total of 20.22 mg-at NH_4-N $m^{-2}d^{-1}$ was regenerated each day. The phytoplankton uptake of ammonium at the time was estimated to be 17.5 mg-at $m^{-2}d^{-1}$ (Whitledge and Conway, 1977), resulting in an excess of 16% of ammonium produced.

Discussion

Nutrient regeneration in a stable environment such as the euphotic zone in an oceanic gyre is vital to maintaining productivity (Eppley, Renger, Venrick, and Mullin, 1973). The regenerated nutrients are thought to cycle through the nutrient-phytoplankton-zooplankton chain several times befor they are lost from the upper mixed layer in the form of sinking or swimming particles. The regeneration lengthens the time and possibly the spatial scales of the original nutrient input. The same processes occur in upwelling areas even though strong upwelling brings enormous quantities of new nutrients into the euphotic zone. At what space and time scale would all of the upwelled nutrients be stripped from the water column? Observations on the *Anton Bruun* cruise 15 to Peru in 1966 showed that in as little distance as 40 km (3.5 days) essentially all upwelled nutrients were converted to plant material (Ryther *et al.*, 1971). Thus the regeneration of nutrients in an upwelling area is an important process that can extend the upwelling effect to longer

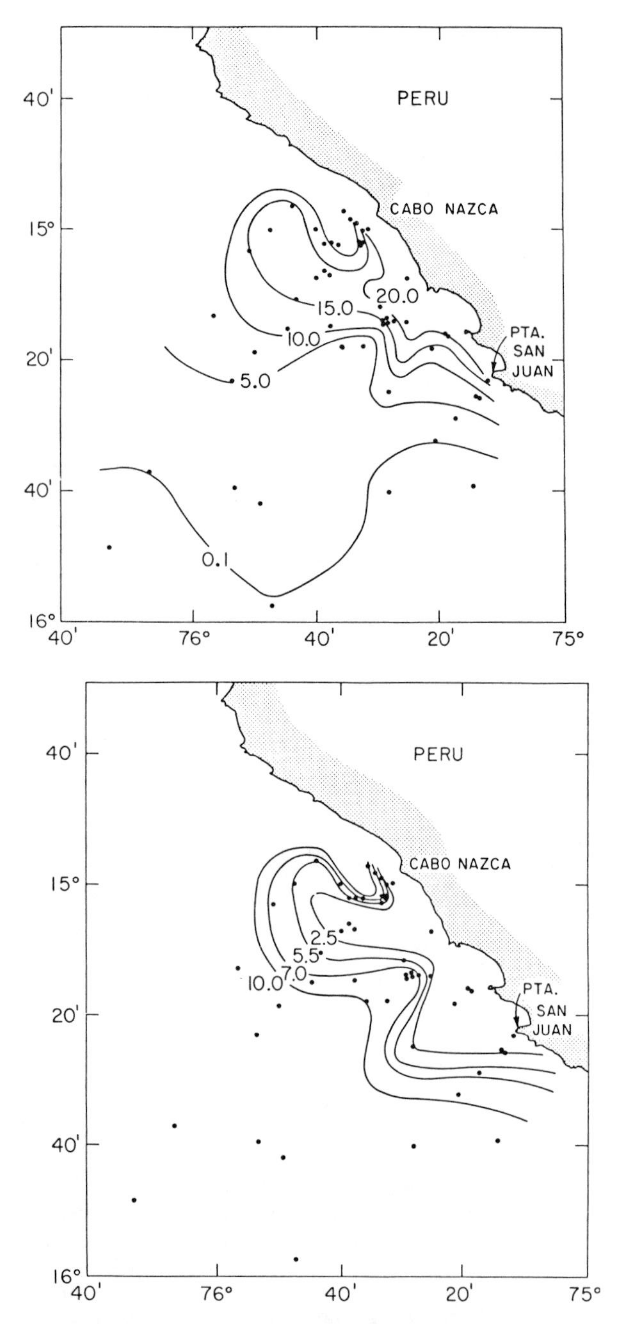

Fig. 6. a) Nitrate concentrations (μg-at/l) at 3 m and b) chlorophyll concentration (μg/l) integrated from 100 to 10% light depths on stations near 15°S off Peru in April and May, 1969 (Walsh, Kelley, Dugdale, and Frost, 1971).

spatial and temporal scales. This was demonstrated on the *Anton Bruun* cruise when a drogue was placed near an upwelling center and was followed from the high nitrate-low chlorophyll water to low nitrate-high chlorophyll water (Fig. 5). The ammonium increased during the latter part of the drogue study. The increase apparently occurred because dissolved silicon was depleted and the predominantly diatom population became silicon limited (Dugdale and Goering, 1970). On most occasions when silicon is not limiting, high concentrations of ammonium are not observed because it is used as rapidly as it is recycled.

As the newly upwelled water enters the euphotic zone the principal form of nitrogen in the water is nitrate (Fig. 6a), which the phytoplankton start to utilize to produce more phytoplankton biomass. As the newly upwelled water enters the euphotic zone there is usually a strong advective flow alongshore (often equatorward) and a smaller advective flow offshore. This flow of nutrients and phytoplankton gives the appearance of a plume of chlorophyll (Walsh, Kelley, Dugdale, and Frost, 1971) if the origin is a point source (Fig. 6b). The plume is in effect defined by the isopleths of nutrient concentrations as they decrease or phytoplankton as they increase while they are advected. The largest zooplankton biomass appears about 20 km offshore as shown by the net samples (Table 5) and a vertically integrating zooplankton pump (Walsh *et al.*, 1980). The maximum in zooplankton biomass is near the fringes of the plume, although some zooplankton are in the freshly upwelled water and they increase in numbers (along with the chlorophyll content) until the nutrients are reduced to low levels. The outer boundary of the plume is the region where regeneration is important in maintaining phytoplankton growth because upwelled nutrients are depleted and only regenerated nutrients remain. Eventually the ecosystem loses all of the inorganic nitrogen because a fraction is lost at each step in the food chain as a result of biological inefficiencies or loss by sinking and swimming particles.

Simulation models constructed to mimic nutrient-phytoplankton-zooplankton interactions in various ecosystems (Dugdale and Whitledge, 1970; Walsh and Dugdale, 1971; Walsh, 1975, 1977) have incorporated ammonium uptake as an integral part of the nutrient-phytoplankton relationship. If ammonium is omitted and phytoplankton growth is simulated using only nitrate, the resulting standing crop of phytoplankton never reaches observed concentrations. Thus even in areas with small concentrations of ammonium, phytoplankton growth is enhanced by ammonium uptake.

In terms of the percentage of phytoplankton production that utilizes regenerated nitrogen, upwelling areas have rates comparable to most other coastal and estuarine systems that do not have appreciable upwelling. The upwelling areas discussed previously have 42 to 72% of total nitrogen productivity being supplied by regeneration (Table 10). The range of values implies that regeneration must be a rather stable and constant feature of upwelling ecosystems. The upwelling areas investigated are highly productive and rank among the highest in terms of fish production.

TABLE 10. Percent Regenerated Productivity in Marine Ecosystems and Percentage of Nitrogen Uptake Originating from Different Animal Groups.

	Regenerated Prod (% of Total Prod)	Zoo-plankton	Nekton	Micro-Nekton	Demersal Fish	Benthos & Sediment	Refer-ence
Upwelling Areas							
NW Africa	72	33	15	-	-	24	Herbland,
Cape Blanc Shelf	61	24	29	-	-	8	LeBorgne,
Cape Blanc Outer Shelf	15	15	-	-	-	-	and Voi-
Cape Timiris Shelf							turiez,'73
Peru							
Shelf	56	19	18	-	-	19	
Off Shelf	42	23	5	-	1	13	
Baja California							
Shelf-March	47	3	1	32	-	11	
Shelf-April	52	16	1	23	-	12	
Coastal Areas							
NE United States							
Narragansett Bay	64	64	-	-	-	-	Martin,'68
Long Island Sound	55-75	43-66	-	-	-	9-12	Harris,'59
NW United States							
Shelf	36	36	-	-	-	-	Jawed,'73
Columbia River Plume	90	90	-	-	-	-	Jawed,'73
Open Ocean							
N. Pacific Gyre	44-83	44-83	-	-	-	-	Eppley *et al.*, '73

Zooplankton may recycle the major fraction of nitrogen in upwelling ecosystems, although zooplankton were not so important in the Peru upwelling ecosystem when the anchoveta stocks were large (Beers, Stevenson, Eppley, and Brooks, 1971; Whitledge and Packard, 1971; Dugdale, 1972). Areas with organically-rich sediments or large numbers of micronekton or benthos also reduce the relative importance of zooplankton regeneration. The importance of zooplankton varies somewhat between the different regions, a useful clue to pathways of material flow through the different upwelling ecosystems. The largest percentage of primary production due to zooplankton nitrogen recycling occurred in the Cape Blanc northwest Africa ecosystem and was slightly smaller off Peru, Cape Timiris off northwest Africa, and Baja California, with a range for the three systems from 3 to 33%. The larger values occurred after the onset of upwelling had been completed for each of the areas while the low value in March for Baja California was observed as the seasonal upwelling was starting. The time lag may be likened somewhat to the spatial lag that produces a plume downstream from an active upwelling center. So after upwelling conditions have stabilized it appears that 20 to 30% of primary productivity is driven by ammonium recycled by zooplankton (Table 10).

For the nearshore summer upwelling areas off Washington and Oregon, Jawed (1973) concluded that ammonium regeneration was not a crucial nitrogen source in the upwelled waters. However no nitrogen uptake rate studies were undertaken so the actual phytoplankton uptake of ammonium could have been substantial because large quantities of ambient ammonium were observed in the water column. Evidently it was presumed that large concentrations of nitrate reduced the importance of ammonium in the ocean.

Of the three upwelling systems, the regeneration of ammonium by nekton is most important in the two areas where commercial quantities of small pelagic nekton exist. Since the decline in the anchoveta biomass, Peru has regeneration rates similar to northwest Africa while the clupeoid derived rates are small off Baja California. Micronekton are important in the Baja California

TABLE 11. Zooplankton and Nekton Ingestion Rates and Percent of Daily Phytoplankton Grazed. Units are g dry wt/m^2 for biomass and mg-at $m^{-2}d^{-1}$ for rates.

	Northwest Africa	Peru	Baja California
Zooplankton			
Zooplankton biomass	3.78	4.41	6.38
Zooplankton growth	0.24	0.28	0.40
Zooplankton excretion	7.82	4.11	6.28
Zooplankton assimilation	8.06	4.39	6.68
Zooplankton ingestion	11.51	6.27	9.54
Nekton			
Nekton biomass	10.55	31.80	7.50
Nekton growth	0.10	0.29	0.47*
Nekton excretion	3.48	4.61	9.40
Nekton assimilation	3.58	4.90	9.87
Nekton ingestion	5.11	7.00	14.10
Total ingestion	16.62	13.27	23.64
Phytoplankton growth-N	23.5	21.7	39.5
% daily phytoplankton-N production grazed	71	61	60

* Calculated using *Pleuroncodes planipes* rates (Blackburn, 1977).

ecosystem because the red crab has taken the herbivore-omnivore role that the clupeoids occupy in the other two upwelling areas. Demersal fish play an insignificant role in the recycling of nutrients according to the best data, but a thorough examination has not been undertaken for any area with large benthic fish populations.

The regeneration of nitrogen from sediments appears to be most important off northwest Africa and Peru although the rates are only slightly smaller for Baja California. Large amounts of organic material are thought to sink to the bottom off Peru, where the sediments have a high organic carbon content. A substantial quantity of organic material may also be lost by sinking off Baja California; however, it may be the principal food source for the older, benthic red crabs.

All protozoa, ciliates, and nauplii were excluded from the mesoplankton studied. There are indications that microplankton may be important in upwelling ecosystems, although there are differences of opinion. Energy budgets for equatorial upwelling regions show that heterotrophic respiration constituted 90% of the total (Sorokin, 1979) and 30 to 90% of the total in coastal upwelling regions (Sorokin, Korsak, Mamaeva, and Kogelschatz, 1980). Setchell and Packard (1979) reported respiration values 10 to 50% as large for the same stations. So the microplankton are a potentially important component that the sampling may have missed.

The importance of open ocean regeneration was studied in the North Pacific gyre where zooplankton was found to produce 44 to 83% of the nitrogen requirements of phytoplankton (Eppley *et al.*, 1973). The methods were comparable to those employed in the upwelling regions. The investigators measured all the dissolved forms of inorganic nitrogen plus urea, phytoplankton biomass, the uptake of ammonium, nitrate, and urea using ^{15}N techniques, zooplankton biomass, and excretion rates. The turnover time for ammonium was 3 to 5 days and the phytoplankton crop was small, so it was concluded that grazing was important in keeping the phytoplankton below maximum possible levels. The remaining 17 to 56% of the nitrogen needed to fulfill phytoplankton requirements was supposed to have come from microzooplankton smaller than 183 μm or from vertical turbulent diffusion. Urea was also considered to be an important nitrogen source.

The regeneration of relatively large quantities of nitrogen by zooplankton and nekton populations implies that these organisms are major consumers of primary (and secondary) productivity. The excretion rates measured for zooplankton and nekton can be used to estimate their ingestion rates if growth and excretion are assumed to be the two main pathways of assimilated nitrogen. An earlier estimate was that in 1969 the Peruvian anchoveta consumed 54 to 61% of the daily phytoplankton nitrogen productivity in the region (Whitledge, 1978). Zooplankton ingestion was not estimated at that time because few biomass or excretion rates were available. The assimilation efficiency of both zooplankton and nekton used in calculations of ingestion (Table 11) is 70%. The growth rates were obtained from literature values of estimated secondary productivity in each of the upwelling areas or from laboratory rates of similar species. The combined zooplankton and nekton ingestion of primary productivity ranged from 60% in Baja California to 71% off northwest Africa, while the Peru (April, 1977) value was 61%. The absolute ingestion rates were quite different from the percentage of productivity grazed. The smallest ingestion rate was estimated to be 13.27 mg-at N $m^{-2}d^{-1}$ for Peru with 53% coming from nekton; northwest Africa was 16.62 mg-at N $m^{-2}d^{-1}$ with 31% of ingestion by nekton. Baja California had a value of 23.64 mg-at N $m^{-2}d^{-1}$ grazed with nekton and micronekton (red crabs) consuming 60% of that amount. So the zooplankton off Peru had ingestion rates of 54 to 66% of the rates in the two other systems, and Peruvian nekton ingestion rates were 50 to 137% of the rates for Baja California and northwest Africa. The values reflect the low primary productivity estimated on the C line off Peru in 1977 and the relatively lower biomass of zooplankton and nekton than in the other upwelling systems. In the zooplankton in the Peru upwelling system no more than 50% of the zooplankton were filter feeders at any location (Flint and Timonin, 1980). In 1977 more of the organic matter produced may have sunk and been incorporated in benthic processes or decomposed, depleting oxygen concentrations to lower than normal values (Dugdale *et al.*, 1977). In 1978 it was estimated that 12% of the material produced by primary production sank to the bottom (Vinogradov, Shushkina, and Lebedeva, 1980); the lowest concentration of oxygen observed was 0.12 ml/l (Bordovsky, Gasarova, Domanov, and Stunghas, 1980). The 1976 observations of anoxic conditions probably were the result of oscillatory current flows and the increased sinking of organic matter, which would probably have to have been larger than 12% of the primary production. J.J. Walsh (personal communication) has estimated the input to have been as much as 50% of the primary production since the anchoveta stocks have declined.

Acknowledgements. This research was supported by National Science Foundation Grant OCE78-05737 as a component of the United States IDOE Coastal Upwelling Ecosystems Analysis (CUEA) program. The analysis was also partially supported by the United States Department of Energy under Contract No. DE-AC02-76CH00016.

References

Armstrong, F.A., M.P.M. Williams, and J.D.H. Strickland, Photo-oxidation of organic matter in seawater by ultraviolet radiation, *Nature, 211*, 481-483, 1966.

Barber, R.T., The JOINT-I expedition of the coastal upwelling ecosystems analysis program, *Deep-Sea Research, 24*, 1-6, 1977.

Beers, J.R., M.R. Stevenson, R.W. Eppley, and E.R. Brooks, Plankton populations and upwelling off the coast of Peru, June 1969, *Fishery Bulletin, 69*, 859-876, 1971.

Blackburn, M., Temporal changes in pelagic biomass of *Pleuroncodes planipes* Stimson (Anomura, Golatheidae) off Baja California, Mexico, *Crustaceana, 23*, 178-184, 1977.

Blackburn, M., Zooplankton in an upwelling area off northwest Africa: composition, distribution, and ecology, *Deep-Sea Research, 26*, 41-56, 1979.

Blackburn, M. and W. Nellen, Distribution and ecology of pelagic fishes studied from eggs and larvae in an upwelling area off Spanish Sahara, *Fishery Bulletin, 74*, 885-895, 1976.

Blackburn, M. and R. Thorne, Composition, biomass, and distribution of pelagic nekton in a coastal upwelling area off Baja California, Mexico, *Tethys, 6*, 28.-290, 1974.

Bordovsky, O.K., A.N. Gusarova, M.M. Domanov, and P.A. Stunzhas, Hydrochemical conditions in the coastal region off the coast of Peru during an *El Aguaje*, in: *Productivity of Upwelling Ecosystems*, R.T. Barber and M.E. Vinogradov (eds.), Elsevier, Amsterdam, 380 pp., 1980.

Boyd, C.M., The biology of a marine decapod crustacean, *Pleuroncodes planipes* Stimpson, 1860, Ph.D. Thesis, University of California, San Diego, 123 pp., 1962.

Codispoti, L.A., and G.E. Friederich, Local and mesoscale influences on nutrient variability in the northwest African upwelling region near Cabo Corbeiro, *Deep-Sea Research, 25*, 751-770, 1978.

Coste, B., and G. Slawyk, Structures de repartitions superficielles des sels nutritifs dans une d'upwelling (Cap Corveiro, Sahara Espagnol), *Tethys, 6*, 123-132, 1974.

Dagg, M., T. Cowles, T. Whitledge, S. Smith, S. Howe, and D. Judkins, Grazing and excretion by zooplankton in the Peru upwelling system during April 1977, *Deep-Sea Research, 27*, 43-59, 1980.

Dugdale, R.C., Chemical oceanography and primary productivity in upwelling regions, *Geoforum, 11*, 47-61, 1972.

Dugdale, R.C., and J.J. Goering, Uptake of new and regenerated forms of nitrogen in primary productivity, *Limnology and Oceanography, 12*, 196-206, 1967.

Dugdale, R.C., and J.J. Goering, Nutrient limitation and the path of nitrogen in Peru current production, *Anton Bruun Report No. 4*, Texas A&M Press, pp. 5.3-5.8, 1970.

Dugdale, R.C., and T. Whitledge, Computer simulation of phytoplankton growth near a marine sewage outfall, *Revue Internationale d'Oceanographie Medicale, 17*, 201-210, 1970.

Dugdale, R.C., J.J. Goering, R.T. Barber, R.L. Smith, and T.T. Packard, Denitrification and hydrogen sulfide in the Peru upwelling region during 1976, *Deep-Sea Research, 24*, 601-608, 1977.

Eppley, R.W., E.H. Renger, E.L. Venrick, and M.M. Mullin, A study of plankton dynamics and nutrient cycling in the central gyre of the north Pacific ocean, *Limnology and Oceanography, 18*, 534-551, 1973.

Flint, M.V., and A.G. Timonin, Trophic structure of mesoplankton in the northern part of the Peruvian coastal region, in: *Productivity of Upwelling Ecosystems*, R.T. Barber and M.E. Vinogradov (eds.), Elsevier, Amsterdam, 380 pp., 1980.

Geynrikh, A.K., Horizontal distribution of copepods in the Peru current regions, *Oceanology, 13*, 94-103, 1973.

Gulland, J.A., The fish resources of the ocean, *FAO Fisheries Technical Paper No. 97*, Rome, 425 pp., 1970.

Haedrich, R.L., M. Blackburn, and J. Brulhet, Distribution and biomass of trawl-caught animals off Spanish Sahara, West Africa, *Matsya, 2*, 38-65, 1976.

Harris, E., The nitrogen cycle in Long Island Sound, *Bulletin of the Bingham Oceanographic Collection*, Yale University, *17*, 31-65, 1959.

Herbland, A., R. LeBorgne, and B. Voituriez, Production primaire et secondaire et régéneration des sels nutritifs dans l'upwelling de Mauritanie, *Centre de Recherches Oceanographiques Abidjan, Documents Scientifiques, 4*, 1-75, 1973.

Huntsman, S.A., and R.T. Barber, Primary production off northwest Africa: the relationship to wind and nutrient conditions, *Deep-Sea Research, 24*, 25-33, 1977.

Jawed, M., Ammonia excretion by zooplankton and its significance to primary productivity during summer, *Marine Biology, 23*, 115-120, 1973.

Judkins, D.C., Vertical distribution of zooplankton in relation to the oxygen minimum off Peru, *Deep-Sea Research, 27*, 475-487, 1980.

Lee, K.C., O.A. Mathisen, and R.E. Thorne, Acoustic observations on the distribution of nekton off the coast of Peru, *Coastal Upwelling Ecosystems Analysis Data Report 59*, Fisheries Research Institute, University of Washington, Seattle, 84 pp., 1979.

Martin, J.H., Phytoplankton-zooplankton relationships in Narragansett Bay. III. Seasonal changes in zooplankton excretion rates in relation to phytoplankton abundance, *Limnology and Oceanography, 13*, 63-71, 1968.

McCarthy, J.J., The uptake of urea by natural populations of marine phytoplankton, *Limnology and Oceanography, 17*, 738-748, 1972.

McCarthy, J.J., and T.E. Whitledge, Nitrogen excretion by anchovy (*Engraulis mordax* and *E. ringens*) and jack mackerel (*Trachurus symmetricus*), *Fisheries Bulletin, 70*, 395-401, 1972.

Packard, T.T., Respiration and respiratory electron transport activity in plankton from the northwest African upwelling area, *Journal of Marine Research, 37*, 711-742, 1979.

Rowe, G.T., C.H. Clifford, and K.L. Smith, Jr., Nutrient regeneration in sediments off Cape Blanc, Spanish Sahara, *Deep-Sea Research, 24*, 57-63, 1977.

Ryther, J.H., D.W. Menzel, E.M. Hulburt, C.J. Lorenzen, and N. Corwin, The production and utilization of organic matter in the Peru coastal current, *Investigacion Pesquera, 35*, 43-59, 1971.

Setchell, F.W., and T.T. Packard, Phytoplankton respiration in the Peru upwelling, *Journal of Plankton Research, 1*, 343-354, 1979.

Smith, K.L., Jr., G.R. Harbison, G.T. Rowe, and C.H. Clifford, Respiration and chemical composition of *Pleuroncodes planipes* (Decapoda: Galatheidae): Energetic significance in an upwelling system, *Journal of the Fisheries Research Board of Canada, 32*, 1607-1612, 1975.

Smith, S.L., Nutrient regeneration by zooplankton during a red tide off Peru, with notes on biomass and species composition of zooplankton, *Marine Biology, 49*, 125-132, 1978.

Smith, S.L., and T.E. Whitledge, The role of zooplankton in the regeneration of nitrogen in a coastal upwelling system off northwest Africa, *Deep-Sea Research, 24*, 49-56, 1977.

Smith, S.L., and T.E. Whitledge, Regeneration of nutrients by zooplankton and fish off northwest Africa, *Rapport et procès-verbaux des rêunions Conseil permanent pour l'exploration de la Mer*, in press.

Sorokin, Y.I., Decomposition of organic matter and nutrient regeneration, in: *Marine Ecology, Vol. 4*, O. Kinne (ed.), John Wiley, New York, pp. 501-616, 1979.

Sorokin, Y.I., M.N. Kasak, T.I. Mamaeva, J.E. Kogelschatz, Primary production and microbial activity in the Peruvian upwelling area, in: *Productivity of Upwelling Ecosystems*, R.T. Barber and M.E. Vinogradov (eds.), Elsevier, Amsterdam, 380 pp., 1980.

Thorne, R.E., O.A. Mathisen, R.J. Trumble, and M. Blackburn, Distribution and abundance of pelagic fish off Spanish Sahara during CUEA expedition JOINT-I, *Deep-Sea Research, 24*, 75-82, 1977.

Vinogradov, M.E., E.A. Shushkina, and L.P. Lebedeva, Production characteristics of plankton communities in coastal waters of Peru, in: *Productivity of Upwelling Ecosystems*, R.T. Barber and M.E. Vinogradov (eds.), Elsevier, Amsterdam, 380 pp., 1980.

Walsh, J.J., A spatial simulation model of the Peru upwelling ecosystem, *Deep-Sea Research, 22*, 201-236, 1975.

Walsh, J.J., A biological sketchbook for an eastern boundary current, in: *The Sea: Ideas and Observations on Progress in the Study of the Seas*, E.D. Goldberg (ed.), John Wiley, p. 923-968, 1977.

Walsh, J.J., and R.C. Dugdale, A simulation model of the nitrogen flow in the Peruvian upwelling system, *Investigacion Pesquera, 35*, 309-330, 1971.

Walsh, J.J., J.C. Kelley, R.C. Dugdale, and B.W. Frost, Gross features of the Peruvian upwelling system with special reference to possible diel variation, *Investigacion Pesquera, 35*, 25-42, 1971.

Walsh, J.J., J.C. Kelley, T.E. Whitledge, J.J. MacIsaac, and S.A. Huntsman, Spin-up of the Baja California upwelling ecosystem, *Limnology and Oceanography, 19*, 553-572, 1974.

Walsh, J.J., T.E. Whitledge, J.C. Kelley, S.A. Huntsman, and R.D. Pillsbury, Further transition states of the Baja California upwelling ecosystem, *Limnology and Oceanography, 22*, 264-280, 1977.

Walsh, J.J., T.E. Whitledge, W.E. Esaias, R.L. Smith, S.A. Huntsman, H. Santander, and B.R. de Mendiola, The spawning habitat of the Peruvian anchovy, *Engraulis ringens, Deep-Sea Research, 27*, 1-27, 1980.

Watson, S.W., Role of bacteria in an upwelling ecosystem, in: *Upwelling Ecosystems*, R. Boje and M. Tomczak (eds.), Springer-Verlag, Berlin, p. 139-154, 1978.

Whitledge, T.E., and R.C. Dugdale, Creatine in seawater, *Limnology and Oceanography, 17*, 309-314, 1972.

Whitledge, T.E., Regeneration of nitrogen by zooplankton and fish in the northwest Africa and Peru upwelling ecosystems, in: *Upwelling Ecosystems*, R. Boje and M. Tomczak (eds.), Springer-Verlag, Berlin, p. 90-100, 1978.

Whitledge, T.E., and H.L. Conway, R/V *T.G. Thompson* cruise 78 (Mescal II-Outfall II) Part II: Productivity, micronekton biomass, current meters and drogue velocities, *Coastal Upwelling Ecosystems Analysis Data Report 37*, 143 pp., 1977.

Whitledge, T.E., and T.T. Packard, Nutrient excretion by anchovies and zooplankton in Pacific upwelling regions, *Investigacion Pesquera, 35*, 243-250, 1971.

Wood, J.D., Nitrogen excretion in some marine teleosts, *Canadian Journal of Biochemistry and Physiology, 36*, 1237-1242, 1958.

WATER EXCHANGE AND THE BALANCE OF PHOSPHATE IN THE GULF OF CARIACO, VENEZUELA

Taizo Okuda

Instituto Oceanografico, Universidad de Oriente
Apartado 94, Cumana, Venezuela

Abstract. The Gulf of Cariaco, east of the Cariaco Basin, is a semi-enclosed basin hydrographically affected mainly by the wind regime of northeastern Venezuela. During periods of stagnation (periods of weak winds from the east-southeast), a strong pycnocline develops in the 40 to 60-m layer. This leads to anoxic conditions and the extraordinary accumulation of phosphate (6 µg-at./l) in the deep waters. Above the pycnocline there is a depletion of phosphate caused by uptake by primary producers. During periods of renewal (strong winds from the east-northeast) the stratification is weakened and vertical mixing takes place. On the basis of observations during 1972 to 1975, the change of phosphate in relation to the stagnation and renewal has been quantified. During the changes from deep-water stagnation to its renewal, the average net increase of phosphate-phosphorus in the layers above 50 m was 187 tons and the average net decrease in the layers below 40 m was estimated to be 482 tons. During the advent of stagnation, phosphate-phosphorus decreased an average of 224 tons in the upper layers and increased by 468 tons in the lower layers.

Introduction

Hydrographic conditions along the northeastern Caribbean coast of Venezuela are influenced principally by variations in upwelling, which are intimately associated with the wind regime. Since the first publication on the hydrochemical conditions in the Gulf of Cariaco by Richards (1960), which was based on observations carried out on board the R.V. *Atlantis*, the circulation and biochemical characteristics of the gulf have been studied by many workers (Gade, 1961a, 1961b; Kato, 1961; Griffiths and Simpson, 1967; Simpson and Griffiths, 1967; Benitez-Alvarez, 1974; Okuda, Benitez-Alvarez, Bonilla, and Cedeno, 1978a; Okuda, Garcia, Gamboa, and Fernandez, 1978b).

The present paper is a quantitative interpretation of the phosphate balance in relation to stagnation and renewal based on observations from 1972 to 1975 by personnel of the Instituto Oceanografico, Universidad de Oriente, Cumana, Venezuela.

Gulf of Cariaco

The Gulf of Cariaco (Fig. 1) joins the Cariaco Basin through an entrance about 5.5 km wide. The gulf is about 62 km long in the east-west direction and has a maximum width of 15 km in the north-south direction. The gulf has a surface area of 642 km^2 and its volume is estimated to be about 31.5 km^3 (Fig. 2). The average depth is about 50 m; the central basin is about 90 m deep, i.e., deeper than the entrance (60 to 70 m) The surface waters (0 to 30 m) represent about 55% of total volume, while the waters below a depth of 60 m contain only about 9% of the total volume (Fig. 2).

The Gulf of Cariaco is a semi-enclosed basin in an arid region where evaporation far exceeds precipitation and runoff. The annual precipitation ranges from 220 to 570 mm, in comparison with a rate of evaporation on the order of 2,180 to 2,720 mm per year for the years 1967 to 1975 (data compiled from the Meteorological Station, Instituto Oceanografico, Universidad de Oriente). The Manzanares River empties near the entrance of the gulf and has an annual discharge of about 6 x 10^6 m^3. However, the runoff seems to have no significant influence on the hydrochemical conditions of the gulf, probably because it discharges to the northwest, slightly away from the entrance of the gulf. Some small streams along the south coast also empty into the gulf, but they have little effect on the hydrochemical conditions.

A definite periodic variation of wind speed and direction was observed from 1972 to 1975. There were weak winds from the east-southeast from August to November or December and strong E-NE winds from December or January to June or July (Fig. 3). There were also two distinct seasonal trends of air temperature (mean values for every 10 days), i.e., a range from 24.8 to 25.9°C from December to March with slightly higher temperatures, almost 26.0 to 26.8°C, from the end of 1972 to the beginning of 1973. On the other hand,

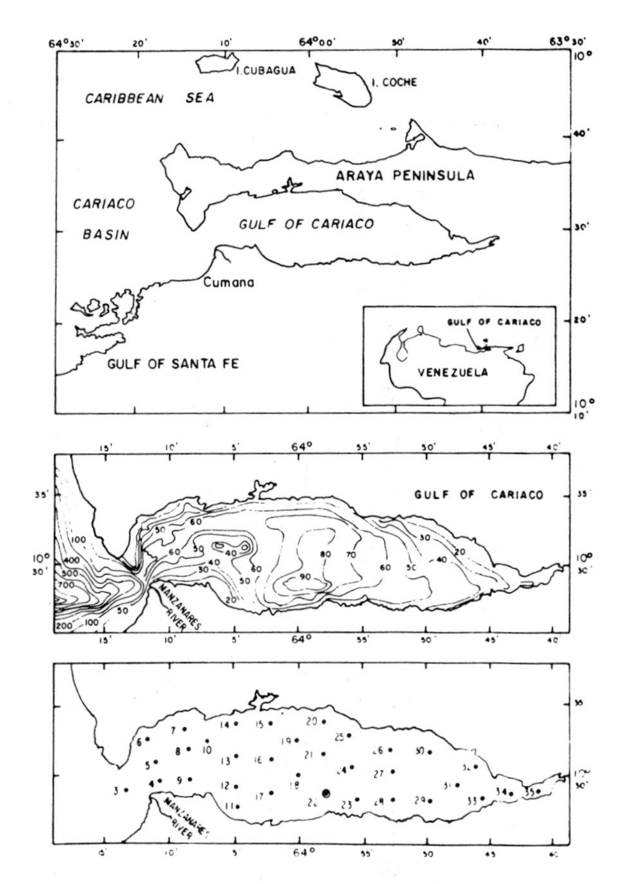

Fig. 1. Map of Gulf of Cariaco showing bathymetric contours in meters and hydrochemical stations. Weekly observations were carried out in Sta. 22 during the years 1972-1975.

temperatures were relatively high (25.8 to 27.8°C) from April to November (Fig. 4).

Hydrographic Conditions

Hydrographic and chemical conditions in the Gulf of Cariaco are affected mainly by water exchanges with the open sea, which in turn de-

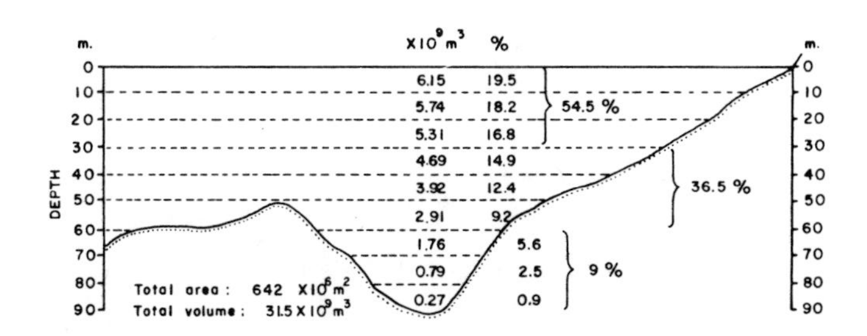

Fig. 2. Volume of different depth zones in Gulf of Cariaco.

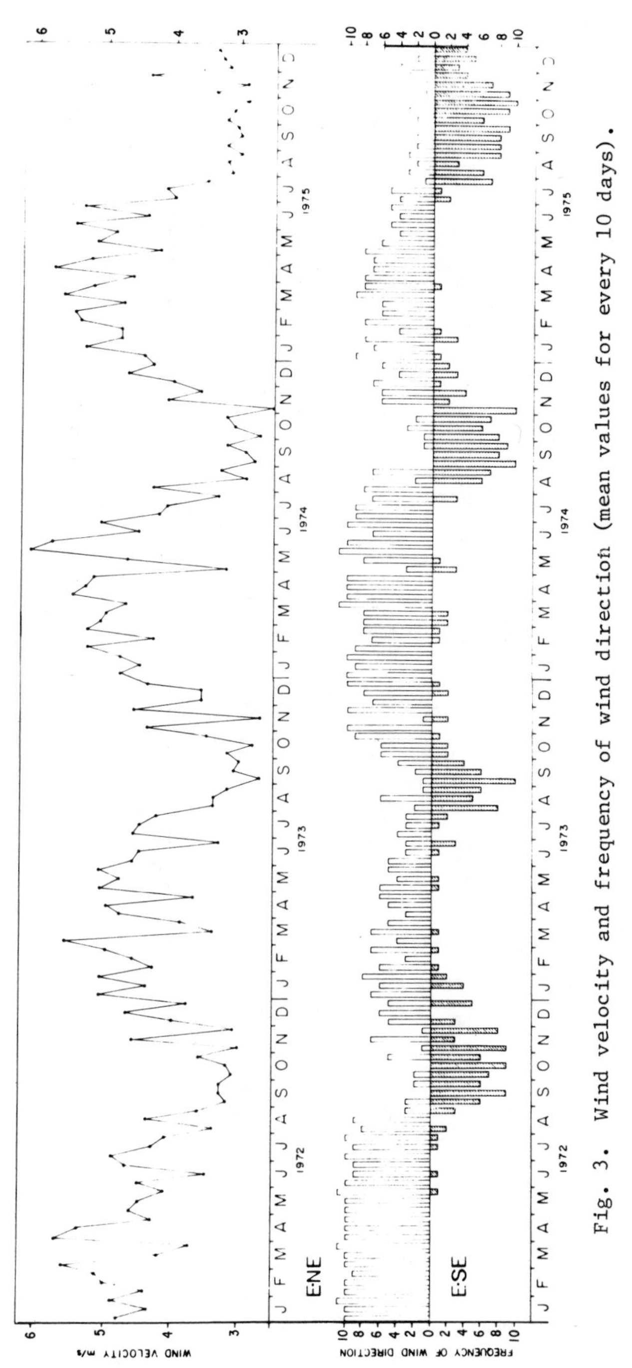

Fig. 3. Wind velocity and frequency of wind direction (mean values for every 10 days).

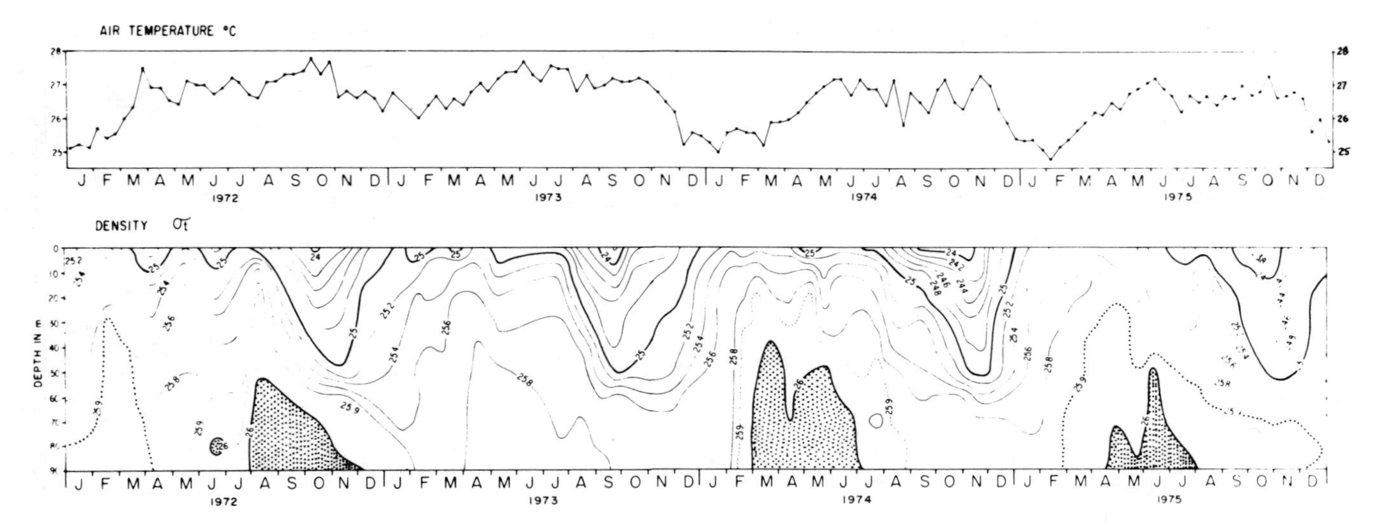

Fig. 4. Seasonal variation of air temperature (mean values for every 10 days) and water density (sigma-t).

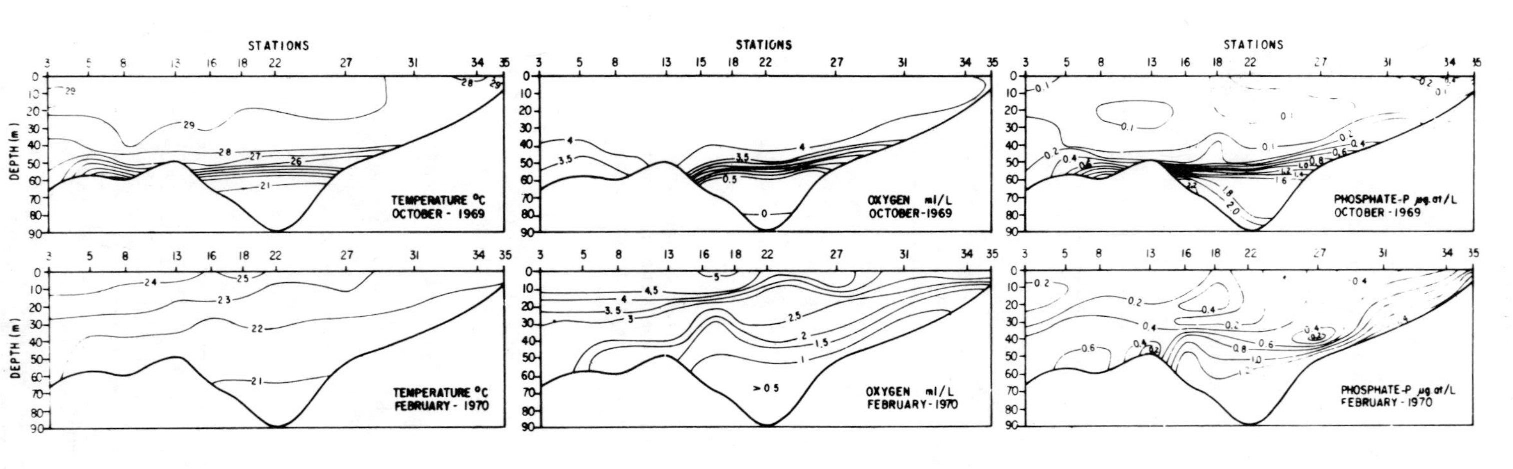

Fig. 5. Vertical distribution of temperature, oxygen, and phosphate in the east-west section.

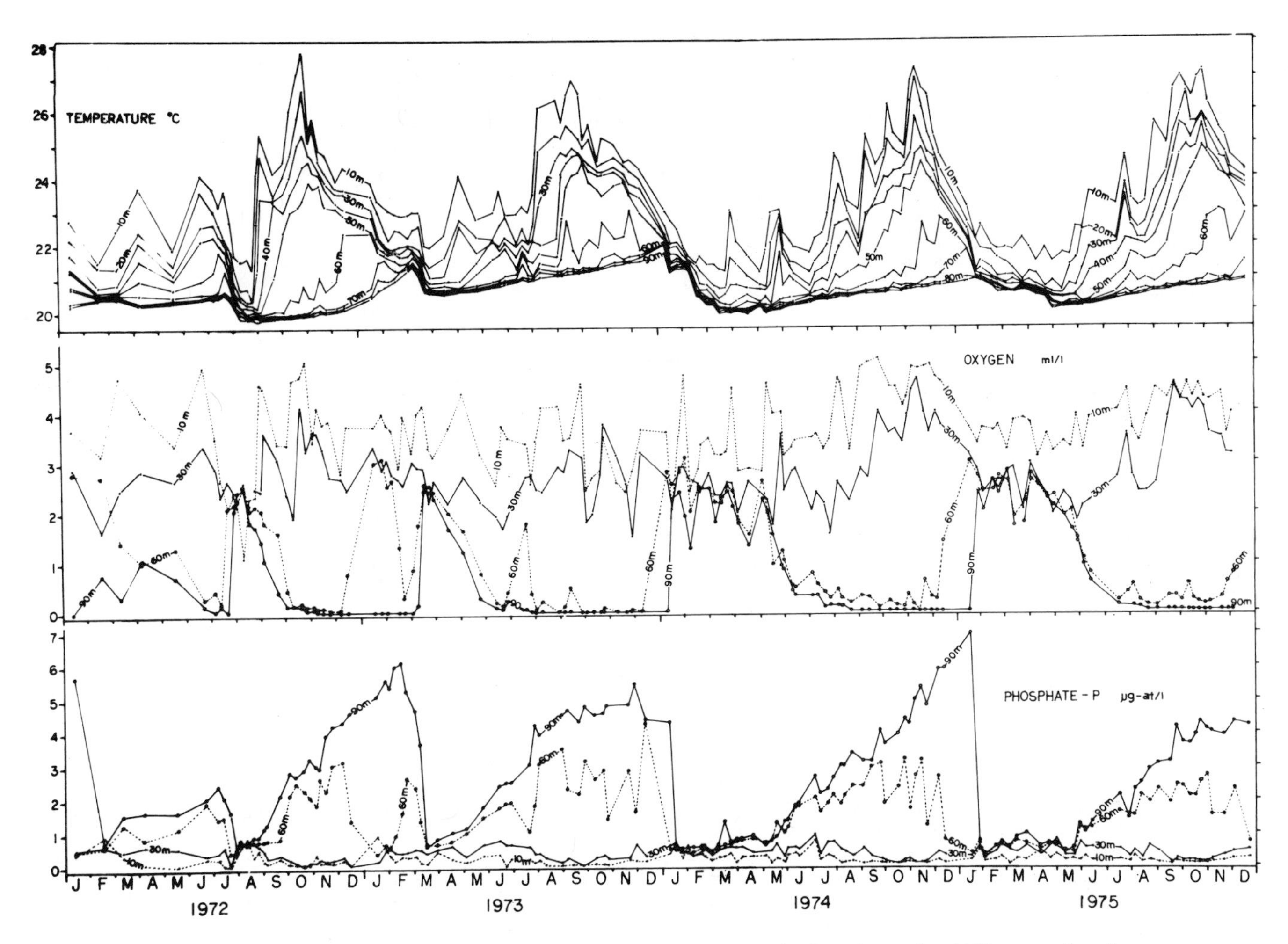

Fig. 6. Seasonal variations of temperature, oxygen, and phosphate in different levels.

pend not only on the local wind fluctuation but also on the large-scale variation of upwelling off the northeast coast of Venezuela. Prevailing winds from the east-northeast enhance the outflow of surface waters from the gulf and the inflow of the sub-surface waters from the Cariaco Basin to the intermediate and deep layers of the gulf.

Vertical and seasonal variations of salinity in the gulf are relatively less than those of temperature, and density variations depend mainly on the temperature. It is of interest to note that the air and water temperatures and water density differed in 1973 from the other years (Fig. 4). Namely, during active water renewal, sigma-t in the deep water was more than 26, except in 1973, when the density in the deep layers was less than sigma-t 26. The differences in 1973 were probably because of relatively weaker upwelling in that year.

Two markedly different hydrographic regimes exist during periods of weak winds from the east-southeast, which are unfavorable for upwelling and favor water stagnation and periods of strong winds from the east-northeast, which are favorable for upwelling and water renewal.

A pronounced pycnocline develops between 40 and 60 m that prevents vertical mixing between surface and deep waters and leads to the formation of an anoxic environment below the pycnocline. Discontinuities of the other variables, such as temperature, dissolved oxygen, and nutrients, develop near the depth of the pycnocline (Fig. 5).

Above the pycnocline, there is a significant warming and increase of dissolved oxygen because of the more restricted water exchange with the open Caribbean. Low density water (σ_t < 24.5) in the upper layers results from the warming and heavy rainfall. Okuda *et al.* (1978a) observed that the water below 80 m warmed slightly (0.7 to 0.9°C) during the period of stagnation. During the period the water column has wide ranges of temperature (almost 20 to 28°C) and of dissolved oxygen (0 to 5 ml/l). On the other hand, the variables during renewal show no such wide ranges because of vertical mixing.

It is of interest to observe (Fig. 6) that sea-

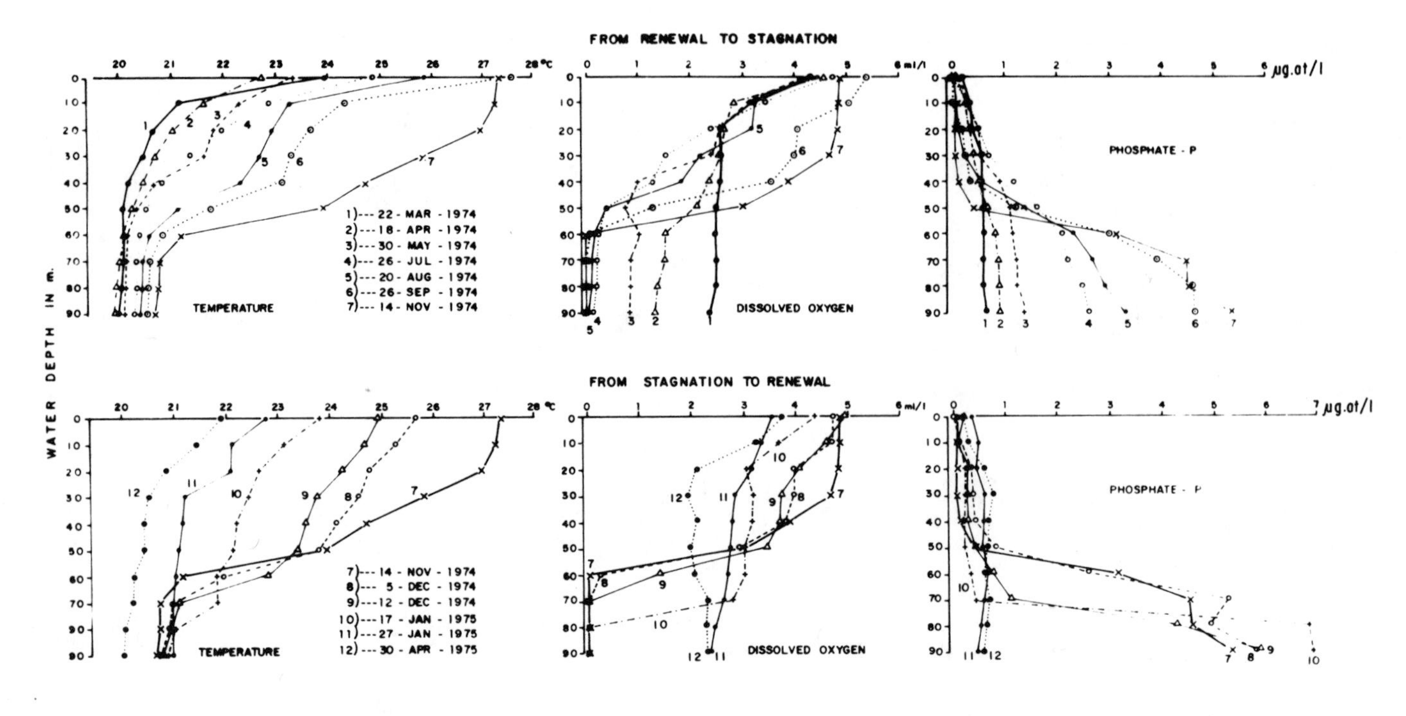

Fig. 7. Seasonal variation on the vertical distribution of temperature, oxygen, and phosphate.

sonal variations of water temperatures of the surface layers reach their maximum between September and November and then decrease gradually until the beginning of water renewal (January to February). On the other hand, temperatures at 70 to 90 m increase continuously from the beginning of the renewal period to the end of stagnation; between 60 and 90 m, decreases of dissolved oxygen and increases of phosphate are gradual during stagnation in contrast to their abrupt changes during renewal periods. The almost linear increase of phosphate in the bottom water (90 m) during stagnation is remarkable, as are the variations in dissolved oxygen and phosphate (Fig. 7).

During strong E-NE winds, the deep waters of the gulf move along the bottom towards the inner part of the gulf, following the penetration of sub-surface waters from the Cariaco Basin. The deep water then moves upwards, reaching the surface. Subsequently, with continuing E-NE winds, the upwelled water moves westward. When the winds are from a direction favorable for surface outflow, even if of low intensity, water exchange may occur within the upper layers, and the degree of water exchange depends not only on the intensity and direction of wind but also on the prevailing density gradients.

During strong winds from the east-northeast, a well-developed pycnocline that was formed during stagnation disappears as a result of water exchange. This results in relatively weak stratification of chemical variables.

Nutrient Regime

There is a great difference in the nutrient concentrations in the deep waters of the Gulf of Cariaco during periods of stagnation and of renewal. The development of a pycnocline during stagnation results in the accumulation of nutrients, with phosphate concentrations up to approximately 6 μg-at./l in the layers below the pyc-

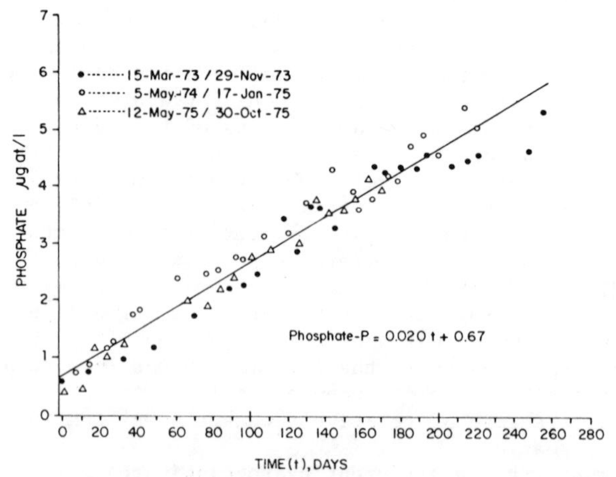

Fig. 8. Increase of phosphate-P in the deep waters during the period of water stagnation (Phosphate-P: mean values in 80-90 m levels).

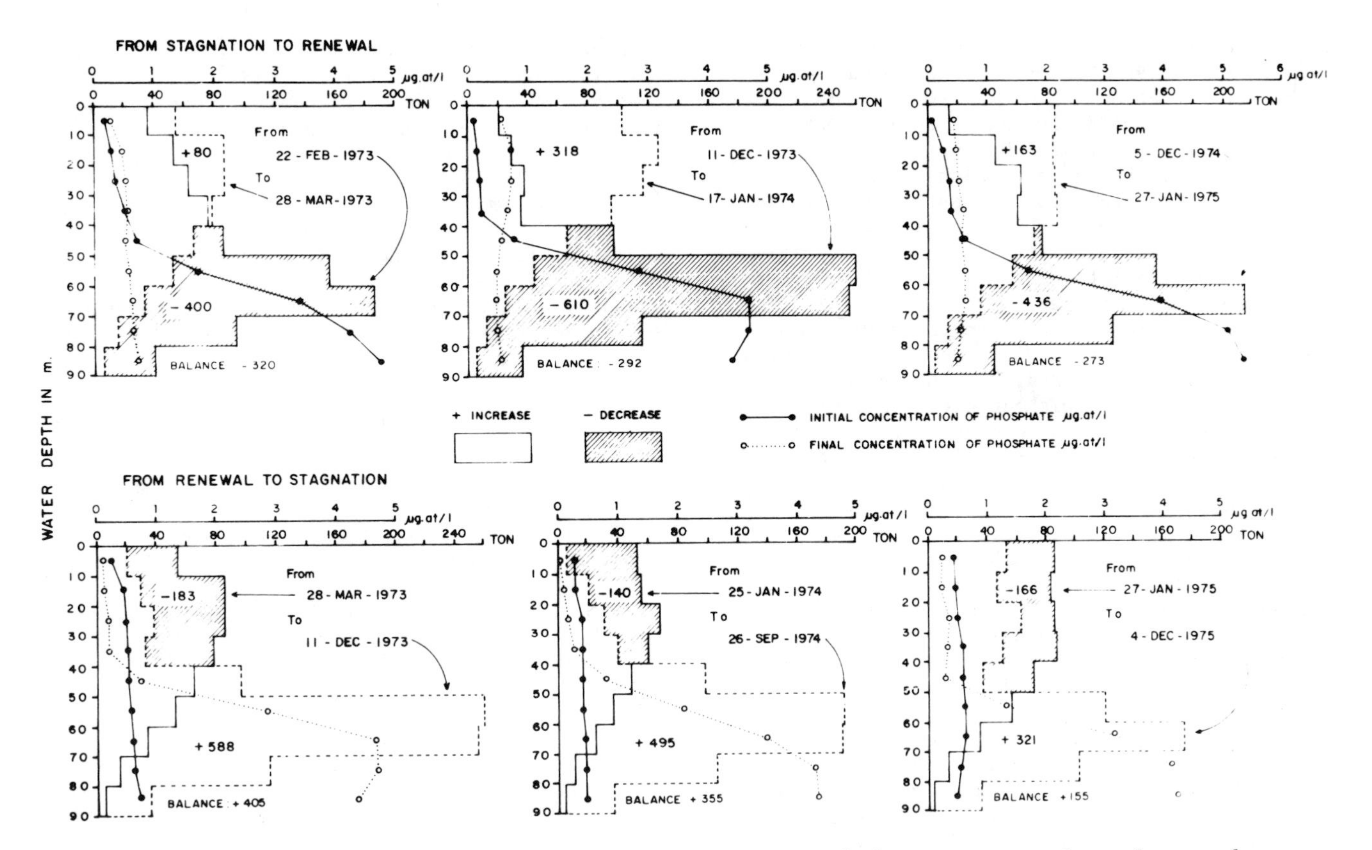

Fig. 9. Quantitative changes of phosphate during the periods between stagnation and renewal.

nocline (Figs 6 and 7). The accumulation rate of phosphate in the bottom waters below 80 m was estimated to be 0.6 mg $m^{-3}day^{-1}$ during the period of stagnation (Fig. 8). Above the pycnocline nutrients were depleted.

Okuda *et al.* (1978b) observed seasonal variations of inorganic nitrogen in the gulf from 1972 to 1975 and reported as follows: During stagnation, nitrate and nitrite were present in very low concentrations in the whole water column. There was an accumulation of ammonia in the deep waters of more than 25 µg-at./l. The concentrations of nitrate in the inflowing waters entering the deep layers of the gulf were 8 to 10 µg-at./l at the beginning of water renewal (January - March), then increased to about 16 µg-at./l with the development of stagnation in the deep water. During the development of stagnation, the concentration of nitrate suddenly decreased to almost nil (in July), and the oxygen content decreased to 0.2 to 0.5 ml/l. Such conditions continued until the next period of deep water renewal. High nitrite concentrations (up to 6 µg-at./l) were observed just before the increase (in February) and just after the decrease (in June and July) of nitrate. High ammonia concentrations were encountered only in the anoxic waters during sulfate reduction under well-developed stagnation (mostly from August to December).

With the onset of strong winds from a direction favorable for upwelling, the deep nutrient-rich waters are pushed up to the surface by the intrusion of sub-surface waters from the Cariaco Basin. The displacement enriches the surface layers in nutrients. However, a considerable amount of the deep water displaced to the surface flows out of the gulf. Therefore, when the deep water of the gulf is replaced by sub-surface waters from the Cariaco Basin, the concentrations of phosphate in the deep waters are decreased to 0.4 to 0.6 µg-at./l.

As the sub-surface waters of the Cariaco Basin contain relatively high amounts of nutrients, their inflow to the gulf may contribute to the nutrient enrichment of the upper layers, which tend to lose nutrients by photosynthetic activity. The upper layers of the gulf appear to receive a continuous supply of nutrients from the open sea, even during periods of stagnation, except during the most intense stagnation period (generally September to October).

Phosphate Balance

Fig. 6 shows quite different patterns of seasonal variation of phosphate at various depths (10, 30, 60, and 90 m). Throughout most of the year, the concentrations of phosphate are generally lower at 10 and 30 m than in the deeper layers (60 and 90 m). On the other hand, phosphate concentrations at 60 and 90 m do not differ greatly from those at 10 and 30 m during the period of water renewal. However, with the onset of stagnation, the concentrations of phosphate in both the layers start to increase gradually and continue to increase at 90 m until the next period of water renewal. At 60 m the concentrations vary irregularly between September and November after reaching their maximum.

For the relationship between the mean values of phosphate in the water column at 35 hydrochemical stations and those at Sta. 22, the observed values were distributed without great deviation from the line that indicates an equal value of the two parameters (Okuda *et al.*, 1978b). For this paper, the variation of the quantity of phosphate-phosphorus was estimated from the weekly observations at Sta. 22 (Fig. 9). In the whole water column of the gulf, the decrease in phosphate-phosphorus from stagnation to renewal was estimated to be 273 to 320 tons, while the increase was 155 to 405 tons in the renewal to stagnation period. During both periods the division between increase and decrease of phosphate-phosphorus was at 40 m, during the period from renewal to stagnation in 1975, when the division was at the 50-m level. During the period from stagnation to renewal, the net increase of phosphate-phosphorus was 80 to 318 tons (187 tons average) in the layers above 40 m and the net decrease was 400 to 610 tons (482 tons average) in the layers below 40 m. On the other hand, during the period from renewal to stagnation, phosphate decreased 140 to 183 tons (224 tons average) in the upper layers and increased 321 to 588 tons (468 tons average) in the lower layers.

Despite the fact that the volume of water in the 60 to 90-m layer is only 9% of the total volume of the gulf, high accumulations of phosphate in the layer (273 to 383 tons) during the period of stagnation appear to play a significant role in the phosphate budget. The deep layers (60 to 90 m) appear to be an effective phosphate trap during the period of stagnation.

During the period from stagnation to renewal, the increase of phosphate in the upper layers is much less than the decrease in the lower layers. The decrease may be attributable to the outflow of significant amounts of phosphate during water renewal, as well as to the uptake of phosphate by phytoplankton.

In the period from renewal to stagnation, the increase of phosphate in the deep layer is 1.9 to 3.5 times the decrease in the upper layers. The difference may be the result of the continuous inflow of open ocean water, which is relatively rich in phosphate, even though its flow rate is not so high.

Waters flowing into the gulf originate in a sub-surface layer of the Cariaco Basin in which an accumulation of organic detritus has been observed (Okuda and Benitez-Alvarez, 1974; Okuda, Bonilla, and Garcia, 1974), so that part of the increase of phosphate in the deep layers during stagnation may arise from the decomposition of organic detritus transported into the gulf.

Acknowledgements. The author is indebted to Dr. E.K. Ganesan for reviewing the manuscript, to the staff of the Chemical Oceanographic Department of the Instituto Oceanografico, Universidad de Oriente, for their generous cooperation, and to Mr. Jorge Sanchez for drafting assistance.

References

Benitez-Alvarez, J., Golfo de Cariaco, In revision de los datos oceanograficos en el Mar Caribe Suroriental, especialmente el margen continental de Venezuela, *Cuadernos Azules, 15*, 110-124, 1974.

Gade, H.G., On the hydrographic conditions in the Gulf of Cariaco during the months from May to November 1960, *Boletin del Instituto Oceanografico, Universidad de Oriente, 1*, 21-46, 1961a.

Gade, H.G., Further hydrographic observations in the Gulf of Cariaco, Venezuela, The circulation and water exchange, *Boletin del Instituto Oceanografico, Universidad de Oriente, 1*, 359-395, 1961b.

Griffiths, R.C., and J.G. Simpson, Temperature structure of Gulf of Cariaco, Venezuela, from August 1959 to August 1961, *Investigaciones Pesqueras, Serie Recursos y Explotacion Pesqueros, 1*, 116-169, 1967.

Kato, K., Oceanochemical studies on the Gulf of Cariaco, I. Chemical and hydrographical observations in January, 1961, *Boletin del Instituto Oceanografico, Universidad de Oriente, 1*, 49-72, 1961.

Okuda, T., and J. Benitez-Alvarez, Condiciones hidrograficas de las capas superiores en la Fosa de Cariaco y areas adyacentes durante la epoca lluviosa, *Boletin del Instituto Oceanografico, Universidad de Oriente, 13*, 147-162, 1974.

Okuda, T., J. Benitez-Alvarez, J. Bonilla, R., and G. Cedeno, Caracteristicas hidrograficas del Golfo de Cariaco, Venezuela, *Boletin del Instituto Oceanografico, Universidad de Oriente, 17*, in press, 1978a.

Okuda, T., J. Bonilla, R., and A.J. Garcia, Algunas caracteristicas bioquimicas en el agua de la Fosa de Cariaco, *Boletin del Instituto Oceanografico, Universidad de Oriente, 13*, 163-174, 1974.

Okuda, T., A.J. Garcia, B.R. Gamboa, and E. Fernandez, Variacion estacional de P-phosphato y nitrogeno inorganico en el Golfo de Cariaco, Venezuela, *Boletin del Instituto Oceanografico, Universidad de Oriente, 17*, in press, 1978b.

Richards, F.A., Some chemical and hydrographic

observations along the north coast South America, I. Cabo Tres Puntas to Curacao, including the Cariaco Trench and the Gulf of Cariaco, *Deep-Sea Research, 7*, 163-182, 1960.

Simpson, J.G., and R.C. Griffiths, Density distribution in the Gulf of Cariaco, eastern Venezuela, *Investigaciones Pesqueras, Serie Recursos y Explotacion Pesqueros, 1*, 305-327, 1967.

A COMPARISON OF DISTRIBUTIONS OF SUSPENDED MATTER IN THE PERU AND NORTHWEST AFRICAN UPWELLING AREAS

Gunnar Kullenberg

Institute of Physical Oceanography
University of Copenhagen
Haraldsgade 6, 2200 Copenhagen N
Denmark

Abstract. Observations of suspended matter observed by a light scattering technique in the northwest African and Peru upwelling areas are discussed. The distributions have an overall similarity but differ in details. The distributions off northwest Africa have more pronounced front-like features than off Peru, but the Peru distributions are much more stratified.

Introduction

A light scattering technique has been used to observe the distribution of suspended matter in the upwelling areas off northwest Africa and Peru (Table 1). For positions see Kullenberg (1978) and Hafferty, Codispoti, and Huyer (1978).

Scattering of light in the sea is caused by water itself, by dissolved salts, and by suspended matter. In the relatively particle-rich waters of upwelling areas the scattering by suspended matter dominates. The scattering exhibits only a slight wave-length dependence and in the present work observations in the red part of the spectrum, $\bar{\lambda}$ = 655 nm, are used as a measure of the total suspended matter, including organic and inorganic material.

An *in situ* light scattering meter capable of measuring the volume scattering function, i.e., the angular distribution of the scattered light between scattering angles 7° and 168° from the incoming beam was used (Kullenberg, 1978). Observations at a fixed angle of 20° will be used, given in relative units. A continuous scattering profile was obtained by lowering the instrument to about 1000 m. The volume scattering function was observed at discrete levels when hauling back. All signals were recorded on board in analog form.

The scattering measurements were insofar as possible carried out in conjunction with conductivity-temperature-depth (CTD) measurements and water sampling for chemical analysis.

Objectives

The overall aim of this work was to investigate the distributions of suspended matter in relation to characteristic features of the upwelling regions as related to dynamics and biological and chemical processes. An intercomparison of the distributions of suspended matter in the two upwelling areas, observed with the same equipment and using the same procedure, will be made to identify similarities and differences and to attempt an interpretation on the basis of other observed features.

To give an idea of the hydrographic structure during the observations, examples of profiles of salinity and temperature are shown in Fig. 1. The distances between the stations in the sections were similar, i.e., between 5 and 20 km, for both areas at corresponding distances from the coast. Over the shelf and slope the station distances were generally 5 to 10 km.

The scattering values in the figures are given in relative units at a 20° angle. They are relative to the dark value obtained when measuring the volume scattering function at fixed depths, whereby the detector rotating around the scattering volume also cuts the light source at 180° angle, thus defining the dark signal (Kullenberg, 1978). The relative values can be converted to absolute units, i.e., m^{-1}, str^{-1}, by multiplication by the factors 8.5×10^{-4}, 1.3×10^{-4}, and 2×10^{-5} for the relative values 130, 100, and 70, respectively. The factors have been calculated using the calibration technique of Kullenberg (1978) and Kullenberg and Berg Olsen (1972). The calibration factor varies with the signal due to the logarithmic amplification used in the instrument.

Northwest African Observations

Observations in three sections (Table 1) discussed in an earlier paper (Kullenberg, 1978) will be compared with the Peru observations. The

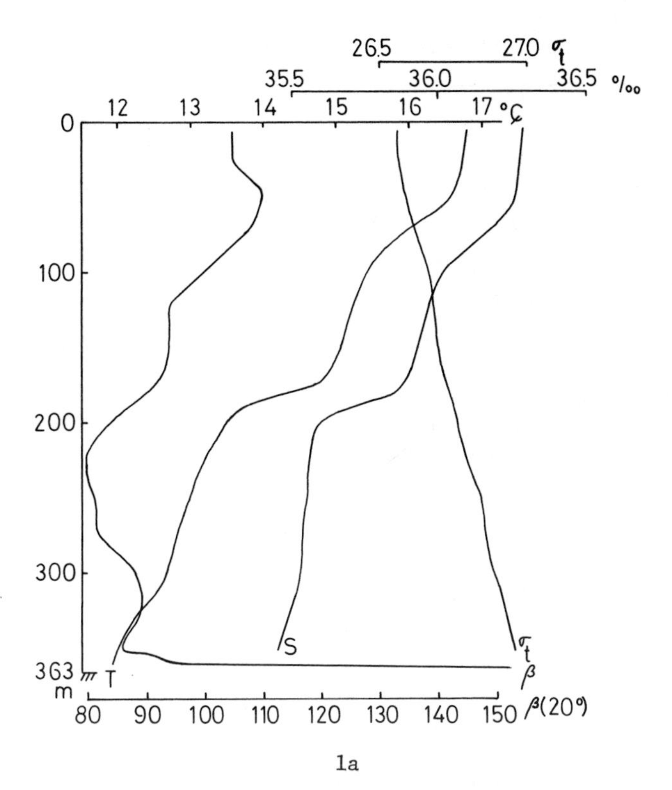

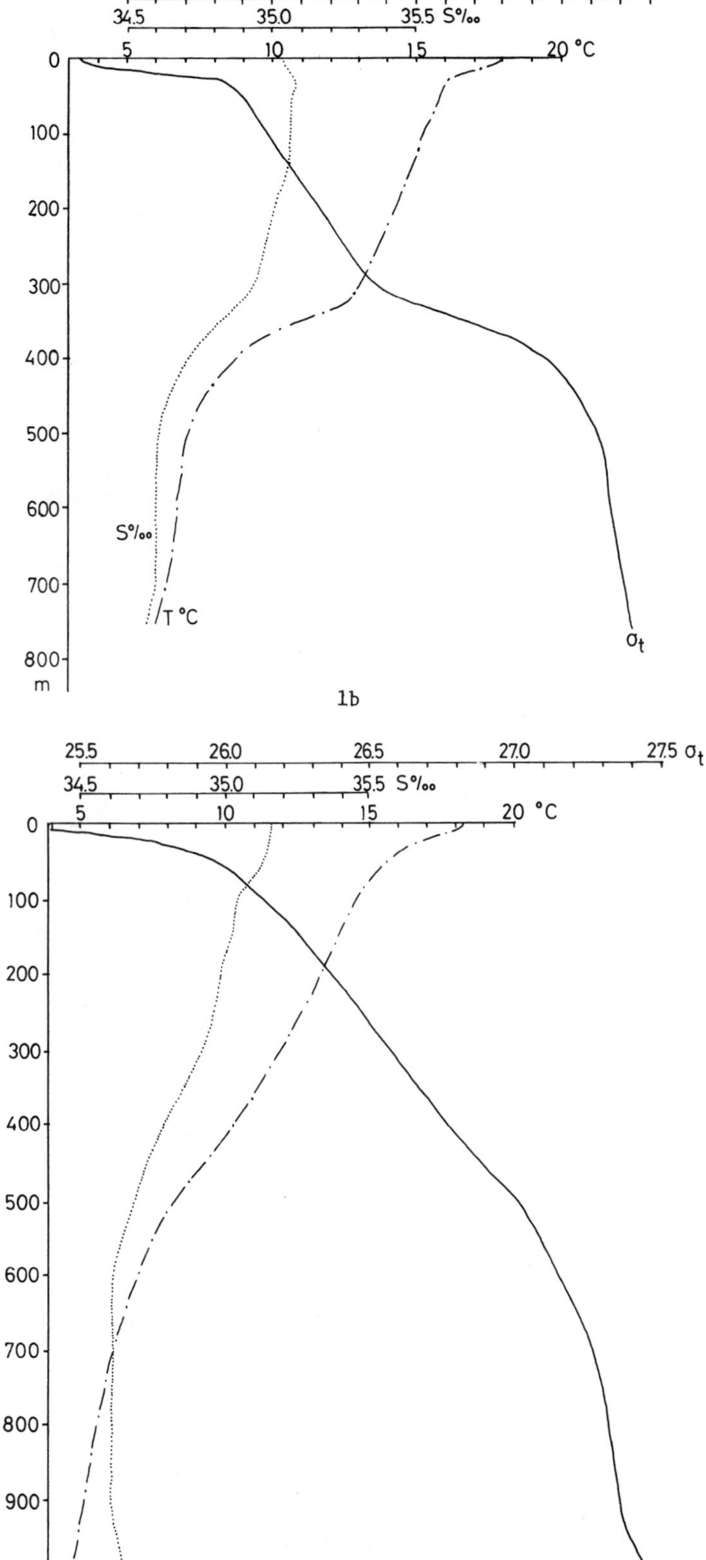

Fig. 1. a) Profiles from the northwest African area, *Discovery* Sta. 8811/1, 22°44.2'N, 17°12.4'W, 3 March 1975; b) Hydrographic (CTD) profiles from the Peru area, *Melville* Sta. 413, 15°15.3'S, 75°39.0'W, 13 May 1977; c) Hydrographic (CTD) profiles from the Peru area, *Melville* Sta. 458, 05°00.0'S, 81°34.0'W, 22 May 1977.

longest section extended 160 km from the coast, the width of the shelf being about 70 km. During the cruise the scattering meter could be lowered only to about 450 m. The observations will be summarized here, and reference is made to Kullenberg (1978) for further details. An example of a northwest African section is given in Fig. 2.

In all the sections there was a 10- to 20-m thick layer of low scattering at approximately 20- to 40-m depth over the shelf. Over the shelf break and slope the layer deepened to just above the bottom (see Fig. 3) and was present down to about 250 m. Close to the bottom there was a layer with high scattering. Over the shelf the thickness of the layer was about 20 m, over the shelf break it was less than 5 m, increasing to 10 to 15 m over the slope. The transition from the low-scattering layer just above was very sharp. There were generally no continuous layers of high scattering extending out from the shelf break or slope. In one section there was a scattering maximum at the slope stations between 150 and 230 m, which, however, was not present at the stations further offshore. There was also a maximum centered at 330 m at one of the slope sta-

TABLE 1. Characteristics of the Sections

Date	Stations	End Positions	Length (km)
1975	*Discovery*		
February 28	8782-8785	22°43.9'N, 17°04.1'W 22°40.8'N, 16°51.6'W	20
February 28 - March 1	8785-8793	22°40.8'N, 16°51.6'W 22°54.0'N, 17°58.4'W	115
March 3 - March 4	8811-8815	22°40.5'N, 17°07.0'W 22°48.0'N, 17°28,1'W	45
1977	*Melville*		
May 12-13	C, 409-419	15°02.0'S, 75°26.0'W 16°20.5'S, 76°45.8'W	208
May 14-16	C, 420-432	15°47.7'S, 76°10.8'W 15°02.1'S, 75°26.5'W	110
May 18-19	H, 435-448	10°39.6'S, 79°44.2'W 9°58.2'S, 78°16.0'W	226
May 21-22	P, 449-462	4°59.4'S, 83°05.8'W 5°0.00'S, 81°09.0'W	225

tions, but it was not found at the next offshore station. In another section there was a 10- to 20-m thick layer with relatively high scattering centered at 240 m at stations outside the slope, but the layer was not present at the slope stations. Patches of high scattering were also observed at offshore stations at depths between 100 and 150 m.

The surface layer over the shelf was 10 to 30 m thick, with increasing thickness over the shelf break and slope. Over the central parts of the slope the surface layer with nearly uniform scattering was generally 70 to 100 m thick, and it was sometimes over 100 m thick. Further offshore the thickness decreased to 40 to 50 m. Patches of high scattering 10 to 30 m thick were observed over the shelf and offshore. The transition from the deep surface layer over the central slope zone in the offshore direction had a front-like structure, extending over less than 5 to 10 km.

The minimum scattering was found in the subsurface waters, generally starting about 80 to 90 km from the coast at depths around 300 m. A dome of low-scattering water appeared to be present over the outer parts of the slope.

The meteorological conditions during the observations were favorable for upwelling, with northerly winds of speeds between 13 and 24 knots.

The Peru Observations

Sections occupied at three different latitudes, CUEA lines C, H, and P during leg 4, May 1977, of the R.V. *Melville* CUEA cruise will be discussed (Table 1, Figs 4 and 5). The C line was occupied three times and to illustrate variations two distributions are presented (Figs 6 and 7). During 12 to 14 May the scattering in the green part of the spectrum, $\bar{\lambda}$ = 520 nm, was measured. However, since the wave-length dependence of the particle scattering is weak the basic properties of the distribution can be compared with the other distributions, when the scattering in the red was measured. The width of the shelf is very different at the C and P lines than at the H line. Pertinent features of the sections will be summarized, referring to the figures for details.

In the wide-shelf section (H line, Fig. 4) there was a weakly scattering layer over the shelf at depths between 40 and 110 m. The minimum deepened slightly over the shelf break, but it was present across the section at 100 to 150 m.

A distinct bottom layer of high scattering about 20 m thick was present over the shelf (see also Fig. 9). Over the slope the bottom layer was also evident with scattering maxima extending

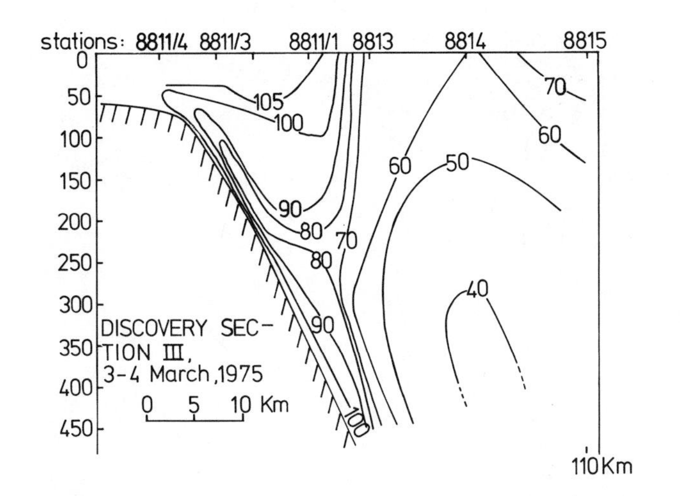

Fig. 2. *Discovery* section 3 to 4 March 1975, northwest African area.

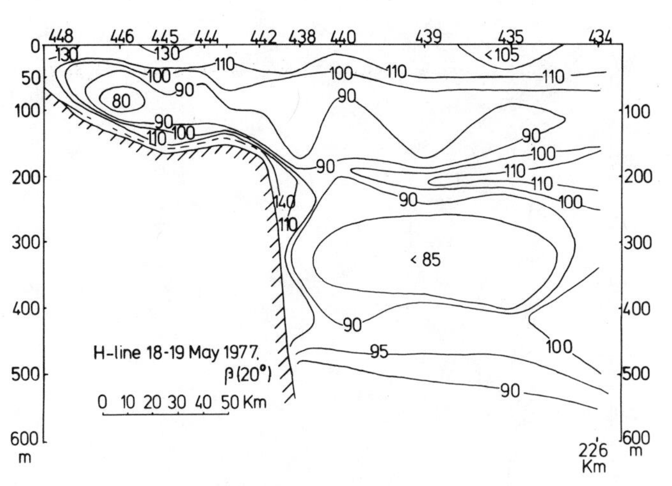

Fig. 4. The H line distribution 18 to 19 May. Numbers are scattering at 20°, relative units, in the red.

out from the slope centered at 240 and 420 m. Over the whole section there were layers of scattering maxima centered at 200 and at 450 m. They were separated by a thick layer with the same low scattering as in the scattering minimum at 80 m depth over the central parts of the shelf. At the outermost stations the scattering minimum was considerably less pronounced than at the central stations.

The surface layer was 20 to 40 m thick over the whole section. Patches of high and low scattering occurred and in the top 50 m there were also layers of scattering maxima of varying thickness.

The northernmost section (P line, Fig. 5) has an intermediate shelf width, less than half that of the H line. There was a marked scattering minimum over the shelf and slope centered at 50 m (see also Fig. 9). Further offshore the scattering minimum was deeper, between about 100 and 350 m, with an absolute minimum in the section observed at 150 m at Sta. 451.

Over the shelf there was a bottom layer of high scattering about 10 m thick, becoming thicker over the break and slope. Several layers of high scattering which appeared to extend out from the

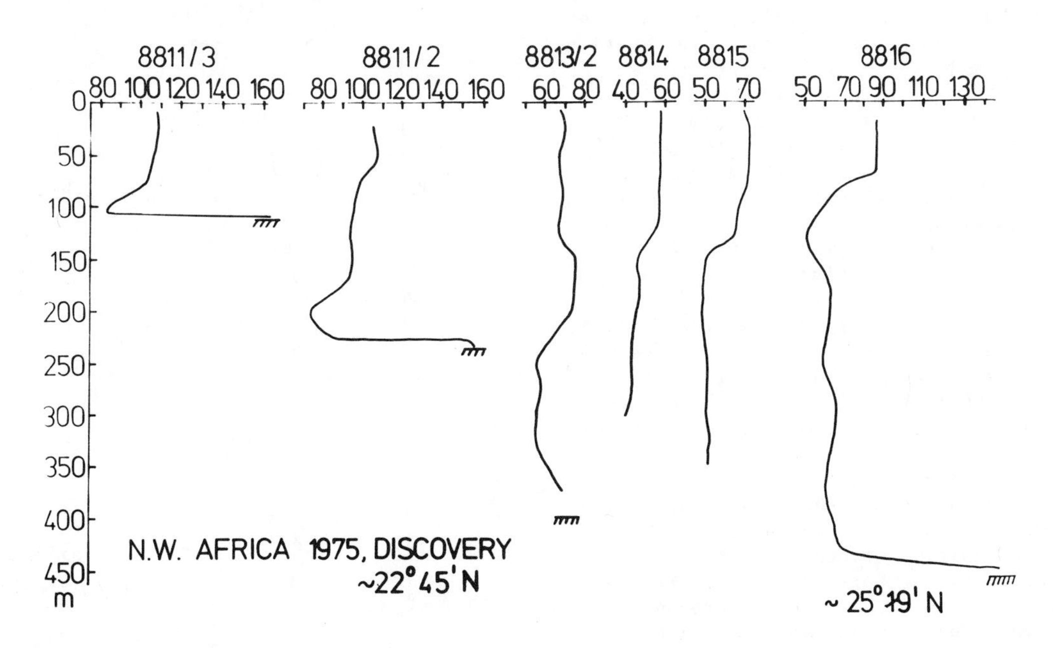

Fig. 3. Profiles of light scattering, 20° scattering angle, relative units, at different positions in the northwest African section and one profile from a slightly more northerly position.

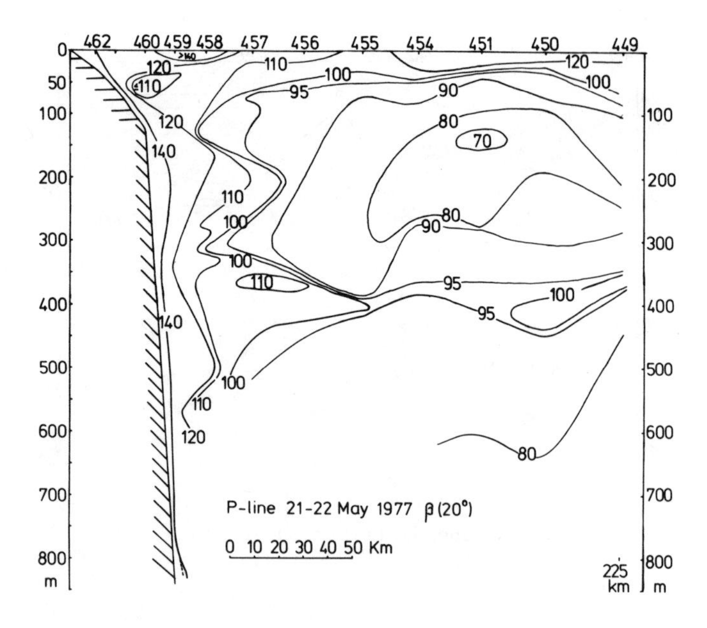

Fig. 5. The P line distribution 21 to 22 May. Numbers are scattering at 20°, relative units, in the red.

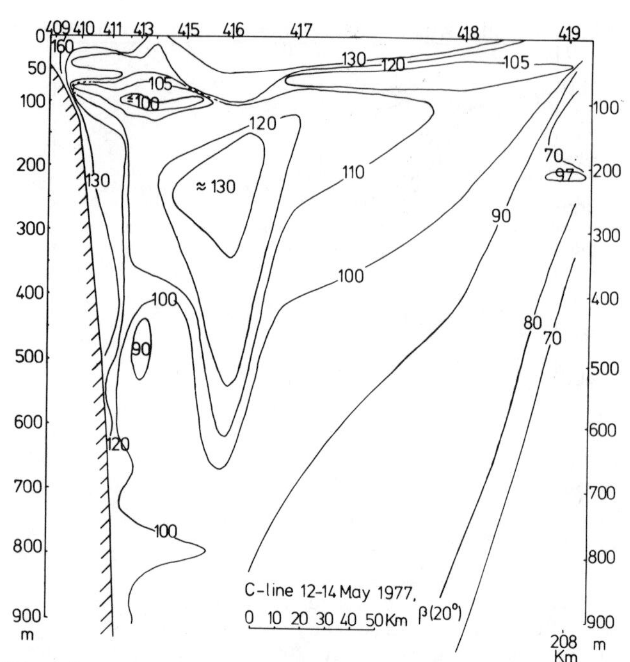

Fig. 6. The C line distribution 12 to 14 May. Numbers are scattering at 20°, relative units, in the green ($\bar{\lambda}$ = 520 nm).

slope, the most prominent ones centered at 170 to 200 m and at about 400 and 500 m, were observed. At the 200-m level there was a scattering minimum further offshore, the high scattering layer extending about 50 km out from the slope. At the 400-m level the high scattering layer was present over the whole section, being 30 to 80 m thick. The lowest scattering values in the layer were observed in the central parts of the section. Around the 500-m level there were internal maxima at several of the offshore stations.

The surface layer was only about 20 m thick in large parts of the section. In the top 50 m there were several internal maxima and minima of the scattering. Although there were internal scattering minima at most stations at different depths, there was no connected minimum scattering layer across the whole section, as was present across the H section.

At the C line the shelf is very narrow. In the section occupied 12 to 14 May (Fig. 6) there was an internal scattering minimum centered around 100 m over the shelf and slope, but it was centered around 70 to 50 m in the outer half of the section. The minimum layer extended, however, as a connected feature across the whole section.

Over the shelf and slope there was generally a bottom layer of high scattering about 30 m thick. It was not so sharply defined as in the H section. At the slope stations there was a scattering maximum between 150 and 500 m. It was present across the whole section but was limited to depths between 140 and 330 m further offshore. A core of very high scattering, reaching surface layer values, was observed between 200 and 300 m

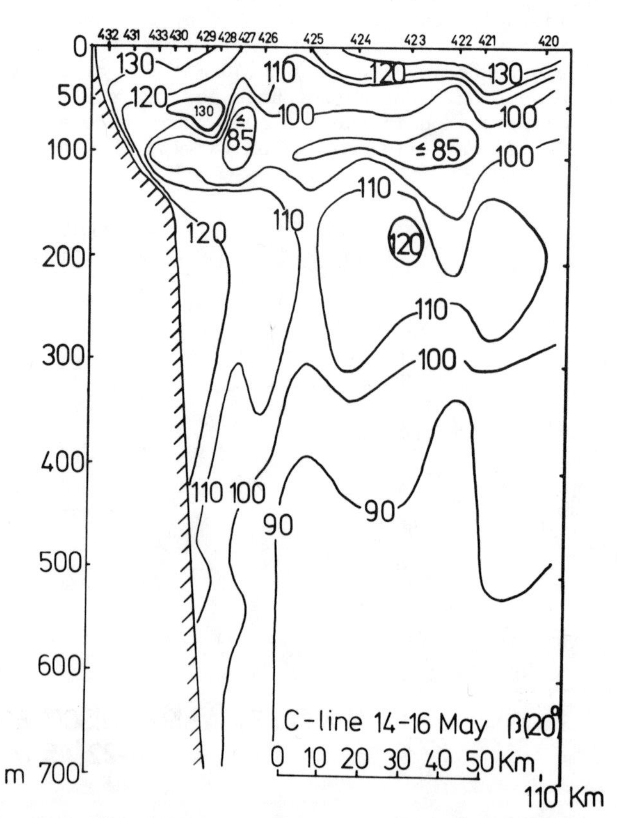

Fig. 7. The C line distribution 14 to 16 May. Numbers are scattering at 20°, relative units, in the red ($\bar{\lambda}$ = 650 nm).

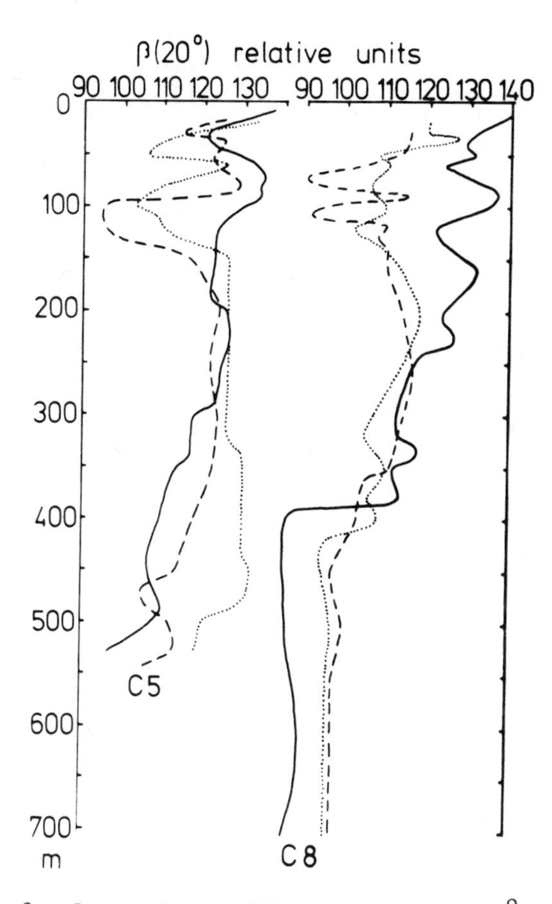

Fig. 8. Scattering profiles at Stas C-5 (15° 12'S, 75°34'W) and C-8 (15°18'S, 75°42'W).
C-5: full drawn 9 May, dotted 12 May, dashed 15 May
C-8: full drawn 8 May, dotted 13 May, dashed 15 May

at Stas 415 and 416. This absolute internal scattering maximum was quite clearly disconnected from the slope. At the outermost station the internal maximum was reduced to a 30-m thick layer at about 200 m.

Centered at 800 m there was a high scattering layer apparently extending out from the slope about 35 km. The layer was separated from the above maximum by a minimum, present at four stations just off the slope. The absolute internal maximum mentioned above extended to about 650 m at Sta. 416, where it formed a marked feature.

Further out the isopleths showed a marked upward slope, varying in the range 5 to 15 x 10^{-3}, slopes quite comparable to those found in frontal zones.

The thickness of the surface layer varied from about 20 m over the shelf break and slope to about 40 m in the central parts of the section. At the outermost station the scattering was markedly less than at the next inshore station.

In the section 14 to 16 May (Fig. 7), which was only half as long as the outer section, there was also a scattering minimum centered around 100 m. It extended across the whole section with an indication of shoaling to about 70 m at the outermost station.

Over the shelf and shelf break there was a 10- to 20-m thick bottom layer with high scattering, but again it was not so sharply defined as in the H section. An internal scattering maximum was present between 140 and about 400 m over the slope. Further out the maximum was between about 150 and 300 m. At the slope stations there was another internal maximum between 500 and 600 m, extending about 15 km from the slope. In the center of the section there was a core of weakly scattering water below about 400 m.

The surface layer was generally about 20 m thick. Centered about 45 km from the coast there was a low scattering zone near the surface with front-like transitions towards high scattering water both inshore and offshore.

Another C section, 70 km long, occupied 8 and 9 May (Kullenberg, 1978a), showed similar features. However, the bottom layer of high scattering was generally not present over the slope, or if it was present it was very thin. Layers of high scattering where present in the section at depths varying between 150 and 600 m. Features apparently extending out from the slope were, however, not a distinct characteristic.

The meteorological conditions during the observations were variable often with low winds or calm. Winds were from S to SE with speeds between 10 and 20 knots.

Comparisons and Interpretations

There are a number of overall similarities in the structure of the distributions in all the sections. The minimum over the shelf is a common feature although it is slightly deeper off Peru than off Africa. It is also found over the slope, but there it is deeper off Africa than

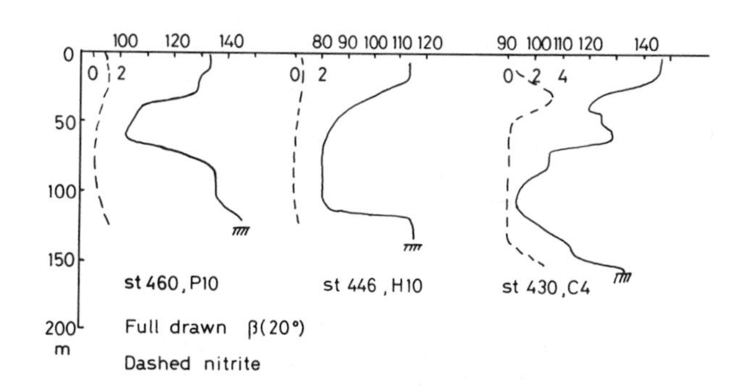

Fig. 9. Scattering and nitrite profiles from one shelf station at each Peru line. Dashed lines are nitrite given in µg-at/l (data from Hafferty *et al.*, 1978), and full drawn lines are scattering at 20°, relative units, scale on top of the abscissa.

off Peru. The bottom layer of high scattering has a similar thickness over the shelf, but it is thinner over the shelf break and slope off Africa than off Peru. There are indications that the layer may not always be present off Peru. The relatively high scattering with patchiness in the surface layer is common to both areas. The similarities are most obvious when comparing the wide-shelf Peru section with the African sections.

Despite the overall similarities the differences are striking. The wide-shelf Peru section is much more stratified than the sections off Africa. The layers apparently extending out from the Peru slope are much less evident over the African slope. The prominent deep, relatively well-mixed surface layer observed in a zone over the African slope is not present in the Peru section, which shows no such front-like structure.

The Peru sections all have a relatively layered structure, including high scattering layers extending out from the slope at different depths. However, the layers do not generally extend in a connected way across the complete section. In several cases relatively higher scattering values were observed in the offshore parts of such layers than in the central parts of the section.

The distributions along the Peru sections differ markedly. On the C line the intermediate scattering maximum along the slope is much thicker than on the H and P lines. On the H line the layer is about 100 m thick at the slope, it decreases to about 50 m over most of the section, and is centered slightly deeper than the shelf break. On the P line the corresponding feature only extends over about 1/3 of the section. In both the sections a deeper high scattering layer extended across the sections, at around 400 m in the P section and slightly deeper in the H section. In the C section the layer was not observed.

The surface layer distributions also differ among the Peru sections. On the C line there are indications of a front-like zone outside the slope about 25 km from the shelf break; they are not so marked as in the African sections, where they were observed 15 to 20 km from the shelf break. The H and P sections show no such features.

Although the basic structure of the two C sections is the same, they differ in detail. The prominent maximum centered 75 km from the coast in the section occupied 12 to 14 May is not prominent in the later section. Although time variability quite clearly occurs the basic structure is often remarkably similar, as suggested by the sequence of scattering profiles observed at Stas C-8 and C-5 (Fig. 8).

The changes in scattering values between the C-sections (Figs 6 and 7) were about 10 to 25% and similar variations are shown in the consecutive profiles (Fig. 8). Such variations are comparable to the differences in scattering observed between the Peru and northwest African areas. However, the interest in the intercomparison between the areas is centered on the distributions and their possible variation, not on the absolute values of the scattering.

The C line distributions of 12 to 14 May are the only ones showing steeply sloping isopleths in the outer parts of the section. Unfortunately the other C-sections do not extend so far out, and the same feature is not evident in the other sections although they do extend as far out.

To interpret the distributions, topography, meteorological conditions, currents, density distributions, and biological and biochemical processes should be taken into account. A basic question is where do the particles come from? Possible sources are atmospheric fall-out, production in the sea, and erosion on the slope. The suspended matter is advected by currents while sinking. Locomotive plankton can migrate and thus counteract sinking and possibly also advection by moving between different current regimes. Particle sinking rates can vary considerably, depending upon size, shape, and composition, falling in the range of 1 to 100 m/d for the large size fraction (Riley, 1970). Garfield, Packard, and Codispoti (1979) found that sinking rates between 1 and 5 m/d fitted their electron transport system (ETS) respiration versus depth distribution. The advective velocities can clearly be much higher.

The light scattering intensity also depends upon the size, shape, and composition of the particles. These parameters can all be expected to vary in time and space. Observations in different oceanic water masses show that the volume scattering function at 20° scattering angle can vary more than one order of magnitude (Kullenberg, 1974). Across frontal zones the properties of the suspended matter can vary considerably (Kitchen, Zaneveld, and Pak, 1978). In upwelling areas such variations can be expected. However, considering the weak frontal development in the Peru area, the horizontal variations of the particle properties across the sections are expected to be fairly limited.

The observations in northwest Africa suggest that there was no erosion on the slope except possibly during current pulses (Kullenberg, 1978). The mean currents at the Peru C line (Brink, Smith, and Halpern, 1978) cannot cause erosion and probably cannot maintain the settling material in suspension. The current systems in both upwelling areas are basically similar. However, the Peru observations show that strong current pulses, with velocities 25 m from the bottom at the shelf break of over 25 cm/s, occur and last up to several days. The pulses are often associated with poleward flow and a much weaker offshore flow component. Such a pulse occurred 12 May, starting about 11 May. The scattering profiles observed at Sta. C-5 (Fig. 8), near the current meter moorings Ironwood, Lagarta, and Lobivia (Brink *et al.*, 1978) show marked variations, with 20 to 30% increases in the scattering at 300 to 500 m from 9 May to 12 May,

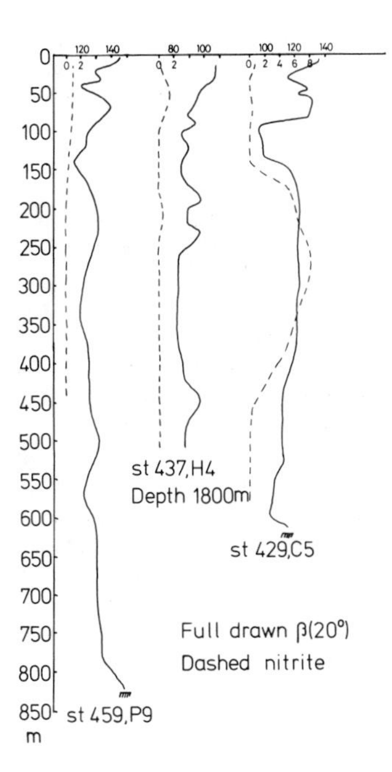

Fig. 10. Scattering and nitrite profiles from one mid-section station at each Peru line. Notations as in Fig. 9.

followed by a decrease to 15 May. At 200 m, at mooring Lagarta, the alongshore current was about 25 cm/s on 12 May and 1 to 4 cm/s on 8 to 10 May. The offshore flow was strong on 12 and 13 May at 100 to 200 m but it was onshore or zero at 500 m on 12 May. It seems likely that the variations of the scattering at C-5 were related to the current event. At Sta. C-8 there was also a considerable variation in the whole scattering profile over the same time period (Fig. 8).

Considering the current strength and direction at 500 m, observed at the Lagarta mooring, it does not appear that an offshore flow of particles below 400 m in the section can explain the scattering variations. The scattering variation below 400 m is not much more than 10%. Instrumental variations have been found to be about ± 5% (Kullenberg, 1978). A real scattering increase is, however, supported by the simultaneous increase of phaeopigment content in the water column below 400 m, with a marked increase both from 8 to 13 May and again from 13 to 15 May (Hafferty, Codispoti, and Huyer, 1978). It may also be noted that the stability across the relevant part of the water column decreased slightly between 8 and 13 May. The density difference between 500 and 200 m was 0.74×10^{-3} g/cm^3 on 8 May and 0.65×10^{-3} g/cm^3 on 13 May. The change is certainly not large, but it suggests a certain amount of internal mixing. The increased particle content below 400 m may have been due to alongshore advection and sinking or a combination of both. It does not appear likely that it can be explained by an offshore transport of particles only, but it might be explained by an offshore transport further to the north combined with alongshore advection and sinking. An assumed sinking rate of 100 m in 30 days yields about 5° latitude displacement with an alongshore advection of 20 cm/s. Such a velocity is not typical over a period of 30 days; more typical is a velocity of 5 to 10 cm/s. The onshore-offshore velocity averaged over a 30-day period is normally well below 5 cm/s, except possibly in the near-surface layer, and it is often below 1 cm/s. It appears that sinking must be regarded as a factor of importance for the particle distribution.

On the other hand the strong current pulse can most likely cause resuspension of particles along the slope, or at least prevent settling out of the material, thus explaining the observed bottom layer of high scattering present in both sections along the slope. It is interesting to note that on 7 and 8 May the currents were generally much weaker, less than 5 cm/s in near-bottom layers (Brink *et al.*, 1978). Thus the general lack of a near-bottom scattering layer at the slope stations in the section of 8 and 9 May (Kullenberg, 1978a) appears to conform with the current distribution in time and space.

The scattering isopleths in the Peru area often show a gentle slope upwards from the inshore towards the offshore parts. This may be related to the offshore extent of the high primary production regime and the upwelling regime in general. It appears that the very high primary production zone is limited to inshore of about 150 km, at least at the C lines (see Garfield *et*

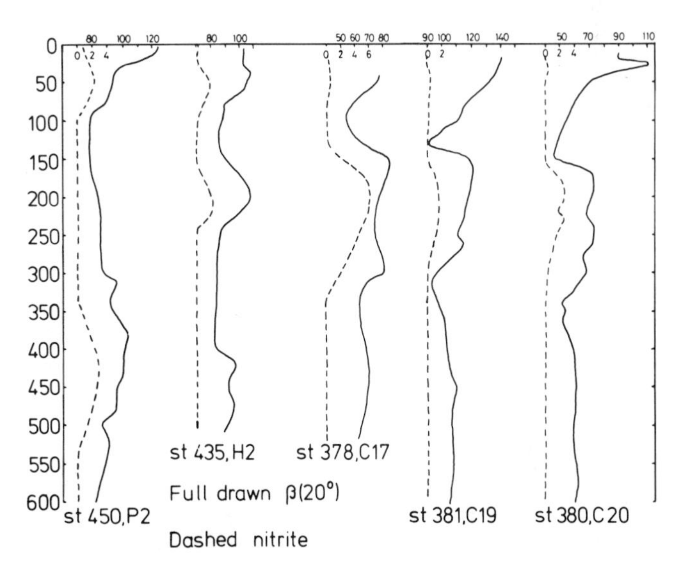

Fig. 11. Scattering and nitrite profiles from outer parts of each Peru section; several profiles from C line to show development of internal maxima. Notations as in Fig. 9.

al., 1979, Hafferty *et al.*, 1978).

The relatively more layered structure observed in the Peru area than in the African area may be related to primary production in the Peru area being higher than in the African area, to differences in stability conditions in the top 100 m, and to the current events discussed above. This aspect requires further study.

Considering the expected particle sinking rates and current velocities, the internal scattering minimum consistently centered around 100 m or somewhat shallower suggests a continuous supply of relatively particle-poor water. One would think that this could come from offshore, but such a hypothesis does not appear to be supported by the current data. At mooring Lagarta, for instance, the current had an onshore component around 100 m on 8 and 10 May but a marked offshore component on 12 May. The mean flow also had an offshore component (Brink *et al.*, 1978). The question requires more study.

Nitrite profiles (bottle data) observed in conjunction with the scattering profiles are included in Figs 9, 10, and 11. The nitrite layers generally coincide with internal scattering maxima, but scattering maxima can occur without nitrite being present in the layer. The possible connection between the features needs further study.

Acknowledgements. I want to acknowledge support from the CUEA program, both during the field phase outside Peru and afterwards, and from the joint IOS, Wormley and IFM, Kiel program in 1975. Financial support has further been given by the Danish Natural Science Research Council under grant Nos. 511-3855, 511-3562, and by NATO Science Committee Special Program Panel on Marine Sciences under grant MS SRG7.

Special thanks are due to R.L. Smith, L.A. Codispoti, and to H. Hundahl for encouragement, cooperation, and help during the observations.

References

Brink, K.H., R.L. Smith, and D. Halpern, A compendium of time series measurements from moored instrumentation during the MAM 77 phase of JOINT-II, Reference 78-17, CUEA Technical Report 45, 72 pp., 1978.

Garfield, P.C., T.T. Packard, and L.A. Codispoti, Particulate protein in the Peru upwelling system, *Deep-Sea Research, 26*, 623-639, 1979.

Hafferty, A.J., L.A. Codispoti, and A. Huyer, Data Report 45, JOINT-II, R/V *Melville* legs I, II, and IV, R/V *Iselin* leg II. Reference M78-48, University of Washington, Department of Oceanography, Seattle, WA 98195, 779 pp., 1978.

Kitchen, J.C., J.R.V. Zaneveld, and H. Pak, The vertical structure and size distributions of suspended particles off Oregon during the upwelling season, *Deep-Sea Research, 25*, 453-468, 1978.

Kullenberg, G., Observed and computed scattering functions. In *Optical Aspects of Oceanography*, N.G. Jerlov and E. Steemann-Nielsen (eds.), Academic Press, 25-49, London, 1974, 494 pp., 1974.

Kullenberg, G., Light scattering observations in the northwest African upwelling region, *Deep-Sea Research, 25*, 525-542, 1978.

Kullenberg, G., Light scattering observations in frontal zones, *Journal of Geophysical Research, 83*, 4683-4690, 1978a.

Kullenberg, G., and N. Berg Olsen, A comparison between observed and computed light scattering functions, Report No. 19, Institute of Physical Oceanography, University of Copenhagen, 28 pp., 1972.

Riley, G.A., Particulate organic matter in the sea, *Advances in Marine Biology, 8*, 1-118, 1970.

A NOTE ON SHORT-TERM PRODUCTION AND SEDIMENTATION IN THE UPWELLING REGION OFF PERU

Klaus von Bröckel

Institut für Meereskunde, Düsternbrooker Weg 20,
2300 Kiel 1, West Germany

Abstract. On a section close to 15°S off the Peruvian coast ^{14}C-uptake and sedimentation were determined at eight stations between March 23 and 28, 1978. The general upwelling situation is described in terms of physical (temperature, salinity), chemical (nitrate, nitrite, ammonia, oxygen), and biological (particulate organic carbon, particulate organic nitrogen, chlorophyll *a*, ^{14}C-uptake) variables. Average sedimentation rate was approximately 13% of primary production expressed as particulate organic carbon, with maximal and minimal values of 18.9 and 6.7%, respectively.

Introduction

The only energy source for benthic communities not exposed to light is organic material produced within the euphotic zone. Levels of primary production in the upwelling area off the Peruvian coast, about 300 to 500 g C $m^{-2}yr^{-1}$ (Ryther, 1969; Cushing, 1971) with daily values up to 11 g C $m^{-2}d^{-1}$ (Ryther *et al.*, 1971), are among the highest reported for marine ecosystems. To some extent, the organic material is used by herbivorous and omnivorous zooplankton and fishes (for example the Peruvian anchoveta, *Engraulis ringens*) to build up their biomass and to support their metabolic functions. As a result of their activities these organisms contribute to the remineralization of organic material in the upper part of the water column. On the other hand, part of the particulate organic material settles out of the euphotic zone and eventually sinks to the sediment surface. The sedimentation can be accelerated several-fold when phytoplankton cells are "packed" in fast-sinking pellets or fecal casts (Smayda, 1971).

The occurrence of very low oxygen concentrations not far below the euphotic zone (Dugdale, 1972; and my own measurements) and of large bottom areas with anoxic, organic-rich sediments (Jordan, personal communication) indicate that a large amount of organic material is incorporated into the sediments. Due to problems involved in sediment trap methodology, there have been no direct sedimentation rate measurements within upwelling ecosystems up to now. The main difficulty is caused by water movements relative to moored traps (Bröckel, 1975; Gardner, 1977; Staresinic, Rowe, Shaughnessey, and Williams, 1978). Using newly developed free-drifting sediments traps, as described by Staresinic *et al.* (1978), the problem seems to be solved. In this paper, data on short-term sedimentation rates will be compared with primary production rates within the context of a general description of the Peruvian upwelling area during the fall of 1978.

Material and Methods

Sampling procedure. Sampling was carried out during a 12-day cruise of R.V. *Knorr* 73-3 (Woods Hole Oceanographic Institution) off the Peruvian coast. Eight stations (a to h) were occupied between March 23 and 28, 1978 (Fig. 1, Table 1).

Water samples from discrete depths were taken with 5-ℓ Niskin bottles. Sedimentation rates were estimated from samples from two traps on three occasions and once with one trap (Table 2) close to stations a, d, and h.

Sediment traps. Two multi-sample sediment traps, consisting of a funnel (40 cm max. diameter, 64 cm deep) with a lid (opening 20-cm diameter) covered by a grid (Zeitzschel, Diekmann, and Uhlmann, 1978) were used. Each trap was mounted on the frame of a free-drifting sediment trap developed by Staresinic *et al.* (1978) and set below the euphotic zone at a water depth of 30 m from late evening to early morning (Table 2).

Analyses of collected material. Material collected in the traps was filtered through 300-μm gauze to break up fecal casts, thus providing a better suspension. Unlike sediment trap samples from other regions, no large zooplankton (>300 m) was found in the gauze. Subsamples for the analyses described below were obtained with a Folsom plankton splitter; fractions split ranged from 1/4 to 1/16 of the initial sample volume.

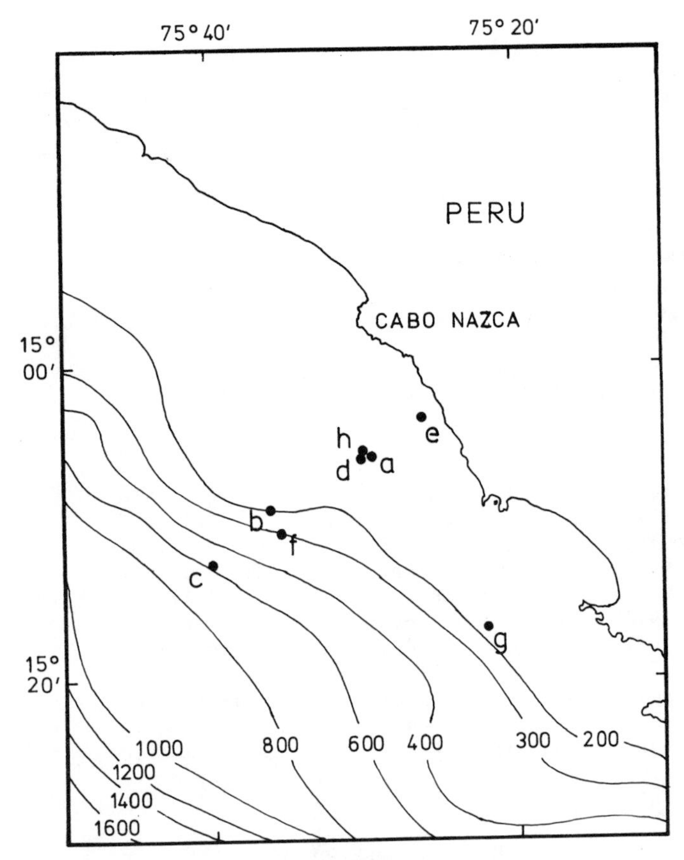

Fig. 1. Locations of Stas a to h in the area of investigation (depth in fathoms; 1 fathom = 1.83 m).

Particulate organic nitrogen (PON) and particulate organic carbon (POC) were measured with a CHN-analyzer (Hewlett-Packard, 185 B) on precombusted Whatman GF/C filters, which were treated with dilute HCl before measurement to remove particulate inorganic carbon. Samples for counting of phytoplankton were preserved with Lugol's solution.

TABLE 1. Depth of Stations and Date of Sampling

Station	Station Depth (m)	Sampling Day	Time (Hour)
a	110	23/March	8:50
b	330	24/March	11:25
c	1010	25/March	10:10
d	110	26/March	8:50
e	62	26/March	11:20
f	530	26/March	16:00
g	178	28/March	9:30
h	110	28/March	13:50

Other data were kindly provided by J. Kogelschatz, Duke University Marine Laboratory: temperature (determined with reversing thermometers); salinity (measured with a salinometer); chlorophyll *a*, nitrate, nitrite, ammonia, and ^{14}C-uptake (all estimated according to Strickland and Parsons, 1968). ^{14}C-uptake was measured as "simulated *in situ*" over a period of 24 h.

Results

To give a general impression of the area investigated, physical, chemical, and biological variables measured from the water surface down to 100 m on Sta. g will be discussed in detail. All other stations showed a qualitatively similar picture; only the depths of some maxima and minima, as well as absolute values, differed somewhat between stations.

Salinity and temperature. Salinity and temperature values (Fig. 2a) indicate a homogeneous water column down to 100 m. The low surface temperature was typical of the whole region during the sampling period. Average surface temperature at Stas a to h was $16.7 \pm 0.54°C$, thus demonstrating, together with the homogeneous water column, the effect of intensive upwelling in the area of investigation between March 23 and 28. As a comparison, a "non-upwelling" station was studied about 20 nautical miles to the west on March 27; it showed a surface temperature of 22.4°C and a distinct thermocline at about 40 m.

Nitrogen and oxygen. In the uppermost 20 m low ammonia concentrations (Fig. 2b) were detected; they increased below the euphotic zone, between 30 and 40 m, and decreased again towards the bottom. This probably reflects the effect of nitrogen remineralization by zooplankton and fishes within and below the productive zone (Whitledge and Packard, 1971; Dugdale and Goering, 1970). Nitrite values, low throughout the water column, showed a steady decrease with increasing water depth. Nitrate concentration showed a minimum of 14 µg-at/l within the euphotic zone (Fig. 2b). Thus, nitrogen did not seem to be a limiting nut-

TABLE 2. Period of Sedimentation Trap Deployment

Sediment Trap	Date and Time of Deployment	Date and Time of Recovery
II	23 March 22:30 h	24 March 6:45 h
I	24 March 21:20 h	25 March 6:30 h
II	24 March 21:40 h	25 March 6:40 h
I	25 March 20:10 h	26 March 7:00 h
II	25 March 20:20 h	26 March 6:30 h
I	27 March 22:15 h	28 March 8:30 h
II	27 March 22:00 h	28 March 8:15 h

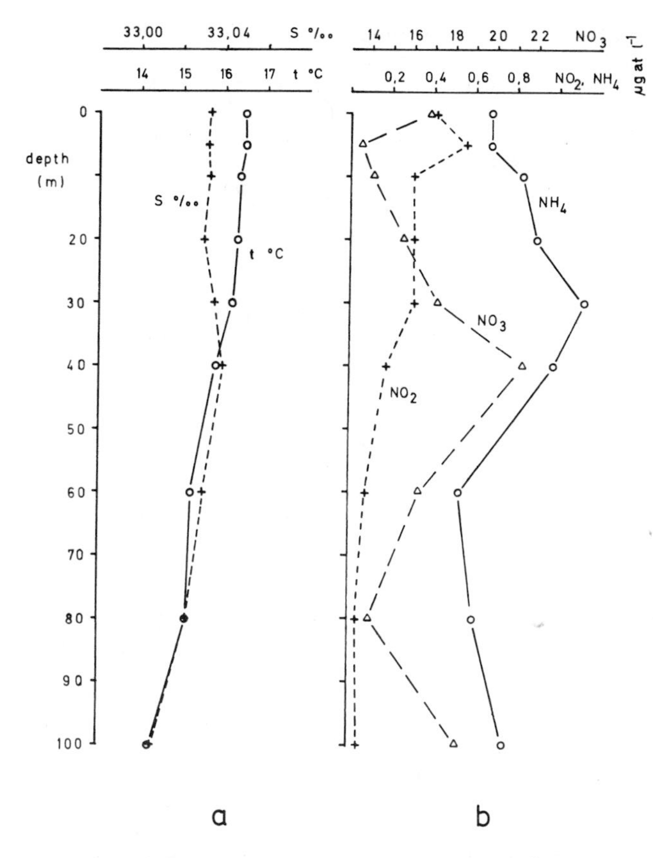

Fig. 2. a: Distribution of temperature (T °C) and salinity (S °/oo) at Sta. g, b: Distribution of nitrate (NO_3, µg-at/l), nitrite (NO_2, µg-at/l) and ammonia (NH_4, µg-at/l) at Sta. g.

rient for phytoplankton growth. Dugdale (1972), describing nutrient conditions in the same general area, concluded that silica is probably the nutrient that limits phytoplankton production in the upwelling region off Peru.

A remarkable decrease in dissolved oxygen concentration was found below 40 m (Fig. 3b). Values near zero were found at a depth of 80 m. Oxygen depletion was observed at all stations; oxygen concentrations below 1 ml/l were measured between 30 and 80 m. A similar oxygen distribution for the region was mentioned by Dugdale (1972), thus demonstrating the high amount of oxygen consumed during remineralization of organic material.

Chlorophyll a, particulate organic carbon and ^{14}C-uptake. The distributions of chlorophyll *a* and particulate organic carbon in the water column were rather similar (Fig. 3a). All were uniform down to about 40 m with values decreasing at depth. Microscopic examination revealed that the phytoplankton community at Sta. g, as well as the other stations, consisted mainly of centric diatoms (*Thalassiosira* spp., *Coscinodiscus* spp., *Skeletonema costatum*, *Chaetoceros* spp., and others) and pennate diatoms (*Pleurosigma* spp., and small *Nitzschia* spp.).

Primary production was restricted to the upper 30 m (Fig. 3b) at Sta. g; at Stas a to f and h the depths of euphotic zone were somewhere between 20 and 30 m.

Integrated values of chlorophyll *a* and particulate organic carbon concentrations for the uppermost 100 m (Sta. e, 55 m) and of ^{14}C-uptake within the euphotic zone are shown in Fig. 4. The same data, as well as particulate organic nitrogen concentrations and C/N ratios, are presented in Table 3. ^{14}C-uptake is considered to approximate net primary production and hence is sometimes referred to as primary production.

There is no obvious relation between the concentrations of observed variables and station depth. Besides a higher chlorophyll *a* value at Sta. a and a higher ^{14}C-uptake at Sta. f, all variables were rather uniform throughout Stas a to h, indicating similar conditions over the whole area of investigation.

Sedimentation. Sedimentation rate, expressed as particulate organic carbon and particulate organic nitrogen in mg/m^2 (12 h)$^{-1}$ from sunset to sunrise, as well as C/N ratios for sedimented material, are given in Table 4. For periods of duplicate sedimentation measurements, the mean values are presented. Single values as well as a comparison between two different trap shapes are given elsewhere. The highest values measured were between March 25 and 26, about 320 and 45 mg POC and PON m^{-2} (12 h)$^{-1}$, respectively. The lowest rates, about 130 and 19 mg POC and PON m^{-2} (12 h)$^{-1}$, respectively, were found from March 27 to 28.

Staresinic (personal communication) measured sedimentation rates in the same region some weeks earlier with a similar free-drifting sediment trap and found sedimentation rates during the night to be about 40% higher than during the day. From the data presented in Table 4 and assuming a 40% higher sedimentation between sunset and sunrise than during the day, sedimentation for the 24-h period was calculated for each station. The results, together with daily primary production (when there was more than one productivity measurement during a single day the mean value was calculated), are given in Table 5.

In terms of particulate organic carbon, the average sedimentation rate was about 13% of the organic material produced by primary production (range: 18.9 to 6.7%). Most of the material collected in the sediment traps was in form of fecal casts, most probably originating from anchoveta. Microscopic examination of the material showed that the pellets contained phytoplankton cells, mainly diatoms, in every state of breakdown. Smayda (1971) investigated the sinking velocity of phytoplankton and of zooplankton fecal pellets; velocities of zooplankton fecal pellets ranged from about 50 to several hundred meters per day. The fecal casts found in the

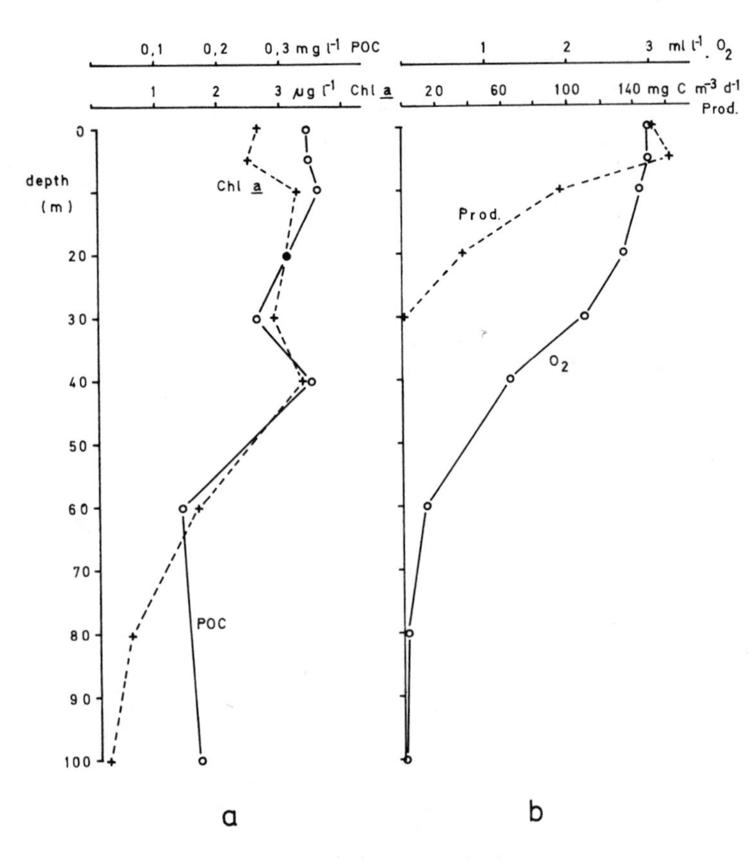

Fig. 3. a: Distribution of chlorophyll *a* (chl *a*, µg/1) and particulate organic carbon (POC mg/1) at Sta. g, b: Distribution of ^{14}C-uptake (Prod., mg C $m^{-3}d^{-1}$) and dissolved oxygen (O_2, ml/1) at Sta. g.

sediment traps are much bigger and thus have a higher velocity. Staresinic *et al.* (1978) found the still-water terminal velocity of similar material collected off Peru in 1977 to be over 800 m/d. One can thus assume that the time lag between the uptake of particulate organic material by fishes and the collection of part of it in the form of fecal casts is small.

Discussion

Few direct sedimentation measurements in relation to primary production are available from the literature. This is partly due to difficulties involved with sediment trap methodology, which hamper direct comparison between actual sedimentation and material collected in moored sediment traps (Bröckel, 1975; Gardner, 1977; Staresinic *et al.*, 1978). In this study most of the problems are believed to have been avoided by using a free-drifting sediment trap.

For comparison, some relationships between sedimentation and primary production rate from the literature are presented in Table 6. The value found in this study, approximately 13% sedimentation in relation to ^{14}C-uptake, is in the lower range of the data. The lowest rates, 7 to 8%, were reported by Ansell (1974) for sedimentation during summer months in a Scottish sea loch. The only value for the Peruvian upwelling area was calculated by Dugdale and Goering (1970); according to them, about 36% of a phytoplankton population was renewed daily. Of this 20% was taken up through grazing and 16% was lost by sinking and mixing. That means that nearly 45% of the organic material produced daily is lost through sinking and mixing. The value is considerably higher than the one we estimate. Keeping in mind that Dugdale and Goering calculated the sedimentation only for one station and that our value resulted from measurements over just a 6-day period, both indicate the range possible for short-term sedimentation rates and show the necessity for more

TABLE 3. Integrated values for a 100-m Water Column of Chlorophyll *a*, Particulate Organic Carbon, and Particulate Organic Nitrogen, Integrated ^{14}C-uptake, and the Ratio of POC to PON (C/N) for Stas a to h.

Station	Chlorophyll *a* (g/m^2)	^{14}C-uptake (g C $m^{-2}d^{-1}$)	Part. Org. Carbon (g/m^2)	Part. Org. Nitrogen (g/m^2)	C/N
a	0.415	2.71	-	-	-
b	0.160	3.22	28.41	4.35	6.53
c	0.279	3.17	27.04	4.69	5.79
d	0.230	2.59	27.50	4.36	6.31
e	0.323	4.06	19.83	3.21	6.17
f	0.222	5.69	28.99	4.29	6.75
g	0.204	2.28	24.17	3.71	6.51
h	0.244	2.81	-	-	-

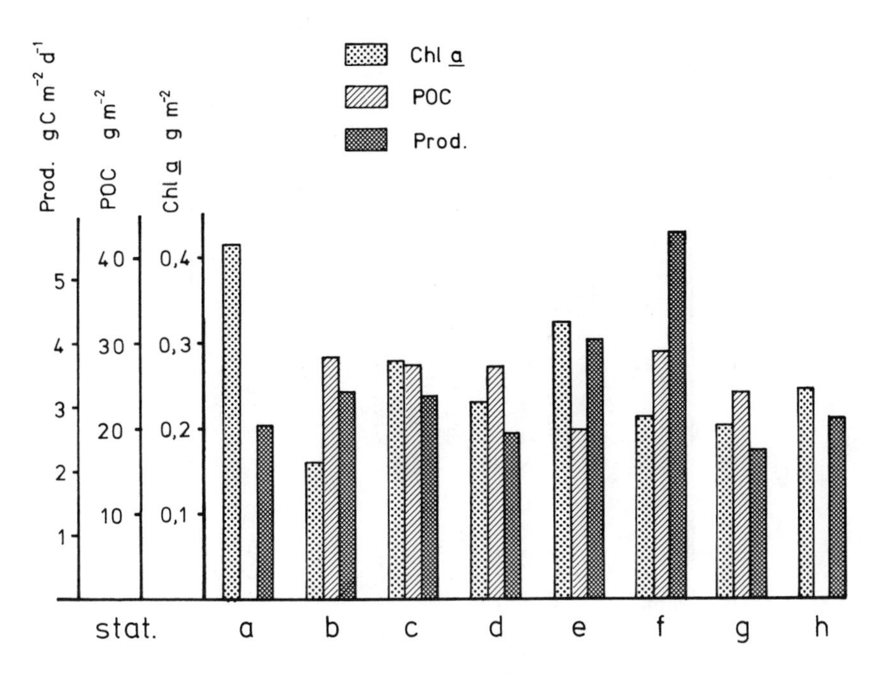

Fig. 4. Integrated values of chlorophyll *a* (chl *a*, g/m^2), particulate organic carbon (POC, g/m^2), and ^{14}C-uptake (Prod., g C $m^{-2}d^{-1}$) at Stas a to h for a 100-m water column (Sta. e, 55m)

detailed investigations over longer periods and wider areas.

Ryther *et al.* (1971) studied the production and cycling of organic material off the Peruvian coast over 5 days within a parcel of newly-upwelled water. Average phytoplankton production amounted to 10 g C $m^{-2}d^{-1}$. This enormous primary productivity resulted at the end of the 5-day period in a standing stock of particulate organic carbon of only about 10 g C/m^2 (integrated for a 150-m water column). Ryther *et al.* found no explanation for the difference between the very high production and the comparably low increase in standing stock. In our study, most of the material collected in the sediment traps was fecal casts. It seems possible that the "loss" of particulate organic material observed by Ryther *et al.* (1971) was due to anchoveta grazing and hence to an accelerated sedimentation of fecal casts, which could not be detected.

Although the sedimentation rate of 13% of primary production found here is not very high, the total amount of particulate organic material reaching the sediments is, compared to other regions, high. Estimates give an annual primary production for the Peruvian upwelling area of somewhere between 300 and 500 g C $m^{-2}yr^{-1}$ (Ryther, 1969; Rowe, 1971). Assuming an average sedimen-

TABLE 4. The Measured Sedimentation Rate Expressed as Particulate Organic Carbon and Particulate Organic Carbon and Particulate Organic Nitrogen in mg m^{-2} $(12\ h)^{-1}$ from Sunset to Sunrise and C/N-ratios of Sedimented Material.

Date	Part. Org. Carbon (mg m^{-2} 12 h^{-1})	Part. Org. Nitrogen (mg m^{-2} 12 h^{-1})	POC/PON
23-24 March	319.6	36.1	8.86
24-25 March	190.6	24.5	7.77
25-26 March	365.7	44.5	8.21
27-28 March	150.9	18.6	8.10
		$\bar{x}$:	8.24

TABLE 6. Values of the Ratio of Sedimented to Produced Matter

Sedimentation in % of Primary Production	Area of Investigation	Method Used	Author
7 - 8	Loch Etive & Creran, Scotland	Moored sediment traps	Ansell, 1974
13	Upwelling area off Peru	Floating sediment traps	this study
20	Eckernförde Bay, Baltic Sea	Moored sediment traps and calculations	Zeitzschel, 1965
30	Loch Ewe, Scotland	Moored sediment traps	Steele and Baird, 1972
20 - 40	Long Island Marsh; USA	Calculation from sediment data	Armentano and Woodwell, 1975
40	Shallow water ecosystem Venezuela	Moored sediment traps	Edwards, 1973
45	Upwelling region off Peru	Calculations	Dugdale and Goering, 1970
50	Departure Bay, B.C., Canada	Moored sediment traps	Stephens, Sheldon, and Parsons, 1967
60	Eckernförde Bay, Baltic Sea	Moored sediment traps and calculations	Bröckel, 1978

TABLE 5. The Calculated Amount of Sedimented and Produced Particulate Organic Material and the Ratios of Sedimented Material in a % of Particulate Material Produced

Date	1 Sedimented Material (POC) ($g\ m^{-2}d^{-1}$)	2 Produced Material ($g\ C\ m^{-2}d^{-1}$)	Ratio of 1:2 (%)
23 March	0.511	2.71	18.9
24 March	0.401	3.22	12.5
25 March	0.445	3.17	14.0
26 March	0.583	4.11	14.2
27 March	0.241	3.59	6.7
28 March	0.241	2.55	9.5
	$\bar{x}$: 0.404	$\bar{x}$: 3.23	$\bar{x}$: 12.7

tation rate of 13% throughout the whole year, between 39 and 65 $g\ C\ m^{-2}yr^{-1}$ would be deposited in the sediment.

Acknowledgements. This study was supported by the Deutsche Forschungsgemeinschaft. I am indebted to the Woods Hole Oceanographic Institution and Dr. S.W. Watson, chief scientist, for the opportunity to participate on *Knorr* Cruise, 73-3. I thank N. Staresinic and H. Clifford for helping with the traps and especially J. Kogelschatz (Duke University Marine Laboratory) for supplying most of the physical, chemical, and biological data.

References

Ansell, A.D., Sedimentation of organic detritus in lochs Etive and Creran, Argyll, Scotland, *Marine Biology, 27,* 263-273, 1974.

Armentano, T.V., and G.M. Woodwell, Sedimentation rates in a Long Island marsh determined by ^{210}Pb dating, *Limnology and Oceanography, 20,* 452-456, 1975.

Bröckel, K. von, Der Energiefluß im pelagischen Ökosystem vor Boknis Eck (Westl. Ostsee), Ph.D. Thesis, University of Kiel, 96 pp., 1975.

Bröckel, K. von, An approach to quantify the energy flow through the pelagic part of the shallow water ecosystem off Boknis Eck (Eckernförde Bay), *Kieler Meeresforschungen,* Sonderheft 4, 233-243, 1978.

Cushing, D.H., Upwelling and fish production, *Advances in Marine Biology, 9,* 255-334, 1971.

Dugdale, R.C., Chemical oceanography and primary productivity in upwelling regions, *Geoforum, 11,* 47-61, 1972.

Dugdale, R.C., and J.J. Goering, Nutrient limitation and the path of nitrogen in Peru Current production, *Anton Bruun Reports, 5,* 3-8, 1970.

Edwards, R.R.C., Production ecology of two Caribbean marine ecosystems II. Metabolism and energy flow, *Estuarine and Coastal Marine Science, 1,* 319-333, 1973.

Gardner, W.D., Fluxes, dynamics, and chemistry of particulates in the ocean, Ph.D. Thesis, M.I.T./W.H.O.I. Joint Program in Oceanography 1977, 402 pp., 1977.

Rowe, G.T., Benthic biomass in the Pisco, Peru upwelling, *Investigacion Pesquera, 35,* 127-135, 1971.

Ryther, J.H., Photosynthesis and fish production in the sea, *Science, 166,* 72-76, 1969.

Ryther, J.H., D.W. Menzel, E.M. Hulbert, C.J. Lorenzen, and N. Corwin, The production and utilization of organic matter in the Peru coastal current, *Investigacion Pesquera, 35,* 43-59, 1971.

Smayda, T.J., Normal and accelerated sinking of phytoplankton in the sea, *Marine Geology, 11,* 105-122, 1971.

Staresinic, N., G.T. Rowe, D. Shaughnessey, and A.J. Williams III, Measurement of the vertical flux of particulate organic matter with a free-drifting sediment trap, *Limnology and Oceanography, 23,* 559-563, 1978.

Steele, J.H., and I.E. Baird, Sedimentation of organic matter in a Scottish sea loch, *Memorie dell'Istituto Italiano di Idrobiologia, Suppl., 29,* 73-88, 1972.

Stephens, K., R.W. Sheldon, and T.R. Parsons, Seasonal variations in the availability of food for benthos in a coastal environment, *Ecology, 48,* 852-855, 1967.

Strickland, J.D.H., and T.R. Parsons, A practical handbook of seawater analysis, *Fisheries Research Board of Canada, Bulletin, 167,* 311 pp., 1968.

Whitledge, T.E., and T.T. Packard, Nutrient excretion by anchovies and zooplankton in Pacific upwelling regions, *Investigacion Pesquera, 35,* 243-250, 1971.

Zeitzschel, B., Zur sedimentation von Seston, eine produktionsbiologische Untersuchung von Sinkstoffen und Sedimenten der Westlichen und Mittleren Ostsee, *Kieler Meeresforschungen, 21,* 55-80, 1965.

Zeitzschel, B., P. Diekmann, and L. Uhlmann, A new multisample sediment trap, *Marine Biology, 45,* 285-288, 1978.

NATURAL AND MAN-MADE RADIOCARBON AS A TRACER FOR COASTAL UPWELLING PROCESS

Stephen W. Robinson
U.S. Geological Survey
345 Middlefield Road
Menlo Park, California 94025
U.S.A.

Abstract. Radiocarbon enters the ocean by gas exchange of carbon dioxide at the surface. Radioactive decay produces a vertical profile with radiocarbon concentration decreasing with depth and makes the radiocarbon content of marine bicarbonate suitable as a tracer for the upwelling process. Because of the slow exchange with atmospheric carbon dioxide, radiocarbon is a more conservative tracer than temperature, salinity, or nutrient content over the annual upwelling cycle.

Large seasonal variations in radiocarbon in surface water attributed to coastal upwelling were observed at Half Moon Bay, California in 1978 and 1979. The shells of West Coast mollusks that lived before 1950 have significantly less radiocarbon than those from coasts not affected by upwelling.

Introduction

Since the pioneering studies of Broecker, Gerard, Ewing, and Heezen (1960) in the Atlantic and of Bien, Rakestraw, and Suess (1960) in the Pacific, the use of natural and man-made radiocarbon as a tracer of ocean mixing processes has been recognized as a powerful tool in chemical and physical oceanography (Broecker, 1974). Over the time scale of the annual upwelling cycle, radiocarbon is an almost conservative tracer in the near-surface waters affected by upwelling. We have examined the utility of radiocarbon for coastal upwelling studies with a two-year time series of surface water radiocarbon determinations at Half Moon Bay, California (37.5°N) and a 137-km offshore transect along 38.0°N, as well as analyses of the shells of molluscs that lived prior to 1950. The surface water measurements show marked seasonal variations indicating the presence of upwelled water at the surface during the summer months, and the pre-1950 mollusks have significantly lower radiocarbon concentrations than mollusks from regions not affected by upwelling.

Radiocarbon in the Pacific Ocean

Radiocarbon (^{14}C) with a half-life of 5730 years is produced naturally at high altitudes by cosmic ray bombardment of the atmosphere. Concentrations in nature are reported as $\Delta^{14}C$, which is the deviation in parts per thousand (o/oo) of the sample's $^{14}C/^{12}C$ ratio from that of the 1950 atmosphere (Stuiver and Polach, 1977). In addition to naturally produced radiocarbon, atmospheric nuclear weapons testing has added significantly to the global radiocarbon inventory since the mid 1950's; radiocarbon in the northern hemisphere atmosphere reached twice the 1950 concentration in late 1963 (Nydal, 1966).

Inorganic carbon dissolved in the oceans dominates that part of the global carbon cycle in exchange with the atmosphere on time scales shorter than the half-life of radiocarbon, comprising about 94% of the total carbon in the system. Radiocarbon is transferred into the marine carbon reservoir by gas exchange across the ocean-atmosphere interface at a global average rate of 15 to 25 moles of carbon $m^{-2} yr^{-1}$ (Broecker, Peng, and Stuiver, 1978). Prior to 1950, the transfer of ^{14}C into the oceans by gas exchange was approximately in balance with radioactive decay. The balance resulted in $\Delta^{14}C$ values of about -45 o/oo in the surface waters and about -235 o/oo below a depth of 1300 m in the Pacific. In the open Pacific, the vertical distribution of radiocarbon is a result of the interplay between vertical eddy diffusion, advection, and radioactive decay, termed the diffusion-advection model (Craig, 1969; Broecker, 1974). Profiles of $\Delta^{14}C$ in the northeast Pacific are reasonably well approximated in the upper 100 m by the diffusion-advection model with vertical diffusivity, K_v = 1260 m^2/yr (0.4 cm^2/s) and upward advection of 1.5 m/yr (Fig. 1). This value of K_v in the upper thermocline agrees with the findings of Sarmiento *et al.* (1976) that the vertical diffusivity varies inversely with the density gradient and that of Broecker (1974), who suggested a K_v of 12,600 m^2/yr below 1000 m in the northeast Pacific. To

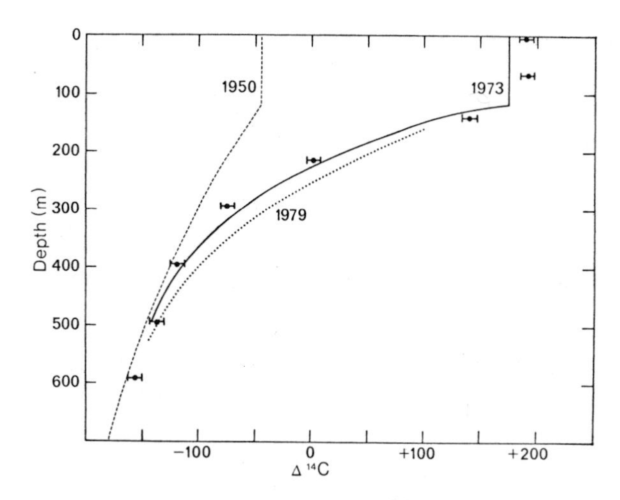

Fig. 1. Vertical profiles of $\Delta^{14}C$ from model calculation with K_v = 1260 m^2/yr, w = 1.5 m/yr, E = 21 moles $m^{-2}yr^{-1}$, and mixed layer thickness of 120 m for 1950, 1973, and 1979. The measured profile at GEOSECS Station 201 (34°N, 128°W; Ostlund *et al.*, 1979) is shown for comparison with the 1973 calculated profile.

model the evolution of the radiocarbon profile in response to the excess atmospheric radiocarbon produced in nuclear weapons testing, an atmospheric exchange rate of 21 moles $m^{-2}yr^{-1}$ and an annual average mixed layer thickness of 120 m are added to the diffusion-advection model. Diffusion from the mixed layer into the thermocline is reduced in the model by 2/3 to approximate the inhibiting effect of the strongly developed summer pycnocline. Results of the calculation are shown in Figs 1 and 2. In the absence of offshore radiocarbon data for the period of this study (1978-1979), the model will be used in the interpretation of the effect of coastal upwelling on the reported measurements.

Data from Pre-1950 Molluscs

To study the radiocarbon distribution in coastal waters prior to the onset of nuclear testing in the atmosphere it is necessary to analyze a proxy recorder of marine inorganic carbon isotopic composition such as coral (Druffel and Linick, 1978) or mollusc shell (Mangerud and Gulliksen, 1975). Carbon in these materials is believed to be in isotopic equilibrium with dissolved inorganic carbon (Mook and Vogel, 1968). The radiocarbon content of inter-tidal mollusc shells from the west coast of North America (Fig. 3) shows a marked depression from the open ocean value of -45 o/oo obtained in the U.S.G.S. laboratory from a Hawaiian mollusc shell, taken to represent a non-upwelling regime. Five measurements from the central California coast (33 to 38°N) gave an average $\Delta^{14}C$ value of -81 o/oo, which corresponds to a depth of 210 m on the pre-1950 open ocean radiocarbon profile of Fig. 1. The effect causes radiocarbon dates on mollusc shells from the West Coast to give dates 600 to 900 years too old and for accurate dating an

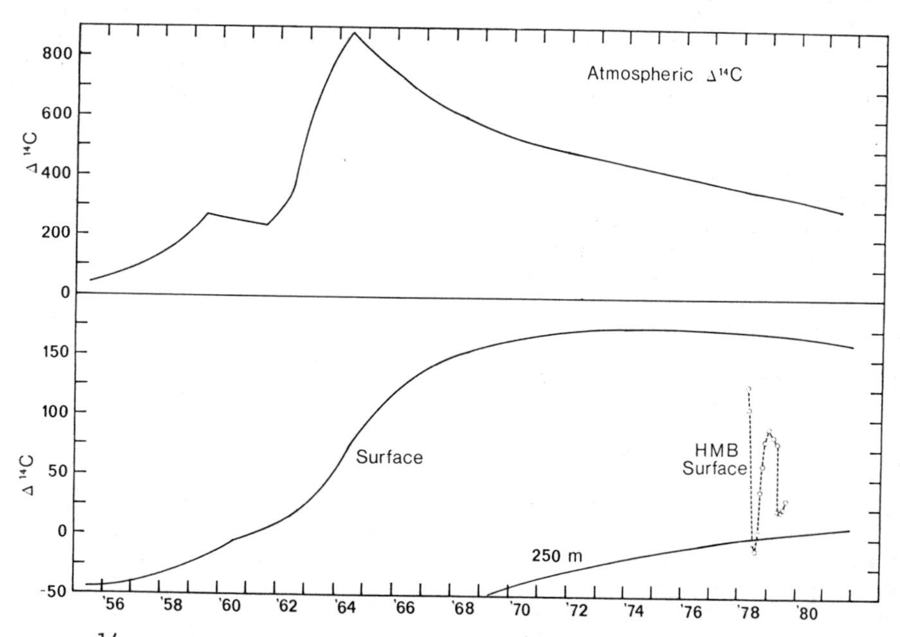

Fig. 2. Variation of $\Delta^{14}C$ in the atmosphere from analysis of tree-ring wood by Cain and Suess (1976) and calculated variation of ocean surface and at 250-m depth, with results of Half Moon Bay surface water analyses (--o--).

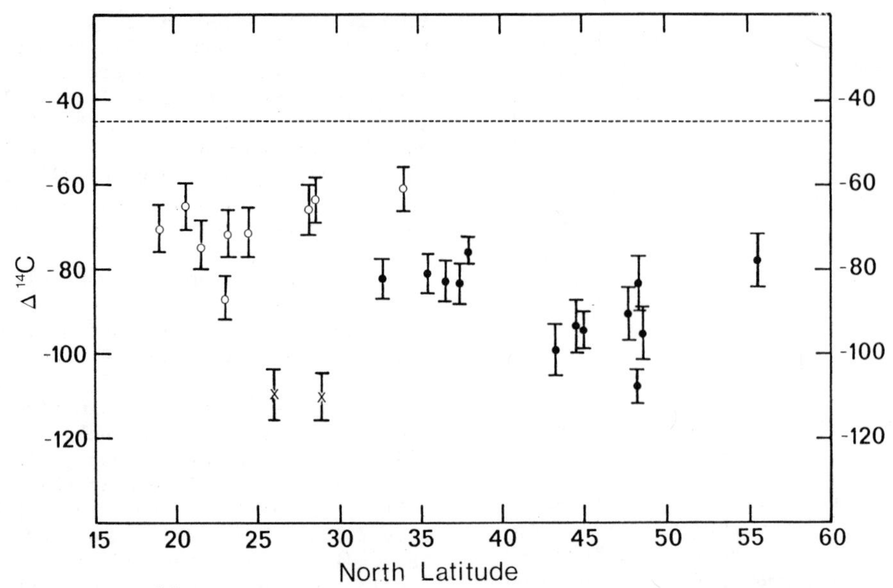

Fig. 3. Reconstruction of $\Delta^{14}C$ in pre-1950 west coast dissolved inorganic carbon from analysis of museum specimen intertidal mollusc shells. Included are results from Baja California (o) and the Gulf of California (x) calculated from data in Berger, Taylor, and Libby (1966). The four points near 48°N are from northern Puget Sound. For comparison, the result for a Hawaiian shell (-45 o/oo) is shown by dashed line.

appropriate correction must be applied (Robinson and Thompson, 1979).

Direct Surface-Water Measurements

Seasonal variations in $\Delta^{14}C$ associated with upwelling have been monitored at Half Moon Bay, California (37.5°N) since April 1978. Water samples (90 l) were collected at about one-month intervals at the surface in Pillar Point Harbor. The dissolved inorganic carbon was extracted as carbon dioxide, and the radiocarbon activity was measured by gas proportional counting to a precision of at least ± 5 o/oo (Robinson, 1979). Tidal mixing of the harbor occurs with a time constant of about one day and does not have a significant damping effect on short-term variations. On two occasions in the summer of 1978 samples were collected simultaneously on the beach just outside the harbor, and their radiocarbon values did not differ significantly from the harbor sample. From the fairly smooth seasonal variation of the results (Fig. 4), it appears that short wind events do not produce significant variations in $\Delta^{14}C$ at the coast. In addition to the time series of Half Moon Bay samples, an offshore transect of three samples was collected on 31 August 1978 by the U.S. Coast Guard Cutter *Yocona*. The transect extended to 137 km from the coast and is supplemented by a Half Moon Bay sample collected five days later. In Fig. 5 the radiocarbon results of the transect are compared with the estimated 1978 $\Delta^{14}C$ at GEOSECS (Geochemical Ocean Sections) Station 201 (34°N, 128°W; Ostlund, Brescher, Oleson, and Ferguson, 1979).

Discussion

The Half Moon Bay time series of surface water analyses (Fig. 4) shows the general features to be expected from a summer upwelling regime. An abrupt drop in the radiocarbon content in April or May from a late winter maximum indicates the

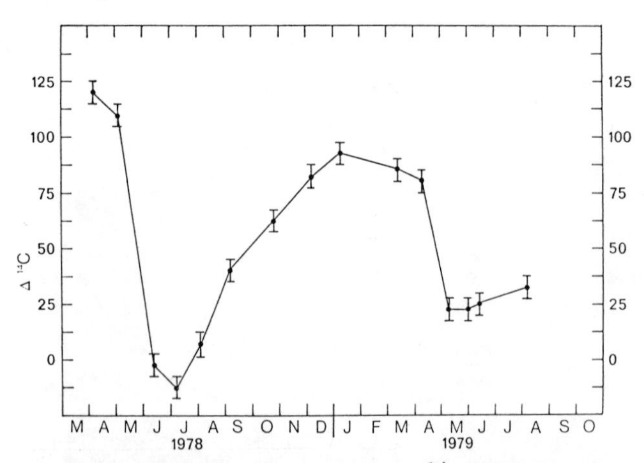

Fig. 4. Seasonal variations of $\Delta^{14}C$ in surface water samples from Pillar Point Harbor, Half Moon Bay, California (37.5°N). The May 1978 sample was collected in Monterey Bay (36.6°N).

presence of upwelled water at the surface. After two to four months of upwelling, $\Delta^{14}C$ begins to climb gradually toward another late winter maximum. The summer $\Delta^{14}C$ minima observed in 1978 and 1979 correspond to average source water depths of 260 and 230 m, respectively, on the calculated radiocarbon profiles for those years. These water depths occur about 45 km offshore of our sampling site. In the Half Moon Bay record, the winter $\Delta^{14}C$ maxima are still 40 to 80 o/oo below the expected surface values far from the coast. The difference indicates a persisting negative radiocarbon anomaly along the coast, maintained either by occasional winter upwelling events or the slowness of horizontal mixing. Another feature of the record is the marked difference between the seasonal signatures in 1978 and 1979, the former year having both a higher winter maximum and a lower summer minimum. This effect appears to be correlated with the meteorologically derived upwelling index of Bakun (1973 and personal communication) for the period of our measurements. The monthly average upwelling indices for 1978-1979 are shown in Fig. 6. The high winter 1978 $\Delta^{14}C$ values seem to be related to the onshore flow indicated by the negative indices of that winter. The upwelling indices show that 1978 upwelling was more intense and occurred over a longer period than in 1979, which may account for the lower $\Delta^{14}C$ minimum during the summer of 1978.

The offshore transect (Fig. 5) shows the distribution of radiocarbon at approximately the end of the upwelling season, though at the coast roughly half of the relaxation toward the winter maximum had already occurred by 5 September. However, the transect probably shows the maximum seasonal upwelling effect in the offshore waters. Fig. 5 also shows the calculated surface $\Delta^{14}C$ for an open ocean station (Fig. 2b) plotted at the offshore distance of GEOSECS Station 201 (615 km). If offshore flow had continued throughout the upwelling season without subsequent sinking of the upwelled water, a strong effect on $\Delta^{14}C$ would have been observed hundreds of kilometers from the coast. The transect indicates that the upwelling circulation cell was confined to within about 120 km from the coast. Seasonal variations at the 137-km station are probably much smaller than at the coast, as indicated by the similarity of $\Delta^{14}C$ with that observed at the coast in April 1978 after a winter of negative upwelling indices and presumed onshore flow. Between 140 km of shore and the open ocean, then, the radiocarbon distribution is the product of horizontal and vertical mixing. In this discussion the effect of the southern transport of the surface water during upwelling is neglected. This may be justified by the small variation with latitude shown in the pre-1950 mollusc shell analyses (Fig. 3).

It is possible to reinterpret the data on pre-1950 molluscs (Fig. 3) in the light of the data now available on the seasonal variations in $\Delta^{14}C$ produced by coastal upwellings. As a result, the

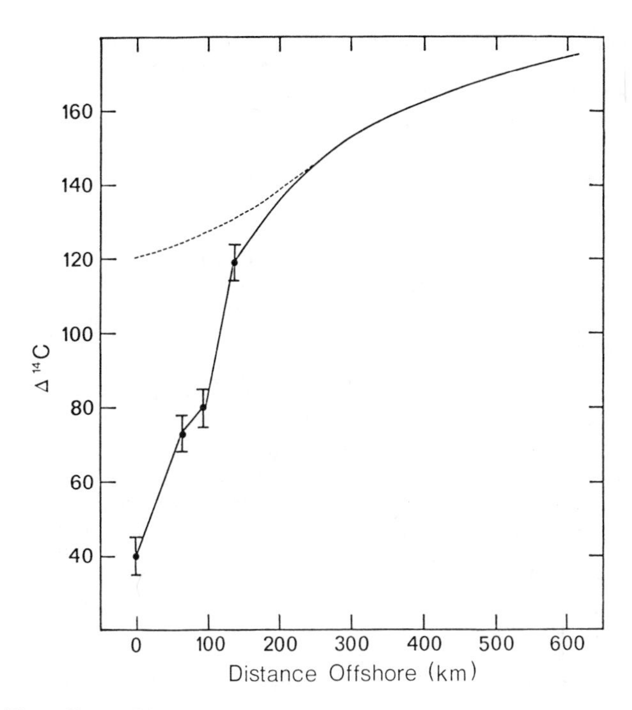

Fig. 5. Offshore transect of 31 August 1978 along 38.0°N. Coastal data point was at 37.5°N on 5 September 1978 and the model calculation is plotted at 615 km, the distance of GEOSECS Station 201. Dashed curve indicates hypothetical late winter situation.

average ^{14}C content of the shell depends on the seasonal distribution of shell growth. In the simplest approximation

$$\Delta_{shell} = f\,\Delta_{upwelling} + (1-f)\Delta_{winter}, \quad (1)$$

where f is the fraction of growth during the summer months. The Half Moon Bay seasonal record

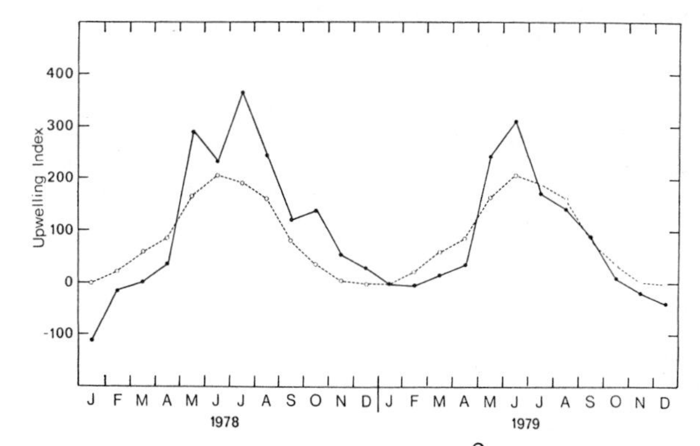

Fig. 6. Upwelling indices for 39°N (Bakun, 1973 and personal communication). Twenty-year average 1948-1967 (--o--) and monthly indices for 1978 and 1979 (—•—).

shows that the winter $\Delta^{14}C$ at the coast reaches about 2/3 of the way between the summer minimum and the open ocean surface value. Therefore, for the pre-1950 shells

$$\Delta_{winter} = 1/3\ \Delta_{upwelling} - 30 \qquad (2)$$

so that

$$\Delta_{upwelling} = 3\ \frac{\Delta_{shell} + 30(1-f)}{2f + 1} \qquad (3)$$

Reasonable values of f lie in the range of 0.7 to 0.9, yielding upwelling source depths of 260 and 240 m, respectively, on the 1950 $\Delta^{14}C$ profile of Fig. 2. The values are in excellent agreement with the conclusions based on the 1978 and 1979 summer minima in the Half Moon Bay data.

There are no data on the distribution of radiocarbon in the waters of the Gulf of California. However, the two pre-1950 shells (Fig. 3) show a lower $\Delta^{14}C$ (-110 o/oo) than the west coast shells. The difference indicates either a deeper source of upwelled water (ca. 450 m) or a generally lower radiocarbon concentration in the waters of the Gulf of California. Intense upwelling in the gulf is known to cause high organic productivity and laminated sediments (Schrader *et al.*, 1980), the accurate radiocarbon dating of which will require an understanding of the upwelling effects.

References

Bakun, A., Coastal upwelling indices, west coast of North America, 1946-71, NOAA Technical Report NMFS SSRF-671, 103 pp., 1973.

Bien, G.S., N.W. Rakestraw, and H.E. Suess, Radiocarbon concentration in Pacific Ocean water, *Tellus, 12,* 436-443, 1960.

Berger, R., R.E. Taylor, and W.F. Libby, Radiocarbon content of marine shells from the California and Mexican West Coast, *Science, 153,* 864-866, 1966.

Broecker, W.S., *Chemical Oceanography,* Harcourt, Brace, Jovanovich, New York, 214 p., 1974.

Broecker, W.S., R. Gerard, M. Ewing, and B.C. Heezen, Natural radiocarbon in the Atlantic Ocean, *Journal of Geophysical Research, 65,* 2903-2931, 1960.

Broecker, W.S., T.-H. Peng, and M. Stuiver, An estimate of the upwelling rate in the equatorial Atlantic based on the distribution of bomb radiocarbon, *Journal of Geophysical Research, 83,* 6179-6186, 1978.

Cain, W.F., and H.E. Suess, Carbon 14 in tree rings, *Journal of Geophysical Research, 81,* 3688-3694, 1976.

Craig, H., Abyssal carbon and radiocarbon in the Pacific, *Journal of Geophysical Research, 74,* 5491-5506, 1969.

Druffel, E.M., and T.W. Linick, Radiocarbon in annual coral rings of Florida, *Geophysical Research Letters, 5,* 913-916, 1978.

Mangerud, J., and S. Gulliksen, Apparent radiocarbon ages of recent marine shells from Norway, Spitsbergen, and Arctic Canada, *Quarternary Research, 5,* 263-273, 1975.

Mook, W.G., and J.C. Vogel, Isotopic equilibrium between shells and their environment, *Science, 159,* 874-875, 1968.

Nydal, R., Variations in C14 concentration in the atmosphere during the last several years, *Tellus, 18,* 271-279, 1966.

Ostlund, H.G., R. Brescher, R. Oleson, and M.J. Ferguson, GEOSECS Pacific radiocarbon and tritium results (Miami), Tritium Laboratory Data Report #8, 14 pp., University of Miami, Rosenstiel School of Marine and Atmospheric Sciences, Miami, Florida, 1979.

Robinson, S.W., Radiocarbon dating at the U.S. Geological Survey, Menlo Park, California, in *Radiocarbon Dating,* edited by R. Berger and H.E. Suess, p. 268-273, University of California Press, Berkeley, 1979.

Robinson, S.W., and G. Thompson, Radiocarbon corrections for marine shell dates with application to prehistory of the southern Northwest Coast, Program and Abstracts, Annual Meeting, Society for American Archaeology, Vancouver, B.C., April 1979, p. 68.

Sarmiento, J.L., H.W. Freely, W.S. Moore, A.E. Bainbridge, and W.S. Broecker, The relationship between vertical eddy diffusion and buoyancy gradient in the deep sea, *Earth and Planetary Science Letters, 32,* 357-370, 1976.

Schrader, H., K. Kelts, J. Curray, D. Moore, E. Aguayo, M.P. Aubry, G. Fornari, J. Gieskes, J. Guerrero, M. Kastner, M. Lyle, Y. Matoba, A. Molina-Cruz, J. Niemitz, J. Rueda, A. Saunders, B. Simoneit, and V. Vaquier, Laminated diatomaceous sediments from the Guaymas Basin slope (Central Gulf of California): 250,000-year climate record, *Science, 207,* 1207-1209, 1980.

Stuiver, M., and H.A. Polach, Discussion: reporting of ^{14}C data, Radiocarbon, *19,* 355-363, 1977.

Upwelling, Primary Production, and Phytoplankton

PHOTOSYNTHETIC PARAMETERS AND PRIMARY PRODUCTION OF PHYTOPLANKTON POPULATIONS OFF THE NORTHERN COAST OF PERU

W.G. Harrison and T. Platt

Fisheries and Oceans, Marine Ecology Laboratory,
Bedford Institute of Oceanography, Dartmouth, N.S. Canada

R. Calienes and N. Ochoa

Instituto del Mar del Peru, Esquina Gamarra y
General Valle, Apartado 3734, Callao, Peru

Abstract. A comparison of results from light-saturation experiments and standard productivity measurements of natural phytoplankton populations from the coastal waters off northern Peru revealed that dense, actively growing populations were confined close inshore and were highly variable over small space and time scales. Variability in maximum specific production rate (P_m^B), low NO_3/SiO_3 concentration ratios and high C:N particulate and assimilation ratios suggested that production in the region was probably nitrogen limited during the study period. Production and nutrient characteristics of some "brown" and "blue" water patches suggested that conditioning of upwelled water may have been necessary for production to be enhanced. Population adaptation to prevailing light conditions was rapid and at least partially under endogenous control.

Introduction

Primary production off the northern coast of Peru varies significantly over seasonal time scales and shows considerable spatial variance within seasons (Guillen, Calienes, and Rondan, 1977). Nutrients and light (variations in mixed layer depth) have been considered the most important environmental regulators based on results from primary productivity survey data.

Platt and Jassby (1976) proposed a new approach for studies of environmental control of primary production processes based on changes in the fundamental relationship between photosynthesis and light (light-saturation curve) under varying environmental conditions. Changes in the parameters characterizing the *P-I* (photosynthesis-irradiance) relationship for natural phytoplankton populations can be related to changes in environmental properties (light, nutrients, temperature, etc.) and the strength of this covariance can be used to assess the relative importance of environmental properties in regulating photosynthesis.

Light-saturation experiments were done on a number of occasions coincident with conventional hydrography-productivity stations off the northern coast of Peru with the objective of comparison of results to evaluate environmental control of primary production using the two independent approaches.

Methods

Light-saturation experiments. Photosynthesis (^{14}C uptake) was measured using natural phytoplankton populations taken from depths selected according to surface light penetration; in most cases, the 50 and 5% light levels were used. Most light-saturation experiments were run in duplicate for each depth using incubators of the design described by Jassby and Platt (1976). Twenty 100-ml sample bottles (two of which were dark) were each inoculated with 5 μCi ^{14}C-bicarbonate and exposed for 4 h to a range of light levels (0 to 200 W/m^2) supplied by filtered, artificial illumination (GTE sylvania PAR 150 lamp). Samples were then filtered onto membrane filters (Millipore R HA) and analyzed for radioactivity by scintillation spectrometry. A number of experiments were done using a modified incubator (Platt, Callegos, and Harrison, in press), which increased the light range to 700 W/m^2. Data from both incubator systems were pooled to construct a single light-saturation curve for each experiment, described by the equation (Platt *et al.*, in press).

$$P^B = P_s^B \left(1-e^{-\frac{\alpha I}{P_s^B}}\right) e^{-\frac{\beta I}{P_s^B}} . \qquad (1)$$

In this equation, the superscript B indicates normalization to chlorophyll *a*, and P^B is photo-

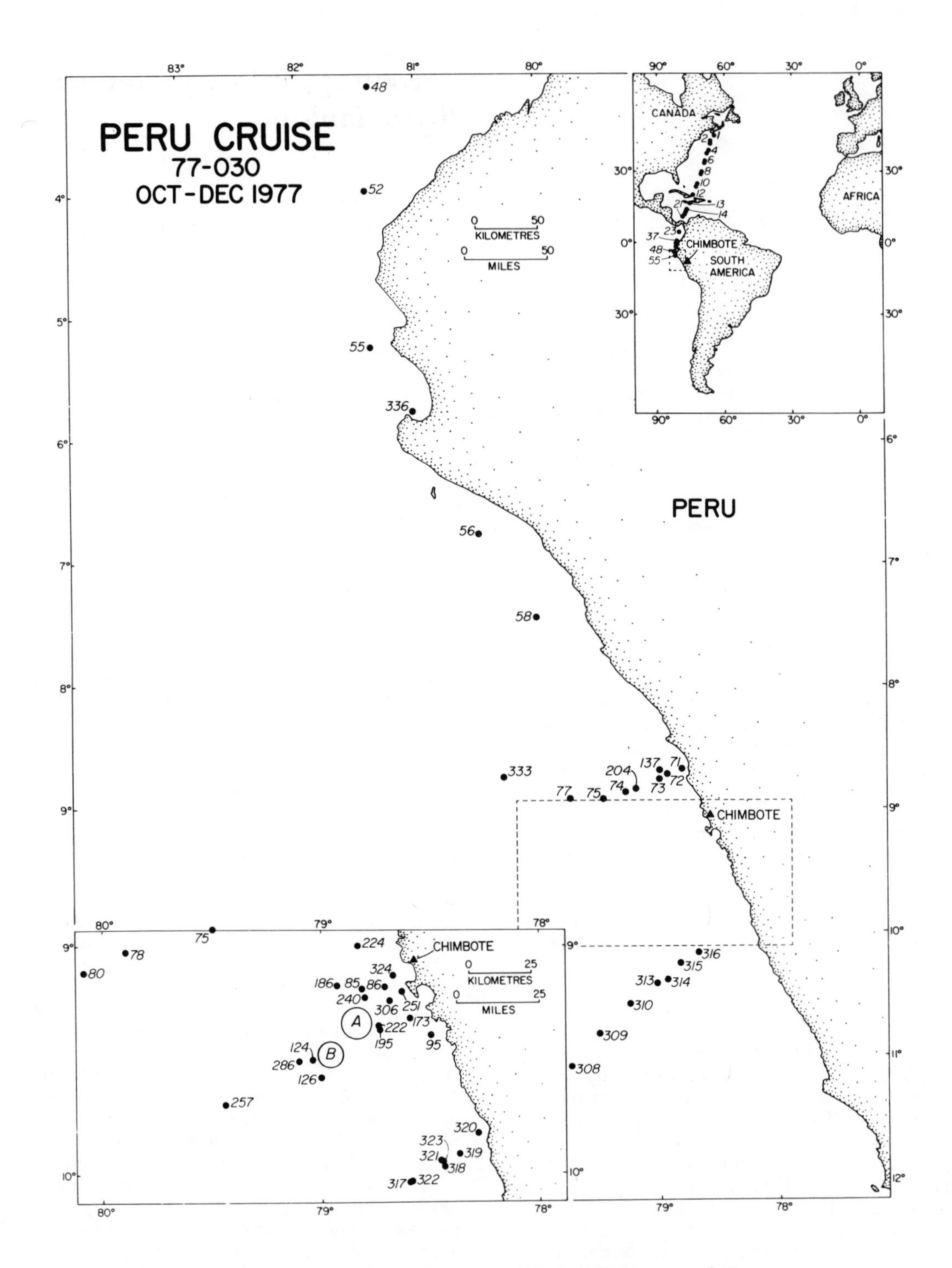

Fig. 1. Station locations, 1977 ICANE Peru cruise.

TABLE 1. Summary Statistics of Photosynthetic and Environmental Parameters, Stations 71-250, Peru, 1977, α = mg C(mg Chl)$^{-1}$h^{-1}W^{-1}m^{2}; P^B_m = mg C(mg Chl)$^{-1}$h^{-1}; $I_k = P^B_m/\alpha$; Chl, POC, PON = mg/m^3; T = ^oC; I_o = daily radiation (langleys) at sea surface; I_z = radiation at sample depth (langleys); NO_3, PO_4, SiO_3, NH_3 = mg-at/m^3.

Parameter	Mean	Range	C.V.	Number of Observations
α	0.14	0.01 - 0.33	50	35
P^B_m	4.12	0.18 - 9.90	70	35
I_k	29	4 - 66	58	35
Chl	11.61	0.11 - 110.01	207	35
POC	846	122 - 3957	106	19
PON	117	27 - 688	85	19
T	17.4	17.2 - 20.1	5	35
I_o	391	222 - 530	26	35
I_z	76	6 - 172	66	35
NO_3	4.59	0.07 - 21.68	122	34
PO_4	1.12	0.59 - 2.35	47	34
SiO_3	3.76	0.37 - 25.38	165	34
NH_3	0.49	0.01 - 2.39	104	34

C.V. = coefficient of variation ($s/\bar{x} \cdot 100$) in percent.

synthetic rate. The three parameters that characterize the P-I relationship are: 1) α, slope of the curve at low light levels, 2) P^B_m, maximum photosynthetic rate at light saturation, and 3) β, a photoinhibition parameter. From these, two additional parameters, $I_k(\equiv P^B_m/\alpha)$ and $I_b(\equiv P^B_s/\beta)$, were derived. I_k is Talling's (1957) photoadaptation parameter and I_b is an analogue for the photoinhibited light range.

Standard productivity experiments. Samples were collected from six depths (100, 50, 25, 10, 5, and 1% light penetration levels) based on euphotic depth estimates of 3X secchi reading. Photosynthesis (^{14}C uptake), $^{15}NO_3$, and $^{15}NH_4$ uptake were measured for each depth using the basic methodology of Strickland and Parsons (1972) for the carbon and of Dugdale and Goering (1967) for nitrogen. Duplicate light and single dark bottles (250 ml) were inoculated with 5 µCi ^{14}C-bicarbonate for photosynthesis measurements. At the end of incubation, particles were concentrated on glass-fiber filters (Reeve Angel 984H) and radioactivity was determined by liquid scintillation.

Nitrogen productivities were run in 1-ℓ pyrex bottles, each inoculated with 0.1 µg-at. of ^{15}N enriched substrate. Particles were collected on glass fiber filters at the end of incubation and later analyzed for ^{15}N enrichment by emission spectrometry (Jasco, Inc.) using procedures outlined by Fiedler and Proksch (1975). Samples were incubated for 24 h in deck "simulated" *in situ* incubators cooled with near-surface seawater. Time-course experiments and comparison of initial nutrient concentrations with 24-h consumption rates showed that nutrient depletion in the bottles was not a serious problem.

Analytical. Nutrients (NO_3, NH_4, SiO_3, PO_4) were determined for each sample using standard methodology (Strickland and Parsons, 1972). Chlorophyll *a* was measured fluorometrically (Holm-Hansen, Lorenzen, Holmes and Strickland, 1965) and particulate organic carbon (POC) and nitrogen (PON) by the method of Sharp (1974). Microscopic analyses of phytoplankton were made on formalin (2%)-fixed samples using the Utermöhl sedimentation technique. Daily total radiation and P-I

TABLE 2. Offshore Gradient in Photosynthetic Parameters (50% Light Level), Peru, 1977. Units as in Table 1.

Station	Distance From Shore (km)	Chl	NO_3	SiO_3	α	P_m^B
71	6	13.2	1.57	25.38	0.14	6.96
137	29	9.8	0.09	1.56	0.14	7.56
204*	58	98.1	0.36	1.06	0.16	9.59
75	92	2.9	9.61	3.19	0.15	2.09
78	142	0.4	11.55	3.58	0.12	2.24

* "Red Tide" - *Mesodinium rubrum*

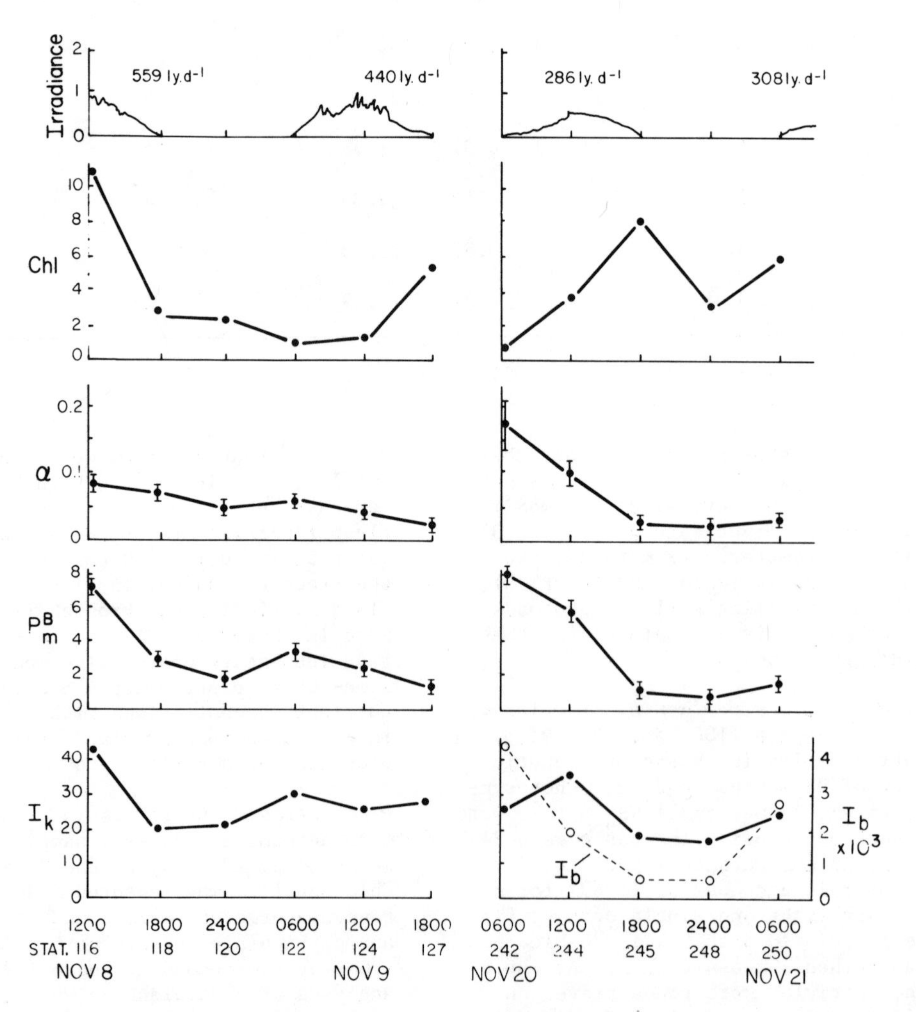

Fig. 2. Diurnal variations in photosynthetic parameters and phytoplankton biomass from two drift stations. Units as in Table 1.

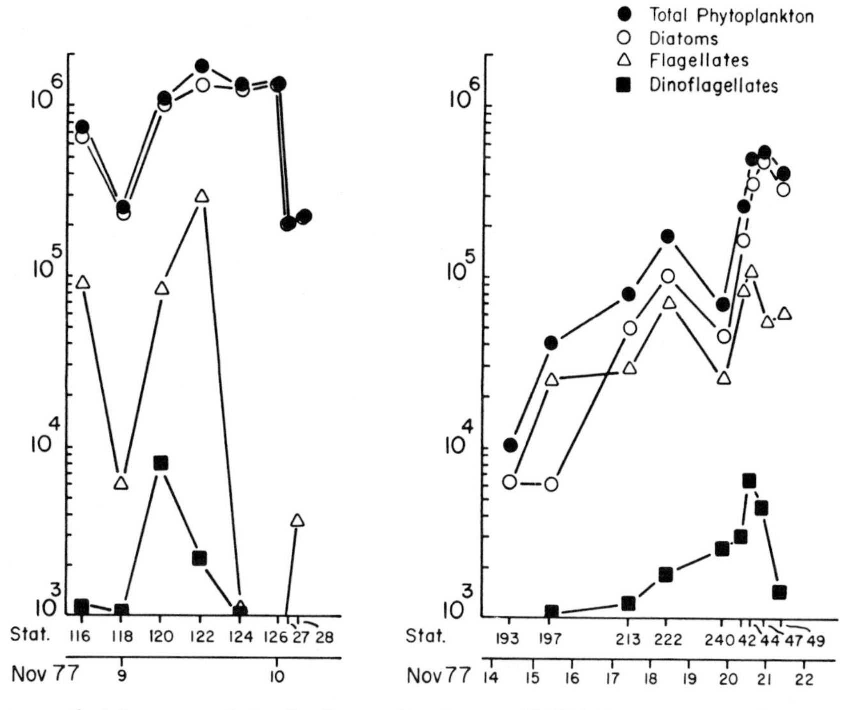

Fig. 3. Average euphotic zone phytoplankton abundance, ICANE Peru cruise, 8-22 November, 1977.

experimental light measurements were made using an Eppley-40 junction pyranometer, corrected to PAR (photosynthetically active radiation). The more conventional irradiance units (langleys) for sunlight measurements are used in this paper for ease of comparison with other studies. Energy units (e.g., Watts) are now in more common use in photosynthesis-light experimental work and have been used in this paper also. According to Strickland's review (1958), and approximate conversion would be 700 W/m^2 = 1 ly/min.

Sampling. Most samples were collected in an area centered at 9°20'S lat., 78°45'W long., covering a region approximately 250 km^2 (Fig. 1). Experiments covered a period of approximately three weeks (November) during the 1977 ICANE cruise (Doe, 1978).

Results and Discussion

Light-saturation experiments. A summary of the photosynthetic and environmental parameters used in analysis of the light-saturation experiments is presented in Table 1.

Systematic variations in photosynthetic parameters were evident from the depth and seaward extent of sampling. On the average, P_m^B(and I_k) were highest in the near-surface samples and decreased with depth. The parameter α was highest and least variable at depth. A clear onshore-offshore gradient was also apparent (Table 2) with highest values of P_m^B and α inshore. Chlorophyll *a* and SiO_3/NO_3 ratios showed a similar trend. The offshore decrease was more pronounced in P_m^B than in α. Highest values in the offshore transect were observed at 58 km from shore in a "red-water" patch dominated by the ciliate *Mesodinium rubrum*.

Samples taken during two drift stations revealed significant changes in photosynthetic parameters and phytoplankton biomass over a 24 to 36-h period (Fig. 2). During the first experiment, greatest changes in photosynthetic parameters were associated with biomass (Chl) variations. For both experiments, maxima for each 24-h period (beginning at 0600 h) were at daybreak and minima were at midnight. Biomass-related changes did not appear to be species dependent, as the relative composition of the phytoplankton remained fairly constant throughout both experiments (Fig. 3). By normalizing the photosynthesis-light curves to maximum observed photosynthetic rate (Fig. 4), a diurnal pattern in variations was more evident. Furthermore, an endogenous rhythm was also apparent from the pronounced changes in P-I curves prior to the onset of sunrise and sunset. The diurnal nature of the photosynthesis-light relationship was much more evident in I_b than in I_k (ref. Fig. 2).

Simple correlation analysis revealed that α and P_m^B were significantly correlated, r = 0.45, (Fig. 5) although less strongly than we have observed in earlier studies (Platt and Jassby, 1976). High correlation between P_m^B and I_k, r = 0.81, suggested that P_m^B variations accounted for

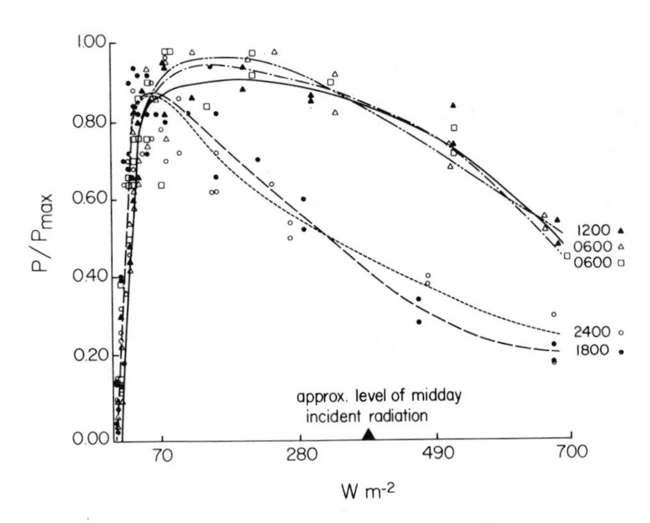

Fig. 4. Photosynthesis-light curves (normalized to P_{max}) for second drift station samples, 20-21 November, 1977.

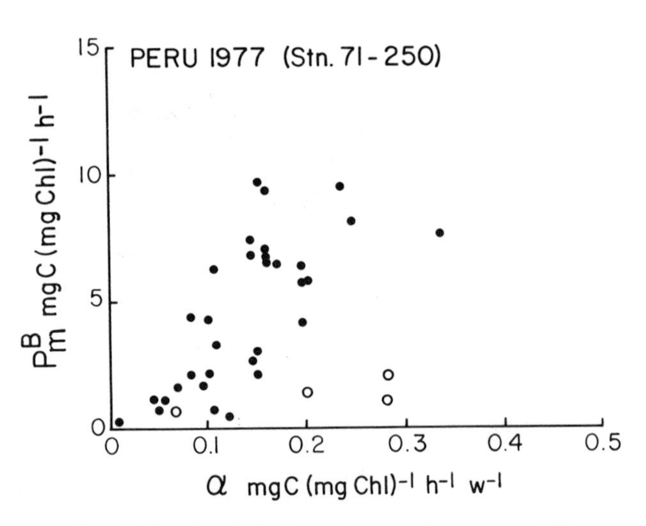

Fig. 5. Relationship between maximum specific production rate (P^B_m) and photosynthetic efficiency below light saturation (α) during ICANE Peru cruise. Open circles are stations 193 and 197.

most of the I_k variations. The parameters P^B_m and α (dependent variables) were further analyzed in relation to environmental variables (independent variables) using multiple regression statistics. Temperature, chlorophyll *a*, total inorganic nitrogen (TIN = $NO_2 + NO_3 + NH_4$), SiO_3, and light (surface and at depth for day of or day previous to sampling) were examined. TIN accounted for most of the variation in P^B_m (25%) while Chl accounted for an additional 12%. SiO_3, temperature, and light variations combined accounted for only an additional 5%. For α, all independent variables taken together explained only 10% of the variation.

The effects of deep-water nutrient enrichment on photosynthetic parameters of near-surface populations was investigated early in the study (Table 3). For two of the three experiments, P^B_m was noticeably depressed. The effects on α were less apparent.

Comparison with standard productivities. Based on the P-I analyses, nitrogen and plankton biomass were important in explaining observed variations in primary production over the spatial and temporal extent of the study. Standard hydrography-productivity observations during the same period reinforced these general conclusions. For stations where P-I analyses and *in situ* production measurements were made simultaneously, there was a clear correlation between P^B_m and euphotic zone production (Fig. 6). Highest biomass and water-column production stations were close inshore, were related to the depth of the nutricline (Fig. 7), and were characteristically nitrogen deficient. A plot of all NO_3 and SiO_3 concentrations revealed a non-linear relationship with NO_3 decreasing more rapidly than SiO_3 in the upper part of the euphotic zone (Fig. 8). Some of the most inshore stations had extremely high

TABLE 3. Effects of Deep-Water Enrichment on Photosynthetic Parameters of Near-surface Phytoplankton, Peru, 1977. Units as in Table 1.

Station	T	Chl	NO_3	SiO_3	α	P^B_m
52	22.5	0.83	0.52	1.63	0.09	2.27
52(m)	19.5	0.89	8.66	9.73	0.11	1.15
56	18.0	10.83	5.17	0.81	0.06	2.13
56(m)	17.8	6.25	12.18	10.19	0.12	2.51
58	18.0	1.78	5.48	3.71	0.10	1.54
58(m)	17.5	0.89	16.03	14.29	0.09	1.20

(m) = 50:50 mixture of near surface and subsurface (45 to 100 m) water.
T = temperature of samples at beginning of experiment.

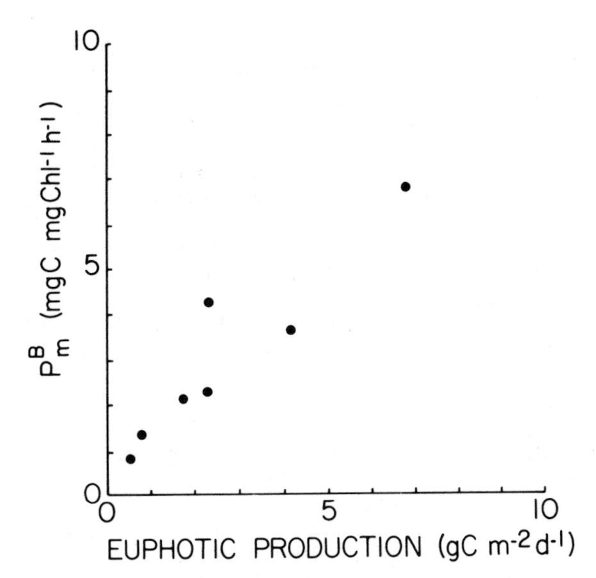

Fig. 6. Relationship between maximum specific production rate (P_m^B) at 50% light level and total euphotic zone daily production rate.

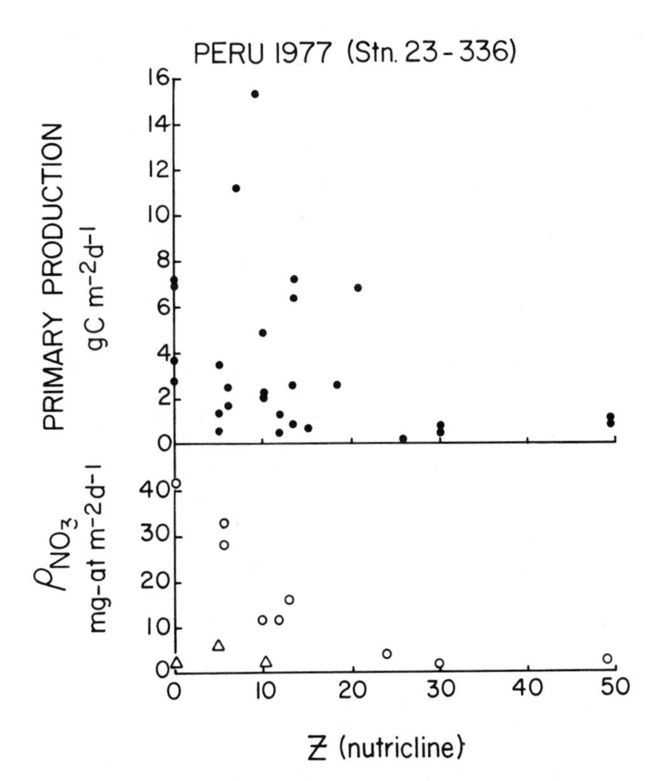

Fig. 7. Relationship between total euphotic zone carbon and nitrate productivities and depth of nutricline. Open triangles are stations where NO_3 concentrations were low at all euphotic depths.

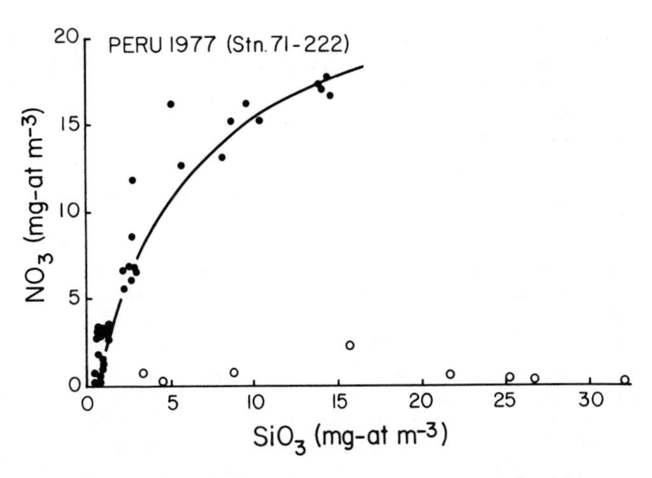

Fig. 8. Relationship between nitrate and silicate concentrations, ICANE Peru cruise. Open circles are stations 71 and 95.

SiO_3 values relative to NO_3. The C:N compositional and assimilation ratios also reflected nitrogen limitation. Inshore, C:N (atomic) ratios were greater than 8 in the particulate matter and C:N assimilation ratios (euphotic zone integrated) were greater than 20. Further offshore, particulate and assimilation ratios were closer to the "Redfield ratio" of 7.

The populations above and below the mixed layer appeared to adapt rapidly to prevailing light conditions. The difference in near-surface and deep-population I_k values increased with the density gradient separating the populations (Fig. 9). This resulted primarily from diverging P_m^B values

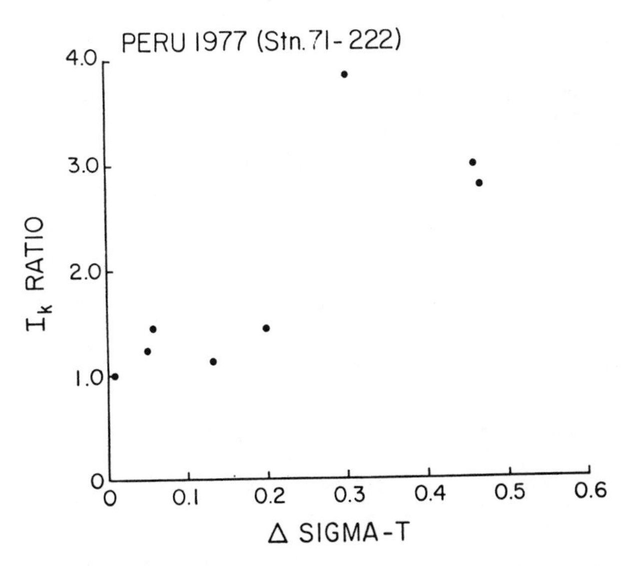

Fig. 9. Relationship between light adaptation (ΔI_k) and water-column stability ($\Delta\sigma_\tau$), ICANE Peru cruise.

although the deep samples on the average also exhibited higher photosynthetic efficiencies (α).

Large variations in P_m^B off Peru are considered typical (Huntsman and Barber, 1977) and were significantly related to biomass variations during our study. Relative species composition was surprisingly constant despite order-of-magnitude changes in total abundance (ref. Fig. 3) and therefore probably did not account for the observed P_m^B - biomass covariance. The "blue" and "brown" water patches observed by Strickland, Eppley, and deMendiola (1969) could have relevance to our observations, particularly if observed low P_m^B values of the "blue" water and P_m^B depression after deep-water enrichment (ref. Table 3) were a consequence of a water "conditioning" requirement (Barber, Dugdale, MacIsaac, and Smith, 1971).

Most studies off Peru have clearly shown or implied the importance of SiO_3 in limiting primary production (e.g., Ryther, Menzel, Hulbert, Lorenzen, and Corwin, 1971; Dugdale, 1972; Dugdale, Jones, MacIsaac, and Goering, 1979). Zentara and Kamykowski (1977) have proposed a general SiO_3 deficiency in the southern hemisphere relative to nitrogen. Our data suggest that this may not always be the case and, in fact, some of the NO_3-SiO_3 plots in the lower southern latitudes used by Zentara and Kamykowski also suggest a similar non-linear relationship with both nutrients converging on the origin. Based on seasonal production studies off northern Peru, O. Guillen (pers. comm.) has concluded that either NO_3 or SiO_3 can become limiting.

Conclusions

Variations in light-saturation curves from natural phytoplankton populations and observations from standard hydrography-productivity stations suggested that areas of high primary production off the northern coast of Peru during November 1977 were extremely variable, principally confined to close inshore (within 10 km), and controlled primarily by the supply of nitrogen to the euphotic zone. Favorable physiological condition of the populations, as indicated by high P_m^B values, was associated with the areas of highest biomass, "brown" water. "Blue" water areas (low biomass, relatively high nutrient concentrations), although of similar species composition, showed less favorable growth conditions through significantly lower P_m^B. Populations appeared to adapt to prevailing light conditions by changes in their photosynthetic efficiency (α), maximum production rate (P_m^B), and sensitivity to inhibiting light intensities. At least part of the adjustment mechanism appeared to be regulated endogenously over diurnal time scales, particularly the light-inhibiting properties.

References

Barber, R.T., R.C. Dugdale, J.J. MacIsaac, and R.L. Smith, Variations in phytoplankton growth associated with the source and conditioning of upwelling water, *Investigacion Pesquera, 35*, 171-193, 1971.

Doe, L.A.E., Project ICANE, a progress and data report on a Canada-Peru study of the Peruvian anchovy and its ecosystem, Bedford Institute of Oceanography Report Series BI-R-78-6, 211 pp., 1978.

Dugdale, R.C., Chemical oceanography and primary productivity in upwelling regions, *Geoforum, 11*, 47-61, 1972.

Dugdale, R.C. and J.J. Goering, Uptake of new and regenerated forms of nitrogen in primary productivity, *Limnology and Oceanography, 12*, 196-206, 1967.

Dugdale, R.C., B.H. Jones, Jr., J.J. MacIsaac, and J.J. Goering, Interactions of primary nutrient and carbon uptake in the Peru current: a review and synthesis of laboratory and field results, *Coastal Upwelling Ecosystems Analysis Technical Report 51*, 40 pp., 1979.

Fiedler, R. and G. Proksch, The determination of nitrogen-15 by emission and mass spectrometry in biochemical analysis: a review, *Analytica Chimica Acta, 78*, 1-62, 1975.

Guillen, O., R. Calienes, and R. de Rondan, Medio ambiente y producion primaria frente al area Pimentel-Chimbote, *Instituto de Mar de Peru Boletin, 3*, 108-159, 1977.

Holm-Hansen, O., C.J. Lorenzen, R.W. Holmes, and J.D.H. Strickland, Fluorometric determination of chlorophyll, *Journal du Conseil, Conseil International pour l'Exploration de la Mer, 30*, 3-15, 1965.

Huntsman, S.A. and R.T. Barber, Primary production off northwest Africa: the relationship to wind and nutrient conditions, *Deep-Sea Research, 24*, 25-33, 1977.

Jassby, A.D. and T. Platt, Mathematical formulation of the relationship between photosynthesis and light for phytoplankton, *Limnology and Oceanography, 21*, 540-547, 1976.

Platt, T. and A.D. Jassby, The relationship between photosynthesis and light for natural assemblages of coastal marine phytoplankton, *Journal of Phycology, 12*, 421-430, 1976.

Platt, T., C.L. Gallegos, and W.G. Harrison, Photoinhibition of photosynthesis in natural assemblages of marine phytoplankton, *Journal of Marine Research*, in press, 1980.

Ryther, J.H., D.W. Menzel, E.M. Hulburt, C.J. Lorenzen, and N. Corwin, The production and utilization of organic matter in the Peru coastal current, *Investigacion Pesquera, 35*, 43-59, 1971.

Sharp, J.H., Improved analysis for "particulate organic carbon and nitrogen" from seawater, *Limnology and Oceanography, 19*, 984-989, 1974.

Strickland, J.D.H., Solar radiation penetrating the ocean. A review of requirements, data and methods of measurement, with particular reference to photosynthetic productivity, *Journal of Fisheries Research Board of Canada, 15*, 453-493, 1958.

Strickland, J.D.H. and T.R. Parsons, A Practical Handbook of Seawater Analysis, *Bulletin No. 167*, 311 pp., Fisheries Research Board of Canada, 1972.

Strickland, J.D.H., R.W. Eppley, and B.R. de Mendiola, Phytoplankton populations, nutrients and photosynthesis in Peruvian coastal waters, *Boletin Instituto del Mar del Peru, 2*, 1-45, 1969.

Talling, J.R., Photosynthetic characteristics of some freshwater plankton diatoms in relation to underwater radiation, *New Phytologist, 56*, 29-50, 1957.

Zentara, S.J. and D. Kamykowski, Latitudinal relationships among temperature and selected plant nutrients along the west coast of North and South America, *Journal of Marine Research, 35*, 321-337, 1977.

UPWELLING OFF CHIMBOTE

O. Guillén

and

R. Calienes

Instituto del Mar del Peru, Lima, Peru

Abstract. Using data from cruises off Chimbote, Peru between 1964 and 1978 the mean annual and seasonal horizontal and vertical distributions of temperature, salinity, density, phosphate, dissolved oxygen, geostrophic flow, Ekman transport, chlorophyll *a*, and primary productivity are presented. Upwelling takes place off Chimbote in fall and winter with the source water from 50 to 100-m depths. The mean width of the upwelling area is 50 km, and waters brought to the surface are an extension of the Cromwell Current. The upwelling velocity calculated from isopycnal displacement was 3.9×10^{-4} cm/s for fall and 1.5×10^{-3} cm/s for winter.

The circulation is dominated by anticyclonic and cyclonic eddies that govern the regional dynamics; they are apparently the result of offshore flow of cool, upwelled water and the surface intrusion of warm, oceanic water. The eddies produce a meander-like surface circulation.

The mean primary production of the area between 1964 and 1978 was 402 g C $m^{-2}yr^{-1}$; the mean concentration of chlorophyll *a* in the euphotic zone was 23 mg/m^2. During the abnormal conditions of *El Niño* of 1972 the primary production was one-third of the mean and a decrease in the biomass of anchovy occurred at the same time in the Chimbote area.

Introduction

Of the upwelling areas along the Peruvian coast, the center off Chimbote at 9°S latitude is the most important because it supports a major portion of the Peruvian anchovy fishery. Characteristics of the area have been described by Zuta and Guillén (1970) and several studies of the environment, plankton, and primary production have shown the seasonal variations (Guillén, 1971, 1973; Guillén and Mendiola, 1971; Santander, 1974; Guillén and Rondan, 1974; Guillén and Calienes, 1980). Interrelations of changes in the environment to aggregation and dispersion of anchovy schools were studied by Guillén, Calienes, and Rondan (1969). To place the oceanographic work in perspective for fisheries management, estimates of biomass of the second and third trophic level were made from the primary production of the area (Guillén, Calienes, and Rondan, 1977). In this paper new information on the variability of properties within the ecosystem is given together with additional estimates of the second and third trophic level yields.

Results

The data on temperature, salinity, dissolved oxygen, and phosphate for the area between 08°30' to 09°30'S have been taken from cruises during 1964 to 1978. Air temperature, atmospheric pressure, wind speed and direction are from Zuta, Rivera, and Bustamante (1978) and were recorded at Chicama and Chiclayo. The Ekman transport was calculated according to the equation

$$E = \frac{\tau_y}{f} \quad (1)$$

where f is the Coriolis force and τ_y is the alongshore component of wind stress.

Winds off Chimbote are generally weak and predominately from the southeast, favorable for upwelling. During the upwelling season the winds fluctuate in speed from 2.0 to 2.5 m/s (Zuta and Guillén, 1970). The atmospheric pressure is higher from August to October, with a maximum in October and an annual amplitude of 5.0 mb.

The bottom topography and the coastline have an influence on coastal upwelling (Mooers and Allen, 1973). The coastline off Chimbote (Fig. 1) is oriented 135° (SE) and the isobaths are parallel to it. The widest shelf is between 7 and 10°S, where the great anchovy fishery is located. The 200-m isobath reaches its maximum offshore extension of approximately 120 km off Chimbote.

The sea-surface temperature distribution (Fig. 2) shows cold water tongues or eddies that extend away from the coast and represent the upwelled

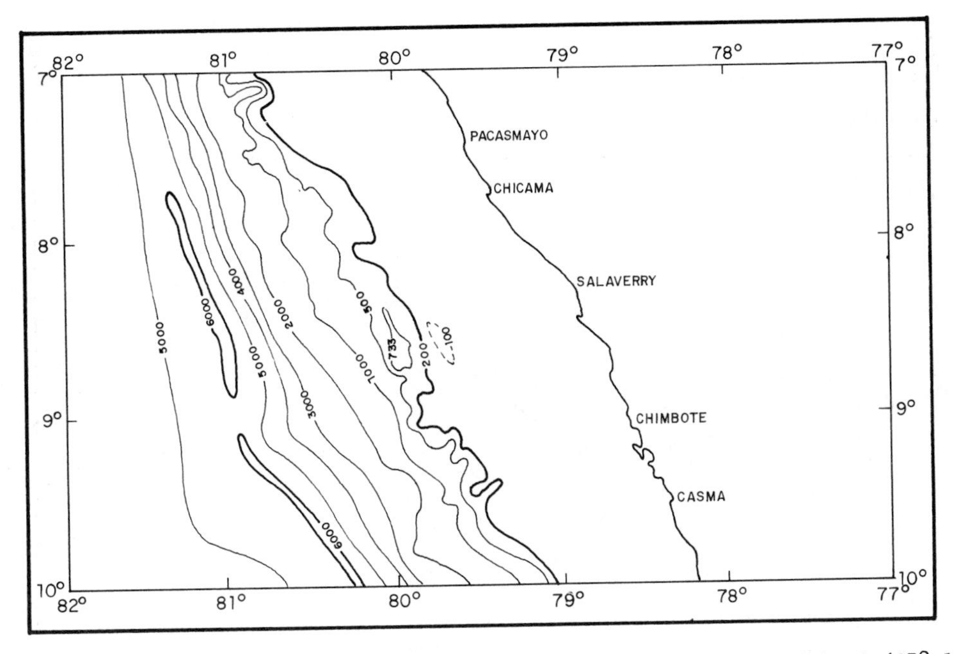

Fig. 1. Topographic conditions (contours in fathoms) of the coastal band (07°-10°S).

water. The offshore extent of the cold water varies in time and space and sets the boundary of the ecosystem. In the same way, the flow of the oceanic waters towards the coast is evident with variations depending on the geostrophic circulation. The monthly distribution of temperature (Fig. 3) shows that cooling occurs from March to October, with an amplitude of 3 to 5°C.

The greatest seasonal changes of the different variables (Fig. 4) are in the 0- to 50-m layer, especially in the 0- to 20-m layer, which corresponds to the euphotic zone. Below 100 m the changes are small and regular. When the mean vertical distributions of physical-chemical parameters are compared with other regions, differences are evident between Chimbote and other upwelling areas. Density in the austral winter increases from 25.6 on the surface to 26.2 at 100 m in Chimbote, from 25.5 to 26.1 in Paita, from 25.9 to 26.3 in San Juan (Zuta *et al.*, 1978), and in Callao from 25.8 to 26.2 (Guillén, 1966).

The 0- to 100-m layer in Chimbote has oxygen

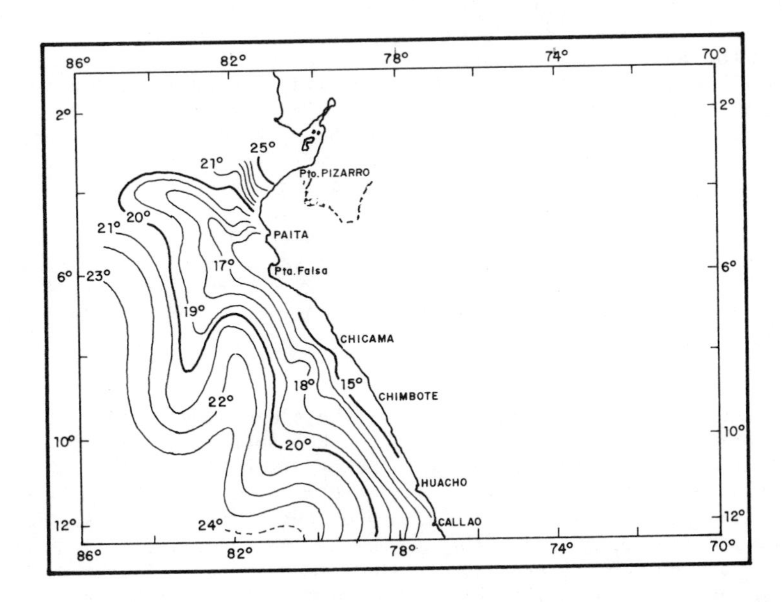

Fig. 2. Sea surface temperature (°C), cruise 7303 (May 6 - June 12, 1973)

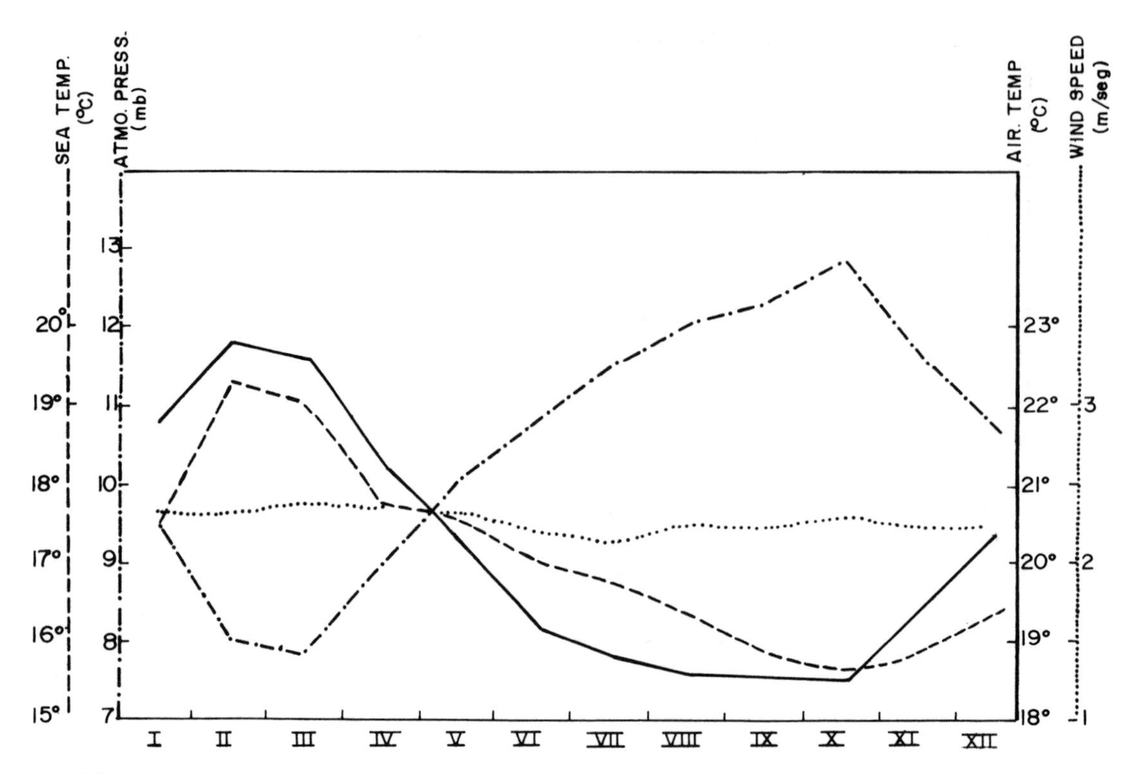

Fig. 3. Monthly means of sea level atmospheric pressure, wind speed, surface air temperature, and sea surface temperature.

values of 0.25 to 0.75 ml/l, values that are less than at Paita (1.0 to 1.5 ml/l) and overlap those at San Juan (0.5 to 1.0 ml/l). The pycnocline is better developed at Paita than in the other areas. The oxygen concentration in the waters of the upper 50 m is highest in autumn and winter due to the influence of the waters of the extension of the Cromwell Current (Equatorial Undercurrent) (White, 1969). In the 50- to 150-m layer oxygen concentrations are rather uniform for most of the seasons with values below 0.5 ml/l, except for the austral autumn when increased oxygen is observed. Below the 50- to 150-m layer, the oxygen content decreases to 0.25 ml/l. Offshore from Callao the oxygen minimum layer near the coast rises to 20 m with values <0.25 ml/l, while further from the coast the minimum layer is below 50 m.

Fig. 5 shows the mean monthly variation of temperature, salinity, density, dissolved oxygen, and phosphate in the water column from 0 to 100 m. The temperature distribution is similar to that of density, with great changes in the superficial layer down to 50 m, especially in the upper 20 m. The changes in salinity were due to the influence of the superficial equatorial waters. Changes of oxygen from May to September may also be due to the influence of the extension of the Cromwell Current during the upwelling period. Waters of high temperature and low density appear from February to June, and waters of low temperature and high density appear from May to October, the period of main upwelling.

Water is upwelled from above 50 m in May, but from 150 m in September with densities of 25.4 and 25.8 (σ_t), respectively, reaching the surface within 20 to 40 miles of the coast.

The average upwelling speed calculated from isopycnal displacement was 3.9 x 10^{-4} cm/s for autumn and 1.5 x 10^{-3} cm/s for winter, with an annual average speed of 0.9 x 10^{-4} cm/s. The mean Ekman transport was 13 kg $cm^{-1}s^{-1}$ and assuming an upwelling extension of 60 km we have an upwelling speed of 1.86 m per day, twice the rate calculated from the displacement of isopycnals (0.83 m per day). The Ekman transport average of 1.3 m^3/s per meter of coast is similar to that calculated by Richmann and Smith (1980) for the same area.

Another important characteristic of the upwelling system is the stratification. Off Oregon density is strongly stratified while off northwest Africa the stratification is weak (Huyer, 1976). Moderate stratification was observed off Chimbote although it depends on the season and the wind fluctuations.

The seasonal variation of the isopycnals off the Chimbote area (Fig. 6) shows that divergence takes place between 50 and 75 m within 40 miles of the coast. In abnormal years (Figs 7 and 8) when the phenomenon known as *El Niño* occurs as in 1972, the upwelling is less and the divergence of the isopycnals is deeper with the 25.8 isopycnal deeper than in normal years. In cold years such as 1964 it was more superficial (Fig. 8).

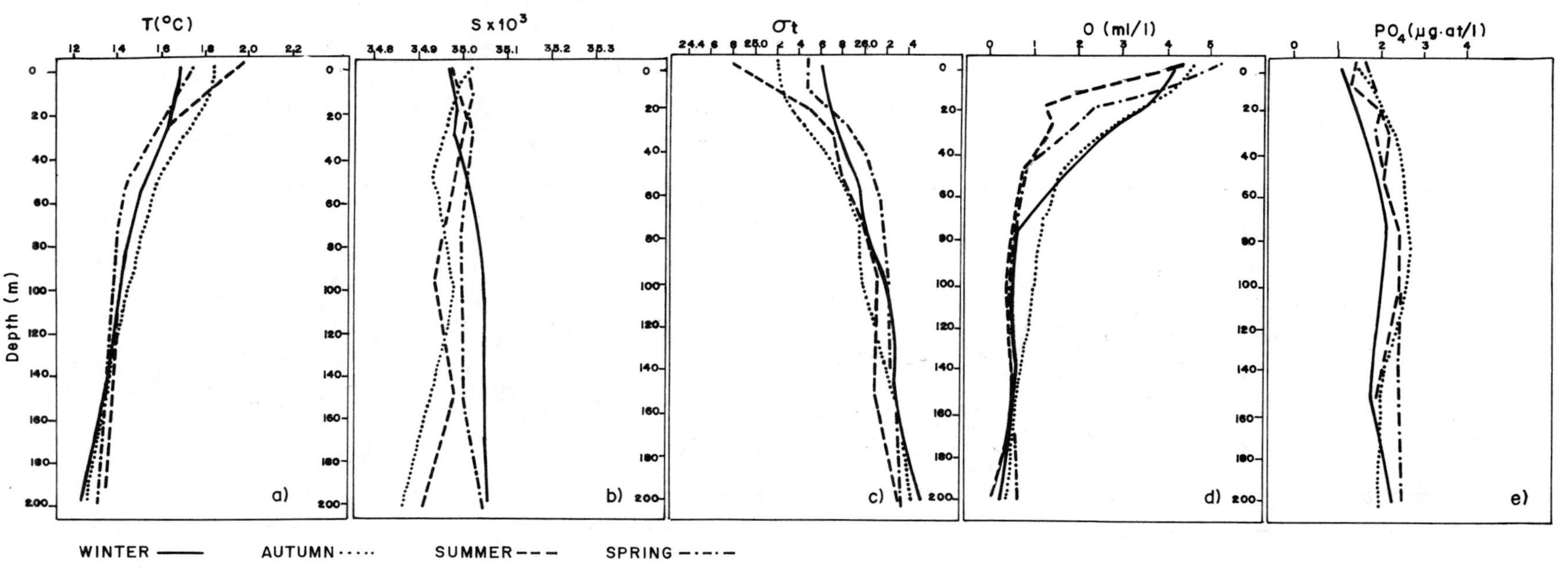

Fig. 4. Vertical profiles of seasonal mean a) temperature, b) salinity, c) density, d) oxygen, e) phosphates for the coastal areas of Chimbote.

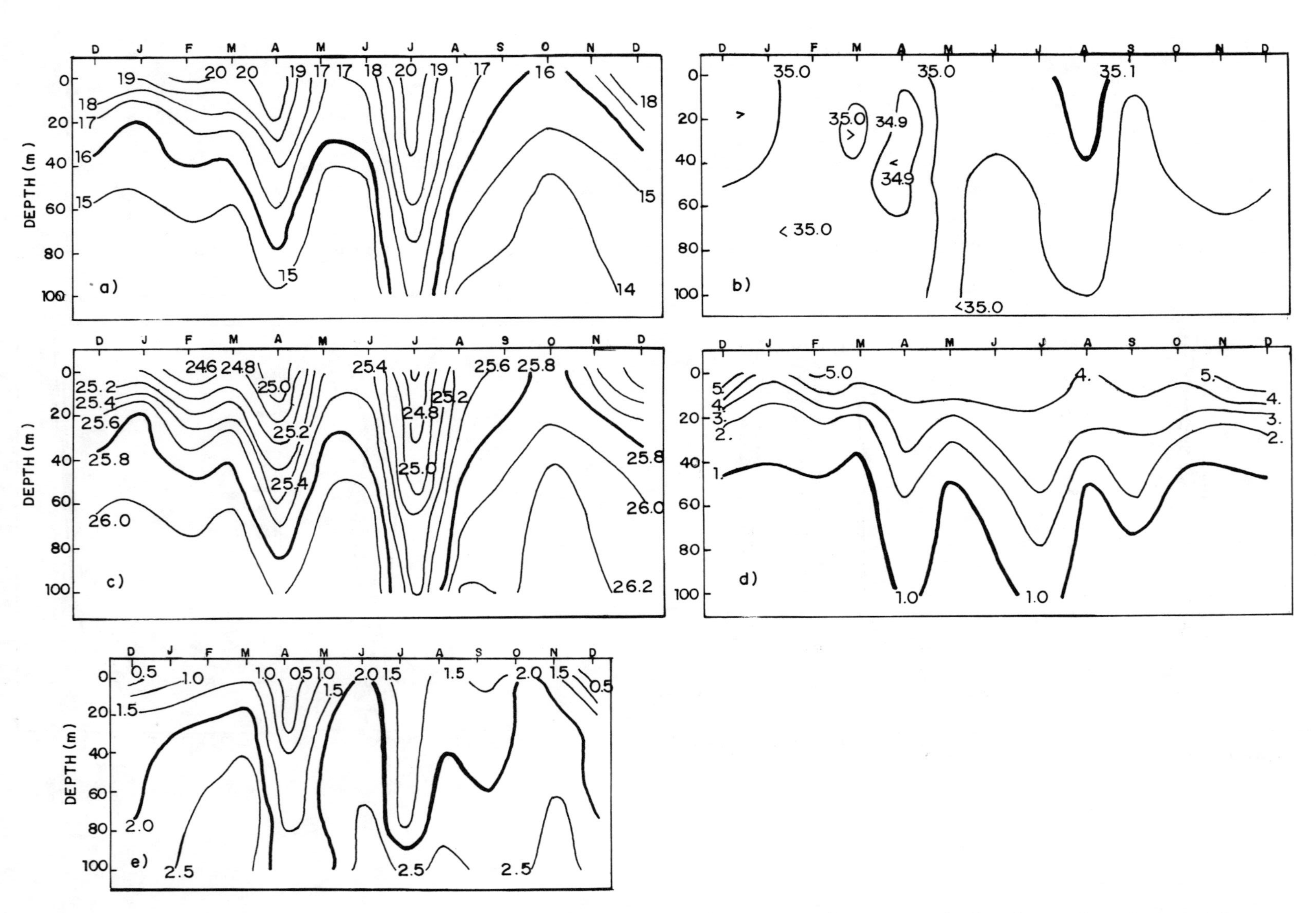

Fig. 5. Monthly mean distribution of a) temperature, b) salinity, c) density, d) oxygen, e) phosphates for the coastal area off Chimbote.

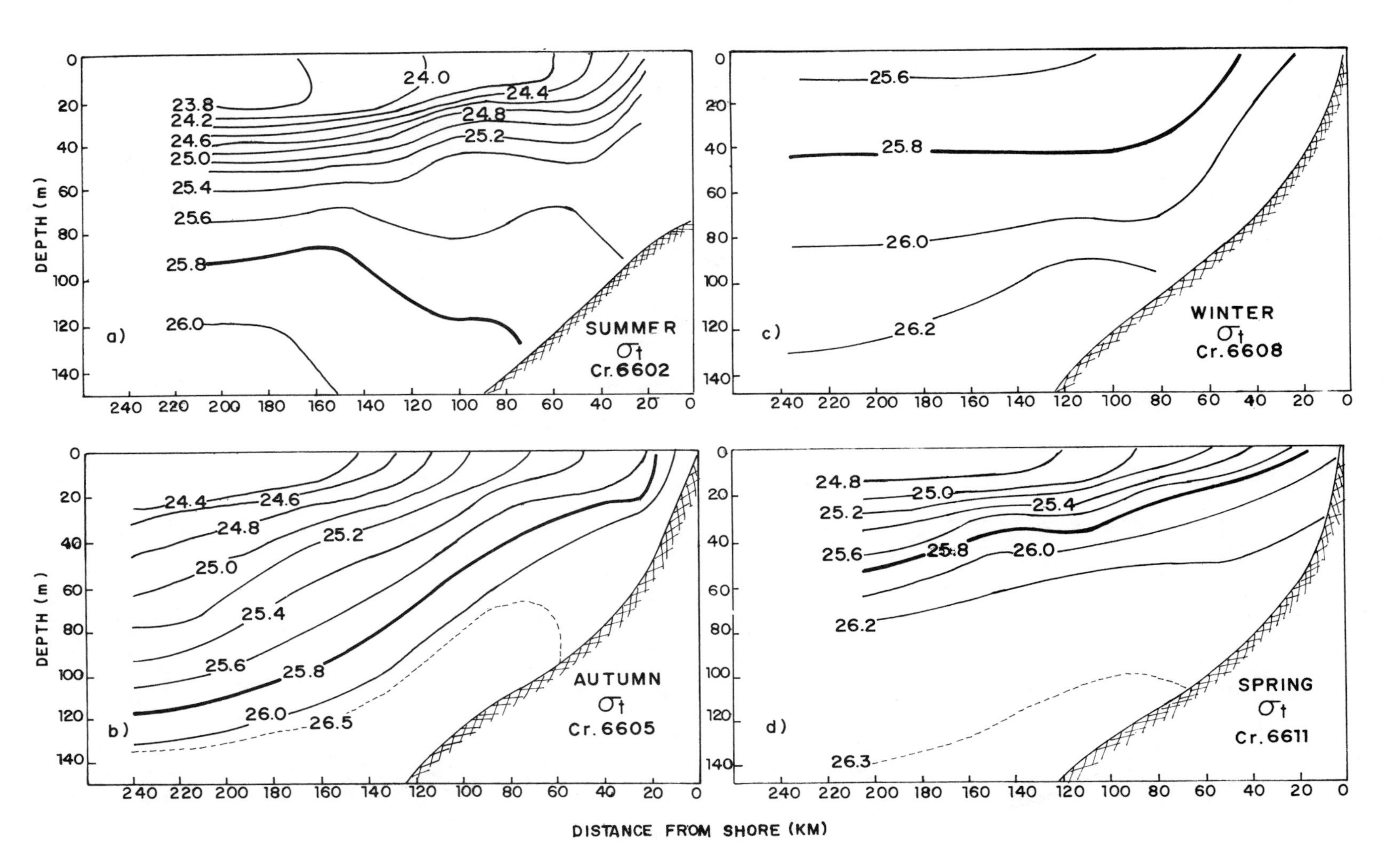

Fig. 6. Seasonal distribution of density during 1966: a) summer, b) autumn, c) winter, and d) spring.

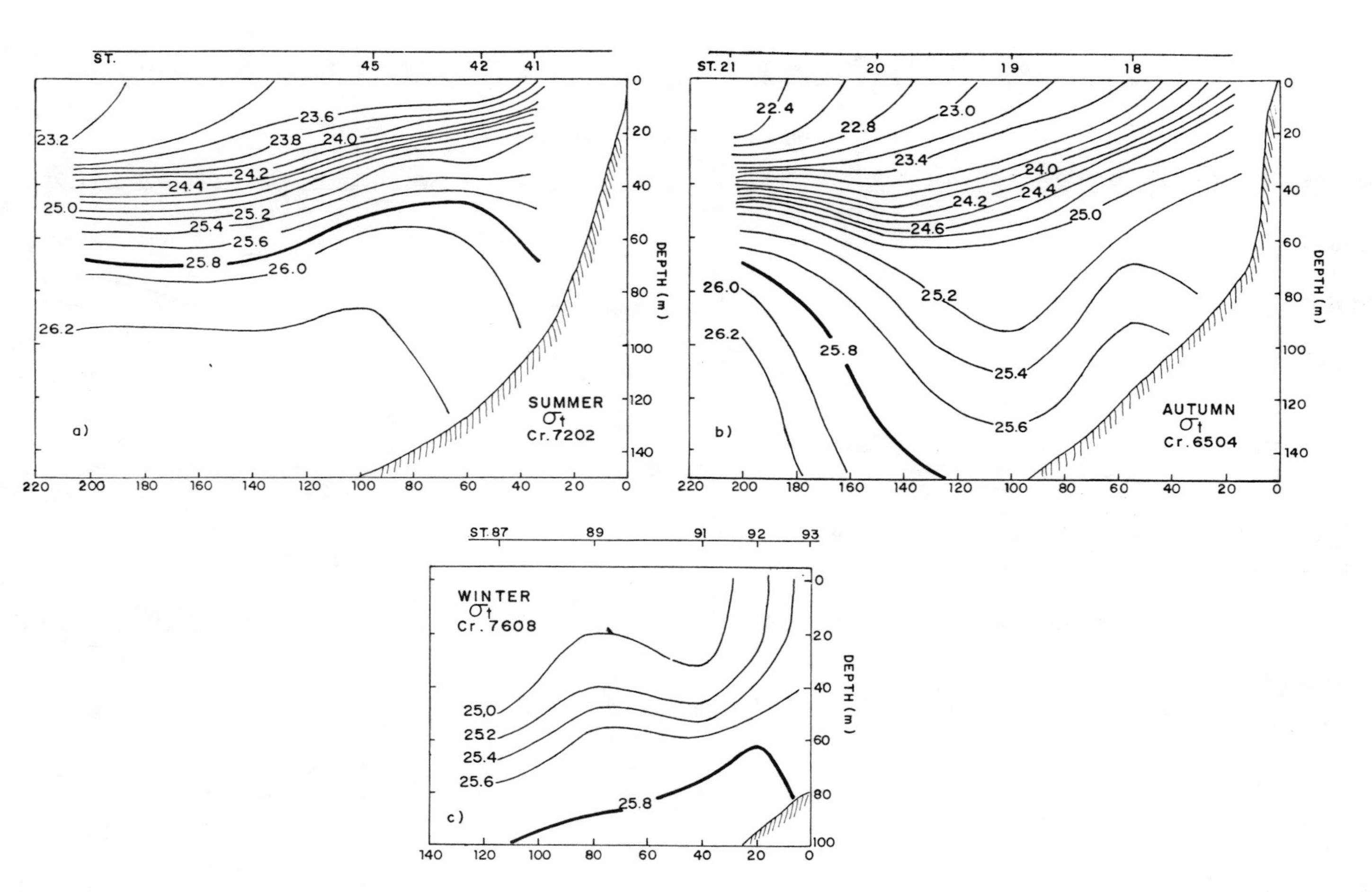

Fig. 7. Distribution of density during *El Niño* phenomenon: a) summer, b) autumn, and c) winter.

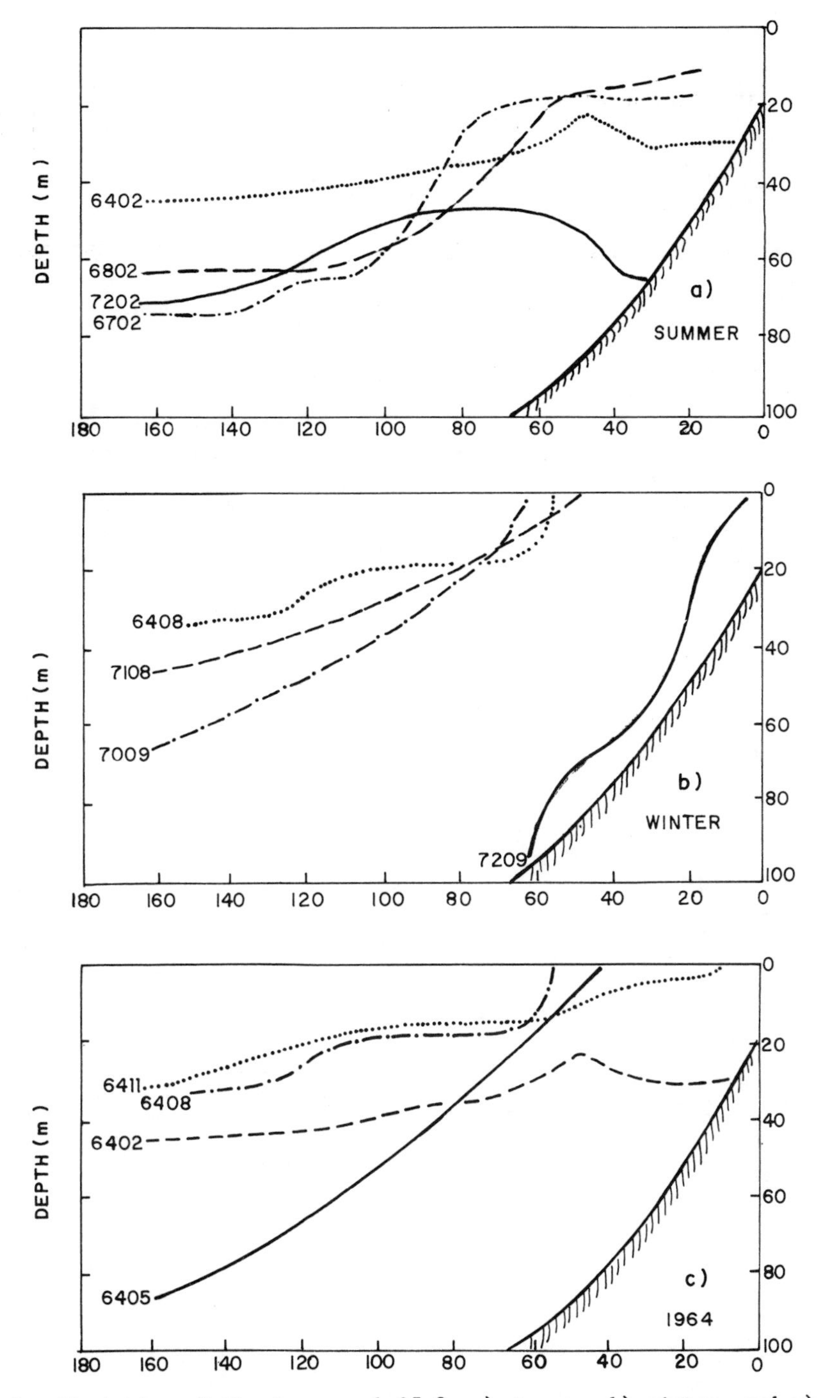

Fig. 8. Variation of the isopycnal 25.8: a) summer, b) winter, and c) 1964.

The average primary production at the surface was 71 mg C $m^{-3}d^{-1}$ and for the water column within the euphotic layer, 1.1 g C $m^{-2}d^{-1}$, giving 402 g C $m^{-2}yr^{-1}$. The average chlorophyll concentrations at the surface and in the euphotic zone were 2.63 and 23 mg/m^2, respectively. The productivity index (mg C/mg Chl/d) in the euphotic zone was 48, similar to that found by Guillén (1973) for the same area. In order to know the interrelations between nutrients, chlorophyll, and carbon we have used the C:N:P relations given by Redfield, Ketchum, and Richards (1963) and the 35:1 carbon: chlorophyll ratio given by Ryther and Menzel (1965). Using the average chlorophyll *a* for the euphotic zone a production of 294 g C/m^2 was calculated; this is lower than that

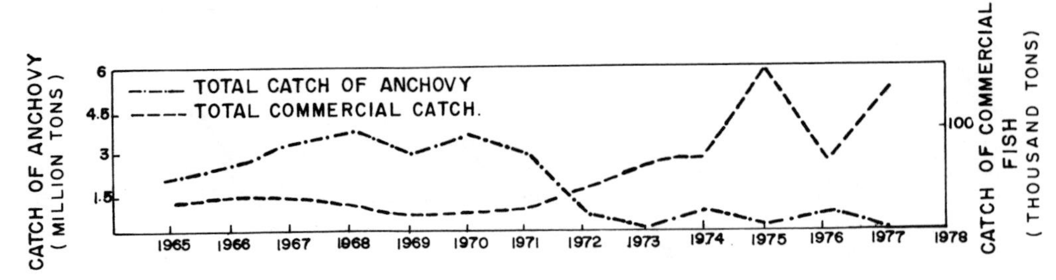

Fig. 9. Catches of anchovy and commercial fishes.

measured in the area. This is because the assumed C:Chl ratio (35) is lower than the productivity index (48) measured off Chimbote. Considering an average euphotic zone 25 m deep, the average contents of phosphate and nitrate in the layer were 43 and 400 μg-at/m^2, respectively. If we assume that these amounts were taken up by the phytoplankton of normal atomic ratio, a carbon production of 53 and 33 g C/m^2, respectively, would result.

Productivity and its Relationship to Fisheries

A common method of estimating the productive potential of an ocean area is through food chain dynamics using phytoplankton biomass (chlorophyll, primary production, and assumed percentages of transfer efficiency. Jordan (1976) and other authors point out that the major food of the Peruvian anchovy is phytoplankton. Using the mean primary production and the area of upwelling, we have calculated the biomass of the second and third trophic levels following the method of Schaeffer (1965) and Cushing (1969). Assuming that the upwelling area is 2.5×10^{10} m^2 and the primary production is 402 g C $m^{-2}yr^{-1}$, the total production is 1×10^7 ton C/yr. Using an efficiency of 10%, assuming that live weight is ten times the carbon weight and considering anchovy as the herbivore that consumes 80% of the production, we calculate an anchovy production of 8×10^6 ton/yr for the Chimbote area.

Secondary production of 0.6×10^6 ton C/yr was calculated assuming a mean zooplankton abundance of 0.8 ml/m^3 (settled volume) in a 100-m water column.

The mean tertiary production was calculated assuming that 1% of the primary production and 10% of the secondary production was transferred, giving an annual tertiary production of 1.25×10^6 tons wet weight. This predicted tertiary production is within the values calculated for commercial non-anchovy fish stocks by Mejia, Samamé, and Esquerre (1974).

During *El Niño* of 1972 (Guillén and Calienes, 1980) positive anomalies of temperature up to 6°C were present due to the intrusion of waters from the equatorial region. The intrusions were characterized by high temperature, very low salinity, low nutrient content, and low productivity. The water had a marked effect on living resources, especially in the decrease of anchovy (Fig. 9). Figs 10a and 10b show a good correlation between the catch of commercial fishes and anchovy along the entire Peru coast and that obtained in the area off Chimbote. The distribution of superficial temperature at Chimbote shows *El Niño* clearly (Fig. 11), so we suggest that the physical parameters in Chimbote could be used to predict the total fishery using the known physical → chemical → productivity → fish stock relations.

Discussion

The hydrographic conditions show the greatest changes in the 0- to 50-m layer, especially the 0- to 20-m layer, which is similar to other upwelling areas such as Paita and San Juan (Zuta *et al.*, 1978), Punta Falsa, and Callao (Zuta and Guillén, 1970). However, there are variations in salinity due to the upwelled water. The upwelled waters off Chimbote come from an extension of the Cromwell Current during most of the year and are characterized by higher oxygen contents than other waters at that depth. The pycnocline is less well developed in Chimbote than off Paita or San Juan, where there are strong oxyclines. The oxygen minimum that most strongly developed from October to March is probably associated with the flow towards the south of the Peru-Chile Undercurrent (Wooster and Gilmartin, 1961).

Off northwest Africa the isopycnals ascend from 200 m at a distance of 80 km off the coast (Barton, Pillsbury, and Smith, 1975), while in Oregon they ascend from 150 m at 60 km off the coast (Barstow, Gilbert, and Wyatt, 1969). Off Chimbote the winter ascent came from 120 m at 140 km off the coast, while in autumn isopycnals rose from about 70 m depth at 60 km off the coast. Winds are stronger and much more variable in strength and direction off northwest Africa and Oregon, while off Chimbote winds vary with a mean of 2.2 to 2.4 m/s during the year and are always from a southeasterly direction favorable for upwelling.

The average upwelling speed calculated from the ascent of isopycnals is 3.9×10^{-4} cm/s for autumn and 1.5×10^{-3} cm/s for winter. The rates are similar to the value found by Zuta *et al.* (1978) for Paita and lower than those calculated in the same way for Callao (7×10^{-4} cm/s) by Guillén

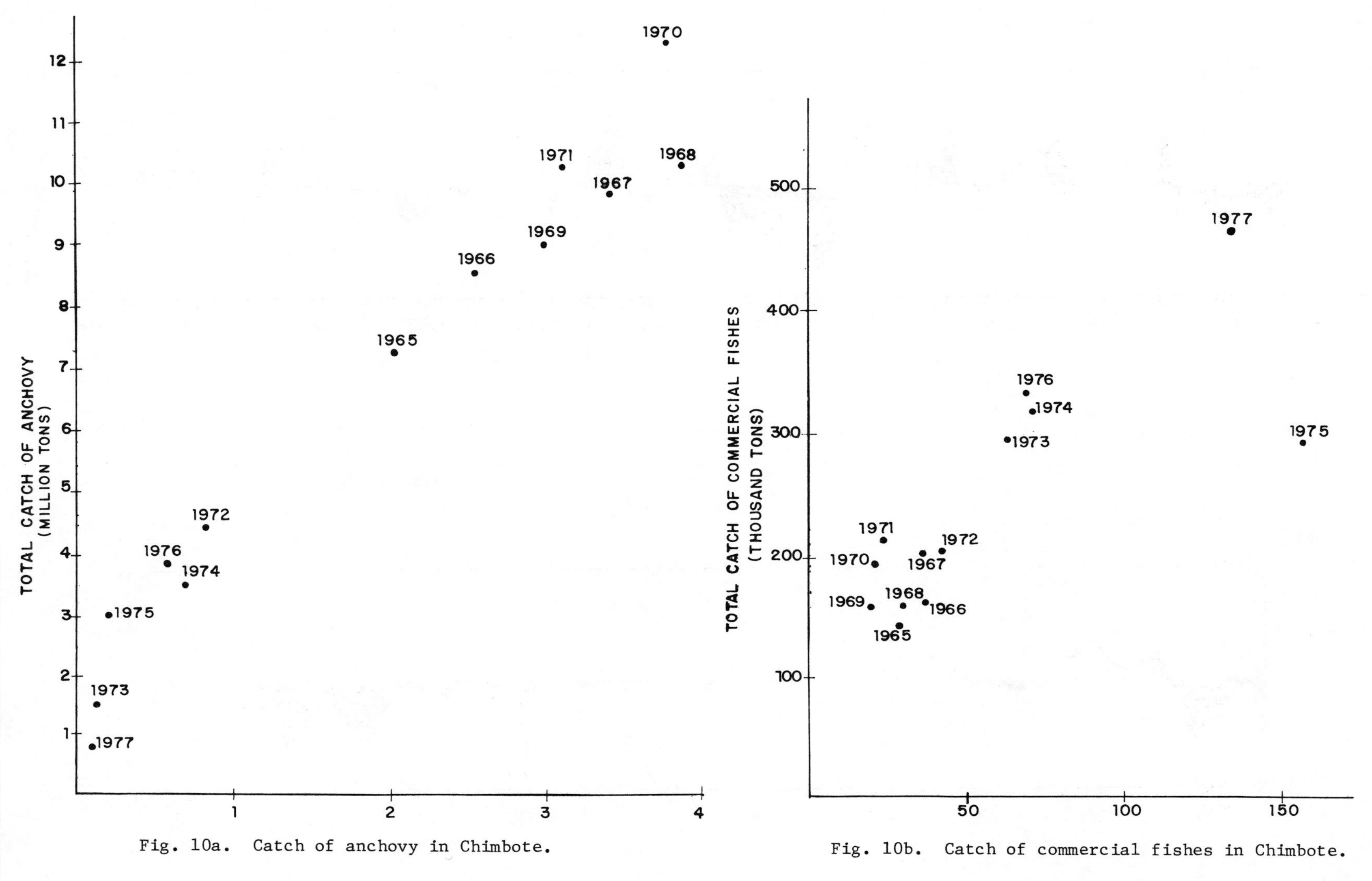

Fig. 10a. Catch of anchovy in Chimbote.

Fig. 10b. Catch of commercial fishes in Chimbote.

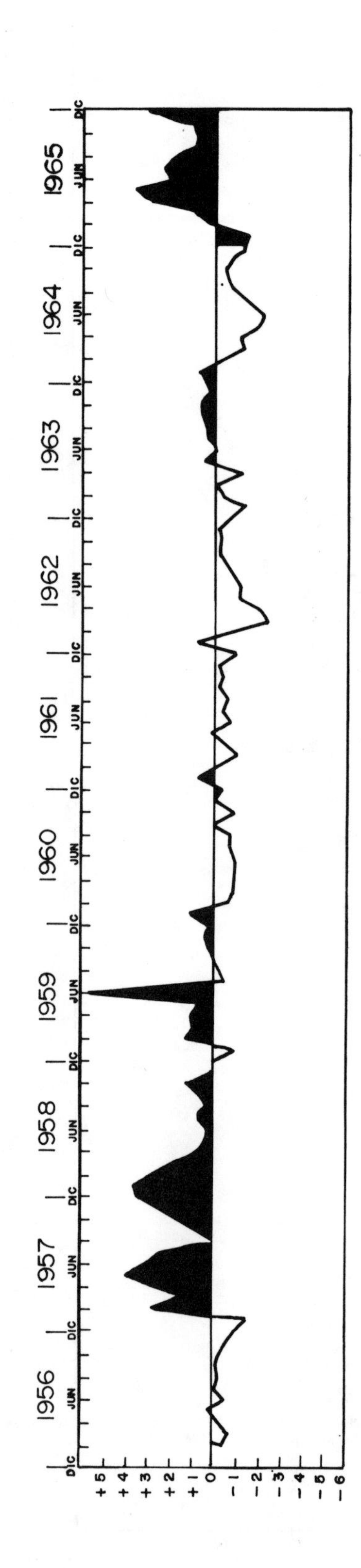

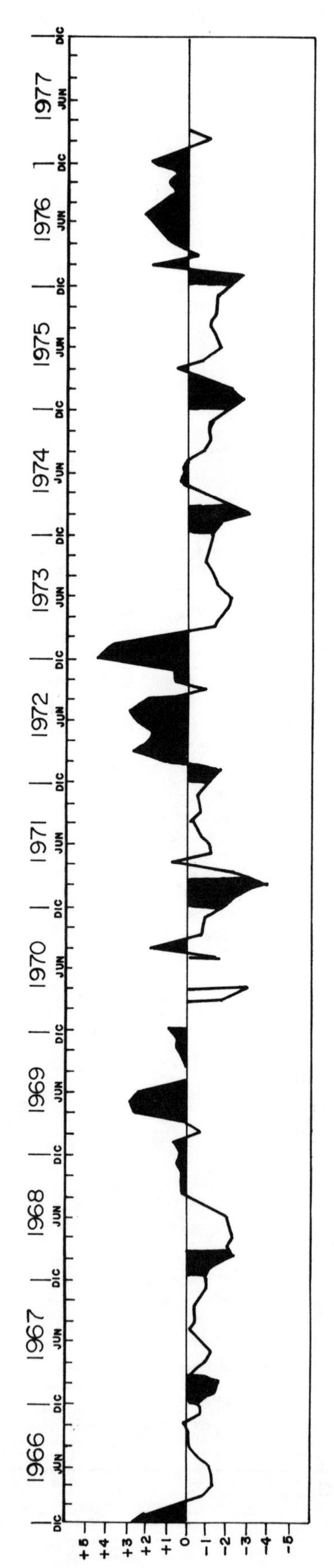

Fig. 11. Mean annual variation of temperature in Chicama.

(1966) and for the upwelling of Oregon, Baja California, and northwest Africa (Table 2).

The upwelling area is variable, depending on season and the predominant hydrographic characteristics of the area. The average width of the upwelling zone is 60 km, similar to that of Paita and San Juan (Zuta *et al.*, 1978). Similarly, the isopycnal divergence depth (50 to 70 m) was similar to that found by Wyrtki (1963) and Zuta *et al.* (1978) for other areas. The upwelling depth is similar to Paita and deeper in winter than off Callao and San Juan. Another characteristic of the Chimbote region (Table 1) is that the cooling period lasts eight months (March to October), longer than for other areas.

Dugdale (1972) and Guillén *et al.* (1977) suggested that the limiting element in productivity along the Peru coast could be silicon, but Harrison and Platt (1979) found that for the Chimbote area in November 1977 nitrogen was the limiting element. Assuming a mean upwelling velocity of 1.3 m/day and a phosphate concentration of 2.0 mg-at/m^3 at 50 m, we have an upwelled flux of 2.6 mg-at/$m^{-2}day^{-1}$. Assuming this concentration was distributed in the 25-m euphotic zone and there was a loss of 50% of the upwelled flux, there is 0.05 µg-at PO_4-P $l^{-1}day^{-1}$ for biological uptake. Assuming a ratio of C:P of 41:1, we have a production equivalent of 1.6 g C $m^{-2}day^{-1}$, somewhat higher than the mean production measured in the area of 1.1 g C $m^{-2}day^{-1}$. The average production of the area was 402 g C m^{-2} yr^{-1}, equivalent to a usage of 330 and 4820 µg-at $m^{-2}yr^{-1}$ of phosphate and nitrate, respectively. With a normal growth rate, only 48 and 30 days would be needed to achieve this production. During *El Niño* of 1972, productivity was 0.39 g C $m^{-2}day^{-1}$, one-third of the average for the area (Guillén *et al.*, 1977).

If anchovy biomass is related to the amount of phytoplankton food (in terms of chlorophyll) required for the maintenance of the fish, and assuming that zooplankton graze 25% of the phytoplankton food (Beers, Stevenson, Eppley, and Brooks, 1971), the dependence of anchovy biomass on available food is clearly evident (Fig. 12). Mathisen, Thorne, Trumble, and Blackburn (1978) suggested that food availability influences species by limiting the larval stages and not the adults.

Light is not limiting in the upwelling region off Chimbote; the amount of light is rarely below the critical level of limiting productivity (Guillén, 1973; Guillén and Calienes, 1980). Similarly, nutrients are not limiting, but silica and nitrate most frequently regulate productivity. There is evidence of the significance for productivity of the relationship of the euphotic zone to the mixed-layer depth (Guillén and Calienes, 1980). Dugdale and Goering (1970) and Whitledge (1972, 1978) pointed out that ammonia may supply half of the daily requirements of nitrogen of phytoplankton with the other half supplied by upwelled nitrate. If this is the case, then in

TABLE 1. Mean Euphotic Zone Properties of the Four Major Upwelling Centers Along the Coast of Peru

Area	T (°C)	S (o/oo)	O_2 (ml/l)	PO_4 µg-at/l	SiO_4 µg-at/l	NO_3 µg-at/l	P:N:Si	Chl*a* mg/m^3	Prod. mg C m^{-3} /day^{-1}	C/Chl mg C $(mgChl)^{-1}$ day^{-1}	Euphotic Zone (m)	Mixed Layer (m)	Total Prod. g C m^{-2} day^{-1}
Paita: 5°-6°	18.90	34.89	4.40	1.45	11.65	9.40	1:6.5:8.0	1.81	66.80	36.9	22.3	13.3	0.35
Chimbote: 8°-9°	18.70	35.01	4.70	1.35	8.45	8.00	1:5.9:6.3	3.28	68.63	20.9	19.1	10.8	1.34
Callao: 12°-13°	17.90	35.04	4.90	1.47	8.62	6.27	1:4.3:5.9	3.43	75.20	21.9	18.7	14.5	0.53
San Juan: 15°-16°	16.80	34.99	4.20	1.72	11.35	11.0	1:6.4:6.6	3.65	154.7	42.4	29.4	15.3	2.58

TABLE 2. Seasonal Variability in the Upwelling Areas off Peru

Area	Isopycnal (σ_t)	Upwelling Depth (m)	Average Winds (m/s)	Month	Speed (cm/s)	Cooling Period
Paita:[1]						
Autumn	25.3-25.5	50	4.9[7]	May	4.6×10^{-4}	April - August (5 m)
Winter	25.8-26.1	150		Sept.	2.2×10^{-3}	
Mean		100			Mean 1.3×10^{-3}	
Chimbote:						
Autumn	25.4-25.6	50	4.6	May	3.9×10^{-4}	March - October (8 m)
Winter	25.8-26.0	120		Sept.	1.5×10^{-3}	
Mean		85			Mean 0.9×10^{-3}	
Callao:[2]						
Autumn	25.5-25.8	40	3.2[7]	May	6×10^{-4}	April - September (6 m)
Winter	25.6-25.9	60		Sept.	3.8×10^{-3}	
Mean		50			Mean 2.2×10^{-3}	
San Juan:[1]						
Autumn	25.6-25.8	40	4.8[7]	June	2.1×10^{-3}	April - September (6 m)
Winter	25.8-26.1	70		Aug.	4.0×10^{-3}	
Mean		55			Mean 3.1×10^{-3}	
Oregon:[3]					$2-0.5 \times 10^{-2}$	
Oregon:[4]					$0.2-7.0 \times 10^{-3}$	
Northwest Africa:[5]					10^{-1}	
Baja California:[6]					$5.4-7.9 \times 10^{-3}$	

1 Zuta *et al.* (1978)
2 Flores (1979)
3 Halpern (1976)
4 Smith, Pattullo, and Lane (1966)
5 Barton, Huyer, and Smith (1977)
6 Stevenson (1974)
7 Zuta and Guillén (1970)

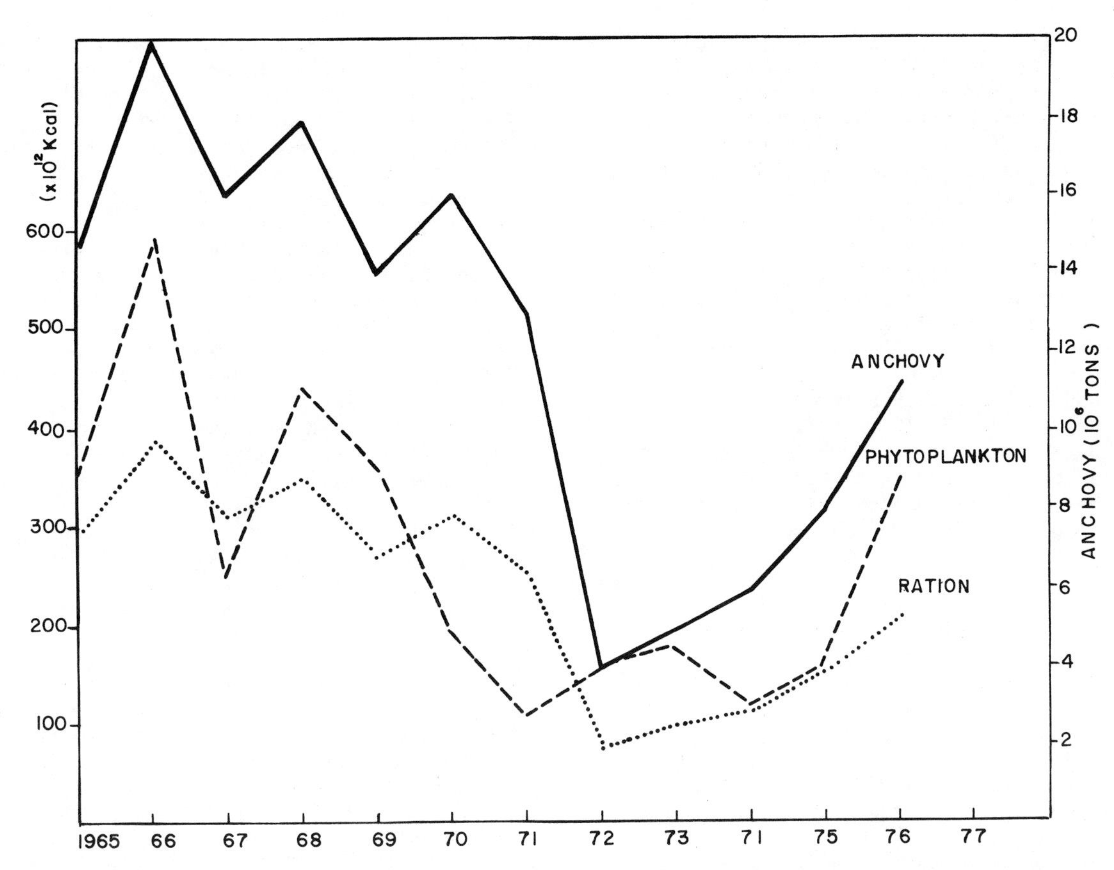

Fig. 12. Anchovy biomass (tons), phytoplankton food (Kcal) and ration (Kcal) for the period 1965-1977.

Peru anchovies influence primary production through excretion of ammonia. According to Whitledge (1978) anchovy and zooplankton off Peru contribute 59 and 41%, respectively, of the ammonia required by phytoplankton production, but Harrison and Platt (1979) reported that 90% of the nitrogen assimilated by phytoplankton near the coast of Chimbote is ammonia. Both results emphasize the importance of ammonia regeneration and suggest that the decrease in primary production in the last few years could be caused by the decreased anchovy stocks, decreasing the ammonia supplied by excretion (Guillén and Calienes, 1980).

References

Barstow, D., W. Gilbert, and B. Wyatt, Hydrographic data from Oregon waters, Department of Oceanography, Oregon State University, Ref. No. 69-6, 84 pp., 1969.

Barton, E.D., A. Huyer, and R.L. Smith, Temporal variations observed in the hydrographic regime near Cabo Corveiro in the northwest African upwelling region, February - April, 1974, *Deep-Sea Research, 24*, 7-24, 1977.

Barton, E.D., R.D. Pillsbury, and R.L. Smith, A compendium of physical observations from JOINT-I, vertical section of temperature, salinity, and sigma-t from R.V. *Gillis* data and low pass filtered measurements of wind and currents, School of Oceanography, Oregon State University, Ref. No. 75-17, 60 pp., 1975.

Beers, J., M. Stevenson, R. Eppley, and E. Brooks, Plankton populations and upwelling of the coast of Peru, June 1969, *Fishery Bulletin, 69*, 859-876, 1971.

Cushing, D.H., Upwelling and fish production, *FAO Fisheries Technical Paper, 84*, 40, 1969.

Dugdale, R.C., Chemical oceanography and primary productivity in upwelling regions, *Geoforum, 2*, 47-61, 1972.

Dugdale, R.C., and J.J. Goering, Nutrient limitation and the path of nitrogen in Peru Current production, in: *Anton Bruun Report, 4*, Texas A&M Press, 5.3-5.8, 1970.

Flores, R., Regimen hidrico del afloramiento en el area de Callao, CUEA Project Workshop, Corvallis, Oregon, 1979 (unpublished document).

Guillén, O., Variación de los fosfatos en el Callao como medida de la producción primaria, I. *Seminario Latino-Americano sobre el Océano Pacífico Oriental, Universidad Nacional Mayor de San Marcos, Lima, Peru*, 192-198, 1966.

Guillén, O., The "*El Niño*" phenomenon in 1965 and its relations with the productivity in coastal Peruvian waters, in: *Fertility of the Sea,*

J. Costlow (ed.), Gordon and Breach Science Publishers, New York, *Vol. 1*, 157-185, 1971.

Guillén, O., Carbon/chlorophyll relationships in Peruvian coastal waters, in: *Oceanography of the South Pacific 1972*, Comp. R. Fraser, New Zealand National Commission for UNESCO, Wellington, 373-385, 1973.

Guillén, O., R. Calienes, and R. de Rondán, Contribución al estudio del ambiente de la anchoveta (*Engraulis ringens* J.), *Boletin del Instituto del Mar del Peru, 2*, 49-76, 1969.

Guillén, O., and B. Mendiola, Primary productivity and phytoplankton in the coastal Peruvian waters, in: *Fertility of the Sea*, J. Costlow (ed.), Gordon and Breach Science Publishers, New York, *Vol. 1*, 157-185, 1971.

Guillén, O., and R. de Rondán, Productividad de las aguas costeras frente al Peru, I Parte: Medio ambiente y producción primaria, *Informe Especial del Instituto del Mar del Peru, 148*, 1-35, 1974.

Guillén, O., R. Calienes, and R. de Rondán, Medio ambiente y producción primaria en el area Pimental-Chimbote, *Boletin del Instituto del Mar del Peru, 3*, 107-159, 1977.

Guillén, O., and R. Calienes, Biological productivity and *El Nino*, *Fisheries Management and Environmental Uncertainties: El Niño* and the anchovy, Wiley Interscience, 600 pp., 1980.

Halpern, D., A compilation of wind, current, and temperature measurements near Spanish Sahara during March and April 1974, Pacific Marine Environmental Laboratory/NOAA, University of Washington Technical Report, 10, 100 pp., 1976.

Harrison, G., and T. Platt, Primary production and nutrient fluxes off the coast of Peru, Workshop of ICANE Project, Lima, Peru, 1979.

Huyer, J., A comparison of upwelling events in two locations: Oregon and northwest Africa, *Journal of Marine Research, 34*, 531-546, 1976.

Jordán, R., Biología de la anchoveta, Parte I: Resumen del conocimiento actual, Actas Reunión de Trabajo sobre el fenómeno coincido como *El Niño*, Guayaquil, Ecuador, *FAO Informes de Pesca, 85*, 359-399, 1976.

Mattisen, O.A., R.E. Thorne, R.J. Trumble, and M. Blackburn, Food consumption of pelagic fish in an upwelling area, in: *Upwelling Ecosystems*, R. Boje and M. Tomczak (eds.), Springer-Verlag, New York, pp. 111-123, 1978.

Mejía, J., M. Samamé, and M. Esquerre, Informe sobre los stocks disponibles de peces de consumo en la costa peruana, *Informe Interno del Instituto del Mar del Peru, 5*, 1-15, 1974.

Mooers, C.N.K., and J.S. Allen, Final report of the Coastal Upwelling Ecosystems Analysis Summer 1973 Theoretical Workshop, School of Oceanography, Corvallis, Oregon State University, 1973.

Redfield, A.C., B.H. Ketchum, and F.A. Richards, The influence of organisms on the composition of sea water, in: *The Sea, 2*, M.N. Hill (ed.), Wiley, New York, pp. 26-77, 1963.

Richmann, J.G., and S.L. Smith, On the possible enhancement of oxygen depletion in the coastal waters of Peru between 6°S and 11°S, *Journal of Marine Research*, in press.

Ryther, J.H., and D.W. Menzel, On the production, composition, and distribution of organic matter in the Western Arabian Sea, *Deep-Sea Research, 12*, 199-209, 1965.

Santander, H., Cosecha estable del zooplancton, *Informe Especial del Instituto del Mar del Peru, 148*, 1-9, 1974.

Schaeffer, M.B., The potential harvest of the sea, *Transactions of the American Fisheries Society, 94*, 123-128, 1965.

Smith, R.L., J.G. Pattullo, and R.K. Lane, An investigation of the early stage of upwelling along the Oregon Coast, *Journal of Geophysical Research, 71*, 1135-1140, 1966.

Stevenson, M., Preliminary results from the coastal upwelling ecosystems analysis cruise - Mescal II, Proceedings of the Twenty-Fourth Tuna Conference, October 1-7, 1973, 1974.

White, W.B., The Equatorial Undercurrent, the South Equatorial Countercurrent, and their extension in the South Pacific Ocean east of the Galapagos Islands during February - March, 1967, *Technical Report, Texas A&M University, Department of Oceanography, Ref. 69-41*, 74 pp., 1969.

Whitledge, T.E., Excretion measurements of nekton and the regeneration of nutrients near Punta San Juan in the Peru upwelling systems derived from nekton and zooplankton excretion, Ph.D. Thesis, University of Washington, Seattle, 115 pp., 1972.

Whitledge, T.E., Regeneration of nitrogen by zooplankton and fish in the northwest Africa and Peru upwelling ecosystems, in: *Upwelling Ecosystems*, R. Boje and M. Tomczak (eds.), Springer-Verlag, New York, pp. 90-100, 1978.

Wooster, W. and M. Gilmartin, The Peru - Chile Undercurrent, *Journal of Marine Research, 19*, 97-122, 1961.

Wyrtki, K., The horizontal and vertical field of motion in the Peru Current, *Bulletin of the Scripps Institution of Oceanography 8*, 313-346, 1963.

Zuta, S. and O. Guillén, Oceanografía de las aguas costeras del Peru, *Boletin del Instituto del Mar del Peru, 2*, 157-324, 1970.

Zuta, S., T. Rivera, and A. Bustamente, Hydrographic aspects of the main upwelling areas off Peru, in: *Upwelling Ecosystems*, R. Boje and M. Tomczak (eds.), pp. 235-257, 1978.

COMPOSITION AND DISTRIBUTION OF PHYTOPLANKTON IN THE UPWELLING SYSTEM OF THE GALAPAGOS ISLANDS

Roberto Jimenez *

Instituto Oceanografico de la Armada
P.O. Box 5940
Guayaquil, Ecuador

* Present address: Instituto Nacional de Pesca
P.O. Box 5918
Guayaquil, Ecuador

Abstract. During a cruise to the Galapagos Islands in November, 1978, phytoplankton was collected at all hydrographic stations occupied by the BAE *Orion* of the Oceanographic Institute of Ecuador. The upwelling in the Galapagos region caused by the Equatorial Undercurrent affects the horizontal and vertical distribution of phytoplankton in a large area of the eastern Pacific. The highest concentrations (8,000 x 10^3 cells/l) were just west of Isabela Island, especially in Elizabeth Bay. The numbers decreased north and south of the upwelling area. The highest primary production (1.433 mg C $m^{-2}d^{-1}$) and chlorophyll content (102 mg/m^2) were also in Elizabeth Bay and decreased to the north and south. The phytoplankton composition had special characteristics related to hydrographic conditions. The diatoms *Thalassiosira* spp., *Nitzschia* spp., and *Chaetoceros* spp. were important components of the phytoplankton in the upwelling region, while microflagellates predominated north and south of the equator. The coccolithophorids were well represented in the area and included *Emiliania huxleyi*, *Cyclococcolithus fragilis*, *Gephyrocapsa oceanica*, and *Calyptrosphaera oblonga*. The distribution of the principal phytoplankton components is examined in relation to the hydrography.

Introduction

The Galapagos Islands are a volcanic archipelago, rising from the sea floor on the equator about 600 miles west of Ecuador. Oceanographic and biological information about the Galapagos region has been increasing in recent years. Pak and Zaneveld (1973) analyzed the Equatorial Undercurrent that flows eastward through the islands, as well as the Equatorial Front associated with upwelling of the Undercurrent (Pak and Zaneveld, 1974). Houvenaghel (1978) described the oceanographic conditions among the islands in relation to seasonal patterns. Additional physical and biological information is available for west of the Galapagos from the 17th cruise of R.V. *Akademik Kurchatov*, which focused on equatorial upwelling (Sorokin, Pavel'eva, and Vasil'eva, 1975; Sorokin, Sukhanova, Konovalova, and Pavel'eva, 1975; Vinogradov and Semenova, 1975; Finenko and Lauskaia, 1975). Physical and biological data have also been reported for east of the Galapagos in relation to the occurrence of the equatorial upwelling (Jimenez, 1978). The presence of the Equatorial Front between the islands and the continent have been described by Enfield (1976) and the distribution of plankton associated with the front by Bonilla (1979). Systematic phytoplankton records in some Galapagos coastal stations have been presented by Hendey (1972), Marshall (1972), and Grillini (1975). This report describes the composition, distribution, and abundance of phytoplankton in relation to the hydrographic characteristics east and west of the Galapagos Islands.

Methods

Phytoplankton samples were collected at hydrographic stations occupied by the research vessel *Orion* of the Oceanographic Institute of Ecuador in November, 1978. Sampling was on two meridional transects from 1°04'N to 2°00'S; the transect through the eastern part of the Galapagos was at 89°42'W and the one west of the islands was at 91°43'W (Fig. 1).

Water samples were collected at depths which correspond to 100, 50, 30, 16, and 1% of the incident light and in offshore stations at greater depths. A quanta meter, constructed at the Institute of Physical Oceanography in Copenhagen, Denmark, was used for the determination of light levels at local noon. Samples of 50 ml fixed with Lugol's solution were examined using combined

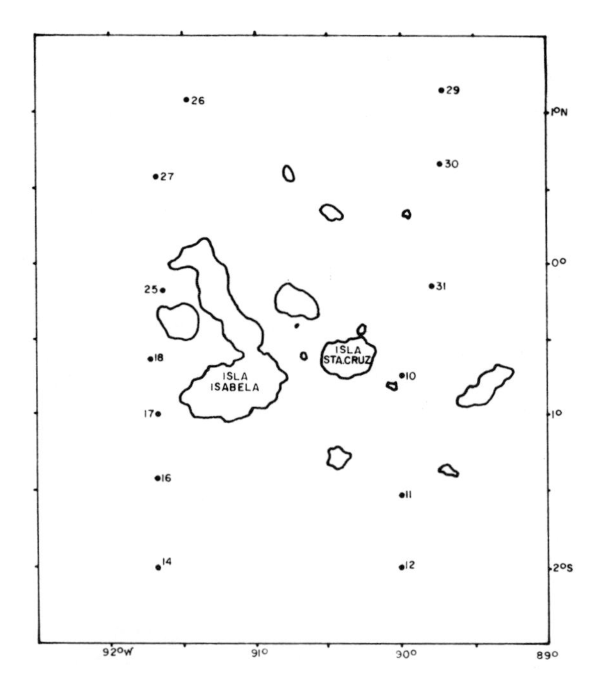

Fig. 1. Location of the meridional transects west and east of the Galapagos Islands.

sedimentation chambers in an inverted microscope. Total cells in a 1-cm^2 chamber bottom were scanned under high (400x) magnification to include smaller organisms. Some forms, especially of naked flagellates, could not be identified to species and were recorded as microflagellates or small dinoflagellates.

Margalef (1978) pointed out that many phytoplankton species-level names refer to larger groups or collections of species rather than to one entity appropriate for taxonomic work, such as the *Nitzschia "delicatissima"* complex. With the inverted microscope it is impossible to identify to species level all the *Thalassiosira* species that occur, but samples were also examined by scanning electron microscopy and different small *Thalassiosira* were identified as being *T. lineata, T. oestrupii, T. subtilis, T. fere-lineata, T. tenera,* and *T. minuscula.* Hasle (1975) suggested that certain differences in colony shape and cell shape and size offer possibilities for distinguishing between species, so living phytoplankton were examined at sea with a light microscope. Observations of living colonies of *Thalassiosira* spp. showed that the colonies ranged in size from 5,000 cells/colony to a few cells in each colony. The scanning electron misroscope also revealed the presence of small cells of thecate and athecate dinoflagellates.

On the transects through the eastern and western parts of the archipelago temperature, salinity, nutrients, chlorophyll, phytoplankton, zooplankton net catches, and, at some stations, primary production with ^{14}C, were observed.

Results

Hydrographic Aspects

The Equatorial Undercurrent is a fast sub-surface current flowing eastward to the Galapagos Islands at the equator (Knauss, 1960). In November, 1978, the distribution of temperature and oxygen at 93°W indicated that the axis of the Equatorial Undercurrent west of the islands was slightly displaced south of the equator. East of the main islands the current was observed in two narrow zonal bands. The bands are not symmetrical with respect to the equator, and it is possible that the locations of the branches vary either seasonally or as a result of other large-scale dynamic conditions (Pak and Zaneveld, 1973). Pak and Zaneveld (1973) pointed out that a complicated distribution of water properties results from the intense mixing caused by the strong subsurface shear between the surface water and the Equatorial Undercurrent flowing in the opposite direction. There is a large velocity shear, and the interruption of flow by the islands causes eddies of various sizes. Intense mixing takes place in an area where there is a convergence of surface water from the north and south. As the two water masses have distinctive characteristics, a strong front is formed. The front contains irregularities of various sizes resulting from intrusions of one water type into another. Houvenaghel (1978) has shown by oceanographic properties that upwelling occurs throughout the different parts of the Galapagos area and indicated the dependence of upwelling in the Galapagos on the Equatorial Undercurrent.

The distribution of water properties west of the Galapagos shows that in 1978 the salinity maximum that characterizes the core of the Equatorial Undercurrent (Stevenson and Taft, 1971) is displaced south of the equator. The displacement is also indicated by the temperature minimum at 100 m with a pronounced thermostad south of the equator.

The Equatorial Undercurrent, with its high-velocity core, has intense vertical mixing in the region of the thermocline. The lines of constant sigma-t, temperature, and oxygen in the meridional section west of the Galapagos bulge upward above the thermocline and bend downward below the thermocline. The core of the Undercurrent is homogeneous with respect to density and temperature. The Equatorial Undercurrent is indicated by the layer of relatively well mixed water with a maximum spreading of isotherms, isohalines, and isopleths of sigma-t and oxygen. In contrast to the considerable homogeneity in the vicinity of the current, there are steep vertical gradients just north and south of the current. Similar hydrographic features were shown by Barber and Ryther (1969) in data collected at five stations between 2°S and 2°N at 92°W during April, 1966. It is evident that in 1978 at stations 24 and 25 water from the core of the Equatorial

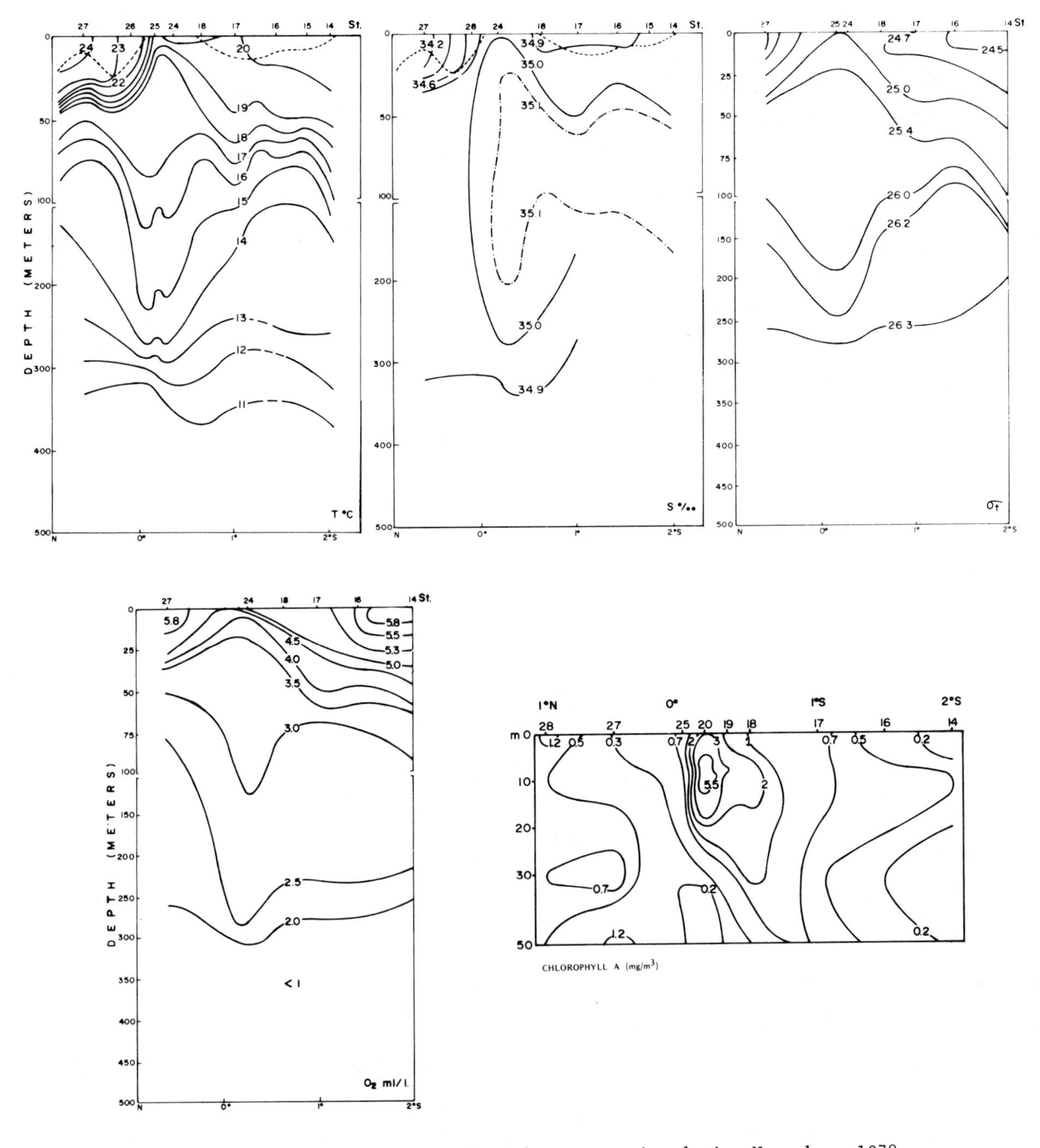

Fig. 3. Water masses in the Galapagos region during November, 1978.

Undercurrent upwelled, the mixed layer reaching the surface. Meanwhile, north of the equator, Tropical water was present at the surface, creating the characteristic strong front that is evident in the vertical and horizontal temperature gradients.

In the eastern part of the Galapagos archipelago the water properties change drastically. Tropical Surface water was present along most of the north-south transect (Fig. 3), and the vertical profiles show a well stratified water column (Fig. 4). The salinity maximum was south of 1°S but separated from the surface by a well stratified water column. The salinity maximum tended

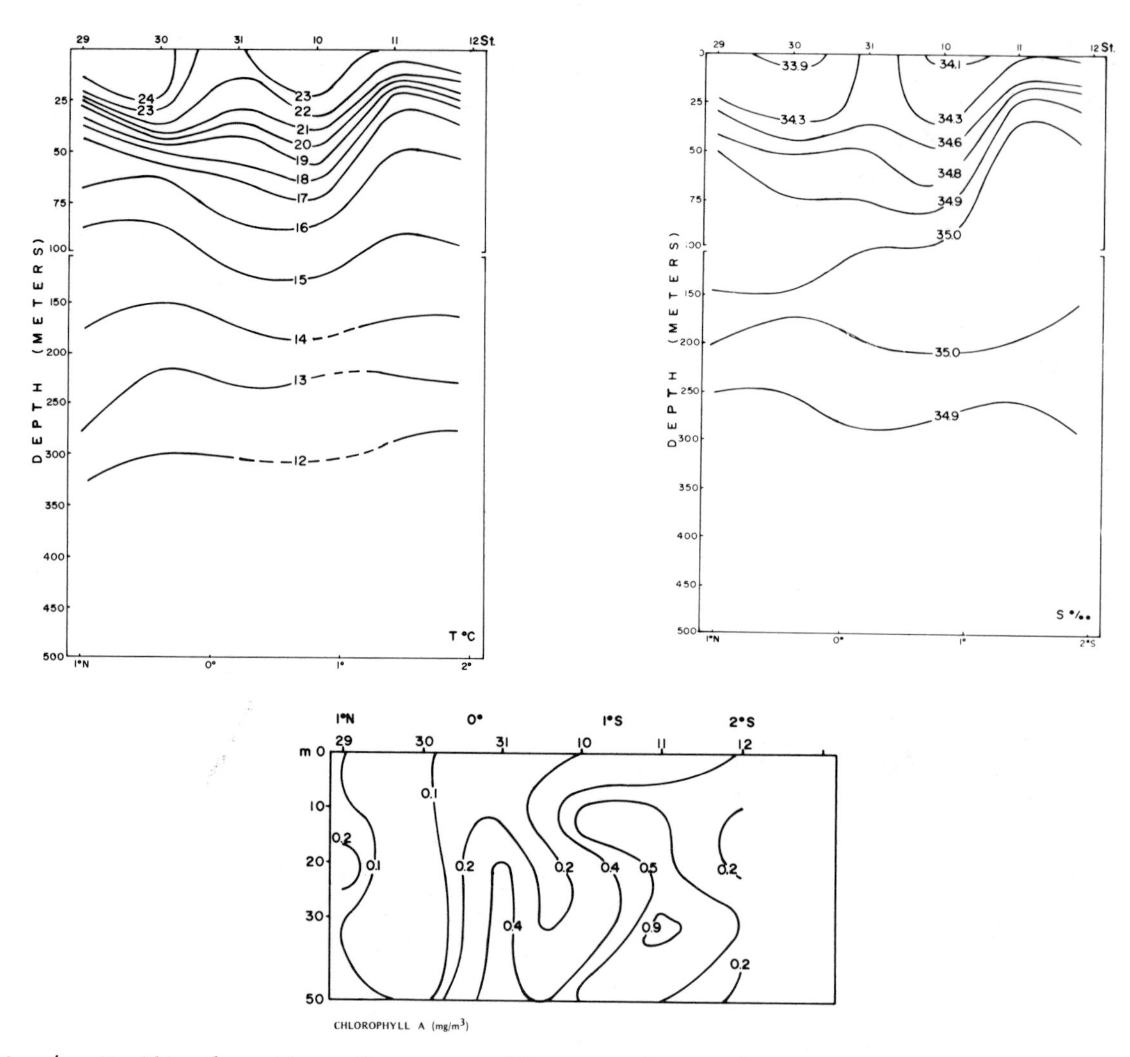

Fig. 4. Meridional sections of oceanographic properties in the eastern part of the Galapagos.

to bend downward north of 1°S. A more detailed description of the water masses present in the area is given by Arcos (1980) and Moreano (1980).

Biological Aspects

The biological characteristics west of the Galapagos show evidence of upwelling in terms of cooling, nutrient content, and increases in productivity caused by the fertilization of surface waters. The highest primary production (from 830 to 1433 mg C $m^{-2}day^{-1}$) was recorded at stations 18, 19, and 20 south of the equator. At station 24, where the ascending water was more evident, the primary production decreased to 277 mg C $m^{-2}day^{-1}$. The relatively poor ability of the most recently upwelled water at the equator to support phytoplankton growth was reported by Barber and Ryther (1969). Enhanced biological production in these upwelled equatorial waters occurs only after they have been at the surface for some time.

The chlorophyll section across the equator west of the Galapagos shows high concentrations (1.0 to 5.5 mg/m^3) at Stas 18, 19, and 20 south of the equator. North and south of this area, chlorophyll values range from 0.7 to 0.2 mg/m^3. Characteristic values of 0.5 mg/m^3 or greater extend from the upwelling center area to 2°S. On the equator there is a sharp decrease in chlorophyll where the core of the Equatorial Undercurrent upwells. There is a slight increase in chlorophyll north of the latter area.

The north-south chlorophyll section through the eastern part of the Galapagos (Fig. 4) is distinctly different from that to the west. Between 1 and 2°S there is a chlorophyll maximum of 0.5 mg/m^3 in the sub-surface water. North of this area, where the salinity maximum was present in the well-stratified surface water, there is a

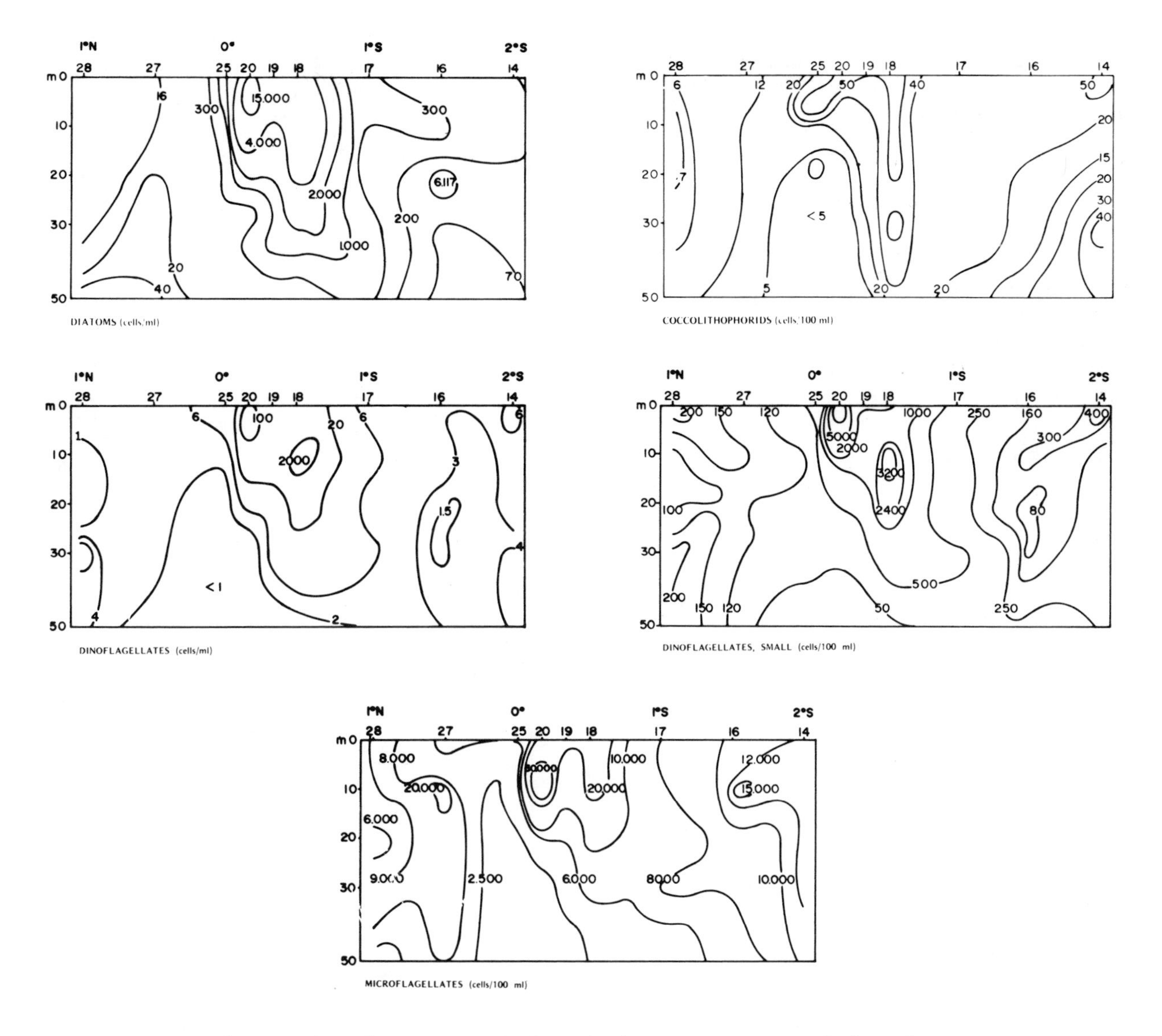

Fig. 5. Meridional sections of phytoplankton groups west of the Galapagos.

marked decrease in the concentration of chlorophyll. Values ranged between 0.1 and 0.2 mg/m^3, typical of surface waters in the eastern tropical Pacific.

Distribution of Phytoplankton

The distribution of phytoplankton west of the Galapagos, where the main upwelling region develops, is similar to distributions of phytoplankton in other upwelling areas, such as northwest Africa (Margalef, 1978). Some of the phytoplankters were of the same genera or species that dominate in other upwelling regions, and diatoms were the dominant group (Fig. 5). They were closer to the core of the upwelling with high cell concentrations corresponding specifically to the *Thalassiosira* spp. and to the area between 0° and 1°S. South of this area the concentration of diatoms decreased to the range of 200 to 300 cells/ml. A significant decrease in density of diatoms was noted north of the equator in the Tropical Surface waters (16 cells/ml). A small increase was evident beneath the thermocline in the sub-surface waters. This was probably caused by the pronounced convergence in the area and the sinking of surface blooms at the convergence.

The coccolithophorids were also abundant in the surface waters affected by upwelling with concentrations between 40 and 50 cells/ml, but they were almost completely absent from the core of the undercurrent where the temperature was below 18°C (Fig. 5). Where recently upwelling water was detected (Sta. 25), concentrations less than

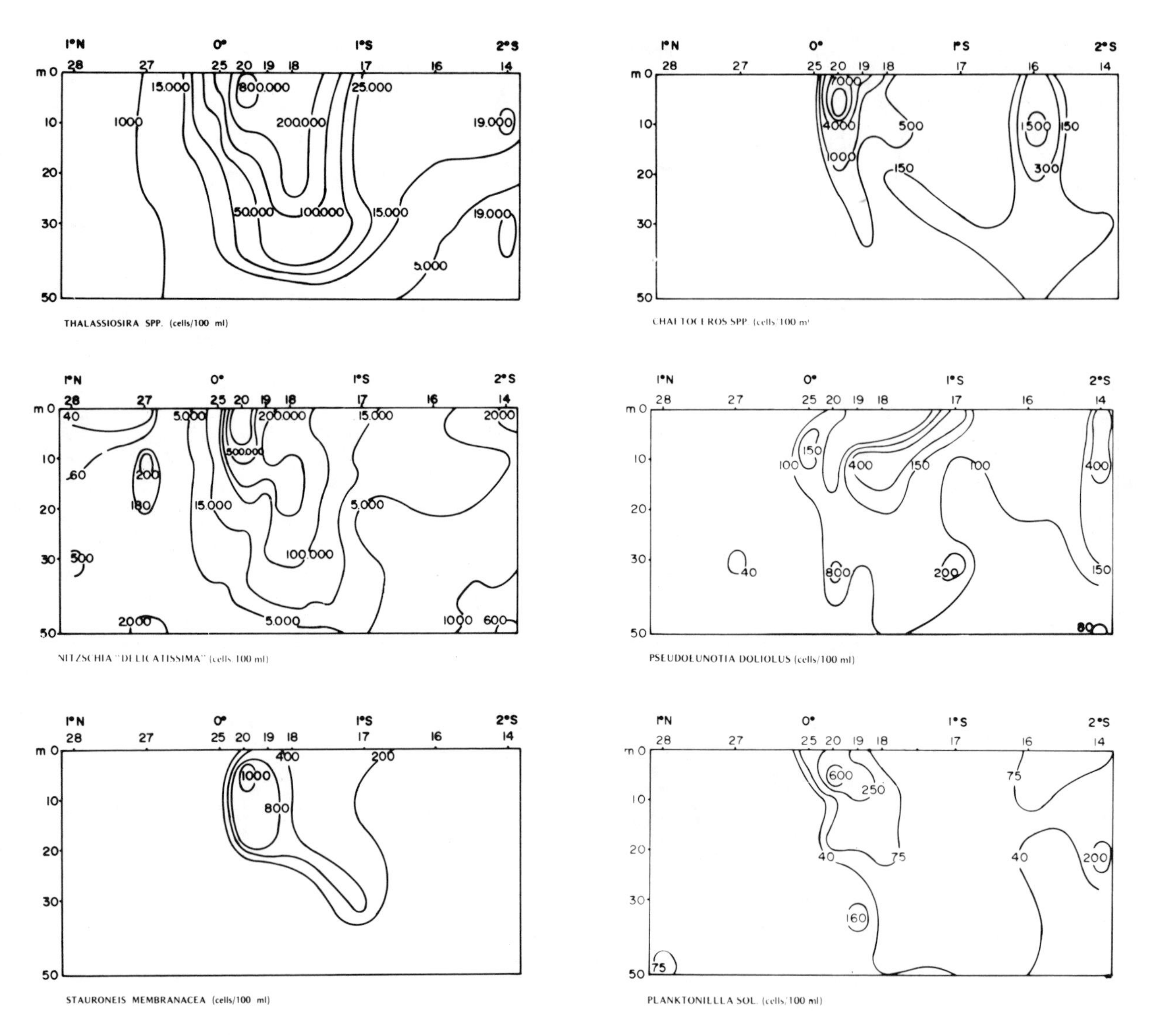

Fig. 6. Meridional sections of phytoplankton species west of the Galapagos.

5 cells/ml were recorded. This observation confirms that the cold Equatorial Undercurrent does not favor the presence of coccolithophorids, as was suggested by Okada and Honjo (1973), who encountered pronounced decreases in coccolithophorid densities at 100 to 125 m in the Central Pacific. The coccolithophorids were relatively abundant between the equator and 2°S with a maximum of 20 cells/ml; they were scarce north of the equator.

The dinoflagellates (Fig. 5) were less abundant overall than the diatoms and the coccolithophorids but had a density maximum close to the upwelling area with concentrations between 20 and 2,000 cells/ml. North and south of the area the dinoflagellates were scarce with concentrations between 2 and 6 cells/ml. Like the other phytoplankton groups, dinoflagellates were poorly represented in the surfacing water of the Equatorial Undercurrent with less than 1 cell/ml. This characteristic corroborates the poor ability of phytoplankton to develop in the recently upwelled water and to survive in the sub-surface core of the Undercurrent.

Using the inverted microscope 130 species of phytoplankton were identified. The basis for the selection of several species discussed in this paper has been its dominance in the upwelling region west of the Galapagos. The dominant diatoms in the area were the *Thalassiosira* spp., *Chaetoceros* spp., *Nitzschia "delicatissima"*, *Pseudoeunotia doliolus*, *Planktoniella sol*, and *Stauroneis membranacea* (Fig. 6). Less important diatom species were the *Rhizosolenia* spp., including *R. bergonii*, *R. acuminata*, *R. stolterfothii*, *R. styliformis*, *R. delicatissima*, *R. alata*, *R.*

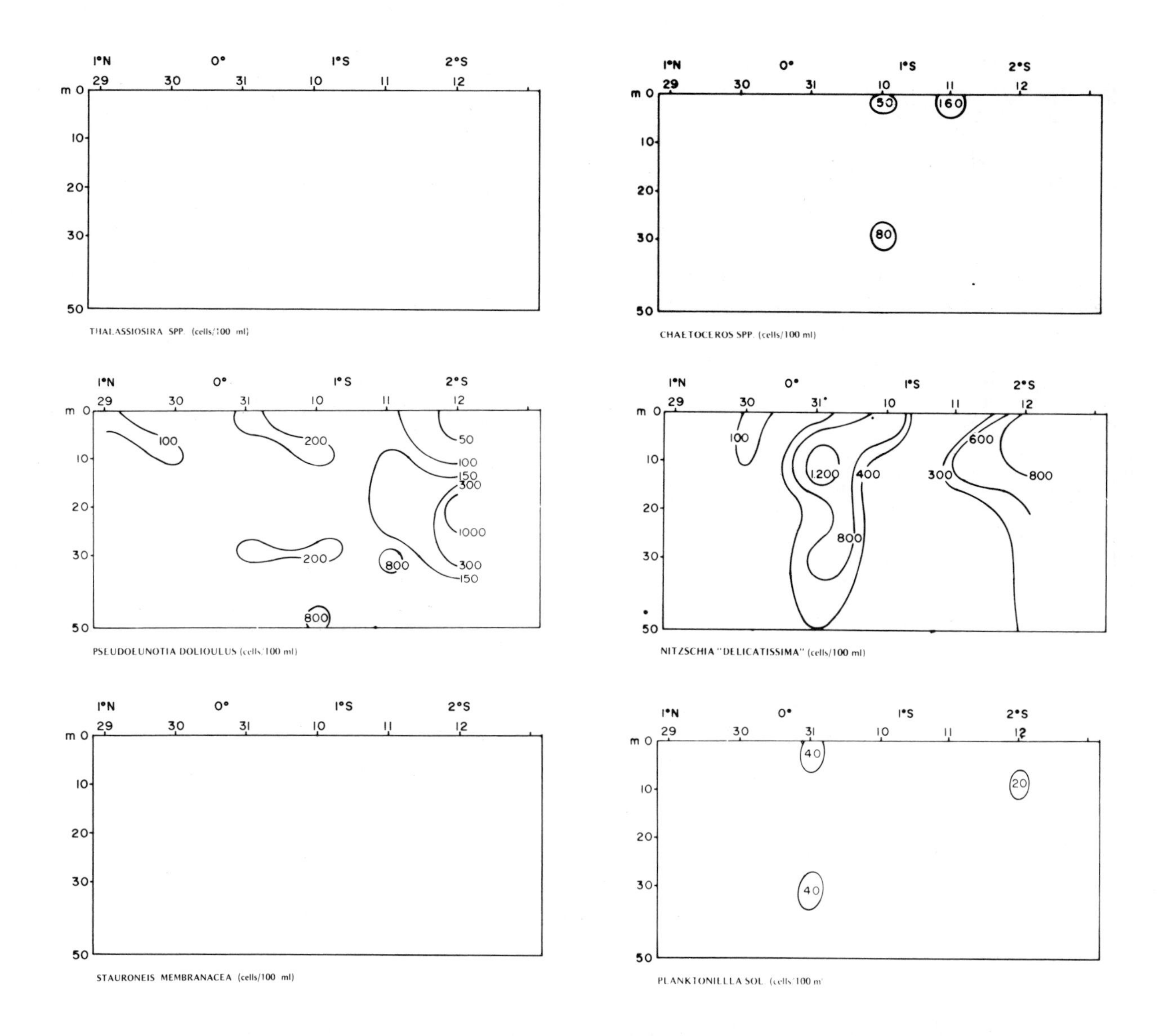

Fig. 7. Meridional sections of phytoplankton species in the eastern part of the Galapagos.

robusta, *R. hebetata*, and *R. setigera*. Although always present, low concentrations were recorded of *Coscinodiscus* species including *C. crenulatus*, *C. excentricuslineatus*, *C. radiatus*, *C. thorii*, *C. asteromphalus*, *C. curvatulus*, and *C. nitidus*. Also present were *Thalassiothrix frauenfeldii*, *T. gibberula*, *T. longissima*, *T. delicatula*, *Thalassionema elegans*, *T. nitzschioides*, *Asteromphalus heptactis*, and *A. elegans*. Colonies of *Phaeocystis* were recorded only west of the Galapagos with a maximum of 600 colonies/100 ml.

Many of the 70 to 80 known *Thalassiosira* species are present in upwelling areas. Large colonies of *T. partheneia*, *T. diporocyclus*, and *T. miniscula* have been found in upwelling areas along the African and American west coasts. Although they are not restricted to upwelling areas, their colony formation is obviously favored by the environmental conditions in such areas (Hasle, 1975). Large *Thalassiosira* spp. colonies like *T. partheneia* contribute up to one-third of the total primary production in the northwest Africa upwelling region (Elbrachter and Boje, 1978). It has been demonstrated that they contribute importantly to the food web in an upwelling current system as well as in oceanic waters where the colonies can drift (Elbrachter, 1978). Characteristic of *Thalassiosira* spp. is colony formation by means of one or more connecting central threads embedded in a mass of mucilage. Other colony-forming genera such as *Phaeocystis* and *Chaetoceros* have been reported in upwelling areas (Margalef, 1978).

In the Galapagos the distribution of *Thalassio-*

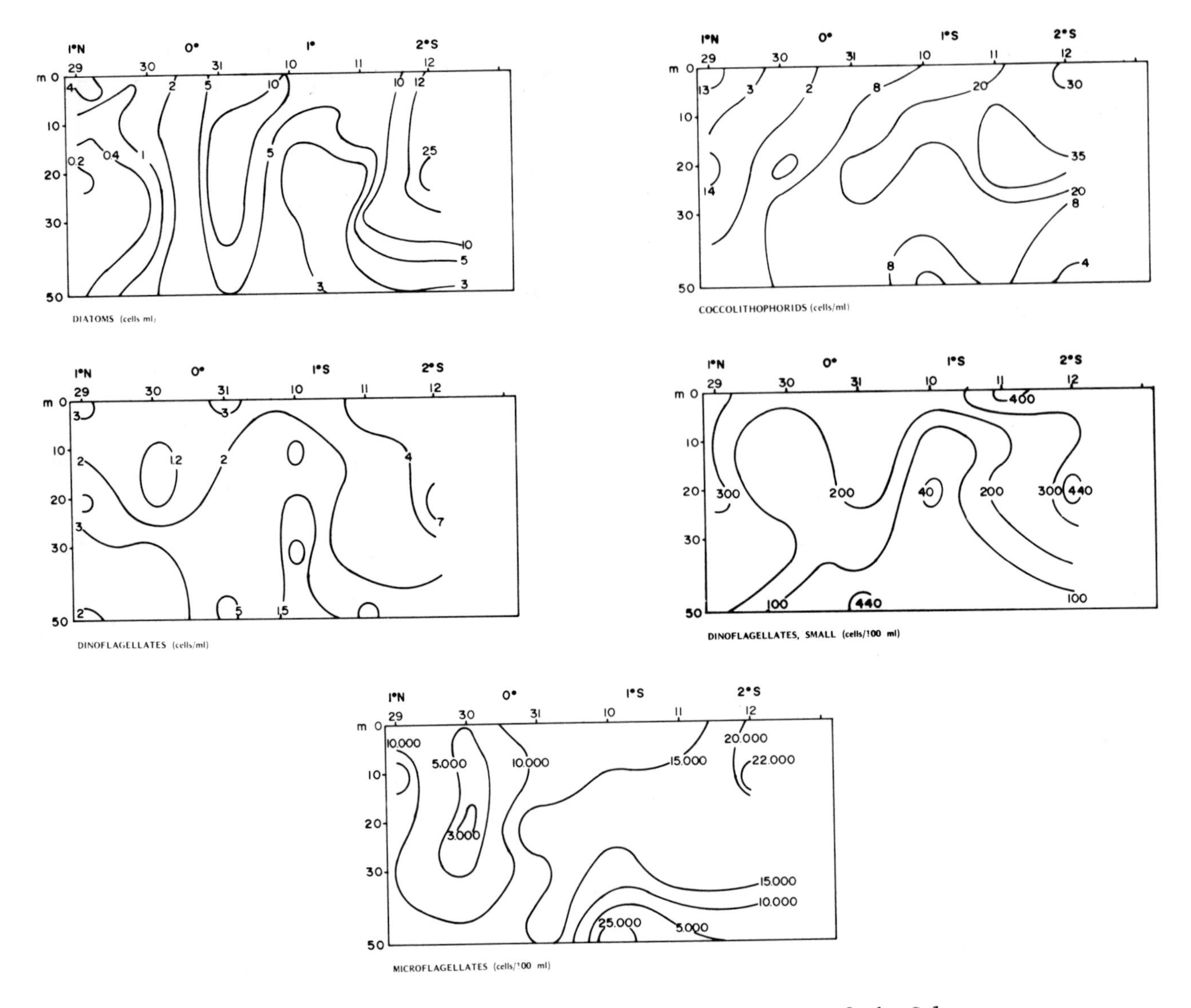

Fig. 8. Meridional sections of phytoplankton groups east of the Galapagos.

sira spp. (Fig. 6) appears to be directly related to upwelling. Maximum concentrations were in the upper layers between 0° and 1°S. Concentrations decreased south of the upwelled water. North of the equator there was a sharp decrease of these diatoms and no cells of the genus were present in the Tropical Surface water. Almost no *Thalassiosira* spp. were present east of the Galapagos and this characteristic emphasizes the dependence of these diatoms on upwelling. Nevertheless, *Thalassiosira* spp. have been observed east of the Galapagos in other years when hydrographic conditions were different.

The *Chaetoceros* spp., including *Ch. peruvianus*, *Ch. lorenzianus*, and *Ch. radicans*, were located in the main upwelling area, specifically in the upper photic layers, but they tended to occur deeper to the south. Certain parcels of water were free of *Chaetoceros* and no cells were present north of the equator or in the core of the Equatorial Undercurrent itself (Fig. 6). Most of the *Chaetoceros* found at the time of our cruise were large individual cells. Unpublished data confirm the presence of small cells of *Chaetoceros*, like *Ch. gracilis* and *Ch. perpusillus*, in high concentrations (2,190 cells/ml) when there is stronger upwelling west of the Galapagos Islands.

Large concentrations of *Nitzschia "delicatissima"* (500,000 cells/100 ml) were present in and around the upwelling area. Concentrations decreased to the south and sharply to the north of the equator (Fig. 6). Margalef (1978) suggested that the dominance of *N. "delicatissima"* in ascending water results because of its ability to grow faster at high concentrations of nutrients.

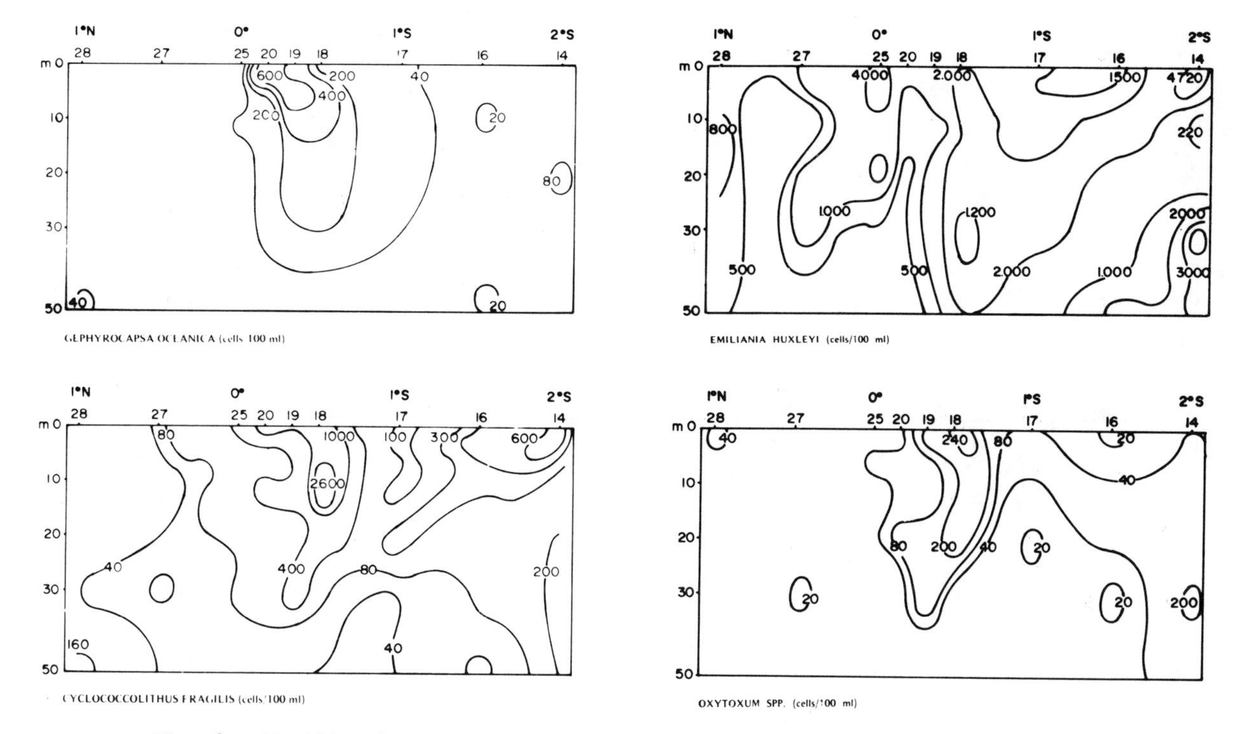

Fig. 9. Meridional sections of phytoplankton species west of the Galapagos.

East of the Galapagos *N. "delicatissima"* (Fig. 7), although in lower densities, was present where the isotherms and isohalines rise in the surface layers. Other *Nitzschia* spp. such as *N. closterium* (= *Cylindrotheca closterium* (Ehrb.) Reimann and Lewin), *N. bicapitata,* and *N. gaarderi,* were present in lower concentrations.

Pseudoeunotia doliolus was present in and around the upwelling area (Fig. 6), with maximum densities in the upwelling zone at 10 to 15-m depth, and far from the upwelling zone both 1 and 2°S in the surface layers. Almost no cells were present between the equator and 1°N on the transect west of the Galapagos. This diatom was present in greater concentrations at 2°S east of the islands, but in deeper layers (Fig. 7). This may be associated with the extension of the undercurrent east of the islands.

Stauroneis membranacea was restricted to the upwelling zone but was not abundant in the strong upwelling region of the Undercurrent core at Sta. 25 (Fig. 6).

Planktoniella sol (Fig. 6) was present in the upwelling area and the drift of the upwelled water to the south produced the highest densities. Fewer individuals and a more erratic distribution of this species were encountered east of the Galapagos. Although *P. sol* is a tropical species widely distributed in tropical and subtropical waters (Semina, 1977), it is clear from the data that the density is dependent on enrichment by upwelling.

Small dinoflagellates were present in large numbers (Fig. 5) at the edge of the zone of strong upwelling between the equator and 1°S, but concentrations were lower in the upwelling center. Between 1 and 2°S concentrations decrease but rise again slightly in the southernmost part of the transect. East of the Galapagos, small dinoflagellates are dispersed throughout the area with a small increase in areas divergent from the equator (Fig. 8). The hydrographic differences east and west of the Galapagos explain the differences in the distributions of the small dinoflagellates. This was also true for *Oxytoxum* spp. (Fig. 9). The distribution of this group of species was closely related to the position of upwelled water but further south. *Oxytoxum* spp. and large dinoflagellates actually appear to have their distribution centers south of the diatom distribution center at Sta. 20, perhaps because of a successional feature; almost no cells were present north of the front where Tropical Surface water was present. East of the main islands (Fig. 10) the numbers tended to increase south of the equator to a maximum at 2°S.

Dinoflagellates of the following species were also identified in low concentrations: *Protoperidinium divergens, P. grande, P. oceanicum, Gonyaulax polygramma, G. catenella, G. fratercula, Ceratium azoricum, C. pentagonum* var. *tenerum, C. furca, Blepharocystis splendor maris, Goniodoma polyedricum, Dinophysis ovum,* and *Diplopsalis* sp.

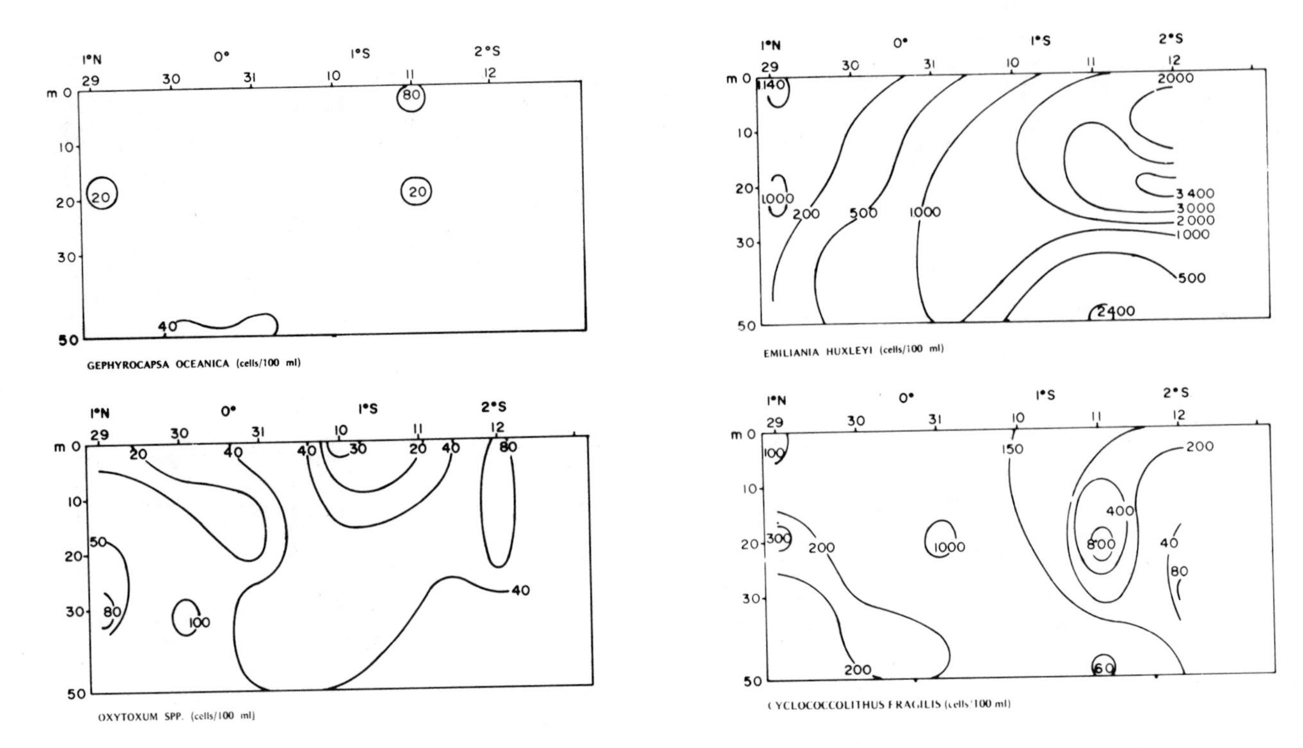

Fig. 10. Meridional sections of phytoplankton species in the eastern part of the Galapagos.

Although the coccolithophorid *Emiliania huxleyi* (Fig. 9) is common in and around the upwelling region, there was a marked decrease of these cells in deeper, cooler, water. A more homogeneous distribution was observed where there was no injection of warmer or cooler water into the euphotic zone. East of the Galapagos the maximum concentrations of these cells (Fig. 10) were at 2°S with densities decreasing to the north in Tropical Surface water.

Cycloccolithus fragilis (Fig. 9) was found in large numbers around areas of moderate upwelling, but a sharp decrease in density was present in areas of strong upwelling and north of the equator. The distribution is somewhat similar to that of *E. huxleyi* east of the islands (Fig. 10).

Gephyrocapsa oceanica, a dominant species of the South Equatorial Current, like *E. huxleyi*, was also largely restricted to the upwelling zone between 0° to 1°S west of the Galapagos (Fig. 9). Few cells of the species were present north or south of this area.

Although the dominant above mentioned coccolithophorids were quantitatively more important, a large group of species was recorded: *Cyclococcolithus leptoporus*, *C. pelagicus*, *Helicosphaera carteri*, *H. hyalina*, *Anthosphaera robusta*, *A. quatricornu*, *Acanthoica quatrospina*, *Halopappus adriaticus*, *Calciopappus caudatus*, *Acanthoica acanthifera*, *Calyptrosphaera oblonga*, *C. sphaeroidea*, *C. pirus*, *Cristallolithus hyalinus*, *Thorosphaera flabellata*, *Umbilicosphaera sibogae*, *Umbellosphaera hulburtiana*, *Scyphosphaera apsteinii*, *S. prolongata*, *S. histrica*, *S. pirus*, *Ophiaster hydroideus*, *Syracospheara binodata*, *Coccolithus pelagicus*, *Discophaera tubifer*, *Caneosphaera molischii*, *Coronosphaera mediterranea*, and *C. binodata*. Most of these species have been recorded in the central Pacific by Hasle (1960a).

The greatest concentration of microflagellates (Fig. 5) was in the surface layers of the main upwelling region while in the deep layers of the area the lowest concentration of microflagellates were found. North and south of the equator the density of microflagellates at depth increased. From this distribution it appears that the cooler water of the Equatorial Undercurrent limits the growth of these species. East of the islands the stratified water column supports relatively larger concentrations of microflagellates (Fig. 8) than the west. However, east of the main islands the large numbers were encountered at temperatures below 19°C and at depths exceeding 25 m. Within the blue-green algae, only a few filaments of *Trichodesmium* sp. were recorded at Sta. 31.

Discussion

Previous authors have emphasized the dominance of coccolithophorids over the diatoms in the central Equatorial Pacific (Hasle, 1960a). Under the abnormal conditions of *El Niño* 1976, the area west of the Galapagos was dominated by coccolithophorids and flagellates (Jimenez, 1977). It seems from the present investigation that the upwelling of the Equatorial Undercurrent favors the dominance of diatoms over the coccolithophorids not only close to the Galapagos but also at some distance from the islands (0°17'N, 93°59'W) (Marshall, 1972). In one surface sample in August, 1968, Marshall reported mainly the diatoms *Nitzschia "delicatissima"*, *N. pacifica*, *Planktoniella sol*, etc., although large numbers of the coccolithophorids *Emiliania huxleyi* and *Gephyrocapsa oceanica* were present. Also, Sorokin *et al.* (1975b) reported a maximum in the number of diatoms of the genera *Nitzschia*, *Thalassiosira*, and *Thalassiothrix* in equatorial cross sections west of 97°W. A significant biomass of the minute colonial alga *Phaeocystis* was also present. Nevertheless, the coccolithophorids dominated in a station at 115°W followed by the diatoms and peridinin dinoflagellates. In general, the biomass of phytoplankton gradually decreased from east to west.

In November, 1978, the area of strong upwelling west of the Galapagos was dominated by diatoms, especially small diatoms such as the colonial or chain-like forms of the genera *Thalassiosira*, *Nitzschia*, and *Chaetoceros*. Populations of small dinoflagellates and coccolithophorids were also abundant in the area. Microflagellates were more abundant north and south of the equator, where an increase of these cells occurred in the less productive waters.

If we compare the November, 1978, observations of primary production and density of phytoplankton immediately west of the Galapagos with the high values reported by Sorokin *et al.*(1975b) from 300 miles west of the islands between the equator and 2°S, it is evident that enrichment by equatorial upwelling occurs in a zonal band from the islands west to the Sorokin *et al.* (1975b) site. In the area south of the equator and east of the Galapagos the species composition and distribution of phytoplankton in November, 1978, seemed to be related to the extension of the Undercurrent into deeper layers. The highest densities of phytoplankton were on the southern edge of the upwelled water; the lowest concentrations were in the warmest waters north of the equator and east of the Galapagos Islands. The water masses as well as the current pattern appear to determine the vertical and horizontal distribution of the species of phytoplankton (Fig. 3). This work tends to confirm the relatively poor ability of newly upwelled water to support large phytoplankton populations. This may be partly due to the lack of organic conditioning in the nutrient-rich, newly upwelled water (Barber and Ryther, 1969; Sorokin *et al.*, 1975), which lengthens the lag time of phytoplankton growth.

This work permits the main factors controlling the variability in composition and distribution of phytoplankton around the Galapagos to be identified. Principal among the factors are: (1) the surfacing of the Equatorial Undercurrent west of the islands, (2) the position of the center of upwelling, (3) the poleward extent of the upwelling effect on the surface, (4) the boundary of the upwelled Equatorial Undercurrent water and the Tropical Surface water north of the equator, and (5) the dispersion by the equatorial current regime east of the Galapagos Islands.

References

Arcos, F., A dense patch of *Acartia levequei* (Copepoda, calanoida) in upwelled equatorial undercurrent water around the Galapagos Islands, *This Volume*, 1980.

Barber, R.T., and J.H. Ryther, Organic chelators: factors affecting primary production in the Cromwell Current upwelling, *Journal of Experimental Marine Biology and Ecology*, *3*, 191-199, 1969.

Bonilla, D., Distribución de los quetognatos en las aguas de las Islas Galapagos, *Acta Oceanografica del Pacifico*, in press, 1979.

Elbrachter, M., Distribution and ecological significance of the plankton diatom *Thalassiosira partheneia*, paper presented at the Symposium on the Canary Current: Upwelling and Living Resources, International Council for the Exploration of the Sea, Las Palmas, Canary Islands, 1978.

Elbrachter, M., and R. Boje, On the ecological significance of *Thalassiosira partheneia* in the northwest African upwelling area, in: *Upwelling Ecosystems*, R. Boje and M. Tomczak (eds.), Springer-Verlag, New York, 24-31, 1978.

Enfield, D.B., Oceanografia de la region norte del frente equatorial: aspectos fisicos, in: *Reunion de Trabajo sobre el Fenomeno Concido como "El Niño"*, 229-334, Guayaquil, Ecuador, 1976.

Finenko, Z.Z., and L.A. Lanskaia, The growth rate of phytoplankton in the equatorial region of the Pacific Ocean, *Academy of Science of the USSR Trudy P.P. Shirshov Institute of Oceanology*, *102*, 123-130, 1975.

Grillini, C.R.L., Contributo alla valutazione della biomassa fitoplanctonica e della producttivita marina primaria nell'arcipelago delle Galapagos, Publicato a cura del Museo Zoologico dell'Universita de Firenze, 1-22, 1975.

Hasle, G.R., Species of the diatom genus *Thalassiosira* in upwelling areas, paper presented at the Third International Symposium on Upwelling Ecosystems, Kiel, Federal Republic of Germany, August 25-28, 1975, Unpublished manuscript.

Hasle, G.R., Plankton coccolithophorids from the subantarctic and equatorial Pacific, *Nytt Magasin for Botanikk*, *8*, 77-88, 1960a.

Hasle, G.R., Phytoplankton and ciliate species from the tropical Pacific, *Skrifter Utgitt av Det Norske Videnskaps-Akademi i Oslo, I. Matematisk Naturvidenskapelig Klasse*, *2*, 1-50, 1960b.

Hendey, N.I., Some marine diatoms from the Galapagos Islands, *Nova Hedwigia*, *22*, 371-422, 1972.

Houvenaghel, G.T., Oceanographic conditions in the Galapagos Archipelago and their relationships with life on the islands, in: *Upwelling Ecosystems*, R. Boje and M. Tomczak (eds.), Springer-Verlag, New York, 181-202, 1978.

Jimenez, R., Phytoplankton biomass and composition west of the Galapagos Islands, Ecuador, *ERFEN Bulletin, 1*, No. 2, 1977.

Jimenez, R., Mise en evidence de l'upwelling equatorial a l'est des Galapagos, *Cahiers ORSTOM, serie Oceanography, 16*, 137-155, 1978.

Knauss, J.A., Measurements of the Cromwell Current, *Deep-Sea Research, 6*, 265-286, 1960.

Margalef, R., Phytoplankton communities in upwelling areas: The example of NW Africa, *Oecologia acta, 3*, 97-132, 1978.

Marshall, H.G., Phytoplankton composition in the southeastern Pacific between Ecuador and the Galapagos Islands, *Proceedings of the Biological Society of Washington, 85*, 1-38, 1972.

Moreano, H.R., The upwelling at Elizabeth Bay and Bolivar Channel, Galapagos Islands, paper presented at the International Symposium on Coastal Upwelling, Los Angeles, California, Feb. 4-8, 1980, Abstract.

Okada, H., and S. Honjo, The distribution of oceanic coccolithophorids in the Pacific, *Deep-Sea Research, 20*, 355-374, 1973.

Pak, H., and J.R. Zaneveld, The Cromwell Current on the east side of the Galapagos Islands, *Journal of Geophysical Research, 78*, 7845-7859, 1973.

Pak, H., and J.R. Zaneveld, Equatorial front in the eastern Pacific Ocean, *Journal of Physical Oceanography, 4*, 570-578, 1974.

Semina, H.J., Distribution of indicator species of planktonic algae in the world ocean, *Oceanology, 17*, 573-579, 1977.

Sorokin, Y.I., E.B. Pavel'eva, and M.I. Vasil'eva, The productivity and trophic role of the bacterio-plankton in the region of equatorial divergence, *Academy of Science of the USSR, Trudy P.P. Shirshov Institute of Oceanology, 102*, 184-198, 1975a.

Sorokin, Y.I., I.N. Sukhanova, G.V. Konovalova, and E.B. Pavel'eva, Primary production and phytoplankton in the area of equatorial divergence in the eastern part of the Pacific Ocean, *Academy of Science of the USSR Trudy P.P. Shirshov Institute of Oceanology, 102*, 108-122, 1975b.

Stevenson, M., and B.A. Taft, New evidence of the equatorial undercurrent east of the Galapagos Islands, *Journal of Marine Research, 29*, 103-115, 1971.

Vinogradov, M.E., and T.N. Semenova, A trophic characterization of pelagic communities in the equatorial upwelling, *Academy of Science of the USSR Trudy P.P. Shirshov Institute of Oceanology, 102*, 232-237, 1975.

SHORT TIME VARIABILITY OF PHYTOPLANKTON POPULATIONS IN UPWELLING REGIONS - THE EXAMPLE OF NORTHWEST AFRICA

Dolors Blasco

Bigelow Laboratory for Ocean Sciences,
West Boothbay Harbor, Maine 04575, U.S.A.

Marta Estrada

Instituto de Investigaciones Pesqueras, Barcelona, Spain

Burton H. Jones *
Bigelow Laboratory for Ocean Sciences,
West Boothbay Harbor, Maine 04575, U.S.A.

* Present Address: Department of Biological Sciences
Allan Hancock Foundation
University of Southern California
Los Angeles, California,
90007, U.S.A.

Abstract. During the JOINT I program, phytoplankton samples were collected in the Cabo Blanco upwelling region from 8 to 25 March, 1974. Principal component analysis reveals a banded structure in the phytoplankton composition across the shelf. Each band is characterized by a different phytoplankton assemblage. Comparison of the phytoplankton and hydrographic data indicates a clear relationship between the width and the position of the bands and the intensity of the upwelling regime. The analysis also made it possible to associate the sequence of the phytoplankton assemblages with the different stages of decay of the upwelling circulation.

Introduction

Until the last decade the importance of the variability of upwelling phenomena has not been fully acknowledged. Since then, three major time scales of significance in upwelling processes have been recognized: the seasonal scale, months to years; the upwelling "event scale", days to weeks; and the diel scale, hours to a day (Walsh *et al.*, 1977). The event scale has been the one toward which most observational effort has been directed in recent years. The result has been a series of papers describing the relationship that exists between the local wind field and the distribution of the physical and chemical properties (Barton, Huyer, and Smith, 1977; Codispoti and Friederich, 1978; Huyer, Smith, and Pillsbury, 1974; Huyer, 1976) and between the fluctuation in the wind velocities and the current regime (Halpern, 1976; Brink, Halpern, and Smith, 1980). It has also become evident that marked changes observed in the distribution of phytoplankton biomass, primary productivity, and assimilation numbers are closely related to wind events (Huntsman and Barber, 1977; Small and Menzies, 1981; Jones and Halpern, 1981). On this basis physical processes, such as advection, dispersion, and mixing, have been invoked to explain the relationship. However, the exact mechanisms by which these physical processes cause the observed variability in the biological properties are not fully understood. For instance, how much of the changes observed in the assimilation number are due to physiological adaptations of a phytoplankton assemblage subject to different environmental conditions and how much is due to changes in the species composition is not yet known.

A previous study of the taxonomic composition of the northwest African upwelling region from March to May, 1974 (Blasco, Estrada, and Jones, 1980) showed that variations in phytoplankton composition during the period were related to a change of the upwelling water source and to a general change of the character of the upwelling. The study also suggested that the variability of the physical forcing functions at a shorter time scale had a prominent effect on the phytoplankton species distribution. The object of the present paper is to analyze further the effect of the variability of the hydrographic field during the development of an upwelling event on the phytoplankton composition and distribution. The basis for the analysis is the phytoplankton data collected during the JOINT-I expedition in March, 1974, from a section at 21°40'S off the northwest African coast. During this period, upwelling along the northwest African coast was characterized by a banded structure (i.e., surface isopleths of the variables were parallel to the coast) (Mittelstaedt, Pillsbury, and Smith, 1975). Temporal variations of the upwelling were characterized by events, i.e., periods of strong upwelling-favorable wind separated by periods of weak upwelling-unfavorable wind (Barton *et al.*, 1977). Periods for the upwelling events were on the order of 7 to 10 days. The general pattern of upwelling structure was that during periods of weak to moderate upwelling-favorable wind, the coldest, most nutrient-rich water occurred near the coast. With stronger wind, the center of the cold, nutrient-rich water migrated offshore toward the shelf break. Weak or upwelling-unfavorable wind resulted in the cessation of upwelling and warming of the near-surface water over the shelf. To illustrate the evolution of the hydrographic fields during an upwelling event, five temperature cross-shelf sections are shown (Fig. 1). More detailed hydrographic descriptions are given by Barton *et al.* (1977) and by Codispoti and Friederich (1978).

Materials and Methods

The phytoplankton data used in this paper were collected during the JOINT I program of the Coastal Upwelling Ecosystems Analysis Group in the upwelling region of northwest Africa during March to May, 1974. Samples of 100 ml of seawater were taken with 5-ℓ Niskin bottles at standard depths down to 200 m at all hydrographic stations (40 stations, 247 samples) during Leg 1 (9 to 24 March) of the cruise. Samples were preserved with Lugol's solution and phytoplankton cells were counted and identified by the Utermöhl inverted microscope technique (Margalef, 1973). Cells were identified to species when possible; if not, they were assigned to a genus or to a group. Physical and chemical variables were measured at all stations and depths (Friebertshauser *et al.*, 1975). The current meter data were taken from the report of Barton, Pillsbury, and Smith (1975).

Principal Component Analysis (PCA) (Legendre and Legendre, 1979) was used to describe the variability of the phytoplankton data and to define changes in phytoplankton composition. The approach, the method, and the system of analysis were described and discussed by Estrada and Blasco (1979) and Blasco *et al.* (1980).

More than 200 species were identified, but because a large number of zero abundance values in the data may distort the analysis, only 104 species have been considered. This selection includes all the species present in more than 20 samples. The phytoplankton data were logarithmically transformed ($x \rightarrow \ln(x + 1)$, x = cells/50 ml). Scores of the principal components (PC) were calculated for each sample from the matrix of factor loadings and the values of the standardized original variables (Cooley and Lohnes, 1971).

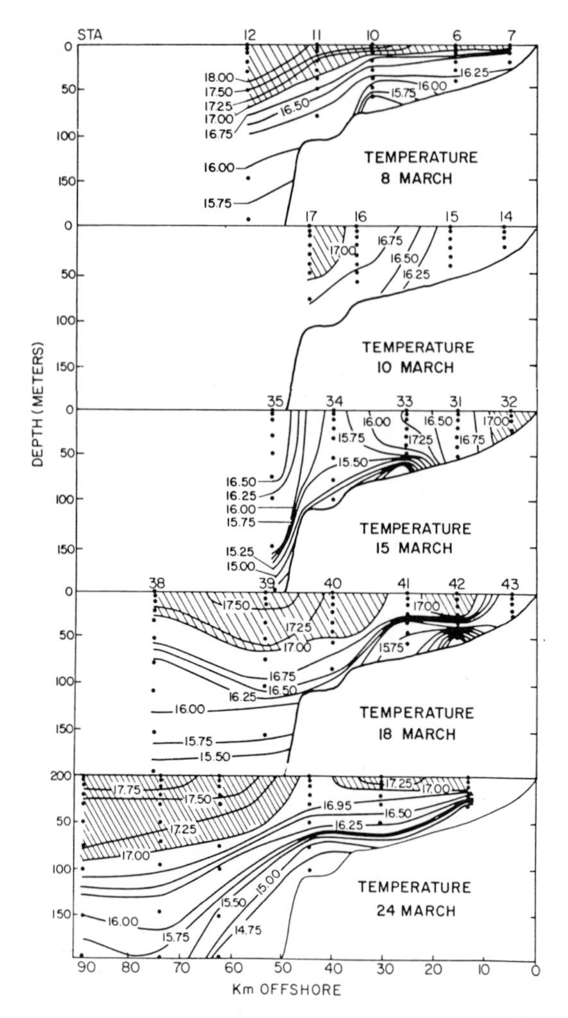

Fig. 1. Temperature sections along the 21°40'N line during the development of the upwelling event from March 8 to 24.

Results

Major trends of the phytoplankton variability

The first three components extracted account for 15.8, 12.7, and 5.3%, respectively, of the total phytoplankton variability. Comparison of the distribution of the first principal component (PC I) scores (Fig. 2) and the temperature distribution (Fig. 1) shows the PC I is related to the upwelling phenomena. High negative scores of PC I are observed in the deep, cold water and they increase gradually as the upwelled water

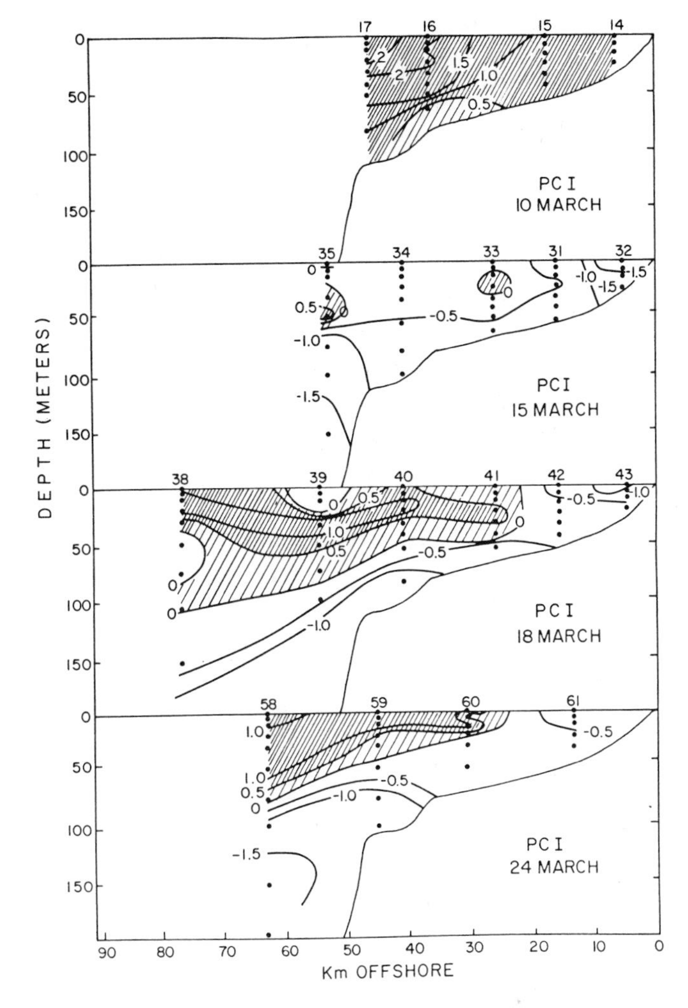

Fig. 2. Cross-shelf distribution of the first principal component at the 21°40'N line during the development of an upwelling event.

approaches the surface and moves offshore. During the period of strong upwelling (15 March, Fig. 2) low PC I scores occur over the entire study area. Most species have positive correlation with PC I, which indicates that this component also reflects total cell and species abundance, in agreement with previous observations (Blasco, 1971; Blasco *et al.*, 1980).

The second principal component (PC II) (Fig. 3) also seems to be related to the upwelling intensity. Positive scores characterized most of the shelf during the periods of stronger upwelling (10 and 15 March, Figs 1 and 3). However, a major difference from the distribution of PC I is that PC II shows no continuity between the deep, cold core of upwelling water and the colder water over the shelf. Rather, the highest scores of PC II are inshore and close to the bottom. This and the negative scores at Sta. 34 (Fig. 3, 15 March) where strong upwelling occurred but vertical mixing did not reach the bottom, suggest that this component represents the influence of the shelf. Strengthening this suggestion is the fact that benthic diatoms have the highest positive correlation with this component (see Fig. 6).

The significant difference in the distribution of the scores of the third principal component (PC III) during the two relaxation periods, 18 and 24 March (Figs 4 and 5), indicates that in contrast to PC I and PC II, PC III is not directly related to the upwelling intensity. Furthermore, when the score distribution is compared with the cross-shelf structure of the longshore flow, it is observed that in general high positive PC III scores are associated with the southward current that flows along the shelf while the negative values are associated with northward flowing waters poleward undercurrent at depth, offshore countercurrent at surface (Mittelstaedt *et al.*, 1975).

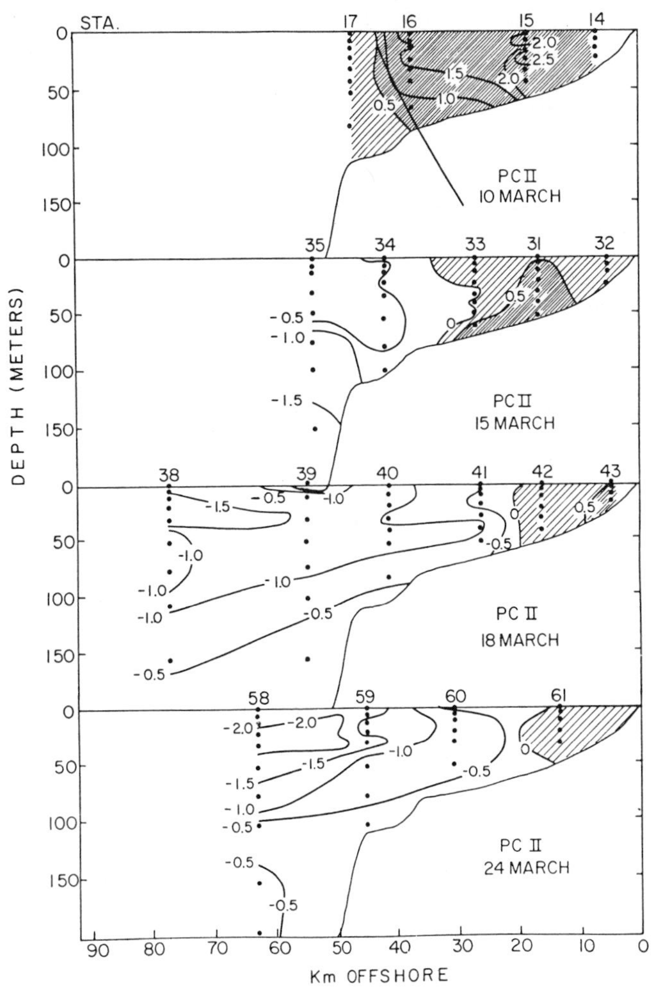

Fig. 3. Cross-shelf distribution of the second principal component at the 21°40'N line during the development of an upwelling event.

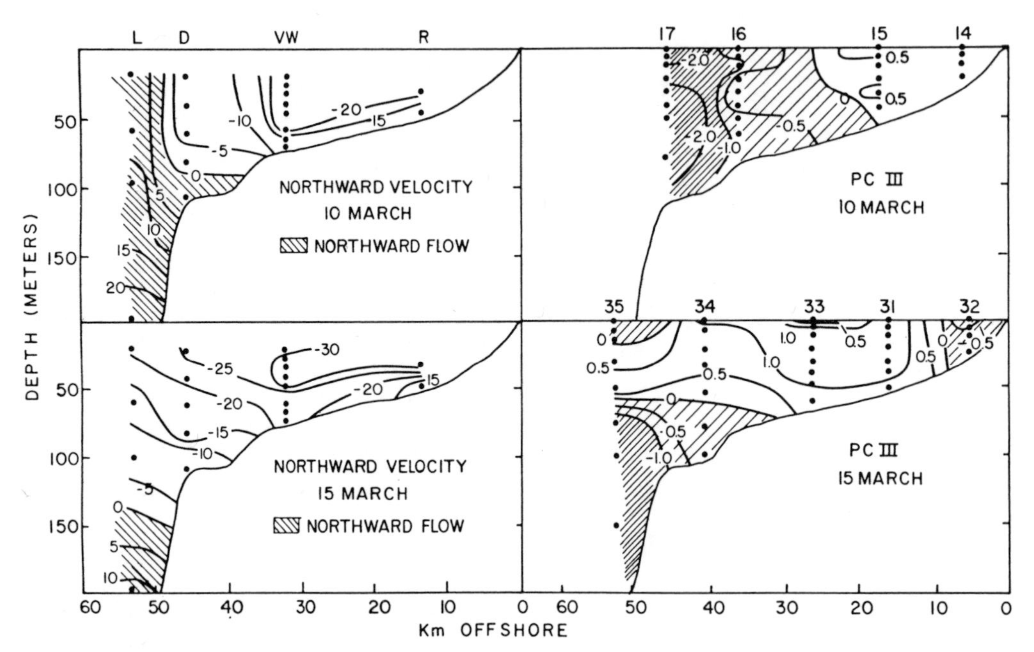

Fig. 4. Cross-shelf distributions of the third principal component and alongshore velocity (cm/s) at the 21°40'N line on 10 and 15 March.

Phytoplankton distribution during an upwelling event

Based on the projection of the phytoplankton species on the space determined by the first (upwelling of deep water), second (influence of the shelf), and third (alongshore flow) principal components (Fig. 6), we have distinguished six different phytoplankton assemblages (Table 1). In defining the groups, only the more frequent species have been considered. The distribution of each of the groups and their numerical contri-

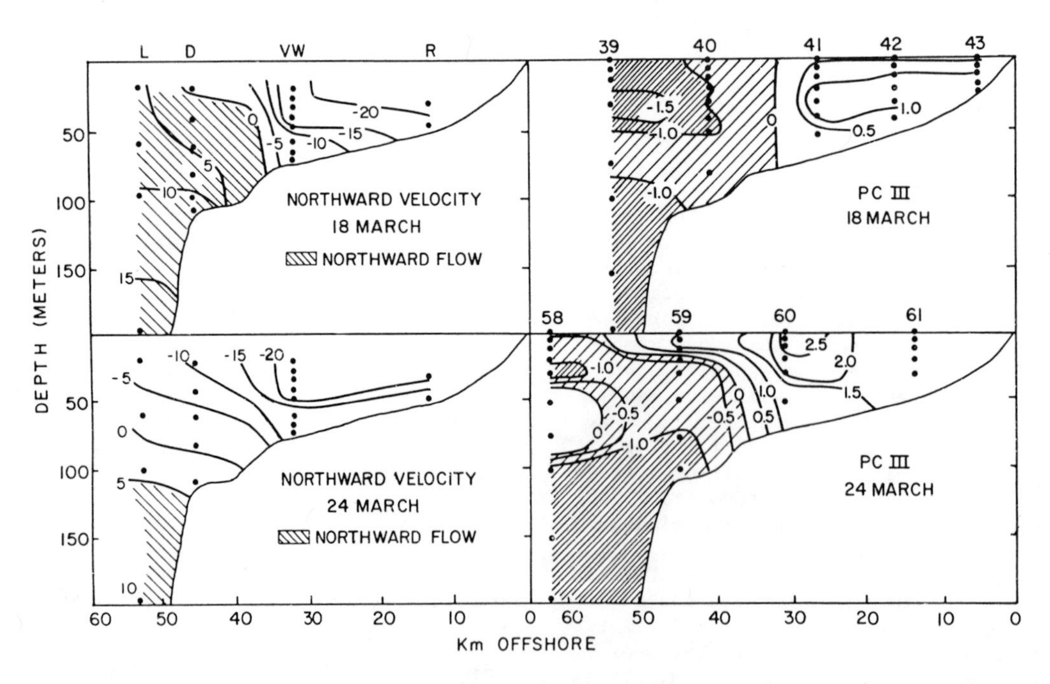

Fig. 5. Cross-shelf distributions of the third principal component and alongshore velocity (cm/s) at the 21°40'N line on 24 March.

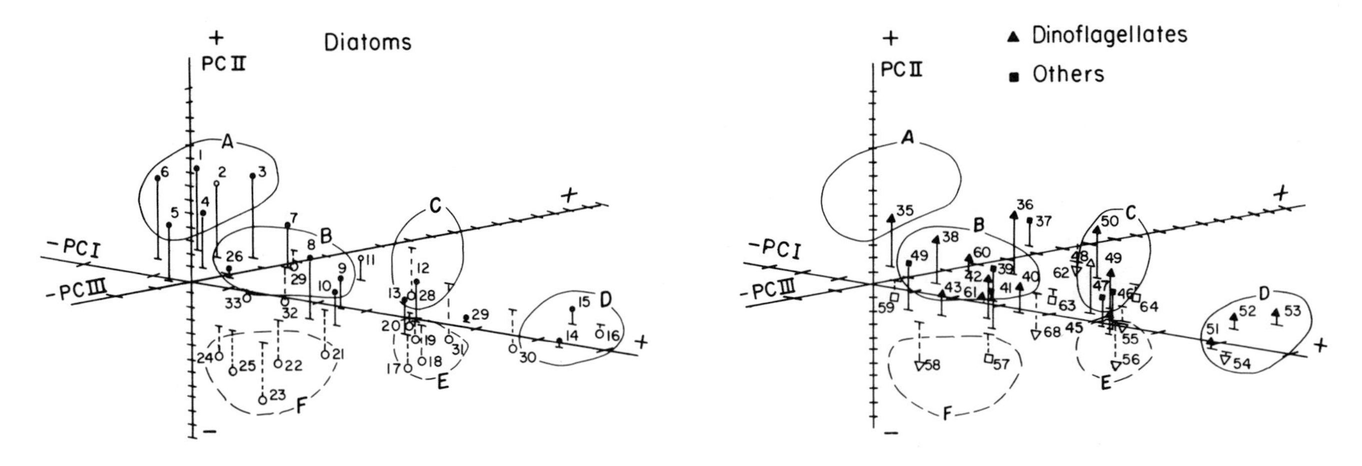

Fig. 6. Relative position of the most frequent species in space dimensioned by the first three principal components. Left - diatoms. Right - dinoflagellates and other flagellates. The identification numbers are given in Table 1.

bution to the total phytoplankton during the development of an upwelling event is presented in Fig. 7. Group A is characteristic of the innermost part of the shelf and its domain extends over the shelf only during periods of strong upwelling. In contrast, groups E and F are characteristic of the offshore region, and their presence and abundance over the shelf are associated with the relaxation periods. Group D seems to have been the most important phytoplankton assemblage during this study, especially during the periods of strong upwelling. Groups B and C are not so easily explained. The positive correlation that the species of these groups have with the second component suggests they are neritic groups. However, this fact does not appear clear in their cross-shelf distribution. Another enigma is why groups B and C had such a high abundance on 10 March. One may speculate that these groups become predominant in the inshore region during the long periods of weak upwelling (Fig. 1, 8 March) and that the distribution observed on 10 March is a remnant of the phytoplankton community that existed in the area before. However, the lack of phytoplankton data previous to 10 March does not permit assessing the validity of this suggestion.

The observed change of the percentage contribution of group D to the total phytoplankton seems to be caused by the relative increase or decrease of the other groups rather than by a change in the species abundance of group D. A point that should be made is that the total sum of the percentage contribution of all the presented groups is around 50 to 60% of the total cell number, and the rest is accounted for by small flagellates and dinoflagellates less than 8 μm in diameter. However, the relatively small size of these organisms versus the larger size of the species included in the groups (Table 1) suggests that, in terms of biomass, these groups played an important role in the phytoplankton composition during this period.

Discussion

This study has shown that the phytoplankton presents a quasi-permanent cross-shelf pattern and that from inshore to offshore a sequence of different phytoplankton assemblages can be identified. According to the relative position of these assemblages with respect to the first three principal components and to their cross-shelf distribution (Figs 6 and 7), the order would be A, B, C, D, E, and F. Although the parallelism is not complete, the assemblages and the ordination can be related to the stages of the succession proposed by Margalef (1978a), and most of the discrepancies can be attributed to the differences in the time and space of the sampling. Margalef's study was based on data collected during a synoptic survey over a large area and no stations were occupied over the shelf. In contrast, most of our data are from repeated sections over the shelf. It has also been shown that the assemblages changed in position relative to the coastline depending upon the strength of upwelling, suggesting strong advective and diffusive influence on their distribution. The importance of physical transport in controlling phytoplankton distribution in upwelling regions has been expressed before (Small and Menzies, 1981; Huntsman and Barber, 1977; Huntsman, Brink, Barber, and Blasco, 1980). The new and striking point here is the existence of different phytoplankton assemblages that seem to be associated with the different stages of the upwelling circulation. Margalef (1978a) suggested that the phytoplankton of the upwelling region of northwest Africa could be described as a sequence of populations developing around the upwelling core, which expand and contract according to the sea-

TABLE 1. Phytoplankton assemblages defined by the projection of the species in the space determined by the first, second, and third principal component (Fig. 6).

Group A		Group B		Group C	
1.	*Paralia sulcata*	8.	*Thalassionema nitzschioides*	12.	*Eucampia zodiacus*
2.	*Pleurosigma* sp.	9.	*Rhizosolenia stolterfothii*	13.	*Lauderia annulata*
3.	*Stephanopyxis turris*	10.	*Chaetoceros affinis*	45.	*Diplopsalis asymmetrica*
4.	*Thalassiosira eccentrica*	38.	*Dinophysis argus*	46.	*Cryptomonas baltica*
5.	*Actinoptychus senarius*	39.	*Calyptrosphaera* sp.	47.	*Coccolithus huxleyi* (= *Emiliana huxleyi*)
6.	*Diploneis* sp.	40.	*Prorocentrum micans*	48.	*Pyrophacus steinii*
35.	*Pyrocystis lunula*	41.	*Helicosphaera carteri*	49.	*Peridinium depressum* (= *Protoperidinium*)
		42.	*Peridinium globulus* (= *Protoperidinium*)		
7.	*Trigonium alternans*	43.	*Exuviella baltica*		
36.	*Dinophysis ovum*				
37.	*Coccolithus pelagicus*	44.	*Coccolithus leptoporus*	50.	*Prorocentrum scutellum*
		11.	*Rhizosolenia robusta*		
		65.	*Ceratium falcatiforms*		

Group D		Group E		Group F	
14.	*Thalassiosira* sp.	17.	*Nitzschia closterium* (= *Cylindrotheca closterium*)	21.	*Nitzschia pungens*
15.	*Rhizosolenia imbricata*			22.	*Chaetoceros socialis*
16.	*Thalassiosira rotula*	18.	*Chaetoceros didymus*	23.	*Planktoniella sol*
51.	*Ceratium kofoidii*	19.	*Dactyliosolen mediterranea*	24.	*Thalassiothrix frauenfeldii*
52.	*Torodinium robustum*	20.	*Hemiaulus sinensis*	25.	*Thalassiothrix mediterranea*
53.	*Ceratium furca*	55.	*Polykrikos kofoidii*	57.	*Mesodinium rubrum*
54.	*Ceratium fusus*	56.	*Spirodinium spirale*	58.	*Prorocentrum obtusidens*

Species not belonging to any group			
26.	*Actinocyclus subtilis*	33.	*Coscinodiscus oculus-iridis*
27.	*Bacteriastrum* sp.	59.	*Gephyrocapsa* sp.
28.	*Rhizosolenia delicatula*	60.	*Pseudophalacroma nasutum*
29.	*Guinardia flaccida*	62.	*Peridinium brochii* (= *Protoperidinium*)
30.	*Rhizosolenia fragilissima*	63.	*Syracosphaera pulchra*
31.	*Rhizosolenia bergonii*	64.	*Dictyocha fibula*
32.	*Thalassiosira partheneia*	65.	*Gonyaulax fragilis*

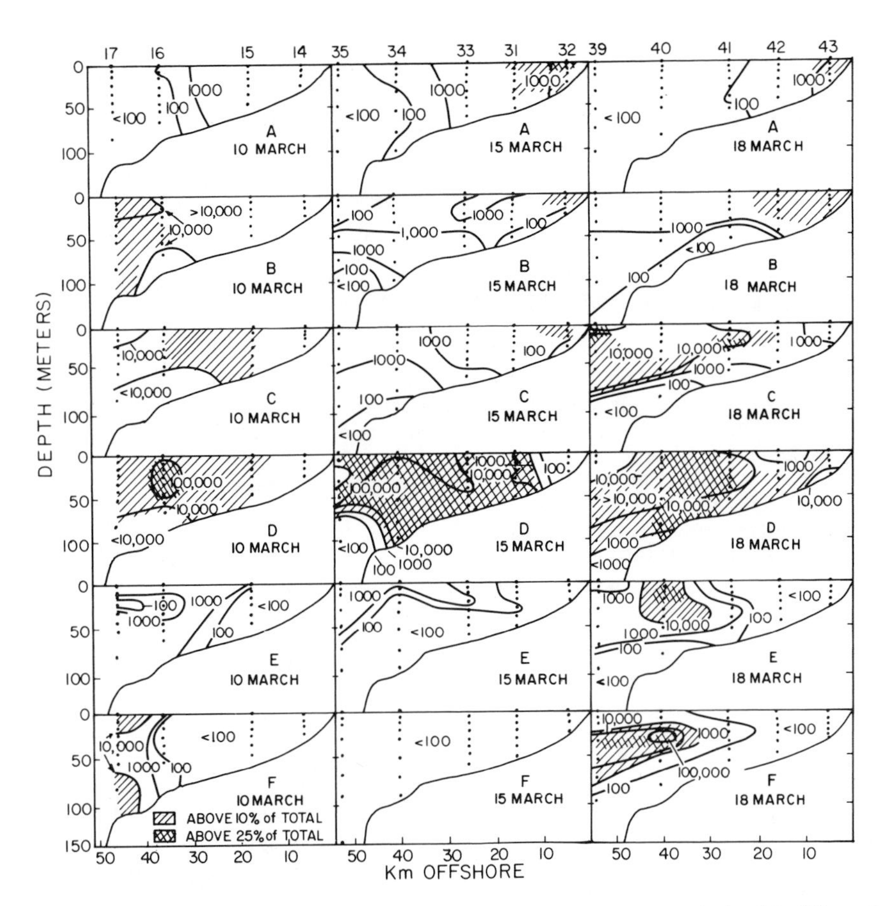

Fig. 7. Distribution of the phytoplankton assemblages defined in Table 1 (cells/l) and their percentage contribution to the total phytoplankton population during an upwelling event.

son. Our data tend to confirm his view and indicate that this interpretation can be extended to shorter time scales. The introduction of this concept provides new light for the interpretation of biomass - specific productivity distribution in this (Huntsman and Barber, 1977; Jones and Halpern, 1981) and other upwelling regions. On the other hand, it raises two interesting questions: one is the persistence of these assemblages as separate entities in such a dynamic system as the northwest Africa upwelling region during this period; the second one is what are the mechanisms that select the groups?

A reduced exchange across the current-countercurrent boundary might be invoked to answer the first question, but the argument is valid only for maintaining the sharp changes of phytoplankton composition in the offshore region. It cannot explain the existence of different groups in the north-to-south flow. A possibility is to assume a cross-shelf gradient of the factor or factors that control the phytoplankton composition, and an alongshore continuity of this gradient. The banded structure of the hydrographic and physical variables during the period (Mittelstaedt *et al.*, 1975; Barton *et al.*, 1977) suggests that the alongshore continuity is a reasonable assumption. Therefore, the only unknowns are the variables that drive the species succession.

In the classical concept of an upwelling ecosystem, the distribution of inorganic nutrients has been proposed to play a major role in the selection of the species. During the period of our study, surface nitrate concentrations oscillated around 5 μg-at/l to 2.5 μg-at/l. The other nutrients also showed high concentrations and their distribution was similar to that of nitrate. Thus, it is reasonable to conclude that nutrients did not play a major role.

Another possibility for the existence of these groups is Margalef's hypothesis that physical processes alone basically control phytoplankton composition (Margalef, 1978b). He argued that decay of external energy has been one of the main agents in phytoplankton selection leading to the evolution of many life forms in phytoplankton. The conclusion is that, in upwelling regions, a sequence of phytoplankton populations would result from the gradient that exists in the physical environment as a consequence of the dissipation of turbulence and decay of the upwelling circulation. The lack of knowledge of the cellular characteristics of the species (sinking rate, growth rates, light requirements, etc.) and the lack of phytoplankton data in the alongshore direction do not permit a full exploration of the above possibility, but the data provide a basis for some speculation.

During the study period the near-shore region was characterized by strong vertical mixing, which resulted in high levels of re-suspended sediments in the water column and, in general, reduced light penetration (Milliman, 1977; Huntsman and Barber, 1977). Under such conditions and given the upwelling circulation with surface flow offshore and deep onshore flow, one can envision that species such as those included in group A (Table 1), with relatively low maximum growth rates and high sinking rates, would be selected. Because of the light limitation, high maximum growth rates would not give a competitive advantage. On the other hand, high sinking rates would prevent the organisms from being washed out of the system.

At mid-shelf, due to a relative increase in the vertical stability and an increase of the shelf depth, re-suspension of sediments did not occur. In general, the environment became less light limited, but the onshore-offshore circulation did not change markedly (Barton *et al.*, 1977). One might expect, in this region, that phytoplankton species would have higher growth rates, although relatively high sinking rates would still be an advantage, at least during certain periods. For example, during strong upwelling, high sinking rates would permit the cells to reach the deep onshore flow and be re-seeded into the system. However, during relaxation periods, the loss of cells to depth would be high and prevent these species from becoming dominant. The distribution of group D and the changes observed in their contribution to the total plankton community during the development of an upwelling event (Fig. 7) fits well with the above description.

The existing data are inappropriate to evaluate fully Margalef's hypothesis. Thus, our inability to explain some of the group distributions or the species composition of the groups is not surprising. In addition, zooplankton grazing may play a role in the selection and maintenance of the groups. The only way that grazing might explain the existence of the different phytoplankton assemblages would be if zooplankton species present a similar cross-shelf structure and zooplankton distribution patterns contract and expand according to the upwelling strength. There is some evidence that this may occur (Colebrook, 1977; Smith *et al.*, 1980), but the existing observations are neither from this area nor the same time and space scales.

Although this study has raised a number of questions which, with the existing data cannot be answered, we think that the recognition of the questions is a step toward understanding phytoplankton in the upwelling regions. This study should help define the relative importance of the different environmental variables to the phytoplankton distribution, and to optimize the space and time scales of their measurements in future phytoplankton research.

Acknowledgements. We thank N. Breitner, J. Garside, J. Botta, J. Rollins, Peg Colby, and Teve MacFarland for their technical assistance and Vicki Jones for editing the manuscript. We also would like to express our appreciation to T.T. Packard and R. Margalef for the encouragement and useful comments throughout this work. Financial support was provided by the Office of International Decade of Ocean Exploration under National Science Foundation grant no. 75-23718 A01 (CUEA-12). Contribution no. 80016, Bigelow Laboratory for Ocean Sciences.

References

Barton, E.D., R.D. Pillsbury, and R.L. Smith, A compendium of physical observations from JOINT-I, vertical sections of temperature, salinity, and sigma-t from R/V *Gillis* data and low pass filtered measurements of wind and currents, Oregon State University, School of Oceanography, Ref. 75-17, 60 pp., 1975.

Barton, E.D., A. Huyer, and R.L. Smith, Temporal variation observed in the hydrographic regime near Cabo Corbeiro in the northwest African upwelling region, February to April, 1974, *Deep-Sea Research, 24*, 7-24, 1977.

Blasco, D., Composicion y distribucion del fitoplancton en la region de las costas peruanas, *Investigaciones Pesqueras, 35*, 61-112, 1971.

Blasco, D., M. Estrada, and B. Jones, Relationship between phytoplankton distribution and composition and the hydrography in the northwest African upwelling region near Cabo Corbeiro, *Deep-Sea Research, 27(10A)*, 799-823, 1980.

Brink, K.H., D. Halpern, and R.L. Smith, Circulation in the Peru upwelling system near 15°S, *Journal of Geophysical Research, 85*, 4036-4048, 1980.

Codispoti, L.A., and G.E. Friederich, Local and mesoscale influences on nutrient variability in the northwest African upwelling region near Cabo Corbeiro, *Deep-Sea Research, 25*, 751-770, 1978.

Colebrook, J.M., Annual fluctuations in biomass of taxonomic groups of zooplankton in the California Current, 1955-59, *Fishery Bulletin, U.S., 75*, 357-368, 1977.

Cooley, W.W., and P.R. Lohnes, *Multivariate Data Analysis*, Wiley, 364 pp., 1971.

Estrada, M., and D. Blasco, Two phases of the phytoplankton community in the Baja California upwelling, *Limnology and Oceanography, 24*, 1065-1080, 1979.

Friebertshauser, M.A., L.A. Codisptoi, D.D. Bishop, G.E. Friederich, and A.A. Westhagen,

JOINT-I hydrographic station data R.V. *Atlantis-II* Cruise 82, Coastal Upwelling Ecosystems Analysis Data Report 18, 243 pp., 1975.

Halpern, D., Structure of a coastal upwelling event observed off Oregon during July 1973, *Deep-Sea Research, 23,* 495-508, 1976.

Huntsman, S.A., and R.T. Barber, Primary production off northwest Africa: The relationship to wind and nutrient conditions, *Deep-Sea Research, 24,* 25-34, 1977.

Huntsman, S.A., K.H. Brink, R.T. Barber, and D. Blasco, The role of circulation and stability in controlling the relative abundance of dinoflagellates and diatoms over the Peru shelf, *This Volume,* 1980.

Huyer, A., A comparison of upwelling events in two locations: Oregon and northwest Africa, *Journal of Marine Research, 34,* 531-546, 1976.

Huyer, A., R.L. Smith, and R.D. Pillsbury, Observations in a coastal upwelling region during a period of variable winds (Oregon coast, July 1972), *Tethys, 6,* 691-404, 1974.

Jones, B.H., and D. Halpern, Biological and physical aspects of a coastal upwelling event observed during March-April 1974 off northwest Africa, *Deep-Sea Research, 28A*(2), 1981.

Legendre, L., and P. Legendre, *Ecologie Numerique,* Masson, Paris, 254 pp., 1979.

Margalef, R., Fitoplancton marino de la region de afloramiento del NW de Africa, *Investigaciones Pesqueras,* Resultados Expediciones Cientificas B.O. *Cornide, 2,* 65-95, 1973.

Margalef, R., Phytoplankton communities in upwelling areas, The example of N.W. Africa, *Oecologia aquatica, 3,* 97-132, 1978a.

Margalef, R., Life forms of phytoplankton as survival alternatives in an unstable environment, *Oceanologica Acta, 1,* 493-509, 1978b.

Milliman, J.D., Effects of arid climate and upwelling upon the sedimentary regime off southern Spanish Sahara, *Deep-Sea Research, 24,* 95-103, 1977.

Mittelstaedt, E., R.D. Pillsbury, and R.L. Smith, Flow patterns in the northwest African upwelling area, Results of measurements along 21° 40'N during February-April 1974, JOINT-I, *Deutsche hydrographische Zeitschrift, 28,* 145-167, 1975.

Small, L.F., and D.W. Menzies, Patterns of primary productivity and biomass in a coastal upwelling region, *Deep-Sea Research, 28A*(2), 1981.

Smith, S.L., K.H. Brink, H. Santander, T.J. Cowles, and A. Huyer, The effect of advection on variations in zooplankton at a single location near Cabo Nazca, Peru, *This Volume,* 1980.

Walsh, J.J., T.E. Whitledge, J.C. Kelley, S.A. Huntsman, and R.D. Pillsbury, Further transition states of the Baja California upwelling ecosystem, *Limnology and Oceanography, 22,* 264-380, 1977.

SEASONAL PHYTOPLANKTON DISTRIBUTION ALONG THE PERUVIAN COAST

B. Rojas de Mendiola

Instituto del Mar del Peru, Lima, Peru

Abstract. During a 10-year study more than 2,000 phytoplankton samples were collected from the entire coast of Peru and analyzed. In general, diatoms were the most abundant group of organisms in all seasons. Predominant species were *Rhizosolenia delicatula, Skeletonema costatum, Thalassiosira subtilis, Thalassionema nitzschioides,* and several species of the genus *Chaetoceros*. Dinoflagellates and flagellates were observed frequently during summer. The mean distribution of the phytoplankton concentration during the 10 years shows the existence of several centers with higher cell densities along the coast, coinciding with the areas of more intense and persistent upwelling. Four major centers have been identified: Pimentel (∿6°S), Chimbote (∿9°S), Callao (∿12°S), and Tambo de Mora-Pisco (∿15°S); and two minor centers, Talara (∿4°S) and Ilo (∿17°S). The relative importance of each center seems to change according to the season. The highest phytoplankton concentration tended to be in the northern part of the coast during fall and winter and in the south through spring and summer.

Introduction

It is well known that the upwelling region off the Peru coast supports one of the largest fisheries in the world, and that several of the fish species that compose these fisheries (*Engraulis ringens* J., *Sardinops sagax sagax* J., *Ethmidium maculata chilcae* H.) are phytophagous organisms. Thus, it is not surprising that the phytoplankton distribution and composition in the region have been subject of numerous scientific investigations.

It has been argued in recent years (Walsh, 1978) that because fish are usually able to migrate over large areas, a knowledge of the large scale phytoplankton distribution and composition is essential in order to understand fish distribution and success. However, most of the studies (Gunther, 1936; Landa, 1953; Barreda, 1957; Rojas de Mendiola, 1958, 1963a, b, 1966, 1979; Strickland, Eppley, and Rojas de Mendiola, 1969; Blasco, 1971; Semina, 1971; Sukhanova, Konovalova, and Rat'kova, 1978; Guillen, Rojas de Mendiola, and Izaguirre de Rondan, 1972, 1973) have been restricted to a small area or to a short time period. The present paper describes the phytoplankton distribution and composition along the entire Peru coast from samples collected during a 10-year study. The main object is to provide general large-scale information that has been lacking, and in addition, to serve as a reference basis for previous and future small scale (time and space) phytoplankton studies in the region.

Materials and Methods

For the present study more than 2,000 water samples (Table 1) for phytoplankton analysis were collected in Nansen bottles at 10-m depth during the following cruises: BAP *Bondy* 1961, 1962, and 1963; BAP *Unanue*, 1964, 1965, 1966, 1967, and 1968, and *SNP*-1, 1969 and 1970. Until 1966 the samples were preserved with Lugol's solution, and thereafter with neutralized 4% formalin. Phytoplankton cells were counted and identified by the Utermöhl inverted microscope technique (Utermöhl, 1958) and often individual organisms were transferred to a compound microscope for further identification.

Results

Phytoplankton species composition and distribution

The list of all the species identified during the study is given in Table 2. A total of 216 species have been observed in the region: 134 diatoms, 65 dinoflagellates, 13 coccolithophorids, 2 silicoflagellates, 1 heterochloridale, and 1 euglenoid. Diatoms were the dominant phytoplankton organisms among the entire Peruvian coast, especially within the 60-mile zone. The most abundant and frequent species were *Schroederella delicatula, Thalassiosira subtilis, Skeletonema costatum, Thalassionema nitzschioides, Nitzschia delicatissima*, and several species of *Chaetoceros*. These species alone often accounted for more than 75% of the total phytoplankton (Fig. 1), and at times they reached cell densities as high as 4 x

TABLE 1. Total number of stations occupied at each degree of latitude between 1961 and 1970. In parenthesis total number of years sampled.

Degrees Latitude	3-4	4-5	5-6	6-7	7-8	8-9	9-10	10-11	11-12	12-13	13-14	14-15	15-16	16-17	17-18	18-19
Winter		46 (7)	28 (6)	58 (6)	49 (7)	88 (6)	66 (9)	44 (6)	51 (8)	41 (8)	21 (6)	24 (4)	25 (5)	33 (5)	37 (4)	18 (4)
Spring	2 (1)	31 (6)	26 (6)	37 (5)	61 (7)	50 (6)	57 (6)	40 (6)	48 (7)	62 (6)	25 (6)	34 (5)	46 (6)	41 (5)	35 (5)	18 (4)
Summer	20 (3)	21 (4)	28 (5)	41 (5)	26 (6)	26 (5)	47 (6)	45 (5)	59 (6)	65 (7)	8 (1)	4 (3)	6 (3)	12 (3)	1 (1)	3 (2)
Fall	1 (1)	10 (3)	4 (2)	13 (4)	9 (4)	10 (3)	20 (5)	26 (6)	99 (8)	131 (8)	47 (7)	15 (6)	21 (4)	30 (5)	17 (4)	10 (4)

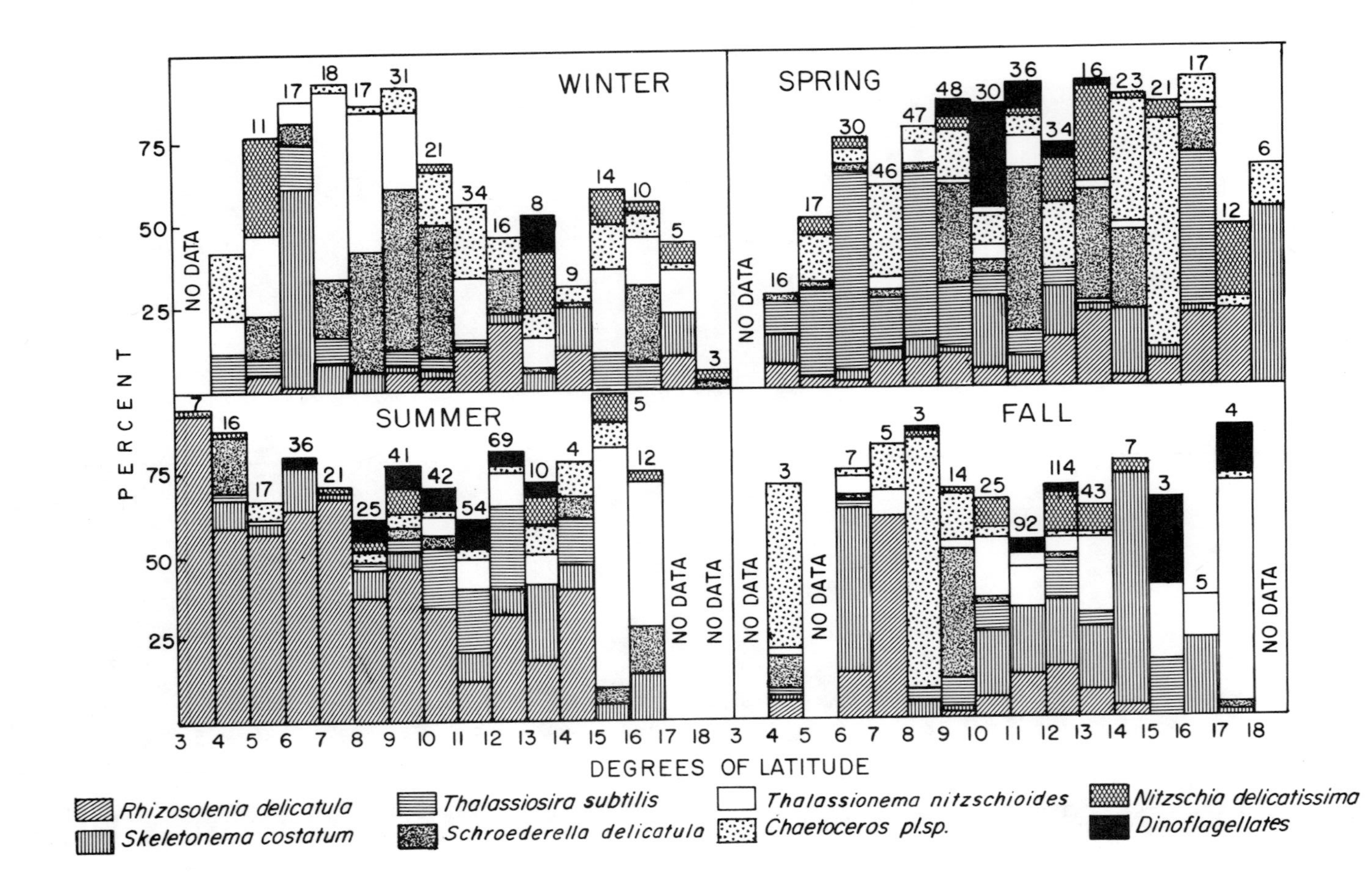

Fig. 1. Mean seasonal and latitudinal variation of the most dominant phytoplankton species. Numbers above the columns indicate number of samples used in calculating the means.

TABLE 2. Phytoplankton species identified during 1961-1970

BACILLARIOPHYTA (DIATOMEAE)

Achnanthes longipes Agardh
Actinocyclus curvatulus Janisch
  octonarius var *octonarius* (Ehrenberg) Hendey
  octonarius var *crassus* (Wm. Smith) Hendey
  octonarius var *tenellus* (de Brébisson) Hendey
Actinoptychus senarius (Ehrenberg) Ehrenberg
  splendens (Shadbolt) Ralfs in Pritchard
Amphora hyalina Kützing
Asterionella japonica Cleve
Asteromphalus flabellatus (Brébisson) Greville
  heptactis (de Brébisson) Ralfs in Pritchard
Bacillaria paxillifer (O.F. Müller) Hendey
Bacteriastrum delicatulum Cleve
  hyalinum Lauder
Biddulphia alternans (Bailey) Van Heurck
  aurita (Lyngbye) de Brébisson
  longicruris Greville
  mobiliensis (Bailey) Grunow ex Van Heurck
  rhombus (Ehrenberg) Wm. Smith
Cerataulina pelagica (Cleve) Hendey
Chaetoceros affinis Lauder
  anastomosans Grunow
  armatum West
  atlanticus Cleve
  brevis Schütt
  chilensis Krasske
  coarctatum Lauder
  compressus Lauder
  concavicorne Mangin
  constrictus Gran.
  convolutum Castracane
  curvisetum Cleve
Chaetoceros danicus Cleve
  debilis Cleve
  decipiens Cleve
  didymus Ehrenberg
  eibenii (Grunow) ex Van Heurck
  gracilis Schütt
  holsaticus Schütt
  lauderi Ralfs
  lorenzianus Grunow
  messanensis Castracane
  parallelis Boden
  pelagicus Cleve
  perpusillus Cleve
  peruvianus Brightwell
  pseudocurvisetum Mangin
  radicans Schütt
  simile Cleve
  sociale Lauder
  subsecundus (Grunow) Hustedt
  subtile Cleve
  teres Cleve
  tetras Karsten
  vanheurcki Gran
  vistulae Apstein
Climacodium frauenfeldianum Grunow
Climacosphenia moniligera Ehrenberg
Corethron criophilum Castracane
Coscinodiscus centralis Ehrenberg
  concinnus Wm. Smith
  crenulatus Grunow
  curvatulus Grunow in Schmidt
  decrescens Grunow in Schmidt
  eccentricus Ehrenberg
  granii Gough
  lineatus Ehrenberg
  marginatus Ehrenberg
  oculus iridis Ehrenberg
  perforatus Ehrenberg
  radiatus Ehrenberg
  wailesii Gran and Angst
Coscinosira polychorda Gran = *Thalassiosiras angusti* lineata (A. Schmidt) Fryxell and Hasle
Cyclotella striata (Kützing) Grunow
Ditylum brightwellii (West) Grunow
Eucampia cornuta (Cleve) Grunow
  zoodiacus Ehrenberg
Grammatophora angulosa Ehrenberg
  marina (Lyngbye) Kutzing
  oceanica Ehrenberg
Guinardia flaccida (Castracane) Peragallo
Hemiaulus hauckii Grunow ex Van Heurck
  sinensis Graville
Hemidiscus cuneiformis Wallich
Lauderia borealis Gran
Leptocylindrus danicus Cleve
  minimus Gran
Licmophora lyngbyei (Kützing) Grunow ex Van Heurck (*L. abbreviata* Agardh)
Lithodesmiun undulatum Ehrenberg
Nitzschia americana Hasle
  bicapitata Cleve
  closterium (Ehrenberg) Wm. Smith = *Cylindrotheca closterium* (Ehrb.) Reimann and Lewin
  delicatissima Cleve
  longissima (Brébisson) Ralfs
  pacifica Cupp
  prolongatoides Mangin
  pungens Grunow
  seriata Cleve
Planktoniella sol (Wallich) Schütt
Paralia sulcata (Ehrenberg) Cleve (*Melosira sulcata*) (Ehrenberg) Kützing
Pleurosigma normanii Ralfs in Pritchard
Pseudoeunotia doliolus (Wallich) Grunow
Rhizosolenia alata Brightwell
  bergonii H. Peragallo
  calcar avis Schultze
  castracanei Peragallo
  delicatula Cleve
  fragilissima Bergon
  hebetata f. *semispina* (Hensen) Gran
  imbricata Brightwell
  robusta Norman
  setigera Brightwell
  stolterfothii Peragallo
  styliformis Brightwell
Roperia tessellata (Roper) Grunow ex Van Heurck
Schroederella delicatula (Peragallo) Pavillard
Stephanopyxis costata (Greville) Hustedt (*Skeletonema costatum* (Greville) Cleve)

turris (Greville) Ralfs in Pritchard
Streptotheca tamesis Shrubsole
Striatella unipunctata (Lyngbye) Agardh
Synedra undulata (Bailey) Gregory
Thalassionema bacillaris (Heiden) Kolbe
nitzschioides (Grunow) Hustedt
Thalassiosira aestivalis Gran and Angst
decipiens (Grunow) Jörgensen
gravida Cleve
mendiolana Hasle
nordenskioldii Cleve
rotula Meunier
subtilis (Ostenfeld) Gran
Thalassiothrix delicatula Cupp
frauenfeldii Grunow
longissima Cleve and Grunow
mediterranea Pavillard

DINOFLAGELLATAE

Amphidinium acutissimum Schiller
Ceratium arietinum Cleve
breve (Ostf et Schmidt) Schröder
buceros (Zacharias) Schriller
dens Ostf et Schmidt
furca (Ehrenberg) Claparede et Lachmann
fusus (Ehrenberg) Dujardin
horridum Gran
lineatum (Ehrenberg) Cleve
longipes (Bailey) Gran
massiliense (Gourret) Jörgensen
pulchellum Schröder
reflexum Cleve
tripos (Muller) Nitzsch
Cladopyxis brachiolata Stein
Dinophysis acuminata Claparede et Lachmann
acuta Ehrenberg
caudata Saville - Kent
ovum Schütt
tripos Gourret
Diplopsalis lenticula Bergh
Exuviaella aequatorialis Hasle
compressa (Bailey) Ostenfeld
marina Cienkowski
Glenodinium gymnodinium Penard (Schiller, 1937)
Goniaulax polygramma Stein
spinifera (Claparede et Lachmann) Diesing
Gymnodinium flavum Kofoid and Swezy
lohmanii Paulsen
splendens Lebour
Gyrodinium fusiforme Kofoid and Swezy
Oxytoxum caudatum Schiller
longiceps Schiller
Peridinium (= Protoperidinium) brochi Kofoid and Swezy
claudicans Paulsen
conicum (Gran) Ostenfeld and Schmidt
crassipes Kofoid
depressum Bailey
diabolus Cleve
divergens Ehrenberg
excentricum Paulsen
globulus Stein
granii Ostenfeld
latum Paulsen *(Peridinium)*
leonis Pavillard
longispinum Kofoid
mariaelebourae Paulsen
minutum Kofoid
monacanthum Broch
murrayi Kofoid
oceanicum Vanhöffen
pellucidum (Bergh) Schütt
pentagonum Gran
peruvianum Balech
punctulatum Paulsen
spiniferum Schiller
trochoideum (Stein) Lemmermann
Podolampas palmipes Stein
Porella perforata (Gran) Schiller
Pronoctiluca spinifera (Lohmann) Schiller
Prorocentrum gracile Schütt
micans Ehrenberg
Pyrocystis lunula Schütt
pseudonoctiluca Thomson
Pyrophacus horologicum Stein

COCCOLITHOPHORACEAE

Acanthoica quattrospina Lohmann
Anthosphaera robusta Lohmann
Calciosolenia cf. *murrayi* Gran in Murray and Hjort
Emiliania huxleyi (Lohmann) Kamptner
Corisphaera gracilis Kamptner
Michaelsarsia splendens Lohmann
Ophiaster hydroideus Lohmann
Pontosphaera inermis Lohmann
nana Kamptner
steueri Kamptner
Rhabdosphaera clavigera Murray and Blackman
stylifera Lohmann
Syracosphaera pulchra Lohmann

SILICOFLAGELLATES

Dictyocha fibula Ehrenberg
Dictyocha speculum Ehrenberg

CRYPTOMONADALES

Chilomonas marina (Braarud) Halldal (*Bodo marina* Braarud)

OTHER GROUPS

Olisthodiscus luteus Carter
Phaeocystis pouchetii (Hariot) Lagerheim
Eutreptiella gymnastica Throndsen

OTHER FORMS IDENTIFIED DURING 1976

BACILLARIOPHYTA (DIATOMEAE)

Chaetoceros costatus Pavillard
Dactyliosolen mediterraneus Peragallo
Detonula confervacea (Cleve) Gran
Rhizosolenia styliformis var. *longispina* Hustedt

Amphiprora sp.
Gyrosigma sp.
Navicula sp.

DINOFLAGELLATAE

Ceratium azoricum Cleve
Ceratium buceros f. *molle* (Kofoid)
Ceratium macroceros (Ehrenberg) Cleve
Gymnodinium sp.
Noctiluca miliaris Suriray
Oxyphysis oxytoxoides Kofoid
Gymnodinium sp.

CRYPTOMONADALES

Monada sp.

10^6 cells/l. Dinoflagellates were second in importance during the period, the most frequent species being *Prorocentrum micans, Protoperidinium pellucidum, Gymnodinium splendens, Ceratium furca,* and *Prorocentrum gracile.* Within the 60-mile region their presence was more variable than the diatoms, but they were always observed at the offshore stations.

Species of other phytoplankton groups were observed much less frequently. Nevertheless, in the near-shore region during the summer months, dense blooms of *Olisthodicus luteus* and *Eutroptiella gymnastica* offten occurred.

Most of the species mentioned above could be observed in the area throughout the year, however, their contribution to the total phytoplankton showed some seasonal pattern (Fig. 1). For example, *Schroederella delicatula* and *Thalassionema nitzschioides* tend to be the dominant species during winter, while in summer *Rhizosolenia delicatula* appears to dominate.

The genus *Chaetoceros* was abundant during the study, but the predominant species changed with the area and the season of the year. In the summer stations *Ch. radicans* predominated between 2 to 12°S latitude while *Ch. debilis* and *Ch. socialis* were the predominant species south of 12°S. During the fall *Ch. convolutus* and *Ch. curvisetus* were the most frequently observed between 4 and 11°S. *Ch. radicans, Ch. affinis*, and *Ch. didymus* were the most abundant species during the winter months, occupying the regions from 2 to 6°S, from 8 to 12°S, and 16 to 19°S, respectively. Higher numbers of *Chaetoceros* species were observed in spring, yet each species still showed regional distribution. *Ch. subtile* was usually observed from 2 to 6°S, *Ch. radicans* occupied the region from 6 to 9°S. South of 9°S the most abundant were *Ch. peruvianus, Ch. debilis*, and *Ch. didymus*, and south of 12°S the dominant species were *Ch. decipiens* and *Ch. curvisetus.*

Other diatom species that also had a clear seasonality in their distribution, although their contribution to the total phytoplankton population was not too important, were *Rhizosolenia alata* and *Guinardia flaccida* during the summer; in the fall *Rhizosolenia fragilissima, Nitzschia* (*epifita*), and *Roperia tesselata*; in winter *Rhizosolenia setigera* and *Corethron hystrix*; and spring, *Dactyliosolen mediterranea, Leptocylindrus minimus, Thalassiosira mendiolana,* and *Coscinodiscus eccentricus.*

Dinoflagellate species were, in general, more frequently observed in the summer and fall samples, and they became much rarer during winter. It is also interesting to note (Fig. 1) that they never contributed significantly to the total phytoplankton population north of 8°S.

During winter and summer, a coherence in the phytoplankton composition seems to exist along the entire Peru coast (Fig. 1). In contrast, during spring and fall, marked local patchiness is evident. Another difference that appears to exist between the seasons is that fewer species accounted for most of the phytoplankton population during winter and summer than during fall and spring.

Total phytoplankton distribution.

The seasonal distribution of the mean concentration of total phytoplankton during the 10 years of the study is shown in Figs 2 and 3. Generally during winter, phytoplankton cell densities were lower than during the other seasons. However, relatively high cell concentration appears to extend farther offshore. In contrast, during the other seasons and especially during summer, cell concentrations were higher but usually were restricted to a narrower region close to the coast and decreased quite drastically offshore.

The phytoplankton were not uniformly distributed in parallel bands along the coast, but rather in a series of discrete centers with higher cell densities, and the position of the centers with respect to the coast seems to be quite constant in time. On this basis four major centers can be identified: Pimentel (∿6°S), Chimbote (∿9°S), Callao (∿12°S), and Tambo de More - Pisco (∿14°S); and two minor centers, Talara (∿4°S) and Ilo (∿17°S). It is also interesting that although the centers with high cell concentrations seem to persist throughout the year, the relative importance of each changes according to the season. During fall and winter, the highest phytoplankton concentrations tended to be in the northern part of the coast and to move southward in spring and summer.

Discussion

The list of identified species presented in this paper agrees with previous taxonomic studies from the region (Gunther, 1936; Rojas de Mendiola; 1958; Strickland *et al.*, 1969; Blasco, 1971), but

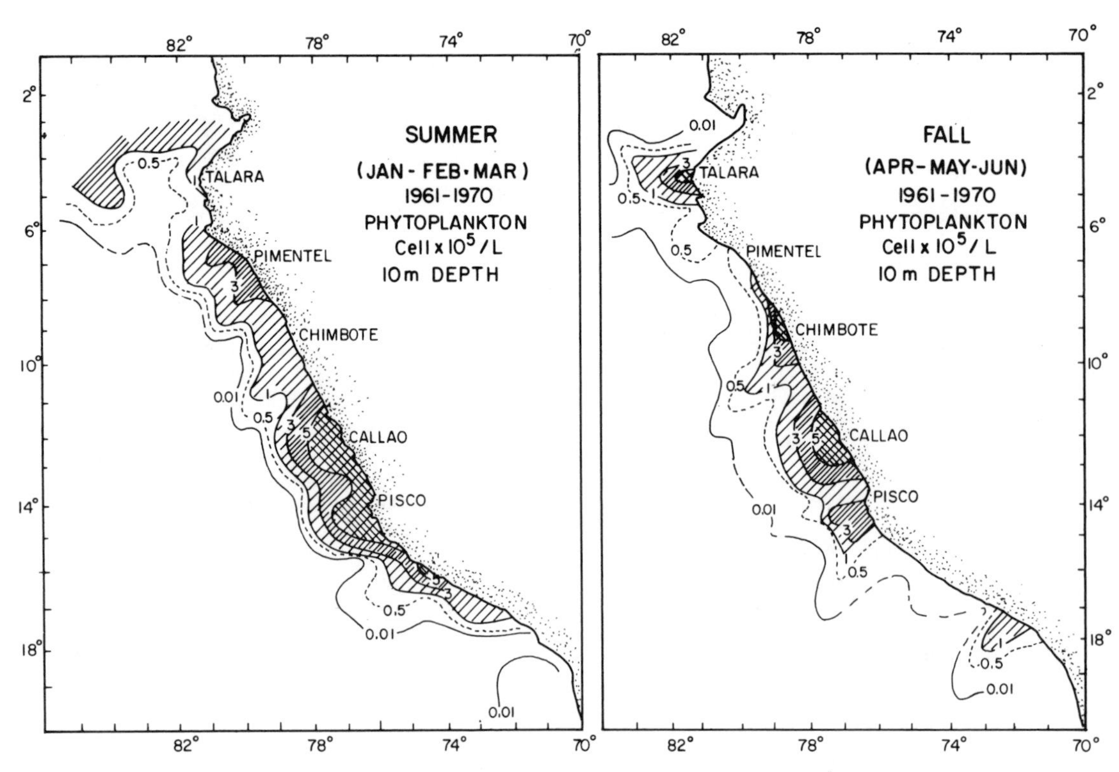

Fig. 2. Mean horizontal distribution of total phytoplankton cell densities along the Peruvian coast during summer and fall, based on samples collected during a 10-year study from 1961 to 1970.

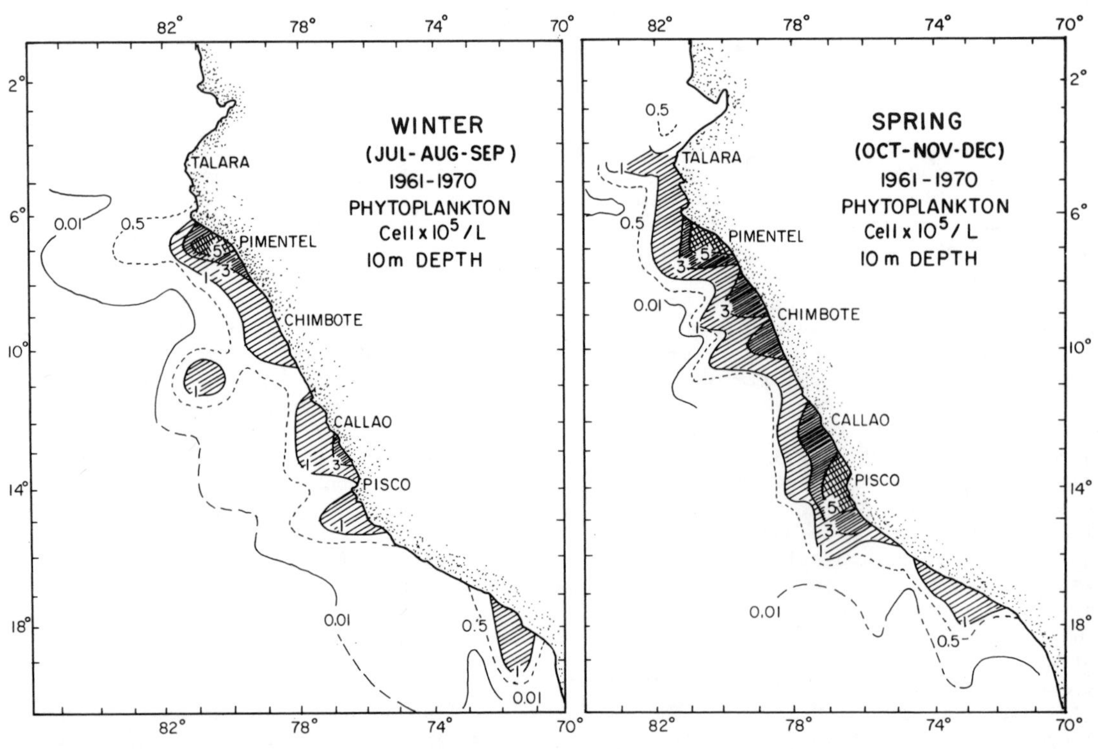

Fig. 3. Mean horizontal distribution of total phytoplankton cell densities along the Peruvian coast during winter and spring based on samples collected during a 10-year study from 1961 to 1970.

it adds a few species. The dominant species during this study have also been previously reported as dominant organisms of the Peruvian region (Strickland *et al.*, 1969; Ryther *et al.*, 1970; Guillen *et al.*, 1972, 1973; Cowles, Barber, and Guillen, 1977; Sukhanova *et al.*, 1978). All are typical "upwelling" diatoms, and therefore the relatively low spatial and seasonal variability observed is not surprising because upwelling occurs along the Peruvian coast throughout the year (Zuta and Guillen, 1970; Wooster and Sievers, 1970). The mean phytoplankton distribution pattern is similar to the mean chlorophyll and temperature distributions, and the centers with maximum cell concentration coincide with the areas where upwelling is more persistent and intense (Zuta and Guillen, 1970).

Although marked differences exist in the intensity and extent of the upwelling between winter and summer (Zuta and Guillen, 1970), the general patterns of distribution and composition in the two seasons are similar. It may be that environmental variability by itself plays a major role in determining composition and distribution, and that during winter and summer upwelling is in a relatively more steady state than during the transient periods of spring and fall.

A question that concerns the results of this study is how representative is a mean value of 10 year's data in describing the Peru upwelling region, where there may be significant inter-annual variability (Zuta and Guillen, 1970). The main object of this paper has been to provide a general description of the phytoplankton in the Peruvian upwelling region, to serve as a reference basis for other small scale (time and space) studies, and to assist in planning future research. As such, the data and approach used seem to be adequate. However, the question of inter-annual variability remains. It is not possible definitely to answer this question with the data at hand, but some analysis of the data from this point of view seems to be appropriate and necessary.

The best known inter-annual variabilities of the Peru upwelling region are the hydrographical and biological changes associated with the warm temperature anomalies called *el Niños*. Since 1960, four such warmer periods have been recognized: 1965, 1969, 1972 (Walsh, 1978), and 1976 (Ramage, 1975). In 1965, 1969, and 1972 (Guillen *et al.*, 1973; Blasco, 1971; Rojas de Mendiola and Estrada, 1976), phytoplankton species composition did not differ markedly from the species composition presented in this paper, nor did a change in the dominant species or in the general pattern of the phytoplankton distribution along the coast seem to occur. The most important difference was that during these periods high phytoplankton cell densities were restricted to the most near-shore region. In 1976, *Gymnodinium splendens*, a common dinoflagellate species of Peruvian waters, occurred in much larger quantities than usual, producing red tide along the entire Peru coast from March to May, 1976 (Rojas de Mendiola, 1979). The dramatic extent of the bloom and the fact that anchovy growth and spawning was seriously reduced during this same period (Valdivia, 1978) has been the basis of much speculation on the importance of inter-annual variability at the phytoplankton composition level (Barber, in press) and specifically questions the results of this study. It should be said, however, that concurrently with *Gymnodinium splendens*, the two other predominant species were *Rhizosolenia delicatula* and *Thalassionema nitzschioides*, both typical dominant species during the summer season in the Peru upwelling region (Fig. 1).

Acknowledgements. I would like to thank the Instituto del Mar del Peru for its continuing support of this research and Dr. Dolors Blasco for her helpful review and criticism of the manuscript. I would also like to express my appreciation to Noemí Ochoa, Olga Gomez, and Emira Autonietti for their technical assistance.

References

Barber, R.T., Variations in biological productivity along the coast of Peru: the 1976 dinoflagellate bloom, *Science*, in press, 1980.

Barreda, M., El plancton de la Bahia de Pisco, *Boletin Compania Administradoro Guano, 33*, 1957.

Blasco, D., Composicion y distribucion del fitoplancton en la region de afloramiento de las costa Peruanas, *Investigacion Pesquera, 35*, 61-112, 1971.

Cowles, T.J., R.T. Barber, and O. Guillen, Biological consequences of the 1975 *el Niño*, *Science, 195*, 285-287, 1977.

Guillen, O., B. Rojas de Mendiola, and R. Izaguirre de Rondan, Primary productivity and phytoplankton in coastal Peruvian waters, in: *Fertility of the Sea, Vol. I*, Gordon and Breach, New York, pp. 157-185, 1972.

Guillen, O., B. Rojas de Mendiola, and R. Izaguirre de Rondan, Primary productivity and phytoplankton in the coastal Peruvian waters, *Oceanography of the South Pacific 1972*, Comp. R. Frazer, New Zealand National Commission for UNESCO, Wellington, pp. 387-395, 1973.

Gunther, E.R., A report on oceanographical investigations in the Peru Coastal Current, *Discovery Reports, 13*, 107-276, 1936.

Landa, A., Analysis de muestras diarias de fitoplancton superficial en Chimbote, *Boletin Cientifico Compania Administradoro Guano, 1*, 1953.

Ramage, C.S., Preliminary discussion of the meteorology of the 1972-1973 *El Niño, Bulletin American Meteorological Society, 56*, 234-242, 1975.

Rojas de Mendiola, B., Breve estudio sobre la variacion cualitativa anual del plancton superficial de la Bahia de Chimbote, *Boletin Com-*

pania Admisistradoro Guano, 34, 7-17, 1958.
Rojas de Mendiola, B., Analysis cuantitativos y cualitativos del fitoplancton en el area de pesca Supe - Cerro Azul, *Informe Interno Instituto de los Recursos Marinos, 38,* 10 pp., 1963a.
Rojas de Mendiola, B., Estudios preliminares sobre los distribuciones del fitoplancton en Noviembre, en el área del Callao - Cabo Blanco, *Informe Interno Instituto de los Recursos Marinos, 34,* 10 pp., 1963b.
Rojas de Mendiola, B., Variacion estacional de fitoplancton en una área fija del Callao, *1er Congres Nacional de Biologia,* 87-89, Lima, 1966.
Rojas de Mendiola, B., Red tide along the Peruvian coast, in: *Toxic Dinoflagellate Blooms,* D.L. Taylor and H.H. Seliger (eds.), Elsevier-North Holland, Amsterdam, pp. 183-190, 1979.
Rojas de Mendiola and M. Estrada, El fitoplancton en el area de Pimentel Verano de 1972, *Investigacion Pesquera, 40,* 463-490, 1976.
Ryther, J.H., D.W. Menzel, E.M. Hulburt, C.J. Lorenzen, and N. Corwin, The production and utilization of organic matter in the Peru coastal current, R.V. *Anton Bruun* Reports, *Scientific Results of the Southern Pacific Experiments, 4,* 12 pp., Texas A&M Press, 1970.
Semina, H.J., Distribution of plankton in the south eastern Pacific, *Trudy Instituta Okeanologii Akademii Nauk, 89,* 43-59, 1971.
Strickland, J.D., R.W. Eppley, and B. Rojas de Mendiola, Poblaciones de fitoplancton, nutrientes y fotosíntesis en aguas costeras peruanas, *Boletin Instituto del Mar del Peru, 2,* 4-12, 1969.
Sukhonova, I.N., G.V. Konovalova, and T.N. Rat'kova, Phytoplankton numbers and species structure in the Peruvian upwelling region, *Oceanography, 18,* 72-76, 1978.
Utermöhl, H., Zur Vervollkommung der quantitativen Phytoplankton-methodic, *Mitteilungen der internationalen Vereinigung fur theoretischie und angewandte Limnologie, 9,* 38 pp., 1958.
Valdivia, E., The anchoveta and *el Niño, Rapports Proces verbaux Reunion Conseil International Exploration Mer, 173,* 196-202, 1978.
Walsh, J.J., The biological consequences of interaction of the climatic, *el Niño,* and event scales of variability in the eastern tropical Pacific, *Rapports Proces verbaux Reunion Conseil International Exploration du Mer, 173,* 182-192, 1978.
Wooster, W.S. and V. Sievers, Seasonal variations of temperature, drift and heat exchange in the surface waters off the west coast of South America, *Limnology and Oceanography, 15,* 595-605, 1970.
Zuta, S. and O. Guillen, Oceanografia de las aguas costeras del Peru, *Boletin Instituto del Mar del Peru, 2,* 161-245, 1970.

THE ROLE OF CIRCULATION AND STABILITY IN CONTROLLING THE RELATIVE ABUNDANCE OF DINOFLAGELLATES AND DIATOMS OVER THE PERU SHELF

S.A. Huntsman

Duke University Marine Laboratory, Beaufort, North Carolina 28516

K.H. Brink

School of Oceanography, Oregon State University, Corvallis, Oregon

R.T. Barber

Duke University Marine Laboratory, Beaufort, North Carolina 28516

D. Blasco

Bigelow Laboratory for Ocean Sciences, West Boothbay Harbor, Maine 04575

Abstract. During a six-week study, the dinoflagellate abundance at C-3, a mid-shelf station off the Peruvian coast, fluctuated widely. At times of maximum cell density (400 cells/ml) dinoflagellates comprised up to 80% of the combined dinoflagellate-diatom population. Our data suggest that after their initial advection into the study site, the dinoflagellates persisted because of stability in the euphotic zone and reseeding by the subsurface poleward undercurrent. The increased stability that favored dinoflagellates did not exclude diatoms at C-3. Diatoms could utilize the available silica and regenerated nitrogen to maintain growth rates adequate to counteract losses from sinking and grazing.

Introduction

The occurrence of "red tides", dense blooms of dinoflagellates, in stratified, nutrient-poor water is well documented (Conover, 1954; Ryther, 1955; Margalef, 1956; Rounsefell and Nelson, 1966). Explanations for the success of dinoflagellates under these conditions involve 1) the use of vertical migration to maintain dinoflagellates near the surface in the absence of turbulence, 2) the ability of dinoflagellates to utilize low nutrient concentrations, or to take up nutrients at depth during vertical migration, or 3) the monopolization of light by dinoflagellates through the formation of dense surface canopies (Ryther, 1955; Margalef, 1978; R.C. Dugdale, personal communication). Because in many cases even dense dinoflagellate blooms contain normal concentrations of diatoms, the mechanisms for the success of dinoflagellates clearly do not result in the exclusion of diatoms. Evaluation of the factors controlling the relative abundance of the two groups may explain the persistence of diatoms in dinoflagellate blooms.

During the first half of 1976, an extensive dinoflagellate bloom dominated by *Gymnodinium splendens* occurred off the coast of Ecuador and Peru. The role of increased static stability in the euphotic layer in initiating the dinoflagellate dominance and the abnormal hydrographic conditions (particularly episodes of extensive denitrification) coincident with its appearance have been described by Barber (in press) and by Dugdale, Goering, Barber, Smith, and Packard (1977). Although chlorophyll *a* concentrations exceeded 400 μg/l in some patches, such extreme levels were not encountered at an inshore study site located over the continental shelf at 15°S (Fig. 1). Here, chlorophyll *a* concentrations were less than 40 μg/l, observed dinoflagellate densities did not exceed 500 cells/ml, and dinoflagellates never comprised over 80% of the combined diatom-dinoflagellate population. We will compare fluctuations in the relative density of dinoflagellates and diatoms to changes in hydrographic and physical conditions at the study site and try to evaluate the factors responsible for the persistent coexistence of the two phytoplankton groups throughout the 6-week study.

Instrumentation and Methods

The study was carried out on the R.V. *Alpha Helix* during two legs of a cruise in 1976. The first leg ran from March 28 through April 11, the second from April 17 through May 4. The princi-

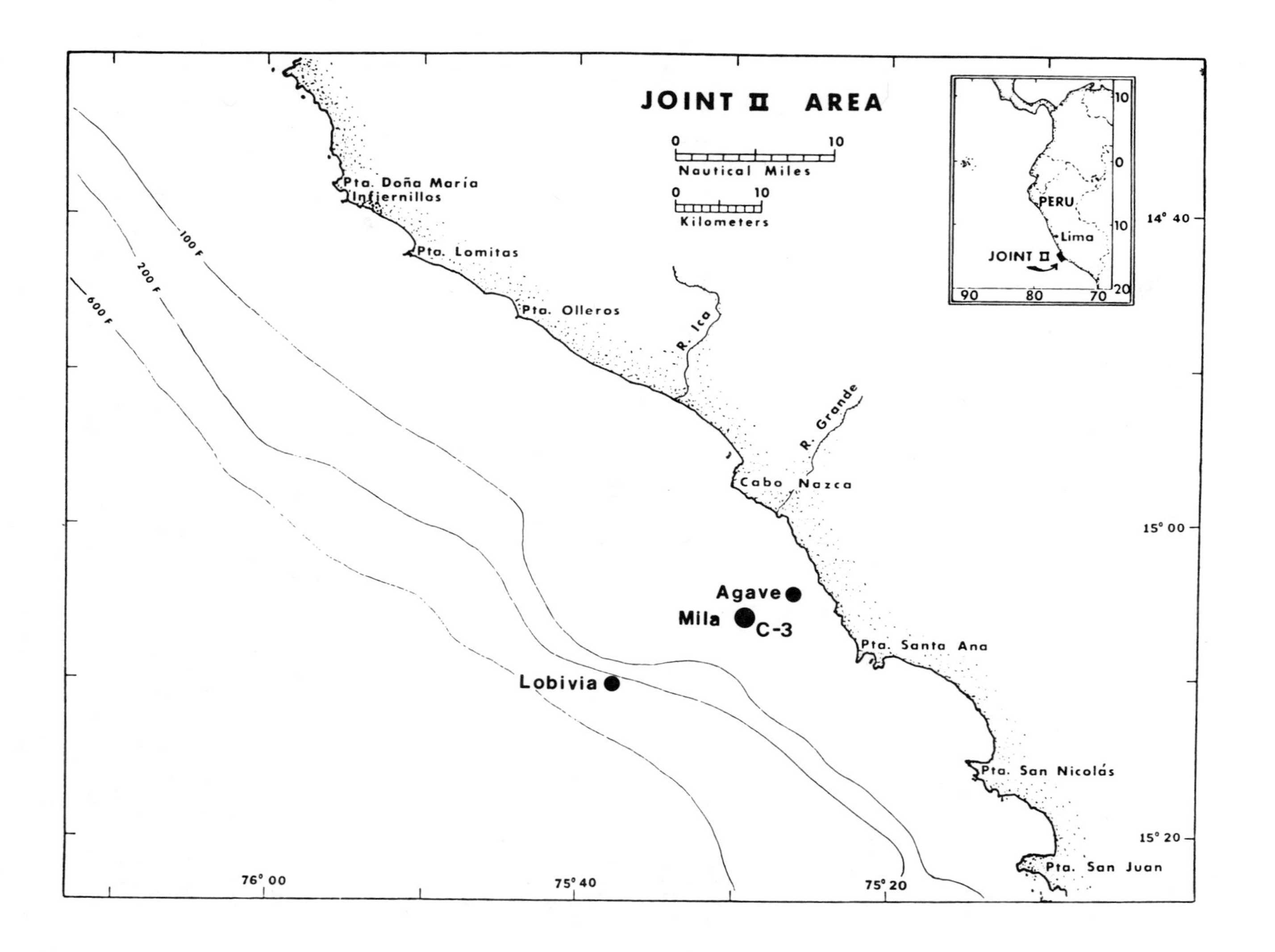

Fig. 1. Study area off the coast of Peru. C-3 is the location of the time series. Agave, Mila (at C-3), and Lobivia identify the position of current meter moorings.

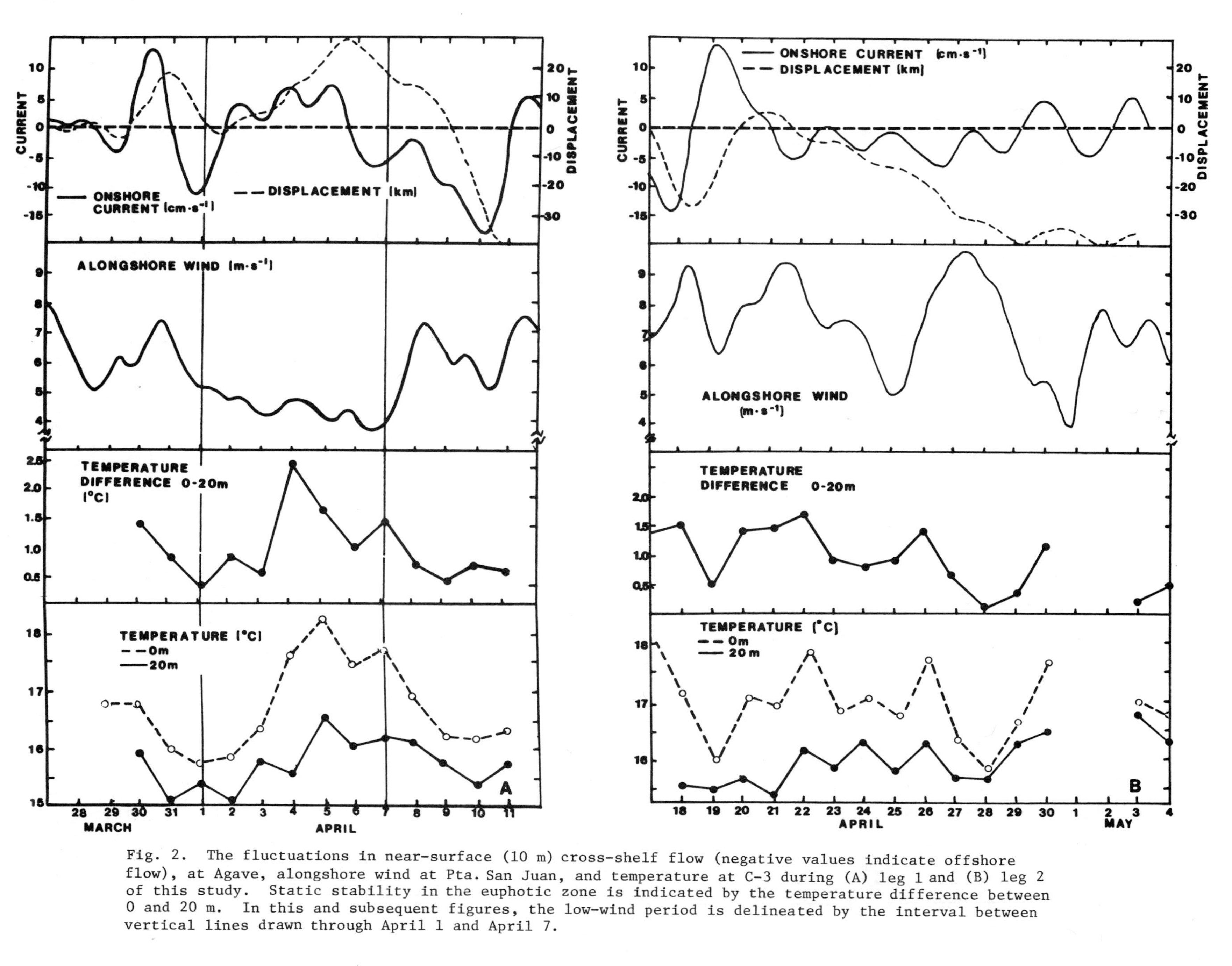

Fig. 2. The fluctuations in near-surface (10 m) cross-shelf flow (negative values indicate offshore flow), at Agave, alongshore wind at Pta. San Juan, and temperature at C-3 during (A) leg 1 and (B) leg 2 of this study. Static stability in the euphotic zone is indicated by the temperature difference between 0 and 20 m. In this and subsequent figures, the low-wind period is delineated by the interval between vertical lines drawn through April 1 and April 7.

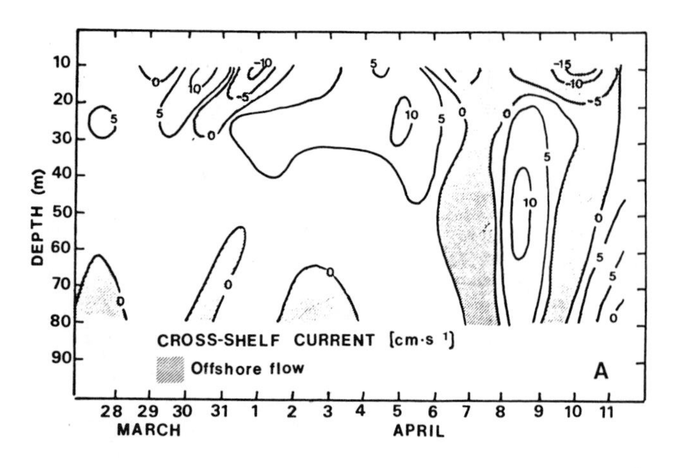

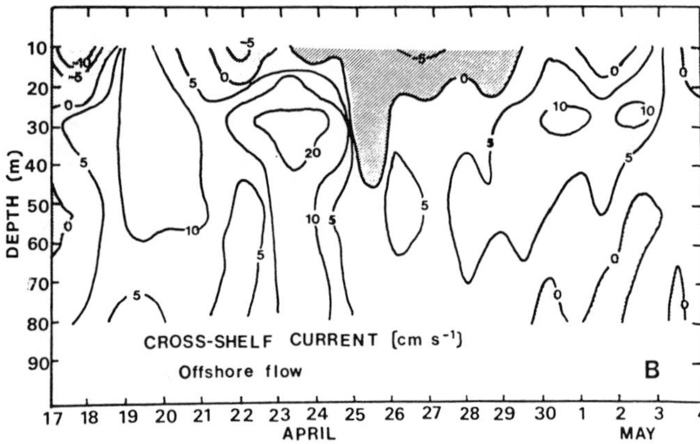

Fig. 3. The depth distribution of cross-shelf current at C-3 during (A) leg 1 and (B) leg 2 of this study, based on records from current meters at 10 m (Agave) and at 28, 53, and 78 m (Mila). Total water depth was 123 m. Reference axis is rotated 30.5° toward the NE-SW to lie perpendicular to the coast.

pal study site was a mid-shelf station ("C-3" in Fig. 1), at 15°05'S and 75°31'W in 123 m of water. Samples were collected daily between 0800 LST and 0900 LST. Nutrients and temperature were collected from discrete depths to 100 m and have been reported by MacIsaac *et al.* (1979). Nutrients were determined as described by Packard, Dugdale, Goering, and Barber (1978). Phytoplankton samples were preserved and identified according to the method of Blasco (1977). Carbon fixation was measured at 100, 50, 30, 15, 5, and 1% light depth by a modification of the ^{14}C technique by Huntsman and Barber (1977). Chlorophyll *a* was determined from the same depths by the SCOR-UNESCO method (1966). The productivity index was determined from the hourly carbon fixation rate per unit chlorophyll *a* (mgC $mgChl^{-1}hr^{-1}$) integrated over the euphotic zone.

Transects were made across the shelf at frequent intervals during which water from 2 m was sampled almost continuously for temperature and chlorophyll fluorescence (Blasco, 1979). On a few of the transects nutrients were also determined. Other data were supplied by occasional hydrographic transects along which vertical profiles of nutrients and temperature were obtained across the shelf.

Three current meter moorings were deployed across the shelf (Fig. 1): one, "Agave", was near-shore in 40 m of water, the second "Mila", was at C-3, the third, "Lobivia", was beyond the shelf break in 656 m. Each was instrumented with two to five current meters and temperature recorders (Enfield, Smith, and Huyer, 1978). Only the inshore array had a near-surface meter shallow enough (10 m) to be in the apparent surface Ekman layer under most circumstances. Because of the importance of this layer to our discussion, we will use data from this meter to represent flow patterns in the surface waters at C-3. As comparisons made a year later showed that the near-surface cross-shelf flow was strongly correlated at the two locations (Brink, Halpern, and Smith, 1980), we feel that the assumption is justified.

Winds were continuously sampled at a meteorological station at Punta San Juan (Fig. 1) and have been decomposed into components tangential and normal to the coast (Stuart, Septseris, and Nanney, 1976). (Sea surface temperature maps made by aircraft over-flights of the region are available in the same report.) Both current and wind data were low-pass filtered with a half-power point of 47 h to remove internal wave, tidal, sea breeze, and other high-frequency variations of short periods.

Results

The physical regime

Throughout the cruise period, the low-pass filtered alongshore wind velocity component was equatorward (upwelling favorable) and variable in strength (Fig. 2). Despite these conditions, the classical upwelling pattern of near-surface offshore flow and deeper onshore flow was observed only occasionally (Fig. 3) and then in conjunction with stronger alongshore winds. It is possible that the 10-m current meter was too deep in the water always to be in the shallow region of offshore flow, but comparable more intensive measurements at a nearby location made during March-May of 1977 indicate that an instantaneous classical upwelling circulation often does not exist (e.g., Brink *et al.*, 1980). The Ekman concept of wind-driven upwelling suggests that fluctuations in near-surface cross-shelf currents should be negatively correlated with changes in the alongshore wind. The relation does seem to have held during much of our observation period, although notable exceptions occurred, e.g., March 27-31 and April 8-12 (Fig. 2).

The local wind stress also causes mixing in the upper water column. The hydrographic observa-

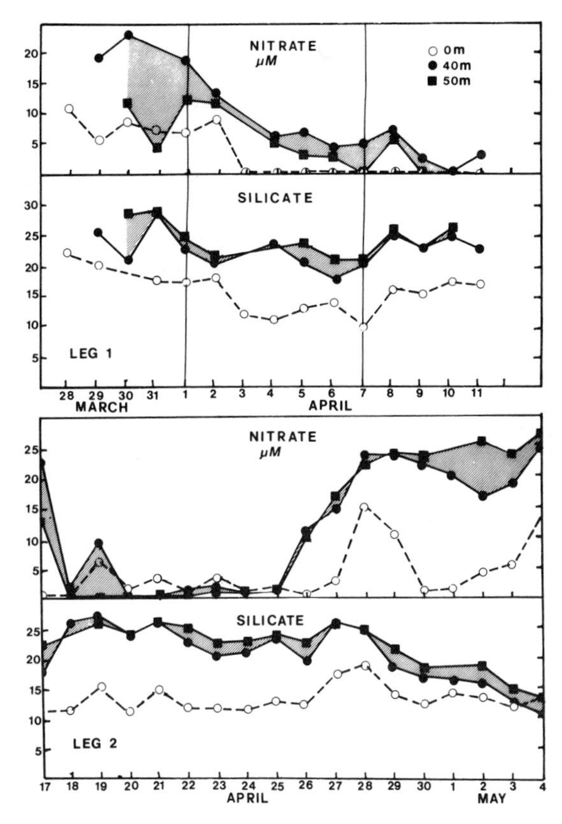

Fig. 4. The fluctuations in nitrate and silica at the surface and in the core of the upwelling source water (40 to 50 m) at C-3 during the cruise.

tions were inadequate to determine a mixed layer depth (temperature was normally not observed between the surface and 20 m), but a reasonable indicator of near-surface stratification can be obtained from the temperature difference between 0 and 20 m. The difference is a good measure of the static stability in the euphotic zone, normally 10 to 20 m deep, because temperature and density are highly correlated in the region (see, e.g., Brink, Gilbert, and Huyer, 1979). Fig. 2 indicates that there is a correlation between alongshore wind component (which is generally representative of the wind speed) and the nearshore static stability. Low winds tend to allow surface heating to strengthen stratification (e.g., April 1-7), while strong winds tend to cause mixing and weaken stratification (April 26-28). Again, the correlation is not perfect, possibly because the surface mixed layer rarely reaches 20 m depth, so that the 0 to 20-m temperature difference may not completely represent the surface mixing regime.

In much of the following, it will be desirable to discuss effects due to cross-shelf advection. Perhaps the simplest way to approach the phenomenon is to assume a simple translation model of advection (Smith, Brink, Santander, Cowles, and Huyer, 1980). In this case, it is assumed that a given cross-shelf distribution of the advected quantity is displaced onshore or offshore by the flow. Thus, a time integral of onshore flow ("displacement", Fig. 2) is calculated, and this distance along with an observed cross-shelf distribution can be used to predict the quantity at a point as a function of time. If the cross-shelf distribution is monotonic, then the fluctuations in the quantity and the displacement would be correlated. The procedure assumes that the currents are constant over all space and ignores boundary effects, *in situ* processes, mixing, and alongshore advection. Although the procedure is approximate, it appears to work tolerably well under some circumstances (e.g., Smith *et al.*, 1980.

Nutrients

Nitrate was abnormally low throughout the water column at C-3 resulting from the upwelling of denitrified water from beyond the shelf break (Dugdale *et al.*, 1977). Nitrate concentrations in the upwelling source water (centered at 40 to 50 m at C-3) decreased from 20 to 3 μM during the first leg, were high at the onset of leg 2, and decreased to less than 2 μM from April 19 to 25, then abruptly increased to nearly normal levels (10 to 20 μM) on April 26 and remained high for the final week of the cruise (Fig. 4). At the surface, nitrate was present in relatively high amounts (6 to 11 μM) until April 3 (Fig. 4). On that date, which coincided with the second day of onshore flow, nitrate in the upper 20 m abruptly dropped to undetectable levels and remained so for the duration of the leg despite restoration of near-surface offshore flow on April 8. Surface nitrate concentrations recovered on April 18 and 19, increasing from 0.1 to 6 μM in the euphotic zone following a strong upwelling event. For the following week, surface nitrate concentrations ranged from 1 to 4 μM, then dropped briefly to 0.2 μM on April 26, coincident with a brief surge of poleward flow that extended into the surface layer. On April 28 the nitrate throughout the water column increased to between 16 (surface) and 22 (below 40 m) μM, values typical of those found in the region in cruises of previous years. Surface nitrate concentrations dropped coincidently with the return of stratification on April 30, then steadily recovered reaching 13 μM by the end of the cruise on May 4.

Nitrite concentrations qualitatively reflected those of nitrate, but were generally much lower. In the euphotic zone, nitrite exceeded 1 μM only on April 18, 19, and 23, the maximum value being 3 μM on April 19.

Ammonia in the surface water was consistently between 0.2 and 0.5 μM, exceeding 1 μM only on March 28 to 30. Although levels in excess of 1 μM were observed in the upwelling source water

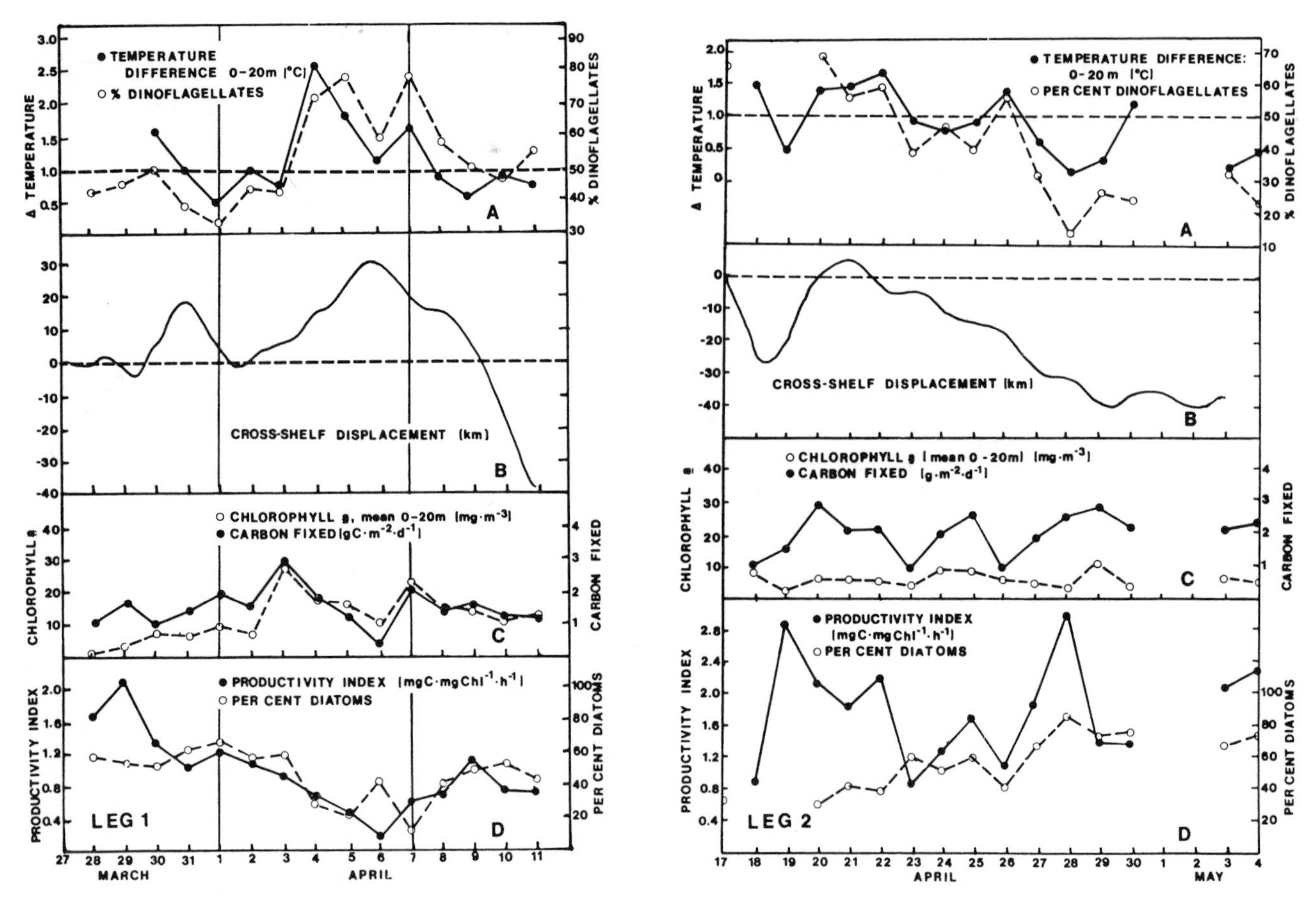

Fig. 5. (Panels A, B) The relationships between the dinoflagellate fraction and stability in the upper 20 m at C-3 or cross-shelf displacement at Agave (10 m). (Panels C, D) A comparison of chlorophyll *a*, primary production, and productivity index in the euphotic zone with diatom fraction at C-3.

during periods of stratification (April 9 to 25), ammonia concentrations in the euphotic zone C-3 did not increase during the period.

Silica concentrations increased in the denitrified water (Dugdale *et al.*, 1977) and was abundant at all times, generally exceeding 20 μM in the upwelling source water and always in excess of 10 μM in the water of the euphotic zone at C-3 (Fig. 4).

Phytoplankton

At the beginning of the cruise diatoms dominated the population and chlorophyll *a* was low, although it increased gradually during the period of abundant nitrate from March 28 through April 2. Onshore flow began on April 1, followed by nitrate depletion and a sharp increase in chlorophyll *a* on April 3. Both diatoms and dinoflagellates increased in numbers, although dinoflagellates contributed a disproportionate share of the chlorophyll due to their large cell size (Blasco, 1979). Dinoflagellates soon dominated the phytoplankton numerically, comprising up to 80% of the cell numbers from April 4 through April 7 (Fig. 5).

The changes in the dinoflagellate component of the phytoplankton correlated with the cross-shelf displacement of water in the upper layer. However, there was also a good correlation between composition and stability of the water column; dinoflagellate dominance in both legs generally occurred when the temperatures at the surface and at 20 m differed by more than 1°C (Fig. 5).

Upwelling resumed vigorously on April 6, and dinoflagellates remained at about 50% of the population for the remainder of the first leg. Except for one day, April 26, dinoflagellates did not exceed 50% of the population after April 22. Although the generally decreasing dinoflagellate fraction again reflected offshore displacement, it is apparent from Fig. 5 that during the second leg changes in stability gave a better correla-

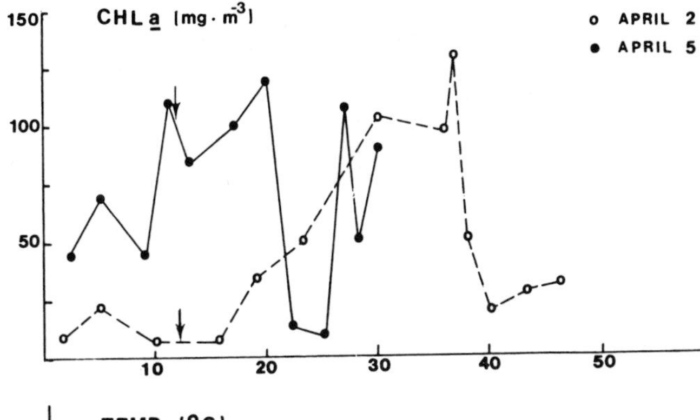

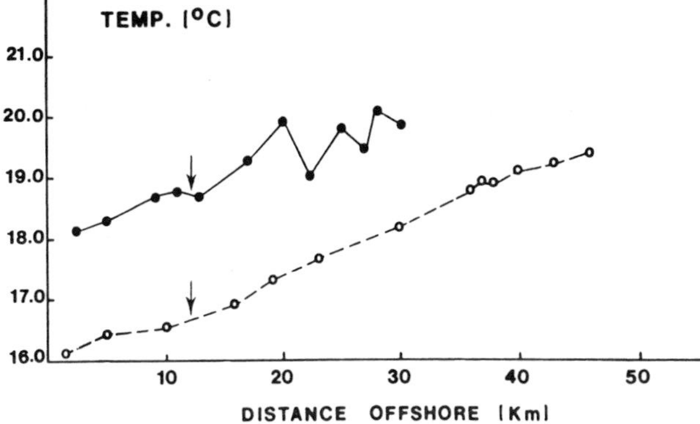

Fig. 6. The change in the cross-shelf distribution of near-surface (2 m) temperature and chlorophyll *a* before and after the response of the surface layer to reduced winds. Transects were made between 1200 and 1445 LST. The position of C-3 is indicated by arrows.

tion with the fluctuations seen in the dinoflagellate fraction.

Primary Production

During leg 1, carbon fixation and chlorophyll *a* in the euphotic zone showed similar trends (Fig. 5) except during the period of highest dinoflagellate density, when carbon fixation dropped to a minimum. During leg 2 carbon fixation was generally higher than it had been on leg 1 while chlorophyll *a* was lower, thus the average productivity index was greater during leg 2 than during leg 1. The changes in productivity index for the euphotic zone appeared to reflect the diatom fraction of the phytoplankton (Fig. 5) during both legs, except during April 20 to 22. Maxima in the value of the index occurred on the dates of the two upwelling events during the second leg, April 19 and 28th. We have no phytoplankton data for the first of these events, but the second coincided with a maximum in the diatom component. Both maxima coincided with minima in the water column stability.

Discussion

During the first week of the study conditions over the shelf appeared unfavorable for the accumulation of high dinoflagellate concentrations. The persistent upwelling-favorable winds produced enhanced nutrient and apparently turbulence levels that favored diatoms.

The data were inadequate to identify the roles of growth and grazing in controlling the species composition of phytoplankton at C-3. However, the low productivity index and extremely rapid rise in dinoflagellates between April 2 and 3 coincident with a period of onshore flow suggest that the bloom observed between April 3 and April 8 was carried onshore to C-3. This view is supported by the frequent observation of surface patches of high chlorophyll beyond the shelf on line maps made prior to April 3 and over the shelf on later maps (e.g., Fig. 6). In the offshore waters inorganic nitrogen (nitrate, nitrite, and ammonia) levels were consistently below 0.5 μM and the silica concentration was around 10 μM in contrast to 15 to 20 μM in the surface waters over the shelf (MacIsaac *et al.*, 1979). This reduced but adequate silica concentration suggests that diatoms were present and actively

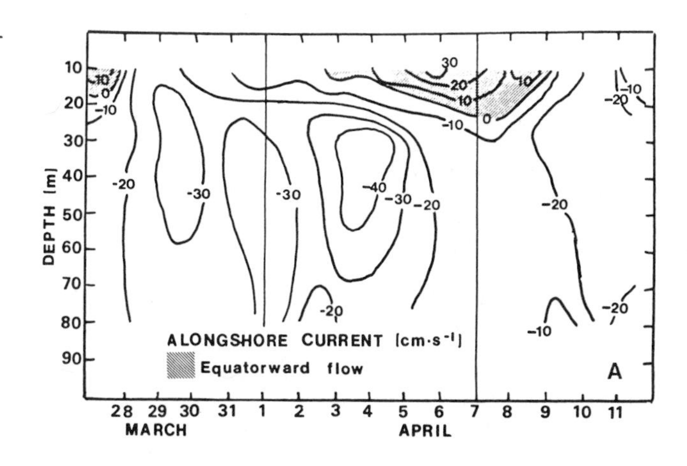

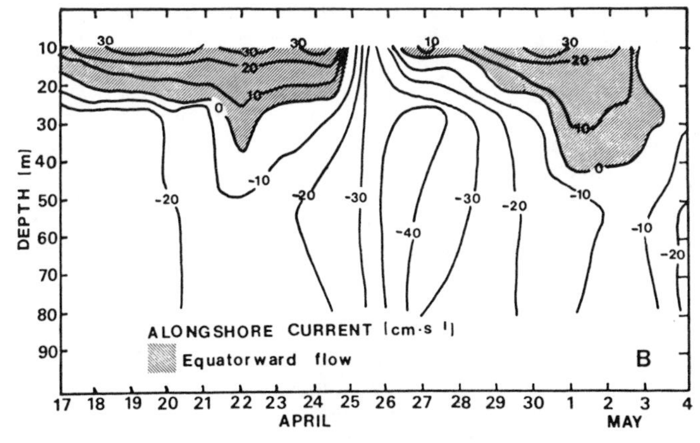

Fig. 7. The depth distribution of alongshore flow at C-3 during (A) leg 1 and (B) leg 2. The reference axis is rotated 30.5° toward the NW-SE orientation to be parallel to the coast.

growing and that silica limitation of diatom growth was not responsible for the success of the dinoflagellates.

At C-3, diatoms persisted at about 100 cells/ml even during the bloom of dinoflagellates. Thus, as was suggested by Blasco (1979), the bloom we observed did not approach monospecific composition. Although *G. splendens* dominated the biomass due to its large size, its density at C-3 was never sufficient to reduce the depth of the euphotic zone, normally about 20 m, to less than 6 m (MacIsaac *et al.*, 1979). While sampling times did not coincide with the mid-day surface maximum of dinoflagellates (Blasco, 1979), it is clear that the dinoflagellate patches we observed did not develop into "red tides" in the traditional sense. Two factors may have contributed to this condition. First, mixing, although relatively weak, was probably sufficient to prevent the accumulation of extremely high populations at the surface. Based on measurements made in 1977 using near-surface current meters, the wind stress measured during the dinoflagellate bloom should have generated a mixed layer approximately 8 m deep. While mixing is unfavorable to migrating dinoflagellates, it would help maintain the diatoms in the euphotic zone. Second, the correlation of the productivity index with the diatom fraction suggests that even during the period of maximum dinoflagellate density, diatoms were responsible for most of the photosynthesis and growth. The high half-saturation constants reported by Eppley, Rogers, and McCarthy (1969) for nitrate (K_s = 3.8 μM) and ammonium (K_s = 1.1 μM) uptake by nitrogen-depleted cultures of *G. splendens* support the hypothesis that the organisms may not be able to compete successfully against diatoms for nitrogen at low concentrations of the nutrient. Although there was almost no measurable nitrogen in the euphotic zone, ammonia excreted by the increased biomass of zooplankton present during the weak-wind period of April 1 to 7 (Smith, 1978) may have been rapidly taken up, providing the nitrogen necessary for diatom productivity.

As the near-surface stability decreased at the end of leg 1, the dinoflagellate fraction declined but remained at a higher fraction of the phytoplankton than they had been before April 3. The continued low levels of nitrogen, which limited diatom growth, and the relatively weak mixing may have allowed the dinoflagellates to persist. In addition, the presence of a poleward current at 10 m directly beneath the wind-driven layer may have prevented the dinoflagellates from being carried northward, out of the system (Fig. 7) (Smith and Barber, 1980). By downward migration at night, dinoflagellates could presumably move into the poleward-flowing layer compensating, to a greater or lesser degree, for their equatorward displacement in the surface layer during the day.

During leg 2, the gradual decline of dinoflagellates may be ascribed to the fairly steady upwelling conditions. Offshore flow tended to carry the cells out of the system, while the lower near-surface stability created conditions unfavorable for successful competition with diatoms. The increased thickness of the equatorward-flowing layer (Fig. 7) prevented dinoflagellates from reaching the return flow during their migratory cycle, further reducing their concentration. In addition to providing a re-seeding mechanism as we have described, the poleward flow appears to have contained an initial seed population of dinoflagellates. Near-surface poleward flow on March 28 to April 3 was accompanied by increasing numbers of dinoflagellates, particularly *G. splendens*, and the brief poleward pulse of April 26 coincided with a 60% increase in dinoflagellate cell numbers. Although the association of poleward flow with dinoflagellate increase did not appear to exist during the poleward flow of April 10 and 11, the period also demonstrated extremely strong offshore flow. The fact that dinoflagellates remained at about 50% of the population suggested again that during April 10 and 11 they were being resupplied to the surface layer by the sub-surface poleward current. This suggests that the region equatorward of 15°S generally had higher dinoflagellate concentrations.

In conclusion, there is little evidence to suggest that the April 3 increase in dinoflagellates at C-3 was due to *in situ* growth rather than advection. The maximum recorded growth rate for cultures of *G. splendens* is 0.6 doublings per day (Thomas, Dodson, and Linden, 1973). However, between April 2 and 3 the density of *G. splendens* at C-3 increased from 24 to 200 cells/ml, a change that, if due to growth, would require three doublings per day. In addition, the productivity index, which reflects photosynthetic activity, bore an inverse relationship to the dinoflagellate fraction during the first leg, dropping close to zero during the periods of greatest dinoflagellate density. Only the period from April 20 to 22 demonstrated a divergence from this relationship, when dinoflaglates dominated the phytoplankton and the productivity index was high. As this was the only time during their dominance that nitrate was present at measurable levels, it may represent a situation similar to that described by MacIsaac (1978) in which dinoflagellates in the presence of adequate nitrate exhibited nitrogen uptake rates similar to those of diatoms; however, the situation clearly did not persist. We therefore return to the overall conclusion that the appearance of the dinoflagellate bloom at C-3 was a result of advective processes, while its persistence in the surface waters was made possible by moderately high stability in the euphotic zone (Barber, 1980) coupled with a subsurface poleward undercurrent that, through the vertical migration of the dinoflagellates, counteracted advective losses in the equatorward flow at the surface. The failure of the system to become highly stratified or silica depleted permitted diatoms to persist by utilizing the small amounts of nitrogen available.

Acknowledgements. This work was supported by NSF grants OCE75-23722, OCE78-03380, and OCE75-23718A01 (Coastal Upwelling Ecosystems Analysis). Early discussions with Sharon Smith and Adriana Huyer helped focus the objectives of this paper.

References

Barber, R.T., Variations in biological productivity along the coast of Peru: the 1976 dinoflagellate bloom, *Science*, in press.

Blasco, D., Red tide in the upwelling region of Baja California, *Limnology and Oceanography, 22*, 255-263, 1977.

Blasco, D., Changes of the surface distribution of a dinoflagellate bloom off the Peru coast related to time of day, in: *Toxic Dinoflagellate Blooms*, D.L. Taylor and H.H. Seliger (eds.) Elsevier/North-Holland, New York, pp. 209-214, 1979.

Brink, K.H., W.E. Gilbert, and A. Huyer, Temperature sections along the C-line over the shelf off Cabo Nazca, Peru, from moored current meters, 18 March - 10 May, 1977, *CUEA Technical Report 49, Ref. 79-2,* 78 pp., School of Oceanography, Oregon State University, Corvallis, 1979.

Brink, K.H., D. Halpern, and R.L. Smith, Circulation in the Peruvian upwelling system near 15°S, *Journal of Geophysical Research, 85*, 4036-4048, 1980.

Conover, S.A.M., Observations on the structure of red tides in New Haven Harbor, Connecticut, *Journal of Marine Research, 13*, 145-155, 1954.

Dugdale, R.C., J.J. Goering, R.T. Barber, R.L. Smith, and T.T. Packard, Denitrification and hydrogen sulfide in the Peru upwelling region during 1976, *Deep-Sea Research, 24*, 601-608, 1977.

Enfield, D.B., R.L. Smith, and A. Huyer, A compilation of observations from moored current meters, Vol. 12, Wind, currents and temperature over the continental shelf and slope off Peru during JOINT-II, *CUEA Data Report 70, Ref. 78-4*, 347 pp., School of Oceanography, Oregon State University, Corvallis, 1978.

Eppley, R.W., J.N. Rogers, and J.J. McCarthy, Half-saturation constants for uptake of nitrate and ammonium by marine phytoplankton, *Limnology and Oceanography, 14*, 912-920, 1969.

Huntsman, S.A., and R.T. Barber, Primary production off northwest Africa: the relationship to wind and nutrient conditions, *Deep-Sea Research, 24*, 24-33, 1977.

MacIsaac, J.J., Diel cycles of inorganic nitrogen uptake in a natural phytoplankton population dominated by *Gonyaulax polyedra, Limnology and Oceanography, 23*, 1-9, 1978.

MacIsaac, J.J., J.E. Kogelschatz, B.H. Jones, Jr., J.C. Paul, N.F. Breitner, and N. Garfield, JOINT-II R.V. *Alpha Helix* productivity and hydrographic data March-May, 1976, *CUEA Data Report 48, Technical Report 2-79*, 324 pp. Bigelow Laboratory for Ocean Sciences, W. Boothbay Harbor, Maine, 1979.

Margalef, R., Estructura y dinamica de la "purga de mar" en la Ria de Vigo, *Investigacion Pesquera, 5*, 113-134, 1956.

Margalef, R., Life-forms of phytoplankton as alternatives in an unstable environment, *Oceanologica Acta, 1*, 493-509, 1978.

Packard, T.T., R.C. Dugdale, J.J. Goering, and R.T. Barber, Nitrate reductase activity in the subsurface waters of the Peru Current, *Journal of Marine Research, 36*, 59-76, 1978.

Rounsefell, G.A., and W.R. Nelson, Red tide research summarized to 1964 including annotated bibliography, *U.S. Fish and Wildlife Service Special Scientific Report Fisheries, No. 535*, 85 pp., 1966.

Ryther, J.H., Ecology of autotrophic marine dinoflagellates with reference to red water conditions, in: *The Lumincescence of Biological Systems*, F.H. Johnson (ed.), American Association for the Advancement of Science, Washington, D.C., pp. 387-414, 1955.

SCOR-UNESCO, Determination of photosynthetic pigments in seawater. Report of SCOR-UNESCO working group 17, *Monographs on Oceanographic Methodology, 1*, 69 pp., 1966.

Smith, S.L., Nutrient regeneration by zooplankton during a red tide off Peru, with notes on biomass and species composition of zooplankton, *Marine Biology, 49*, 125-132, 1978.

Smith, S.L., H.K. Brink, H. Santander, T.J. Cowles, and A. Huyer, The effect of advection on variations in zooplankton observed at a single location over the Peruvian shelf, Abstracts of the IDOE International Symposium on Coastal Upwelling, University of Southern California, Los Angeles, February 4-8, 1980, American Geophysical Union, Washington, D.C. (Abstract only), 1980.

Smith, W.O., and R.T. Barber, Phytoplankton sinking and vertical migration in the Peru upwelling system at 15°S, in: *This Volume*.

Stuart, D.W., M.A. Spetseris, and M.M. Nanney, Meteorological data - JOINT II: March, April, May 1976, *CUEA Data Report 35, Ref. FSU-CUEA-Met-76-2*, 70 pp., Department of Meteorology, Florida State University, Tallahassee, 1976.

Thomas, W.H., A.N. Dodson, and C.A. Linden, Optimum light and temperature requirements for *Gymnodinium splendens*, a larval fish food organism, *Fishery Bulletin, U.W., 71*, 599-701, 1973.

THE ROLE OF CIRCULATION, SINKING, AND VERTICAL MIGRATION
IN PHYSICAL SORTING OF PHYTOPLANKTON
IN THE UPWELLING CENTER AT 15°S

Richard T. Barber

Duke University Marine Laboratory
Beaufort, North Carolina 28516, U.S.A.

Walker O. Smith, Jr.

University of Tennessee
Knoxville, Tennessee 37916, U.S.A.

Abstract. To test the idea that coastal upwelling circulation can physically sort phytoplankton according to sinking rate or vertical migration speed, we used the observed cross-shelf circulation in a trajectory model to predict the displacement of phytoplankton. Trajectories were calculated using the currents observed on 20 to 22 March, 1977 at the upwelling center at 15°S on the coast of Peru. The trajectories show that during this period particles sinking with velocities characteristic for phytoplankton were carried offshore or to the sediments, while a particle with the migratory behavior of the photosynthetic ciliate *Mesodinium rubrum* was physically maintained over the shelf and slope for the three-day period. The model trajectories account for the aggregations of *M. rubrum*, the relative absence of diatoms, and the low biomass of motile microflagellates; the prevalence of these three conditions throughout March and April, 1977 at 15°S suggests that physical sorting was occurring throughout that period.

--o--

During coastal upwelling nutrient-rich water is advected to the surface and the phytoplankton assemblages that develop are frequently dominated by large forms such as chain-forming diatoms and motile dinoflagellates. Short and efficient food webs result when these large primary producers are grazed directly by phytophagous fishes. Ryther (1969) argued that two variables, primary production and trophic efficiency, act additively to produce the high fish yields that characterize upwelling regions. The hypothesized increase in transfer efficiency in the upwelling food web has been a major explanation of enhanced yields from upwelling, so examination of the processes contributing to this efficiency is relevant for understanding such ecosystems. An explanation of why relatively large phytoplankton usually predominate in upwelling species assemblages would contribute to this understanding because size is important in food web energetics (Banse and Mosher, 1980).

Studies of the effect of size and motility on phytoplankton suspension (Smayda, 1970; Semina, 1972; Bienfang, 1979) have considered the consequences of vertical losses from the euphotic zone; in coastal upwelling, horizontal as well as vertical losses are important. Phytoplankton that rapidly sink out of the euphotic zone to the shelf sediments may be resuspended by energetic upwelling events (Blasco, Estrada, and Jones, 1980), but horizontal displacement in the Ekman flow offshore beyond the shelf break irreversibly eliminates smaller phytoplankton that sink more slowly from the upwelling circulation (Malone, 1975). This occurs because the onshore, subsurface flow that provides source water to the upwelling is confined to a relatively narrow band along the coast (Smith, 1980). In March and April, 1977 at 15°S the zone of onshore flow (positive in the u-direction) was only 20 km wide (Fig. 4 in Brink, Halpern, and Smith, 1980). A phytoplankter that moves downward out of the offshore-flowing Ekman layer before being carried 20 km offshore will be entrained in the onshore flow and recycled through the upwelling circulation; the rate of downward movement is critical, however, because of the narrow zone where the congruent flow originates. The ironic aspect of this conceptual model is that fast sinking (or downward migration) is required for a phytoplankter's physical maintenance in the upwelling circulation. The process would favor large, fast-sinking phytoplankton species and eliminate small forms, thereby providing a simple explanation of the predominance of large phytoplankton in coastal upwelling.

The question that prompted our use of a trajec-

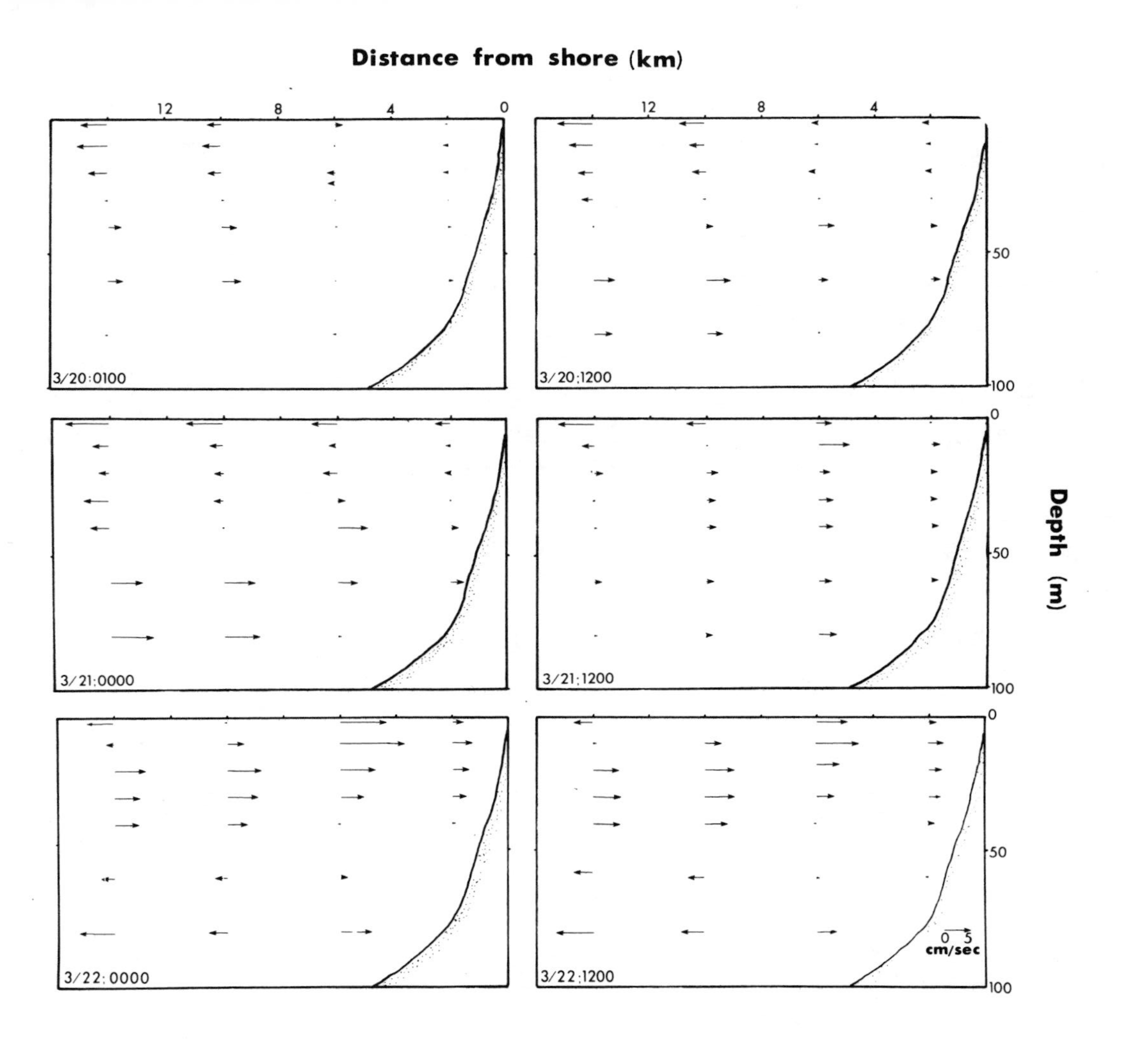

Fig. 1. Onshore-offshore component of current velocity for the period March 20 to 22, 1977. Initially the pattern was a classical upwelling circulation with the surface layers moving offshore and the source water flowing onshore at intermediate depths. A strong reversal occurred at 1200 on March 21 with a change to surface onshore currents.

tory model was not why large phytoplankton predominate in upwelling areas but the desire to explain the opposite observation: Why were characteristic upwelling species and population densities found in previous and subsequent years not present in March and April, 1977 at the 15°S upwelling center? The upwelling center at 15°S has been studied in March and April by expeditions in 1966, 1969, 1976, 1977, and 1978, so there is a great deal of information on the species composition, biomass, and productivity during that season of the year. In 1966 (Ryther *et al.*, 1971), 1969 (Walsh, Kelley, Dugdale, and Frost, 1971), and 1978 (Barber, 1981) high-biomass diatom assemblages were present with chlorophyll *a* concentrations from 2 to 20 mg/m^3; in 1976 a high biomass of *Gymnodinium splendens* was present at 15°S (Huntsman, Brink, Barber, and Blasco, 1980). In March and April, 1977 the phytoplankton conditions in and around the 15°S upwelling center were distinctly different from the other years. Phytoplankton abundance was lower by an order of magnitude, but the specific productivity was high (Barber, 1981). Large diatoms or dinoflagellates were not abundant, aggregations of the photosynthetic ciliate *Mesodinium rubrum* occurred, and motile microflagellates were present as a low biomass background population (Sorokin and Kogelschatz, 1979). In 1977 there were high biomass phytoplankton populations north and south of the 15°S region (Boyd and Smith, 1980), but phytoplankton conditions at the 15°S upwelling center were quite distinct from other years; in this analysis we attempt to explain the cause of the 1977 conditions.

The continuous observations from moored current meters from March and April, 1977, at 15°S on the coast of Peru make it possible to test the idea that physical sorting affects the type of phytoplankton that can accumulate and reach high population densities in upwelling centers. The physical sorting we treat in this report is independent of biological processes that might favor one size class of phytoplankton over another. If there are size-dependent physiological processes, the processes simultaneously, but independently, alter the size composition as the physical sorting occurs. Trajectories were calculated using flow predicted by the O'Brien, Smith, and Heburn (1980) continuity model, which in turn used the observed currents for March 20 to 22. Plate III shows the location of the continuously recording current meter arrays; surface layer flow was measured at two shelf locations (PS and PSS) and intermediate and bottom flow were measured at five stations (PS, PSS, I, M, and A) (Brink *et al.*, 1980). Current velocities were linearly interpolated in the x-z plane at 10 m x 2-km intervals and the upwelling velocities were calculated at these points by assuming three-dimensional continuity. The position of a phytoplankton particle at each hourly time step includes horizontal and vertical displacements by the water and movement from sinking or active migration. Three sinking velocities were selected (0, 0.5, and 1.0 m/h) to span the range observed by Smayda (1970).

During March through May, 1977 there were episodic, but very dense, surface accumulations of *M. rubrum*, an autotrophic ciliate capable of vertical migrations up to 8 m/h (Smith and Barber, 1979). The March 20 to 22 period was chosen for detailed analysis because it was the first period of *M. rubrum* abundance during the 1977 current meter measurements. Since *M. rubrum* makes active diel migrations from 0 to 40 m, we hypothesized that onshore transport at intermediate depth might negate offshore displacement in the surface Ekman layer; the ciliate might therefore be more efficiently retained in an upwelling zone than even a fast-sinking particle such as a large centric diatom. The diel vertical migration of *M. rubrum* used in the calculations was that observed during JOINT-II in March to May, 1977 (Smith and Barber, 1979): from 0000 to 0600 h at 40 m; 0600 to 1100 h migrating up to surface at 8 m/h; 1100 to 1500 h at 0 m; 1500 to 2000 h migrating down to 40 m at 8 m/h; and 2000 to 2400 h at 40 m.

The flow pattern on March 20 was similar to the mean flow of the March through May period (Smith, 1980): surface offshore flow in the upper 25 m and onshore flow at intermediate depths from 40 to 60 m (Fig. 1). On March 21 the surface offshore flow strengthened considerably and extended down to 40 m; a strong reversal occurred during the last hours of 21 March and by 0000 on March 22 flow was onshore throughout the upper 30 m within 8 km of the coast.

In our model, non-sinking particles initially at the surface and 12 km offshore were rapidly moved offshore out of the upwelling circulation (Fig. 2). However, if a non-sinking particle (or motile phytoplankter such as a microflagellate) started 6 km off the coast rather than 12 km, it was retained in the surface waters on the shelf during the entire March 20 to 22 period (Fig. 2). Particles sinking at 0.5 m/h were carried offshore out of the system if they started at the surface 12 km offshore; however, if started at the surface 6 km offshore, their final position was 3.5 km offshore at 56 m (Fig. 2), and they were retained on the shelf below the 22-m thick euphotic layer but in the convergent flow. Particles sinking at 1 m/h, near the maximum sinking speed for living phytoplankton (Smayda, 1970; Bienfang, 1979), were rapidly transported offshore out of the shelf circulation when started at the surface 12 km offshore. When started at 6 km offshore, they were not transported offshore, but were carried to the bottom after 69 h. At 12 km offshore on March 19 the upper water column was mixed to a depth of 16 m (Brink, Halpern, Huyer, and Smith, 1981), and this mixing was used as an additional initial condition (Fig. 2). Particles sinking at 1 m/h starting from the bottom of the mixed layer and 12 km offshore were retained over the shelf at the end of three days at a depth of 60.5 m. A strong upwel-

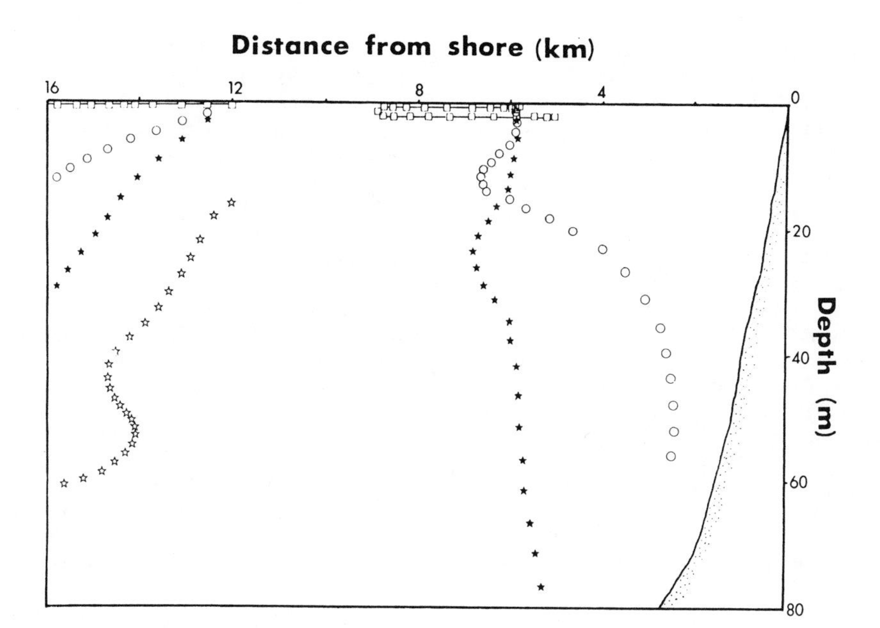

Fig. 2. The trajectories of particles sinking at different rates during March 20 to 22. The symbols are: non-sinking particles □, sinking at 0.5 m/h 0 and at 1.0 m/h * , and a particle sinking at 1 m/h from the bottom of the mixed layer ☆. Each symbol represents a 3-h time increment. The non-sinking particle symbols are slightly displaced vertically for clarity.

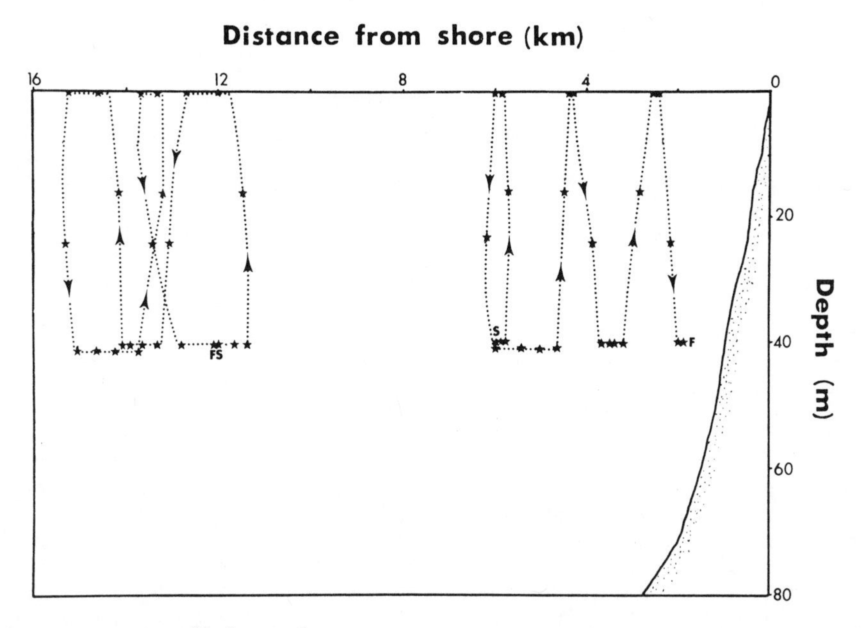

Fig. 3. The trajectory of *Mesodinium rubrum* during March 20 to 22. S = starting point, F = finishing point. *M. rubrum* maintains itself at the shelf-break location with nearly no net movement and is displaced inshore slightly when started at 6 km offshore.

ling event would entrain water from that depth and location into the subsurface onshore flow, so phytoplankton could be advected inshore and upwelled back to the surface layer.

The model suggests that the diel vertical migration pattern of *M. rubrum* was an effective mechanism for maintaining a population in the upwelling water on 20 to 22 March (Fig. 3). Starting 12 km offshore at 40 m at midnight (0000) on March 20 the population was first displaced offshore but later carried onshore. The calculated net cross-shelf movement after 62 h is only 0.06 km. Starting 6 km offshore, *M. rubrum* were transported to a final location 1.9 km offshore, a net inshore movement of 4 km. This analysis of trajectories indicates that during the March 20 to 22 period *M. rubrum* could have been maintained in the upwelling system from a starting position 12 km offshore.

Dense *M. rubrum* aggregations with 200 to 1000 mg/m^3 chlorophyll were observed by ship and aircraft surveys 12 to 20 km offshore on 21 and 22 March and aggregations were present 6 km offshore on March 22 (Smith and Barber, 1979). This is in substantial agreement with the outcome of the model. The spatial retention by circulation and migration, together with the high specific growth rates observed (16.8 mg C (mg chl $a)^{-1}h^{-1}$; Smith and Barber, 1979), could account for the sudden appearance of dense aggregations. Similarly, the observed fluctuations in flow could account for the sudden disappearance of the aggregations.

M. rubrum is ubiquitous in coastal and offshore waters and in upwelling regions dense blooms may occur over very large areas (Ryther, 1967; Barber, White, and Siegelman, 1969; Packard, Blasco, and Barber, 1978). It has not been clear how the blooms or aggregations suddenly appeared and disappeared. The model shows that in the 15°S upwelling center the diel vertical migration of *M. rubrum* can effectively negate strong offshore displacement in the surface Ekman layer and maintain a "standing wave" of high population density.

Margalef (1978) argued that the kinetic energy of circulation in upwelling waters contributes to the quantity of energy captured in primary production because the wind-driven circulation provides a kinetic-energy subsidy to non-motile phytoplankton that is as important as the radiant energy provided by the sun. Our results support this idea but with some caveats. One caveat is that during the period we studied the subsidy was exploited by motile phytoplankton such as the autotrophic ciliate and microflagellates. The unusually deep (40 m) offshore-flowing surface layer on 21 March did not provide a subsidy that would help non-motile diatoms stay in the upwelling system, but the active ciliate could exploit it, and to a lesser degree motile microflagellates could also. During 20 to 22 March, 1977 the kinetic energy of circulation was apparently too strong to be exploited by diatoms. The ability of motile phytoplankton to maintain themselves in the upwelling circulation appears similar to the phenomenon reported by Peterson, Miller, and Hutchinson (1979), in which certain species of copepods are maintained in the food-rich but narrow inshore band of water by the flow pattern. The adaptations of motile phytoplankton are much less precise than those reported for copepods, but even partial recycling together with high specific growth rates can explain the "red-water" aggregations of the ciliate that appear and disappear so rapidly in upwelling regions (Ryther, 1967; Packard *et al.*, 1978).

The model also helps to explain how strong cross-shelf zonation of organisms is maintained in the strong surface flow. Particles with either diatom or microflagellate sinking characteristics were advected offshore out of the upwelling circulation region when started 12 km from shore, but when started 6 km from shore the non-sinking particle was maintained in the euphotic layer inshore over the shelf and the sinking particles moved to the sediments or into the convergent onshore flow at 40 to 60 m. Phytoplankton starting from a position 6 km from shore had the potential of being re-seeded into the upwelling center but those starting 12 km offshore did not; an inshore "cell" with distinct biological characteristics could be generated and maintained with these trajectories.

The model when operated with the currents observed on 20 to 22 March, 1977 did not sort for large, fast-sinking forms and, indeed, these forms were rare during that period. This model result and field observation suggest that the circulation observed in 1977 was not typical of the other years when the 15°S site was studied (Ryther *et al.*, 1971; Walsh *et al.*, 1971; Huntsman *et al.*, 1980; Barber, 1981). The model trajectories account for the aggregations of *M. rubrum*, the persistence of low biomass populations of microflagellates, and the low density of diatoms. The prevalance of these three conditions throughout March and April, 1977 at 15°S suggests that physical sorting similar to that in the model occurred throughout that period.

The two-layer flow regime of coastal upwelling can potentially recycle phytoplankton through the upwelling centers several times, helping to produce the high accumulations characteristic of coastal upwelling. The trajectory model supports the idea that this occurs and, further, the model results suggest that temporal variations in the circulation enable phytoplankton with different sinking and migration characteristics to predominate at different times in the coastal upwelling system of Peru. Variations in physical sorting arising from variations in the circulation appear to account for some of the biological variability that is observed in upwelling. Circulation, sinking, and vertical migration are certainly not the only determinants of phytoplankton abundance and species composition in upwelling, but the trajectory calculations suggest that physical sorting of phytoplankton plays a more important role in regulating species composition and abundance than had previously been appreciated.

Acknowledgements. This report is a contribution of the Coastal Upwelling Ecosystems Analysis (CUEA) program, NSF Grant Number OCE75-23722. George Heburn of Florida State University carried out the hourly interpolations from the 6-h current data. David Halpern (Pacific Marine Environment Laboratory, Seattle) and Robert L. Smith (Oregon State University) made the current meter observations that were the data for the continuity circulation model of James O'Brien of Florida State University.

References

Banse, K. and S. Mosher, Adult body mass and annual production/biomass relationships of field populations, *Ecological Monographs, 50*, 355-379, 1980.

Barber, R.T., Variations in biological productivity along the coast of Peru: The 1976 dinoflagellate bloom, *Science*, in press, 1981.

Barber, R.T., A.W. White, and H.W. Siegelman, Evidence for a cryptomonad symbiont in the ciliate, *Cyclotrichium meunieri, Journal of Phycology, 5*, 86-88, 1969.

Bienfang, P.K., A new phytoplankton sinking rate method suitable for field use, *Deep-Sea Research, 26*, 719-729, 1979.

Blasco, D., M. Estrada, and B. Jones, Phytoplankton distribution and composition in the northwest Africa upwelling region near Cabo Corbeiro, *Deep-Sea Research, 27*, 799-821, 1980.

Boyd, C.M., and S.L. Smith, Interactions of plankton populations in the surface waters of the Peruvian upwelling region: a study in mapping, IDOE International Symposium on Coastal Upwelling, Los Angeles, California, Feb. 4-8, 1980, Abstract.

Brink, K.H., D. Halpern, and R.L. Smith, Circulation in the Peruvian upwelling system near 15°S, *Journal of Geophysical Research, 85*, 4036-4048, 1980.

Brink, K.H., D. Halpern, A. Huyer, and R.L. Smith, Near-shore circulation near 15°S: the physical environment of the Peruvian upwelling system, in: *Productivity of Upwelling Ecosystems*, R.T. Barber and M.E. Vinogradov (eds.), Elsevier, Amsterdam, in press, 1981.

Huntsman, S.A., K.H. Brink, R.T. Barber, and D. Blasco, The role of circulation and stability in controlling the relative abundance of dinoflagellates and diatoms over the Peru shelf, *This Volume*, 1980.

Malone, T.C., Environmental control of phytoplankton cell size, *Limnology and Oceanography, 20*, 490, 1975.

Margalef, R., Life-forms of phytoplankton as survival alternatives in an unstable environment, *Oceanologica Acta, 1*, 493-509, 1978.

O'Brien, J.J., R.L. Smith, and G.W. Heburn, Estimation of vertical velocities over the continental shelf at 15°S on the coast of Peru, Motion picture and manuscript available from Mesoscale Air-Sea Interaction Group, Florida State University, Tallahassee, Florida, 32306, 1980.

Packard, T.T., D. Blasco, and R.T. Barber, *Mesodinium rubrum* in the Baja California upwelling system, in: *Upwelling Ecosystems*, R. Boje and M. Tomczak (eds.), Springer-Verlag, New York, pp. 73-89, 1978.

Peterson, W.T., C.B. Miller, and A. Hutchinson, Zonation and maintenance of copepod populations in the Oregon upwelling zone, *Deep-Sea Research, 26A*, 467-494, 1979.

Ryther, J.H., Occurrence of red water off Peru, *Nature, 214*, 1318-1319, 1967.

Ryther, J.H., Photosynthesis and fish production in the sea, *Science 166*, 71-76, 1969.

Ryther, J.H., D.W. Menzel, E.M. Hulburt, C.J. Lorenzen, and N. Corwin, The production and utilization of organic matter in the Peru coastal current, *Investigacion Pesquera, 35*, 43-59, 1971.

Semina, H.J., The size of phytoplankton cells in the Pacific Ocean, *Internationale Revue der gesamten Hydrobiologie u Hydrographie, 57*, 177-205, 1972.

Smayda, T.J., The suspension and sinking of phytoplankton in the sea, *Annual Review of Oceanography and Marine Biology, 8*, 353-414, 1970.

Smith, R.L., A comparison of the structure and variability of the flow field in three coastal upwelling regions: Oregon, northwest Africa, and Peru, *This Volume*, 1980.

Smith, W.O., Jr., and R.T. Barber, A carbon budget for the autotrophic ciliate, *Mesodinium rubrum, Journal of Phycology, 15*, 27-33, 1979.

Sorokin, Y.I., and J.E. Kogelschatz, Analysis of heterotrophic microplankton in an upwelling area, *Hydrobiologia, 66*, 195-208, 1979.

Walsh, J.J., J.C. Kelley, R.C. Dugdale, and B.W. Frost, Gross features of the Peruvian upwelling system with special reference to possible diel variation, *Investigacion Pesquera, 35*, 25-42, 1971.

DIATOMS IN SURFACE SEDIMENTS: A REFLECTION OF COASTAL UPWELLING

Gretchen Schuette

and

Hans Schrader

School of Oceanography, Oregon State University, Corvallis, Oregon 97331

Abstract. Recent sediments influenced by overlying strong coastal upwelling off Peru, off South West Africa, and within the central Gulf of California can be differentiated from sediments of adjacent oceanic regimes by diatom floral analysis. Abundant remains of well-preserved diatoms are deposited in each area beneath zones of recurrent high primary productivity. The distribution patterns of *Chaetoceros* spores and valves of other meroplanktic diatom species preserve in the sediments off Peru the sinuous seaward boundary of upwelled coastal waters and may reflect the evolution of the biota influenced by episodic nutrient enrichment. Off South West Africa the initial response to nutrient enrichment is preserved in several nearshore areas by a high abundance in the sediments of valves of *Delphineis karstenii*. An association of large centric diatoms occurs in samples from these areas, and *Chaetoceros* spores, representing a final stage of species succession, occurs in highest abundance on the outskirts of the sediment patches. Laminated sediments of the Gulf of California provide deposits representing single upwelling seasons. Core material available from this area will allow fine resolution documentation of the influence of coastal upwelling on relic diatom abundance, preservation, and species composition.

Introduction

Surface sediment samples from off Peru and off South West Africa and subsurface sediment samples from cores collected in the central Gulf of California were analyzed for their diatom content and diatom species composition. The purpose of the study of three coastal upwelling areas was to identify the possible influence of coastal upwelling on the abundance of opaline diatom valves in underlying sediments and on the species composition of relic diatom assemblages.

The uppermost sediments retrieved at 116 coring stations off Peru were analyzed. Samples may represent from about 7 to 300 years of deposition (see Schuette and Schrader, 1979, for site locations and sampling details). Surface sediment samples from off South West Africa consisted of two sets: 1) 82 diatom-rich inner shelf samples, and 2) 39 offshore southeast Atlantic samples, 16 of which had sufficient preservation of opal phytoplankton to allow counts of species abundance. Samples may represent about 10 years of deposition (see Schuette and Schrader, in press, for site locations and sampling details). Core material from the Gulf of California was available from two sources: 1) DSDP Leg 64 HPC Site 480 (Curray *et al*., 1979), and 2) cores retrieved on the 1979 cruise of the H-1 *Mariano Matamoros* (Schrader, 1979). Samples may represent 0.5 year of deposition.

Slide preparation followed the procedure outlined in Schrader (1974) and Schrader and Gersonde (1978). More than 300 valves were counted per slide.

Abundance of Valves in Sediments

Nutrient enrichment of surface waters off Peru and off South West Africa due to the coastal upwelling phenomenon results in both regions in the intense production of diatoms throughout much of the year (Hart and Currie, 1960; Rojas de Mendiola, 1980 and pers. comm.). In the central Gulf of California, upwelling and non-upwelling seasons alternate in response to changing wind direction. In all three areas, the recurring or quasi-constant high production of diatoms results in a high supply of diatom valves to the sea floor. In coastal areas lacking strong upwelling, diatoms are usually completely dissolved in the water column or at the sediment water interface. A high supply of valves, perhaps supplemented by preservational transfer of valves to deeper sites by fecal pellets, and high pelagic sedimentation rates in fertile areas favor the deposition of sediments rich in well preserved diatom valves (Diester-Haass and Schrader, 1979; Berger, 1976).

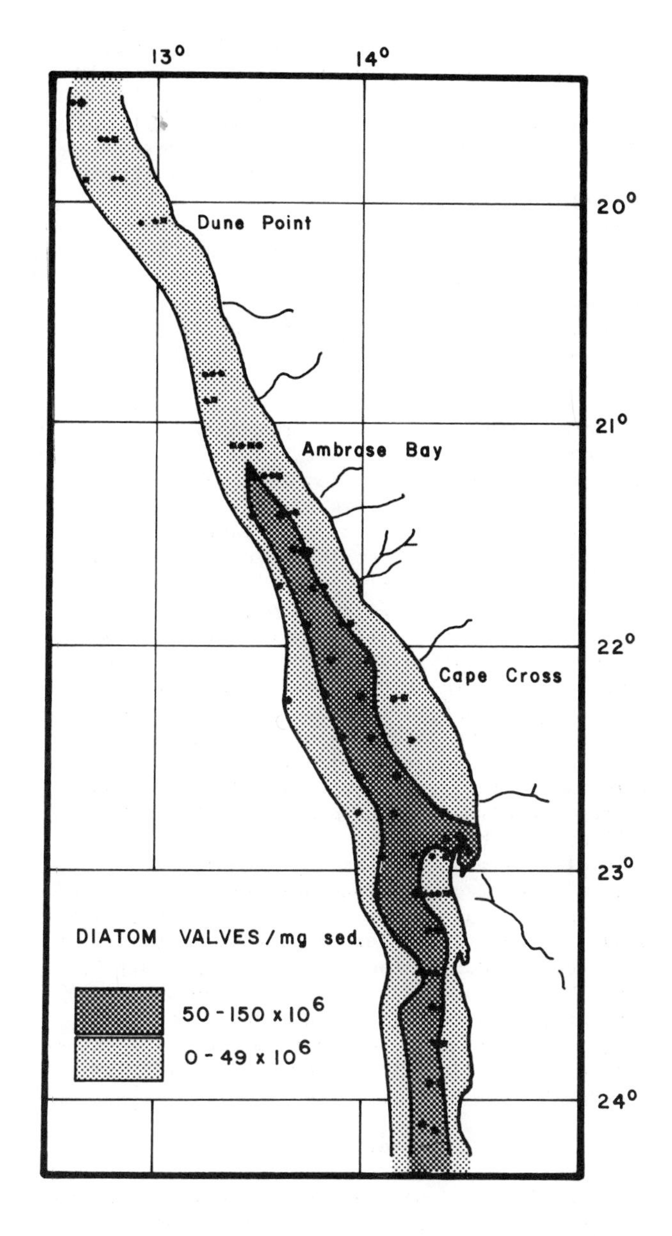

Fig. 1. Numbers of diatom valves per mg of sediment off South West Africa (after Richert, 1976).

Off South West Africa the horizontal distributions of temperature and salinity, especially in winter and spring, show strong gradients and the presence of a conspicuous belt of cold upwelled water adjacent to the coast from about 23°S to 26°S (Stander, 1964). Primary productivity values reach 1600 mgC $m^{-2}day^{-1}$ (Mukhina, 1974). Phytoplankton surveys have characterized the outstanding biological feature of the South West African region as a contrast in densities between "rich" coastal and "poor" oceanic populations of phytoplankton, with densities differing by 3 to 4 orders of magnitude (Hart and Currie, 1960). In their analysis of a characteristically active period of upwelling, Hart and Currie (1960) reported an abrupt drop in numbers of phytoplankton before the shelf edge (about 200-m water depth). In recent survey by the Sea Fisheries Branch of South Africa the richest phytoplankton concentrations were again found in a "so-called belt (although patchy) parallel to the coast" (Cram, 1978).

Sediments rich in diatom remains constitute a record of the nearshore zone of recurrent high primary productivity. A belt of diatomaceous mud extends principally from 19 to about 25°S (Marchand, 1928; Hart and Currie, 1960: Calvert and Price, 1970, 1971; Mukhina, 1974; Bremner, 1980). The mean depth of the seaward edge of the diatomaceous sediments is 130 m; the mean depth of the landward edge is 31 m (Bremner, 1980). Limiting depths are established where frustules are hydrodynamically stable on the landward margin and where they undergo nearly complete dissolution on the seaward margin (Bremner, 1980). Abundance values (numbers of diatom valves/mg sediment) calculated for our samples (Richert, 1976) show three main distribution zones (Fig. 1): 1) the very nearshore (water depths below 80 m) with less than 50 x 10^6, 2) a zone of highest abundance (water depths from 90 to 130 m) with 50 to 150 x 10^6, and 3) an "offshore" zone (water depths about 140 m) with less than 50 x 10^6 valves/mg sediment. This sediment surface distribution with highest values occurring in samples from Walvis Bay and in a band north to about 21°30'S and south past 24°S is well correlated with the reports of high phytoplankton concentrations in surface waters.

Off Peru there is also a sharp contrast between the "coastal" and "oceanic" settings. The coastal upwelling accounts for the recurrent development of high densities of phytoplankton and high primary productivity values nearshore (Guillen and Calienes, 1976; Guillen, Rojas de Mendiola, and Izaguirre de Rondan, 1971, 1973; Zuta, Rivera, and Bustamante, 1978). Correspondingly the sediments reflect the seaward limit of coastal upwelling influence with high abundance of diatom remains in sediments underlying the nearshore region of recurrent high primary productivity (Schuette and Schrader, 1979).

The results of hydrological and biological surveys off Peru during the last two decades give evidence for upwelling centers, sites of recurrent high diatom concentrations or productivity or both (Rojas de Mendiola, 1980; Zuta *et al.*, 1978). In the sediments influenced by this pattern (Fig. 2), there are zones or centers of highest abundance of diatom valves. Off Peru the highest values (>50 x 10^6 diatom valves/g) occur in more or less discrete regions laterally offset from the coast (in addition to an area adjacent to the shore at 12°S). The offshore centers of upwelling influence (water depths 2000 to 3600 m) occur at about 8°S, 13 to 14°S, and 17°S and bear a relationship both to the areas of major concen-

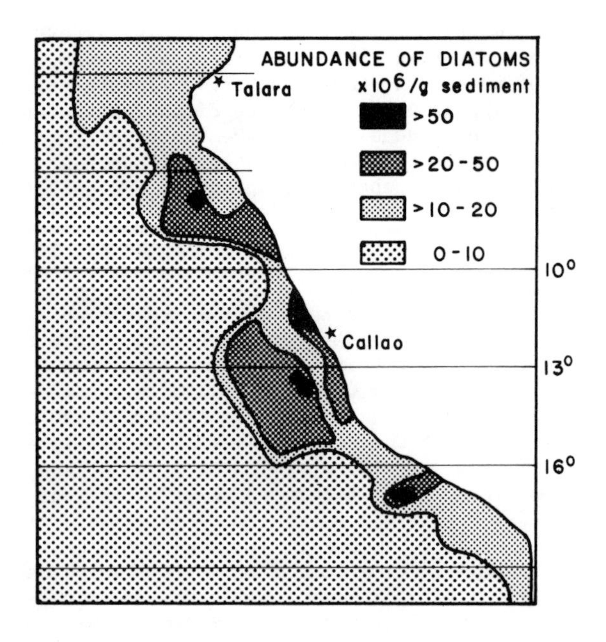

Fig. 2. Numbers of diatom valves per gram of dry sediment off Peru.

trations of phytoplankton where upwelling takes place within 60 miles of the coast (Rojas de Mendiola, 1980) and to observations of tongues of upwelled water extending 70 to 130 miles offshore (Zuta *et al.*, 1978).

Differences between the Peruvian and South West African regions in the offshore boundaries of their sediment records correspond with reported surface water limits of cold, nutrient-rich waters in the respective regions. Both regions are characterized by an eddy-like structure of surface isotherms, but a striking feature of the Peru Current region that distinguishes it from the Benguela Current region is the large apparent extent of the eddies (Hart and Currie, 1960). Fig. 2 suggests that the diatom valve concentrations in the sediments off Peru preserve a reflection of the sinuous seaward boundary of upwelled waters. The initial patch or tongue-like structure of upwelled waters may be preserved in the distribution patterns of various diatom species in sediments off South West Africa (see discussion of distribution patterns below). But the limited offshore extension of the diatom-rich surface waters may restrict sediments rich in accumulated diatom remains into a more band-like pattern off South West Africa.

The small absolute (and relative to the mean) variance in the wind off Peru, where local winds are weak and of short duration (Smith, 1978) contrasts with the wind regime over the inner shelf waters off South West Africa. Nearshore waters are subject here to the effects of coastal winds superimposed upon the processes induced by the trade winds (Hart and Currie, 1960; Stander, 1963). The local winds, because of their variability, may either augment or suppress upwelling (Stander, 1963). The resulting intermittent process may contribute to the high primary productivity of the region by frequent re-initiation of bloom conditions, but it may also contribute to the relatively small seaward extent of nutrient- and diatom-rich waters and, thus, of diatom-rich sediments.

The narrow shelf and steep slope off Peru also contrast with the broad shelf off South West Africa. The higher concentrations of diatoms per gram of sediment off South West Africa may be related to the contrast. Doubtless the shallow depositional environment off South West Africa enhances diatom preservation and thus the abundance of valves in sediments.

Cores collected near Guaymas, eastern-central Gulf of California, contain laminated sediments, some of which are nearly pure diatom oozes (Schrader *et al.*, 1980; Schrader, 1979). Preliminary examination of box core material suggests that the areas of intense seasonal upwelling and phytoplankton concentration have an associated record in the underlying sediments. In addition, the abundance of diatoms may vary from 70 to 80% in one lamina to 15 to 45% in an adjacent lamina. Northwesterly winds trigger coastal upwelling on the eastern side of the Gulf during the dry season (January to May). A pale olive lamina with an excellently preserved diatom assemblage is presumably produced on the Guaymas slope as a result of this seasonal high productivity. Upwelling on this side of the basin ceases during the rainy season (July to September) when winds are from the southeast. Terrigenous material is then washed into the area and contributes to the formation of a darker lamina with moderately preserved diatom assemblages (Schrader *et al.*, 1980).

Preservation

Dissolution of opal may selectively destroy the valves of some diatom species more readily than others and thus increase the relative abundance in sediments of solution-resistant forms (Schrader, 1972). Well preserved diatom frustules dominate the siliceous fraction of nearshore samples off South West Africa. Preservation of those species that dominate the surface waters indicates relatively little transformation of biocoenoses to taphocoenoses (sediment assemblages) in these shallow waters. Dissolution of less strongly silicified forms undoubtedly is responsible for differences between the composition of bottom deposits and the average composition of the plankton (Hart and Currie, 1960), but enrichment in large species in some samples may also reflect surface water proliferation of these forms (see discussion below on characteristic species).

Further offshore, in the low productivity, non-upwelling, oceanic regime, low supply results in many samples barren of opal phytoplankton. All samples are strongly enriched in robust forms that may be rare in surface waters (e.g., *Coscinodiscus nodulifer*).

Off Peru, diverse assemblages are preserved at thousands of meters of water depth. There is however an apparent enrichment of some species in the sediment assemblages. Heavy silicification of the valves of the diatoms accounts for the presence and abundance in sediments of these robust species. The presence even at deep sites of *Skeletonema costatum*, a dissolution-sensitive species, is however testimony to the abundant supply of diatom valves to the sediments in certain locations. Although a common pelagic diatom frequently associated with a coastal flora (Hendey, 1964) and previously reported as a dominant species off Peru (Calienes and Guillen, 1976; Rojas de Mendiola, 1980), *Skeletonema costatum* is rare in our sediment samples. It does occur however in several samples from sites at 3000-m water depth.

In samples from the Gulf of California the preservation of weakly silicified diatoms and even of the delicate bristles of *Chaetoceros radicans* is extraordinary and must bear a relationship to extremely high primary productivity (and to rapid and preservational burial) at these sites in the central Gulf.

Characteristic Species and Species Groups

Species with high relative abundance in our sediment samples were considered characteristic if their distribution in the sediments showed a relationship either to the dynamics of coastal upwelling (insofar as we understand them) or to theories and observations of the biological response to nutrient enrichment. By this criterion, *Chaetoceros* spp. and a set of meroplanktic species characterize coastal upwelling-influenced sediments. In addition, one species (*Delphineis karstenii*) was considered especially diagnostic because of its association with highly productive coastal waters of the present and with deposits representing highly productive ancient environments.

Chaetoceros

Off South West Africa, Hart and Currie (1960) noted the outstanding importance of the *Chaetoceros* group in the rich coastal waters during both seasons studied (autumn and spring of 1950). In spring, the season of active upwelling, the rich coastal flora consisted mainly of *Chaetoceros* spp. associated with *Asterionella* and *Delphineis karstenii* (synonym: *Fragilaria karstenii*, see Fryxell and Miller, 1978). In Kollmer's study (1962, 1963) of data collected in and near Walvis Bay from 1958 to 1960, the nearshore community was characterized mainly by *Chaetoceros* spp., *Delphineis karstenii*, and *Thalassiosira* spp. In the recent surveys by the Sea Fisheries Branch of South Africa, *Chaetoceros* spp. constituted, together with *Delphineis karstenii*, 96 and 87% of the total cell counts during winter and spring months of 1971 and 1972, respectively (Rep. Div. Sea Fish. S. Af. 40, 1972). Again the following year, important diatoms were *Delphineis karstenii* and species of the genus *Chaetoceros* (Rep. Div. Sea Fish. S. Af. 41, 1973). Apparently the richest concentrations of phytoplankton off South West Africa consist mainly of *Delphineis karstenii* and *Chaetoceros* spp. (Cram, 1978). *Chaetoceros* spores occur in all of our sediment samples from the inner shelf off South West Africa and their relative abundance ranges from 18 to 80%. The group has the highest relative abundance of the taxa counted in 72 of the 82 samples from the inner shelf.

Chaetoceros species have been recorded as dominant in phytoplankton off Peru from 4 to 18°30'S (Guillen *et al.*, 1971, 1973). Representatives of the genus are among the dominant species present all year (Guillen *et al.*, 1971, 1973; Rojas de Mendiola, 1980). *Chaetoceros* resting spores contribute abundantly to the surface sediments off Peru (Zhuze, 1972; Schuette and Schrader, 1979), as they contributed in the past to the sedimentation process forming the diatomites of the Pisco formation (Mertz, 1966). In 40 of our 49 samples from the continental margin off Peru *Chaetoceros* resting spores were >30% of the diatom assemblage.

Chaetoceros resting spores may be produced during a *Chaetoceros*-dominated stage of species succession when nutrients are nearly exhausted in the euphotic zone (Guillard and Kilham, 1977; Margalef, Estrada, and Blasco, 1979). Thus, the offshore distribution pattern for these species in the sediments off Peru was interpreted (Schuette and Schrader, 1979) as reflecting the seaward edge of nutrient-replete surface waters. Off South West Africa as well, the *Chaetoceros* pattern of relative abundance may be related to species succession (see discussion of distribution patterns below).

Delphineis

The importance of *Delphineis karstenii* in the phytoplankton off South West Africa was suggested above. The species is characterized as a coastal "pioneer" (Rep. Div. Sea Fish. S. Af. 39, 1971), i.e., a characteristic member of the first seral stage of an ecological succession (presumably initiated by nutrient enrichment of surface waters). Hart and Currie (1960) considered *Delphineis karstenii* as the most strictly coastal species of diatoms encountered in the plankton off South West Africa. Its occurrence in sediments off this coast seems to be related to its restricted occurrence within 40 miles from shore (Rep. Div. Sea Fish. S. Af. 38, 1970)(see discussion below). *Delphineis karstenii* occurs in all but one of our surface sediment samples and ranges from <1 to 49% relative abundance.

Off Peru *Delphineis karstenii* is rare in our samples, but the occurrence is still noteworthy because the species may be a specific indicator of productive coastal waters. The genus was defined by Andrews (1977) and the species he assigned to the genus seem to have fluorished in Miocene shallow shelf environments of the eastern U.S. *Delphineis karstenii* (Fryxell and Miller, 1978) is the only recognized living taxon in the genus so far.

Meroplanktic species

Meroplanktic species are the final group of species we consider to be characteristic of coastal upwelling. Meroplanktic diatoms are species that either produce a resting spore or possess a sedentary stage or dormant phase in their life cycle (Smayda, 1958) and prefer turbulent nearshore waters.

Off South West Africa *Actinocyclus octonarius*, a meroplanktic species, contributes importantly to sediments. Its distribution in the sediments is correlated with that of *Coscinodiscus perforatus* (correlation, 0.87), with *Coscinodiscus asteromphalus* (correlation, 0.69), and *Cosdinodiscus gigas* (correlation, 0.62). *C. asteromphalus* and *C. gigas* are among the largest species present in the sediments off South West Africa.

Mukhina (1974) records the occurrence of *A. octonarius* in suspension samples, but attributes the accumulation of the species in sediments to the coarse silicification of its valves. Hart and Currie (1960) also suggest that the greater proportion of strongly silicified centrics found in sediment samples compared to plankton samples was due to rapid solution of less silicified forms. However, a set of large-celled *Coscinodiscus* species were recorded by recent Sea Fisheries Branch surveys as repeatedly forming continuous belts and patches between 20 and 24°S (I. Kruger, personal communication). The data at least present the possibility that the correlated distributions in the sediments of several large centrics may represent a reflection of surface water assemblages.

The meroplanktic species group is represented in sediments off Peru by *Cyclotella striata/stylorum*, *Actinocyclus octonarius*, and *Actinoptychus senarius*. None of the species is known to dominate living phytoplankton assemblages off Peru. *Coscinodiscus perforatus* occurs also in sediments off Peru (Zhuze, 1972; Schuette and Schrader, 1979), and in our study from that area its distribution is correlated with the abundant meroplanktic species *Cyclotella striata/stylorum*.

Evidence from the Gulf of California suggests that a sediment flora including a significant proportion of large-diameter diatoms may be determined by factors in addition to selective dissolution. Round (1967) proposes that active mixing may transport large forms off the sediments and maintain them in suspension. He cites the frequent records from neritic regions of phytoplankton dominated by large centric diatom species. Round (1967) attributed the abundance of large centrics in sediments of the Gulf of California to the recurrence of blooms of the species in surface waters. *Coscinodiscus asteromphalus*, e.g., forms immense blooms in the gulf and contributes significantly to underlying sediments (Round, 1967, 1968). In some of our samples of individual laminae from cores collected off Guaymas *C. asteromphalus* may form almost monospecific assemblages (Schrader *et al.*, 1979).

Distinct Distribution Patterns of Coastal Upwelling Seabed Diatom Assemblages

The coherent distribution of diatom sediment assemblages distinct from the adjacent "oceanic" regime or that have features that can be interpreted as reflecting the biology or hydrology of the upwelling area provides evidence for the spatial and, in the Gulf of California, temporal limits of the strong influence of coastal upwelling.

Off South West Africa, Hart and Currie (1960) distinguished between a coastal and an oceanic flora. Kollmer (1963) used the terms "neritic" and "oceanic" to indicate the main areas of distribution for two associations distinguished by his monthly records of phytoplankton species in the area off Walvis Bay during 1959 and 1960. Of the species that form his "neritic" association, those preserved in the inner shelf sediment assemblages include *Chaetoceros* species, *Delphineis karstenii*, *Thalassiosira* spp., and *Thalassionema* spp. Kollmer's "oceanic" association is comprised mainly of species not preserved in abundance in the offshore southeast Atlantic samples. In these samples, very different assemblages are formed, mainly of species with robust valves, such as *Coscinodiscus nodulifer*, *Pseudoeunotia doliolus*, *Thalassiosira eccentrica*, *Thalassionema nitzschioides*, and *Roperia tesselata*.

There is a negative correlation between the *Chaetoceros* spp. distribution and that of *Delphineis karstenii* (-0.66) in the sediments off South West Africa. Our samples provide a record of overlapping but distinct distribution patterns for these two dominant species. Principal components analysis of the relative abundance data at the 82 inner shelf sites (for details of the statistical approach see Schuette and Schrader, in press) and rotation of the resulting components provides a factor (accounting for about 11% of the variance in the data) whose pattern illustrates the nearshore distribution of *Delphineis karstenii* (Fig. 4). Positive loadings are associated with *Delphineis karstenii* and with *Paralia sulcata*; negative loadings with the

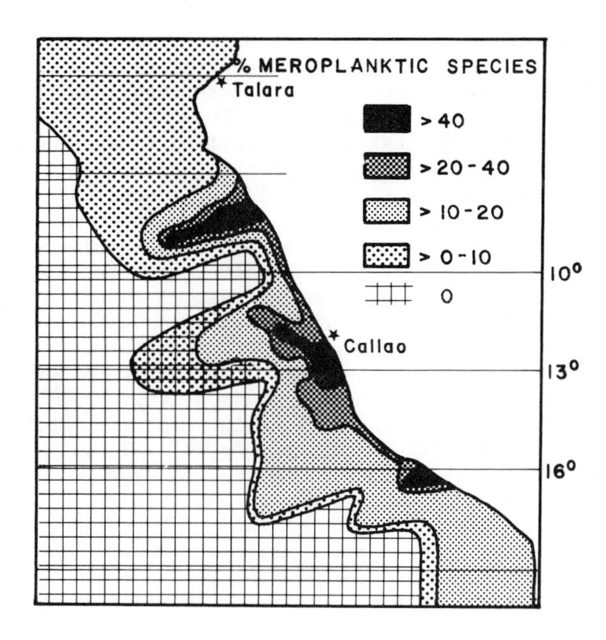

Fig. 3. Relative abundance of the meroplanktic species group in sediment samples off Peru.

Chaetoceros group. Another factor accounting for about 22% of the variance in the data shows a distributional association for the set of large centric diatoms (*Actinocyclus octonarius, Coscinodiscus perforatus, C. gigas, C. asteromphalus*) (Fig. 5). The joint occurrence of these two factors in several nearshore areas, especially those off Ambrose Bay, Cape Cross, and Walvis Bay, suggests the recurrence at these sites of a sequence of dominating diatom assemblages. As productive waters spread from "point" sources at the coast, the biocoenoses to which the sediment assemblages correspond may include the following elements: initial blooms of the pioneer *Delphineis karstenii*, proliferation of large centric forms, and finally widespread formation of *Chaetoceros* resting spores. In other words, there is a recurring biological response to intermittent coastal upwelling and this response is recorded in distinct distribution patterns of diatom taphocoenoses.

Two main complexes of diatoms are evident in the phytoplankton off Peru, one associated with cold coastal waters and one with offshore oceanic waters (Sukanova, Konovalova, and Rat'kova, 1978). Sediment assemblages allow for a distinction between oceanic and coastal taphocoenoses (Schuette and Schrader, 1979). Almost all stations on the continental margin off Peru have greater than 25% relative abundance of *Chaetoceros* resting spores, whereas sediment assemblages offshore are dominated by *Coscinodiscus nodulifer, Nitzschia marina, Pseudoeunotia doliolus,* and species of *Asteromphalus*. (These oceanic assemblages may "intrude" nearshore where *Chaetoceros* spores are in less abundance.)

A high abundance of meroplanktic species also distinguishes the productive coastal region off Peru from incursions of "oceanic" assemblages and the distribution of the abundant meroplanktic species reflects the sinuous seaward boundary of the upwelled coastal waters. The highest relative abundance of the component is in relatively

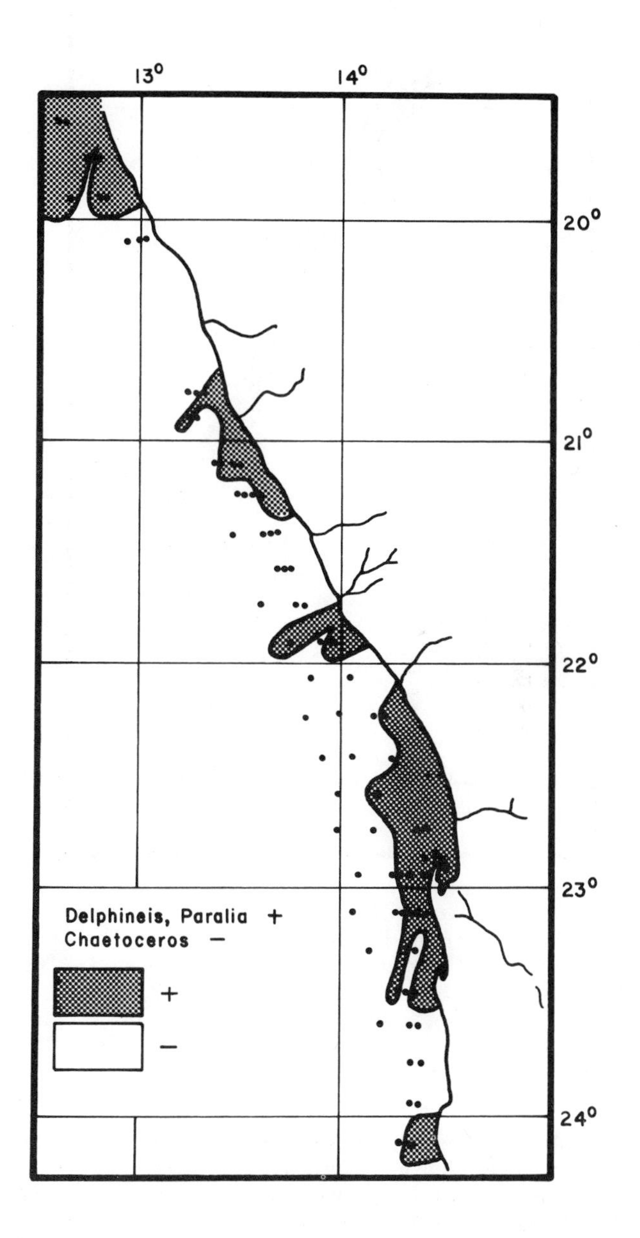

Fig. 4. Distribution off South West Africa of a factor associated with the *Chaetoceros* resting spores (negative loadings), the *Delphineis* complex, and *Paralia sulcata* (positive loadings).

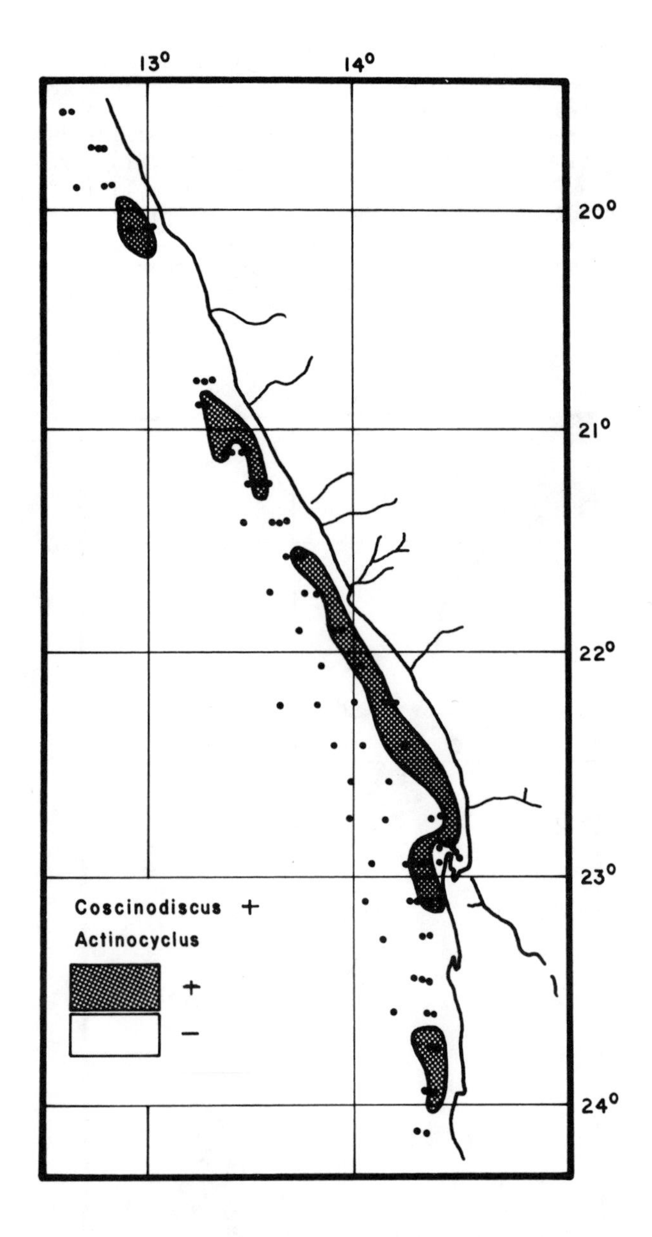

Fig. 5. Distribution off South West Africa of a factor correlated with several large centric diatoms: *Actinocyclus octonarius*, *Coscinodiscus perforatus*, *C. gigas*, and *C. asteromphalus*.

nearshore waters (between ∿25-75 miles from shore and in water depths less than 2500 m) (Fig. 3). *Chaetoceros* resting spores on the other hand have their peak abundance further offshore and thus at greater water depths. Off Peru the last two stages of a "seral" response to nutrient enrichment of surface waters may be recorded in the meroplanktic and *Chaetoceros* group distributions in the sediments. The initial bloom portion of the record perhaps was not sampled, is not preserved in deep waters, or does not deliver a large signal to the sediments under the more constant upwelling conditions off Peru.

Rhythmites indicative of upwelling in the central Gulf of California contain abundant vegetative cell remains of such delicate species as *Synedra indica*, *Skeletonema costatum*, and *Chaetoceros radicans*, among others; whereas adjacent laminae, presumably formed under less nutrient-rich "oceanic" conditions, contain in abundance the robust frustules *Coscinodiscus nodulifer*, *Pseudoeunotia doliolus*, and *Coscinodiscus asteromphalus*, among others. Complete documentation of diatom species composition in a long series of laminae containing alternating upwelling and non-upwelling diatom assemblages will allow further characterization of the influence of coastal upwelling on diatom sediment composition.

In summary, comparison of the South West Africa data with the analogous data from the Peru upwelling area suggests similar features in both regions. Evidence for centers of coastal upwelling off Peru is preserved in areas of high abundance of diatom valves per gram of sediment, in the distribution of large meroplanktic species, and -- still abundant further offshore -- in the distribution of *Chaetoceros* spores. Strong, quasi-constant winds may be implicated in the offshore distribution of diatom taphocoenoses off Peru. Intermittent winds and a shallow depositional environment off South West Africa confine the sediment record of coastal upwelling near shore and impose a bandlike structure on the pattern of abundance of diatom valves in sediment samples. Here, the pioneer flora -- dominated by *Delphineis karstenii* -- is preserved in abundance in discrete nearshore areas. The distribution reflects the patch or bloom structure of the initial response to nutrient enrichment. An association of large centrics occurs at these sites, and *Chaetoceros* spores, representing a final stage of species succession, occur in highest abundance on the outskirts of the sediment patches. The opportunity residing in the laminated sediments of the Gulf of California is to document the variability in the seasonal, interannual, and longer term record of the patterns of abundance and distribution that characterize the influence of coastal upwelling.

Acknowledgements. This study was supported by National Science Foundation Grant OCE 77-20624 and ATM 7919458. We thank J.M. Bremner and N. Kipp for providing some of the sediment samples. Samples were also provided by the core repositories of the Lamont-Doherty Geological Observatory, Oregon State University, and the Deep Sea Drilling Project. Special thanks are due L. Hutchings and I. Kruger for providing data and publications from South West Africa and to Paul Loubere for his advice on the computer programming.

References

Andrews, G.W., Morphology and stratigraphic significance of *Delphineis*, a new marine diatom genus, *Nova Hedwigia, Beiheft, 54*, 243-260, 1977.

Berger, W.H., Biogenous deep sea sediments: Production, preservation, and interpretation, in *Chemical Oceanography, 5*, J.P. Riley and R. Chester (eds.), pp. 265-388, Academic Press, London, 1976.

Bremner, J.M., Physical parameters of the diatomaceous mud belt off South West Africa, *Marine Geology, 34*, M67-M76, 1980.

Caliences, R. and O. Guillen, The seasonal cycle of phytoplankton and its relation to upwelling in Peruvian coastal waters (abstract), *CUEA Newsletter, 5(1)*, 8, 1976.

Calvert, S.E. and N.B. Price, Minor metal contents of recent organic-rich sediments off South West Africa, *Nature, 227*, 593-595, 1970.

Calvert, S.E. and N.B. Price, Recent sediments of the South West African shelf, in *The Geology of the East Atlantic Continental Margin*, F.M. Delaney (ed.), pp. 171-185, Institute of Geological Sciences, London, Report 70/16, 1971.

Cram, D.L., A role for the Nimbus-G coastal-zone colour scanner in the management of a pelagic fishery, *Fisheries Bulletin South Africa, 11*, 1-9, 1978.

Curray, J.R., D.G. Moore, *et al.*, Leg 64 seeks evidence on development of basins, *Geotimes, July*, 18-20, 1979.

Diester-Haass, L. and H.-J. Schrader, Neogene coastal upwelling off northwest and southwest Africa, *Marine Geology, 29*, 39-53, 1979.

Fryxell, G.A. and W.J. Miller III, Chain-forming diatoms: Three araphid species, *Bacillaria, 1*, 113-136, 1978.

Guillard, R.L. and P. Kilham, The ecology of marine planktic diatoms, in *The Biology of Diatoms*, D. Werner (ed.), pp. 372-469, University of California Press, Botanical Monographs, 13, Berkeley, 1977.

Guillen, O. and R. Calienes, Seasonal variation of productivity and its relation to upwelling in Peruvian coastal waters (abstract), *CUEA Newsletter, 5(1)*, 9, 1976.

Guillen, O., B. Rojas de Mendiola, and R. Izaguirre de Rondan, Primary productivity and phytoplankton in the coastal Peruvian waters, in *Fertility of the Sea, 1*, J.D. Castlow (ed.), pp. 157-185, Gordon and Breach, New York, 1971.

Guillen, O., B. Rojas de Mendiola, and R. Izaguirre de Rondan, Primary productivity and phytoplankton in the coastal Peruvian waters, in *Oceanography of the South Pacific 1972*, R. Fraser (ed.), pp. 405-418, New Zealand National Commission for UNESCO, Wellington, 1973.

Hart, T.J. and R.J. Currie, The Benguela Current, *Discovery Report 31*, 123-298, University Press, Cambridge, 1960.

Hendey, N.I., *An Introductory Account of the Smaller Algae of British Coastal Waters, Part V: Bacillariophyceae*, Her Majesty's Stationary Office, London, 317 pp., 45 plates, 1964.

Kollmer, W.E., The pilchard of South West Africa (*Sardinops ocellata*): The annual cycle of phytoplankton in the waters off Walvis Bay 1958, South West Africa Administration Marine Research Laboratory Investigational Report, 4, 1-44, Verenigde Pers, Beperk Windhoek, 1962.

Kollmer, W.E., The pilchard of South West Africa (*Sardinops ocellata* Pappe): Notes on zooplankton and phytoplankton collections made off Walvis Bay, South West Africa Administration Marine Research Laboratory Investigational Report, 8, 1-78, John Meinert (Pty.) Ltd., Windhoek, 1963.

Marchand, J.M., The nature of sea-floor deposits in certain regions of the west coast, *Fish and Marine Biology Survey, Union of South Africa, Rep. No. 6*, pp. 1-11, Pretoria, 1928.

Margalef, R., M. Estrada, and D. Blasco, Functional morphology of organisms involved in red tides, as adapted to decaying turbulence, in *Toxic Dinoflagellate Blooms*, D. Taylor and H. Seliger (eds.), pp. 89-94, Elsevier-North Holland, Inc., New York, Amsterdam, Oxford, 1979.

Mertz, D., Mikropaläontologische und sedimentologische Untersuchung der Pisco-Formation Südperus, *Paleontolographica, Abt. B. 118(1-3)*, 1-51, 1966.

Mukhina, V., Diatoms in the suspension and surface sedimentary layer of the shelf of South Western Africa, in *The Micropaleontology of Oceans and Seas*, A.P. Zhuse (ed.), pp. 94-105, Nauka, Moscow, (in Russian), 1974.

Report of the Division of Sea Fisheries South Africa 38, 12, Sea Fisheries Branch, Cape Town, 1970.

Report of the Division of Sea Fisheries South Africa 39, 24, Sea Fisheries Branch, Cape Town, 1971.

Report of the Division of Sea Fisheries South Africa 40, 26, Sea Fisheries Branch, Cape Town, 1972.

Report of the Division of Sea Fisheries South Africa 41, 23, Sea Fisheries Branch, Cape Town, 1973.

Richert, P., Relationship between diatom biocoenoses and taphocoenoses in upwelling areas off West Africa (abstract), 4th Symposium on Recent and Fossil Diatoms, Oslo, 1976.

Rojas de Mendiola, B., Variations of phytoplankton along the Peruvian Coast, *This Volume*, 1980.

Round, F.E., The phytoplankton of the Gulf of California, Part I, Its composition, distribution, and contribution to the sediments, *Journal of Experimental Marine Biology and Ecology, 1*, 76-97, 1967.

Round, F.E., The phytoplankton of the Gulf of California, Part II, The distribution of phytoplankton diatoms in cores, *Journal of Experimental Marine Biology and Ecology, 2*, 64-86, 1968.

Schrader, H.-J., Kieselsäureskelette in Sedimenten

des ibero - marokkanischen Kontinentalrandes und angrenzender Meeresgebiete, "*Meteor*" *Forschungsergebnisse, Reihe C, v. 8*, 10-36, 1972.

Schrader, H.-J., Proposal for a standardized method of cleaning diatom bearing deep-sea and land-exposed marine sediments, *Nova Hedwigia, Beiheft 45*, 403-409, 1974.

Schrader, H., Cruise report: Baja Vamonos 79, Sept. 7 - Sept. 30, 1979, Oregon State University Data Report 78, Reference 79-15, 1979.

Schrader, H.-J. and R. Gersonde, Diatoms and silicoflagellates, *Utrecht Micropaleontological Bulletin, 17*, 129-176, 1978.

Schrader, H., K. Kelts, J. Curray, D. Moore, *et al.*, Laminated diatomaceous sediments from the Guaymas basin slope (central Gulf of California): 250,000 year climate record, *Science, 207*, 1207-1209, 1980.

Schuette, G. and H. Schrader, Diatom taphocoenoses in the coastal upwelling area off western South America, *Nova Hedwigia, Beiheft, 64*, 359-378, 1979.

Schuette, G. and H. Schrader, Diatom taphocoenoses in the coastal upwelling area off South West Africa, *Marine Micropaleontology*, in press.

Smayda, T.J., Biogeographical studies of marine phytoplankton, *Oikos, 9*, 158-191, 1958.

Smith, R.L., Physical oceanography of coastal upwelling regions. A comparison Northwest Africa, Oregon, and Peru, Paper presented at the Symposium on the Canary Current: Upwelling and Living Resources, International Council for the Exploration of the Sea, Las Palmas, Canary Islands, 1978.

Stander, G.H., The pilchard of South West Africa (*Sardinops ocellata*): Temperature: Its annual cycle and relation to wind and spawning, South West Africa Administration Marine Research Laboratory, Investigational Report, 9, 1-57, Verenigde Pers, Beperk, Windhoek, 1963.

Stander, G.H., The pilchard of South West Africa (*Sardinops ocellata*): The Benguela current off South West Africa, South West Africa Administration Marine Research Laboratory, Investigational Report, 12, 1-122, Windhoek, 1964.

Sukhanova, I.N., G.V. Konovalova, and T.N. Rat'kova, Phytoplankton numbers and species structure in the Peruvian upwelling region, *Oceanology, 18*, 72-76, 1978.

Zhuse, A.P., Diatoms in the surface layer of sediments in the vicinity of Chile and Peru, *Oceanology, 12*, 697-705, 1972.

Zuta, S., T. Rivera, and A. Bustamante, Hydrologic aspects of the main upwelling areas off Peru, in *Upwelling Ecosystems*, R. Boje and M. Tomczak (eds.), pp. 235-257, 1978.

ADVECTION AND UPWELLING IN THE CALIFORNIA CURRENT

Patricio A. Bernal

Departamento de Biología y technología del Mar
P. Universidad Catolica de Chile, Sede Talcahuano
Casilla 127, Talcahuano, Chile

John A. McGowan

Scripps Institution of Oceanography
University of California, San Diego
La Jolla, California 92093, U.S.A.

Abstract. The classical view of the dynamics of epipelagic ecosystems in eastern boundary currents emphasizes the role of coastal upwelling as the major external forcing mechanism. However the physical dimensions of the highly productive area indicate that off California the system is at least 500 km wide. Latitudinal gradients in the distribution of properties suggest that the system responds to both large-scale horizontal advection from the north and local upwelling.

A 21-year time series of biological and physical data from the California Current is reviewed and the ability of indices of upwelling and transport from the north to predict significant, large-scale anomalies in zooplankton biomass is tested. The analyses indicate that changes in zooplankton biomass are uncorrelated with upwelling and that transport from the north is a better predictor of such changes.

Introduction

The goal of studies of coastal upwelling ecosystems is to achieve enough understanding that their response to changes can be predicted (Anonymous, 1973, 1978). It has been known for many years that these coastal regions are cooler, with higher dissolved nutrient concentrations, primary production, and biomass of phytoplankton and zooplankton than more oceanic regions at the same latitudes. The regions support large but variable stocks of commercially important fish and squid.

In studying these or other marine ecosystems, however, the dimensions of the system must be estimated. Boje and Tomczak (1978) pointed out that it is not now possible to decide whether an upwelling region should be described as a complete ecosystem on its own with elements of internal homogeneity, a degree of self regulation, and recognizable boundaries, or whether it cannot be separated from the larger oceanic ecosystem. They further point out that in the course of making such a determination it would be helpful to find some "natural boundaries" of upwelling regions. In our study we use the extensive series of observations of the California Cooperative Fisheries Investigations (CalCOFI) to determine some natural biotic and physical boundaries of the coastal zone where several elements of the ecosystem are enriched; we also analyze a time series from within the enriched system to describe the response of the biological part to the physical part. We then suggest reasons for the observed responses.

Major Features of the System. Fig. 1-a is a vertical section showing the salinity distribution from a point about 74 km offshore to a point well out into the North Pacific central gyre. Shoreward of about 126°W there is a well developed surface low-salinity layer and a shallow salinity minimum. Farther west there appears to be a frontal zone, then a complex area with several salinity minima and maxima and finally, beginning at about 138°W (1570 km offshore), a well developed halocline, underlying high salinity surface waters, appears. The latter structure is typical of the North Pacific central water mass. The nitrate distribution (Fig. 1-b) in the low-salinity area clearly differs from the more oceanic region in that nitrate is systematically and regularly detectable in the euphotic zone and the deeper isopleths of nitrate dome upwards here. The salinity "front" west of Sta. H5 is 820 km offshore. The coastal zone of low-salinity surface water and shallow salinity minimum is a widespread, typical feature of the California Current (Reid, Roden, and Wyllie, 1958). Fig. 2 indicates that the low-salinity

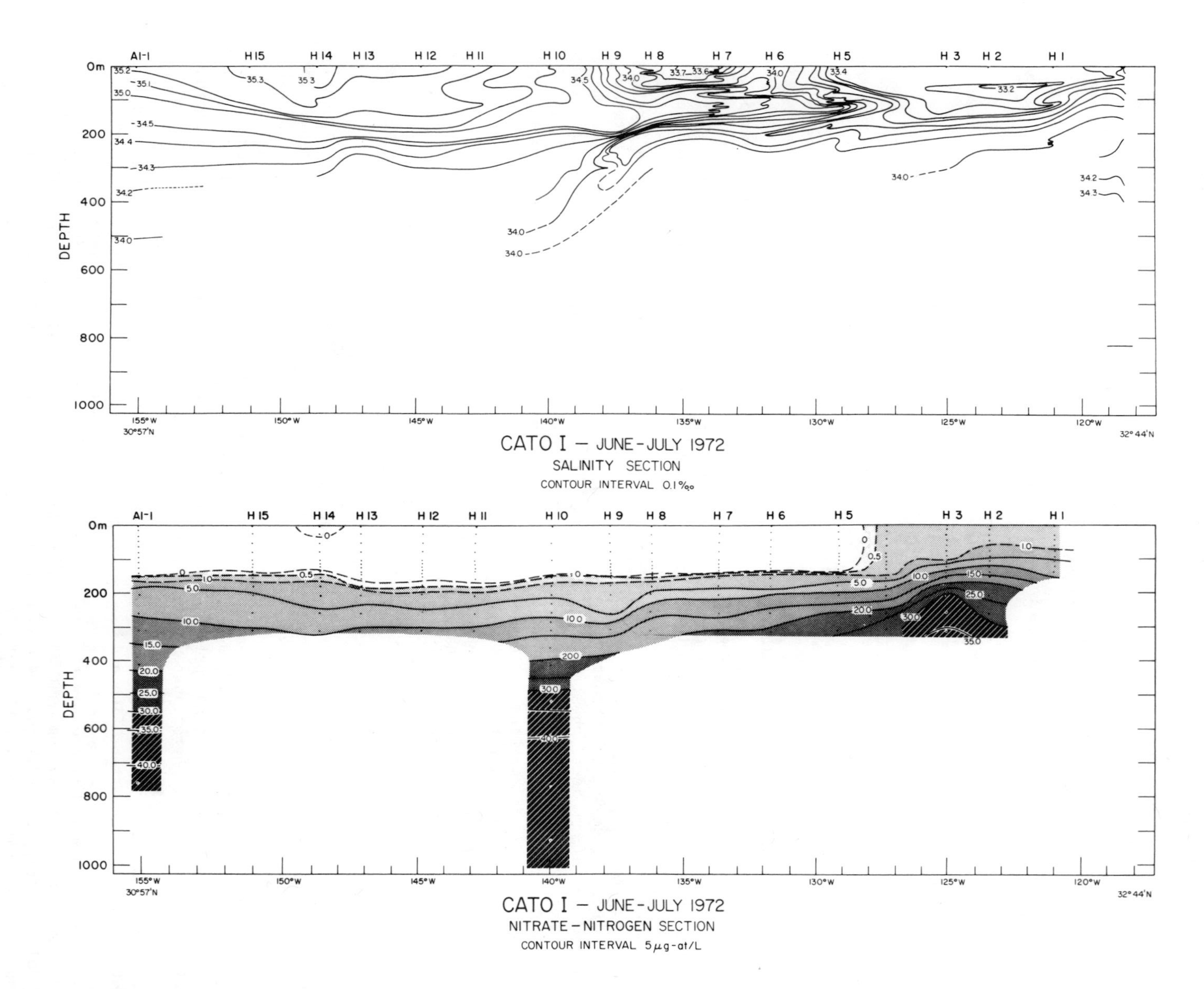

Fig. 1. Contoured salinity (upper panel) and nitrate-nitrogen concentrations (lower panel) on a vertical section between the coast of California and the central gyre of the North Pacific. The depths sampled (black dots) are shown only in the lower panel. H-1 through Al-1 are the station numbers.

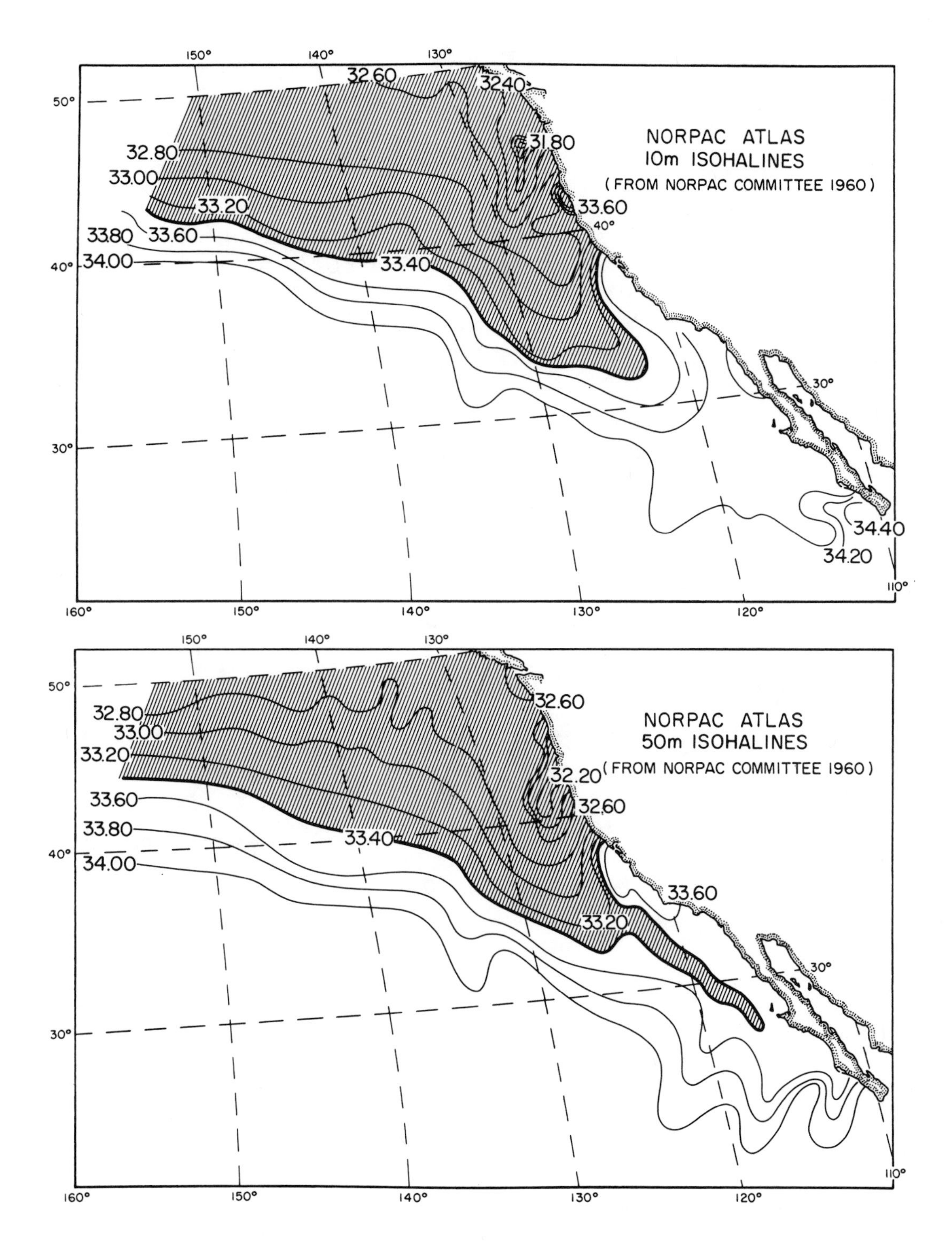

Fig. 2. Contoured salinity concentrations at 10 and 50 m for the eastern North Pacific. Isopleths of salinities greater than 34.00 have been omitted except at the southern end of Baja California. For numbers of stations see NORPAC Atlas (NORPAC Committee, 1960).

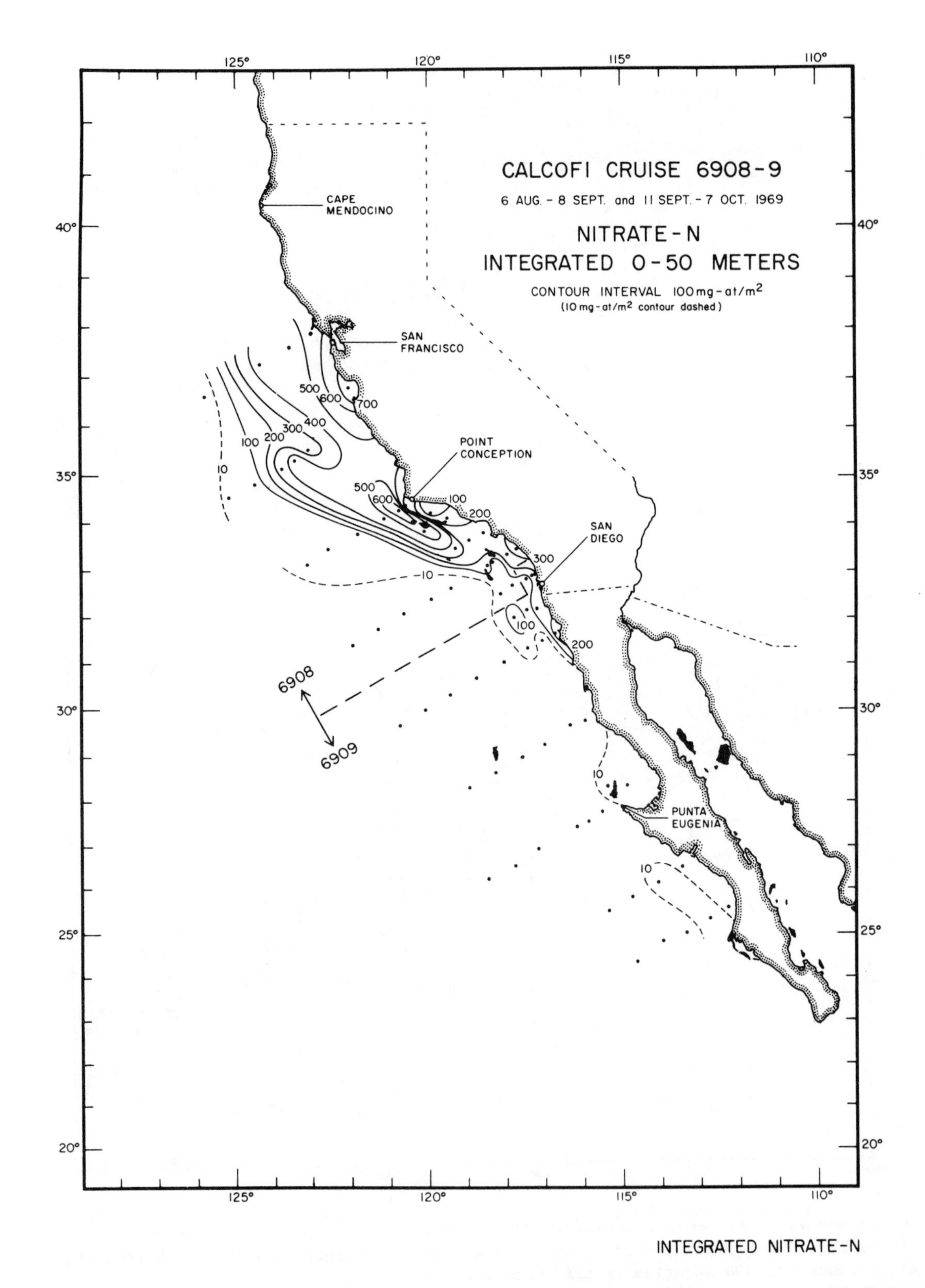

Fig. 3. Integrated 0 to 50-m concentrations of nitrate-nitrogen in the California Current (Thomas and Siebert, 1974).

feature is continuous all the way north to the Subarctic Pacific. The integrated (0 to 50 m) nitrate distribution (Fig. 3) shows both positive and negative offshore gradients and a pronounced negative gradient from north to south. The area of enrichment of nitrate (i.e., >10 mg-at/m^2) is 320 km wide off San Francisco and about 205 km off the southern California bight. Venrick (1979) presented a trans-Pacific vertical section of chlorophyll concentrations at 35°N. It shows high concentrations of chlorophyll in the upper 80 m from a point near the American coast to about 130°W, where a deep (>100 m) chlorophyll maximum, typical of the North Pacific central gyre, is present. The sharp chlorophyll front in her Fig. 2. coincides with the salinity and nitrate fronts shown in our Fig. 1. Maps of the distribution of chlorophyll *a* in both winter and spring show the same north-south and onshore-offshore gradients as the nitrate map. The width of the zone of enrichment (>0.05 mg/m^3) is about 670 km off San Francisco and 450 km off the southern California bight (Figs 4-a and 5-a). Maps of macrozooplankton biomass have major trends similar to those of nitrate and chlorophyll *a*; although the zooplankton is very patchy, there are clear north-south and offshore gradients of abundance. Off northern California, the outer boundary of the zone of high biomass (>64 ml $10^{-3}m^{-3}$)(displacement volume) is about 650 km offshore in the winter and 700 km in spring (Figs 4-b and 5-b). There are no extensive data on nekton but patterns of abundance of larval fish have been determined. Figs 6-a through 7-b show the distribution of four of the most abundant larvae in the California Current (anchovy *Engraulis mordax*, Pacific hake *Merluccius productus*, jack mackerel *Trachurus symmetricus*, and a mesopelagic fish *Leuroglossus stilbius*). In contrast to the other properties of the ecosystem, these do not show a north-south gradient but tend to be most frequent and abundant south of Point Conception. The boundaries in the area range from 330 to 600 km offshore. The mean temperature of the mixed layer (as indicated by the 10-m temperature) for summers shows cool water near shore and a strong north-south gradient further offshore. In winter, the narrow coastal band of cool water disappears but the offshore north-south gradient persists (Lynn, 1967).

The maps of important ecosystem properties were selected from a much larger data set, but they are representative and can serve to indicate the natural boundaries or dimensions of the coastal area where there are higher nutrient concentrations and increased biomass of phytoplankton, zooplankton, and nekton. While the boundaries are variable and not all property boundaries coincide, it is, nevertheless, evident that the enriched system is some 400 to 500 km wide north of Point Conception and about 260 to 300 km wide off southern California and northern Baja California.

As coastal upwelling appears to be limited to the 20- to 50-km offshore zone (Yoshida, 1955; Allen, 1973; O'Brien *et al.*, 1977) and the dimensions of the enriched system are about an order of magnitude larger, it seems reasonable to ask if variations in the intensity of coastal upwelling cause variations in the biomass of the of the system. The question is further sharpened by the observation that a large amount of cold, low-salinity, nutrient-rich water enters the system, horizontally, from the north (Sverdrup, Johnson, and Fleming, 1942; Reid *et al.*, 1958; Hickey, 1979). That is, is the enriched area of the California Current self-contained and self-regulating, in the sense that variations in the biotic carrying capacity are caused primarily by a local process such as coastal upwelling, or is the carrying capacity also influenced by input from outside the system?

In attempting to answer the question we use the natural boundaries outlined above to define the ecosystem. However as the boundaries are somewhat variable in space and time we have, for the purposes of quantitative analysis, established some more arbitrary boundaries within the system of natural boundaries. There are two ecosystem properties that have been measured on a large spatial (approximately 6.6 x $10^5 km^2$) and long-time (21 years) scale: macrozooplankton biomass and hydrography (temperature and salinity data). Fig. 8 shows the pattern of stations used in our study. The region is divided into four areas on the basis of biogeographic evidence (Bernal, 1979). Each point represents a station where a large (ca. 500 m^3), oblique (0 to 140 m) net tow (1 m diameter, 505 mm mesh) was taken. At each net tow station, a hydrographic cast (0 to 600 m, 18 bottles average) was made for determination of temperature, salinity, and oxygen. The number of stations occupied each month varied and the average per area ranged from 23 to 47.

The outer boundary of the areas were chosen on the basis of the average patterns of biomass normal to the coast at a northern line (line 60) and at a line mid-way (line 90) through the north-south axis of the system (Fig. 9). The two lines were chosen because they started a few kilometers from the coast but often extended more than 1000 km offshore thus including the western boundaries of the system. It is evident that zooplankton abundance peaks strongly about 180 km offshore (point A at Sta. 70 in Fig. 9-a) and then declines to near oceanic levels some 500 km offshore (at Sta. 120). A similar pattern is seen in the fall-winter data from the line, but the biomass is systematically lower in the first 500 km offshore (about 200 ml $10^{-3}m^{-3}$). At line 90 (Fig. 9-b) the pattern is even more bimodal, with peaks at about 150 km (point A, Sta. 45) and 275 km (point B, Sta. 60) offshore, and the decline to near oceanic abundance is between 400 and 500 km offshore. In addition to the average zooplankton abundance, the average integrated transport from 0 to 200 m was calculated

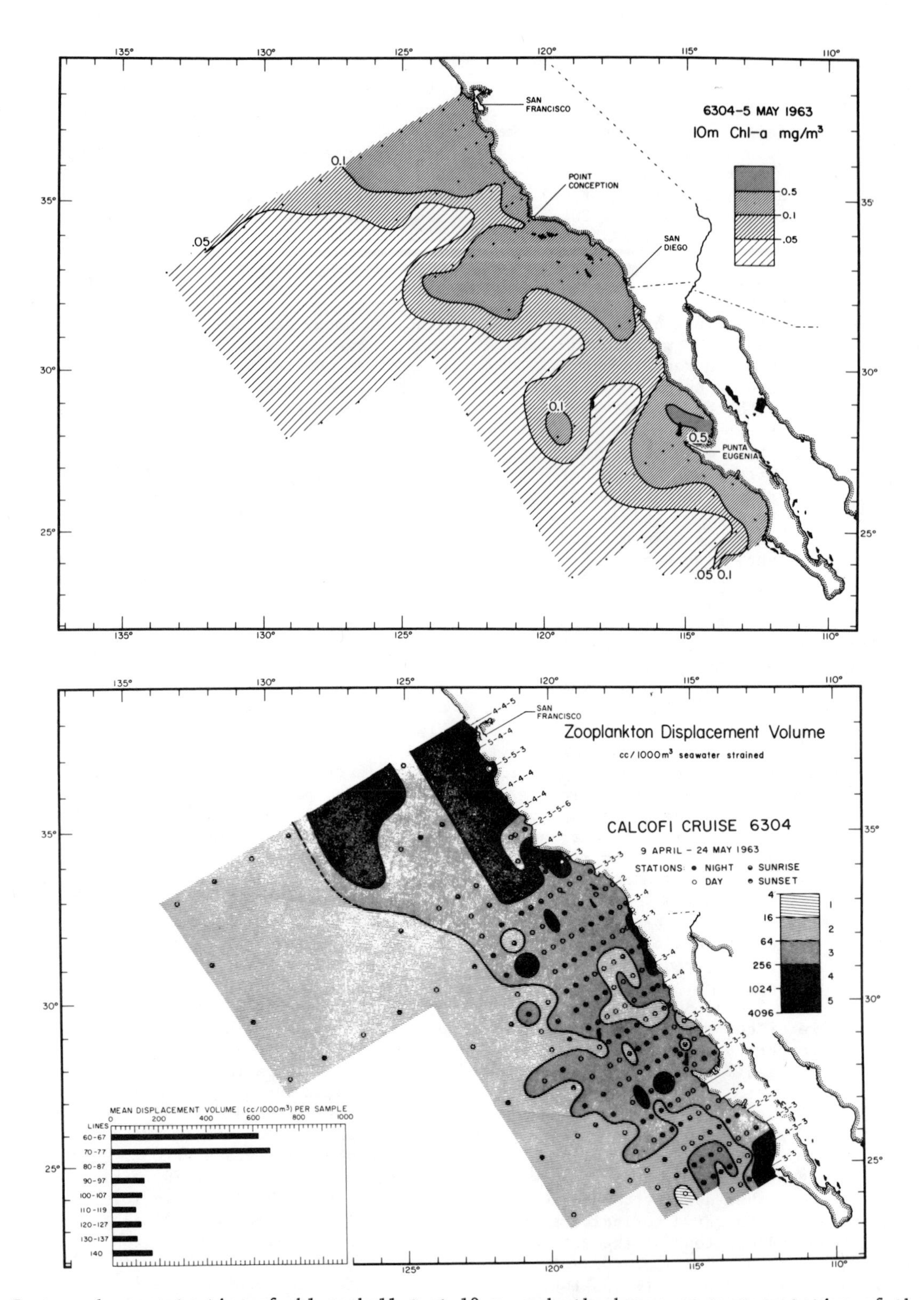

Fig. 4. Contoured concentration of chlorophyll *a* at 10 m, a depth chosen as representative of the mixed layer (upper panel, a). Contoured zooplankton biomass estimates from oblique, integrating net tows (lower panel, b). The numbers to the right of the shoreline represent biomass estimates at special inshore stations occupied during the cruise. These stations are too closely spaced to be shown by station circles. A number index appears to the right of the estimated biomass key. The values for the inshore stations were not used in contouring. The box in the lower left shows the mean estimates of biomass from north (lines 60-67) to south (line 140). Zooplankton data from Smith (1971).

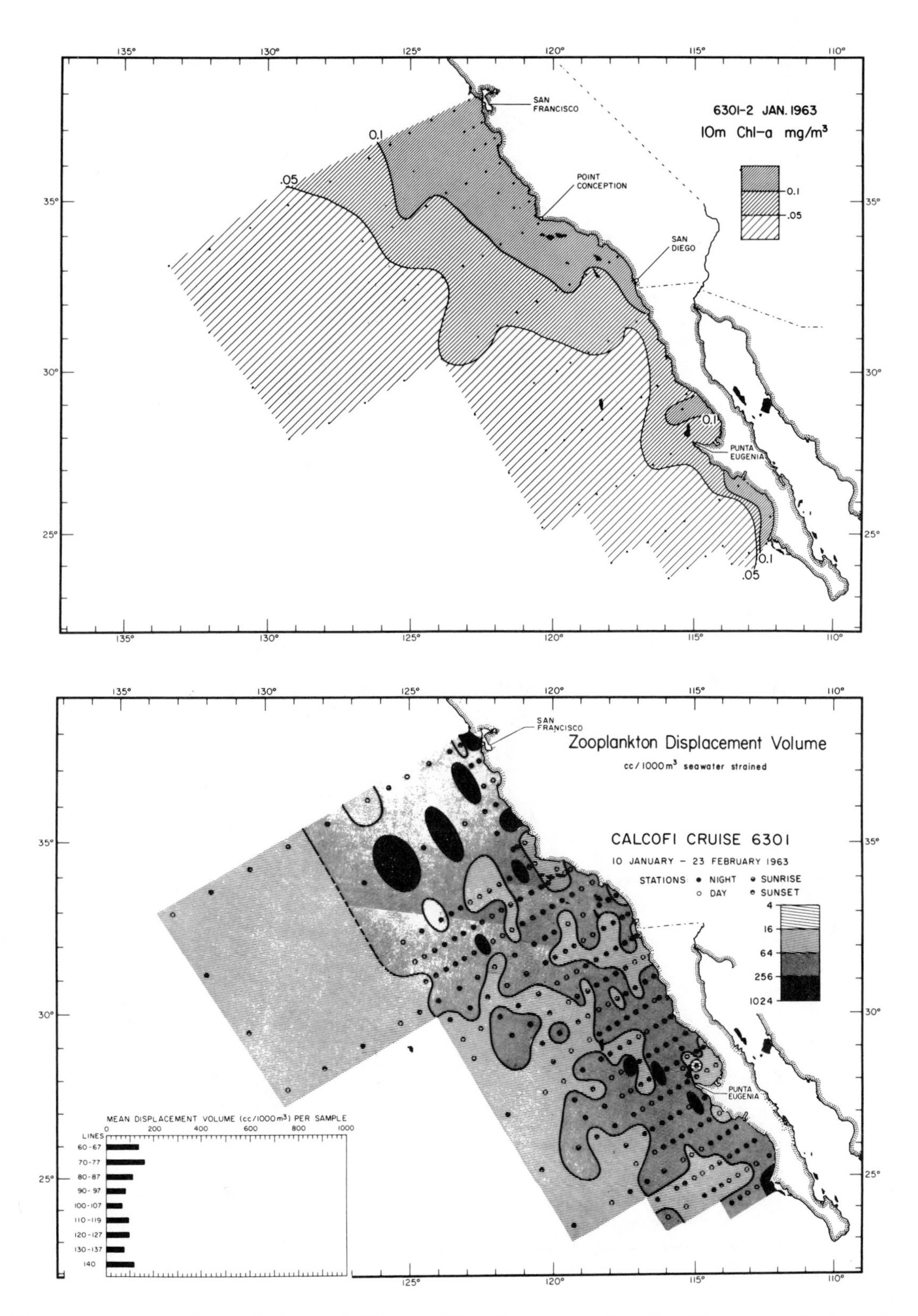

Fig. 5. Winter concentrations of chlorophyll *a* at 10 m (upper panel, a). Winter zooplankton biomass estimates (lower panel, b). Explanation as in Fig. 4.

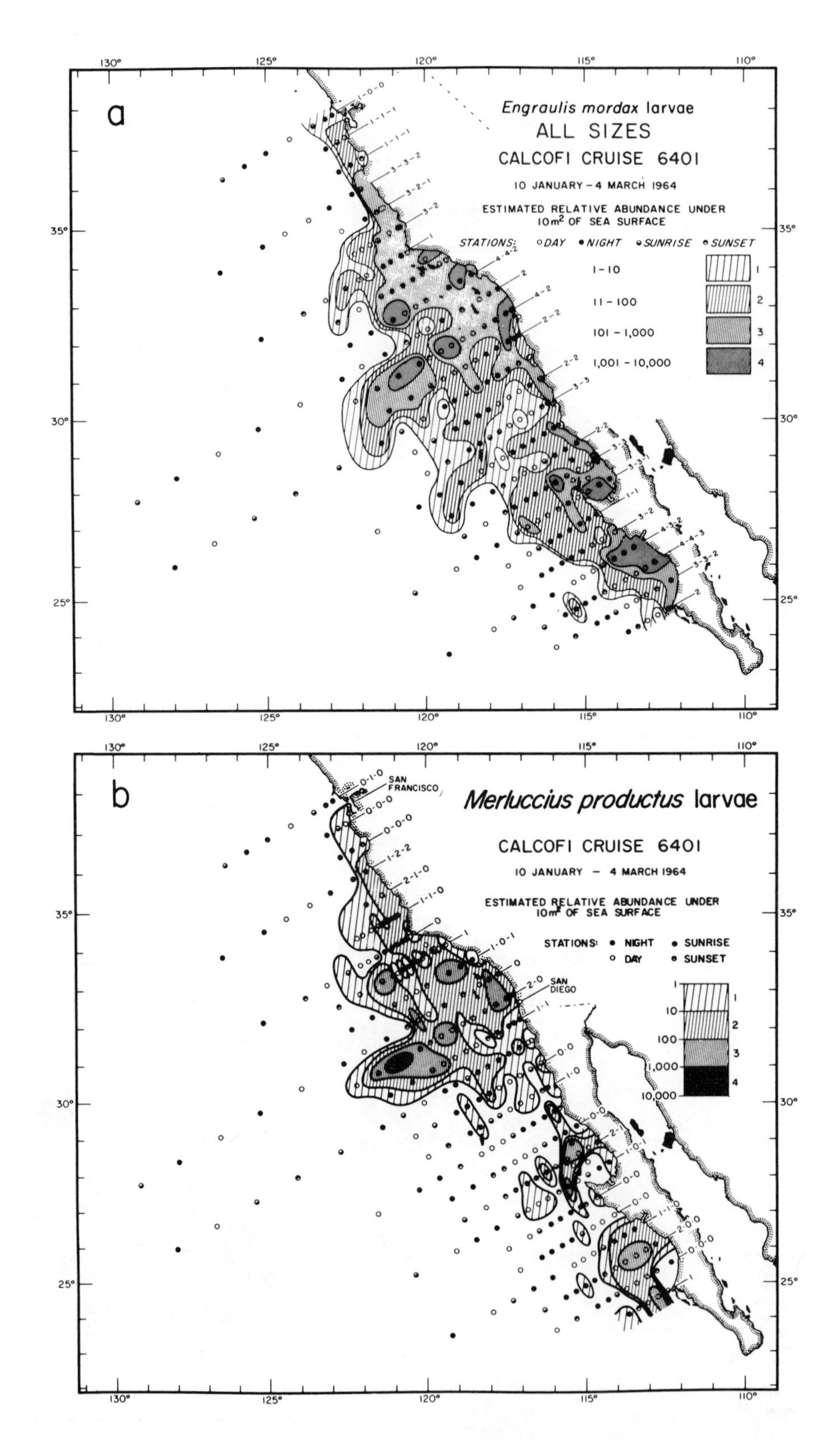

Fig. 6. Contoured estimates of abundance of a) anchovy (*Engraulis mordax*) and b) Pacific hake (*Merluccius productus*) larvae during the peak of the spawning season for the species. Numbers on the shore as in Fig. 4 (from Kramer and Ahlstrom, 1968; Ahlstrom, 1969).

for the lines. The transport represents the cross-shore averages for the months of January and July of all observations available from 1950 to 1978. Inshore of 180 km, the transport is northward at about 0.45 x 10^4 m^3 $s^{-1}km^{-1}$. West of 180 km, the flow is southward at about 1.0 x 10^4 m^3 $s^{-1}km^{-1}$. It is of interest that the first maxima in biomass at 180 and 150 km in both lines (points A in Figs 9-a, 9-b) and the secondary maximum in line 60 (point B Fig. 9-a) coincide with areas where the cross-shore gradient of average transport is also maximum. The gradients of average flow, although not strictly comparable with maximum shear because of the temporal average they represent, nevertheless may represent quasi-permanent features where turbulent mixing might become a three-dimensional process and where the vertical input of nutrients from below could be enhanced. A feature common to all the cross-shore patterns is the relative scales of the zone influenced by coastal upwelling, illustrated by a band with a cross-shore dimension equal to one baroclinic Rossby radius of deformation (R_{ϕ} in Figs 9-a, 9-b). This is, in general, less than 50 km, but the zone of high biomass is several hundred kilometers wide and is well offshore of R_{ϕ}. The seasonal reduction of biomass in the zone of high biomass coincides with a reduction of the intensity of southward flow (minimum in December) and with the relaxation of coastal upwelling (minimum in January, Bakun, 1973). Smith (1971) also documented the inshore-offshore and north-south trends in zooplankton biomass.

Time series analysis. In addition to the seasonal signal, the epipelagic ecosystem of the California Current also undergoes large scale, low-frequency fluctuations. Based on the record of varved sediments of biological origin, Soutar and Isaacs (1969, 1974) showed that the fluctuations seemed to have been the rule rather than the exceptions. The fluctuations are evidence of non-steady-state behavior of the ecosystem in the low-frequency range, presumably in response to forcing at commensurate scales. Colebrook (1977) studied the year-to-year fluctuations in biomass of 17 broad taxonomic categories (total copepods, euphausiids, chaetognaths, decapods, etc.) in the California Current for the period 1955 to 1959, using principal component analysis. His results indicate that changes in biomass show a remarkable coherence among categories and among different geographic localities. In other words, a large part of the year-to-year fluctuation in biomass was common to all geographic areas and to a majority of taxa.

The temporal scales of demographic phenomena relevant to primary and secondary production levels of the ecosystem are much shorter than the characteristic periods of the fluctuations. The scales are of the order of hours to days for phytoplankton doubling times and about two or three months for macrozooplankton. This mismatch of scales and Colebrook's work suggests that the fluctuations represent the forced response of the ecosystem to driving functions that are essentially independent of the internal structure and function of the ecosystem or food chain.

We have analyzed the 21-year record of zooplankton biomass by area (Fig. 8) for the purpose of describing the main features of non-seasonal variability. The techniques used in the field and laboratory to obtain the biomass data were described by Kramer *et al.* (1972). The data were pooled by area and anomalies from the seasonal mean calculated (Bernal, 1979). The four seasonally corrected time series of anomalies from the long term means are shown in Fig. 10. The figure shows that there are large scale, low-frequency trends. Particular years, 1950, 1953, and 1956, had simultaneous maxima in at least two of the areas; years 1958 and 1959 show coherent sets of minima in all four areas.

The main feature of the spectral analyses of data (Bernal, 1979) is that a large fraction of the total variance lies within the low-frequency band (i.e., less than 1.7 x 10^{-3} cycles/d). The north-south trend of biomass is paralleled by an increasing fraction of the total variance clustered in the low-frequency band. This means that as one moves from north to south an increasing amount of the variability is associated with some slow response, large scale process.

The presence of a well defined, low-frequency response of the epipelagic ecosystem of the California Current poses the question: what is the principal forcing mechanism driving the system in this frequency domain? The first part of the answer is provided by a significant negative correlation between sea surface temperatures and zooplankton biomass (Reid *et al.*, 1958; Reid, 1962). However, the correlation by itself, interesting as it is, does not allow one to differentiate between the two causal mechanisms potentially involved with the input of non-regenerative nutrients, namely coastal upwelling and large scale horizontal advection, because both processes lower the sea surface temperature and could have forced the statistical association. We have attempted to resolve this dichotomy by assuming first that variations in coastal upwelling intensity are the dominant process and second that large scale advection (and eventual mixing) of cold, low-salinity, nutrient-rich water from the north is the dominant process. The method used was to perform time-lagged cross-correlation analysis with the time-series of zooplankton biomass as the dependent variable and the index of coastal upwelling (Bakun, 1973, 1975) and an index of horizontal transport from the north as independent variables.

There were two separate cross-correlation analyses. Fig. 11 shows the cross-correlation functions, with time lags from 0 to 18 months, for zooplankton with the Bakun upwelling index for the region. In the left hand graph the index

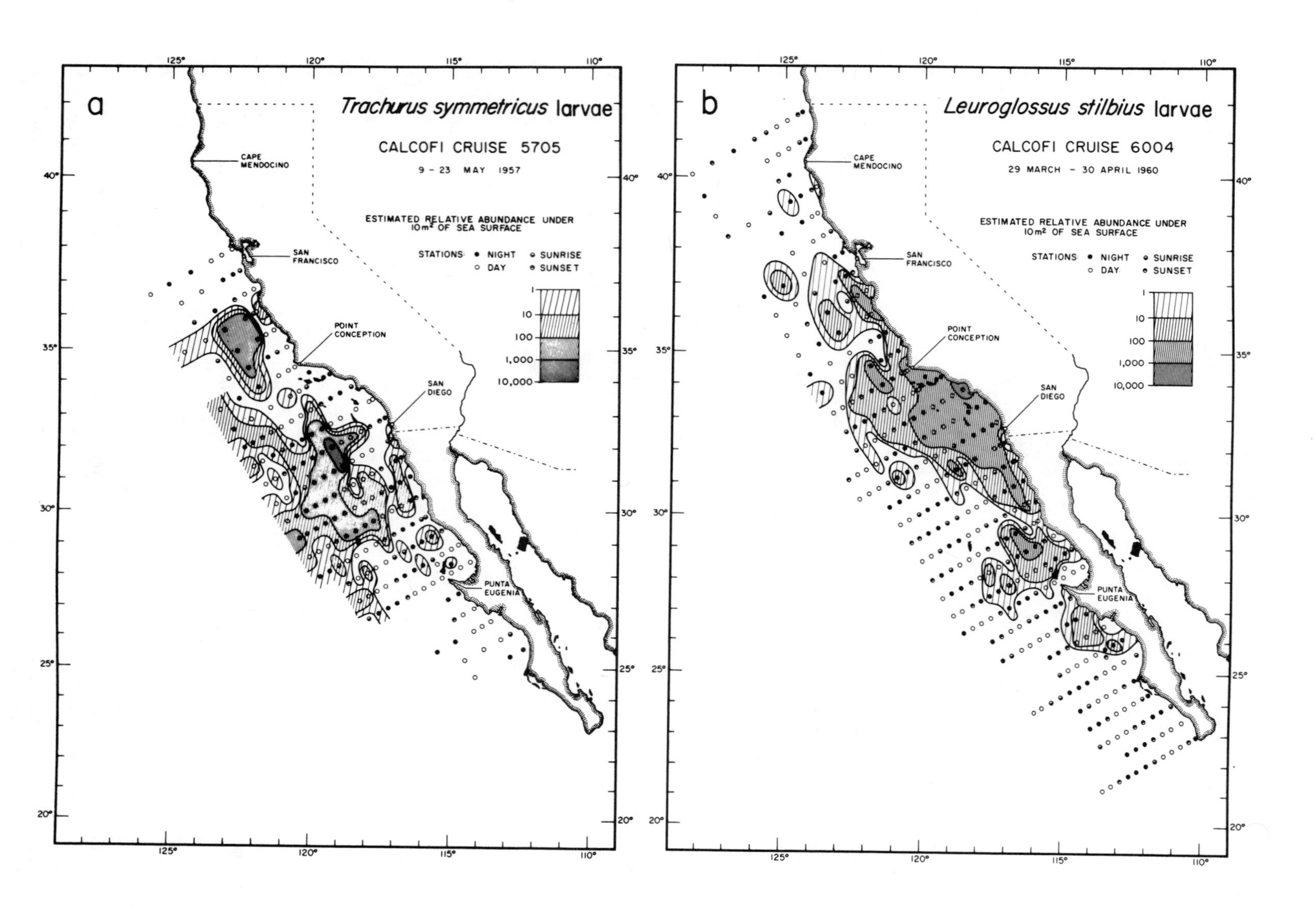

Fig. 7. Estimates of the abundance of a) jack mackerel (*Trachurus symmetricus*) larvae during the peak of the spawning season for the species and b) a typical distribution of the abundance of the larvae of a fish (*Leuroglossus stilbius*) that is mesopelagic as an adult (from Ahlstrom, 1969, 1972).

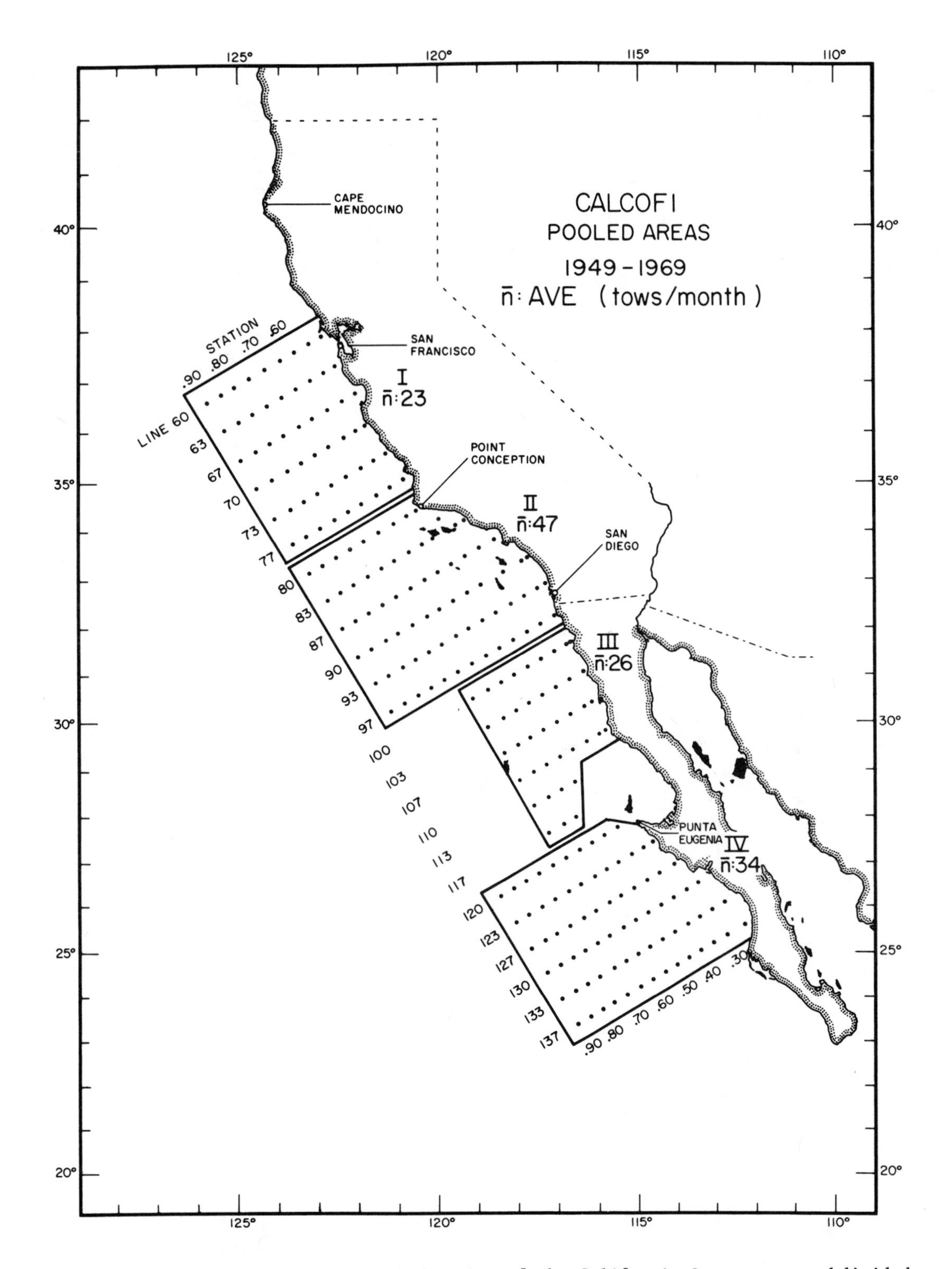

Fig. 8. The four areas into which the enriched region of the California Current was subdivided. Dots indicate the position of CalCOFI stations identified by a line number on the left margin and by a station number on top and bottom. As not all stations within each area were occupied every cruise, the average number of stations occupied per month is given as $\bar{n}$.

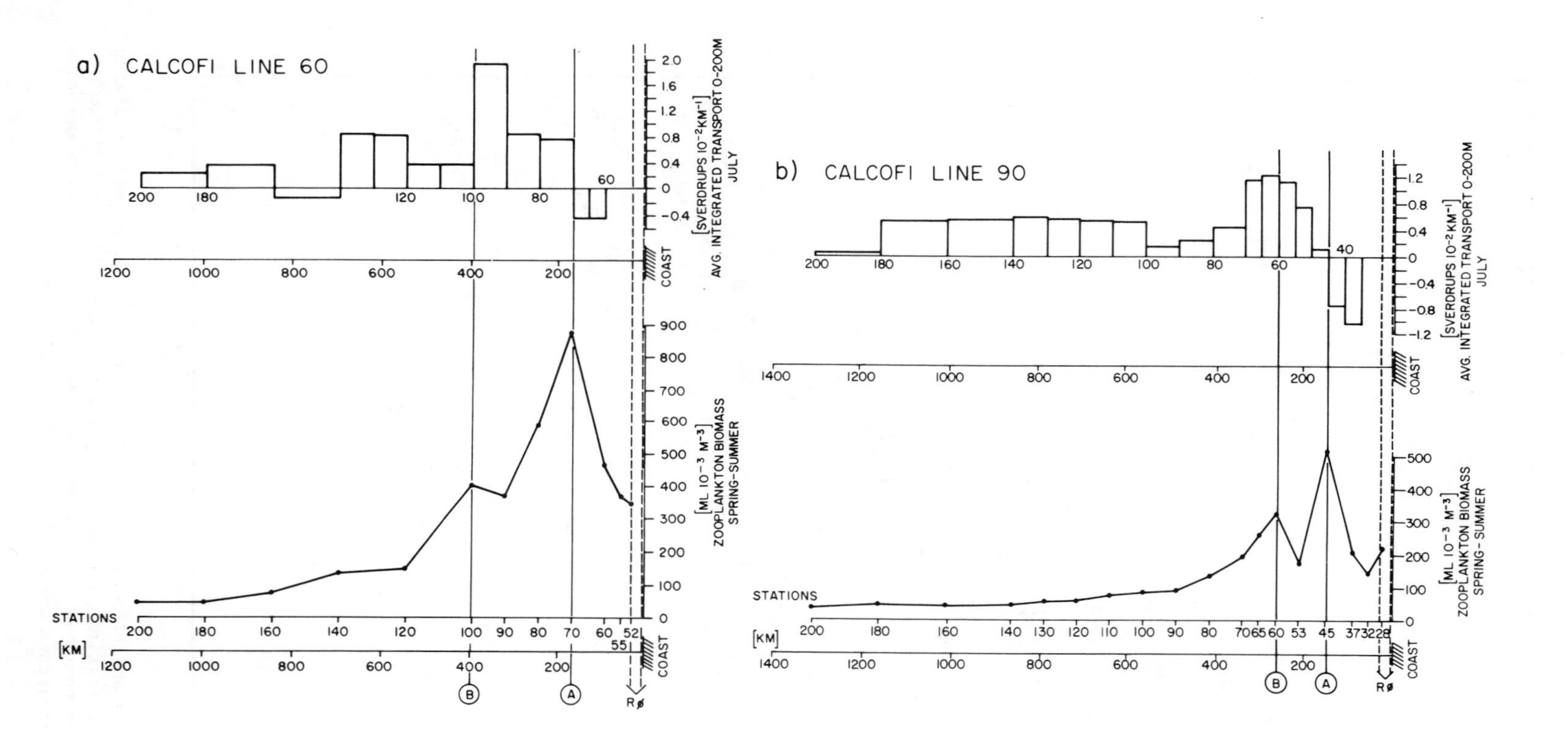

Fig. 9. Cross-shore patterns of mean longshore transport (0 to 200 m) and zooplankton biomass at CalCOFI lines 60 (a) and 90 (b) during spring and summer. Units of transport are Sverdrups (10^6 m^3s^{-1}) per 100 km cross-shore distance, with positive values indicating southward flow. Units for zooplankton are ml of biomass per 1000 m^3 of water filtered by the net. R_ϕ is the distance equivalent to one baroclinic Rossby radius of deformation. Vertical lines A and B label features discussed in the text.

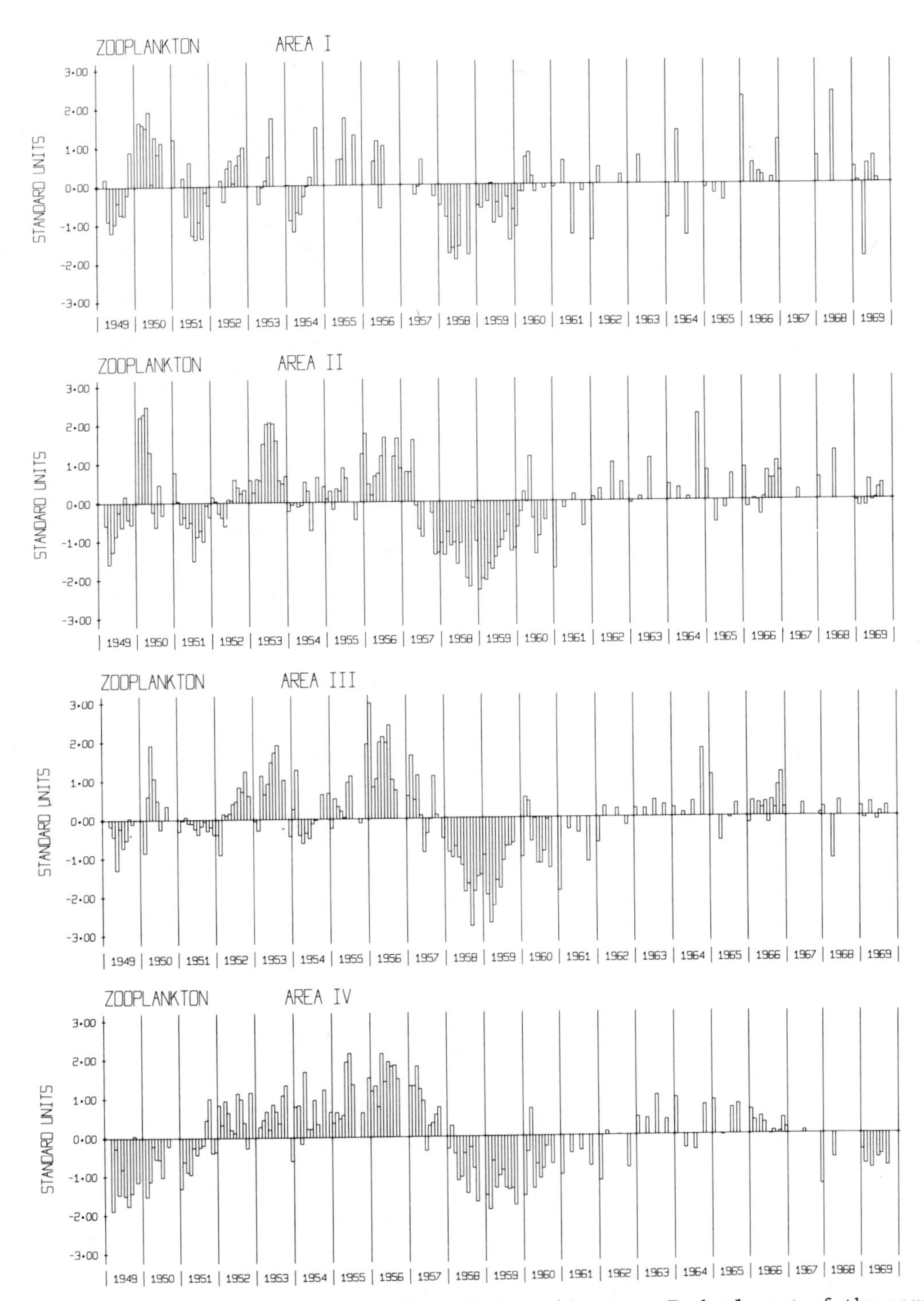

Fig. 10. Seasonally corrected time series of zooplankton biomass. Each element of the series was derived from the average over space of all log-transformed estimates of biomass within each area, a procedure that filters short-term (daily, weekly) and small-scale variability (patchiness). The data are anomalies from the long term (21-year) seasonal mean standardized with their corresponding standard deviations, accordingly the magnitude of 1 standard unit is equal to one standard deviation.

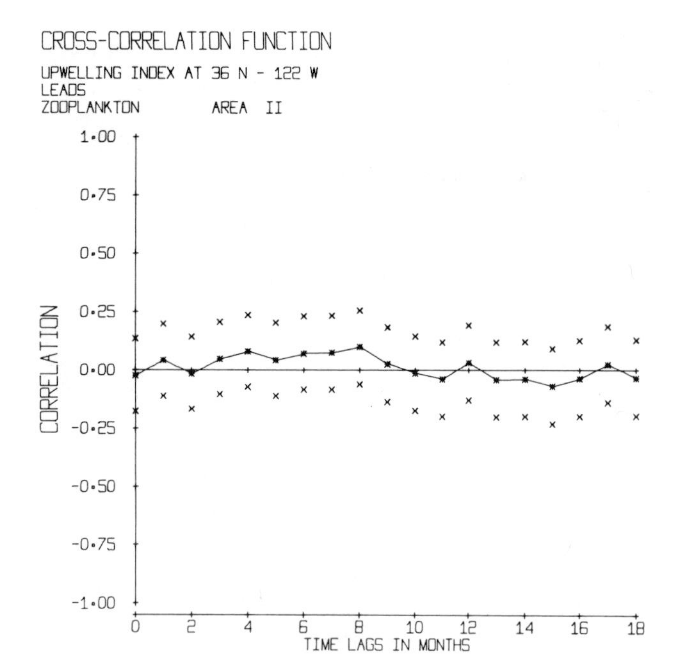

Fig. 11. Time lagged cross-correlation functions from 0 to 18 months. Upper: Bakun's upwelling index in Area I leads zooplankton biomass in Area II. Lower: upwelling index in Area II leads zooplankton biomass in Area II. Crosses above and below the continuous line indicate the 95% confidence limits of the correlation $r(t) \neq 0$ at $P \leq 0.05$.

within area I was used with zooplankton from area II, assuming that macrozooplankton biomass might respond to changes in upwelling intensity taking place upstream. None of the correlation values was significantly different from 0. On the right hand side of the figure are the time-lagged correlations of the zooplankton of area II with the upwelling index from the same area. Only three of 19 correlations depart significantly from zero, and they are negative and small. Correlation coefficients for areas I, III, and IV of zooplankton with upwelling indices centered within the areas did not depart from 0 at any time lags. The inference to be drawn from the statistical tests is that changes in the carrying capacity of the system (anomalies of zooplankton biomass from the long-term mean) are unrelated to changes in the intensity of upwelling as indicated by the Bakun index.

As there is abundant circumstantial evidence that large amounts of cold, nutrient-rich water enter the system from the larger oceanic system to the north and northwest, it is reasonable to attempt the same sort of analyses with measures of horizontal transport from the north. The large tongue of low-salinity water penetrating into the system is an excellent marker of water of northern origin (Fig. 2). We have used the 33.40 isohaline as an index of the degree of penetration of these northern waters. Fig. 12 shows the long term mean position of the isohaline and its southernmost extent for different months of several years. The degree of penetration of this water into the system is highly variable. Using this marker, its variability, and the extensive time series of hydrographic data we have calculated an index of transport from the north. For each month along CalCOFI line 80, the geostrophic transport (in reference to an assumed level of no motion at 500 decibars) and the field of salinity were interpolated from observations into a fixed vertical grid. The sub-set of grid points less than or equal to 33.40 was determined on the salinity field, and the associated transport in the upper 200 m was calculated. The monthly transport estimates, expressed as Sverdrups* per 100 km of cross-shore distance, were also seasonally corrected by subtracting them from their long-term monthly means (1950 to 1978) and standardized with their corresponding standard deviations.

The calculations (Fig. 13) show that there are large positive and negative anomalies from the 21-year mean of transport of water from the north. We have used this information to calculate time-lagged cross-correlations with the anomalies of zooplankton biomass as the dependent variable. The results (Fig. 14) show that in areas downstream from line 80 (areas II and III), anomalies in zooplankton abundance are well correlated with anomalies in transport at time lags from 0 to 7 months (area II) or 0 to 4 months (area III). Area I was not particularly well correlated, with only the 2- and 4-month lags

* One Sverdrup = $10^6 m^3 s^{-1}$

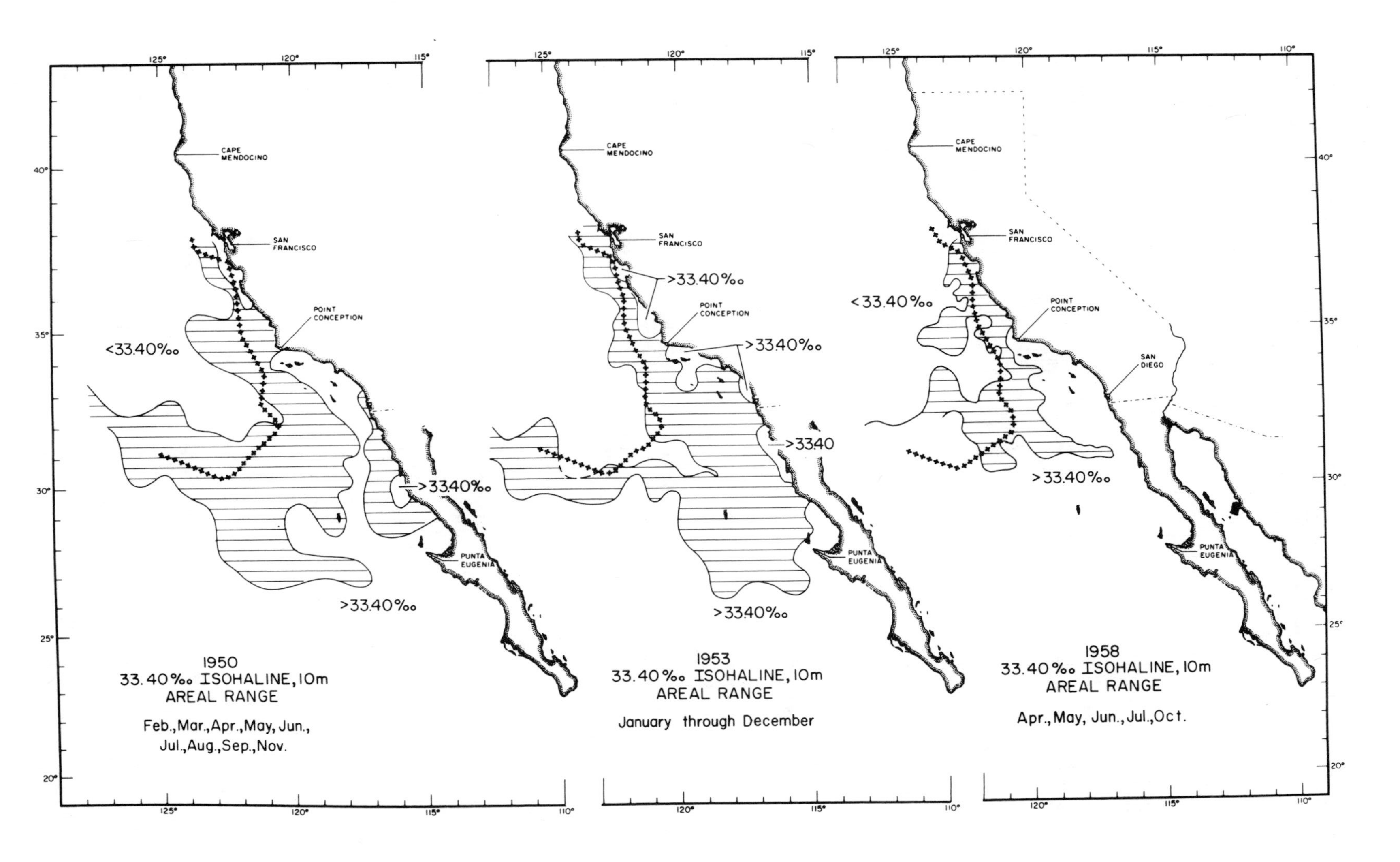

Fig. 12. Inter-annual fluctuations of the 33.40 isohaline at 10 m in the California Current region, 1950, 1953, and 1958. On a given year the "areal range", cross hatched, is the region defined by the maximal northern and southern excursions of the isohaline. This region separates waters that never were more saline than 33.40 to the north from waters that never were less saline than 33.40 to the south. As a reference the long-term average position of the 33.40 isopleth (Lynn, 1967) is outlined by crosses.

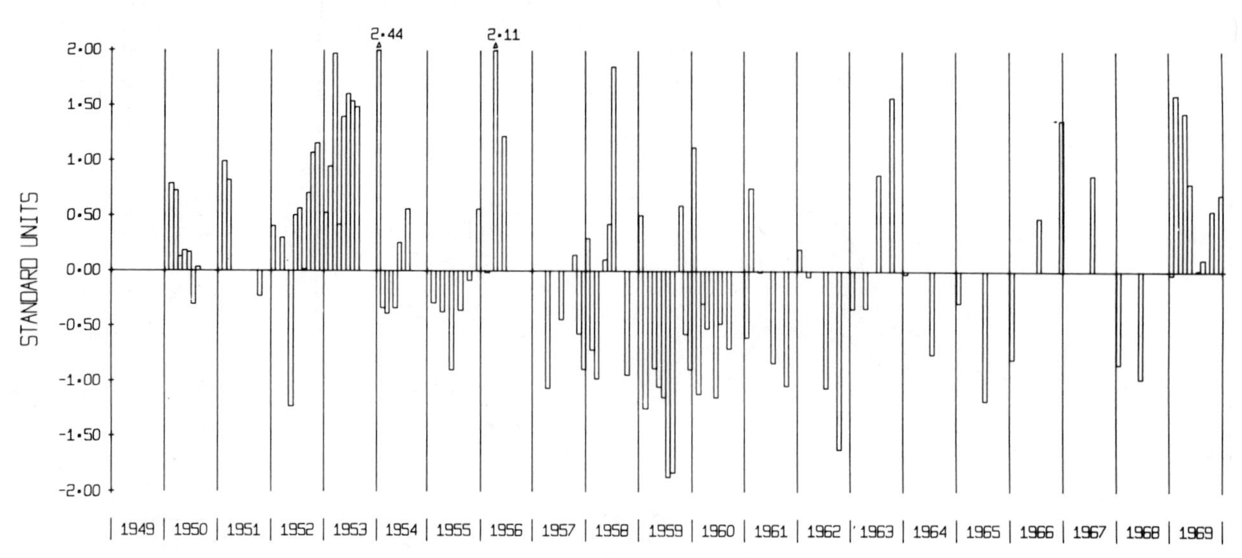

Fig. 13. Seasonally corrected time series of the anomalies of transport from the north across CalCOFI line 80. Units are analogous to those described for Fig. 10.

showing significant positive correlations. Area IV had significant positive correlations at 0- to 5-month lags, but the correlation coefficients were all less than 0.35.

The long-term means of zooplankton abundance and the anomalies from them were based on samples taken at the positions shown in Fig. 8. Most were outside the R_ϕ distance from the coast. It is, therefore, conceivable that coastal upwelling, within this narrow band, could lead to a close inshore zone of high biomass regulated by the local process, but the possibility seems unlikely. Beginning in 1963 many special inshore stations were occupied well inside the R_ϕ distance from the coast (see maps in Smith, 1971). We have shown one example of such a cruise in Fig. 4b. Neither on this cruise nor on the 21 other cruises with special inshore stations were there spatial or temporal trends indicating an additional biomass peak near shore.

Discussion

In view of the strong north-south gradient in zooplankton abundance it could be argued that the positive correlations of abundance and the transport across line 80 are merely a product of advective movement of the high northern biomass. However, studies of species biogeography have shown that area II and III tends to have different species assemblages than area I, north of line 80 (Alvariño, 1964, 1965; Berner, 1967; Brinton, 1967, 1973; Bowman and Johnson, 1973; Fleminger, 1964, 1967; McGowan, 1967, 1968). Area I is frequently dominated by species with subarctic water mass affinities. While specimens of this assemblage can often be found south of line 80, particularly in the colder years, they do not dominate the biomass, and the zooplankton assemblage south of the line is much more diverse than north of it (McGowan and Miller, 1980).

The simplest explanation of our results is that the observed, large, low-frequency signals are the response of the ecosystem to the input of cold, low-salinity, high-nutrient water coming in horizontally from the north as the main body of the California Current. The input of nutrients enhances production. Nutrients obviously enter the system at euphotic zone depths, but those entering below that depth may also be important. For example, shear induced turbulence, cyclonic eddies, island and sea-mount effects, and a number of other mechanisms may all stir and mix the system vertically, thus introducing more nutrients into the euphotic zone.

That the input of water derived from the north should have such a large effect on the productivity of this coastal ecosystem should not be a surprise. Wickett (1967), using a sub-set of the zooplankton data we used, found a positive correlation ($r = 0.84$, $p \leq 0.01$) between southward Ekman transport at 50°N, 140°W and the zooplankton biomass off southern California one year later, suggesting that the zooplankton might be responding to a nutrient input taking place far upstream. The upstream subarctic water is rich in nutrients. A nutrient budget for the subarctic Pacific has been computed (McGowan and Williams, 1973); an excess of inorganic phosphorous of about 0.13 mg-at PO_4 $m^{-2}d^{-1}$ was found. Because

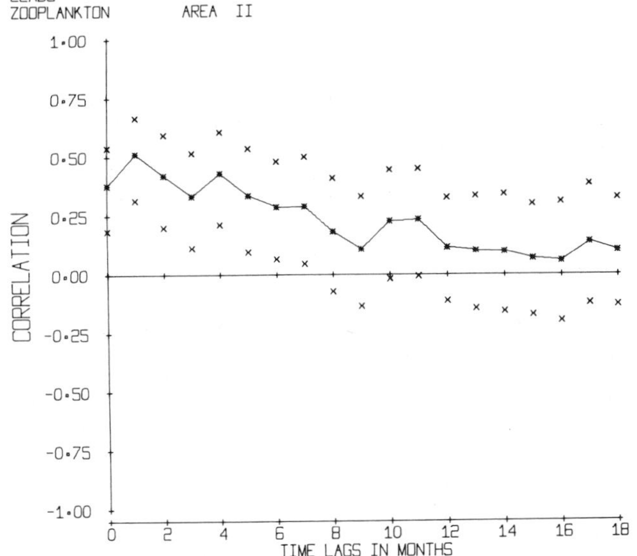

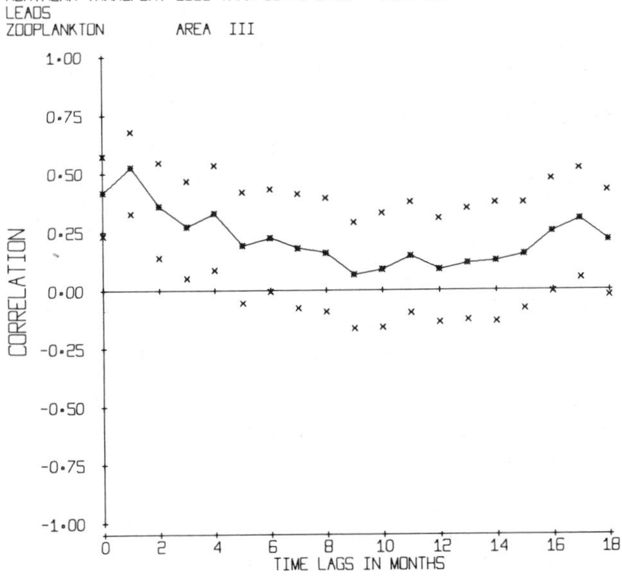

Fig. 14. Time lagged cross-correlation functions from 0 to 18 months. Upper: transport from the north across line 80 leads zooplankton in Area II. Lower: transport from the north across line 80 leads zooplankton in Area III. Confidence limits as in Fig. 11.

phosphorous is not accumulating in the upper layers of the subarctic, it was suggested that the balance must be achieved by a net flux from the subarctic into the California Current. Colebrook (1977) concluded that "whatever influence or influences are responsible for the fluctuations in the plankton either have their origin in the north of the (CalCOFI) survey area or have a greater effect on those categories with northern patterns of distribution". A correlation between cooler periods and high zooplankton abundance in the main body of the California Current was shown by Reid *et al.* (1958) and Reid (1962). The functional relationship here is probably due to higher nutrient levels during cool periods, for Zentara and Kamykowski (1977) showed that, in the eastern North Pacific, nitrate, phosphate, and silicate follow an inverse, monotonic relationship with temperature.

That large-scale, low-frequency variations in transport from the north occur has been independently demonstrated by Chelton (1980), who used a long time series of tide gauge and steric height data to determine variations of sea level on the west coast of North America. His time series of the low-frequency component of steric height is well correlated with the time series of zooplankton anomalies.

That changes in zooplankton abundance are totally uncorrelated with changes in intensity of upwelling is a surprise. The result could be due to some error on our part in the way the upwelling index was used, or it may be a failure of the index itself to indicate upwelling intensity or, as seems more likely, upwelling intensity has little to do with large-scale variations in secondary productivity in the main body of the California Current.

Acknowledgements. This research was supported by the Marine Life Research program, the Scripps Institution of Oceanography part of the California Cooperative Oceanic Fisheries Investigations (CalCOFI), which is sponsored by the Marine Research Committee of the State of California, and by a grant from the Ford Foundation to P.A. Bernal through the University of Chile — University of California Cooperative Program.

References

Ahlstrom, E.H., Distributional atlas of fish larvae in the California Current region: jack mackerel, *Trachurus symmetricus*, and Pacific hake, *Merluccius productus*, 1951 through 1966, *CalCOFI Atlas No. 11*, v-xi, 1-187, 1969.

Ahlstrom, E.H., Distributional atlas of fish larvae in the California Current region: six common mesopelagic fishes: *Vinciguerria lucetia, Triphoturus mexicanus, Stenobrachius leucopsarus, Leuroglossus stilbius, Bathylagus wesethi*, and *Bathylagus ochotensis*, 1955 through 1960, *CalCOFI Atlas No. 17*, v-xv, 1-306, 1972.

Allen, J.S., Upwelling and coastal jets in a continuously stratified ocean, *Journal of Physical Oceanography, 3*, 245-257, 1973.

Alvariño, A., Zoogeografia de los Quetognatos, especialmente de la region de California, *Ciencia (Mexico), 23*, 51-74, 1964.

Alvariño, A., Distributional atlas of Chaetognatha in the California Current region, *CalCOFI Atlas No. 3*, vii-xiii, 1-291, 1965.

Anonymous, International Decade of Ocean Exploration, Progress Report Vol. 2, Living resources program, 36-42, Office for the International Decade of Ocean Exploration, National Science Foundation, Washington, D.C. 20550, 1973.

Anonymous, International Decade of Ocean Exploration, Progress Report Vol. 7, Living resources program, 69-77, National Science Foundation, International Decade of Ocean Exploration Section, Superintendent of Documents, U.S. Government Printing Office, Washington D.C. 20402, 1978.

Bakun, A., Coastal upwelling indices, west coast of North America, 1946-71, U.S. Department of Commerce, NOAA Technical Report, NMFS, SSRF-671, 1-103, 1973.

Bakun, A., Daily and weekly upwelling indices, west coast of North America, 1967-73, U.S. Department of Commerce, NOAA Technical Report, NMFS, SSRF-693, 1-114, 1975.

Bernal, P.A., Large-scale biological events in the California Current *CalCOFI Report 20*, 89-101, 1979.

Berner, L.D., Distributional atlas of Thaliacea in the California Current region, *CalCOFI Atlas No. 8*, vii-xi, 1-322, 1967.

Boje, R., and M. Tomczak, Ecosystem analysis and the definition of boundaries in upwelling regions, In *Upwelling Ecosystems*, R. Boje and M. Tomczak (eds.), 3-11, Springer-Verlag, Berlin, 1978.

Bowman, T.E., and M.W. Johnson, Distributional atlas of calanoid copepods in the California Current region, 1949 and 1950, *CalCOFI Atlas No. 19*, v-vii, 1-239, 1973.

Brinton, E., Distributional atlas of Euphausiacea (Crustacea) in the California Current region, Part I, *CalCOFI Atlas No. 5*, vii-xi, 1-275, 1967.

Brinton, E., Distributional atlas of Euphausiacea (Crustacea) in the California Current region, Part II, *CalCOFI Atlas No. 18*, v-vii, 1-336, 1973.

Chelton, D.B., Low frequency sea level variability along the west coast of North America, Ph.D. Thesis, University of California, San Diego, 1-151, 1980.

Colebrook, J.M., Annual fluctuations in biomass of taxonomic groups of zooplankton in the California Current, 1955-59, *Fishery Bulletin (U.S.)*, *75*, 357-368, 1977.

Fleminger, A., Distributional atlas of calanoid copepods in the California Current region, Part I, *CalCOFI Atlas No. 2*, ix-xvi, 1-313, 1964.

Fleminger, A., Distributional atlas of calanoid copepods in the California Current region, Part II, *CalCOFI Atlas No. 7*, vii-xvi, 1-213, 1967.

Hickey, B.M., The California Current System, hypotheses and facts, *Progress in Oceanography*, *8*, 191-279, 1979.

Kramer, D., and E.H. Ahlstrom, Distributional atlas of fish larvae in the California Current region: northern anchovy, *Engraulis mordax* (Girard), 1951 through 1965, *CalCOFI Atlas No. 9*, vii-xi, 1-269, 1968.

Kramer, D., M.J. Kalin, E.G. Stevens, J.R. Thrailkill, and J.R. Zweifel, Collecting and processing data on fish egg and larvae in the California Current region, U.S. Department of Commerce, NOAA Technical Report, NMFS Circ-370, iii-iv, 1-38, 1972.

Lynn, R.J., Seasonal variation of temperature and salinity at 10 meters in the California Current, *CalCOFI Reports 11*, 157-186, 1967.

McGowan, J.A., Distributional atlas of pelagic molluscs in the California Current region, *CalCOFI Atlas No. 6*, vii-xiv, 1-218, 1967.

McGowan, J.A., The Thecosomata and Gymnosomata of California, *The Veliger* (Supplement), 103-122, 1968.

McGowan, J.A. and C.B. Miller, Larval fish and zooplankton community structure, *CalCOFI Reports, 21*, 29-36, 1980.

McGowan, J.A., and P.M. Williams, Oceanic habitat differences in the North Pacific, *Journal of Experimental Marine Biology and Ecology, 12*, 187-217, 1973.

NORPAC Committee, *Oceanic observations of the Pacific: 1955, the NORPAC Atlas*, University of California Press and University of Tokyo Press, Berkeley and Tokyo, 123 pl., 1960.

O'Brien, J.J., R.M. Clancy, A.J. Clarke, M. Crepon, R. Elsberry, T. Gammelsrød, M. MacVean, L.P. Röed, and J.D. Thompson, Upwelling in the ocean: two- and three-dimensional models of upper ocean dynamics and variability, In *Modelling and Prediction of the Upper Layers of the Ocean*, E.B. Kraus (ed.), 178-228, Pergamon Press, Oxford, 1977.

Reid, J.L., On the circulation, phosphate-phosphorous content and zooplankton volumes in the upper part of the Pacific ocean, *Limnology and Oceanography, 7*, 287-306, 1962.

Reid, J.L., G.L. Roden, and J.G. Wyllie, Studies in the California Current System, *CalCOFI Reports, 6*, 27-57, 1958.

Smith, P.E., Distributional atlas of zooplankton volume in the California Current region, 1951 through 1966, *CalCOFI Atlas No. 13*, v-xvi, 1-144, 1971.

Soutar, A., and J.D. Isaacs, History of fish populations inferred from fish scales in anaerobic sediments off California, *CalCOFI Reports, 13*, 63-70, 1969.

Soutar, A., and J.D. Isaacs, Abundance of pelagic fish during the 19th and 20th centuries as recorded in anaerobic sediments off the Californias, *Fishery Bulletin (U.S.), 72*, 257-273, 1974.

Sverdrup, H.U., M.W. Johnson, and R.H. Fleming, *The Oceans, their Physics, Chemistry and General Biology*, v-x, 1-1087, Prentice-Hall, 1942.

Thomas, W.H., and D.L.R. Siebert, Distribution of

nitrate, nitrite, phosphate, and silicate in the California Current region, 1969, *CalCOFI Atlas No. 20,* vii-x, 2-97, 1974.

Venrick, E.L., The lateral extent and characteristics of the North Pacific Central environment at 35°N, *Deep-Sea Research, 26A,* 1153-1178, 1979.

Wickett, W.P., Ekman transport and zooplankton concentrations in the North Pacific Ocean, *Journal of the Fisheries Research Board of Canada, 24,* 581-594, 1967.

Yoshida, K., Coastal upwelling off the California coast, *Records of Oceanographic Works in Japan, 2,* 8-20, 1955.

Zentara, S.J., and D. Kamykowski, Latitudinal relationships among temperature and selected plant nutrients along the west coast of North and South America, *Journal of Marine Research, 35,* 321-337, 1977.

THE EFFECT OF ADVECTION ON VARIATIONS IN ZOOPLANKTON AT A SINGLE LOCATION NEAR CABO NAZCA, PERU

S.L. Smith

Brookhaven National Laboratory, Upton, New York 11973

K.H. Brink

School of Oceanography, Oregon State University, Corvallis, Oregon 97331

H. Santander

Instituto del Mar del Peru, Apartado 22, Callao, Peru

T.J. Cowles

Woods Hole Oceanographic Institution, Woods Hole, Massachusetts 02543

A. Huyer

School of Oceanography, Oregon State University, Corvallis, Oregon 97331

Abstract. Temporal variations in the biomass and species composition of zooplankton at a single mid-shelf station in an upwelling area off Peru can be explained to a large extent by onshore-offshore advection in the upper 20 m of the water column. During periods of strong or sustained near-surface onshore flow, peaks in biomass of zooplankton were observed at mid-shelf and typically oceanic species of copepod were collected. In periods of offshore flow at the surface, a copepod capable of migrating into oxygen-depleted layers deeper than 30 m was collected. A simple translocation model of advection applied to the cross-shelf distribution of *Paracalanus parvus* suggests that the fluctuations in *P. parvus* observed in the mid-shelf time series were closely related to onshore-offshore flow in the upper 20 m.

Fluctuations in abundance of the numerically dominant copepod, *Acartia tonsa*, were also affected by near-surface flow. The population age-structure suggests that *A. tonsa* was growing at maximal rates, due in part to its positive feeding response to an unusual assemblage of phytoplankton in 1976.

Introduction

Temporal variations in zooplankton at a given location in an upwelling region can occur as a result of both biological and physical processes. Biologically, variations occur on a time scale of weeks associated with the life cycle of certain organisms, and on a diurnal time scale associated with vertical migration. Physically, variations may occur as a result of onshore-offshore and alongshore advection, which generally have a dominant time scale of several days, intermediate to the two dominant biological time scales. We shall examine the temporal variability of zooplankton observed during a 15-day period of daily observations at a single location in the coastal upwelling region off Peru.

The observations of zooplankton were obtained from vertical net hauls made primarily at a single mid-shelf location, called C-3, at about 15°07'S, 75°31'W over the shelf just south of Cabo Nazca (Fig. 1). The hauls were made from R.V. *Alpha Helix* daily from 28 March to 11 April 1976, at about the same time each morning, between 0900 and 1100 LST, to avoid posible aliasing from the diurnal cycle of vertical migration. Live specimens of the dominant species, *Acartia tonsa*, were collected from the net hauls and maintained in the laboratory to study their grazing habits. The offshore distributions of zooplankton were inferred from occasional observations between 1 and 5 April at other locations along the C line (Fig. 1). No direct information is available about the alongshore distribution of zooplankton, but alongshore variations in biological variables are believed to be much less than offshore varia-

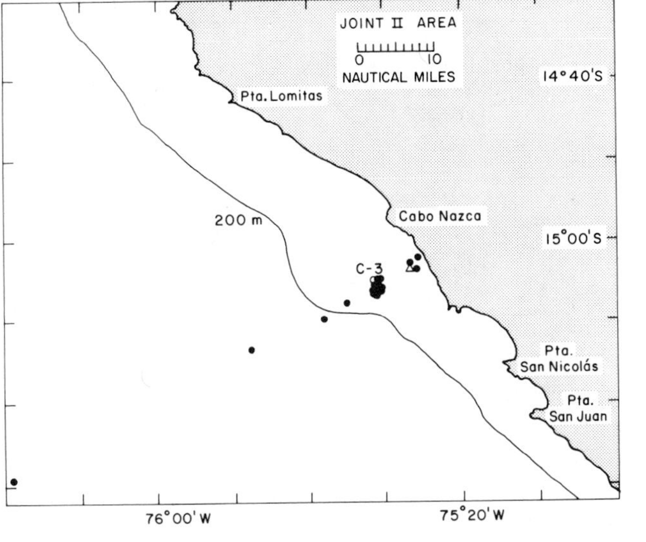

Fig. 1. Location of zooplankton observations from R.V. *Alpha Helix*, 28 March to 11 April 1976. C-3 is the location of the daily time series. Currents were measured at 10 and 29 m at Agave (Δ) and at 28, 53, and 78 m at Mila (O). The wind was measured at a coastal station at Punta San Juan.

tions (Dugdale *et al.*, 1977). Our observations were made during an atypical period in the upwelling regime off Peru. Dinoflagellates dominated the phytoplankton along much of the coast (Dugdale *et al.*, 1977; Rojas de Mendiola, 1979), and denitrification was taking place (Packard, Dugdale, Goering, and Barber, 1978). The prevailing conditions in 1976 have been characterized both as an *aguaje* (Rojas de Mendiola, 1979) and *El Niño* (Wyrtki, 1977).

Currents were observed during the period at two locations over the shelf, at Mila, alongside C-3, and at Agave, inshore of C-3 (Fig. 1). The current meters at Mila were all at depths greater than 25 m, but at Agave one of the current meters was at 10 m, which is usually well within the surface Ekman-like layer (the planned array did not include near-surface measurements). Wind observations were made at a coastal station at San Juan (Fig. 1). Preliminary analysis indicated that the zooplankton data were not correlated with either the alongshore wind or the alongshore currents at Mila or Agave. However, the onshore component of the near-surface current seemed to account for some of the variability in the zooplankton samples. Some of the variability, particularly in the dominant species *A. tonsa* and hence in the biomass, can also be explained in terms of *in situ* growth. We shall demonstrate that onshore-offshore advection alone accounts for the occurrence of most of the species of zooplankton at C-3, while both growth and advection affect the occurrence of the dominant species, *A. tonsa*.

Methods

Zooplankton were collected in vertical hauls of 60-cm diameter Bongo frames fitted with 102-μm mesh nets (S.L. Smith, 1978) and in vertical hauls of a 70-cm diameter Hensen net fitted with a 300-μm mesh net and a 10-mm mesh prefilter in the net mouth to exclude large medusae. At stations with depths less than 100 m, hauls of the Hensen net were from bottom to surface and 50 m to surface, while at deeper stations hauls were from 200 m to surface and from 50 m to surface. One of the deeper catches at each station was frozen and returned to the laboratory where it was washed with distilled water onto a preweighed filter, dried at 60°C for 48 h, and weighed. All other samples were preserved in formalin for determination of displacement volume and the quantitative analysis of species composition. The hauls of the Bongo nets were from 100 or 200 m to the surface, depending on the depth. The collection from one net was poured through sieves of 102, 223, 505, and 1050-μm mesh, dried at 60°C, weighed, and corrected for phytoplankton (see S.L. Smith, 1978, for details). Net hauls at C-3 and C-5 were taken during daylight to remove the effects of diurnal migration.

Grazing rates of *A. tonsa*, the most abundant copepod during this study, were measured in late April, 1976. Adult *A. tonsa* were sorted from samples collected by vertical net tows (1 to 100 m) into beakers containing unfiltered surface seawater. Experimental food conditions were established from ambient chlorophyll *a* concentrations at selected euphotic zone depths. Ambient concentrations were occasionally diluted with filtered seawater to create a range of food concentrations. Two adult *A. tonsa* were placed in 250-ml bottles, each experiment having at least two replicates plus a control. Bottles were held in a light-proof water bath for 24 h. Surface seawater circulated through the bath, keeping temperatures within 1°C of "catch" temperatures, and the polyethylene bottles were gently rolled to keep the food in suspension. A Coulter counter was not available, so grazing rates have been calculated from initial and final chlorophyll *a* concentrations in the control and experimental bottles, using the equations of Frost (1972). Data are presented in terms of carbon, assuming a carbon:chlorophyll ratio of 50 (see Cowles, 1977, for details).

The current and temperature observations were obtained from Aanderaa current meters on subsurface moorings with instruments at 10 and 29 m at Agave, at 28, 53, and 78 m at Mila, and with additional instruments at more distant locations. The time series were low-pass filtered with a half-power point of 47 h to remove tidal and other variations of shorter period. Details of installation and data treatment can be found in Enfield, Smith, and Huyer (1978), and data are discussed more fully by Brink, Allen, and R. Smith (1978) and R. Smith (1978). The San Juan

TABLE 1. Mean Current Meter Data

Position and Instrument Depth		Cross-shelf Velocity Component $\bar{u}$(cm/s)*	Alongshore Velocity Component $\bar{v}$(cm/s)**	Temperature $\bar{T}$(°C)
Agave	10 m	-0.7	9.8	15.59
	29 m	1.8	-3.5	15.12
Mila	28 m	6.3	-19.9	15.50
	53 m	3.6	-22.6	14.65
	78 m	2.3	-19.0	14.36

* Positive (negative) values are for onshore (offshore) flow.
** Positive (negative) values are for equatorward (poleward) flow.

wind data were obtained from a land-mounted anemometer with good exposure to the coastal winds (Stuart, Spetsaris, and Nanney, 1976).

Results

The Flow Field

The mean currents (Table 1) for 39.5 days from 28 March to 6 May, 1976, including the period of daily observations of zooplankton, show that the flow is equatorward and weakly offshore near the surface at Agave, and poleward and onshore at each of the deeper current meters at both locations. The equatorward flow at 10 m at Agave was part of the wind-driven surface current, while the poleward flow elsewhere was part of the undercurrent (Brink *et al.*, 1978). The mean onshore flow at depth and offshore flow at the surface are consistent with an upwelling circulation pattern driven by the mean equatorward alongshore wind stress of 0.82 dynes/cm^2. The mean flow at 10 m at Agave is probably representative of the mean near-surface currents across the whole shelf in the region. Direct observations of both near-surface and sub-surface currents with better vertical resolution at nearby locations in 1977 (Brink, Smith, and Halpern, 1978) substantiate this.

Fluctuations in the currents in the region were associated with fluctuations in the local wind stress and with remotely generated disturbances that appeared to propagate as low frequency, coastally trapped waves (R.L. Smith, 1978). At both Agave and Mila, fluctuations in the alongshore current were not significantly correlated with the local wind stress (Brink, Allen, and R.L. Smith, 1978). The alongshore current fluctuations at all current meters are significantly correlated with each other, and with current observations off Callao (375 km north of Mila and Agave) but with a lag of 2 days due to the wave propagation (R.L. Smith, 1978). Fluctuations in the onshore flow at the deeper current meters are significantly correlated with the alongshore wind stress, with stronger onshore flow when the equatorward wind stress was stronger. Simple Ekman dynamics would lead us to expect that the surface offshore flow was also stronger when the wind stress was stronger, but the onshore component of the current at 10 m at Agave was not significantly correlated with the local wind stress (Brink *et al.*, 1978). Part of the variance in the onshore flow might be due to propagating disturbances. In either process, variations in the onshore current at 10 m at Agave probably represent the near-surface flow across the whole shelf in this region.

Biomass of Zooplankton

Net hauls at C-3 were made from both 50 and 100 m with the 300-μm Hensen net. The time series of settled volume from both hauls (Fig. 2) shows that they captured similar amounts of zooplankton. On 2 and 8 April, jellyfish fragments were abundant; they probably account for the higher volume from 50 m than from 100 m on April 8. Except on 28 March and 1 April, when the 100-m haul had a significantly greater volume than the 50-m haul, the zooplankton at C-3 were concentrated in the upper 50 m.

The concentration of the zooplankton within the upper 50 m may have been due in part to the presence of water with very low oxygen concentrations. The vertical distribution (Fig. 3) of dissolved oxygen observations during this period (MacIsaac *et al.*, 1979) shows values less than 1 ml/l at depths greater than 30 m. These values may be high, because the water showed evidence of denitrification (Packard *et al.*, 1978). If most species of zooplankton avoid such oxygen-deficient water, and there is evidence that some do (Boyd, Smith, and Cowles, 1980; Judkins, 1980), most of them would be concentrated in the upper 30 m. Furthermore, the surface mixed layer was less than 15 m deep throughout the study (MacIsaac *et al.*, 1979). If such near-surface concentration occurred, we would expect the near-surface currents, represented by 10 m at Agave, to have a significant effect on the distribution of zooplankton.

Time series of the biomass of zooplankton at C-3 (Fig. 2) were obtained from the 100-m hauls of each net; estimates from the 300-μm Hensen net are about one-fifth those from the 102-μm Bongo net. Dry weights from two size fractions collected in the 102-μm Bongo net (Fig. 4) show that over the shelf most of the biomass of zooplankton was in the smaller (<505 μm) size category. Variations in the two biomass estimates are qualitatively similar, except for a large discrepancy on April 5, for which the Bongo net estimate is inexplicably low. Comparing day-to-day fluctuations in biomass with the cross-shelf flow measured at 10 m at Agave shows that peaks in biomass generally occurred during onshore flow and that biomass was least during offshore flow (Fig. 2). The largest biomass appeared near the end of the period of sustained onshore flow between 2 and 6 April (Fig. 2).

The qualitative similarity between biomass and onshore flow can be explained by the offshore distribution of zooplankton biomass: for each size range, the biomass increased near the shelf break (Fig. 4). The total biomass of zooplankton smaller than 505 μm had a maximum at the shelf edge, with lower biomass farther offshore. As an additional comparison between the cross-shelf flow and the biomass, we computed a "predicted" time series by assuming that the section shown in Fig. 4 is representative for a particular date and

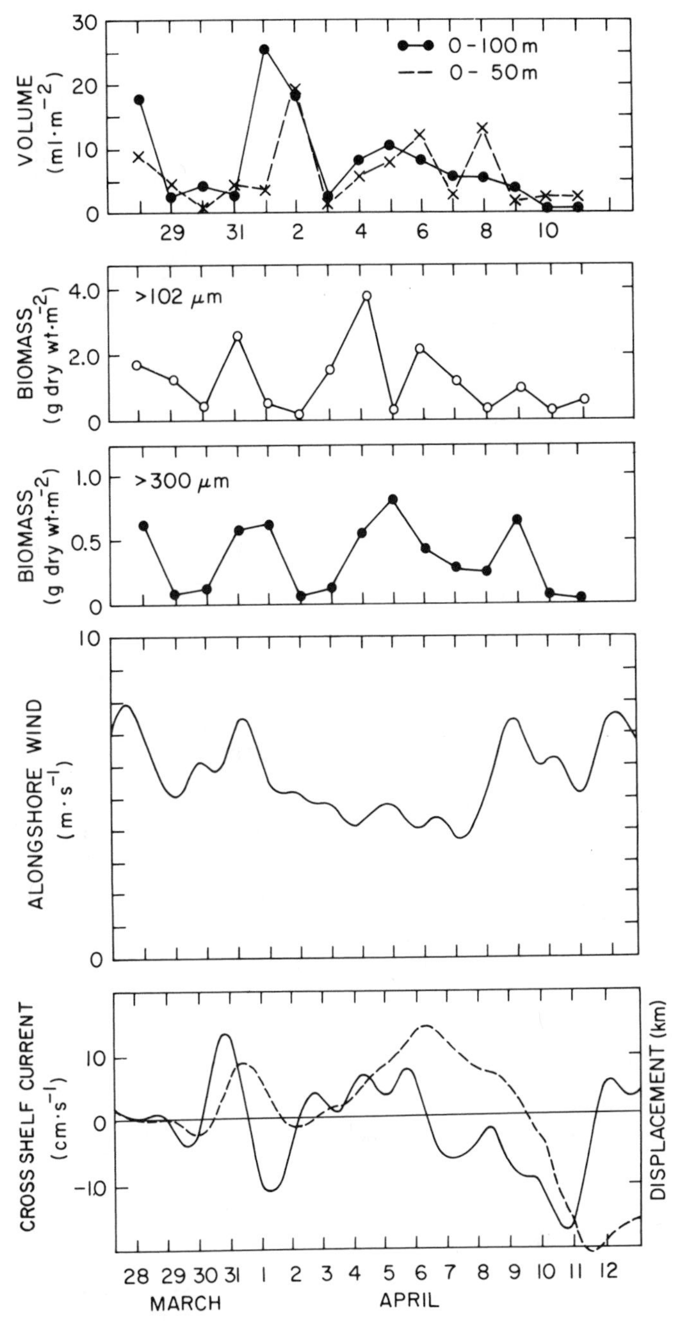

Fig. 2. Settled volume and biomass of the zooplankton at C-3, cross-shelf current at Agave, and wind at San Juan. Settled volume is shown for 50- and 100-m net hauls of the 300-μm Hensen net; biomass as dry weight is shown for 100-m hauls of both 102-μm Bongo net and 300-μm Hensen net. The cross-shelf current (positive for onshore flow) at 10 m at Agave is shown (——) with it times integral (---) depicting displacement of a water mass from its position on 3 April if the currents were homogenous in space.

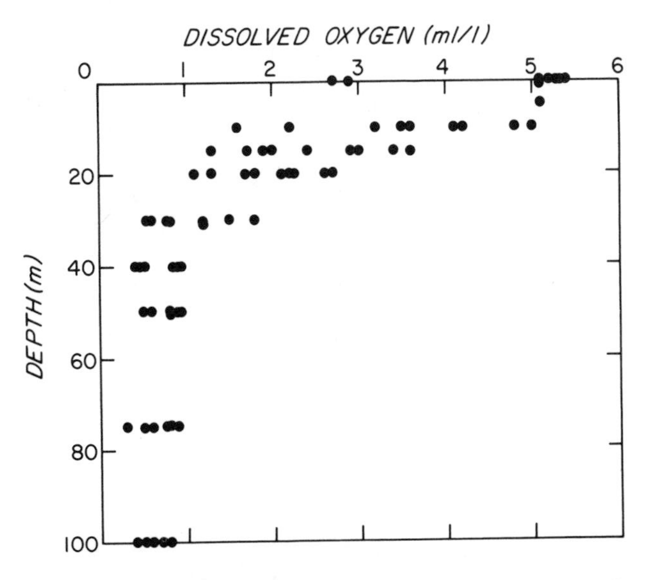

Fig. 3. Vertical profile of the concentration of dissolved oxygen at C-3, from *Alpha Helix* stations 20, 22, 24, 27-29, 31, and 33, occupied daily 3 April to 11 April 1976. Values were determined with a polarographic oxygen electrode (MacIsaac *et al.*, 1979).

that the section is passively displaced onshore or offshore without distortion. We assumed the offshore profile of total biomass (both >102 μm and >300 μm) was representative for 3 April, the date on which the net hauls at the shelf break (C-5) were made. The displacement for each date before and after 3 April was computed from the time integral of the cross-shelf current at 10 m at Agave. The resulting time series of predicted biomass (Fig. 5) does not agree very well with the observed biomass, but it does show the two major peaks in biomass that occurred on 31 March and 4 to 6 April. We repeated the calculation of predicted biomass using cross-shelf currents averaged over several depths but all other combinations failed to show the March 31 peak.

We also used the offshore profiles of biomass for different size fractions, but found no improvement in the predicted biomass. Although agreement is not good, the coincidence of peaks for the two different series of net hauls suggests that the zooplanktonic community as a whole was sensitive to currents in the uppermost part of the water column. The implication is that for

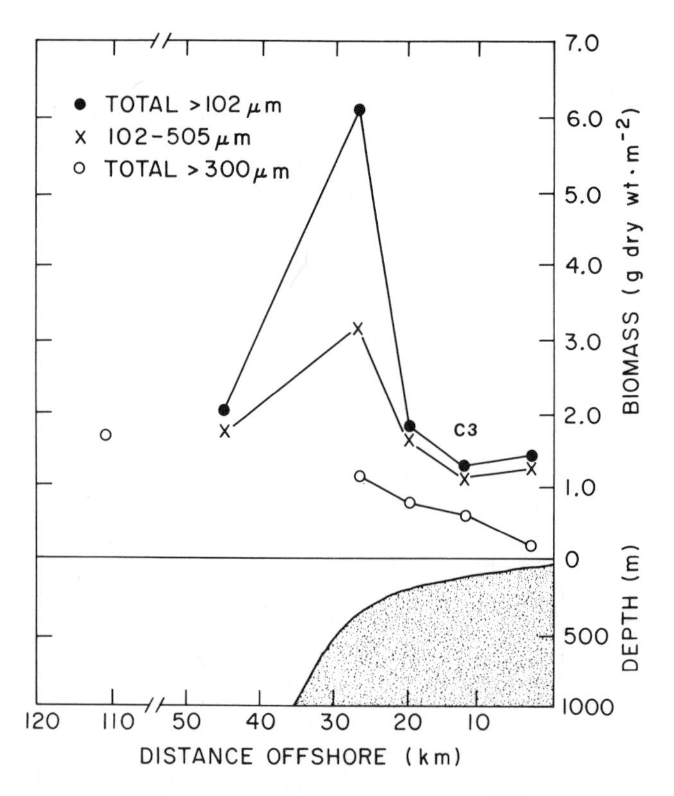

Fig. 4. The offshore distribution of biomass collected in vertical hauls of the 300-μm Hensen (O) and 102-μm Bongo (●) nets between 1 and 5 April, 1976. All hauls except the Hensen net haul at 110 km offshore were made during daylight. Within 5 km of shore, net hauls were made to 50 km; between 5 and 25 km, hauls were to 100 m; farther offshore, hauls were to 200 m.

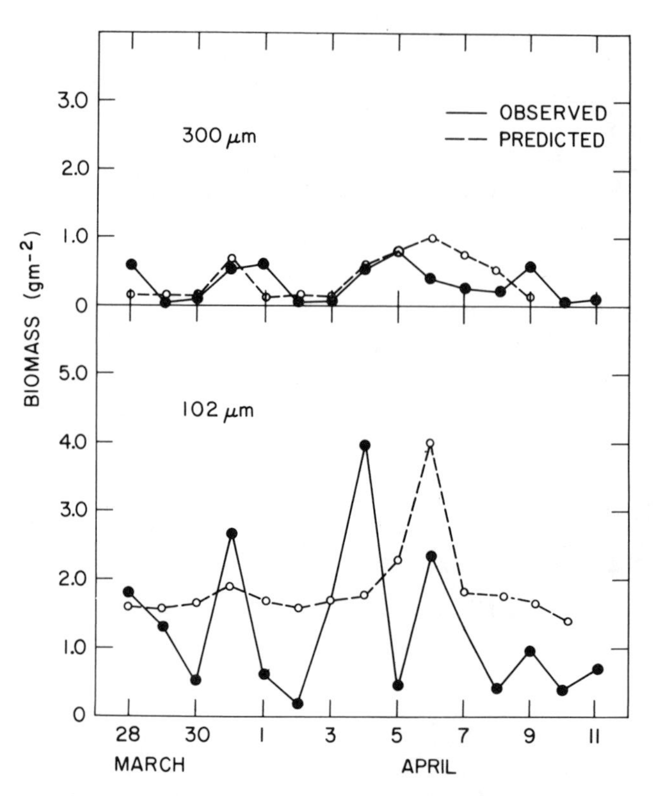

Fig. 5. Time series of the observed and predicted biomass at C-3 for the 102-μm Bongo net (lower panel) and the 300-μm Hensen net (upper panel). Predicted biomass was computed from the onshore (positive) or offshore (negative) displacement of the offshore profile of biomass from its assumed position of 3 April (Fig. 4).

the most part zooplankton were restricted to the layer influenced by surface currents. Any agreement is encouraging because there was bound to be distortion in the pattern, the cross-shelf profile (Fig. 4) was not synoptic, and *in situ* processes were ignored.

Species Composition

Patterns of the abundance and diversity show three groups of species of copepod associated with the continental shelf, continental slope, and deep ocean. The shelf group, composed of *A. tonsa* Dana and *Centropages brachiatus* Dana, decreased sharply in abundance offshore of the shelf break (Fig. 6). The continental slope group, containing *Paracalanus parvus* Claus, *Eucalanus* spp., and *Oncaea* spp., increased gradually in abundance from mid-shelf to 110 km offshore (Fig. 6). The deep-water or oceanic group contained 17 species, which were most abundant between 80 and 110 km offshore; four are shown in Fig. 6.

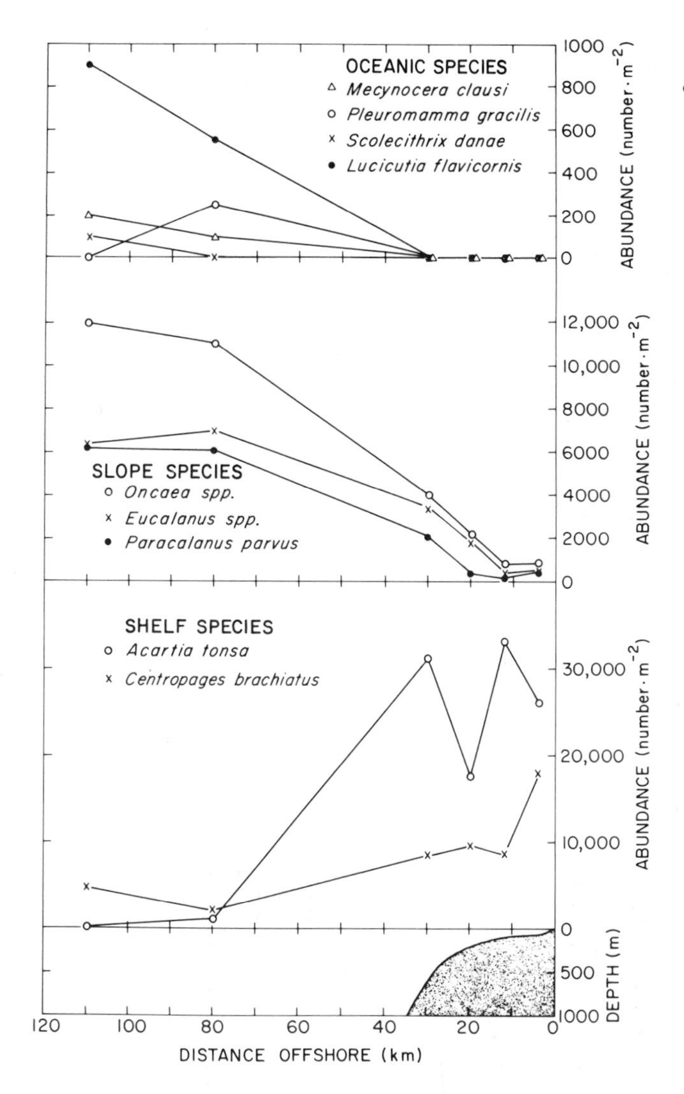

Fig. 6. Offshore profiles of abundance of various species of zooplankton in the oceanic group, the continental slope group, and the continental shelf group. Note the change of scale from one group to another. All abundances are from hauls of the 300-μm Hensen net from 50 m to the surface, except at 77 km from shore, where the 50-m sample was lost and we used the 200-m haul.

Day-to-day fluctuations in the abundance of each group of species at C-3 (Fig. 7) show the effects of onshore-offshore advection more clearly than does biomass. The numerically dominant copepod during this study was *A. tonsa*, a ubiquitous and opportunistic species often found associated with dinoflagellate blooms (Conover, 1956; Seki, Tsuji, and Hattori, 1974; S. Smith, 1978; Morey-Gaines, 1979; Walsh *et al.*, 1974). The species is regularly present in coastal waters off northern Chile (Vidal, 1968) and off Peru (Heinrich, 1973), but it is not abundant in per-

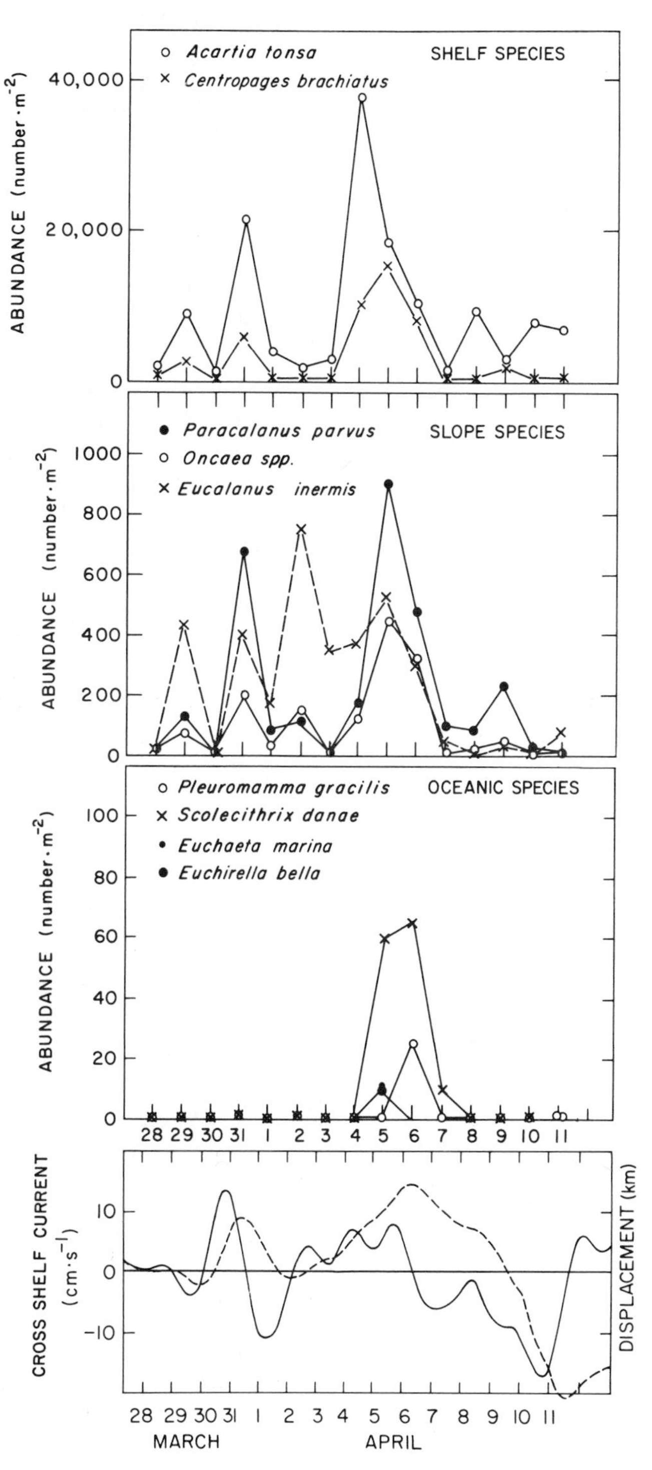

Fig. 7. Time series of the abundance of various species of zooplankton in the continental shelf, continental slope, and oceanic groups, with the cross-shelf component of the current at 10 m at Agave (——) and its time integral (---). All abundances are from the 50-m haul of the Hensen net.

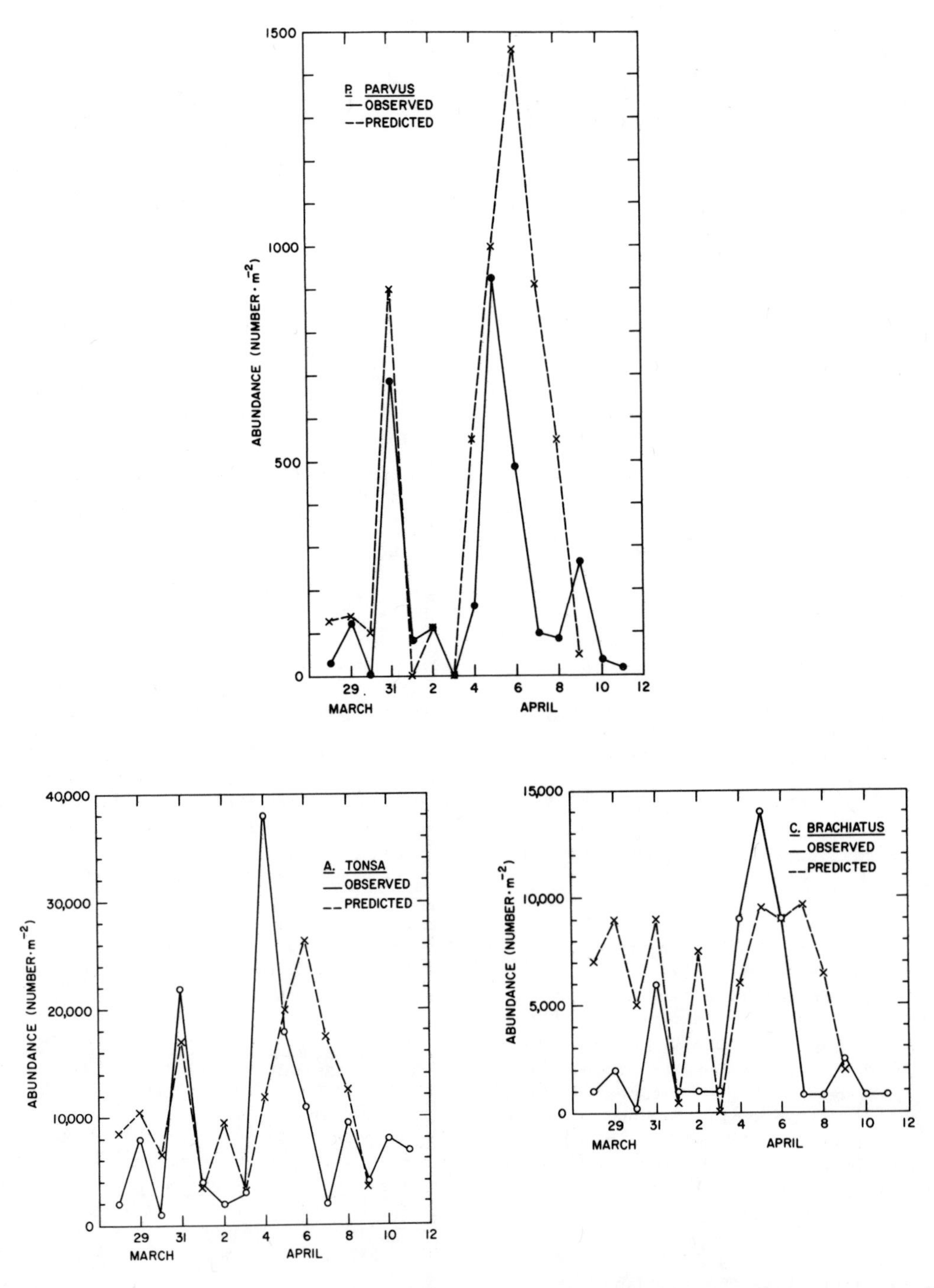

Fig. 8. Time series of the observed and predicted abundance at C-3 of *Paracalanus parvus*, *Centropages brachiatus*, and *Acartia tonsa*. Data are from 50-m hauls of the Hensen net.

iods of normal upwelling (Heinrich, 1973). The maximum abundance of *A. tonsa* occurred during the period of sustained onshore flow (2 to 6 April) with a secondary peak during the relatively brief but strong onshore flow of 30 and 31 March (Fig. 7). Lowest numbers were associated with periods of offshore flow. The next most abundant species, the shelf copepod *Centropages brachiatus*, also had maxima during periods of onshore flow (Fig. 7), suggesting that shelf species accumulate at C-3 (and probably inshore of C-3) during onshore flow.

Two of the slope species, *Paracalanus parvus* and *Oncaea* spp., also had peak abundances during

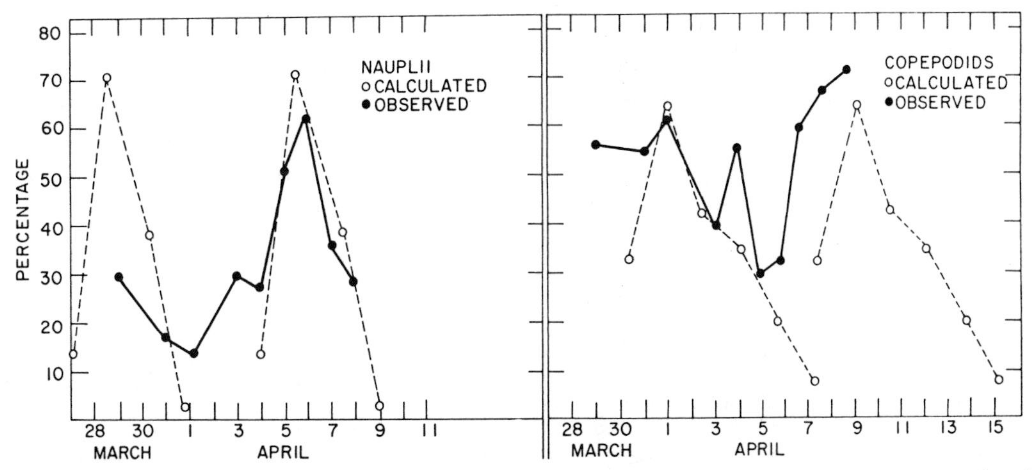

Fig. 9. Observed and calculated percentages of the number of nauplii and copepodids in the *Acartia tonsa* population at C-3. The dashed lines are fractions that would result if *A. tonsa* was growing at the maximum rate for the existing sea-surface temperature based upon laboratory results obtained by Heinle (1966).

periods of onshore flow; during the period of sustained onshore flow (3 to 6 April) they lagged the shelf species by a day. The oceanic species did not appear during or after the short burst of strong onshore flow on 31 March, but they were observed during and immediately after the period of sustained onshore flow (Fig. 7). In addition, other organisms such as *Euphausia mucronata* (Euphausiacea), *Muggiaea atlantica* (Siphonophora), and doliolids, which generally occur seaward of the shelf break, were captured at C-3 during periods of onshore flow. Fig. 7 suggests a lag in abundances of the three groups following onshore flow beginning on 3 April; *A. tonsa* peaked on 4 April, the slope species peaked on 5 April, and the oceanic species peaked on 6 April.

One of the species in the continental slope group, *Eucalanus inermis* Giesbrecht, was most abundant at C-3 on 2 April, in a period of offshore flow at Agave 10 m (Fig. 7). The species was not confined to the upper layer (< 30 m) over the shelf in this area in 1977 (Boyd *et al.*, 1980); it could migrate deeper than 30 m where oxygen concentrations were generally less than 0.1 ml/l. Thus *E. inermis* was probably not restricted to the near-surface layer and is likely indicative of freshly upwelled water being advected offshore.

Since there was a cross-shelf pattern in the species of zooplankton, we computed predicted abundances from the offshore abundance profile and the onshore current data, in the same way we computed predicted biomass described above. In particular, we computed the predicted abundance of *P. parvus*, one species of the continental slope group. We assumed that the offshore distribution of *P. parvus* shown in Fig. 6 was valid for 3 April; its onshore or offshore displacement from that date was determined from the time-integral of the cross-shelf current at 10 m at Agave (other estimates of the cross-shelf current were also used, but the 10-m Agave data again gave the best prediction). The overall agreement of the observed and predicted curves (Fig. 8) is startling considering how simplistic the calculation is: ignoring the effect of the nearby coast, assuming onshore currents are the same in all space, neglecting alongshore effects, and assuming that zooplankton are conservative, passive tracers. We conclude that advective effects are extremely important in determining the distribution of *P. parvus*, and that as shallow currents give by far the best prediction, its population is concentrated in the upper 10 to 20 m of the water column. The second conclusion is supported by the observation that generally 0 to 50 and 0 to 100-m casts show similar patterns in *P. parvus* counts. On 7 and 8 April large numbers of *P. parvus* were collected below 50 m while few were collected above 50 m, resulting in an over-prediction of the *P. parvus* population. Similar calculations for *C. brachiatus* and *A. tonsa* show general agreement between observed and predicted abundances at C-3 as well (Fig. 8) though both patterns have over-predictions on 28 and 29 March and 7 and 8 April. The over-prediction on 7 and 8 April suggests that on those days much of the zooplanktonic community may have been distributed deeper in the water column than on other days of the time series.

The dominant species: Acartia tonsa

Up to this point we have presented evidence that mass and species composition of zooplankton passing a fixed point (C-3) on the continental shelf off Peru in 1976 were associated with the onshore flow measured at 10-m depth near shore. That advection was not the only factor determining the abundance of zooplankton at C-3 becomes clear when we consider the biological characteristics of *A. tonsa*. Analyses of daily collections with the 102-μm mesh nets from C-3 for the percent composition of nauplii and copepodids of *A. tonsa* show three peaks in copepodid fractions and two in the fraction of nauplii (Fig. 9). Using the laboratory data of Heinle (1966) and the sea-surface temperatures prevailing during this time series, the maximal growth rate of *A. tonsa* could produce an egg-to-egg cycle of 11 to 13 days with the age structure shown as a dashed line in Fig. 9. Comparing the age structure of an *A. tonsa* population resulting from the maximum growth rate calculated from laboratory data with the observed population age structure suggests that most of the observed fractions of nauplii and copepodids in the *A. tonsa* population at C-3 could result from the introduction of new nauplii and copepodids on two different days followed by maximum growth (Fig. 9). The days on which new nauplii would have to have been advected into the C-3 area are 27 March and 4 April 1976, days which are near the end of periods of onshore flow.

The ingestion rate of *A. tonsa* increased linearly with food concentrations (Fig. 10). The ingestion experiments were over a period of 7 days (22 to 28 April, 1976) when the composition of phytoplankton was changing from about 50 to about 85% diatoms (D. Blasco, personal communication). Comparison of the results in Fig. 10 with phytoplankton composition data indicate that shifts in the relative abundance of diatoms and dinoflagellates had little or no effect on the measured ingestion rate of *A. tonsa*. The phytoplankton were dominated by diatoms or *Gymnodinium splendens* throughout March and April 1976 (D. Blasco, personal communication), and mean chlorophyll *a* for the euphotic zone between 28 March and 11 April 1976 ranged from 2.5 to 50.5 mg/m^3

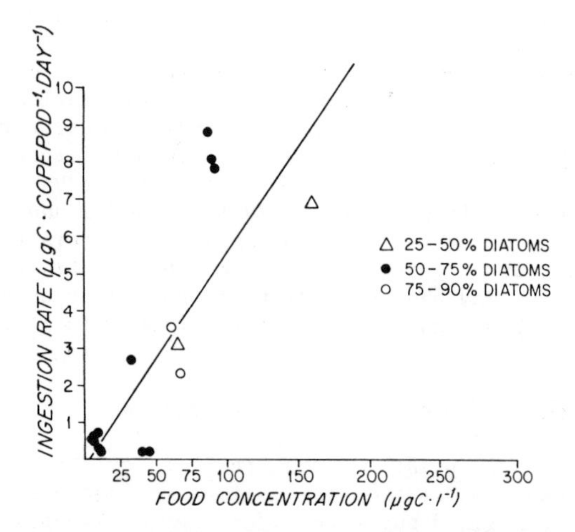

Fig. 10. Ingestion rates of *Acartia tonsa* feeding on phytoplankton assemblages containing 25 to 50, 50 to 75, and 75 to 90% diatoms. Data are from grazing experiments conducted on the *Alpha Helix* during 22 to 28 April 1976.

(MacIsaac *et al.*, 1979), indicating that adequate plant food was available throughout the time series. As food for *A. tonsa* was abundant and ingested (Fig. 10), and our estimates of the dates of new introductions of nauplii have a reasonable foundation, we suggest that advection and a potentially maximal growth rate could operate together to create the large numbers of *A. tonsa* observed.

Discussion

Previous studies of advection and zooplankton in areas of upwelling have not demonstrated explicitly the relationship between daily fluctuations in the cross-shelf patterns in species of zooplankton and onshore-offshore advection (Blackburn, 1979; Peterson, Miller, and Hutchinson, 1979). The combination of current meters moored at mid-shelf and near-shore at 15°S, cross-shelf sampling of zooplankton, and daily biological measurements of the entire mid-shelf water column have allowed us to demonstrate that the species composition and biomass of zooplankton at mid-shelf were closely related, on a daily basis, to onshore-offshore displacement of approximately the upper 20 m of the water column, as recorded somewhat inshore of the mid-shelf station. We have been able to describe short-term advective effects on the zooplankton at mid-shelf and to separate those effects from the effect of *in situ* growth by the numerically dominant copepod in the area. The considerable variability on short time scales (daily) observed during this study suggests that studies of seasonal means, such as that of Peterson *et al.* (1979), neglect a time scale that may be important in understanding the dynamics of upwelling systems.

We observed no effects that we can attribute to alongshore flow, although the alongshore current is much stronger than the cross-shelf current over the Peruvian shelf (Brink *et al.*, 1978). At Mila, adjacent C-3, the alongshore flow is stronger than the cross-shelf flow by a factor of at least three for the mean, and a factor of two for the standard deviation. Nearer shore, at Agave, the differences are even greater: a factor of four or more for the means and a factor of three for the standard deviations. If alongshore gradients were weaker than offshore gradients by a factor of two or three, alongshore currents still would have about as great an effect in producing local results at C-3 as cross-shelf currents, because the advective effect due to a component is proportional to velocity times the gradient in the given direction. Because our results can be explained by cross-shelf currents and cross-shelf biological profiles, we conclude that the alongshore gradients are even weaker. The presence of the dinoflagellate *Gymnodinium splendens* along the entire coast of Peru during this period (Rojas de Mendiola, 1979; Dugdale *et al.*, 1977) also suggests that alongshore biological gradients were weak.

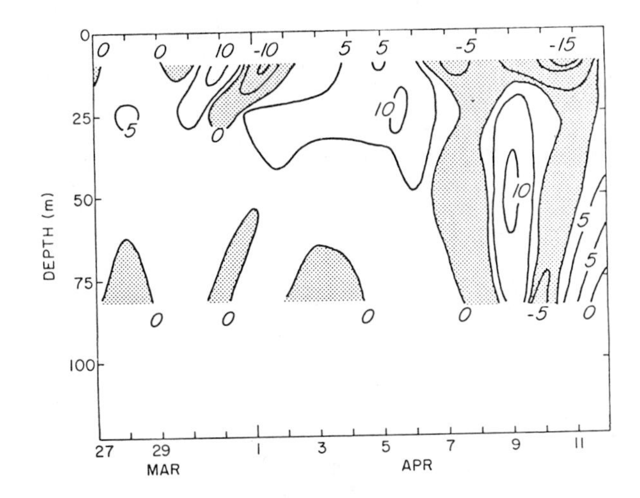

Fig. 11. Observed mid-shelf onshore (positive) and offshore (negative) flow at C-3, 28 March to 11 April, 1976. Data are from 10 m at Agave and 28, 53, and 78 m at Mila.

We would not expect similar results in regions where alongshore gradients are as sharp as offshore gradients, or for zooplankton species that undergo extensive diel vertical migration. In upwelling regions, the cross-shelf component of flow is strongly depth-dependent, with offshore flow limited to a rather thin surface layer; the compensating onshore flow may occur along the bottom as it appears to off northwest Africa (Mittelstaedt, Pillsbury, and Smith, 1975) or in the interior as it does off Oregon (Huyer, 1976). Thus lateral advection of a migrating organism would depend upon the vertical structure of the currents. The upwelling circulation off Peru was generally three layered: the surface offshore flow was confined to approximately the upper 20 m, the compensating onshore flow was maximum at mid-depth (about 60 m at C-3), and there was weak offshore flow along the bottom (Brink *et al.*, 1978). Even though an upwelling circulation did not dominate the cross-shelf currents during the period of our study, there was significant vertical structure in the cross-shelf currents (Fig. 11), which would subject migrating zooplankton to opposing lateral flows during a daily cycle. Our conclusion that the abundance of zooplankton at C-3 is strongly affected by the near-surface current implies that there was little or no vertical migration and that the zooplankton were concentrated in the upper layer. The conclusion that there was little vertical migration is consistent with the dominance of *A. tonsa* and the presence of oxygen-depleted water below 30 m; *A. tonsa* seldom exhibits diel migratory behavior even in oxygenated waters (Conover, 1956) and the lack of sufficient oxygen probably inhibits any tendency of many other species to migrate downward (Judkins, 1980).

In many coastal upwelling regions the currents fluctuate with a time scale of several days, while most species of zooplankton have a life cycle of several weeks. The numerical dominance of *A. tonsa* during this study may have been related to the unusual temporal scales for processes affecting *A. tonsa*. The dominant period of onshore advection was about 6 days, while that for maximum reproduction and growth of *A. tonsa* was about 13 days (Heinle, 1966). Thus, the advective time scale and the time scale for maximum rates of production for *A. tonsa* are of the same order, suggesting comparable importance of the two processes. As a food source of dinoflagellates is known to enhance egg production by *A. tonsa* (Morey-Gaines, 1979) and dinoflagellates, particularly *Gymnodinium splendens*, were unusually abundant off Peru in 1976 (Dugdale *et al.*, 1977), the numerical dominance of *A. tonsa* was apparently the result of its ability to grow and reproduce rapidly on the unusual phytoplankton assemblage, its tendency not to undergo diel vertical migration into deeper, oxygen-depleted water, and the effect of a favorable advective regime.

Acknowledgements. This study was supported by the Coastal Upwelling Ecosystems Analysis Program of the Office for the International Decade of Ocean Exploration of the National Science Foundation (NSF Grants OCE 78-03380, OCE 78-03381, OCE 78-05737-A01 and OCE 78-03043) and by the Instituto del Mar del Peru. Robert L. Smith provided the 1976 current meter data, and David Stuart provided data on winds. Susan A. Huntsman and Dolors Blasco participated in early discussions of data that led to this paper.

References

Blackburn, M., Zooplankton in an upwelling area off northwest Africa: composition, distribution and ecology, *Deep-Sea Research, 26A*, 41-56, 1979.

Boyd, C.M., S.L. Smith, and T.J. Cowles, Grazing patterns of copepods in the upwelling system off Peru, *Limnology and Oceanography, 25*, 583-596, 1980.

Brink, K.H., J.S. Allen, and R.L. Smith, A study of low-frequency fluctuations near the Peru coast, *Journal of Physical Oceanography, 8*, 1025-1041, 1978.

Brink, K.H., R.L. Smith, and D. Halpern, A compendium of time series measurements from moored instrumentation during the MAM '77 phase of JOINT-II, CUEA Technical Report 45, Ref. 78-17, School of Oceanography, Oregon State University, Corvallis, 72 pp., 1978.

Conover, R.J., Oceanography of Long Island Sound, IV. Biology of *Acartia clausi* and *A. tonsa*, *Bulletin of the Bingham Oceanographic Collection, 15*, 156-233, 1956.

Cowles, T.J., Copepod feeding in the Peru upwelling system, Ph.D. Thesis, Duke University, Durham, 173 pp., 1977.

Dugdale, R.C., J.J. Goering, R.T. Barber, R.L. Smith, and T.T. Packard, Denitrification and hydrogen sulfide in the Peru upwelling region during 1976, *Deep-Sea Research, 24*, 601-608, 1977.

Enfield, D.B., R.L. Smith, and A. Huyer, A compilation of observations from moored current meters, Volume XII: wind, currents and temperature over the continental shelf and slope off Peru during JOINT-II, March 1976 - May 1977, CUEA Data Report 70, Ref. 78-4, School of Oceanography, Oregon State University, Corvallis, 343 pp., 1978.

Frost, B.W., Effects of size and concentration of food particles on the feeding behavior of the marine planktonic copepod *Calanus pacificus*, *Limnology and Oceanography, 17*, 805-815, 1972.

Heinle, D.R., Production of the calanoid copepod, *Acartia tonsa*, in the Patuxent River estuary, *Chesapeake Science, 7*, 59-74, 1966.

Heinrich, A.K., Horizontal distribution of copepods in the Peru Current region, *Oceanology, 13*, 94-103, 1973.

Huyer, A., A comparison of upwelling events in two locations: Oregon and northwest Africa, *Journal of Marine Research, 34*, 531-546, 1976.

Judkins, D.C., Vertical distribution of zooplankton in relation to the oxygen minimum off Peru, *Deep-Sea Research, 27A*, 475-487, 1980.

MacIsaac, J.J., J.E. Kogelschatz, B.H. Jones, Jr., J.C. Paul, N.S. Breitner, and N. Garfield (eds.), JOINT-II R.V. *Alpha Helix* productivity and hydrographic data, March-May 1976, CUEA Data Report 48, 324 pp., 1979.

Mittelstaedt, E., R.D. Pillsbury, and R.L. Smith, Flow patterns in the northwest African upwelling area, *Deutsche Hydrographische Zeitschrift, 28*, 145-167, 1975.

Morey-Gaines, G., The ecological role of red tides in the Los Angeles - Long Beach Harbor food web, in *Toxic Dinoflagellate Blooms*, D.L. Taylor and H.H. Seliger (eds.), Elsevier North Holland, Inc., pp. 315-320, 1979.

Packard, T.T., R.C. Dugdale, J.J. Goering, and R.T. Barber, Nitrate reductase activity in the subsurface waters of the Peru Current, *Journal of Marine Research, 36*, 59-76, 1978.

Peterson, W.T., C.B. Miller, and A. Hutchinson, Zonation and maintenance of copepod populations in the Oregon upwelling zone, *Deep-Sea Research, 26*, 467-494, 1979.

Rojas de Mendiola, B., Red tide along the Peruvian coast, in *Toxic Dinoflagellate Blooms*, D.L. Taylor and H.H. Seliger (eds.), Elsevier North Holland, Inc., pp. 183-190, 1979.

Seki, H., T. Tsuji, and A. Hattori, Effect of zooplankton grazing on the formation of the anoxic layer in Tokyo Bay, *Estuarine and Coastal Marine Science, 2*, 145-151, 1974.

Smith, R.L., Poleward propagating perturbations in currents and sea levels along the Peru coast, *Journal of Geophysical Research, 83*, 6083-6092, 1978.

Smith, S.L., Nutrient regeneration by zooplankton during a red tide off Peru, with notes on biomass and species composition of zooplankton, *Marine Biology, 49*, 125-132, 1978.

Stuart, D.W., M.A. Spetseris, and M.M. Nanney, Meteorological data - JOINT-II: March, April, May 1976, CUEA Data Report 34, Ref. FSU-CUEA-MET-76-2, Department of Meteorology, Florida State University, Tallahassee, 69 pp., 1976.

Vidal, J., Copepodos calanoideos epipelagicos de la expedicion Marchile II, *Gayana, 15*, 1-98, 1968.

Walsh, J.J., J.C. Kelley, T.E. Whitledge, J.J. MacIsaac, and S.A. Huntsman, Spin-up of the Baja California ecosystem, *Limnology and Oceanography, 19*, 553-572, 1974.

Wyrtki, K., Advection in the Peru Current as observed by satellite, *Journal of Geophysical Research, 82*, 3939-3942, 1977.

THE ZOOPLANKTON IN AN UPWELLING AREA OFF PERU

Haydee Santander Bueno

Instituto del Mar del Peru, Lima, Peru

Abstract. In autumn 1976 and 1977, interdisciplinary multi-ship studies were made in the area off Pisco to San Juan, Peru, to integrate biological, physical, and chemical aspects of the upwelling ecosystem in the area. Zooplankton was one of the biological elements considered. This report describes the general results relative to the zooplankton community including biomass, composition of the zooplankton groups, and some dominant organisms, principally copepods. In profiles perpendicular to the coast there were low biomass values near shore, increasing to the shelf break. Biomass patterns were similar for the copepods and other holoplanktonic organisms. The pattern is altered for short periods by local environmental or biological processes. In 1976 and 1977, the levels of primary production were similarly low, but the zooplankton biomass was higher in 1977 and the specific composition also differed in the two years. In 1976, the dominant organisms were euryphagous and in 1977 carnivorous. A common feature during the two years was the reduced numbers of phytophagous copepods, indicating a lower level of production in the period.

Introduction

Many factors affect the specific composition and biomass fluctuations of the zooplankton. One way to understand the changes and the interacting processes behind them is to know first the basic biological processes of the community elements and then to take up the development processes within the communities as well as functional characteristics of the ecosystem, such as upwelling. The impact of zooplankton on phytoplankton through grazing and its influence on the abundance and distribution of fish is quite variable, depending on specific zooplankton composition, specific biological characteristics, and on the multiple effects of the physical environment, that is the concentration or dispersion of zooplankton populations by the hydrodynamics.

With the aim of contributing to the knowledge of zooplankton population structure in an area of upwelling, biomass, species composition, and some ecological characteristics of the zooplankton with special reference to the dominant group, the copepods, were studied in April of 1976 and 1977. The area studied, 15°S, is characterized by permanent and intense coastal upwelling (Zuta and Guillén, 1974) and in contrast, is an area of little or no fish spawning compared with other coastal upwelling areas to the north (Santander and S. de Castillo, 1979).

Upwelling, as evident from temperatures in the area, developed similarly in both years, although the temperatures were slightly lower in 1976. Oxygen was deficient in both years. Concentrations below 1 ml/1 occurred between 20 and 30 m but occasionally also at 10 m in 1976 on the continental shelf. In 1977 the low concentrations occurred even shallower (Packard, Dugdale, Goering, and Barber, 1978).

The surface phytoplankton in 1976 was dominated by *Gymnodinium splendens* and in 1977 by *Mesodinium rubrum*, and by surface chlorophyll concentrations from 2.5 to 50.5 mg/m^3 in the first year (MacIsaac *et al.*, 1979) and 1.3 to 19.2 mg/m^3 in the second year (Packard *et al.*, 1978).

Methods and Material

All samples were from vertical tows from 50 m to the surface; at most of the stations there were also tows from 100 or 200 m as depth permitted. In 1976 a net with a 0.70-m diameter mouth was used and a Bongo net was used in 1977; both had 300-μm mesh. One of the samples from each of the deeper hauls was frozen on board and later used for dry weight biomass estimations after being dried at 60°C for 48 h; the remaining samples were preserved in formalin for qualitative and quantitative analysis, using up to eight subsamples, depending on volume.

From March 28 to April 11, 1976, the sampling was between 15°00' and 15°29'S and 75°22 and 76°03'W and consisted of 83 samples from 31 stations, mostly along a profile called the C line. In 1977 (April 21 to 27), seven lines similar to and including the C line were occupied between 14°35' and 15°50'S, including 31 stations (Fig. 1).

Results and Discussion

In April, 1976 the zooplankton larger than 300μ on the C line had a rather low biomass, between

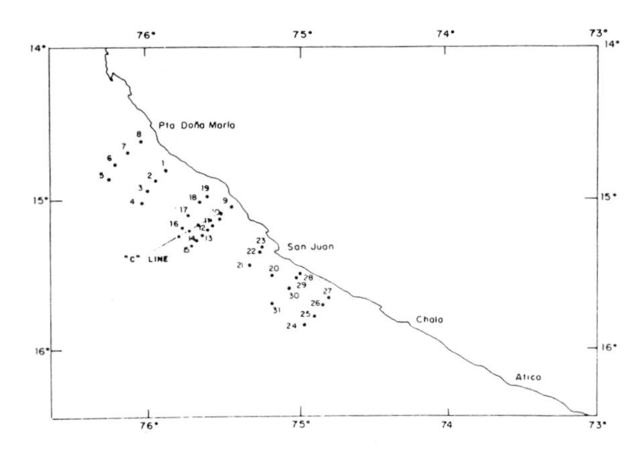

Fig. 1. Location of zooplankton stations, JOINT II, 1977 *Cayuse* Cruise, 21-27 April.

0.01 and 3.07 g/m^2, the lower values on the shelf and up to 20 km from the coast (Fig. 2). Near shore the copepods dominated in numbers (Fig. 3) followed by larvae of polychaetes, brachiopods, cirripeda and other crustacea. Radiolaria were not quantified because of their massive occurrence. From March to May all the area studied had abundant hydromedusas at different depths. Some holoplanktonic larger organisms like siphonophors and chaetognaths were common beyond 77 km

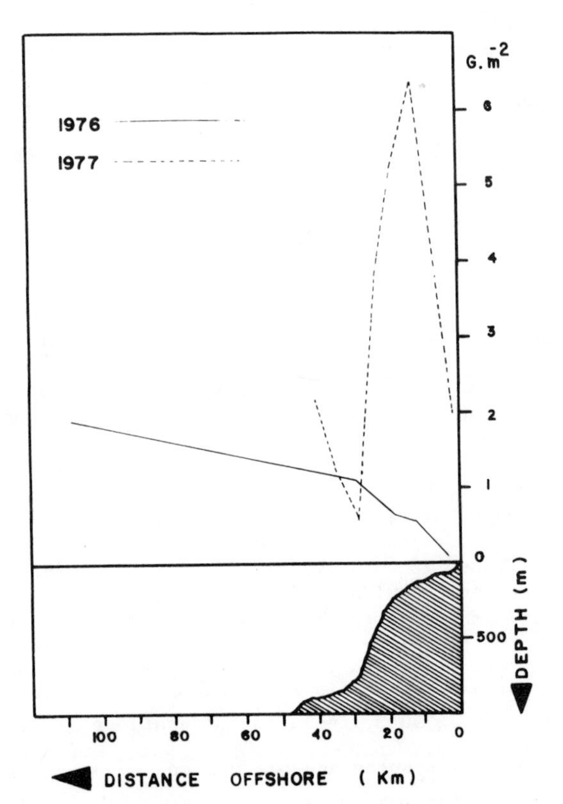

Fig. 2. Distribution of zooplankton biomass at 15°00'S in April 1976 and April 1977.

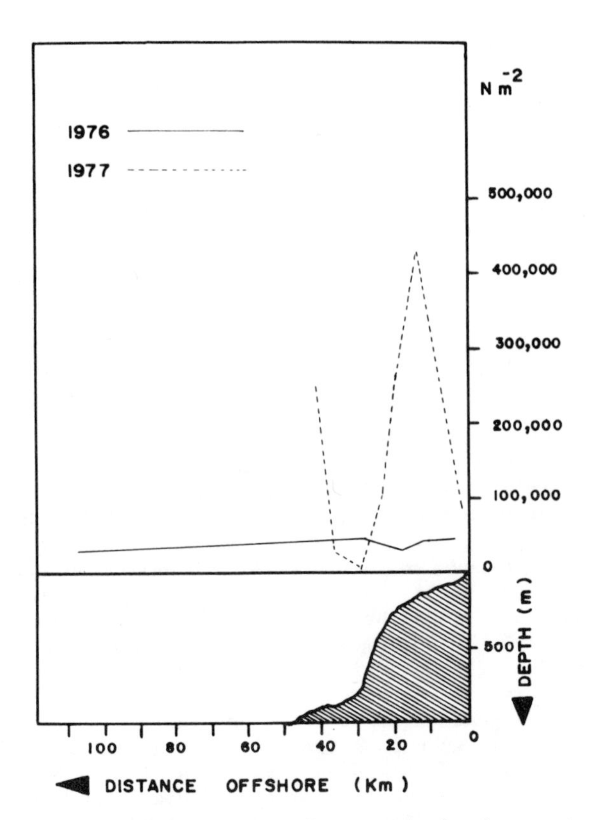

Fig. 3. Distribution of copepod abundance at 15°00'S in April 1976 and April 1977.

from the coast but not on the shelf area. The same can be said of the euphausids; they occurred where the depth was more than 1000 m (Fig. 4). Of the 28 species of copepods, the dominant one was *Acartia tonsa*. This is a euryphagous species, 0.95 to 1.30 mm long, typically from neritic areas. It can live on a variety of diatom and flagellate food (Conover, 1956), and it can regulate its vertical migration. These two qualities give it an advantage relative to other species, permitting it to live in the surface layer and avoid the oxygen-depleted water below 30 m, and to feed on one of the components of its varied diet. About seven species were present near the coast and of them about 95% of the individuals were euryphagous (Fig. 5). Farther than 30 km from the coast the species were notoriously more diverse where the oxygen values increased (Fig. 6); 52% were phytophagous, 19% euryphagous, and 29% carnivorous; these categories made up 36, 31, and 33% of the copepods 108 km from the coast. If the area is subdivided into neritic, slope, and oceanic sub-areas, each species tended to occupy only one of the sub-areas. Among the more noticeable neritic species were (Fig. 5) *A. tonsa* and *Centropages brachiatus*. There were about 17 species in the oceanic area, beyond 50 km off the coast (Fig. 6).

Two communities were also observed in the zoo-

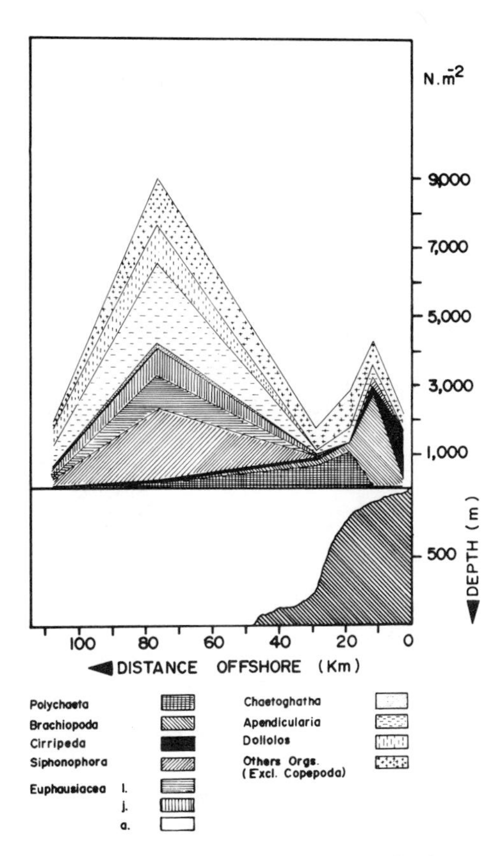

Fig. 4. Composition of the zooplankton at 15°00'S in April 1976.

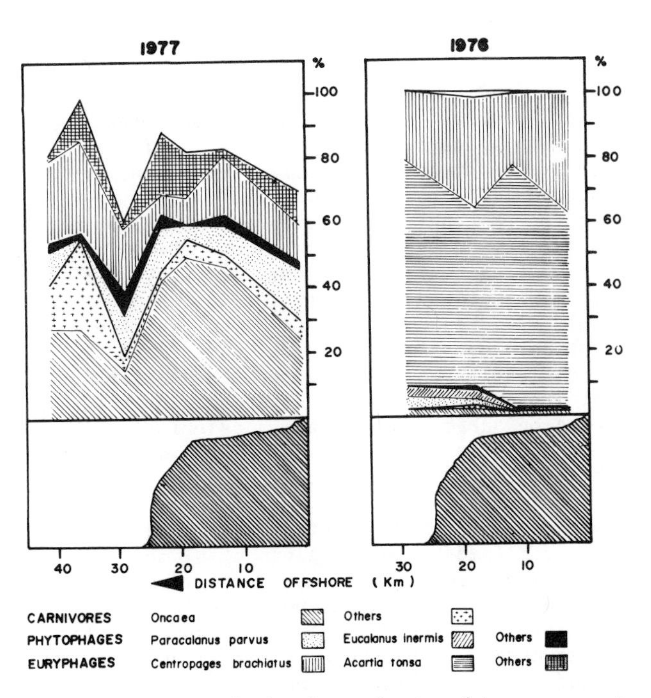

Fig. 5. Copepod abundance by trophic groups at 15°00'S in April 1976 and April 1977.

plankton population, a neritic one up to 20 km offshore, composed of meroplanktonic organisms and copepods. The oceanic community was composed primarily of larger zooplankters, holoplanktonic, carnivorous, and phytophagous copepods and, farther (50 km) from the coast the proportion of copepods in the trophic groups was similar to that in the neritic community, but the species composition differed. In 1977, the biomass varied between 0.6 and 6.39 g/m^2 (Fig. 2); as in 1976 the biomass was low at the station nearest to the coast, it was greater at mid-shelf, decreased on the slope, and increased again farther offshore. Although in general the values were higher than those of 1976, low concentrations were found at localities north and south of the C line (Fig. 7), corresponding to upwelling foci with temperatures below 16 and 15.5°C at the surface and at 20 m, respectively (Fig. 8) (Huyer and Gilbert, 1979). The two major concentrations seem to have originated by a local process at the prolongation of a tongue of water warmer than 19.0°C (Fig. 9) and of salinity lower than 35.05°/oo at the surface. In the area between 14°35' and 15° 50'S the lower biomass values tended to coincide with the lower temperatures associated with centers of upwelling. The zooplankton groups present in 1977 were mostly the same as in 1976, dominated by copepods (Fig. 3) near the coast with high frequencies of bivalves, polychaetes, brachiopods, and ostracod larvae (Fig. 10); however, the radiolaria (Acanthaires) and hydromedusae present in large concentrations in 1976 were almost

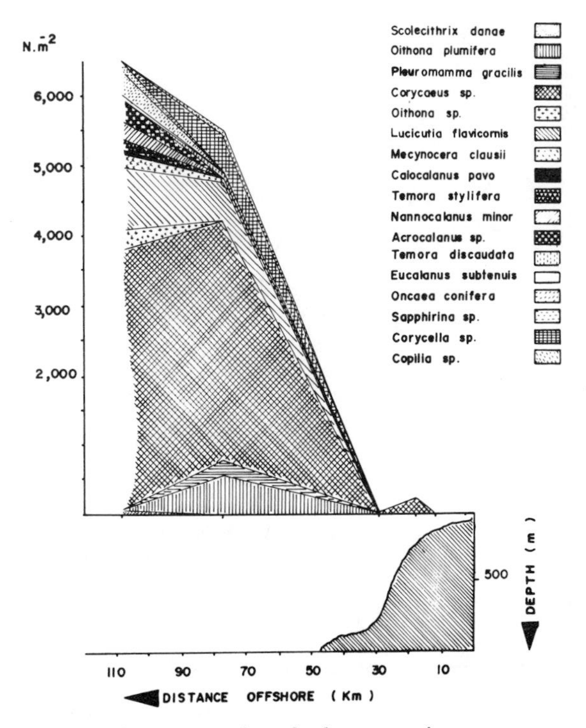

Fig. 6. Copepods of the oceanic area.

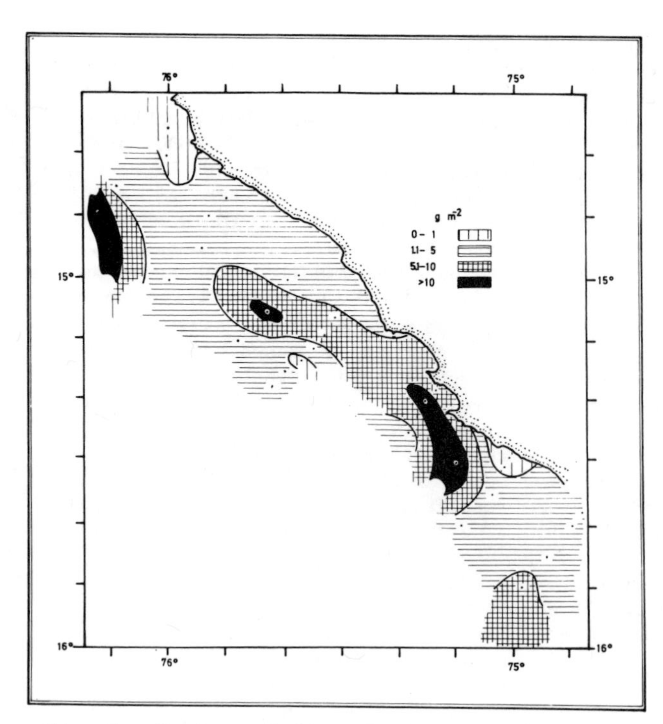

Fig. 7. Horizontal distribution of zooplankton biomass in April 1977.

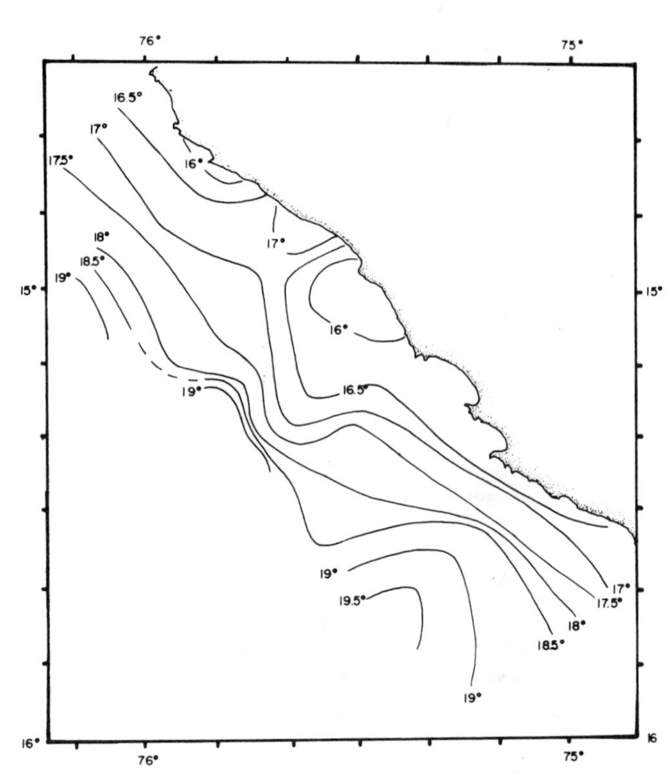

Fig. 9. Temperature (^{o}C) at the surface, 17-20 April, 1977.

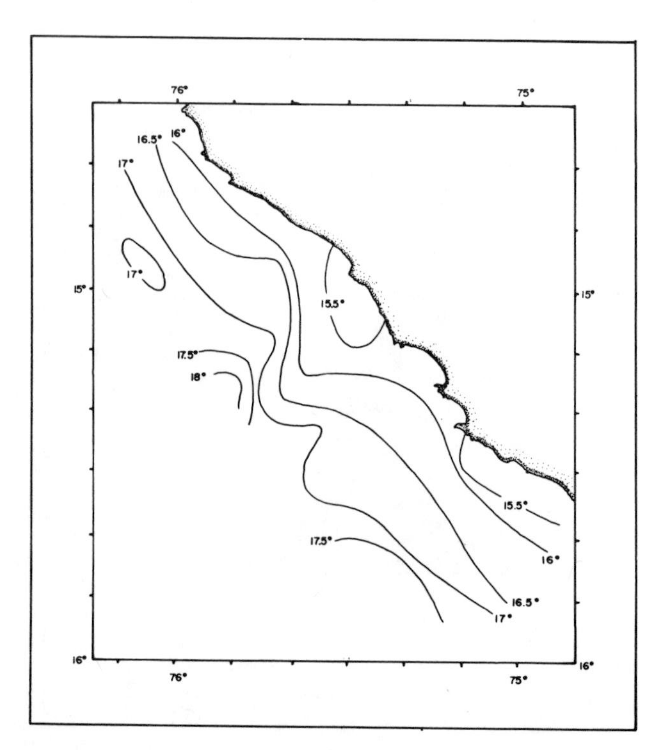

Fig. 8. Temperature (^{o}C) at 20 m, 17-20 April, 1977.

absent in 1977. On the slope the principal groups were the siphonophors, bivalves, foraminifera, and radiolaria; farther from the coast the proportion of euphausids and siphonophors was greater, as in 1976.

The year 1977 was similar to 1976 in the number of copepod species. On the C line the dominant species was the carnivorous *Oncaea* sp., with an average frequency of 33 and 50% of the carnivorous copepods (Fig. 5).

While the phytophagous species were less abundant close to and far from the coast, the euryphagous species tended to be more abundant farther from the coast. The carnivorous copepods dominated all the area, and *Oncaea* sp. was the dominant species. The horizontal distribution followed closely the zooplankton biomass distribution; its presence was characteristic of the greater zooplankton concentration associated with an extension of water warmer than 19°C that occurred between two plumes of upwelling. On the other hand, large concentrations of phytophagous species were associated with the upwelling areas and characterized by lower values of biomass. The euryphagous copepods group was larger than the phytophagous group in these upwelling areas.

As in 1976 along the C line the meroplanktonic organisms and the copepods were distributed near the coast in 1977, while the holoplankters were beyond the slope. The species composition of copepods differed in the two years and on average

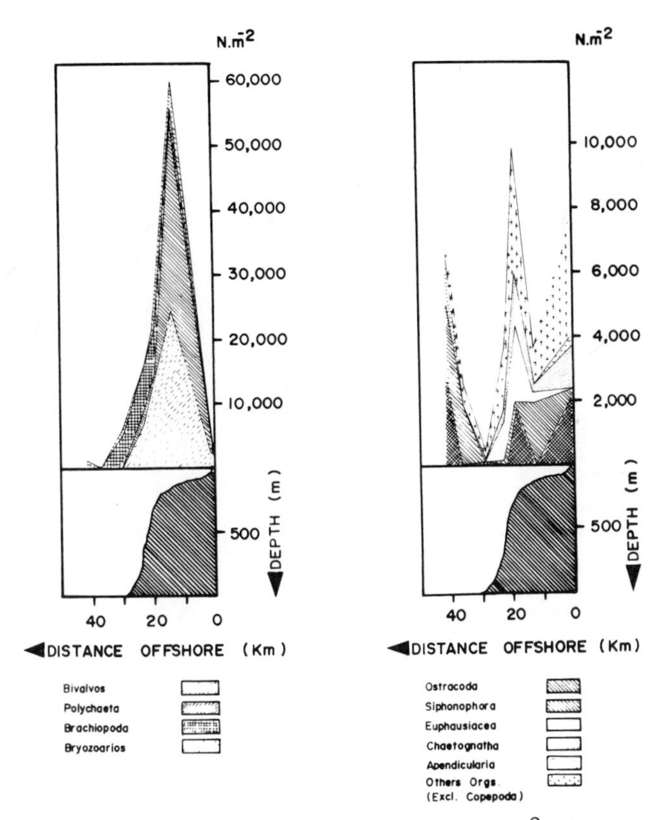

Fig. 10. Zooplankton composition at 15°00'S in April 1977.

the carnivorous species dominated out to 40 km in 1977. From 1964 to 1969 the zooplankton biomass in the upper 50 m within 110 km of the coast was estimated from volumetric measurements. The more consistent biomass characteristics observed along the coast and through the year were the higher values occurring mainly between 4 and 6°S and south of 14°S (Fig. 11). These areas are characterized by a narrow shelf, around 18 to 28 km, while the rest of the shelf is wider, up to 120 km at 9°S.

One outstanding feature of the zooplankton distribution is that at 15°S the groups of larger organisms such as euphausids, chaetognaths, siphonopors, etc., dominate, especially from the slope outward, and probably determine the local increase of zooplankton biomass near the slope as it occurred in 1976. In April, 1977 this increase, plus local modifications of the species composition, could have arisen from local dynamic processes resulting in oceanic waters closing up to the coast, with the additional effect of increasing the mortality of neritic species by predation from the invaders.

The upwelling process itself is also active in modifying the large-scale pattern. Samyshev (1973) noted that the accumulation of zooplankton is a result of the rate and strength of the upwelling. When upwelling is weak, the tendency is for organisms to accumulate in the nearshore areas; when strong, the tendency is for transport to other regions.

Adaptations of organisms to extreme conditions are not known. A direct relation between upwelling centers and large concentrations of zooplankters is not immediately apparent. On the other hand few zooplanktonic organisms are able to live under oxygen-deficient conditions. All these factors together with the dominance of *G. splendens* in the surface layer, the layer occupied by the zooplankton, mean that the community had to live in a narrow range of three variables on the shelf, temperature, oxygen, and food. The conditions led to the situation in 1976, i.e., few species on the shelf dominated by euryphagous species, which seem to be resistent organisms because of their feeding habits; the particular species in 1976 had an adaptable migratory behavior. Diversification of species would on the contrary be expected with greater distance from the upwelling center, more diverse phytoplankton species, more abundant oxygen, and greater depth allowing for more vertical migration. But these conditions did not prevail in either 1976 or 1977, when productivity was low (Barber *et al.*, 1978), although the upwelling was permanent. The dominance of euryphages in 1976, of carnivores in 1977, and the absence of phytophages on the shelf

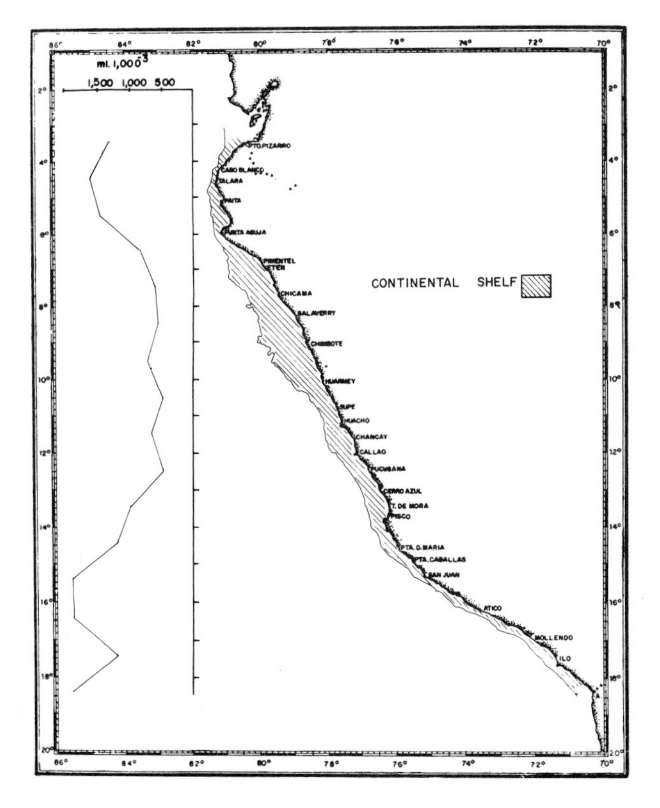

Fig. 11. Latitudinal variation of zooplankton biomass, 1964-1969.

in both years, would indicate a relaxation of phytoplankton development in April, 1977 or a long period of low production in 1976, with dominance of phytoplankton in a narrow size range on the surface layers.

Acknowledgements. This study was supported by the Instituto del Mar del Perú and by the Coastal Upwelling Ecosystems Analysis Program. G. Luyo and M. Véliz analyzed some of the zooplankton samples.

References

Barber, R.T., S.A. Huntsman, J.E. Kogelschatz, W.O. Smith, B.H. Jones, and J.C. Paul, Carbon, chlorophyll and light extinction from JOINT II 1976 and 1977, *CUEA Data Report 49*, 476 pp., 1978.

Conover, R.J., Oceanography of Long Island Sound, VI. Biology of *Acartia clausi* and *A. tonsa*, *Bulletin of the Bingham Oceanographic Collection, 15*, 156-233, 1956.

Huyer, A. and W.E. Gilbert, Vertical sections and mesoscale maps of temperature, salinity and sigma-t off the coast of Peru July to October 1976 and March to May, 1977, *CUEA Technical Report, 60*, 162 pp., 1979.

MacIsaac, J.J., J.E. Kogelschatz, B.H Jones, Jr., J.C. Paul, N.S. Breitner, and N. Garfield (eds.) JOINT II R/V *Alpha Helix* productivity and hydrographic data March-May, 1976, *CUEA Data Report, 48*, 324 pp., 1979.

Packard, T.T., R.C. Dugdale, J.J. Goering, and R.T. Barber, Nitrate reductase activity in the subsurface waters of the Peru Current, *Journal of Marine Research, 36*, 59-76, 1978.

Samyshev, E.V., Relationships between trophic copepod groups in the zooplankton of the Gulf of Guinea, *Hidrobiological Journal, 9*, 25-32, 1973.

Santander, M.H. and O.S. de Castillo, El ictioplancton de la costa Peruana, *Boletin Instituto del Mar del Peru, 4*, 69-112, 1979.

Zuta, S. and O. Guillén, Oceanografia de las aguas costeras del Peru, *Boletin Instituto del Mar del Peru, 2*, 157-324, 1970.

SHORT TERM VARIATION IN THE VERTICAL DISTRIBUTION OF SMALL COPEPODS OFF THE COAST OF NORTHERN PERU

S.L. Smith

Brookhaven National Laboratory, Oceanographic Sciences Division, Upton, New York 11973

C.M. Boyd

Department of Oceanography, Dalhousie University, Halifax, Nova Scotia, Canada

P.V.Z. Lane

Brookhaven National Laboratory, Oceanographic Sciences Division, Upton, New York 11973

Abstract. Vertical profiles of chlorophyll *a*, oxygen, density, and copepods were observed during November 1977 near 9°S off Peru. The majority of three groups of copepod, the Oncaeidae, the Oithonidae, and small calanoids, remained above the depth (∿30 m) where concentrations of oxygen became less than 0.5 ml/l both day and night. Centers of population of all three groups were in or below the pycnocline at all times. In daytime all three groups accumulated at depth, while at night all three groups showed some dispersion throughout the upper 30 m with statistically significant separation in the layers of Oncaeidae and small calanoids. Small calanoids were always higher in the water column than the Oncaeidae at night. The rather small, daily vertical excursions by the Oncaeidae and small calanoids exposed them to mean onshore, poleward flow by day and mean offshore, equatorward flow at night.

Introduction

The association of vertical structure, layering, in the distribution of zooplankton with variables in the water column that might cause or strongly influence that structure has been described in several areas of the world's oceans. The variables most frequently observed to correlate with layering of zooplankton are density (Banse, 1964; Longhurst, 1976; Boyd, 1973; Zalkina, 1977), oxygen (Longhurst, 1967; Antezena-Jerez, 1978), potential food (Anderson, Frost, and Peterson, 1972; Vinogradov, 1972; Hobson and Lorenzen, 1972; Mullin and Brooks, 1972), and light (summarized by Raymont, 1963). The variables are usually not independent of one another, which complicates any analysis of vertical profiles for meaningful, possibly causative, relationships. Nevertheless, experiments continue because in the organization of marine ecosystems, space is a resource that could be partitioned by swimming metazoans. The trophic structure of a marine system could vary considerably as a function of the living space occupied by an organism or group of organisms and temporal patterns in the ways that space is used.

Off Peru feeding patterns of the herbivorous copepod *Calanus chilensis* were modified according to food availability. In areas of high concentrations of food, *C. chilensis* fed evenly day and night, a pattern it shared with the omnivorous *Centropages brachiatus*. When food became scarce, *C. chilensis* fed primarily in daytime and ranged to greater depths while *C. brachiatus* fed primarily at night (Boyd, Smith, and Cowles, 1980). Thus when food was relatively scarce, the two species partitioned their time of feeding, and *C. chilensis* broadened its living space in the vertical.

In the present study we sampled zooplankton in the upper 35 m of the water column at 5-m intervals with the ship anchored for 42 h at 9°20'S, 78°51'W and with the ship following a drogue set at 10 m for 32 h near 9°28'S, 78°58'W (see Plate IV). The trajectory of the drogue was roughly a circle 6 km in diameter. The study was conducted in November, 1977, from C.S.S. *Baffin* as part of the Canada-Peru study of the ecosystem of the Peruvian anchovy (I.C.A.N.E.; Doe, 1978). We shall describe vertical distribution of organisms, chlorophyll *a*, and oxygen and examine the data from the anchor station for evidence of vertical structure in the distribution of organisms and the influence of the pycnocline on vertical structure.

Methods

Zooplankton were sampled each 5 m of the upper 35 m by a reciprocating pump attached to a 3.2-cm internal diameter hose. Water was discharged from the hose on deck through a plankton net of 158-μm aperture mesh. Each depth was sampled for 5 min, and the flow rate of the pumping system, which averaged 80 l/min, was measured by allowing the outflow to fill a large container of known volume. Collections were preserved in 5% neutral formalin. All copepods and nauplii in each sample were identified to genus or group and counted. The configuration of the pump and hose were such that the copepods were generally in poor condition, precluding any reliable identification of stages in the life cycle. We have chosen three categories for analysis: species of *Oncaea*, species of *Oithona*, and small calanoids. The categories contain both adults and late copepodids, with the calanoid category being *Paracalanus parvus* largely.

The water column was sampled each 4 h for vertical profiles of salinity, temperature, oxygen, chlorophyll *a*, nutrients, and zooplankton. Temperature and salinity observations were made using a Guildline 8705 digital conductivity, temperature, and depth probe (CTD) with a 500-decibar pressure sensor. Water samples for dissolved oxygen, nutrients, and chlorophyll *a* determinations were obtained from a hydrocast after each CTD cast. Dissolved oxygen was determined by the Winkler method (Strickland and Parsons, 1968) with an uncertainty of 0.2 ml/l at oxygen concentrations below 0.5 ml/l. Chlorophyll *a* was measured fluorometrically as described by Lorenzen (1966). Wind speed and direction data were recorded daily by ship's officers. Further details and data can be found in Doe (1978).

Results

Hydrographic and nutrient data suggest that both the anchor and the drogue studies were not within zones of active upwelling. Mean sea-surface temperatures were 18.0 ± 0.2°C during the anchor study and 17.6 ± 0.4°C during the drogue study while average surface nitrate concentrations were 1.47 and 0.35 μg-at/l, respectively. Winds paralleled the coast just before and during both studies with speeds averaging 13 ± 5 knots during the anchor study and 11 ± 4 knots during the drogue study.

During the anchor study, oxygen concentrations were less than 2.5 ml/l at the surface and generally less than 0.3 ml/l below 30 m (Fig. 1). Chlorophyll *a* was less than 5.0 mg/m^3 and without structure in the upper 40 m (Fig. 1). The abundance of copepods in general was sharply reduced above 10 m and below 30 m at all sampling times (Fig. 2), the lower boundary corresponding to the depth where concentrations of oxygen became less than 0.3 ml/l. At night the depth of maximum abundance of total organisms was 20 m with species of *Oncaea* and small calanoids composing that layer (Fig. 2). In daytime (0950 and 1500 h), the composition of the layer of maximum abundance was the same, but it occurred at 25 m (Fig. 2). Throughout the anchor study species of *Oithona* were less abundant than the *Oncaea* spp. or calanoids and more evenly distributed throughout the upper 35 m (Fig. 2). As there was no vertical structure in chlorophyll *a*, the layers observed in the copepods do not correlate with layers of phytoplankton.

During the drogue study, higher and more variable surface concentrations of oxygen were observed (Fig. 1), but below 30 m concentrations were less than 0.3 ml/l. Vertical distribution of chlorophyll *a* had subsurface maxima exceeding 10 mg/m^3 at depths of 10 to 25 m in several samples (Fig. 1). The depth of the maximum total number of copepods was 15 m day and night, with small calanoids dominant numerically (Fig. 3). When a subsurface maximum in chlorophyll *a* was observed, peaks in abundance of copepods were observed just beneath it generally.

We converted the absolute numbers of each profile of copepods to relative abundance (percent) in 35 m in order to smooth effects of horizontal patchiness and examine the vertical distributions of the copepods throughout each time series with respect to centers of populations and dispersal, concentrations of oxygen, and movement of the pycnocline (*sigma-t* = 25.35 to 25.75). The effect of the concentration of oxygen less than 0.5 ml/l on distributions is clear. For the most part the lower contour enclosing the 10% level of total numbers of each group of copepod is above the depth where oxygen was less than 0.5 ml/l (Figs 4 and 5). During daytime, the population center (relative abundace > 30%) of species of *Oithona* covered approximately 5 to 10 m and varied in depth from 15 to 25 m (Figs 4 and 5), while the population centers of species of *Oncaea*, which spanned approximately 5 m, were at depths of 10 to 30 m. Daytime population centers of small calanoids spanned 5 to 10 m and were at depths of 5 to 27 m. At night, population centers similar to those observed in daytime tended to be maintained but dispersion through the water column occurred, presumably for feeding, and was evident in the wider depth range enclosed by the 10% contours (Figs 4 and 5). In the anchor study, species of *Oithona* show the greatest nighttime dispersion while the small calanoids show the least (Fig. 4). During the drogue study, nighttime dispersion occurred in all three groups, most noticeably in the small calanoids and species of *Oncaea* (Fig. 5). The dispersion at night indicates migration by some individuals in the populations during both the anchor and drogue studies.

The overall pattern in both the anchor and drogue studies has four noteworthy components. Most of the population of all three groups re-

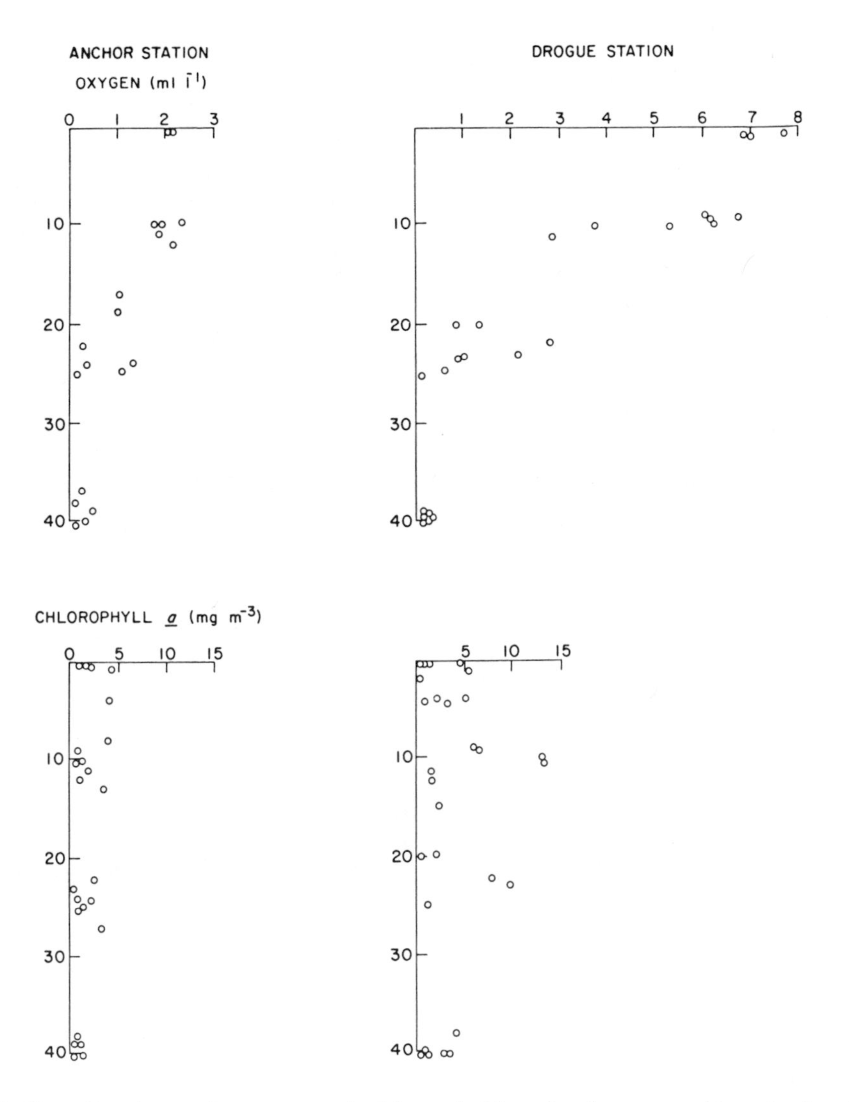

Fig. 1. Vertical distributions of oxygen and chlorophyll *a* in the upper 40 m during the anchor and drogue studies.

mained above the depth where oxygen concentrations were less than 0.5 ml/l day and night. Population centers of all three groups were in or below the pycnocline day and night. In daytime, all three groups accumulated at depth. At night all three groups showed some dispersion throughout the upper 30 m.

The anchor station, because of its more complete coverage of distributions of copepods and water column structure over time and because the sampling platform was fixed, is useful for examining relationships between density structure and vertical distribution of copepods. The depths of maximum abundance of the three groups of copepods from Fig. 2 have been replotted along with depths of *sigma-t* equaling 25.35, 25.55, and 25.75 (Fig. 6). A *sigma-t* of 25.75 was generally the bottom of the pycnocline while the 25.35 was the top. The maximum abundances of the small calanoids and species of *Oncaea* were below the bottom of the pycnocline, with one exception, while the species of *Oithona*, if they had a clear depth of maximum abundance (Fig. 2), realized it without any relationship to the pycnocline (Fig. 6). Thus, while the oxygen minimum layer establishes a lower boundary for the distribution of these copepods, the bottom of the pycnocline seems to establish an upper boundary for the centers of population of small calanoids and species of *Oncaea* (Fig. 6).

Near sunset (1740 h) the population centers of all three groups of copepod move toward the surface (Fig. 4) suggesting cohesive diurnal vertical migrations (Pearre, 1979). However, density profiles show that a different water mass in which the bottom of the pycnocline was nearer the surface was sampled at that time (Fig. 6). The

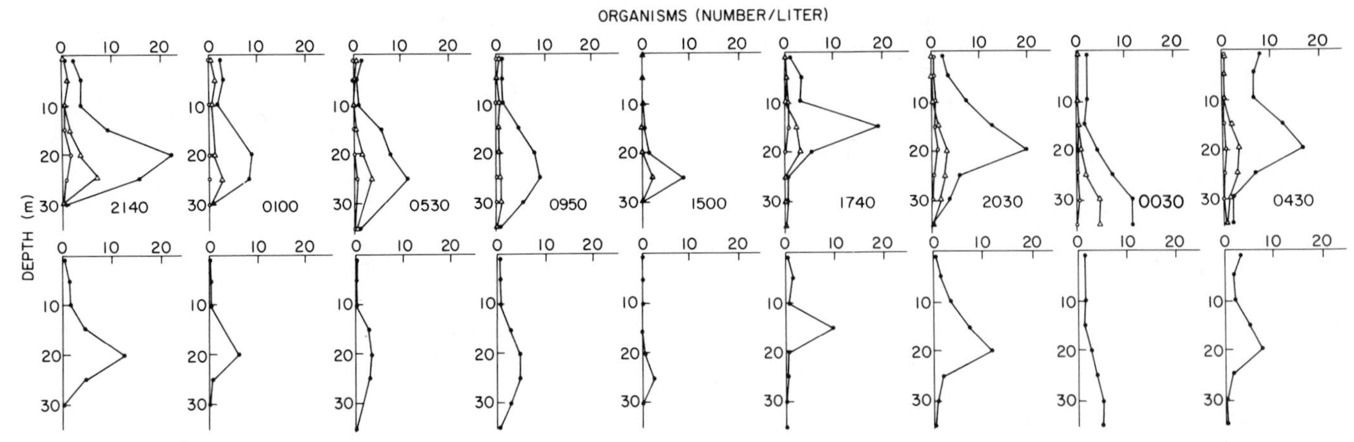

Fig. 2. Profiles of abundance of species of *Oncaea* and *Oithona* and small calanoids collected during the anchor study. In the upper panel, ● represents total organisms, o represents species of *Oithona*, and Δ represents species of *Oncaea*. The bottom panel shows distributions of small calanoids, and time is local standard time.

vertical distribution of the copepods with respect to the pycnocline was similar to that observed at other times in the series suggesting that observed vertical excursions in the centers of population were not the result of cohesive vertical migration but rather a result of the shoaling of the pycnocline from some physical process that carried the copepods with it. If density data had not been available, vertical migration would have been described for the copepods quite incorrectly.

The depth of maximum abundance of species of *Oncaea* is generally deeper than the depth of maximum abundance of small calanoids (Fig. 6), while the depth of maximum abundance of *Oithona* is variable, often occurring at more than one depth. We used the Kolmogorov-Smirnov test (Tate and Clelland, 1957) modified for discrete samples (Conover, 1972), and not as used by Shulenberger (1978), to test whether such layering in the vertical distributions of small calanoids and spe-

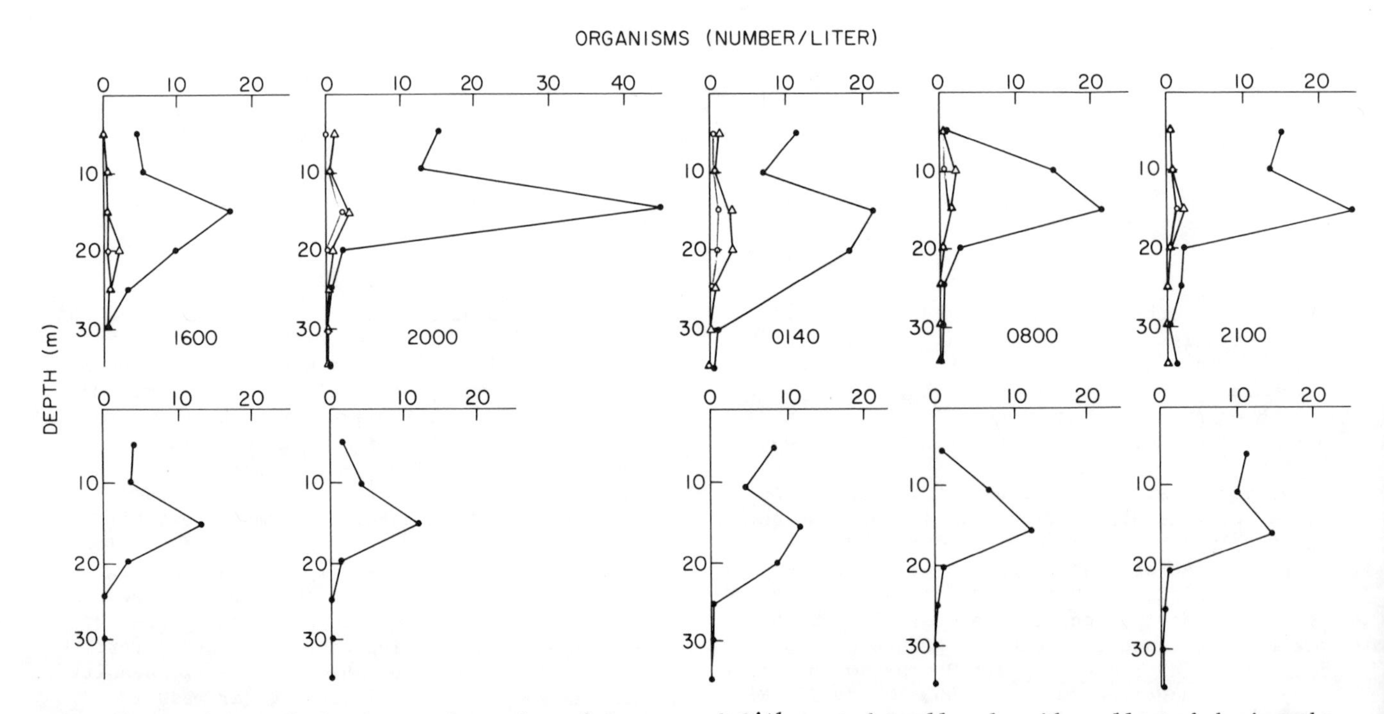

Fig. 3. Profiles of abundance of species of *Oncaea* and *Oithona* and small calanoids collected during the drogue study. Symbols are the same as in Fig. 2 and time is local time.

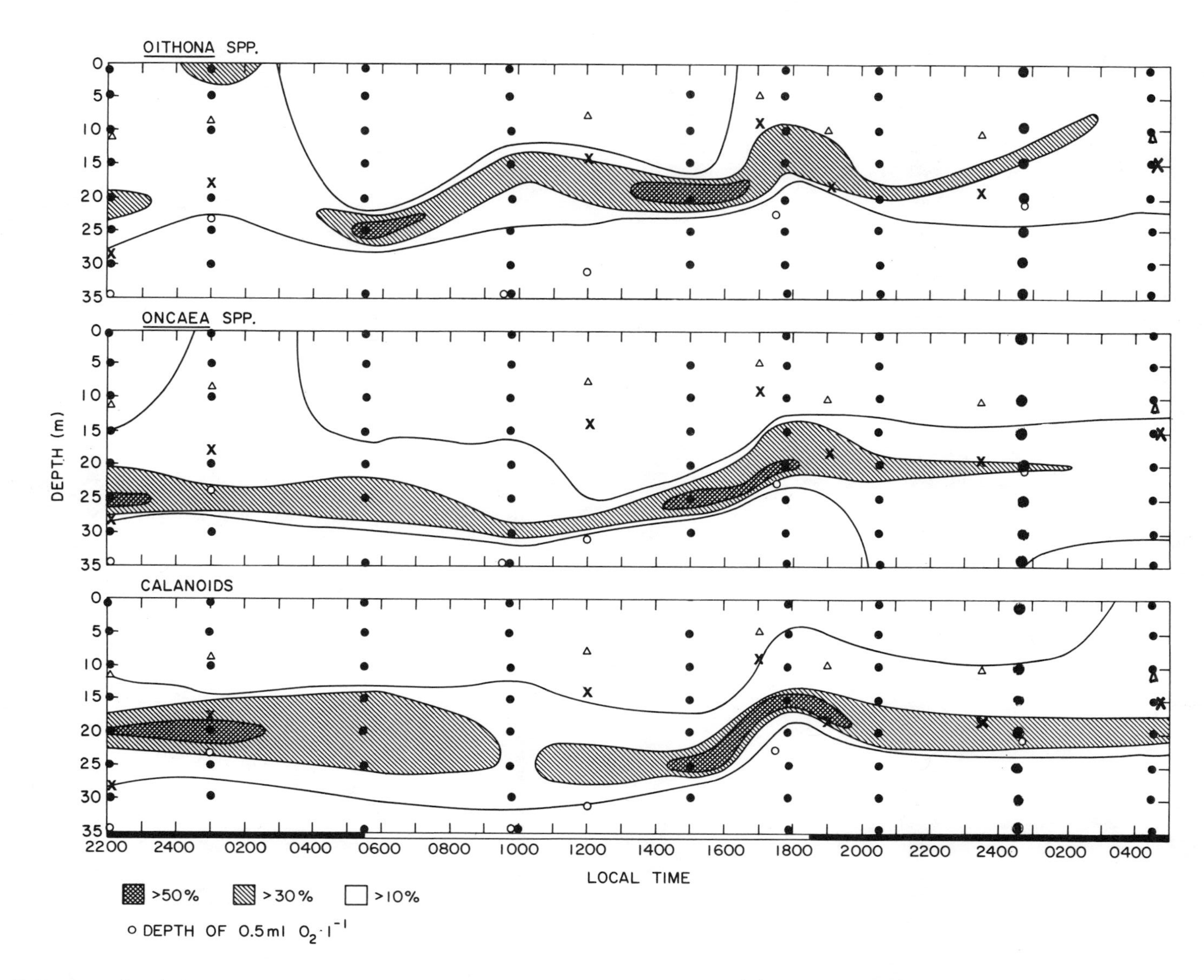

Fig. 4. Relative abundance of three groups of copepod collected during the anchor study. The abscissa is depth in meters. A Δ represents the depth at which σ_t was 25.35, and an X represents the depth at which σ_t was 25.75.

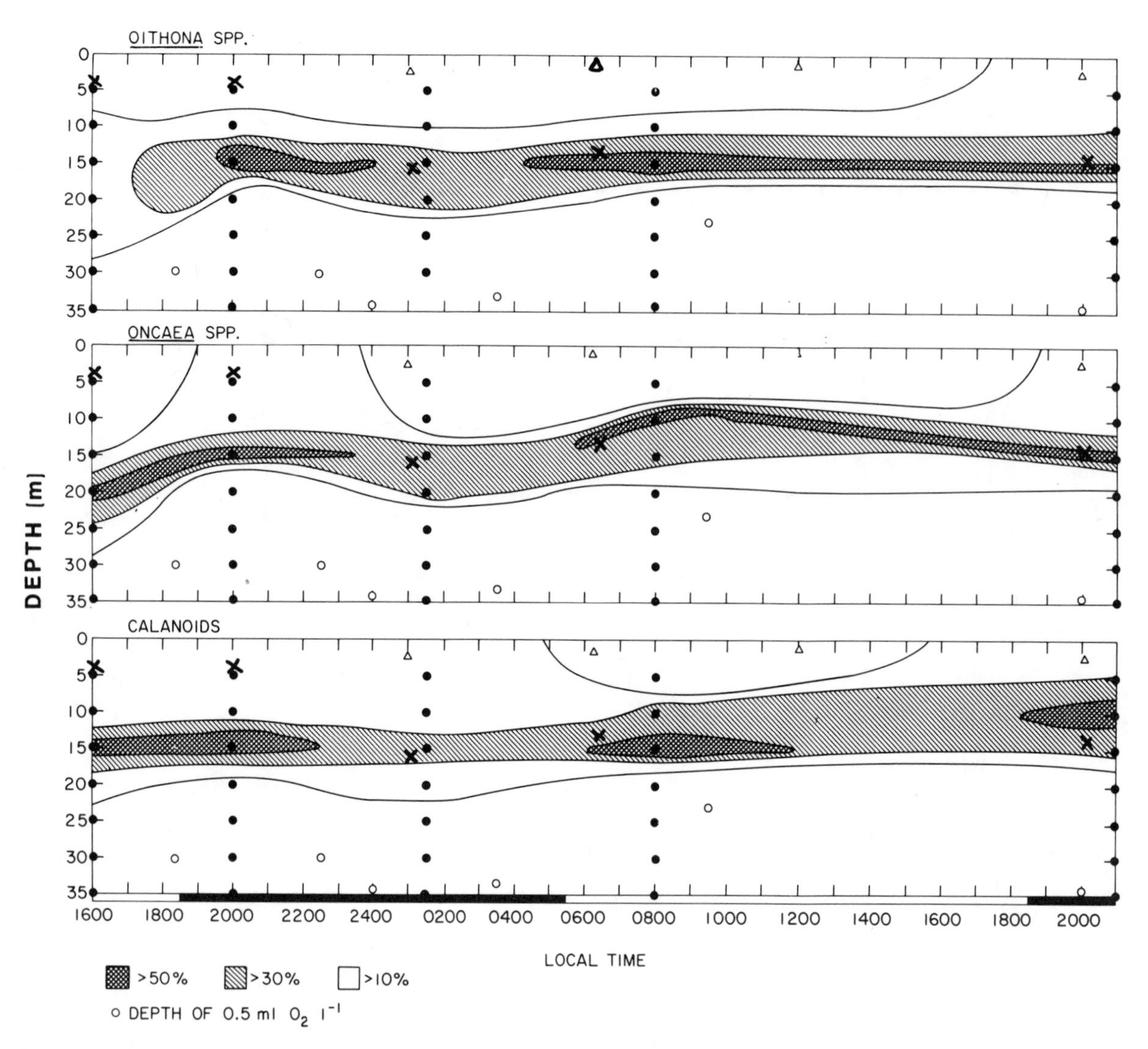

Fig. 5. Relative abundance of three groups of copepod collected during the drogue study. The abscissa is depth in meters. Symbols are the same as in Fig. 4.

cies of *Oncaea* was significant (Fig. 7). In all profiles sampled at night, small calanoids were significantly ($P \leq 0.05$) higher in the water column than were species of *Oncaea* below the pycnocline (*sigma-t* = 25.75; Fig. 7). In daytime (0950 and 1500 h) below the pycnocline, there was no significant difference in depth·frequency. Above the pycnocline at night there were significant depth differences between the two groups, but no consistent pattern was apparent. Early in the series species of *Oncaea* were significantly higher in the water column than were calanoids; late in the series calanoids were higher than species of *Oncaea* (Fig. 7). At all times except that of the first profile, more than 50% of both groups of copepods were at or below the bottom of the pycnocline.

We pooled all data above and below the depth of *sigma-t* = 25.75 by day and by night in order to generalize the patterns. At night above the depth of *sigma-t* = 25.75, small calanoids were significantly higher in the water column than were species of *Oncaea* except at 1 and 5 m (Fig. 8). At night below the pycnocline, small calanoids were significantly higher in the water column at all depths than were species of *Oncaea*. In daytime above *sigma-t* equaling 25.75, species of *Oncaea* were significantly higher in the water column than were small calanoids (Fig. 8), while below *sigma-t* = 25.75 small calanoids were significantly higher than *Oncaea* at depths of 10 to 20 m but not significantly different at depths greater than 20 m. In general, at night small calanoids were significantly higher in the water column than were species of *Oncaea*, while in the daytime there were no significant differences below 20 m, where

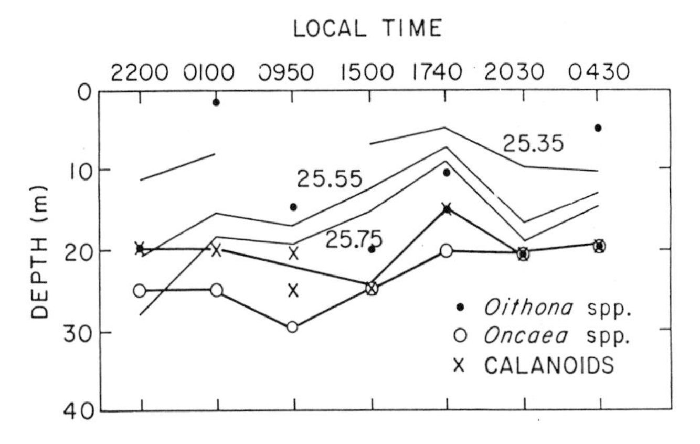

Fig. 6. Depths of three σ_t surfaces (25.35, 25.55, and 25.75) and depths of maximum abundance of three groups of copepod during the anchor study.

the bulk of both groups occurred (Fig. 4). Thus, during the anchor study the daytime accumulation of species of *Oncaea* and small calanoids at depth (Fig. 4) can be confirmed statistically, while dispersion of these groups at night has the feature of statistically significant layering.

The Kolmogorov-Smirnov test applied to the vertical distributions of species of *Oithona* at the anchor station showed that at night *Oithona* was generally significantly higher in the water than small calanoids, while in the day *Oithona* was either higher or not significantly different from the other two groups. The general overall pattern at night was that species of *Oithona* were highest in the water column, small calanoids were below them, and species of *Oncaea* were deepest in the water column. In the day, statistical separation of layers was not so pronounced (Fig. 8) because of the tendency of all three groups to accumulate just above the oxycline (Fig. 4).

When the Kolmogorov-Smirnov test was applied to the results from the drogue study, small calanoids were significantly higher in the water column than species of *Oncaea* except at 0800 and 2000 h, when the vertical distributions were not significantly different. Species of *Oithona* had no consistent vertical distribution with respect to the other two groups during the drogue study.

Discussion

For the most part 35 m covered the depth range inhabited by species of *Oncaea*, *Oithona*, and the small calanoids, because abundances had decreased to nearly zero at 35 m (Figs 2 and 3). The oxy-

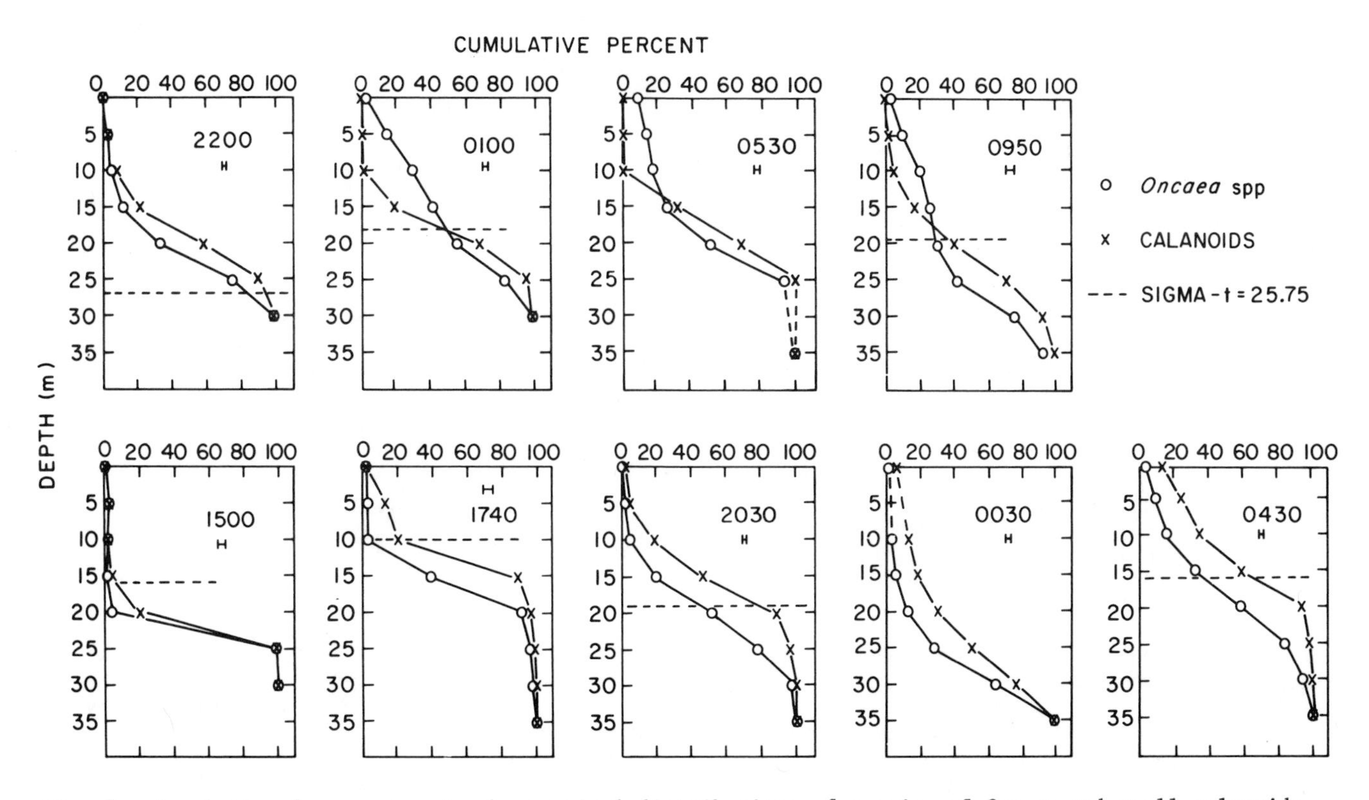

Fig. 7. Cumulative frequencies in the vertical distributions of species of *Oncaea* and small calanoids collected during the anchor study. Dashed line is depth of σ_t equaling 25.75. Time is local time. The 95% confidence interval is shown beneath local time.

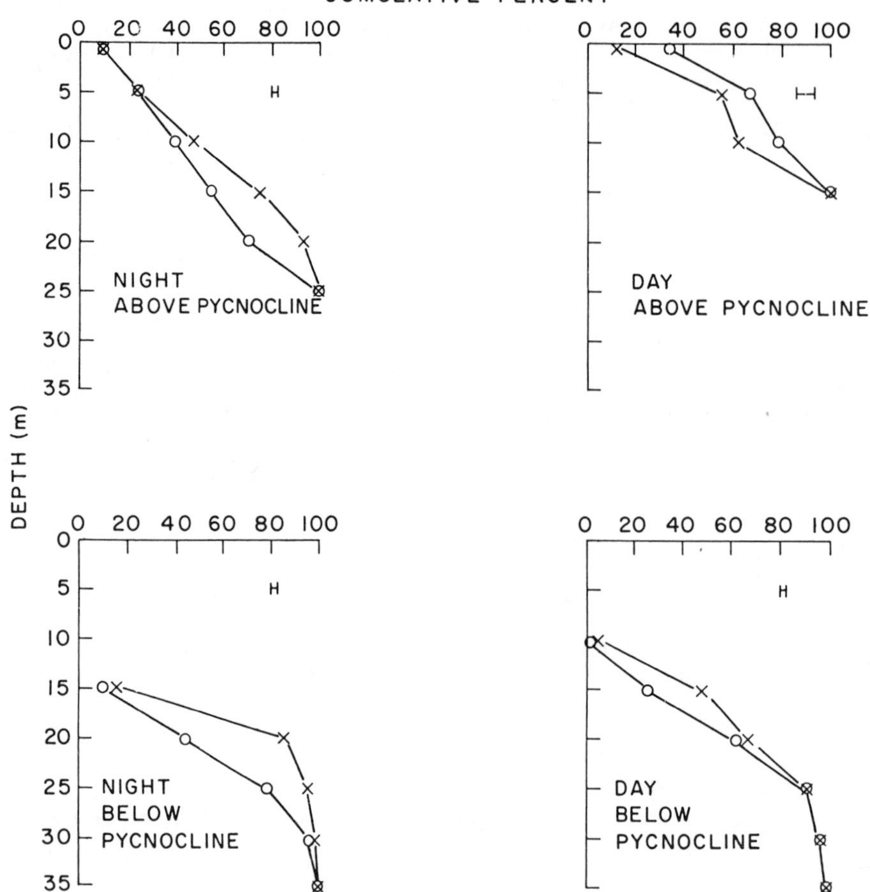

Fig. 8. Cumulative frequencies in the vertical distributions of species of *Oncaea* (O) and small calanoids (X) for pooled data collected during the anchor study. The 95% confidence interval is shown in the upper right-hand corner of each graph.

gen depletion in subsurface waters off Peru establishes a distinct lower boundary for the distribution of many organisms in the water column and causes most zooplankton to inhabit a relatively thin surface layer. Restriction to the upper 35 m maintains zooplankton within currents which may be quite variable in this complex upwelling system (Brink, Halpern, Huyer, and Smith, 1980) and have strong influence on the distribution of zooplankton in the onshore-offshore dimension (Smith *et al.*, this volume). The major features of the mean circulation at 15°S off Peru also occur at 10°S (Brink *et al.*, 1980), suggesting that offshore and equatorward flow was restricted to the upper 20 to 25 m during our study with onshore and poleward flow beneath. Considering the data from the anchor station (Fig. 7) in view of the mean circulation, at night an average of 78% of the small calanoids and 49% of the species of *Oncaea* were in the layer having offshore, equatorward flow. In the day, an average of 70% of the small calanoids and 82% of species of *Oncaea* were in the layer of onshore, poleward flow. Most of the *Oithona* at the anchor station were in the layer of offshore, equatorward flow both day and night. Thus, the rather small vertical excursions by individuals of species of *Oncaea* and the small calanoids within the upper 35 m were sufficient to expose them on a daily basis to a mean circulation tending to maintain them in one place. The pycnocline, which seems to be an upper boundary for the center of population of species of *Oncaea* and small calanoids, would act to keep the centers of population within the layer of transition between onshore and offshore flow or in the layer of onshore flow. Of the three groups examined, only the species of *Oithona* had a vertical distribution suggesting they would be transported out of the area of coastal upwelling ultimately. The vertical distributions and movements of the other two groups would maintain them in coastal waters.

The descriptive study of vertical migration of

zooplankton on small spatial scales is difficult probably because the variance introduced by our ability to make appropriate collections is similar to the variance in the behavior itself. The present study has revealed structure that is persistent and statistically significant over depths of less than 35 m. The patterns, however, are not clear, cohesive (Pearre, 1979) vertical migrations that have been described for copepods and euphausiids that cover several hundreds of meters on a daily basis (summarized by Raymont, 1963). There were three indications of vertical migration in the present study: (1) occasional bimodalities in profiles at night, (2) generally upward dispersion at night when relative abundances are considered, and (3) the average depths (from the statistical analyses) below which 60% of the individuals were found. For small calanoids the depth was 20 m by day and 15 m by night and for *Oncaea* spp. 23 m by day and 18 m by night. The migration of individuals was not synchronous; they rose and sank probably in accordance with hunger and satiation, which are variable with the individual copepod as well as with species (Mackas and Bohrer, 1976). Asynchronous migrations are nonetheless real vertical migrations on the part of individuals. Pearre (1979) suggested that when food is abundant individuals might spend relatively little time feeding at the surface and a large fraction of time at depth once they are satiated. The vertical distributions observed in this study are consistent with that hypothesis.

Studies of patchiness in the distribution of zooplankton are frequently conducted by collecting data at a fixed, shallow depth from relatively long transects. These studies have shown that abundance of zooplankton varies six-fold over 80 to 100 km in the North Sea (Mackas and Boyd, 1979) and off Georges Bank (Denman and Mackas, 1978) with maximum abundances of 60/l at the surface in the North Sea and 30/l off Georges Bank. Physical processes that lift or mix subsurface zooplankton into the layer being sampled can create patches 100 m long in Massachusetts Bay (Haury, McGowan, and Wiebe, 1978). The restriction of most zooplankton to the upper 35 m off Peru and the abundance (up to 20 organisms per liter) and vertical structure of those organisms (for example, Fig. 2) suggests that any patchiness observed at the surface should be correlated unambiguously with either the upwelling circulation or coastally trapped waves (Smith, 1978), which dominate the physical processes off Peru (Brink *et al.*, 1980). The strong and shallow vertical gradients in zooplankton combined with physical processes should dominate any patchiness at the surface in this area. Cross-shelf gradients in numbers of zooplankton off Peru at 15°S are on the order of 1 to 10 organisms per liter over 30 km (Smith *et al.*, this volume) while the vertical gradients observed in this study were 1 to 20 organisms per liter over 30 m.

Previous studies of vertical distribution of Cyclopoida in the equatorial Pacific Ocean have shown that several of the species of *Oncaea* and *Oithona* likely to be in our samples have centers of population (25 to 75%) extending to 150 m (Zalkina, 1970), emphasizing the reduction in living space experienced by the Cyclopoida collected off Peru. In contrast with our results, the Oithonidae tended to occur deeper in the water column (75 to 150 m) than the Oncaeidae (25 to 125 m) with the vertical distributions of the two groups separate by night and overlapping by day (Zalkina, 1970). The compression of most of the zooplankton into a surface layer of 35 m suggests that competitive interactions among zooplankters might be intensified off Peru. We speculate that the significant and small-scale layering of the three groups of zooplankton in the study could be a response to competition. The layering is such that small calanoids, presumed to be grazers, are frequently at the bottom of the pycnocline, and when a subsurface maximum in chlorophyll *a* occurs, they are just beneath it. The Oncaeidae, whose feeding habits are not well known but may be carnivorous, are found consistently beneath the small calanoids. The vertical structure observed in the study may be an important factor in the transfer of organic material within the upwelling system. At the moment, the trophic significance of these small, but numerically dominant (Dagg *et al.*, 1980), copepods is unexplored.

Acknowledgements. A large number of people contributed to the field work of this study which was supported by a Canadian International Development Agency grant to Dalhousie University. We are especially grateful to Dr. D. Mackas, S. Waddell, C. Wirick, and Capt. D. Deere and crew of C.S.S. *Baffin* for their help. National Science Foundation grant OCE 78-05737-A01, part of the Coastal Upwelling Ecosystem Analysis program of I.D.O.E., supported one of us (S.L.S.) during preparation of the manuscript.

References

Anderson, G.C., B.W. Frost, and W.K. Peterson, On the vertical distribution of zooplankton in relation to chlorophyll concentration, in: *Biological Oceanography of the Northern North Pacific Ocean*, A.Y. Takenouti (ed.), Idemitsu Shoten, Tokyo, pp. 341-345, 1972.

Antezena-Jerez, T., Distribution of euphausiids in the Chile-Peru current with particular reference to the endemic *Euphausia mucronata* and the oxygen minimum layer, Ph.D. Thesis, University of California, San Diego, 489 pp., 1978.

Banse, K., On the vertical distribution of zooplankton in the sea, in *Progress in Oceanography*, M. Sears (ed.), *2*, 55-125, 1964.

Boyd, C.M., Small scale patterns of marine zooplankton examined by an electronic *in situ* zooplankton detecting device, *Netherlands Journal of Sea Research*, *7*, 103-111, 1973.

Boyd, C.M., S.L. Smith, and T.J. Cowles, Grazing patterns of copepods in the upwelling system off Peru, *Limnology and Oceanography, 25,* 585-598, 1980.

Brink, K.H., D. Halpern, A. Huyer, and R.L. Smith, Near-shore circulation near 15°S: the physical environment of the Peruvian upwelling system, in *The Bioproductivity of Upwelling Ecosystems,* M.E. Vinogradov and R.T. Barber (eds.), in press, 1980.

Conover, W.J., A Kolmogorov goodness-of-fit test for discontinuous distributions, *Journal of the American Statistical Association, 67,* 591-596, 1972.

Dagg, M., T. Cowles, T. Whitledge, S. Smith, S. Howe, and D. Judkins, Grazing and excretion by zooplankton in the Peru upwelling system during April 1977, *Deep-Sea Research, 27A,* 43-59, 1980.

Denman, K.L. and D.L. Mackas, Collection and analysis of underway data and related physical measurements, in *Spatial Pattern in Plankton Communities,* J.H. Steele (ed.), Plenum Press, New York, 470 pp., 1978.

Doe, L.A.E., Project ICANE: a progress and data report on Canada-Peru study of the Peruvian anchovy and its ecosystem, Report BI-R-78-6, Bedford Institute of Oceanography, Dartmouth, Nova Scotia, 211 pp., 1978.

Haury, L.R., J.A. McGowan, and P.H. Wiebe, Patterns and processes in the time-space scales of plankton distributions, in *Spatial Pattern in Plankton Communities,* J.H. Steele (ed.), Plenum Press, New York, 470 pp., 1978.

Hobson, L.A. and C.J. Lorenzen, Relationships of chlorophyll maxima to density structure in the Atlantic Ocean and Gulf of Mexico, *Deep-Sea Research, 19,* 297-306, 1972.

Longhurst, A.R., Vertical distribution of zooplankton in relation to the eastern Pacific oxygen minimum, *Deep-Sea Research, 14,* 51-63, 1967.

Longhurst, A.R., Interactions between zooplankton and phytoplankton profiles in the eastern tropical Pacific Ocean, *Deep-Sea Research, 23,* 729-754, 1976.

Lorenzen, C.J., A method for the continuous measurement of *in vivo* chlorophyll concentration, *Deep-Sea Research, 13,* 223-227, 1966.

Mackas, D. and R. Bohrer, Fluorescence analysis of zooplankton gut contents and an investigation of diel feeding patterns, *Journal of Experimental Marine Biology and Ecology, 25,* 77-85, 1976.

Mackas, D. and C.M. Boyd, Spectral analysis of zooplankton spatial heterogeneity, *Science, 204,* 62-64, 1979.

Mullin, M.M. and E.R. Brooks, The vertical distribution of juvenile *Calanus* (Copepoda) and phytoplankton within the upper 50 m off La Jolla, California, in *The Biological Oceanography of the Northern North Pacific Ocean,* A.Y. Takenouti (ed.), Idemitsu Shoten, Tokyo, pp. 347-354, 1972.

Pearre, S., Jr., Problems of detection and interpretation of vertical migration, *Journal of Plankton Research, 1,* 29-44, 1979.

Raymont, J.E.G., *Plankton and Productivity in the Oceans,* Pergamon Press, Oxford, 660 pp., 1963.

Schulenberger, E., Vertical distributions, diurnal migrations, and sampling problems of hyperiid amphipods in the North Pacific central gyre, *Deep-Sea Research, 25,* 605-623, 1978.

Smith, R.L., Poleward propagating perturbations in currents and sea levels along the Peru coast, *Journal of Geophysical Research, 83,* 6083-6092, 1978.

Smith, S.L., K.H. Brink, H. Santander, T.J. Cowles, and A. Huyer, The effect of advection on variations in zooplankton at a single location near Cabo Nazca, Peru, *This Volume.*

Strickland, J.D.H. and T.R. Parsons, *A Practical Handbook of Seawater Analysis,* Bulletin 167, Fisheries Research Board of Canada, Ottawa, 293 pp., 1968.

Tate, M.W. and R.C. Clelland, *Nonparametric and Shortcut Statistics,* Interstate Printers and Publishers, Inc., Danville, Illinois, 171 pp., 1957.

Vinogradov, M.E., Vertical stratification of zooplankton in the Kurile-Kamchatka trench, in *The Biological Oceanography of the Northern North Pacific Ocean,* A.Y. Takenouti (ed.), Idemitsu Shoten, Tokyo, pp. 333-340, 1972.

Zalkina, A.V., Vertical distribution and diurnal migration of some Cyclopoida (Copepoda) in the tropical region of the Pacific Ocean, *Marine Biology, 5,* 275-282, 1970.

Zalkina, A.V., Vertical distribution and diurnal migration of Cyclopoida (Copepoda) in the waters of the north trade winds current and Sulu Sea, *Polish Archives of Hydrobiology, 24* (suppl.), 337-362, 1977.

A DENSE PATCH OF *ACARTIA LEVEQUEI* (COPEPODA, CALANOIDA) IN UPWELLED EQUATORIAL UNDERCURRENT WATER AROUND THE GALAPAGOS ISLANDS

Fernando Arcos

Instituto Oceanografico de la Armada, Guayaquil, Ecuador

Abstract. The use of the copepod *Acartia levequei* as an indicator of upwelled Equatorial Undercurrent water is suggested, and the occurrence of a dense patch of *A. levequei* has been related to the hydrography of the region around the Galapagos Islands.

--oo--

The region around the Galapagos Islands is highly productive and characterized by oceanic fronts, strong current shear, and patches of cool upwelled water. The sharp discontinuities in currents, water masses, and organisms have attracted the attention of both physical and biological oceanographers.

Observations west of the Galapagos Islands indicate that the Equatorial Undercurrent (Cromwell Current) becomes much shallower before striking the west side of the archipelago (Pak and Zaneveld, 1973). Houvenaghel's (1974) detailed analyses of the hydrographic conditions in the area showed that the undercurrent surfaces in several locations among the islands. Studies of primary production and chlorophyll by Maxwell (1974) and Jimenez (1979) show the highest values on the western side of the Galapagos, off Isabela and Fernandina islands. The association of chaetognaths and hydromedusae with the current patterns and water masses in the Galapagos was described by Houvenaghel (1974) and Alvarino (1977). Recent work by D. Bonilla related chaetognath abundance with the upwelling observed during the November 1978 cruise of *Orion*. Gueredrat (1971) studied copepod species abundance in relation to the equatorial current system west of and among the islands.

One species of copepod, *Acartia levequei*, reported by Grice (1964) as being abundant in Academy Bay, is apparently a good indicator species of upwelled Equatorial Undercurrent water around the Galapagos. The species is abundant in the upwelled water, ranging up to 19,000 organisms per cubic meter and produces clearly visible discoloration of the water. This paper reports one phase of a study of the zooplanktonic indicator species. The overall goal of the study is to identify organisms that will aid in understanding the circulation, mixing, and upwelling of the complex oceanic region around the Galapagos Islands and the coast of Ecuador. In this report an occurrence of a dense patch of *Acartia levequei* is related to the hydrography of the region.

The samples used for quantitative analysis of *A. levequei* and zooplankton biomass determinations were obtained during November, 1978 aboard the B.A.E. *Orion* in the C.0-2-78 cruise of the Oceanographic Institute of the Ecuadorian Navy (INOCAR) in a program to study equatorial upwelling processes (Fig. 1). Seventeen samples were taken from 10-min horizontal surface tows with a standard net having a mouth opening of 0.27 m^2 and a mesh size of 150 μm (Table 1). Additional samples for the determination of *A. levequei* abundance throughout the archipelago were obtained from ships of opportunity that worked in the region from August, 1975 to November, 1979 (Table 2). For the quantitative analysis of the organisms, the method of abundance classes was used as proposed by Frontier (1969) for ecological interpretations of the zooplankton data.

The area studied, in the eastern tropical Pacific, is limited on the north by 01°23'N, on the south by 02°00'S, on the east by 89°36'W, and on the west by 91°50'W. The hydrographic conditions of the area during November, 1978 were described on the basis of temperature-salinity diagrams of two meridional transects (Fig. 2a and 2b). The water masses in the area are the following (Fig. 3):

(1) The northern half of the archipelago is dominanted by Tropical Surface water from the north. Normally the strong front of the Tropical Surface water is further north in the boreal summer and fall.

(2) In the southern part of the Galapagos archipelago South Equatorial Current water is present and flows around the southwest corner of Isabela Island into Elizabeth Bay, apparently forced by a strong southeasterly wind (Moreano, 1980). At the southeast corner of Isabela Island this water is mixed with Surface Tropical water, and at the southwest corner the South Equatorial water is mixed (by wind) with upwelled

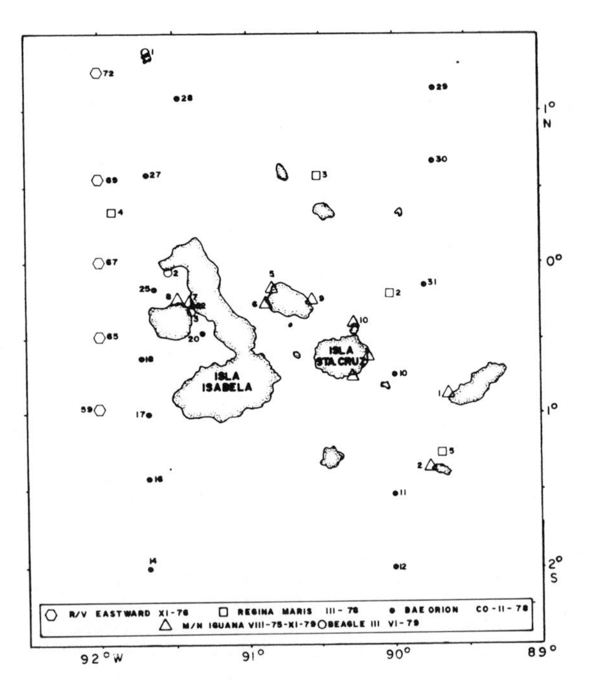

Fig. 1. Chart showing the stations in the Galapagos region where the plankton samples were collected between 1975 and 1979.

Equatorial Undercurrent water. Pak and Zaneveld (1973) found that in November, 1971 the South Equatorial Current was present as a zonal band south of about 01°30'S and east of 91°00'W. In November, 1971 the South Equatorial water extended further north among the islands than in November, 1978.

(3) The hydrographic data show that on the west side of Isabela Island Equatorial Undercurrent water is forced upward by topography causing areas of surface cooling. As it moves to the south this water mixes on the surface with South Equatorial Current water. The hydrographic results (Fig. 3) show clearly that at all of the southern stations, both east and west of the Galapagos, the Equatorial Undercurrent water was present at 75 m beneath the South Equatorial Current water. Equatorial Undercurrent water could surface in topographically driven upwelling among the other islands of the archipelago. Undercurrent water has been identified east of the islands beneath the southern edge of the Equatorial Front and within 110 km of the coast of Ecuador and northern Peru (Stevenson and Taft, 1971), at 84°W by Pak and Zaneveld (1973), at 85°W by Jimenez (1978), and at 82°W at a depth of 75 m by F. Pesantes (personal communication).

Among the Galapagos Islands the variability of plankton abundance and species composition is great throughout the year because the transport, growth, and reproduction of phytoplankton and zooplankton depend on the strength and position of the Equatorial Undercurrent. During November 1978 in the channel between Isabela and Fernandina islands, a large area of brown-red discoloration was observed (Fig. 4). The cause of the discoloration was a dense swarm of the copepod, *A. levequei*. In general copepods are the first zooplankton group to increase after enrichment by

TABLE 1. Station data from the November 1978 cruise.

Station	Date	Time	Latitude	Longitude	Salinity (x 10^3)	Temperature (°C)
10	17-XI-78	21:50	00° 45.0'S	90° 00.0'W	34.150	23.17
10'	19-XI-78	14:30	01° 15.0'S	90° 00.0'W	34.340	22.40
11	19-XI-78	19:06	01° 32.0'S	90° 00.0'W	34.337	22.66
12	20-XI-78	02:00	02° 00.0'S	90° 00.0'W	34.361	22.58
14	20-XI-78	19:15	02° 00.0'S	91° 40.0'W	34.939	21.28
16	21-XI-78	01:50	01° 25.0'S	91° 40.0'W	34.868	20.78
17	21-XI-78	08:10	01° 00.0'S	91° 40.0'W	34.896	19.83
18	21-XI-78	13:45	00° 30.0'S	91° 43.0'W	34.878	20.25
X	21-XI-78	15:15	00° 36.0'S	91° 32.0'W	34.966	20.30
20	22-XI-78	10:45	00° 30.2'S	91° 18.5'W	34.776	21.26
22	22-XI-78	15:05	00° 17.3'S	91° 23.5'W	34.994	21.01
25	23-XI-78	12:30	00° 10.6'S	91° 38.6'W	34.993	19.11
27	23-XI-78	23:20	00° 35.0'N	91° 41.0'W	34.057	24.35
28	24-XI-78	05:15	01° 04.2'N	91° 28.0'W	33.833	25.06
29	24-XI-78	23:35	01° 07.8'N	89° 42.5'W	33.998	24.54
30	25-XI-78	07:05	00° 09.5'N	89° 42.5'W	33.992	24.49
31	25-XI-78	17:30	00° 09.0'S	89° 46.4'W	34.418	22.33

X is the area of water discoloration.

TABLE 2. Station data from cruises between 1975 and 1979.

Cruise	Station	Locality	Date	Time	Latitude	Longitude
	1	Isla San Cristobal			00°53.1'S	89°36.0'W
	2	Punta Suárez			01°21.1'S	89°44.0'W
	3	Academy Bay			00°44.1'S	90°18.0'W
	4	Isla Plaza			00°35.0'S	90°09.1'W
M.N. *Iguana*	5	James Bay	22 cruises between August of 1975 and November of 1979.		00°14.0'S	90°51.0'W
	6	Isla Espumilla			00°11.0'S	90°50.0'W
	7	Caleta Tagus			00°13.0'S	91°22.0'W
	8	Punta Espinosa			00°15.1'S	91°26.1'W
	9	Isla Bartolome			00°17.0'S	90°33.1'W
	10	Isla Seymour			00°24.2'S	90°17.0'W
	59		5-XI-76	02:49	00°58.3'S	92°01.0'W
	65		6-XI-76	02:26	00°29.0'S	91°58.9'W
R.V. *Eastward*	67	West of the Archipielago	7-XI-76	06:17	00°00.1'S	92°00.0'W
	69		7-XI-76	10:34	00°33.0'N	92°00.0'W
	72		7-XI-76	18:50	01°14.8'N	92°02.1'W
	2		13-VIII-78		00°12.0'S	90°01.2'W
R.V. *Regina Maris*	3	Among the Islands.	15-VIII-78		00°34.0'N	90°31.0'W
	4		16-VIII-78		00°19.6'N	91°54.3'W
	5		24-VIII-78		01°15.8'S	89°40.2'W
	1	Isla Wolf	23-VI-79	14:00	01°23.0'N	91°49.5'W
Beagle II	2	North of Isabela Is.	24-VI-79	16:00	00°03.0'S	91°32.0'W
	3	Caleta Tagus	25-VI-79	13:00	00°14.0'S	91°32.5'W

upwelling, with increases in total copepod biomass as well as the development of an herbivorous fauna of species that are particularly well adapted to cold or cool water (Thiriot, 1978). Phytoplankton counts and identification by Jimenez (1980) showed that the most abundant species in the upwelled water were relatively small species of diatoms like *Thalassiosira* spp. and *Nitzschia* spp. This type of phytoplankton is optimum for grazing by *A. levequei*, although on occasion big dinoflagellates such as *Ceratium furca* were found in the mouth appendages of this herbivorous copepod.

A. levequei, from the family Acartiidae, belongs to the subgenus Acanthacartia (Steuer, 1923). The organisms we collected fit the taxonomic descrip-

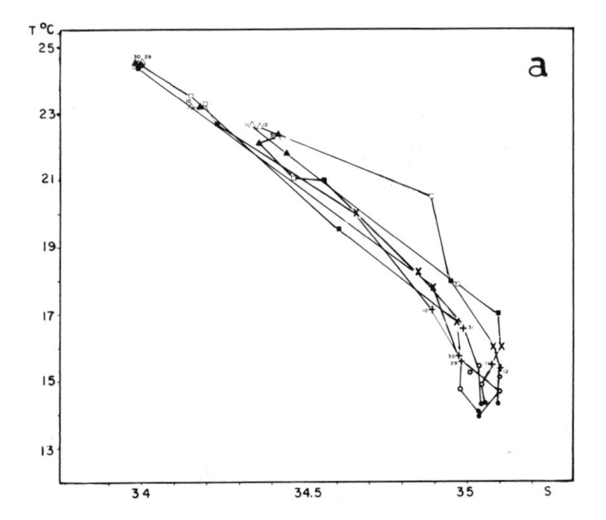

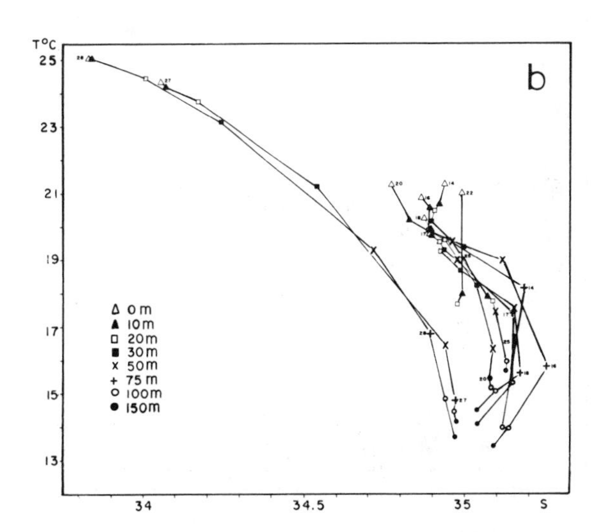

Fig. 2. T-S diagrams of the 150-m water column from stations of the November, 1978 cruise: (a) east of the Galapagos and (b) west of the Galapagos.

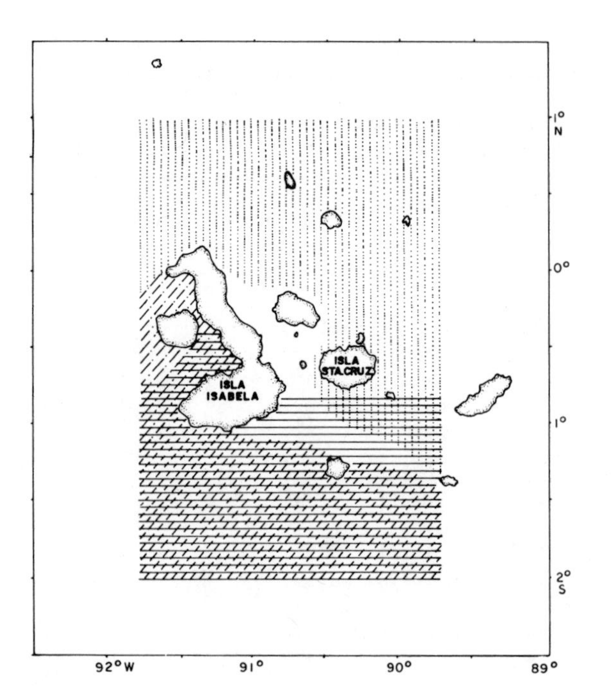

SURFACE TROPICAL WATER

SOUTH EQUATORIAL CURRENT

EQUATORIAL UNDERCURRENT AT SURFACE

EQUATORIAL UNDERCURRENT AT A DEPTH OF 75m

Fig. 3. Water masses in the Galapagos region during November 1978.

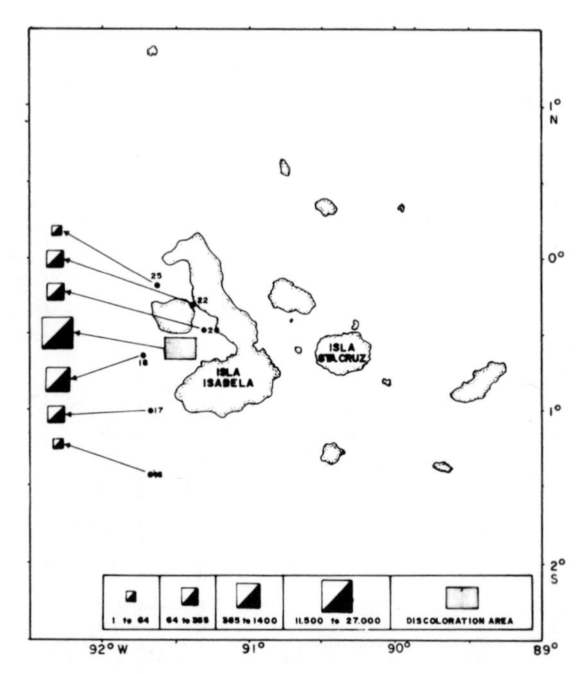

Fig. 4. Numerical distribution of *A. levequei* (in organisms per cubic meter) during the cruise of November, 1978.

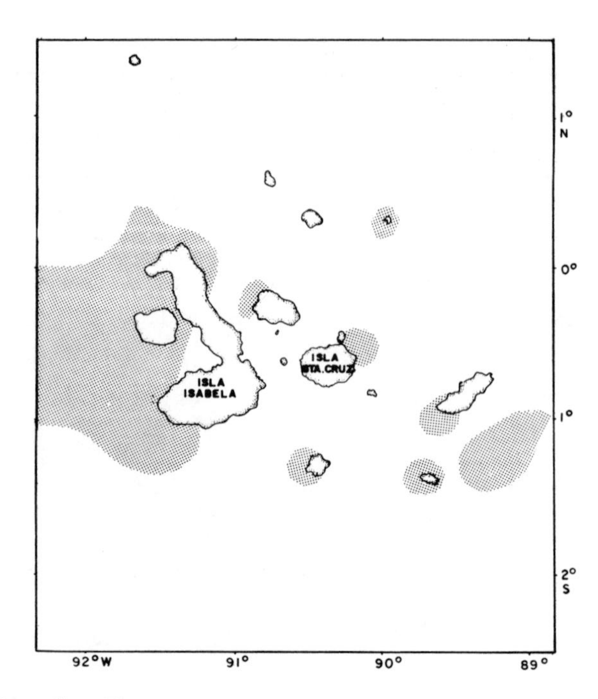

Fig. 5. The areas of most prominent upwelling (shaded areas, from Houvenaghel, 1978).

tion given by Grice (1964). The most important character for its identification is the structure of the fifth pair of feet in both the male and female. The total length of the body was 1.1 to

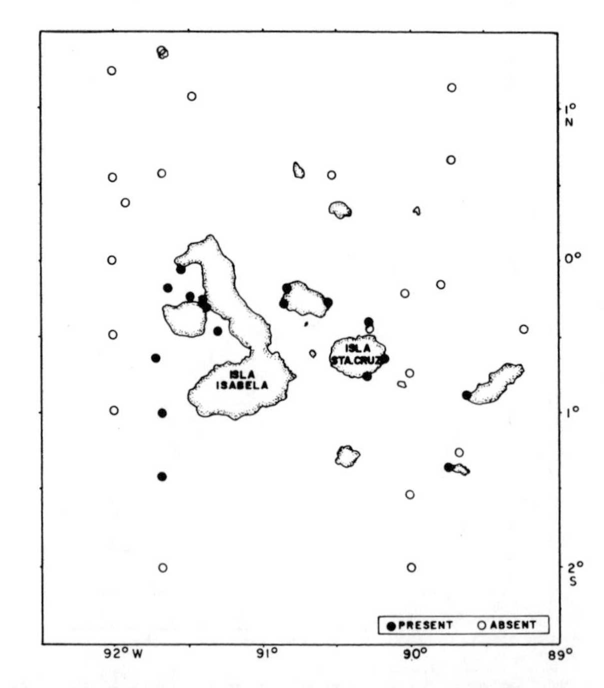

Fig. 6. Chart showing the presence and absence of *A. levequei*.

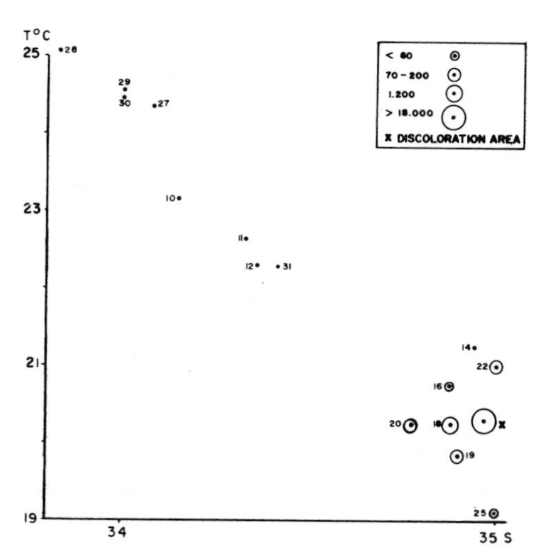

Fig. 7. T-S diagram of the surface waters during November, 1978 related with the densities of *A. levequei* (organisms per cubic meter).

1.2 mm for the males and 1.3 to 1.5 mm for the females. The organisms caught by Grice (1964) measured 1.2 to 1.3 mm for the males and 1.5 to 1.7 mm for the females.

The near-shore habitat of this species (Grice, 1964) is confirmed by its year-round presence in the stations, which were all very close to the islands. *A. levequei* was absent from samples taken by *Regina-Maris* and *Eastward* in offshore waters and from coastal samples taken in the northern half of the archipelago by *Beagle* II at stations where Tropical Surface water was prevalent. But during the November, 1978 cruise in the western part of the Galapagos Islands, the species was found both nearshore and offshore in association with newly upwelled Equatorial Undercurrent water.

Comparing the chart of the most prominent upwelling areas (Fig. 5), described on the basis of temperature by Houvenaghel (1978), with the chart showing the presence or absence of *A. levequei* (Fig. 6), the close relationship between the upwelled waters and the presence of the species is apparent. The T-S diagram (Fig. 7) of the surface waters during November, 1978 and the distri-

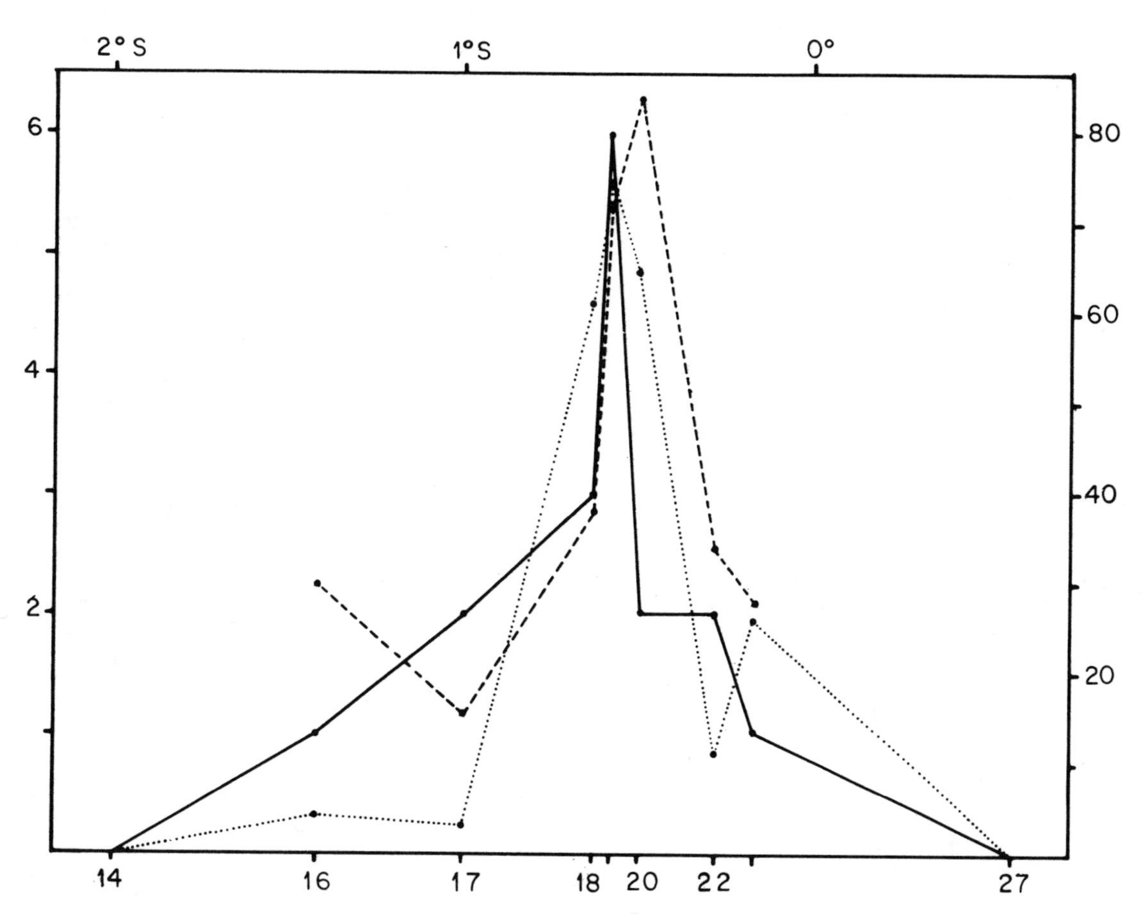

Fig. 8. —— Number of copepods *Acartia levequei* per m^3 expressed in abundance quotations; .... percentage of *A. levequei* in relation to the total number of copepods; ---- percentage of juveniles of *A. levequei* in relation to the adults of the same species.

bution of *A. levequei* abundance show that the species is associated with the lowest temperatures (19.11 to 21.26°C) and highest salinities (34.776 to 34.994). During the 1978 cruise (Fig. 4) the maximum concentration of *A. levequei* in the discolored water was 19,000 organisms per m^3. The species comprised 75% of the copepods in the tow and 73% of the *A. levequei* individuals were juvenile stages. The number of *A. levequei* and their percentage of the total copepod fauna decreased north and south along the transect away from the southwest corner of Isabela Island.

Currents may carry planktonic animals to areas where they can survive but not reproduce (Margalef, 1977). The populations of *A. levequei* in the offshore waters south and west of Isabela Island may be expatriates from the massive patch in the inshore region between Isabela and Fernandina islands where the discoloration was observed (Fig. 4). The rich phytoplankton bloom in the inshore region at 00°40.0'S may have provided a center from which the copepods were dispersed offshore to 00°10.6'S and to 1°25.0'S. The northern boundary of dispersion is clearly limited by the marked front between Equatorial Undercurrent water and Tropical Surface water. The southern boundary extends as far as the presence of Equatorial Undercurrent water is detectable on the surface as a mixture with the South Equatorial water. Verification of the expatriation hypothesis will require analysis of the distribution of *A. levequei* with depth stratified sampling at stations close in both time and space. The present results indicate that *Acartia levequei* should be an indicator species for Equatorial Undercurrent water among the Galapagos archipelago.

Acknowledgements. I thank the Director and Sub-Director of the Instituto Oceanografico de la Armada for making possible this work on indicator species. I am also grateful to Dr. Roberto Jimenez for continuous encouragement and advise and to Mr. Victor Mesias for drafting.

References

Alvarino, A., El zooplancton del Pacífico Ecuatoriano, paper presented at IV Simposio Latinoamericano de Oceanografía Biológica, Guayaquil, Ecuador, 1977.

Bonilla, D., Distribución de los Quetognatos en las aguas de las Islas Galápagos, *Acta Oceanografica del Pacífico,* (in press).

Frontier, S., Sur une méthode d'analyse faunistique rapide de zooplankton, *Journal of Experimental Marine Biology and Ecology, 3,* 18-26, 1969.

Grice, G.D., Two new species of Calanoid copepods from the Galapagos Islands with remarks on the identity of three other species, *WHOI Data Report 1332,* 225-264, Woods Hole Oceanographic Institution, Woods Hole, 1964.

Gueradrat, J.A., Evolution d'une population de copépodes dans le systéme des courants équatoriaux de l'Océan Pacifique, Zoogéographie, écologie et diversité spécifique, *Marine Biology, 9,* 300-314, 1971.

Houvenaghel, G.T., Etude océanographique de l'archipiel des Galapagos et mise en évidence du role des conditions hydrologiques dans la détermination du peuplement des Iles, Doctoral Thesis, Université Libre de Bruxelles, 1974.

Houvenaghel, G.T., Oceanographic conditions in the Galapagos archipelago and their relationships with life on the islands, in: *Upwelling Ecosystems,* R. Boje and M. Tomczak (eds.), Springer-Verlag, New York, 181-200, 1978.

Jimenez, R., Mise en évidence de l'upwelling équatorial a l'est des Galapagos, *Cahiers ORSTOM series Oceanography, 16,* 137-155, 1978.

Jimenez, R., Informe preliminar sobre los pigmentos clorofilicos y la producción primaria en las aguas de las Islas Galápagos, *INOCAR Bulletin,* 1-8, 1979.

Jimenez, R., Composition and distribution of phytoplankton in the upwelling system of the Galapagos Islands, *This Volume,* 1980.

Margalef, R., *Ecologia,* 951 pp., Omega, Barcelona, 1977.

Maxwell, D.C., Marine primary productivity of the Galapagos archipelago, Ph.D. Dissertation, Ohio State University, 167 pp., 1974.

Moreano, H.R., The upwelling at Elizabeth Bay and Bolivar Channel, Galapagos Islands, paper presented at the International Symposium on Coastal Upwelling, Los Angeles, California, 1980.

Pak, H. and J.R. Zaneveld, The Cromwell Current on the east side of the Galapagos Islands, *Journal of Geophysical Research, 78,* 7845-7859, 1973.

Steuer, A., Bausteine zu einer Monographie der Copepodengattung Acartia, *Arbeiten aus dem Zoologischen Institute der Universität Innsbruck, Band I,* Heft 5, 1923.

Stevenson, M.R. and B.A. Taft, New evidence for the Equatorial Undercurrent east of the Galapagos Islands, *Journal of Marine Research, 29,* 103-115, 1971.

Thiriot, A., Zooplankton communities in the West African upwelling area, in: *Upwelling Ecosystems,* R. Boje and M. Tomczak (eds.), Springer-Verlag, New York, 32-61, 1978.

ON THE FOOD OF CALANOID COPEPODS FROM THE NORTHWEST AFRICAN UPWELLING REGION

S.B. Schnack

Institut für Meereskunde an der Universität Kiel
2300 Kiel, West Germany

M. Elbrächter

Biologische Anstalt Helgoland, Litoralstation List
2282 List/Sylt, West Germany

Abstract. At four stations on a horizontal section feeding behavior of adult females of two calanoid copepod species, *Calanoides carinatus* and *Centropages chierchiae*, was studied during a cruise of R.V. *Meteor* to the northwest African upwelling area in January and February, 1977. Two different methods were applied: gut content studies with preserved animals and feeding experiments with naturally-occurring particles as food supply.

In the two inshore stations, where upwelling coincided with small centric diatoms as the major fraction of the phytoplankton assemblages, more than 90% of the copepods examined contained food in their guts. In the two more offshore stations with higher surface temperature (17 to 18°C) and mainly μ-flagellates, dinoflagellates, and large diatoms as available food organisms, the majority of copepods were found with empty guts (*C. carinatus*, 75%, *C. chierchiae*, 62%). Small centric diatoms constituted the major fraction of recognizable food particles in the inshore stations, dinoflagellates in the southeast offshore station.

Particle spectra and feeding behavior differed from station to station. There was evidence for preferential feeding on larger particles and particle concentration peaks in the feeding experiments at the more offshore stations, where total particle concentration was twice that at the inshore stations. Higher filtering rates, however, were observed at the inshore stations.

Altogether, *C. carinatus* and *C. chierchiae* seem to be well adapted to take advantage of the phytoplankton bloom in freshly upwelled water.

Introduction

The interest in physical and biological events in upwelling areas has considerably increased during recent years. The main biological feature of an upwelling process is the increase of phytoplankton standing stock and productivity due to fertilization by cold, nutrient-rich upwelled waters (Boje and Tomczak, 1978). In the northwest African upwelling region, high phytoplankton concentrations were found near the coast, where upwelling is more intense (Estrada, 1978). The high plant concentrations are maintained throughout an upwelling season and induce an increase in zooplankton, especially filter-feeding copepods (Binet, 1973; Thiriot, 1978). The heterogeneity of the phytoplankton in space and time is caused by changing environmental conditions such as aging within upwelled water signalled by increasing surface temperature and decreasing nutrient values. Copepods feeding on natural phytoplankton assemblages have to deal with this mixed distribution of food particles.

Feeding interrelationships at lower trophic levels in upwelling areas have been little studied (Cowles, 1978). The general intent of this study was to determine the relationship between the quantity and quality of the food available and its consumption by copepods. The study was planned with a view to answering the following questions: How do copepods handle the heterogeneous spectra of food particles and do they change their feeding behavior with changing food supply?

To gain an impression on the feeding behavior we followed two different strategies: first, we analyzed the gut contents to gather information on type and amount of food ingested in the natural state (Mullin, 1966). Difficulties with this method are the identification of digested food particles: it is impossible to recognize organisms without indigestible thecae or tests. The method is qualitative rather than quantitative. The second method involves direct measurements of food uptake in feeding experiments and assumes that the feeding habits of copepods are

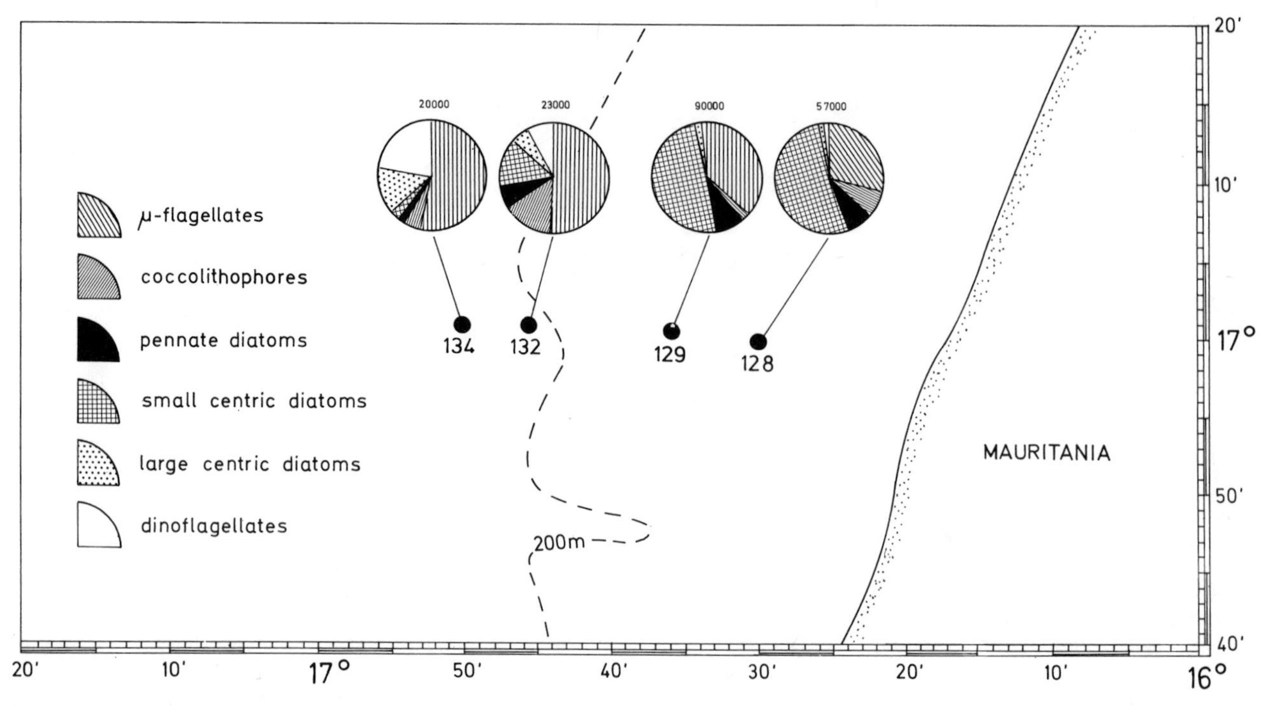

Fig. 1. Positions of the stations and variations of main phytoplankton components. The numbers above the circles indicate the phytoplankton concentration (number of cells per liter).

the same in the laboratory as in the sea.

During a cruise of R.V. *Meteor* in January and February, 1977 studies were carried out with adult females of *C. carinatus* and *C. chierchiae*, two of the most abundant calanoid copepod species in the northwest African upwelling system. This contribution deals with the results from one horizontal section. A more comprehensive description of the hydrographic and biological situation in connection with feeding behavior of copepods will be given elsewhere.

Material and Methods

At four stations along a horizontal section at 17° (Fig. 1) phytoplankton and zooplankton were collected for feeding studies. All samples were taken in the evening (between 19.00 and 20.30 h). For phytoplankton determination, water samples were taken with 30-ℓ Niskin bottles at 10 and 30 m in duplicate. One sample was preserved immediately after sampling with formalin (2% end concentration); the other was screened through a 200-μm sieve to remove large particles and used as food for copepods in the shipboard feeding experiments.

The copepods were collected with nets (mesh size, 300 μm) hauled vertically from 30 m to the surface. The samples were diluted and gently split in two parts, one of which was preserved with formalin (4% end concentration) for later analysis. For the feeding experiments, adult female *C. carinatus* and *C. chierchiae* were selected and placed in beakers containing screened seawater. After 10 to 12 h preconditioning time, healthy, actively swimming females were picked out, and 20 to 25 individuals of each species were added to a 5-ℓ beaker containing 3 ℓ of water with experimental food conditions.

The experiments for each species were run in duplicate. One beaker containing seawater without copepods served as control. All suspensions were stirred with Plexiglass stirrers rotating at 2 rpm (modified after Frost, 1972). The experiments were run in dim light at environmental temperatures (between 15 and 18°C) for 8 h. Before and after each experiment particle counts were made using a Coulter Counter model TA II. Two aperture tubes (140 and 280 μm) were used to measure particle size distribution from 4 to 102 μm mean spherical diameter (Sheldon and Parsons, 1967). Channels 1 to 3 were not used because of noise problems. Samples for microscopic examination were also taken. Filtering rates and average particle concentrations were calculated for each size class using the method described by Frost (1972).

The guts of preserved copepods were dissected, mounted on a glass slide in polyvinyllactophenol, and dyed with lignin pink. The guts were crushed slightly under a cover glass and examined under a microscope at 1000 x magnification.

Results

The location of the four sampling stations is given in Fig. 1. The same figure shows the relative composition of phytoplankton groups as well as the abundance of phytoplankton expressed as cells per litre. At Stas 128 and 129, the phytoplankton assemblage was dominated by small centric diatoms (>53%) and μ-flagellates constituted the second major fraction (>25%). Dinoflagellates and large centric diatoms, such as *Stephanopyxis palmeriana* (Grev.) Grunow, *Rhizosolenia stolterfothii* Pergallo and *R. robusta* Norman were rare (both groups less than 5%), in contrast to the results from the more offshore Stas 132 and 134, where μ-flagellates made up more than 50% of the phytoplankton assemblage. The population of small centric diatoms decreased rapidly (at Sta. 134, <3%) concomitant with an increase in dinoflagellates and large diatoms (22 and 14%, respectively at Sta. 134).

The results clearly indicate that there is a change in phytoplankton composition in an inshore-offshore gradient. The temperature distribution during sampling suggests upwelling with the most intense upwelling closest to the coast (Fig. 2). At Stas 128 and 129 the temperature in the upper 30 m was below 17°C, whereas at Stas 132 and 134 it was warmer. From comparisons with the phytoplankton distribution it is obvious that cold, upwelled waters coincided with small centric diatoms. With increasing surface temperatures towards the offshore stations, the abundance of dinoflagellates and large diatoms increased. The results corroborate those reported by Richert (1975). At the inshore stations, cell numbers per liter were higher than offshore, but the situation was reversed with regard to total particle volumes. The discrepancy was due not only to increasing cell size, but the detritus content also increased from inshore to offshore stations.

At all four stations calanoid copepods dominated by numbers the zooplankton (85 to 88%). *C. carinatus* accounted for 40 to 70% and *C. chierchiae* 20 to 40% of all calanoid species. From the structure of the mouth parts, the copepod *C. carinatus* should be herbivorous, i.e., a filter-feeding species, whereas *C. chierchiae* should be an omnivorous one (Schnack, unpublished data).

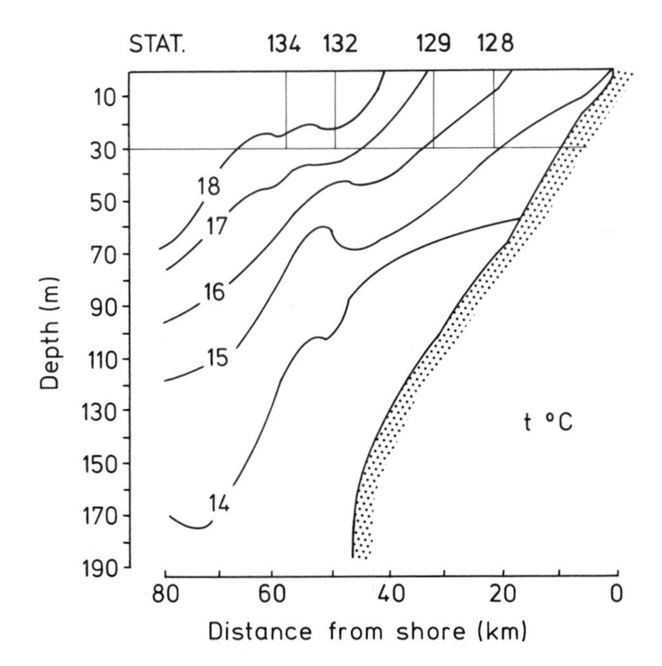

Fig. 2. Distribution of temperature in the section along 17°N. The thin vertical and horizontal lines indicate the zooplankton net hauls (30 to 0 m).

Analysis of the gut contents

The guts of 200 females of *C. carinatus* and *C. chierchiae* were examined from each station (Fig. 3). No great difference is obvious between the two copepod species. At Stas 128 and 129 only a few specimens had empty guts (<8%), and the gut contents of the majority contained unidentifiable as well as identifiable food remnants (Fig. 3a). At Sta. 132, a higher percentage of animals without food were found (about 35%), and at Sta. 134 as much as 75% of the *C. carinatus* and 62% of the *C. chierchiae* contained no food. Diatoms constituted by numbers the major portion of the recognizable food in the guts at Stas 128, 129, and 132. Especially small centric diatoms (such as *Cyclotella* and *Thalassiosira* species, diameter 8 to 25 μm) and small pennate diatoms (type *Navicula*, 18 x 10 μm) were found frequently at Stas 128 and 129. At Sta. 134, dinoflagellates (mainly *Prorocentrum*) constituted the major part of the identifiable food particles (Fig. 3b); μ-flagellates and coccolithophorids were not included in the quantitative account. The μ-flagellates are impossible to recognize in the guts and are possibly part of the unidentifiable mass. Coccolithophorids were found at all stations, but due to the disintegration of shells into single coccoliths, we were not able to recognize all ingested ones. Many of the copepods with food also contained sand grains in the guts: at Stas 128 and 129 about 95%, at Stas 132 and 134 only 70 and 61%, respectively.

Feeding experiments

Particle concentrations (expressed as particle volume) and filtering rates for female *C. carinatus* and *C. chierchiae* relative to particle size for all four stations are shown in Figs 4 and 5. As particle spectra and feeding behavior differed from station to station, the results are presented separately below:

Sta. 128. The particle distribution showed a maximum between 10 and 64 μm with a peak at 20 μm mean spherical diameter (Fig. 4). All size classes were removed by the copepods. *C. cari-*

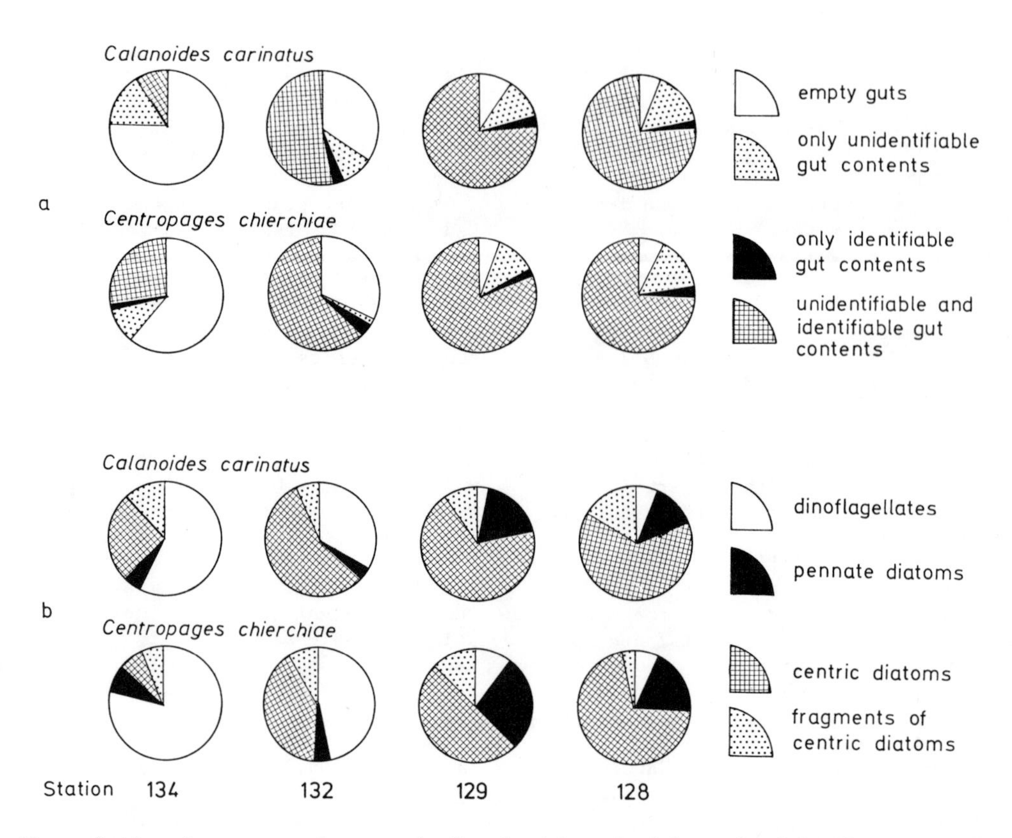

Fig. 3. a) The relative frequency of copepods found with and without food in the guts. Dots indicate percentage of copepods where all particles in the gut were unidentifiable; black, where they were all identifiable and checks denote a mixture of both particle types; b) The relative composition of identifiable food particles by numbers in the guts.

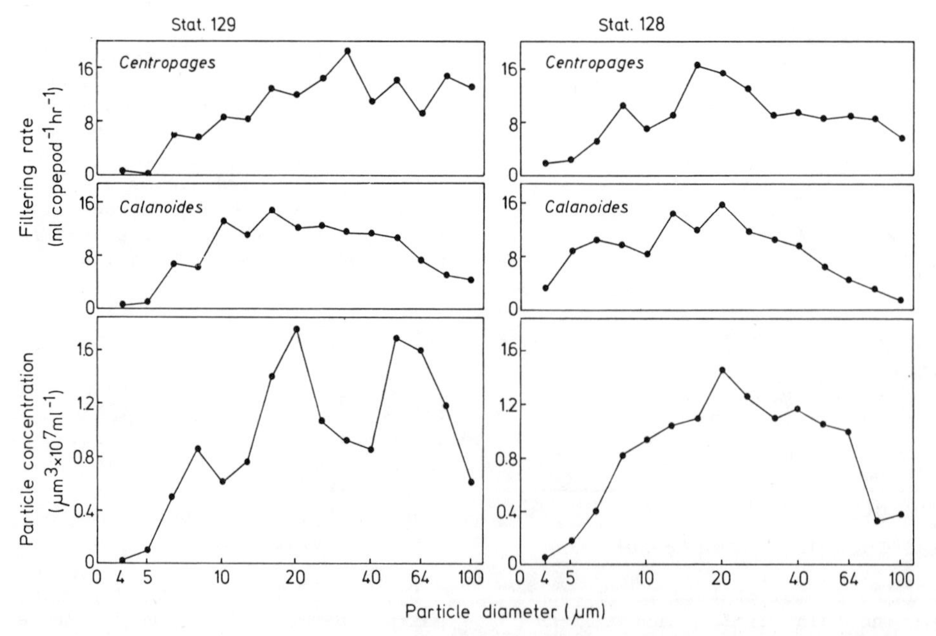

Fig. 4. Relationship between filtering rate and particle size distribution for *Calanoides carinatus* and *Centropages chierchiae* at Stas 128 and 129. Particle size is given in spherical diameter.

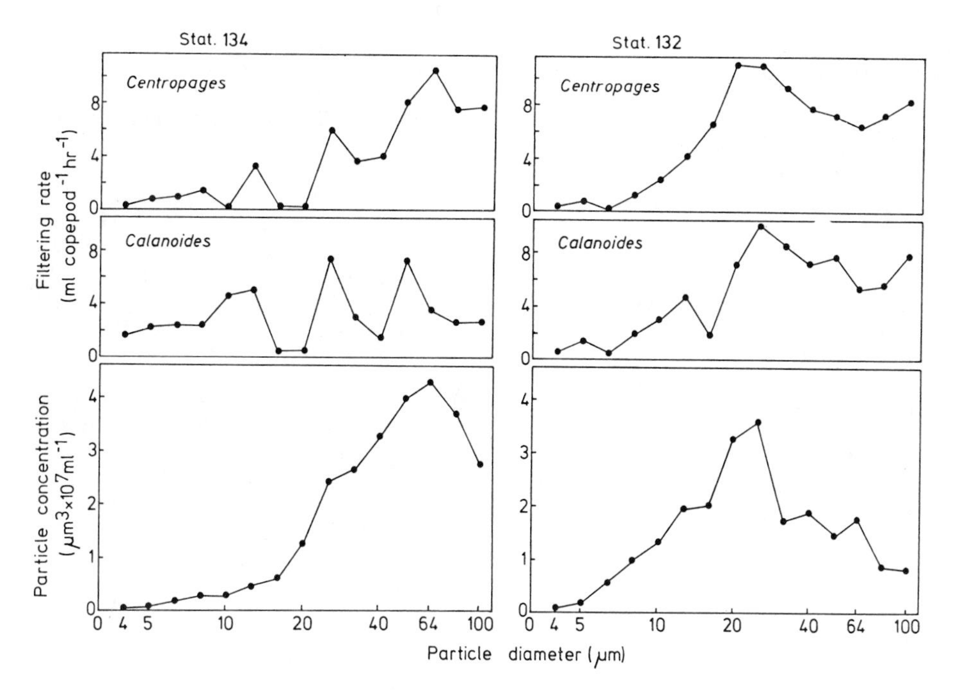

Fig. 5. The same curves as Fig. 4 but for Stas 132 and 134.

natus tended to utilize the smaller particles more efficiently than *C. chierchiae*, whereas the latter fed at higher rates on the larger particles. Both copepod species showed highest filtering rates at or near the particle volume peak.

Sta. 129. A double-peak distribution, with large peaks at 20 and 50 μm was present. A small peak occurred also at 8 μm. Again, *C. chierchiae* grazed more heavily on larger size particles than *C. carinatus*. However, there is no evidence for preferential feeding on the particle peak concentrations.

Sta. 132. A distinct biomass peak for particles of about 20 μm diameter occurred. For both copepods species, filtering rates were higher on particles in the concentration peak as well as larger particles.

Sta. 134. Large particles (>25 μm) were most abundant with a concentration peak at 64 μm. The filtering curves for both species show a zigzag form. It is not clear why both species filtered particles 16 and 20 μm in diameter at such low rates. Again, *C. chierchiae* preferentially fed on the concentration peak and larger particles. In contrast, *C. carinatus* fed on different size classes: 10 to 13, 25, and 50 μm were eaten preferentially and other size classes were fed on at lower rates. The highest filtering rates for *C. carinatus* and *C. chierchiae* were at the inshore stations (18 ml per copepod per hour in contrast to 10 ml per copepod per hour at the offshore stations).

Discussion

Variations in feeding behavior, according to type and concentrations of food particles available, have been reported for different copepod species (Conover, 1978). Evidence for higher filtering rates on larger food particles was observed by Mullin (1963), Richman and Rogers (1969), Paffenhöfer (1971), Frost (1972, 1977), and Gaudy (1974). All the results were based on experiments with only one or a few algal species. Only a few studies have been with naturally occurring food particle spectra. In the studies the animals fed preferentially on particles present in highest concentrations as well as on large particles (Parsons, LeBrasseur, and Fulton, 1967; Parsons, LeBrasseur, Fulton, and Kennedy, 1969; Poulet, 1973, 1974, 1978; Richman, Heinle, and Huff, 1977; Gamble, 1978). In our study, evidence for preferential feeding on large size classes and volume peaks was observed for *C. carinatus* and *C. chierchiae* at Sta. 132, but at Sta. 134 only for the latter copepod species. Cowles (1978) pointed out in his study of feeding of copepods with natural assemblages in the Peru upwelling system that the animals alter their feeding behavior as a function of total food abundance. At high food levels, copepods fed more efficiently on larger, more abundant particles. As food levels decreased, however, Cowles' results indicated that copepods fed more indiscriminately. With one exception, namely *C. carinatus* at Sta. 134, our results are in accord with those of Cowles. However, all particle concentrations in the northwest African upwelling region were much higher than those observed in the Peru upwelling system by Cowles (1978). Hence, no direct comparison of the observed feeding behavior is possible.

The results reported here cannot be explained by purely mechanical feeding behavior as suggested by Boyd (1976), Lam and Frost (1976), Lehman (1976), Nival and Nival (1976), and Frost (1977). Several factors may affect the feeding habits of copepods. According to our gut content studies, species and size composition of phytoplankton may be among the most important factors. In stations with freshly upwelled waters, where small-sized cells with greater surface to volume ratio tend to predominate (Richert, 1975), copepods were found with full guts. However, at stations with aged upwelled waters, where large diatoms and dinoflagellates as well as μ-flagellates dominated, a high percentage of copepods had empty guts. Little is known concerning the causes of this depression of feeding activity, and further investigation of the phenomenon will be necessary to determine why feeding changes to such an extent along an upwelling gradient.

Summarizing, *C. carinatus* and *C. chierchiae* seem to be well adapted to utilizing the phytoplankton blooms formed in freshly upwelled waters. We can thus conclude that the feeding habits of the two species are one precondition for the dominant role they play as secondary producers in the northwest African upwelling area.

Acknowledgements. This work was supported by the Deutsche Forschungsgemeinschaft with the program "Upwelling in the Sea". We would like to thank Dr. E. Mittelstaedt for providing the temperature data and Dr. V. Smetacek for reading the manuscript.

References

Binet, D., Note sur l'évolution des populations de copépodes pelagiques de l'upwelling Mauritanien (Mars - Avril 1972), *Document Scientifique Abidjan, ORSTOM, 4*, 77-90, 1973.

Boje, R., and M. Tomczak, Ecosystem analysis and the definition of boundaries in upwelling regions, in *Upwelling Ecosystems*, R. Boje and M. Tomczak (eds.), Springer-Verlag Berlin, Heidelberg, New York, pp. 24-31, 1978.

Boyd, C.M., Selection of particle sizes in filter-feeding copepods: a plea for reason, *Limnology and Oceanography, 21*, 175-180, 1976.

Conover, R.J., Transformation of organic matter, in *Marine Ecology, Vol. IV, Dynamics*, O. Kinne (ed.), Wiley, Chichester, pp. 221-499, 1978.

Cowles, T.J., Copepod feeding in the Peru upwelling system, *Coastal Upwelling Ecosystems Analysis, Technical Report 36*, 172 pp., 1978.

Estrada, M., Mesoscale heterogeneities of the phytoplankton distribution in the upwelling region off NW Africa, in *Upwelling Ecosystems*, R. Boje and M. Tomczak (eds.), Springer-Verlag, Berlin, Heidelberg, New York, pp. 16-23, 1978.

Frost, B.W., Effects of size and concentration of food particles on the feeding behavior of the marine copepod *Calanus pacificus*, *Limnology and Oceanography, 17*, 805-815, 1972.

Frost, B.W., Feeding behavior of *Calanus pacificus* in mixtures of food particles, *Limnology and Oceanography, 22*, 472-491, 1977.

Gamble, J.C., Copepod grazing during a declining spring phytoplankton bloom in the northern North Sea, *Marine Biology, 49*, 303-315, 1978.

Gaudy, R., Feeding of four species of pelagic copepods under experimental conditions, *Marine Biology, 25*, 125-141, 1974.

Lam, R.K., and B.W. Frost, Model of copepod filtering response to changes in size and concentration of food, *Limnology and Oceanography, 21*, 490-500, 1976.

Lehman, J.T., The filter-feeder as an optimal forager, and the predicted shapes of feeding curves, *Limnology and Oceanography, 21*, 501-516, 1976.

Mullin, M.M., Some factors affecting the feeding of marine copepods of the genus *Calanus*, *Limnology and Oceanography, 8*, 239-250, 1963.

Mullin, M.M., Selective feeding by calanoid copepods from the Indian Ocean, in *Some Contemporary Studies in Marine Science*, H. Barnes (ed.), George Allen and Unwin Ltd., London, pp. 545-554, 1966.

Nival, P., and S. Nival, Particle retention efficiencies of an herbivorous copepod, *Acartia clausi* (adult and copepodite stages): effects on grazing, *Limnology and Oceanography, 21*, 24-38, 1976.

Paffenhöfer, G.A., Grazing and ingestion rates of nauplii, copepodids and adults of the marine planktonic copepod *Calanus helgolandicus*, *Marine Biology, 11*, 286-298, 1971.

Parsons, T.R., R.J. LeBrasseur, and J.D. Fulton, Some observations on the dependence of zooplankton grazing on the cell size and concentration of phytoplankton biomass, *Journal of the Oceanographical Society of Japan, 23*, 10-17, 1967.

Parsons, T.R., R.J. LeBrasseur, J.D. Fulton, and O.D. Kennedy, Production studies in the Strait of Georgia, Part II, Secondary production under the Fraser River plume, February to May, 1967, *Journal of Experimental Marine Biology and Ecology, 3*, 39-50, 1969.

Poulet, S.A., Grazing of *Pseudocalanus minutus* on naturally occurring particulate matter, *Limnology and Oceanography, 18*, 564-573, 1973.

Poulet, S.A., Seasonal grazing of *Pseudocalanus minutus* on particles, *Marine Biology, 25*, 109-123, 1974.

Poulet, S.A., Comparison between five co-existing species of marine copepods feeding on naturally occurring particulate matter, *Limnology and Oceanography, 23*, 1126-1143, 1978.

Richert, P., Die räumliche und zeitliche Entwicklung des Phytoplanktons, mit besonderer Berücksichtigung der Diatomeen, im NW-Afrikanischen

Auftriebsgebiet, Ph.D. Thesis, Universität Kiel, 140 pp., 1975.
Richman, S., and H.N. Rogers, The feeding of *Calanus helgolandicus* on synchronously growing populations of the marine diatom *Ditylum brightwelli*, *Limnology and Oceanography*, *14*, 701-709, 1969.
Richman, S., D.R. Heinle, and R. Huff, Grazing by adult estuarine calanoid copepods of the Chesapeake Bay, *Marine Biology*, *42*, 69-84, 1977.
Sheldon, R.W., and T.R. Parsons, A practical manual on the use of Coulter counter in marine science, Coulter Electronics, Toronto, pp.66, 1967.
Thiriot, A., Zooplankton communities in the West African upwelling area, in *Upwelling Ecosystems*, R. Boje and M. Tomczak (eds.), Springer-Verlag, Berlin, Heidelberg, New York, pp. 32-61, 1978.

RESPONSES OF MACROFAUNA TO SHORT-TERM DYNAMICS OF A GULF STREAM FRONT ON THE CONTINENTAL SHELF

John J. Magnuson
Cynthia L. Harrington

Marine Studies Center
University of Wisconsin-Madison
Madison, Wisconsin 53706

Donald J. Stewart

Fish Division
Field Museum of Natural History
Chicago, Illinois 60605

Gary N. Herbst

Department of Zoology
Hebrew University
Jerusalem, Isreal

Abstract. Benthic and near-bottom fishes, decapod crustaceans, and echinoderms were sampled in two areas through which a Gulf Stream front moved just north of Cape Hatteras along 75°13'W longitude in October 1977. We tested the hypothesis that the frontal aggregations resulted from the response of organisms to the front itself rather than to a geographically fixed feature of the region. The front, 8.5°C and 3.5 ‰ wide, was characterized by horizontal gradients of at least 0.5°C/km and speeds of 30 cm/s. Bottom trawling was conducted day and night in each area (northern and southern) when the front was present and when it was absent. Both areas contained more species and individuals of fish when the front was present. Atlantic croaker, weakfish, and spot left when the front left. Decapods exhibited some response to the front, especially at night. During the day, fewer echinoderms were caught when the front was present — perhaps because the sudden changes in temperature induced burrowing. During the day all three groups were more abundant in the northern than in the southern area in terms of species and individuals. The front itself and more geographically fixed features both influenced distributions.

Introduction

Surface layers of the Gulf Stream, which are relatively high in temperature and salinity but low in nutrients, form a dynamic front with the cooler, more nutrient-rich Virginian water mass on the continental shelf just north of Cape Hatteras. The front may be analogous to those that surround coastal upwelling areas where cool, nutrient-laden waters mix with warmer surface waters. It is well known that plankton and nekton aggregate at oceanic fronts (Uda, 1952, 1966; Hela and Laevastu, 1970; Backus, Craddock, Haedrich, and Shores, 1970), but much of what is known is based either on anecdotal information or on observations from coarse-grained sampling designs that may not allow statistical hypothesis testing. The processes that operate at fronts may be exaggerated near Cape Hatteras because the horizontal thermal gradients are exceptionally sharp (up to 1.2°C/km). The front at Hatteras differs from open-ocean fronts in that it is located in relatively shallow water (e.g., 30 m) and thus may strongly influence behavior and distribution of a wide diversity of benthic as well as pelagic organisms.

During the course of our studies of the Gulf Stream front near Cape Hatteras (Herbst, Weston, and Lorman, 1979; Magnuson, Brandt, and Stewart, 1980) we have observed that certain fish and decapod crustacean species are consistently more abundant along the Gulf Stream front than in reference areas sampled to the north and south. In this paper we test the hypothesis that these apparent frontal aggregations represent a response to the front itself rather than to some geograph-

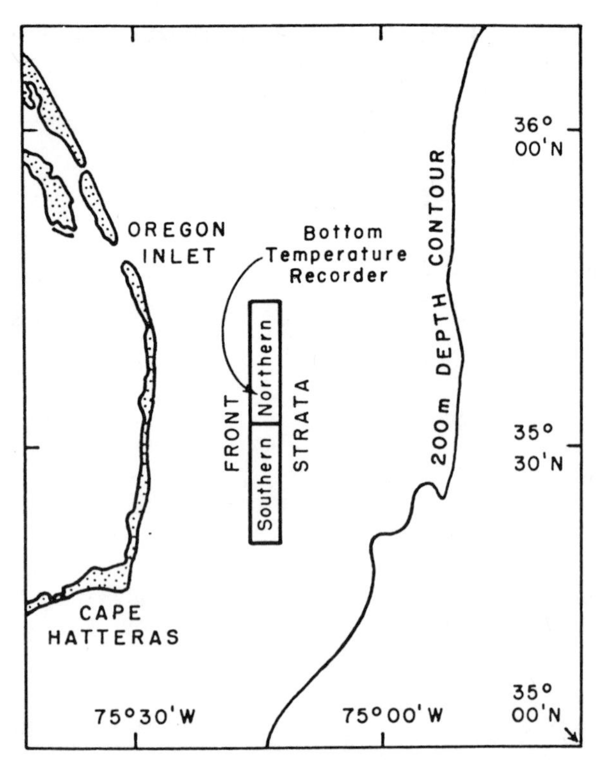

Fig. 1. Map of study site in the western North Atlantic near Cape Hatteras showing sampling areas and location of bottom temperature recorder.

ically fixed feature of the region such as bottom structures. During October 1977, the Gulf Stream front underwent an offshore-onshore excursion that took about 10 days. This allowed us to complete a factorial sampling design with adequate replication in two areas, each with and without the front present. The data provide tests of the foregoing hypothesis.

For a large, fast-swimming fish, a significantly reduced abundance at a given location when the front was absent might be interpreted as migration with the front. Relatively immobile organisms such as echinoderms (sea stars and urchins) might exhibit behavioral responses to the front that alter their catchability and apparent abundance (e.g., burrowing). Analysis of echinoderm catches provides a test of the hypothesis that non-migratory behavioral responses to the front occur. Organisms of intermediate mobility such as decapods (crabs and shrimp) and fishes of small size or special morphology (e.g., eels) could respond to short-term frontal movements by either migration or burrowing. For such organisms our sample design provides a test for behavioral response, but the nature of the response must remain a matter of conjecture.

Structure and Dynamics of the Front

Samples were collected from Duke University's R.V. *Eastward* between 19 and 28 October 1977, on the continental shelf immediately north of Cape Hatteras in the western North Atlantic along 75°13'W longitude (Fig. 1). Depth was typically 31 m (S.D. = 2.5 m). Bottom water temperatures and salinities were determined with a shallow-water mechanical bathythermograph (BT) lowered to touch the bottom and with a Beckman salinometer. During the cruise a waterproof thermograph (Ryan Model J) was positioned 0.5 m above the bottom at 35°35.5'N, 75°13.0'W (Fig. 1).

The bottom temperatures of the front were 25.6°C on the warm edge and 17.1°C on the cold edge, bottom salinities were 35.5 ‰ on the warm edge and 32.0 ‰ on the cold edge. The bottom temperatures recorded on the anchored thermograph changed 1.2 to 2.0°C/h (based on four passes of the front). A speed of 30 cm/s for the front's horizontal movement was estimated from multiple mappings of bottom temperature. The isotherms were nearly vertical (Fig. 2). North-south distances between the 20 and 25°C isotherms were about 9 km at 30 m and 4 km at 10 m on 27 October 1977. The horizontal gradients of 0.5°C/km at the bottom and 1.3°C/km at 10 m are probably underestimates because the cruise track most likely did not cross the front at a right angle. In the areas through which the front passed, bottom water temperatures varied from 16 to 26°C and salinities from 30.5 to 35.5 ‰ during the 10 days of the study (Fig. 3).

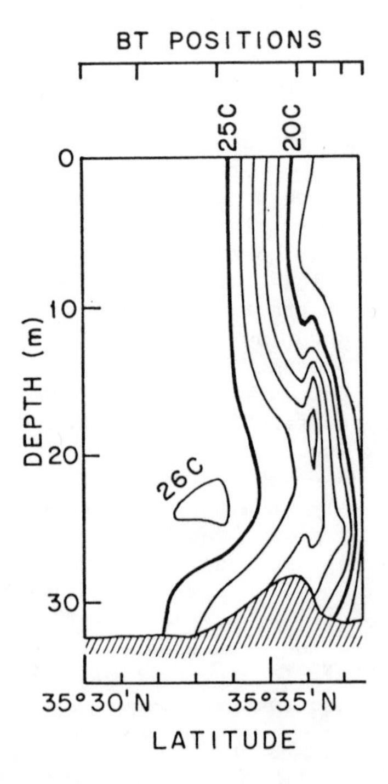

Fig. 2. Temperature profile along 75°13'W of a northward moving front at 35°35'N on October 27, 1977. BT positions are indicated along the top.

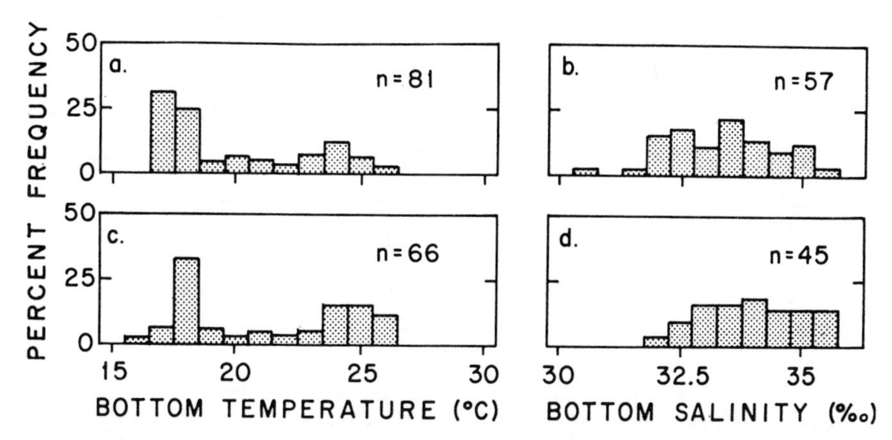

Fig. 3. Relative frequency histograms of bottom temperature and salinity. a and b., northern area; c and d., southern area; n = number of samples.

Methods and Experimental Design

Abundance of benthic and near-bottom macrofauna was estimated along 75°13'W longitude in two geographic areas traversed by the front (labeled southern and northern; see Fig. 1). All animals were collected with 15-min hauls (bottom time) of a 9.1-m semi-balloon otter trawl with a 9.4-m head rope and an 11.6-m foot rope. The nylon netting was 3.8-cm stretch mesh with a cod-end liner of 0.6-cm stretch mesh. Trawling speed was 3 to 4 knots. Fishes, decapod crustaceans, and echinoderms were sorted, identified, and counted on board if possible; otherwise, they were preserved in formalin for identification in the laboratory.

The basic experimental design involved sampling each of the two geographic areas when the front was present and when it was absent (e.g., a 2 x 2 factorial). We first sampled organisms in the southern area as the front was moving southward through it (October 19-21). The front then moved offshore to near the shelf edge for a few days (as determined by an onshore-offshore BT section), and we sampled both southern and northern areas while they were bathed in cold Virginian waters (October 23-24). Finally, the front moved inshore and northward again and we sampled it as it passed

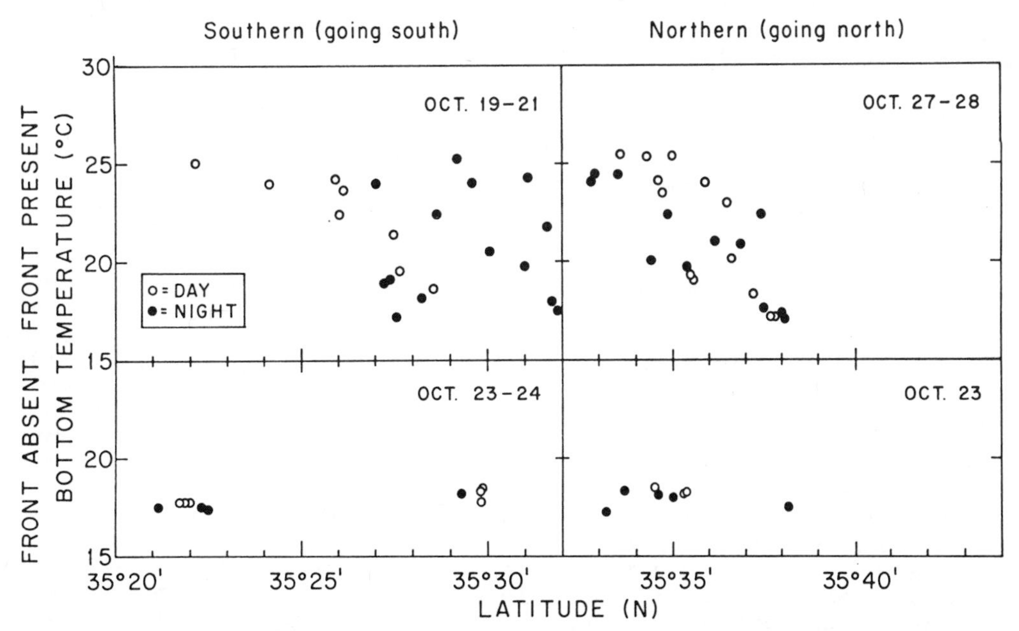

Fig. 4. Bottom temperatures and latitudes of bottom trawl hauls in southern and northern portions of the front area when the front was present or absent in October 1977. Dates of sampling are indicated. When the front was present in the southern portion it was traveling southward, but when in the northern part it was traveling northward.

TABLE 1. Experimental Design (ANOVA)

		Day Samples LATITUDE		Night Samples LATITUDE	
		Southern	Northern	Southern	Northern
FRONT	Present	8	13	14	12
	Absent	6	3	4	5

through the northern area (October 27-28). Bottom temperatures taken before each trawl haul (Fig. 4) illustrate the dynamics of the front as it moved through the study areas and the uniform conditions that prevailed when it moved out.

The experimental design used for analysis of variance (ANOVA) is shown in Table 1 with number of trawl hauls indicated for each cell; Fig. 4 illustrates the latitudinal and thermal range for samples in each cell. Throughout the study we attempted to balance sampling effort between day and night. A preliminary 3-way ANOVA including DAY-NIGHT, LATITUDE (southern-northern) and FRONT (present-absent) revealed significant DAY-NIGHT differences in catches as well as interactions between DAY-NIGHT and the other two factors. To avoid confounding the comparisons of LATITUDE vs. FRONT, we chose to keep day and night samples separate and run only 2 x 2 factorials involving the two factors of primary interest. When this was done, the number of significant interactions was reduced to 1 for the 22 separate ANOVA's that were run. The main effects related to each factor were thus readily interpretable for almost every analysis.

For analyses, catch data were transformed to ln (X + 1) where X is the number per trawl haul of species or individuals in the group (or taxon) being considered. This assumes the data had a negative binomial distribution and the exponent of the negative binomial function (k) has a value of 2 (Taylor, 1953). For presentation in Table 2 and Fig. 5, the means were back-transformed to geometric mean number per trawl haul by taking the antilog and subtracting 1. An ANOVA program (Rummage; Ver 2) supported at the Madison Academic Computing Center was used.

The first 12 ANOVA's (Fig. 5) provide an overview and comparison of how the catch per trawl of fishes, decapods, and echinoderms responded to the presence or absence of the front in the two geographic areas. The number of species per trawl haul and the total number of individuals per trawl haul (all species in an animal group combined) were analyzed according to the foregoing design. These more general analyses were followed by parallel analyses of number per trawl haul for individual species that were contributing to the overall response patterns observed. The 10 individual species analyzed included five fish, three penaeid shrimp, and two sea stars that were chosen because they were relatively aggregated in the front. The fish species considered were bluefish {*Pomatomus saltatrix* (Linnaeus)}, weakfish {*Cynoscion regalis* (Bloch and Schneider)}, Atlantic croaker {*Micropogonias undulatus* {(Linnaeus)}, spot (*Leiostomus xanthurus* Lacepede), and pigfish {*Orthopristis chrysoptera* (Linnaeus)}. Taxonomic undertainty led to the omission of one fish species, scup {*Stenotmus chrysops* (Linnaeus)}, which was also aggregated at the front and thus might have been included. The three shrimp were *Solenocera atlantidis* Burkenroad, *Trachypenaeus constrictus* (Stimpson), and *Sicyonia brevirostris* (Stimpson). The two sea stars were *Luidia clathrata* (Say) and *Astropecten articulatus* (Say). A diverse fauna occurred in the frontal area; 87 species of fishes, 49 decapod crustaceans, and 5 echinoderms were caught. The fish families with the most species were bothids (13), scianids (6), serranids (6), triglids (5), and synodids (4). Twenty-five other families of bony fish were represented as were 5 families of cartilaginous fishes. Sixteen families of decapods were caught in the frontal area including penaeids (10 species), pagurids (5), portunids (5), xanthids (4), and majids (4). Echinoderms were represented by three sea star families and one sea urchin family.

Distribution of Organisms

Our results on total number of species and total number of individuals of fish support our hypothesis that fishes respond to the front itself. More species (night) and more individuals (day and night) were collected per trawl haul when the front was present (Fig. 5). Similarly, the five fish species selected for individual analysis were more abundant when the front was present (day and night)(Table 2). Other evidence suggests that about 60% of the fish species moved back and forth with the front. Of the 87 total fish species caught only 40% were caught both when the front was present and when it was absent. In contrast, 64% of the fish species were caught both in the northern and southern areas.

The results on fishes also suggest a response to some geographically fixed feature. More spe-

TABLE 2. Comparisons of Catch Per Trawl Haul of Selected Species of Fish, Shrimp, and Sea Stars Between FRONT (present-absent) and LATITUDE (southern-northern) in the Frontal Area. Means and statistical significance (indicated by > or <) are for the main effects of the two-way analysis of variance ($p \leq 0.05$). Exceptions are explained in footnotes.

	Geometric Mean Number Per Trawl Haul											
	Day Samples						Night Samples					
	FRONT			LATITUDE			FRONT			LATITUDE		
	Present		Absent	Southern		Northern	Present		Absent	Southern		Northern
Fish[1,2]												
Bluefish	2.5	>	0.0	1.4		3.4	0.0		0.1	0.1		0.0
Weakfish	9.0	>	0.0	1.5	<	22.6	1.6	>	0.0	0.8	<	2.9
Atlantic croaker	34.9	>	0.0	7.2	<	88.1	20.8	>	0.1	5.6	<	86.4
Spot	10.1	>	0.0	1.5	<	26.7	46.9	>	0.0	44.2		49.9
Pigfish	1.6	>	0.0	0.3	<	2.8	1.9	>	0.0	1.7		2.1
Number of Trawl Hauls	21		9	8		13	26		9	14		12
Shrimp												
Solenocera atlantidis	0.5		0.4	0.0	<	1.0	29.0	>	3.3	11.9		9.0
Trachypenaeus constrictus	1.6		0.5	0.0	<	2.8	51.5		17.4	21.7		41.5
Sicyonia brevirostris	0.4		0.4	0.1		0.8	6.0	>*	2.2	3.6		4.0
Sea Stars												
Luidia clathrata	0.5	<	2.6	0.3	<	3.0	1.6		2.2	0.2	<	5.9
Astropecten articulatus	1.2	<	4.3	2.2		2.7	1.9	<	6.1	0.9	<	9.8
Number of Trawl Hauls	21		9	14		16	26		9	18		17

[1] FRONT: Comparisons of fishes between front present and front absent were judged significant if the 95% CI of the mean with the front present did not overlap zero or the 95% CI of the mean number of fish caught when front was absent.

[2] LATITUDE: Comparisons of fishes between northern and southern latitudes based only on catches in front because so few fish were caught when front was absent.

*Significant interaction confounds interpretation.

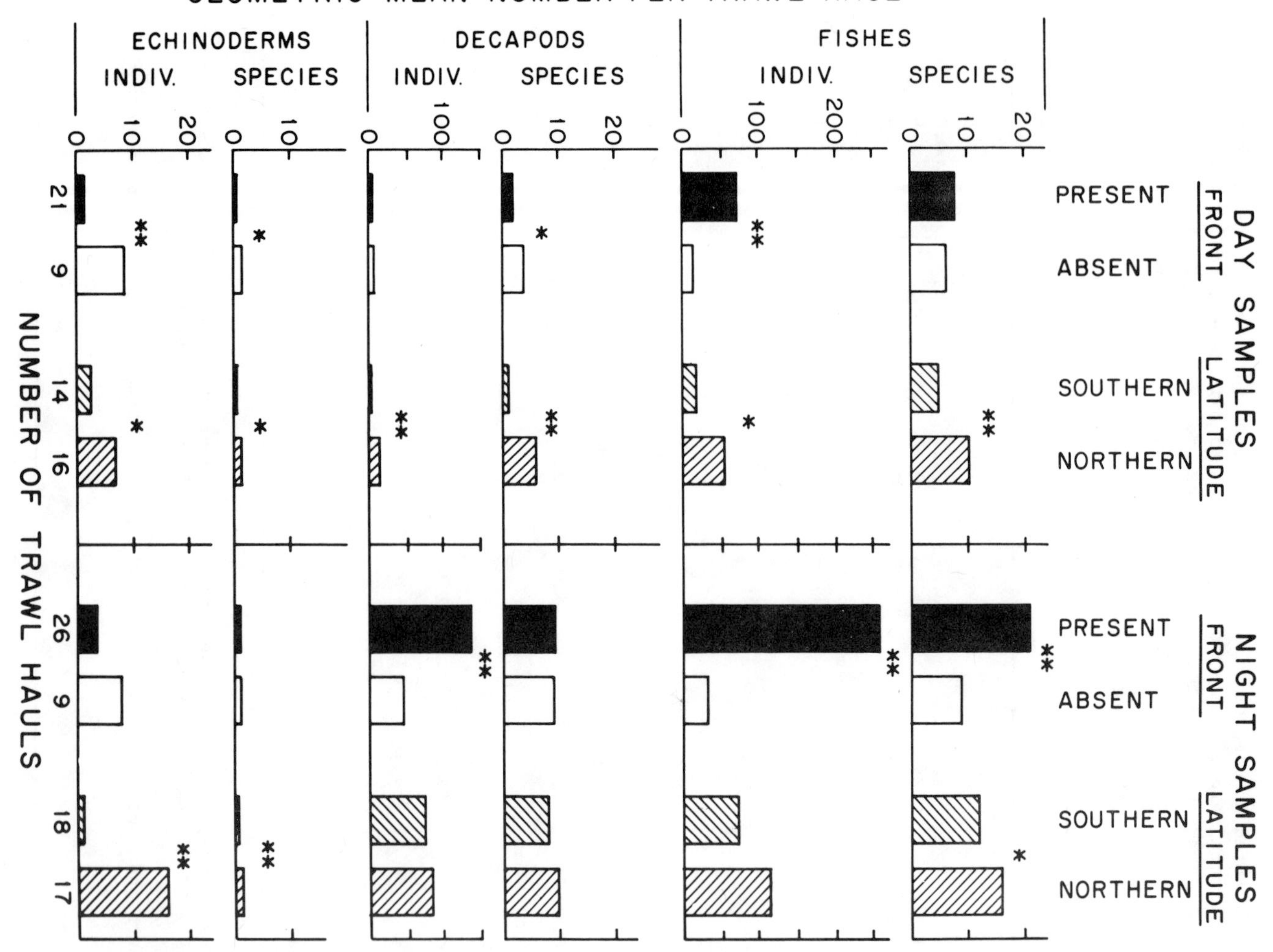

Fig. 5. Main effects of FRONT (present or absent) and LATITUDE (southern and northern) on the geometric mean number of species and individuals per trawl haul of fishes, decapod crustaceans, and echinoderms. Day samples and night samples were analyzed separately. The number of trawl hauls are indicated at bottom. Statistical significance of main effects of FRONT and LATITUDE from 2-way ANOVA are indicated by * $\leq$ 0.05, ** $\leq$ 0.01.

cies (day and night) and more individuals of fish (day) were collected per trawl haul in the northern area (Fig. 5). Some of the selected fish species were more abundant in the northern area (four species in the day analyses, two species in the night analyses)(Table 2).

The response to the front by decapods was not so clear as for fish. More species were caught per trawl haul when the front was absent (day), but more individuals were caught per trawl haul when the front was present (night)(Fig. 5). Two of the three selected species of decapods were more abundant when the front was present at night (Table 2). However, a statistically significant interaction occurred between the two factors for *S. brevirostris* (night) confounding the main effect of FRONT. Some evidence indicates that decapods also respond to some geographically fixed feature. For example, decapods were more abundant in the northern area in the day (both number of species and number of individuals). Two of the three selected species were more abundant in the northern area (day).

Echinoderms may respond behaviorally to the front as well as to geographically fixed features. More species and more individuals of echinoderms were caught per trawl haul when the front was absent (day) (Fig. 5). *Luidia clathrata* was more abundant when the front was absent (day and night) as was *Astropecten articulatus* (day)(Table 2). More species and individuals of echinoderms were caught per trawl haul in the northern area (day and night)(Fig. 5). Both sea star species were more abundant in the northern area at night; in the day only one was more abundant (Table 2).

Discussion of Hypothesis

The influence of the movement of the front on fish distribution was quite clear. The same geographic locations generally contained more species and individuals when the front was present. Species such as Atlantic croaker, weakfish, and spot, which were abundant when the front was present essentially disappeared when the front was absent. These nekton move with the front. The influence of the front on decapod crustaceans was less apparent, but it was evident, especially at night when they are generally more active. The influence of the front dynamics on echinoderms is puzzling. During the day fewer were caught when the front was present. Because these epibenthic organisms are too sedentary to move with the front, their catchability by bottom trawling must have decreased when the front was present. Perhaps rapidly changing temperatures, encountered as the front passes, encourage burrowing in the substrate. Both *Luidia clathrata* and *Astropecten articulatus* are known to burrow (Christensen, 1970; Burla, Ferlin, Pabst, and Ribi, 1972; Ferlin-Lubini and Ribi, 1978; C.L. Harrington, personal observation). Alternatively, when the front was present the trawl may have been so filled with large fish that it was inefficient at catching small or flat-bodied invertebrates.

Distributions associated with geography irrespective of the front were especially apparent for echinoderms, but they were also evident for fish and decapod crustaceans. For fish and decapod crustaceans, the differences between southern and northern parts of the frontal area were more evident during the day than night. Generally the numbers of species and individuals increased to the north in the front area. The trend is in the opposite direction of the general expectations of higher diversity in lower latitudes (see, for example, Friedrich, 1969; and Pielou, 1979). Yet the trends in our analyses of the front were evident for the number of species, total individuals, and selected species of fishes, decapods, and echinoderms. Increases to the north in this region in June 1977 were also reported by our group (Herbst *et al.*, 1979) for decapods from trawl samples and for amphipods from grab samples. The geographic trends in macrofauna just north of Cape Hatteras may well result from longer-term dynamics of the Gulf Stream front in this major zoogeographic transition zone.

There is another potential explanation for the LATITUDE effects. Both day and night samples from the northern area were collected October 27-28 after a storm had passed through the area. The R.V. *Eastward* weather log recorded southeast winds of 40 knots and southeast swells of 4.5 m in the frontal area on October 26, which was followed by increased water turbidity during sampling on October 27-28 (D.J. Stewart, personal observation). The turbid conditions may have stimulated activity among crepuscular or nocturnal species producing the observed differences between southern and northern distributions from day samples. No differences resulting from increased turbidity would be expected from night samples. Our results from night samples are mixed. Some LATITUDE effects remain, most do not. The evidence is strongest for a LATITUDE effect in echinoderms.

In conclusion, the movements of the front have an immediate and direct effect on the distribution of benthic and near-bottom macrofauna. However, even in such dynamic regions, features associated with geographic position and statistically independent of the short-term dynamics of the front may be important. The increases in species and individuals to the north in this dynamic region may be the result of long-term influence of the Gulf Stream front.

Acknowledgements. We thank the crew and the staff of the R.V. *Eastward* and our many colleagues who helped with data collection. Special recognition goes to Paul W. Rasmussen for assisting with computer analyses. The research was supported by a grant from the National Science Foundation (Grant #OCE-8002062).

References

Backus, R.H., J.E. Craddock, R.L. Haedrich, and D.L. Shores, The distribution of mesopelagic fishes in the equatorial and western North Atlantic, *Journal of Marine Research, 28*, 179-201, 1970.

Burla, H., V. Ferlin, B. Pabst, and G. Ribi, Notes on the ecology of *Astropecten aranciacus, Marine Biology, 14*, 235-241, 1972.

Christensen, A.M., Feeding biology of the sea-star *Astropecten irregularis* Pennant, *Ophelia, 8*, 1-134, 1970.

Ferlin-Lubini, V., and G. Ribi, Daily activity pattern of *Astropecten aranciacus* and two related species under natural conditions, *Helgolander Wissenschaftliche Meeresuntersuchungen, 31*, 117-127, 1978.

Friedrich, H., *Marine Biology*, University of Washington Press, Seattle, Washington, 474 pp., 1969.

Hela, I., and T. Laevastu, *Fisheries Oceanography*, Fishing News (Books) Ltd., London, England, 230 pp., 1970.

Herbst, G.N., D.P. Weston, and J.G. Lorman, The distributional response of amphipod and decapod crustaceans to a sharp thermal front north of Cape Hatteras, North Carolina, *Bulletin of the Biological Society of Washington, 3*, 188-213, 1979.

Magnuson, J.J., S.B. Brandt, and D.J. Stewart, Habitat preferences and fishery oceanography, in Bardach, J.E., J.J. Magnuson, R.C. May, and

J.M. Reinhart (eds.), *Fish Behavior and Its Use in the Capture and Culture of Fishes*, International Center for Living Aquatic Resources Management, Manila, in press, 1980.

Pielou, E.C., *Biogeography*, John Wiley and Sons, New York, 351 pp., 1979.

Taylor, C.C., Nature of variability in trawl catches, *Fishery Bulletin, 54*, 145-166, 1953.

Uda, M., On the relation between the variation of the important fisheries conditions and the oceanographical conditions in the adjacent water of Japan, *Journal of the Tokyo University of Fisheries, 38*, 363-389, 1952.

Uda, M., *Advances in Fisheries Oceanography*, No. 1, Japanese Society of Fisheries Oceanography, Tokyo, 25 pp., 1966.

THE EFFECT OF EL NIÑO UPON THE PERUVIAN ANCHOVETA STOCK

D.H. Cushing

Ministry of Agriculture, Fisheries and Food,
Directorate of Fisheries Research,
Fisheries Laboratory,
Lowestoft, Suffolk, NR33 OHT (UK)

Abstract. The biology of the Peruvian anchoveta is described. The spawning season coincides with the period of regression of El Niño. After El Niño of 1965-66 the recruitment to the anchoveta stock was higher than in the years before 1965-66. It is suggested that this was associated with a decrease in the bird population, particularly the juveniles.

Before El Niño of 1972, the anchoveta stock survived El Niño, at least in 1957-58 and 1965-66, because recruitment was not impaired in those years when the fishery was substantial. Since 1972 the anchoveta stock has been replaced to some degree by a sardine population suggesting that the changes in 1972 were to some extent irreversible.

Introduction

The fishery for anchoveta within the Peruvian upwelling system has been the largest fishery in the world and in 1970, 12.28 million tons were landed, mostly through the ports of Chimbote and Callao. In the early stages in the middle fifties catches amounted to less than 100,000 tons, but with increasing demand for fish meal, catches developed as shown in Fig. 1.

During the sixties catches built up by an order of magnitude to the peak catch in 1970. During the late sixties, the catches of anchoveta were dried and pressed into fish meal in such large quantities that the price throughout the world was dominated by the landings in Peru. In 1972, catches dropped to 4.45 million tons and since then, large though the fishery remains, it has not recovered. The initial collapse took place during the major El Niño of 1972.

El Niño is a warm current of trans-equatorial origin that appears every five or ten years off the coast of Peru. Typically the warm current strengthens from the region between the Galapagos Islands and Ecuador and flows south to 5° or to 14°S between February and May. It stabilizes for a period and retreats in August, September, and October, but strengthens again in the following January, February, and March and then regresses again after a total life of about 18 months. Longshore winds persist during the period of El Niño and warm water upwells in the same way as does the cooler water in the normal season, but it is only one-third as productive. Major El Niños appeared in 1891, 1911, 1917, 1925, 1930, 1941, 1957, 1965, 1972, and perhaps in 1976; those in 1891, 1941, and 1972 were very strong.

In any upwelling area there are colonies of guano-producing birds, cormorants, boobies, and pelicans, and in each of the major systems there is a "Cabo Blanco" or "Cap Blanc". One of the first signals of a major El Niño is a premature emigration of cormorants from Peru. Normally they migrate north or south after breeding and return in the southern spring. When El Niño appears in January or February, the cormorants fly to Chile or Ecuador, having abandoned their nests. A reliable and popular signal of a major El Niño is the appearance of starving pelicans on the streets of Guayaquil and Lima.

All three bird species breed on capes and islands off the coast of Peru where there are considerable deposits of guano. It is excavated and exported by the guano companies for fertilizer. Jordán (1967) suggests that there has been an "improvement of nesting sites and protection of various headlands where important centers of nesting were successfully established". Since 1910, the numbers of birds increased from about three million to eight million in the thirties. The numbers fell sharply to four million after the major El Niño of 1940-41 and did not recover until the late forties. Subsequently, numbers continued to rise and reached a peak of twenty million in 1956; the numbers of birds then declined as a consequence of competition with the fishermen and the effects of El Niño, as will be shown below. By 1972-73, the numbers were reduced to about one million. Numbers were estimated from the production of guano (Fig. 2).

It is well known that the anchoveta feeds on phytoplankton during the spawning period in

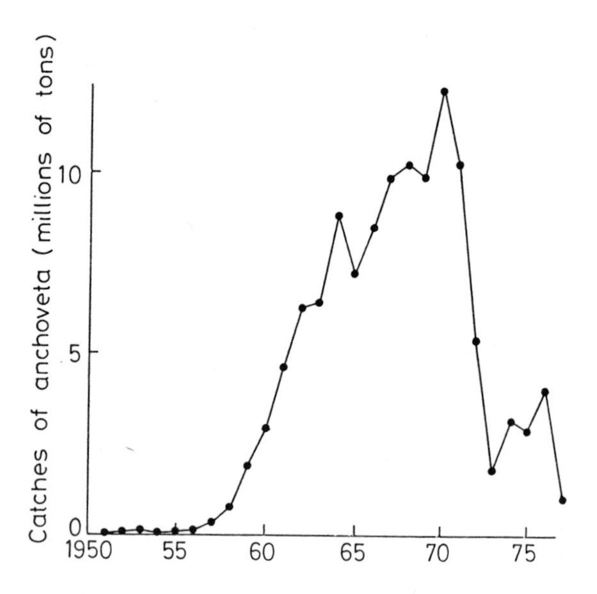

Fig. 1. Catches of anchoveta in millions of tons from 1959-1977.

spring, the period of strong upwelling and high phytoplankton production; because energy is wasted in extensive filtration and because the fish can merely satisfy their maintenance requirements from phytoplankton, it is possible that the fish feed on algae to provide their larvae with subsequent copepod nauplii (Cushing, 1978). Walsh (1977) suggested that the anchoveta feed on zooplankton during the period of weak upwelling and that the anchoveta take 1% of the total productivity, primary and secondary, in strong upwelling and 10% in weak upwelling periods.

The Biology of the Anchoveta

The Peruvian upwelling has been studied by many oceanographers (Gunther, 1936; Wooster, 1961). A detailed account was given by Zuta and Guillen (1970) and by Zuta and Urquizo (1972). Zuta and Guillen tabulated mean temperatures off Punta Falsa, off Puerto Chicama and off Callaõ (Fig. 3) which demonstrate that the upwelling season is most intense from May to November. Zuta and Urquizo (1972) published the average temperature distributions by months for the period 1926 to 1969, which also show the period of intense upwelling. They also show that there appear to be three foci of upwelling: (a) Pimentel to Salaverry in 7 to 8°S, (b) Callaõ and northwards in 10 to 12°S, and (c) Pisco and southwards in 14 to 17°S. Of the three subareas, the latter is most intense. Zuta and Guillen (1970) tabulated primary production in g C $m^{-2}d^{-1}$ by latitude (Table 1). The three upwelling foci tend to be more productive than the waters outside them and from May to November we would expect to find the prominent biological events associated with them.

The anchoveta spawns between August and March with peaks in September and January, as shown in the distributions of maturity stages by months for a number of years (Jordán and Vildoso, 1965; Vildoso and Alegre, 1969; Miñano, 1968a). The same double peak is shown clearly in the seasonal distribution of fat content (Lam, 1968). Extensive egg surveys have been executed in all months of spawning between latitudes 6°S and 16°S and up to 150 nautical miles offshore (Guillen and Vasquez, 1966; Guillen and Flores, 1967; Miñano, 1968b; Zuta and Mejia, 1968; Valdivia and Guillen, 1966; Flores and Elias, 1967; Flores, 1967; Valdivia and Poma, 1969). As the eggs hatch in about two days, they are found close to where they were spawned. Spawning starts in July at a rather low level between 9 and 11°S. In August and September, spawning occurs between 5 and 15°S and up to 150 nautical miles offshore. The interesting point is that the fish spawn so far offshore, far beyond the dynamic boundary at 50 km or so and far beyond the usual area of the fishery (which is within about 30 km of the shore). By November, spawning is restricted to a narrow coastal strip (30 to 50 km offshore). By December and January a northern and southern group of spawners are perhaps distinguishable.

Spawning begins three months after the start of the intense period of upwelling and finishes three months after its end. If upwelling starts in May or June, the first zooplankton generation should appear some six weeks later. The anchoveta larvae may be spawned at the right time to take nauplii from the second zooplankton generation, when numbers might have increased considerably. Similarly the summer spawners (January and February) might rely on a generation of zooplankton that survive after the period of intense upwelling. The Peru coastal current drifts towards the equator at 10 to 30 km/d (Zuta and Guillen, 1970) which means that eggs and larvae could drift out of the area in one to three months. They remain within the upwelling system, perhaps by sinking into the deeper water (above the pycnocline), which then upwells towards the coast.

The growth of the anchoveta has been studied by means of the progress of length modes in time. Saetersdal and Valdivia (1964) showed that there were two length groups of recruits, and tables in Boerema *et al.* (1967) show that fish of 5 cm in standard length appear in two groups in January and May. The first group arises from the spawning in August, September, and October and the second from that in January; the latter grows somewhat more slowly because its initial growth occurs outside the period of major upwelling. Fig. 4 is adapted from Fig. 14 in Saetersdal and Valdivia (1964). The figure displays the course of events for about two and a half years; the periods of upwelling and the spawning periods are shown. On the ordinate is given length in centimeters and across the diagram are shown five important measures of length, (a) 5.5 cm, the length at which the smallest fish have been

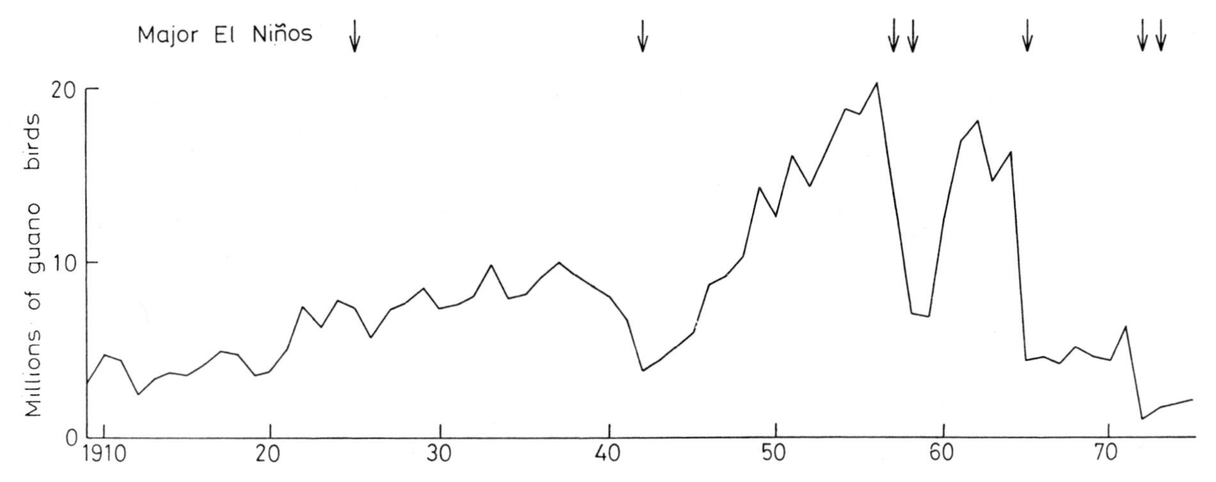

Fig. 2. The numbers of guano birds between 1909 and 1964 (Jordán and Fuentes, 1966).

recorded in the fishery, (b) 8.5 cm, the length at which maturation starts, (c) 11.5 cm, the length at which fish may first spawn, (d) 14.5 cm, the length at which the anchoveta spawn for the second time, and (e) 17.0 cm, the length at which the third spawning might occur (there is no evidence for this in Saetersdal and Valdivia's paper, but by 1974 the 1971 year-class had reached a length of 17 cm (Valdivia, 1976), and tagging experiments have shown that the fish can live for as long as five years). The figure shows the growth curves of the two spawning groups with ages, otolith rings, and stages of maturation and the contribution made by each group to the spawnings shown by vertical arrows. The separation of the two groups is somewhat artificial because as the fish grow older their length distributions tend to merge.

The ages of the fish are estimated from the rings on their otoliths; Vildoso and Chumán (1968) examined otoliths collected between October 1963 and December 1964. Only half were legible but two rings were probably laid down each year. The first ring is laid down one year after spawning and the second about six months later. The sharpest change in ring structure (from opaque to hyaline, or vice versa) occurs in May and November, when the period of intense upwelling starts and ends. This suggests that the biannual rings reflect changes in upwelling. However, in the first year of life checks due to upwelling are not recorded, perhaps because the little fish may have grown predominantly in the cool upwelling water.

In temperate waters, fish spawn during a long season, older fish spawning first and recruits last. The upwelling production cycle resembles that in temperate waters. In Fig. 4 it is implied that the January spawning comprises recruit spawners; although they are abundant, their egg production is low (Miñano, 1968a), and the September spawning is the more prominent, as might be expected if all adults were present. The figure is constructed on the assumption that the anchoveta spawns serially once a year, although the construction demands that the

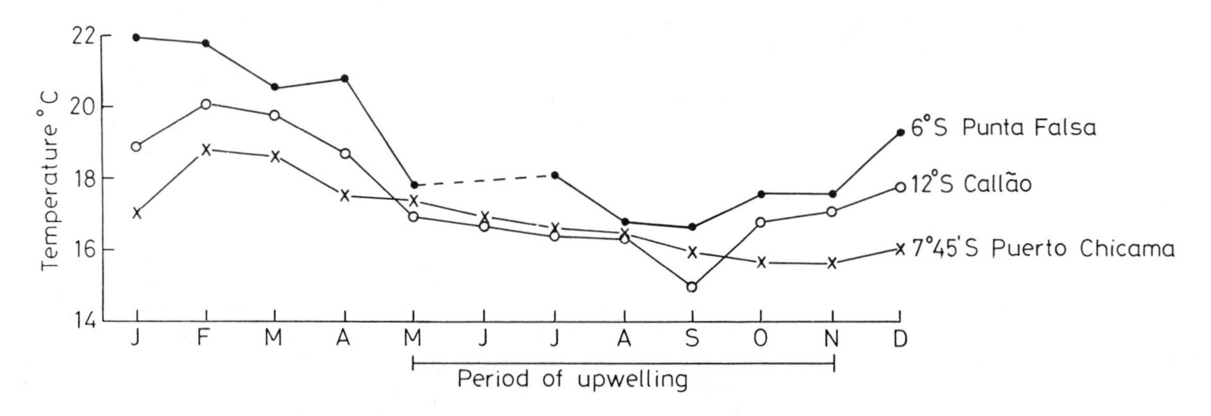

Fig. 3. Duration of the main upwelling system off Peru as shown by the distribution of mean temperatures off Punta Falsa, off Puerto Chicama, and off Callaõ (Zuta and Guillen, 1970).

TABLE 1. Primary Production (g C $m^{-2}d^{-1}$) by Latitude.

Latitude	Primary Production	Upwelling Foci	
4°S	0.20		
5	0.60		0.43 (mean of 4°, 5°, 6°)
6	0.50		
7	0.55		
8	1.30	Pimental-Salaverry	0.92 (mean of 7°, 8°)
9	0.60		0.60 (observation at 9°)
10	0.55		
11	0.65	northward from Callaõ	0.63 (mean of 10°, 11°, 12°)
12	0.67		
13	0.37		0.37 (observation at 13°)
14	0.60		
15	2.30		
16	0.70	southward from Pisco	1.05 (mean of 14° through 17°)
17	0.62		
18	0.58		0.58 (observation at 18°)

recruits spawn in January and again in September or November. The figure subsumes three lines of evidence, on growth, maturation, and age in a consistent manner.

The recruiting fish are first found in the fishery at a length of 5.5 cm, 3 to 5 months after spawning. From the tables given in Boerema *et al.* (1967) the little fish appear at Chimbote and Callaõ (180 miles apart) at about the same time. This implies that they have moved inshore with the upwelled water. Whatever the mechanism, the Peruvian coastal waters are flooded with recruiting fish between January and July, which is why the *peladilla* (small fish) season is closed in February and March and during the winter months.

The biology of the anchoveta is linked to the upwelling region in which it lives. The fish spawn in the period of upwelling across the whole region and even beyond the strict region of coastal upwelling as defined in physical terms. The larvae grow mainly during the upwelling period and the fish start to mature at about nine months and spawn for the first time at about one year of age. They spawn again in their second year and perhaps in their third. The stock maintains itself in the upwelling system, but how it does so is unknown; the adults must drift offshore in the upwelling plumes and perhaps return in the countercurrent, but it is rather deep for such a fish.

The Effect of the Guano Birds on the Stock of Anchoveta

There are three species of birds that feed on the anchoveta, the *guanay* (*Phalacrocorax bougainvilli* L., a cormorant), the *piquero* (*Sula variegata* Tschudi, a booby), and the *alcatraz* (*Pelecanus occidentalis thagus* Molina, a pelican). The cormorants fish in broad bands in daylight and dive to a depth of about 12 m; the booby is less gregarious and dives to depths not greater than 15 m; the pelican fishes both day and night in small flocks and feeds at or just below the surface (Jordán, 1967). All three species breed on the guano islands and headlands in November and the chicks are reared in January and February. After breeding the adults migrate to Chile or Ecuador. Between February and May, the growing juveniles feed on their own and migrate south or north in April, May, or June (Jordán and Cabrera, 1960).

Jordan (1967) estimated that the birds take 430 g/day each of anchoveta of all sizes between 3 and 16 cm (but the larger fish predominated in the samples). Although the guanay can fish up to 30 to 40 miles offshore, they prefer to work inshore near the headlands and islands. It is possible that the growing juveniles feed on the recruiting anchoveta inshore between February and June.

The population (estimated from guano production) increased from 4 million in 1909 to 20 million in 1956; in 1957, due to El Niño, numbers fell to 7 million, but recovered to 18 million by 1962. In 1965 the numbers fell to 4 million and did not recover after El Niño of 1965-66. After El Niño of 1972-73, the population declined to 1 million (Jordán and Fuentes, 1966; Jordán, 1976) (see Fig. 1). The population collapses during a major El Niño because the adult birds abandon their chicks at the onset of El Niño and fly to Ecuador or Chile some time before their normal migration and suffer a high mortality. It is possible that El Niño presses the anchoveta population to the south and the adult birds do not have enough to eat because they do not follow the fish. The population of birds recovered from

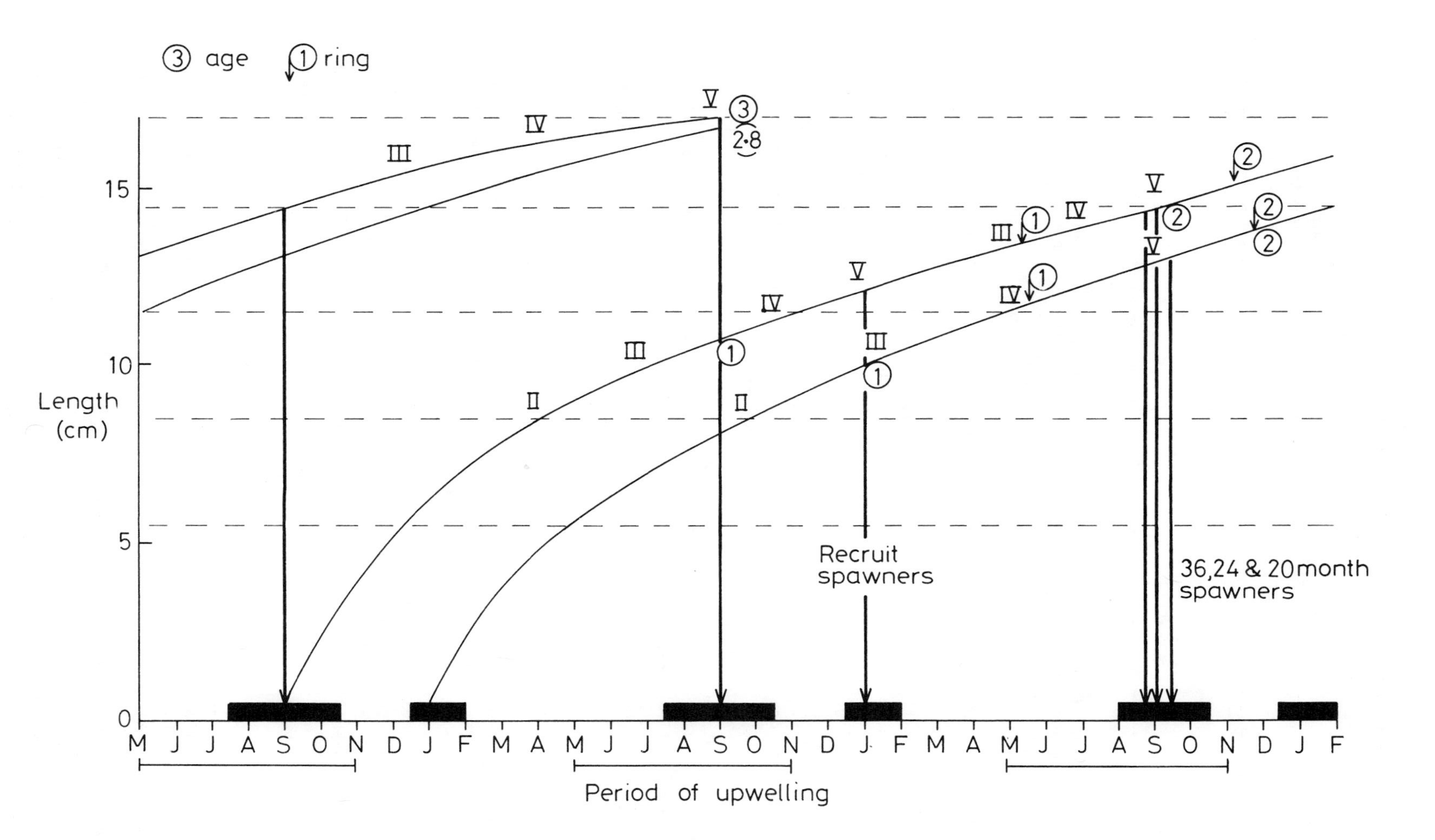

Fig. 4. The growth of anchoveta; periods of spawning (as blocks on the abscissa) and upwelling (as bars below the abscissa) are shown. Four critical standard lengths are given in the diagram: (a) the length at which the smallest fish have been recorded in the fishery, 5.5 cm; (b) the length at which maturation starts, 8.5 cm; (c) the length at which the fish may first spawn, 11.5 cm; (d) the length at which anchoveta spawn for the second time, 14.5 cm; (e) the length at which they may spawn for the third time, 17.0 cm (adapted from Saetersdal and Valdivia, 1964). Ages are indicated as for example ① number of rings on the otolith are indicated, ② maturation stages are given in Roman figures.

TABLE 2. Effect of Bird Predation upon Anchoveta Recruitment

Period	Stock in weight per unit effort	Numbers of recruits per unit effort	Numbers of recruits eaten by birds	Sum of recruits and numbers eaten
1960-1964	583	270 ± 85	146	416
1965-1971	368	438 ± 73	33	468

El Niño of 1957 when the fish catches were less than 2 million tons. In 1965, catches were more than 7 million tons and it is reasonable to suppose that the guanay birds had suffered in competition with the fishermen.

Let us suppose that the young bird population was twice that of the adults and that they eat 300 g/day for 150 days. From 1960 to 1964, the adult bird stock was 15.8 x 10^6 juveniles (2 juveniles per adult), which may have eaten (31.6 x 300 x 150 x 10^6) g, or at 5 g/fish, (31.6 x 300 x 150 x 10^6/5) fish or 0.28 x 10^{12}. During the period, the stock was 13.45 x 10^6 tons or 1.12 x 10^{12} fish (at 12 g/fish) (Anon., 1972); the quantity of recruits eaten may have comprised 25% of the stock. A similar calculation for the period 1965 to 1971 when the number of birds was reduced to 4.98 x 10^6, suggests that the quantity eaten was 0.9 x 10^{12} or 6% of the stock. Anon. (1973) gives a stock-recruitment curve, numbers of recruits per unit of effort and adult stock in weight per unit of effort. From the observations we construct Table 2.

Between the two periods recruitment increased by 63% as the bird stock was reduced after El Niño of 1965-66. The sums of recruits and numbers eaten were not very different, which suggests that the stock-recruitment curve without bird predation is not the dome-shaped one shown in Anon. (1973) but is an asymptotic one; so far as is known dome-shaped stock and recruitment curves are characteristic of some gadoids and some Pacific salmon stocks (Cushing and Harris, 1973) but not engraulids or clupeids. The numbers of recruits eaten were estimated from the percentages of stock as estimated above.

El Niño of 1972-1973

El Niño of 1972-73 was probably the most intense observed since 1891; it started in February 1972, stabilized by March of that year, retreated in August, September, and October; it strengthened again in the following January, February, and March, and died away in May 1973. During the regression in August-October 1972, the sea surface temperature was more than 1°C warmer than the average between 1925 and 1973 (Miller and Laurs, 1975). Wooster and Guillen (1974) noticed some early signals. In March 1971 the equatorial waters between Ecuador and 100°W (the region of the Galapagos Islands) were already warmer, perhaps because the countercurrent had already started to flow. They also observed a patch of warm water offshore in 16 to 18°S in August 1971, as if there had been an offshore southerly flow of transequatorial water in the early months of that year; the pelagic crab *Euphylax dovii* from Central American waters was found in this water mass (Vildoso, 1976).

When the El Niño started in February 1972 the adult birds left Peru and the population was reduced to 1 million and by 1974 that number had not increased. During the period of this El Niño the presence of marine invaders in Peruvian waters was recorded in some detail. They included animals native to the waters of Central America, Mexico, and southern California, but there was one remarkable exotic, the milkfish *Chanos chanos* from Indonesia. It is sometimes recorded in Central American waters, but in 1972-73 it appeared for the first time off Peru (Vildoso, 1976); it may have travelled on the Equatorial Countercurrent. A second event observed in this El Niño was the reduction of primary production by a factor of three (Guillen, 1976). A third event, which happens in any El Niño, was the dispersal of the anchoveta and hence the disruption of the fishery. However, by the autumn of 1972, the shoals had migrated south beyond 15°S and were found close inshore as if they had been pushed south by the warm current (Valdivia, 1976). In the earlier El Niños during the life of the fishery (1957-58 and 1965-66), the fish reassembled after dispersal and neither the stock nor the recruitment appeared to be affected.

In the spring of 1969, the stock was the lowest recorded till then, but with the spawning in January 1970 high recruitment was generated. The spawning in spring 1971 comprised second-time spawners hatched in September 1969 and first-time spawners hatched in January 1970. The year class hatched in spring 1971 was the lowest recorded during the previous history of the fishery, and this was five months before El Niño appeared. The two subsequent year classes hatched during El Niño years of 1972 and 1973 were also low, probably because the primary productivity in the upwelled water was sharply reduced. Either all three year-class strengths were determined after five months of age, or the 1971 year class suffered a history different from its two successors.

In the spring of 1971 the proportion of fish in maturity stage V (nearly ripe) was reduced from a normal 70% to 28.6% at Chimbote and from a normal 60% to 12.8% at Callaõ and at the same time the fat content remained high instead of declining (Valdivia, 1976). In August, September, and October 1971, fish of 13 cm may have comprised second-time spawners (spawned in September 1969) and first-time spawners (spawned in January 1970). The year classes hatched in 1969 and 1970 were the highest recorded (Valdivia, 1976). Let us suppose that growth was density dependent and that as a consequence the usual first-time spawners (spawned in January 1970) did not mature. Then the egg production between August and October 1971 might have been reduced considerably. Such an interpretation would disconnect the observations on maturation from the subsequent El Niño.

In 1974 nearly half the catches comprised large fish, between 15 and 18 cm long, which were survivors of the 1971 year class at three years of age; until 1974 the largest fish observed in the fishery were between 15 and 16 cm and fish three years old had not been seen since 1961 and 1962 (Boerema *et al.*, 1967). The length distribution of the catches in 1974 was drawn from three year-classes, 1971, 1972, and 1973 (Valdivia, 1976); because the year classes of 1971 and 1972 were of the same magnitude in 1974, Valdivia concluded that the 1971 year class had suffered little fishing mortality. Even more remarkable, Jordán (1976) reported the recovery of 14 to 15 cm fish tagged in July 1970 as follows:

1970	1971	1972	1973	1974
4448	442	6	44	27

Not only did the fish live for five years or so, but they were effectively absent from the fishery in 1972, when relatively large quantities of fish were landed. Presumably a fraction of the stock shifted south during the intense El Niño of 1972 and evaded the effects of fishing. If part of the 1970 year class did not spawn in the spring of 1971, the class spawned in 1971 was reduced, but perhaps not so much as was originally thought.

During the period of study of the population dynamics of the anchoveta only one very strong El Niño has appeared, that of 1972-73. In 1965-66 the phenomenon was less intense. In particular the temperatures off Peru in the spring of 1965 during El Niño regression were normal or just above normal (Wooster and Guillen, 1974); in other words some upwelling of cool productive water might have occurred during the major spawning season and hence recruitment of the subsequent year class was sustained. In 1972-73, there was no period of cool water upwelling during the retreat of El Niño, when the main anchoveta spawning takes place.

In the unexploited state, the stock of anchoveta may not be affected much by El Niño, except possibly by enhanced recruitment after the most intense events because predation by young birds may be reduced. Presumably in 1891 and 1940-41 the stock recovered quickly from single recruitment failures if they indeed occurred at all. However an immediate effect of El Niño is to drive the stock south into waters that remain fairly productive. The main conclusion is that in an unexploited state the stock survives El Niño with little loss in numbers or recovers by the following year. But when the stock of anchoveta was exploited, fishermen and birds competed for the fish available. The effect of El Niño reduced the bird population dramatically. We have speculated on the effects on the fish stock particularly upon the 1971 year class. Whatever the mechanism, the sharp fall in the recruitment of three successive year classes (1971, 1972, and 1973) reduced the stock so severely that under continued exploitation it has not recovered.

Events Since El Niño of 1972-73

El Niño was forecast for 1975, but it did not appear, although warm water might have appeared offshore as part of the trans-equatorial flow. In January and February 1976 an El Niño appeared, but by July a normal upwelling system replaced it and lasted through the main spawning season, September, October, and November. However, El Niño of 1976 was remarkable for dense blooms of the dinoflagellate *Gymnodinium splendens*, which produced a dense red tide or aguaje. At the same time large quantities of jellyfish called *malagua* were observed for the first time since 1917. The red tide extended from 5 to 15°S and the chlorophyll concentrations reached 50 to 400 μg/l (in El Niño of 1972-73 the quantities of chlorophyll were as low as 0.2 to 0.6 μg/l). Off San Juan in the south of the region hydrogen sulphide was observed in the compensatory undercurrent. In September, October, and November, no *malagua*, aguaje, or hydrogen sulphide was observed. In 1977 the water was a little warmer than usual as if El Niño had resurged, but there was no red tide or jellyfish (Kesteven *et al.*, 1977). The cause of the red tide in physical terms is unknown unless the trans-equatorial flow was much shallower than usual.

El Niño of 1972 concentrated the adult fish in some cooler areas close to the shore and large catches were made; by September 1972 the stock was reduced to 2 million tons. After 1972, recruitment was much reduced and catches remained as high as 1.5 to 3.5 million tons. However by September 1975 the spawning stock had reached 4 million tons and with a strong recruitment in the spring of that year the total stock (not spawning stock) had risen to 10 to 12 million tons. During the first half of 1976 the growth rate of the anchoveta was reduced, the fat content from March to June was very low, and there was some suggestion that the natural mortality increased. Catches of 4 million tons reduced the spawning stock size to 6 million tons in September 1976, which declined to 2 to 3 million tons by the beginning of 1977, with poor recruitment. A further decrease in stock size occurred when 1 million tons of anchoveta were taken in the fishery. The stock-recruitment relationship shows that the best recruitment is produced from spawning stocks of 10 million tons and if the spawning stocks are reduced below 5 million tons recruitment will be reduced. Hence it would seem reasonable to close the fishery whenever the spawning stock falls below 5 million tons. Another useful rule might be to close the fishery whenever El Niño occurs until it has been established that the recruiting year-class has not been reduced.

Other changes have been taking place. Between 1966 and 1972, sardine catches ranged from 1100 to 6000 tons; in 1973, 130,000 tons were taken and between January and April 1977, 300,000 tons were caught. In 1972 there was a great increase in numbers of sardine larvae in the plankton samples and there were larger catches of mackerel and jack mackerel in 1973. In March 1976 and March 1977 acoustic surveys revealed a stock of 6 million tons predominantly comprised of sardine, mackerel, jack mackerel, and saury. The evidence suggests that stocks of other pelagic fish have risen as the stock of anchoveta has declined. It need not surprise us because the anchoveta stock has suffered from recruitment overfishing since the early months of 1972, and we expect stock switches as the result of recruitment overfishing — as anchovy succeeded sardine after the collapse of the Californian sardine stock.

Conclusion

El Niño, the warm current of trans-equatorial origin that appears from time to time off Peru, is an expression of events in the general atmospheric circulation. It alters the productivity of Peruvian waters. During the period of the fishery, the population of birds that feed on anchoveta has been reduced by a factor of 20, apparently due to competition with fishermen. The numbers of birds have always been reduced by El Niño; Jordán and Cabrera (1960) estimated typical mortalities in El Niño of 1957-58. Their numbers did not recover after two successive El Niños, presumably because there was not enough anchoveta available to the survivors once the fishery took more than 5 million tons each year. Thus the production of guano was reduced as a combined effect of El Niño and the fishery.

El Niño has had a complex effect upon the fishery for anchoveta. There is no doubt that the stock was reduced to low levels by the reduced recruitments in El Niños of 1972-73; it may have been an indirect effect of the earlier El Niños of 1965-66, if recruitment has been enhanced by reduced predation by the birds. But the stock suffered three poor recruiting year classes in succession and thenceforward suffered from recruitment overfishing. As a consequence, sardine stocks and those of other pelagic fish probably increased. Thus the stock of anchoveta and the fishery upon it were both reduced as an indirect effect of El Niño, which reduced recruitment.

References

Anon., Informe sobre la Segunda Reunion del Panel de Expertos en Dinámica de Población de la Anchoveta Peruana, *Boletin Instituto del Mar del Peru, 2, 7*, 377-457, 1972.

Anon., Tercera sesion del Panel de Expertos sobre la Dinámica de la Población de la Anchoveta Peruana, *Boletin Instituto del Mar del Peru, 2, 9*, 525-599, 1973.

Boerema, L.K., G. Saetersdal, I. Tsukayama, J.E. Valdivia, and B. Alegre, Informe sobre los efectos de la pesca en el recurso Peruana de anchoveta, *Boletin Instituto del Mar del Peru, 1, 4*, 136-186, 1967.

Cushing, D.H., Upper trophic levels in upwelling areas, in *Upwelling Ecosystems*, edited by R. Boje and M. Tomczak, pp. 101-110, Springer-Verlag, Berlin, Heidelberg and New York, 1978.

Cushing, D.H., and J.G.K. Harris, Stock and recruitment and the problem of density dependence, *Rapport et Procès-verbaux des Réunions. Conseil Permanent International pour l'Exploration de la Mer, 164*, 142-155, 1973.

Flores, L.A.P., Informe preliminar del Crucero 6611 de la Primavera de 1966 (Cabo Blanco-Punta Coles), *Informe Instituto del Mar del Peru, 17*, 16 pp., 1967.

Flores, L.A.P., and L.A.P. Elias, Informe preliminar del Crucero 6608-09 de Invierno 1966 (Máncora-Ilo), *Informe Instituto del Mar del Peru*, 24 pp., 1967.

Guillen, O., El sistema de la corriente Peruana. Parte I. Aspectos fisicos, *FAO Informes de Pesca, 185*, 243-284, 1976.

Guillen, O., and L.A.P. Flores, Informe preliminar del Crucero 6702 del Verano de 1967 (Cabo Blanco-Arica), *Informe Instituto del Mar del Peru, 18*, 17 pp., 1967.

Guillen, O., and F. Vasquez, Informe preliminar del Crucero 6602 (Cabo Blanco-Arica), *Informe Instituto del Mar del Peru, 12*, 27 pp., 1966.

Gunther, E.R., A report on oceanographical investigations in the Peru coastal current, *Discovery Reports, 13*, 109-276, 1936.

Jordán, R., The predation of guano birds on the Peruvian anchovy (*Engraulis ringens* Jenyns), *Calcofi Reports, XI*, 105-109, 1967.

Jordán, R., Biologia de la anchoveta. Parte I. Resumen del Conocimiento actual, Actas de la Reunion de Trabajo sobre el Fenomeno Conocido como "El Niño". *FAO Informes de Pesca, 185*, 359-399, 1976.

Jordán, R., and D. Cabrera, Algunas resultados de

la anillaciones de Guanay (*Phalacrocorax bougainvillii* L.) efectuadas durante 1939-41 y 1949-53, *Boletin de la Compania Administradora del Guano, 36, 1*, 11-26, 1960.

Jordán, R., and H. Fuentes, La poblaciones de aves guarneras y su situacion actual, *Informe del Instituto del Mar del Peru, 10*, 30 pp., 1966.

Jordán, R., and A.C. Vildoso, La anchoveta (*Engraulis ringens* J.); Conocimiento actual sobre su biologia, ecologia y pesqueria, *Informe Instituto del Mar del Peru, 6*, 52 pp., 1965.

Kesteven, G.L., J.A. Gulland, B. Jones, R. Barber, and L. Boerema, Report of the consultative group convened by the Minister of Fisheries in Peru to advise him on the state of the stocks of anchoveta and other pelagic species and on the courses of action taken for the management of the fishery, July 1977, 17 pp., 1977.

Lam, R.C., Estudio sobre la variacion del contenido de Grasa en la anchoveta Peruana (*Engraulis ringens* J.), *Informe Instituto del Mar del Peru, 24*, 29 pp., 1968.

Miller, F.R., and R.M. Laurs, The El Nino of 1972-1973 in the eastern tropical Pacific Ocean, *Bulletin Inter-American Tropical Tuna Commission, 16, 5*, 401-448, 1975.

Miñano, J.B., Estudio de la fecundidad y ciclo sexual de la anchoveta (*Engraulis ringens* J.) en la zona de Chimbote, *Boletin Instituto del Mar del Peru, 1, 9*, 505-552, 1968a.

Miñano, J.M., Informe preliminar del Crucero 6705-06 del Otoño de 1967 (Cabo Blanco-Ilo), *Informe Instituto del Mar del Peru, 21*, 13 pp., 1968b.

Saetersdal, G., and J.E. Valdivia, Un estudio del crecimiento, tamaño y reclutamiento de la anchoveta (*Engraulis ringens* J.), *Boletin Instituto del Mar del Peru, 1, 4*, 85-136, 1964.

Valdivia, J., Aspectos biologicas del fenomeno "El Niño", 1972-3. Parte II. La poblacion de la anchoveta, Actas Reunion de Trabajo sobre el fenomeno conocido como El Niño. *FAO Informes de Pesca, 185*, 80-93, 1976.

Valdivia, J., and O. Guillen, Informe preliminar del Crucero de Primavera 1965 (Cabo Blanco-Morro Sama), *Informe Instituto del Mar del Peru, 11*, 37-70, 1966.

Valdivia, J., and A. Poma, Informe preliminar del Crucero Unanue 6711 18 Noviembre-20 Diciembre Primavera 1967, *Informe Instituto del Mar del Peru, 29*, 14 pp., 1969.

Vildoso, A. Ch. de, Aspectos biologicos del fenomeno "El Niño" 1972-73. Parte I. Distribucion de la fauna, *FAO Informes de Pesca, 185*, 62-79, 1976.

Vildoso, A.C., and E.D. Chumán, Validez de la lectura de otolitos para determinar la edad de la anchoveta (*Engraulis ringens*), *Informe Instituto del Mar del Peru, 22*, 34 pp., 1968.

Vildoso, A.C., and B. de H. Alegre, La madurez sexual de la anchoveta (*Engraulis ringens* J.) en los periodos reproductivos 1961/8, *Boletin Instituto del Mar del Peru, 2, 3*, 112-125, 1969.

Walsh, J.J., The biological consequences of the interaction of the climatic, El Niño, and event scales of variability in the eastern tropical Pacific, *Rapport et Procès-verbaux des Réunions. Conseil Permanent Internationale pour l'Exploration de la Mer, 173*, 182-197, 1977.

Wooster, W.S., Yearly changes in the Peru Current, *Limnology and Oceanography, 2*, 222-225, 1961.

Wooster, W.S., and O. Guillen, Characteristics of El Niño in 1972, *Journal of Marine Research, 32, 3*, 387-404, 1974.

Zuta, S., and O. Guillen, Oceanografia de las aguas costeras del Peru, *Boletin Instituto del Mar del Peru, 2, 5*, 161,323, 1970.

Zuta, S., and W. Urquizo, Temperatura promedio de la superficie del mar frente a la costa peruana periodo 1928-1969, *Boletin Instituto del Mar del Peru, 8*, 461-519, 1972.

Zuta, S., and J. Mejia, Informe preliminar del Crucero Unanue 6708 24 Agosto-25 de Setiembre Invierno 1967, *Informe Instituto del Mar del Peru, 25*, 23 pp., 1968.

SEASONAL FOOD PRODUCTION AND CONSUMPTION BY NEKTON
IN THE NORTHWEST AFRICAN UPWELLING SYSTEM

R.J. Trumble

Washington Department of Fisheries
Seattle, Washington 98195, U.S.A.

O.A. Mathisen

College of Fisheries, University of Washington,
Seattle, Washington 98195, U.S.A.

D.W. Stuart

Department of Meteorology, The Florida State University,
Tallahassee, Florida 32306, U.S.A.

Abstract. Mesoscale time periods of days to weeks, normally studied by intensive observations in upwelling areas, can adequately depict physical processes and responses of lower trophic levels. Such time scales are too short, however, for the longer-lived nekton. As an example, results of short-term observations of nekton in an upwelling area were extrapolated to a yearly basis by linking them to reconstructed yearly cycles of the physical environment and the primary and secondary production. The conclusion that nekton did not stress their food supply during the JOINT-I expedition to northwest Africa was expanded to apply to other times of the year.

Introduction

Classical upwelling studies have emphasized the response of upwelling to "event"-related phenomena with a time scale of several days to a few weeks. Numbers of phytoplankton generations can pass during an event, and varying intensities of upwelling between events can help define rate-controlling mechanisms. An "event" to long-lived nekton may require an entire yearly cycle. Fish aggregations may not be able to nor need to respond to upwelling on a typical upwelling-event scale, but they may rather integrate short-term changes and interact only to consistent, long-term fluctuations. On an even longer time scale, our experiments may only be observations of transition periods during the evolutionary process.

The JOINT I and II expeditions of the Coastal Upwelling Ecosystems Analysis Program generally encompassed sufficient time to study processes completed during upwelling events. One result of the JOINT I observations was that abundance of nekton was apparently not food-limited, because estimated potential food supply often substantially exceeded daily ration requirements of the nekton (Mathisen, Thorne, Trumble, and Blackburn, 1978). However, one cannot generalize a conclusion on trophic relations between zooplankton abundance and food requirements of fish on the basis of six weeks of observation. One must demonstrate a probable excess food supply during other seasons of the year so that JOINT I zooplankton observations cannot be dismissed as a temporary pulse of food that momentarily exceeded the capacity of the nekton to exploit it fully. This requires that seasonal cycles be constructed for nekton and zooplankton throughout the year.

Published information comparable to that obtained during JOINT I are not available for other times of the year. It is possible, however, to extrapolate from JOINT I observations using other limited data to gain information related to food demands of nekton throughout a year. As much as possible, seasonal cycles for physical changes in the environment for the phytoplankton, zooplankton, and nekton have been constructed, from which one can examine the adequacy of the food base of secondary producers and nekton.

Yearly Physical Cycles

Winds that drive the northwest African upwelling system respond to seasonal variations in the North Atlantic (Azores) anticyclonic high-pressure cell. The cell reaches highest intensities and moves farthest north during summer, when strongest winds occur. Minimum strength and most southerly position occur in winter. The coast of northwest Africa lies under the eastern arm of the cell and experiences the prevailing northerly winds that result. Strength and position of the Azores high-pressure cell during summer prevent most atmospheric disturbances from reaching the African coast and thereby reduce variations in wind speed and direction. Weakening of the pressure cell in winter decreases protection against disturbances.

Wind conditions in the JOINT I area reflect the pattern expected from general atmospheric conditions (Fig. 1). Plotted wind values are monthly averages of twice daily meteorological observations during an 8-yr period at the Nouadhibou (formerly Port Etienne) weather station (U.S. Naval Hydrographic Office, 1952). Weather observed at Nouadhibou agrees closely with weather observed at meteorological buoys and land stations placed during the JOINT I experiment (Stuart, Buchwalter, and Garcia-Meitin, 1975).

The strongest feature of the average wind conditions is a marked consistency through the season. Velocities varied only 50% from maximum to minimum values for 11 months and direction maintained a nearly constant orientation. Maximum average monthly winds occurred from March through July, ranging from 8 to 9 m/s. Most winds blew from the north-northeast (010° to 015°). For the rest of the year, average winds decreased slightly to mostly 6 or 7 m/s and developed a more easterly component (020° to 030°) as velocities dropped. Therefore, wind direction and speed were favorable to upwelling through the year. During the March to July period of strongest average winds, nearly all winds had an upwelling-favorable direction; conversely, unfavorable winds occurred most frequently during periods of weakest wind (Fig. 1).

Another approach is to construct the Ekman flow throughout the year from existing records. Wooster and Reid (1963) and Wooster, Bakun, and McLain (1976) averaged sea-surface temperatures, wind measurements, and wind drift recorded in ships' logs by season and area and calculated a wind stress and Ekman transport (Fig. 2). The Ekman transport should be considered potential upwelling, because other physical forces may overcome and thereby reverse the flow predicted by wind stress. Schultz, Schemainda, and Nehring (in press) presented additional results of five *Alexander von Humboldt* cruises to northwest Africa from 1970 to 1976. The general conclusion is that upwelling occurs through the year near Cap Blanc, although the intensity changes with the seasons. Peak upwelling occurs from March through July with a June peak, while October through January correspond to minimum upwelling.

An oceanographic front separating warm subtropical water from North Atlantic water reaches maximum northerly displacement near Cap Blanc (20 to 21°N) during the summer. Upwelling along northwest Africa is intermittent, even during peak periods. Periods of very strong winds alternate with calm periods on time scales of several days to 1 or 2 weeks.

Yearly Biological Cycles

Primary production cycles are complexly related to the physical processes of upwelling, mixing, and solar irradiation. Contrary to intuitive reasoning, peak primary production does not necessarily occur during peak upwelling (Anderson, 1972; Walsh, 1976).

High wind conditions during JOINT I corresponded to periods of reduced primary productivity along the continental shelf, in spite of a continuous supply of nutrients (Huntsman and Barber, 1977). Strong mixing by the wind carried phytoplankton to depths where the cells became adapted to low light levels with concomittant low photosynthetic efficiency. Stratification of the near-surface waters for several days during intermittent periods of calm permitted phytoplankton to adapt to high levels and to increase productivity rapidly. Nutrient levels declined in response. Nearshore, turbidity from resuspended sediments and aeolian sand reduced the photic zone to shallow depths. Huntsman and Barber observed typical primary production rates of 1 to 3 g C $m^{-2}d^{-1}$ across the continental shelf, except for the turbid coastal zone, and suggested that the values are at or near the yearly maximum. Schultz *et al.* (1978) observed similar maximum rates of 3 to 4 g C $m^{-2}d^{-1}$ in June.

Other periods of high primary production appear likely in the Cap Blanc area. Schultz *et al.* (1978) show high chlorophyll *a* concentrations continuing from June through October, although they specified no primary production values for the period. Østvedt, Blindheim, Føyn, Berge, Smedstad, and Vestnes (1973) reported primary production of 2 to 6 g C $m^{-2}d^{-1}$ off Cap Blanc in November and December, 1972. Such high values may result from sustained, lower level upwelling relative to that observed in late spring and summer. Even though the JOINT I period may not correspond to maximum primary production, the values in March, April, and May were consistently high and would reflect themselves in increased zooplankton production 2 to 3 months later.

Zooplankton extrapolations will be more useful for the present study because zooplankton are the primary forage for fish. During JOINT I, zooplankton densities averaged about 60 g/m^2 (Table 1), almost equal to the nekton density, although zooplankton hardly registered on the echograms during the survey (Thorne, Mathisen, Trumble, and Blackburn, 1977). Often, high catches of zooplankton during acoustic transects would correspond to nearly zero integrated densities. Zooplankton abundance appeared to have been even greater during autumn. Østvedt and Blindheim (1973) and Mathisen, Østvedt and Vestnes (1974) estimated that zooplankton exceeded fish abundance by three to five times, based on echo integration and mid-water trawls during the

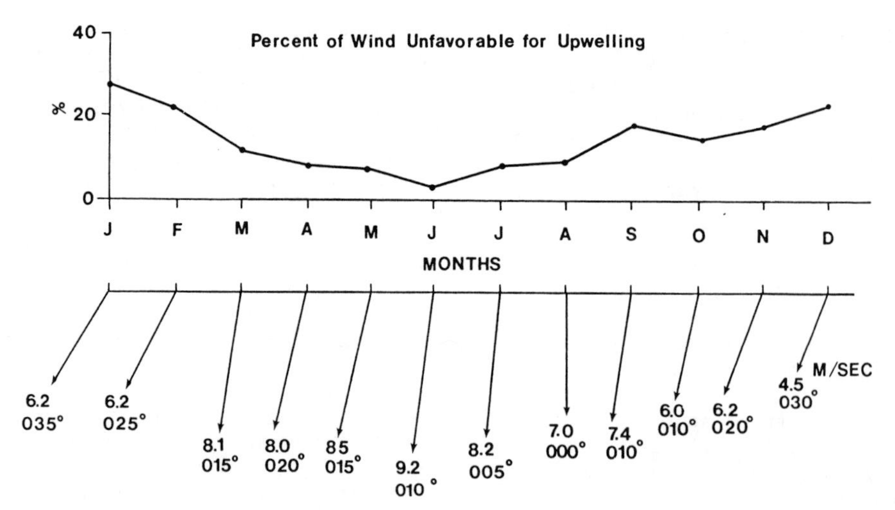

Fig. 1. Wind speed and direction at Nouadhibou (8-yr average).

G.O. Sars cruise off northwest Africa during November, 1972. Since the 120-kHz echosounders used aboard the *G.O. Sars* during the autumn survey were comparable to that used during the JOINT I survey, zooplankton appear to be much more abundant during autumn than during spring.

Flood, Braun, and DeLeon (1976) also provide evidence for high zooplankton density in autumn. Amphioxus larvae (*Branchiostoma senegalense*) off Cap Blanc reached estimated peak abundance of 50 g/m^2 and averaged 10 g/m^2 from August to November. Cephalochordata (including any amphioxus) averaged less than 0.7 g/m^2 during JOINT I (Blackburn, 1979). These larvae constituted up to 65% of the food content of *Scomber japonicus* captured near Cap Blanc in June.

Pavlov (1968) also corroborates the highest zooplankton abundance during autumn. In spring, high-density zooplankton occurred only on the outer continental shelf, but at all depths. Eggs and larval zooplankton were present, as were oceanic species; no amphioxus larvae were collected. In autumn, density remained high on the shelf, but it was equally high or higher in waters well beyond the shelf break. Autumn-collected specimens were much fatter than those of spring. Oceanic species, eggs, and larvae were poorly represented, and amphioxus were abundant.

Upwelling off Morocco begins in spring and brings deep-living (>200 m) zooplankters to the surface in inshore waters. Furnestin and Furnestin (1970) reported peak zooplankton abundance consistently occurring in autumn during six years of study off Morocco (30 to 35°N). Conversely, minimum zooplankton abundance coincided with peak summer upwelling. Furnestin (1976) and Furnestin and Belfquih (1976) relate the Moroccan zooplankton observations to hydrological changes. Concurrently, upwelling flushes nearshore species offshore as the Canary Current surface waters are diverted from the coast. Abundance reaches a minimum in summer during peak upwelling, although the proportion of young zooplankters increases. Upwelling stops during autumn, the Canary Current weakens, surface waters return over the continental shelf, enriched waters mix with normal surface waters, and zooplankton increase in abundance and diversity to a maximum and then decrease slowly through the spring decline.

If timing of zooplankton abundance relative to upwelling for Cap Blanc corresponds to the pattern described for Morocco, then lower zooplankton concentrations would be expected during the late spring and early summer upwelling peak than during the late summer and early fall upwelling lull. Schultz *et al.* (in press) compared zooplankton concentrations in late February and early March with those of June and July and found 2 to 3 times more zooplankton biomass during the earlier period. The composite picture that develops for the Cap Blanc area from the references above is that

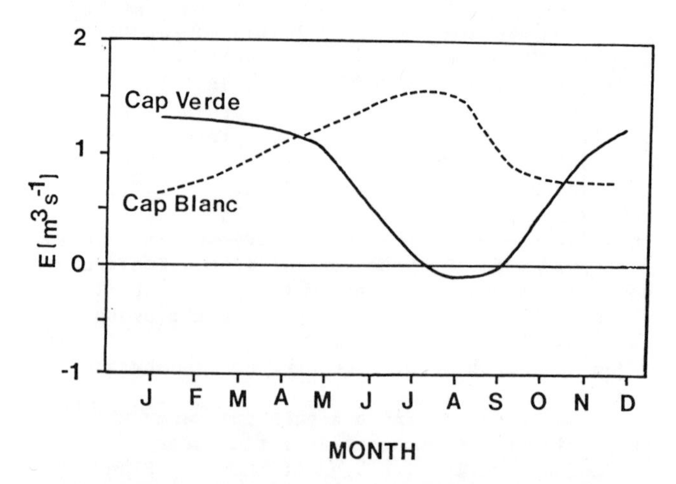

Fig. 2. Schematic average annual cycle of Ekman transport along northwest Africa (after Wooster *et al.*, 1976).

TABLE 1. Mean zooplankton standing stock (g/m^2 surface) observed near Cap Blanc April 6, 1974 through May 6, 1974[1].

Area/Time	1	2 and 3	4
I	38.96	36.32	57.07
II	51.36	43.23	49.37
III	57.86	98.26	118.34

The overall mean is 61.33 g/m^2 with a standard error of 8.86 g/m^2.

[1] Values have been corrected for slight phytoplankton contamination.

spring is a period of relatively high zooplankton biomass, which decreases during summer, possibly from a combination of peak upwelling followed by invading warm water in the north-moving oceanographic front. In autumn, zooplankton reach the highest abundance of the year.

Yearly Nekton Cycles

Pieces of the yearly cycle for fish species at Cap Blanc can be constructed from commercial catch statistics and the results of fish and ichthyoplankton surveys. In the eastern central Atlantic zone of FAO fishery statistics, which includes the JOINT I site, 75% of the catches were composed of four fish types (Anonymous, 1977): clupeids (*Sardina pilchardus* and *Sardinella aurita*) exceeded jacks (*Trachurus* spp. and *Carynx rhonchus*), while redfish (*Sparidae*) and miscellaneous species followed.

Traditionally, the dominant clupeids were *Sardina pilchardus* in northern waters off Morocco and *Sardinella aurita* in more southern waters. However, an apparent species shift during the early 1970's saw a rapid increase in abundance of *S. pilchardus* adjacent to and north of Cap Blanc (Holzlöhner, 1975; Domanevsky and Barkova, 1976) and a growth of the fishery for pilchard beginning in 1971.

Closer examination of available information suggests that conflicting reports of *Sardinella-Sardina* dominance are resolved by seasonal migrations superimposed on the recent surge of *S. pilchardus*. Temperature regimes preferred by sardines would exclude nearly all *S. pilchardus* from the Cap Blanc coast during the warm months of July through September (Fig. 3). Average surface temperatures near the JOINT I area exceeded 22 to 23°C during summer in spite of predominant offshore offshore transport. *S. pilchardus* avoid such warm water, preferring surface temperatures less than 18°C (Furnestin and Furnestin, 1970; Domanevsky and Barkova, 1976). Seasonal migrations along the Moroccan coast by sardines to maintain proper temperature have been reported. Similar migrations appear likely for Saharan sar-

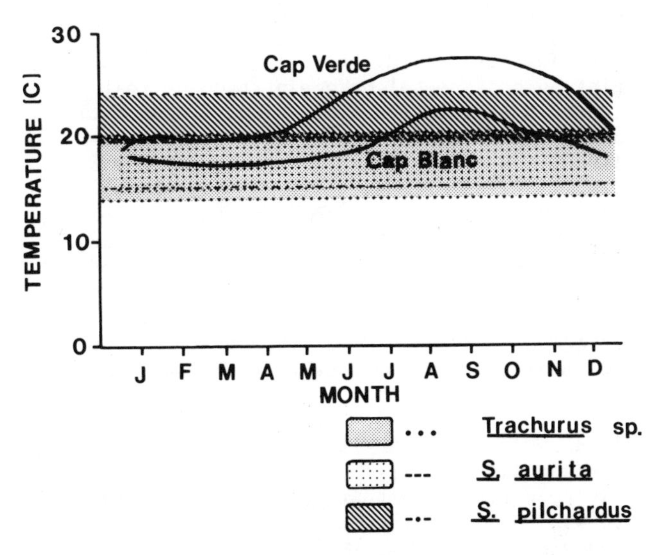

Fig. 3. Schematic average annual cycle of temperature along northwest Africa (after Wooster *et al.*, 1976) with temperature preference of selected species.

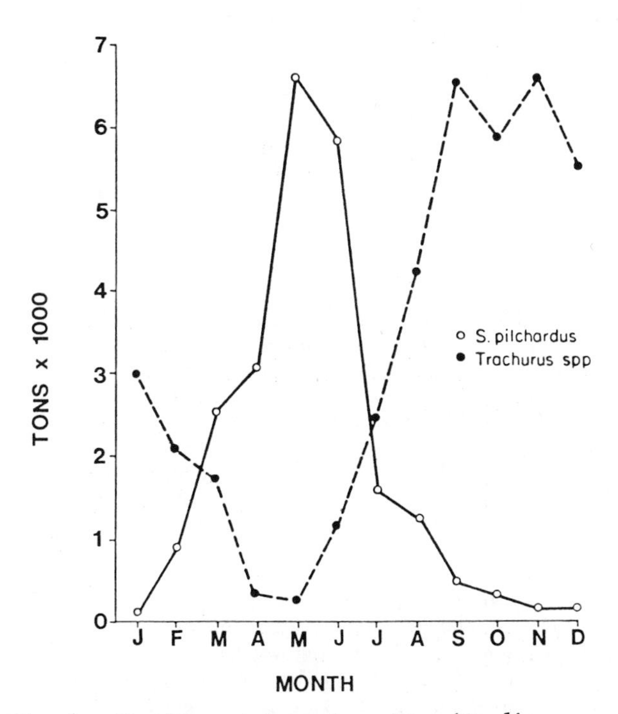

Fig. 4. Monthly catches of sardine (*Sardina pilchardus*) and horse mackerel (*Trachurus* spp.), 1974, by Romania off Rio de Oro (from J.P. Troadec, unpublished data).

dines. The advancing warm front could push sardines considerably north of Cap Blanc during summer.

Catch records of major fishing nations such as Poland show sardines are landed near Cap Blanc only in cool periods and none during the warm months. During the first half of the year, the southern and central sectors receive the heaviest fishing pressure. The fleet shifts to the northern section for the second half of the year. Catch pattern by month for 1974 (J.P. Troadec, personal communication) show peak catches in June and July indicating that the sardines may concentrate as they move north. Romanian trawl landings from 1974 show the same pattern (Fig. 4).

Spawning of the sardines takes place during or shortly after peak zooplankton abundance. The poorly developed larvae have a short point of no return, so presumably the larvae must eat soon after yolk sac absorption or face starvation. High concentrations of food suitable for larval fish would reduce starvation and help offset predation by increasing larval growth.

Horse mackerel also undergoes seasonal fluctuations in availability in the region including Cap Blanc. Temperature preferences for horse mackerel (Blackburn, 1975) are similar to those of sardine (Fig. 3) and may exclude horse mackerel along Rio de Oro from the nearshore waters south of the oceanographic front during warm summer months. Horse mackerel spawning occurs from November through May along Rio de Oro. *Trachurus* spp. have non-functioning mouths, guts, and eyes at hatching, but they develop large mouths capable of catching small zooplankton by the time of yolk sac absorption. In the Pacific, *Trachurus symmetricus* larvae feed heavily on cosmopolitan but uncommon species off the coast of California. Therefore larval *Trachurus* spp. do not seem to be tied to periods of maximum primary production for survival.

The Nature of Limiting Factors

The evidence for seasonal cycles, presented earlier, is largely circumstantial, but at least two seasons (spring and autumn) have been identified in which nekton seem to experience excess food. Zooplankton abundance seemed several times higher in autumn than in spring and therefore enhanced feeding conditions occurred during the fall. But food production in spring and early summer during the JOINT I observations was demonstrated also to exceed demand.

During a third season, summer, the species complex of fish, and possibly zooplankton, shifts as a result of the northern migration of the oceanographic front. Only during winter is food very likely to be in short supply, but winter is also a period of fasting for many fish species.

The nekton biomass at Cap Blanc does not appear to be higher any other time of the year than during JOINT I, and the examined records of yearly wind, Ekman flow, temperature, zooplankton, and fish abundance seem to validate the conclusion that nekton do not stress their food supply. This presumably supports the alternative hypothesis that year class strength is governed primarily by larval survival.

References

Anderson, G.C., Aspects of marine phytoplankton studies near the Columbia River with special reference to a subsurface chlorophyll maximum, in: *The Columbia River Estuary and Adjacent Ocean Waters, Bioenvironmental Studies*, A.T. Pruter and D.L. Alverson (eds.), University of Washington Press, Seattle, pp. 219-240, 1972.

Anonymous, Food and Agriculture Organization of the United Nations, Yearbook of Fish Statistics, Vol. 42, Catches and Landings, 1976, pag. var., 1977.

Blackburn, M., Summary of existing information on nekton of Spanish Sahara and adjacent regions, northwest Africa, CUEA Technical Report 8, 49 pp., 1975.

Blackburn, M., Zooplankton in an upwelling area of northwest Africa: Composition, distribution, and ecology, *Deep-Sea Research, 26*, 41-56, 1979.

Domanevsky, L.N., and N.A. Barkova, Some peculiarities of sardine (*Sardina pilchardus* Walbaum) distribution and spawning along northwest Africa, *International Council for the Exploration of the Sea C.M.* 1976/j:6, 15 pp., 1976.

Flood, P.R., J.G. Braun, and A.R. DeLeon, On the annual production of amphioxus larvae (*Branchiostoma senegalense* Webb) off Cap Blanc, northwest Africa, *Sarsia, 61*, 63-70, 1976.

Furnestin, M.L., Les copepodes du plateau continental marocain et du detroit Canarien, I. Repartition quantitatives, *International Council for the Exploration of the Sea C.M.* 1976/L:8, 5 pp., 1976.

Furnestin, M.L., and M. Belfquih, Les copepodes du plateau continental marocain et du detroit Canarien, II. Les especes au cours d'un cycle annuel dans les zones d'upwelling, *International Council for the Exploration of the Sea C.M.* 1976/L:9, 11 pp., 1976.

Furnestin, J., and M.L. Furnestin, La sardine marocaine et sa peche, migrations trophiques et genetiques en relation avec l'hydrologie et le plancton, *Rapports et Proces-Verbaux des Reunions Conseil Permanent International pour l'Exploration de la Mer, 159*, 165-175, 1970.

Holzlöhner, S., On the recent stock development of *Sardina pilchardus* Walbaum off Spanish Sahara, *International Council for the Exploration of the Sea C.M.* 1975/J:13, 16 pp., 1975.

Huntsman, S.A., and R.T. Barber, Primary productivity off northwest Africa: the relationship to wind and nutrient conditions, *Deep-Sea Research, 24*, 25-34, 1977.

Mathisen, O.A., O.J. Østvedt, and G. Vestnes, Some variance components in the acoustic estimation of nekton, *Tethys, 6*, 303-312, 1974.

Mathisen, O.A., R.E. Thorne, R.J. Trumble, and

M. Blackburn, Food consumption of pelagic fish in an upwelling area, in: *Upwelling Ecosystems*, R. Boje and M. Tomczak (eds.), Springer-Verlag, Berlin, pp. 111-123, 1978.

Østvedt, O., and J. Blindheim, Studies of the abundance and distribution of fish off west Africa, November-December, 1972, *International Council for the Exploration of the Sea C.M.* 1973/J:24, 9 pp., 1973.

Østvedt, O.J., J. Blindheim, L. Føyn, G. Berge, O. Smedstad, and G. Vestnes, Report on a cruise by the R.V. *G.O. Sars* to west Africa, 23 October to 15 December 1972, *International Council for the Exploration of the Sea C.M.* 1973/J:23, 34 pp., 1973.

Pavlov, V. Ya., Plankton distribution in the Cap Blanc region, *Oceanology, 8,* 381-387 (American Geophysical Union Translation), 1968.

Schultz, S., R. Schemainda, and D. Nehring, Seasonal variations in the physical, chemical, and biological features in the CINECA region, in: Rep. of CINECA Symposium on the Canary Current: Upwelling and Living Resources, *Rapports et Proces-Verbaux des Reunions Conseil International pour l'Exploration de la Mer,* in press, 1978.

Stuart, D.W., T. Buchwalter, and J. Garcia-Meitin, JOINT I Meteorological Data, CUEA Data Report No. 30, 88 pp. (Available from the National Technical Information Service, PB 255711/AS), 1975.

Thorne, R.E., O.A. Mathisen, R.J. Trumble, and M. Blackburn, Distribution and abundance of pelagic fish off Spanish Sahara during CUEA expedition JOINT I, *Deep-Sea Research, 24,* 75-82, 1977.

U.S. Naval Hydrographic Office, Sailing directions for the west coasts of Spain, Portugal, and northwest Africa and offlying islands, H.O. Pub. 51 (formerly No. 134), 411 pp., 1952.

Walsh, J.J., Models of the sea, in: *The Ecology of the Seas,* D.H. Cushing and J.J. Walsh (eds.), W.B. Saunders Co., Philadelphia, pp. 388-466, 1976.

Wooster, W.S., and J.L. Reid, Eastern boundary currents, in: *The Sea,* M.N. Hill (ed.), Wiley, New York, pp. 253-280, 1963.

Wooster, W.S., A. Bakun, and D. McLain, The seasonal upwelling cycle along the eastern boundary of the north Atlantic, *Journal of Marine Research, 34,* 131-141, 1976.

THE BENTHIC PROCESSES OF COASTAL UPWELLING ECOSYSTEMS

Gilbert T. Rowe

Oceanographic Sciences Division
Department of Energy and Environment
Brookhaven National Laboratory
Upton, New York 11973, U.S.A.

Abstract. The sediments and benthos on continental margins characterized by upwelling and seasonally-stratified ecosystems are compared. Primary productivity has a major influence on community structure and partitioning of carbon to secondary producers. The highest primary production causes anoxic sediments characterized by sulfate reduction and denitrification, low invertebrate diversity and biomass, and mats of sulfur bacteria with high biomass. Continental margins of lower primary productivity are characterized by aerobic metabolism and invertebrate assemblages of high biomass. In upwelling systems the biomass of the benthos, zooplankton, and fishes is often highest offshore, rather than next to the coast.

Preliminary budgets, based on sediment trap, benthic metabolism, and burial rate data, allow a comparison of the relative importance of the loss or recycling of carbon and nitrogen in each system. Benthic metabolism and nutrient regeneration were important in all sediments studied, even in anaerobic systems that appear to conserve organic matter. Remineralized nutrient and organic carbon concentrations are not in steady state and vary radically over periods of months to years.

Introduction

The benthos is important in the function of coastal ecosystems. It receives a large fraction of the organic matter produced, uses it as energy, converts it to biomass, and finally releases large amounts of inorganic plant nutrients back into the water column. The purpose of my work, as a member of the Coastal Upwelling Ecosystems Analysis Group, has been to quantify these exchanges, to compare their relative importance in different regions, and to draw conclusions about the general significance of biological processes on the bottom in ecosystem function.

Investigations have been made with a number of different collaborators in three different upwelling areas and in the middle Atlantic Bight of the United States, and this paper is a summary comparison of the different systems. Most of the work has been or is in the process of being published in other papers and in a review presenting many specific data and detailed results not repeated here (Rowe, in press). Few technical details or data will be presented, and failures will not be elaborated on, but I shall emphasize constructing carbon budgets for the sediment of each geographic locale, compare system structure and function, and describe some heretofore unrecognized, important dynamic aspects of the bottom as a functioning part of each ecosystem.

Standing Stocks of the Benthos

Quantitative bottom samples taken on a nearly world-wide basis now make possible some broad comparisons of the biomass of what is referred to as macrofauna {retained on sieves ranging in size from 297 μ (for deep-sea assemblages, Hessler and Jumars, 1974) to 2 mm (Mare, 1942)}. Studies specifically in upwelling areas include those by Sanders (1969) off southwest Africa, Nichols and Rowe (1977) off northwest Africa, Gallardo (1963) off Chile, and Frankenberg and Menzies (1968) and Rowe (1971a, b) off Peru. In the regions where primary production in the water column is highest, stocks of benthos are markedly reduced (Chile, Peru, southwest Africa) when the oxygen of the bottom water is reduced to levels below about 5% of saturation. At depths above and below the oxygen minimum, stocks are high (Rowe, 1971a), and at great depths community structure regains an expectedly high number of species (Sanders, 1969).

From general correlations between primary production and benthic stocks (Rowe, 1971b) one would expect the strong positive relationship to extend to upwelling ecosystems, but in general it does not, even in regions without limiting oxygen concentrations. Off northwest Africa, for example, average wet-weight biomass (Table 1, Plate II, from Nichols and Rowe, 1977) was 35 g/m^2 for the entire continental shelf. Primary production, averaging about 1 to 3 g C $m^{-2}d^{-1}$ (Table 1, Huntsman and Barber, 1977), is somewhat higher than that off the mid-Atlantic states of the United States, with an average primary production of about 1 g C $m^{-2}d^{-1}$ (Ryther, 1963; Falkowski, Hopkins, and Walsh, 1980), where recent studies revealed biomass consistently above 100 g/m^2 in many areas (Wigley and Theroux, in press). Indeed, boreal and temperate continental shelves are known to be characterized by biomass levels reaching several hundred grams per square meter (Zenkevitch, 1963, and others), far above those so far seen in upwelling regions.

Off Baja California, Mexico, the upwelling ecosystem is dominated by the red crab *Pleuroncodes planipes* of the family Galatheidae. Most galatheids live in the deep sea and are entirely benthic, but those living in shallow water are primarily pelagic. *P. planipes* is both benthic and pelagic and when it settles to the bottom, which it does for unknown reasons, it totally dominates the community in both biomass and metabolism (Smith, Harbison, Rowe, and Clifford, 1975; Rowe, in press, Table 1). Secondary production estimates can vary widely but probably are about 4.6 g C $m^{-2}y^{-1}$ (Rowe, in press).

Bottom trawls were used to estimate the abundance of large near-bottom organisms (megafauna) off northwest Africa (Haedrich, Blackburn, and Bruhlet, 1976). The highest biomass was found along the continental shelf - slope boundary, presumably because shelf break upwelling is prominent there and considerable organic matter produced on the shelf is exported to less physically dynamic depths (Table 1). Secondary production of this mixed assemblage of organisms is probably less than around 0.2 g C $m^{-2}y^{-1}$ (Rowe, in press), somewhat lower than that of the red crab off Baja California, Mexico, and estimated to be an order of magnitude lower than that of the macrofauna of the same area (Nichols and Rowe, 1977, Table 1).

Smaller size classes of benthos are important components of the sediment communities in upwelling areas. Microbial populations and their dynamics have been studied off northwest and southwest Africa (Watson, 1978), along with total sediment metabolism off northwest Africa (Christensen and Packard, 1977). The abundance of the meiofauna has been compared along several transects extending off northwest Africa and Portugal (Thiel, 1978) and nematode abundance and biomass have been estimated in the 15°S area off Peru (Plate III) by J.A. Nichols (Rowe, in press). The general trend for all these organisms is toward much greater abundances and activities (metabolic) in the top few centimeters of sediment and much higher abundances and activities close to shore, decreasing in an offshore direction and with depth. The meiofaunal nematodes in the oxygen minimum zone off Peru, as with the macrofauna, are less abundant apparently due to low-oxygen stress. Frankenberg and Menzies (1968) and Rowe (1971a) found the restricting effects on macrofaunal-sized nematodes in the oxygen-poor zone was less than the apparent reductions of the remaining macrofauna, but Nichols' data confirm that under the most severe conditions even the meiofaunal nematodes (less than 0.42 mm but larger than 60 micron sieve) are also reduced in abundance (Rowe, in press).

A new discovery in upwelling zones with low oxygen has been a mat of *Thioploca*-like sulfur bacteria on the oxycline's intersection with the sediments. Gallardo (1977) first found it in the Bay of Concepcion in Chile and close inspection of sediments along the Peru coast demonstrated that it extends from Chile to central Peru, often all across the continental shelf (Rowe, in press). Its biomass off Peru has not been measured directly, but a crude displacement volume (Rowe, in press) suggested that it could weigh 1530 g wet weight/m^2, or about 90 g C/m^2. Not only is its productivity unknown, presently it is not even known if it is autotrophic, heterotrophic, or mixotrophic, and it may be an assemblage of a number of functionally different organisms.

Concentration of Organic Matter in the Sediments

The concentration of organic matter in coastal sediments has a pattern that is different from that of biomass. In coastal sediments, even in regions of high primary production, the organic matter concentrations are quite low when the sediment grains are course but when the grains are fine (silt and clay predominate) then organic matter concentrations are high. Evidently in dynamic areas very close to shore, where the bottom is subject to longshore tidal currents and responds to wave action, small particles are resuspended, and winnowing gradually carries finer material offshore, leaving only sand-sized material, relatively free of organic matter. Silt and clay, products of geologic weathering, are carried offshore, along with organic detritus, and supposedly are deposited in physically less dynamic areas (Visher, 1961; Schopf, 1980). Although cross-shelf transects suggest that on the shelf such a pattern of decreasing grain size may be the exception rather than the rule (Emery, 1968), the generally higher concentrations of organic matter on the upper continental slope (Premuzic, 1980) reach even higher concentrations in upwelling regions that have low oxygen concentrations (Demaison and Moore, in press). The latter authors suggest that diagenetic processes in such reducing anaerobic environments allow the preservation of lipid-rich organic matter that leads to fossil fuel deposits. Preservation in such deposits is in apparent contradiction to the view that anaerobic metabolism of such carbon compounds is extensive. Concentrations of organic carbon in these deposits are commonly greater than 4% of the sediment dry weight (Rowe, 1971a, in press; Manheim, Rowe, and Jipa, 1975; Diester-Haass, 1978; Müller and Suess, 1979), much higher than off northwest Africa (<4%) with higher oxygen concentrations (Milliman, 1977; Nichols and

Rowe, 1977; Watson, 1978; Thiel, 1978; Müller and Suess, 1979).

The accumulation of organic matter on the bottom has been estimated in upwelling regions by geologists by multiplying organic matter concentrations by sedimentation rates. Off southwest Africa sedimentation rates reach about 150 cm each 1000 years (Diester-Haass, 1978), about the same as recorded off Peru (66 to 140 cm each 1000 years, Müller and Suess, 1979), or considerably higher than the western continental margin of the United States (10 cm each 1000 years) or northwest Africa (2 to 12.7 cm each 1000 years, Müller and Suess, 1979). Müller and Suess (1979), after compensating for compaction and dilution, calculated that 37.8 and 39.7 g C $m^{-2}y^{-1}$ accumulate on the Peruvian upper continental slope, whereas 1.08 and 1.5 g C $m^{-2}y^{-1}$ accumulate on the northwest African slope.

Accumulation is the input of organic matter minus biological utilization. Traps have been set recently in the water column to catch vertically sinking particles before they reach the bottom. Off Peru (Plate III) three different kinds of traps were used to make such estimates. Rowe (in press) reported on sets of moored arrays of traps and a study by Staresinic (1978) using floating traps in the same region.

An interesting aspect of the preliminary trap data is the variation in time and space, a product both of prevailing biological and physical processes as well as the differences between the methods. Staresinic for example noted yearly (1977 vs. 1978) differences that he believed attributable to the kinds of zooplankton in the water and differences in the depth at which he floated the traps. He thought higher rates at night might be due to migrating zooplankton or wind mixing during the daytime. In the data from moored traps in the area (Rowe, in press), extremely large increases were noted close to the bottom and close to shore in 1976, 1977, and 1978, due to resuspension. The average fluxes of sinking organic carbon, 0.3 and 0.5 g C $m^{-2}d^{-1}$ for moored and floating traps, are important in our general budgets for the bottom, but the abberant near-bottom information gives us some insight into how some potentially important, but unrecognized processes such as resuspension might affect the ecosystem.

Biological Utilization of Organic Matter

Biologists have long recognized that the principal loss of organic matter from an ecosystem is its metabolic conversion to CO_2. Aerobic benthic metabolism, reviewed by Pamatmat (1977), has been measured in many aquatic environments, including upwelling areas (Table 1). Using a presumed stoichiometry between oxygen consumption and carbon dioxide production, called the RQ (for Respiratory Quotient), one can estimate organic matter losses to metabolism, an important part of the budgeting goals of our work. While aerobic metabolism in sand environments is relatively high (northwest Africa and Baja California), comparing well with other shallow areas (Table 1), it appears to be much lower in the mud bottoms where oxygen is low enough to limit invertebrates (e.g., Peru).

The importance of anaerobic metabolism has been recognized by geochemists (Berner, 1971; Goldhaber *et al.*, 1977) and microbiologists (Sorensen, Jorgensen, and Reusbech, 1979). The most important alternative terminal electron acceptors are $SO_4^=$, NO_3^-, and NO_2^-, although others may also be important (Berner, 1971).

Estimates of sulfate reduction were reported by Rowe (in press, Table 1) from an unpublished collaboration with Robert Howarth and C.H. Clifford using labelled-sulfate shipboard incubations to determine sulfate metabolism loss. Consumption of other respiratory substrates such as O_2, NO_3^-, and NO_2^- were not measured in the outer shelf sediment, but the large sulfate losses indicate that metabolic rates are appreciable even though organic levels are extremely high and aerobic respiration is low. Goering and Pamatmat (1971), working in the same area but in a different year, noted significant rates of denitrification in cores incubated on the ship. This also illustrates a pertinent difference between aerobic coastal sediments that produce nitrate (Rowe, Clifford, and Smith, 1977; Florek, 1979; Rowe and Clifford, 1978), and anoxic sediments that consume nitrate, nitrite, and sulfate.

Carbon Budgets for each Upwelling System

Estimates of metabolism, secondary production, accumulation, and particle flux in the water column can be used to construct preliminary budgets (Table 1, Fig. 1) for the bottom in the coastal areas studied. The systems are very different and not all the important variables have been measured. Conceptually the fluxes into and out of the bottom can be represented by Fig. 1. While resuspension by currents has been mentioned and the process is fairly straightforward, bedload transport may not be. By bedload transport is meant, in the geologic sense, the organic matter that is transported horizontally along the bottom as part of the erosion of the total bed. Examples could be sand waves or sediment ripple marks that contain organic detrital particles moving along the bottom with or against the current, or mass sediment movements such as slumps or turbidity currents off the outer continental shelf. Information on these processes is scant and cannot be put in a carbon budget.

Off Baja California we can estimate the production of the macrofauna, of the dominant megafaunal organism, *P. planipes*, and nearshore oxygen demand, but we have no information on burial, export, resuspension or particle sedimentation rate. Off northwest Africa (Plate I), as off Baja California, organic carbon in the sediments is not

TABLE 1. Comparison of Some Sediment Organic Carbon Fluxes (Fig. 1), in g m^{-2} d^{-1}

	Baja California[1]	NW Africa[2]	Peru[3]	Mid-Atlantic Bight[4]	Buzzards Bay[5]	Narragansett Bay[6]
Primary Production	3.9	1.3	∿5	∿1	0.3 to 0.5	0.6
Particle Flux	?	?	0.3 to 0.5	?		0.34 to 0.56
Organic Carbon Accumulation	Low	Low, exports offshore	-0.106	Erosional export, areas as high as -0.07	-0.04	?
Aerobic Respiration	-0.4	-0.2	-0.05 to -0.15	-0.04 to -0.38 (mean ca. -0.11 to -0.24)	-0.26 to -0.45	-0.19 to -0.63
Denitrification	?	At depth in some sediments	Significant, rate unknown	Occasional (spatially seasonal)	?	Occasional (seasonal)
Sulfate Reduction	?	?	-0.25	Occurs, but restricted in time and space	?	?
Secondary Production[7]	-0.019 (macro + megafauna)	-0.018 (macro, meio megafauna)	-0.030 (macro + meiofauna)	-0.027 (macrofauna only)	-0.016 (macrofauna only)	-0.019 to -0.027 (macrofauna only)

1 Data from Smith, Rowe, and Clifford, 1974; Smith, Harbison, Rowe, and Clifford, 1975; Rowe, in press.
2 Data from Huntsman and Barber, 1977; Rowe *et al.*, 1976; Rowe, Clifford, and Smith, 1977; Milliman, 1977; Nichols and Rowe, 1977; Rowe, in press.
3 Data from Barber and Smith, in press; Staresinic, 1978; Rowe, in press; Goering and Pamatmat, 1971; Pamatmat, 1971; Rowe, Clifford, and Howarth, unpub. $SO_4^{=}$ reduction data; Müller and Suess, 1979.
4 Data from Falkowski, Hopkins, and Walsh, 1980; Ryther, 1963; Thomas *et al.*, 1976; Rowe, 1971c; Florek, 1979; Steimle and Stone, 1973.
5 Rowe, Clifford, Smith, and Hamilton, 1975; Florek, 1979; Smith, Rowe, and Nichols, 1973.
6 Nixon, Oviatt, and Hale, 1976; Oviatt and Nixon, 1975.
7 Assumes P/B of 1 for macrofauna and megafauna and P/B of 4 for meiofauna; carbon assumed to be 5% of wet weight; excludes microfauna in all cases. From above sources, plus J. Nichols' unpublished nematode data.

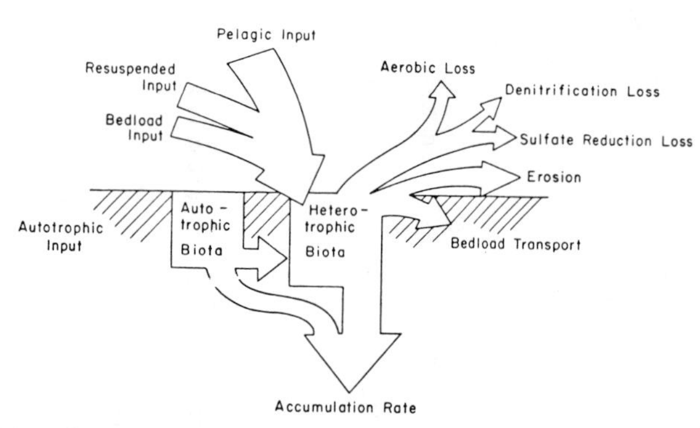

Fig. 1. Sources and sinks of organic matter in a coastal ecosystem. (See Table 1 for available data.)

high, and it is reasonable to believe that considerable export of carbon occurs along shore and down the continental slope via submarine canyons (Koopmann, Sarnthein, and Schrader, 1978). This may also occur off Baja California, but we have no evidence that it does. In addition to total aerobic metabolism, microbial metabolism in the sediment was investigated off northwest Africa (Watson, 1978). Organic carbon concentrations, low on the shelf, increase on the upper slope where grain size decreases, confirming that export from the shelf occurs, even when the carbon does not move down canyons to deeper water. Production of the macrofauna and megafauna were indirectly estimated from the P/B (productivity/biomass) ratios in the literature, but they are subject to large error (Rowe, in press, as discussed in Nichols and Rowe, 1977; Haedrich *et al.*, 1976; Watson, 1978). Little is known of anaerobic metabolism or sedimentation rates in this region.

Considerably more data are available from Peru than the other two upwelling areas because it has been studied more intensively. In the 15°S area (Plate III), metabolic losses, due to sulfate reduction in 1978; denitrification (Goering and Pamatmat, 1971), and oxygen consumption (Pamatmat, 1971), can be estimated. It is not clear that the rates should be added together for the purpose of a budget because they were measured in different years. Nitrate reduction is believed to have been high during the PISCO (1969) expedition to the area (Plate III, Goering and Pamatmat, 1971), while oxygen demand was relatively modest (Pamatmat, 1971, Table 1).

In none of the budgets tabulated do input of organic matter and its consumption or loss agree well. This disagreement must be due to our inability to measure all the variables at the same place and time and also to our misconception that the bottom is more or less in a steady state. One would expect for example that underestimates of consumption would result from not measuring some terminal electron acceptors, or metabolic loss, "net erosion" versus "net deposition" on relatively short time scales, and movement from shallow areas of high wave and tidal energy to greater depths. Such underestimation would generally make for consistent export from shallow to deep water. For example, pelagic inputs in Peru need to be supplemented nearshore or "upstream" to balance the short-term budget, which requires adding resuspended material, bedload, or autotrophic production (Fig. 1), but such variables are difficult to estimate. Subtraction of the pelagic from the pelagic plus resuspended input (the total in the bottom trap collections as reported by Rowe, in press, using the technique of Rowe and Gardner, 1979) increases the input, but erosional export remains unknown, as does the bedload balance. The imbalance is an enigma that may be common to shallow benthic budgets, as indicated by information on Narragansett Bay and Buzzards Bay (Table 1). The possible explanations lie in a general imbalance between geologic inputs and exports, conceptual misunderstandings about the stoichiometry between the oxidation of organic matter and the terminal electron acceptors, or surprisingly high efficiency of the chemolithotrophs in the upper layers of the sediments. Of course, if the sediment systems are not in a steady state on the time scales we are dealing with there is no reason to expect that we will ever achieve a balance other than by chance.

Temporal Variations in Sediment Properties

The dynamics of only a small fraction of all sediment biological processes have ever been quantified, presumably because the dynamics of the benthic biota and environment are inordinately difficult to measure because rates of change are so slow. This presumption of a somewhat steady state makes solving differential equations that model sediment processes easy, but there is mounting evidence that a number of processes not only proceed rapidly but that the processes themselves differ appreciably over periods of just a few months to a few years, and not in "geologic" time scales as often presumed.

One of the most dramatic variations documented in upwelling areas was in nitrate concentrations in surficial sediments along the C line (Plate III). In 1969 Goering and Pamatmat (1971) recorded values of less than 10 μg-at/l, presumably in the pore water of the sediment, and of about 30 μg-at/l in the deep water, with little gradient above the bottom. Although the concentration did not increase with depth into the bottom, the concentration in the sediments in general was well below that in the water, supporting the results of their experiments aboard ship that indicated that denitrification was occurring in the sediments. Then in June 1976, immediately following the *aguaje* (total denitrification followed by sulfate reduction in the water column, Dugdale *et al.*, 1977), pore water nitrates were about 2500 μg-at/l, at about the same depth (equivalent to C 3, 90 to 125-m depth), followed by a decrease in nitrates just deeper in the sediments and no measurable nitrate at C 5, just offshore (240 m). About 9 months later the concentration had decreased to 1300 to 1400 μg-at/l at the same station (100 m deep) and a year after that the value had gone down to about 200 μg-at/l (Rowe, 1979, in press). The most plausible explanation is that microaerophilic nitrifiers took advantage of an abundance of inorganic reduced substrates (energy sources), getting their reducing power by splitting water and oxidizing the energy sources (probably ammonium and sulfide) with something other than free oxygen. The organisms might also be mixotrophs, as opposed to strict obligate chemolithotrophs, thereby being able to utilize the vast amount of organic matter that would have been available at the time. The decrease in nitrate in the pore water by diffusion and denitrification can be interpreted as the return of the system to "normal" conditions where nitrates are low and denitrification is important.

Prior to 1972, before the anchovy fishery suffered *el nino* from which it apparently has never recovered, it was assumed that a large fraction of the primary production was consumed by the anchovy. We can hypothesize that now without them a larger fraction of the phytoplankton sinks to the bottom than prior to 1972, and data from the JOINT II area (Plate III) can be used to test this hypothesis. In nine analyses made in 1969 on the surface sediments between 60- and 2000-m depth (Rowe, 1971a; Manheim *et al.*, 1975), the organic carbon ranged from 1.8 to 4.8%, averaging 3.3%. Far more data are available from the JOINT II expedition (Rowe, 1979) and to prevent biasing a comparison, I lumped the JOINT II values by 100-m depth intervals equivalent to the depths where the nine 1969 values were taken. A total of 27 values, all from the top 1 cm of sediment, were partitioned into nine depth categories. The JOINT II data had a category for 30 m, while the 1969 data did not; the 1969 data had a value at 1500 m, while the JOINT II data did not. The JOINT II average for the nine depths was 5.7% organic carbon, compared to the 3.3% for 1969, using identical techniques.

A doubling of surface sediment organic carbon concentrations is a striking variation, subject, I believe, to considerable question, but some simple calculations suggest that such a change is possible. One would estimate that one cubic centimeter of wet mud, containing about 0.7 g of dry mud, would have contained 0.038 g organic carbon in 1976 to 1978 versus 0.023 g in 1969, making the same assumption of dry to wet weight. This would be equivalent to 0.015 g C/cm^3 in an 8-year period or about 20 g $m^{-2}y^{-1}$. Considering that total primary production is on the order of more than 1000 g C $m^{-2}y^{-1}$ (Barber and Smith, in press), and that we have estimated that about 10% of that product was sinking out of the euphotic zone during JOINT II (Staresinic, 1978; Rowe, in press), such an increase seems entirely possible.

The Impact of Sediment Resuspension

The resuspension of bottom sediments is a common occurrence in nearshore sediments (Fig. 1) and is potentially important in balancing a carbon budget, but its impact on the ecosystem has never been thoroughly treated. In addition to displacement and burial of the benthos and reducing light penetration, resuspension could have an effect on the non-conservative chemical properties of the water column if the sediment has a large biological oxygen demand (BOD). For the phenomenon called *aguaje* to occur (Dugdale *et al.*, 1977), during which all the oxygen, nitrate, and nitrite and some of the sulfate in the water are respired, circulation must be such as to prevent replenishment of oxygen (Barber and Huyer, 1979), and it might also be reasoned that the phenomenon requires an unusually large input of pelagic organic matter. A possible alternative to these requirements could be higher rates of resuspension of bottom sediments with exceptionally high BOD.

Only indirect evidence is available that the *aguaje* was caused by sediment resuspension. The interior flow (poleward undercurrent) was considerably stronger in 1976 than in 1977, and the poleward flow in 1976 had an offshore component along the bottom, whereas in 1977 when conditions were characterized by longshore winds and significant coastal upwelling not seen in 1976, the poleward flow was restricted and often lacked significant offshore flow along the bottom (Brink, Allen, and Smith, 1978; Brink, Halpern, and Smith, 1980). The data from moored sediment traps indicate that nearshore resuspension was much greater in 1976 than in 1977, although the observations were in June of 1976 and not earlier in the year as during the recorded *aguaje* (Rowe, in press). Only organic-rich sediment could have a BOD large enough to affect water column characteristics, but Peru sediments are some of the richest in the world. The high deposition rates for the outer shelf at C 3 to C 5 might suggest that resuspension does not occur there. Resuspension was probably mainly nearshore (C 1 to C 3), resulting in the export of organic-rich matter offshore to the shelf break, where physical energy levels were generally lower.

Acknowledgements. This work was mainly supported as part of the Coastal Upwelling Ecosystems Analysis Program by a series of grants from the IDOE section of the U.S. National Science Foundation, the latest being Grant OCE79-12600, and by the Department of Energy under Contract No. DE-AC02-76CH00016.

References

Barber, R., and A. Huyer, Nitrate and static stability in the coastal waters off Peru, *Geophysical Research Letters, 6*, 409-412, 1979.

Barber, R., and R. Smith, Coastal Upwelling Ecosystems, in: *Analysis of Marine Ecosystems*, A. Longhurst (ed.), Academic Press, in press.

Berner, R., *Principles of Chemical Sedimentology*, 240 p., McGraw Hill, New York, 1971.

Brink, K.H., J.S. Allen, and R.L. Smith, A study of low frequency fluctuations near the Peru coast, *Journal of Physical Oceanography, 8*, 1025-1041, 1978.

Brink, K.H., D. Halpern, and R.L. Smith, Circulation in the Peruvian upwelling system near 15°S, *Journal of Geophysical Research, 85*, 4036-4048, 1980.

Christensen, J.P., and T.T. Packard, Sediment metabolism from the northwest African upwelling system, *Deep-Sea Research, 24*, 331-343, 1977.

Demaison, G.J., and G.T. Moore, Anoxic environments and oil source bed genesis, *Organic Geochemistry*, in press.

Diester-Haass, L., Sediments as indicators of upwelling, in: *Upwelling Ecosystems*, R. Boje and M. Tomczak (eds.), Springer, Berlin, p. 261-281, 1978.

Dugdale, R., J. Goering, R. Barber, R. Smith, and T. Packard, Denitrification and hydrogen sulfide in the Peru upwelling region during 1976, *Deep-Sea Research, 24*, 601-608, 1977.

Emery, K.O., Relict sediments on continental shelves of the world, *American Association of Petroleum Geologists Bulletin, 52*, 445-464, 1968.

Falkowski, P.G., T.S. Hopkins, and J.J. Walsh, An analysis of factors affecting oxygen depletion in the New York Bight, *Journal of Marine Research, 38*, 479-506, 1980.

Florek, R., Measurements of the flux of dissolved inorganic nutrients from marine coastal and shelf sediments, Master of Science Thesis, Wayne State University, Detroit, Michigan, 94 pp., 1979.

Frankenberg, D., and R.J. Menzies, Some quantitative analyses of deep-sea benthos off Peru, *Deep-Sea Research, 15*, 623-626, 1968.

Gallardo, V.A., Notas sobre le densidad de la fauna bentonica en el sublitoral del notre de Chile, *Gayana Zoologica, 8*, 3-15, 1963.

Gallardo, V.A., Large benthic microbial communities in sulphide biota under Peru-Chile subsurface countercurrent, *Nature, 268*, 331-332, 1977.

Goering, J.J., and M.M. Pamatmat, Denitrification in sediments of the sea off Peru, *Investigacion Pesqueres, 35*, 233-242, 1971.

Goldhaber, M.B., R.C. Aller, J.K. Cochran, J.K. Rosenfeld, C.S. Martens, and R.A. Berner, Sulfate reduction, diffusion and bioturbation in Long Island Sound sediments: Report of the FOAM group, *American Journal of Science, 277*, 193-237, 1977.

Haedrich, R.L., M. Blackburn, and J. Bruhlet, Distribution and biomass of trawl-caught animals off Spanish Sahara, West Africa, *Matsya, 2*, 38-46, 1976.

Hessler, R.R., and P.A. Jumars, Abyssal community analysis from replicate box cores in the central North Pacific, *Deep-Sea Research, 21*, 185-209, 1974.

Huntsman, S.A., and R.T. Barber, Primary production off northwest Africa: the relationship to wind and nutrient conditions, *Deep-Sea Research, 24*, 25-33, 1977.

Koopman, B., M. Sarnthein, and H.-J. Schrader, Sedimentation influenced by upwelling in the subtropical Baieda Levrier (West Africa), in: *Upwelling Ecosystems*, R. Boje and M. Tomczak (eds.), Springer, Berlin, pp. 282-288, 1978.

Manheim, F., G. Rowe, and D. Jipa, Marine phosphorite formation off Peru, *Journal of Sedimentary Petrology, 45*, 243-251, 1975.

Mare, M.F., A study of a marine benthic community with special references to the micro-organisms, *Journal of the Marine Biological Association of the United Kingdom, 25*, 517-554, 1942.

Milliman, J.D., Effects of arid climate and upwelling upon the sedimentary regime off southern Spanish Sahara, *Deep-Sea Research, 24*, 95-103, 1977.

Müller, P.J., and E. Suess, Productivity, sedimentation rate, and sedimentary organic matter in the oceans - I. Organic carbon preservation, *Deep-Sea Research, 26A*, 1347-1362, 1979.

Nichols, J.A., and G.T. Rowe, Infaunal macrobenthos off Cap Blanc, Spanish Sahara, *Journal of Marine Research, 35*, 525-536, 1977.

Nixon, S.W., C. Oviatt, and S.S. Hale, Nitrogen regeneration and the metabolism of coastal marine bottom communities, in: *The Role of Terrestrial and Aquatic Organisms in Decomposition Processes*, Blackwell, Oxford, p. 269-283, 1976.

Oviatt, C.A., and S.W. Nixon, Sediment resuspension and decomposition in Narragansett Bay, *Estuarine and Coastal Marine Science, 3*, 201-217, 1975.

Pamatmat, M.M., Oxygen consumption by the sea bed, IV. Shipboard and laboratory experiments, *Limnology and Oceanography, 16*, 536-550, 1971.

Pamatmat, M., Benthic community metabolism: a review and assessment of present status and outlook, in: *The Ecology of Marine Benthos*, University of South Carolina Press, Columbia, p. 89-111, 1977.

Premuzic, E., Organic carbon and nitrogen on the top sediments of world oceans and seas: distribution and relationship to bottom topography, Brookhaven National Laboratory Report, BNL 51804, 100 p., 1980.

Rowe, G.T., Benthic biomass in the Pisco, Peru upwelling, *Investigacion Pesqueres, 35*, 127-135, 1971a.

Rowe, G.T., Benthic biomass and surface productivity, in: *The Fertility of the Sea, Vol. II*, pp. 441-454, Gordon and Breach, New York, 1971b.

Rowe, G.T., The effects of pollution on the dynamics of the benthos of New York Bight, 6th European Symposium on Marine Biology, Rovinj, Yugoslavia, *Thalassia Yugoslavica, 7*, 353-359, 1971c.

Rowe, G.T., Sediment data from short cores during JOINT II off Peru, Coastal Upwelling Ecosystem

Analysis Data Report 65, 57 pp., 1979.
Rowe, G.T., Benthos production and processes off Baja California, northwest Africa and Peru, in: *Bioproductivity of Upwelling Ecosystems*, R. Barber and M. Vinogradov (eds.), in press.
Rowe, G.T., and C.H. Clifford, Sediment data from short cores taken in the northwest Atlantic Ocean, Woods Hole Oceanographic Institution Technical Report WHOI 78-46, 57 pp, 1978.
Rowe, G.T., and W. Gardner, Sedimentation rates in the slope water of the northwest Atlantic Ocean measured directly with sediment traps, *Journal of Marine Research, 37*, 581-600, 1979.
Rowe, G.T., C.H. Clifford, and K.L. Smith, Jr., Nutrient regeneration in sediments off Cap Blanc, Spanish Sahara, *Deep-Sea Research, 24*, 57-63, 1977.
Rowe, G.T., C. Clifford, K. Smith, and P.L. Hamilton, Benthic nutrient regeneration and its coupling to primary productivity in coastal waters, *Nature, 255*, 217-219, 1975.
Rowe, G.T., A.A. Westhagen, C.H. Clifford, J. Nichols-Driscoll, and J. Milliman, JOINT I benthos and sediment station data, R.V. *Atlantis II*, Cruise 82, Coastal Upwelling Ecosystems Analysis Data Report 31, 25 pp., 1976.
Ryther, J.H., Geographic variations in productivity, in: *The Sea*, M.N. Hill (ed.), Interscience, New York, pp. 347-380, 1963.
Sanders, H.L., Benthic marine diversity and the stability-time hypothesis, in: *Diversity and Stability in Ecological Systems*, Brookhaven Symposia in Biology, 11, pp. 71-81, 1969.
Schopf, T.J.M., *Paleooceanography*, Harvard University Press, Cambridge, 341 p., 1980.
Smith, K.L., Jr., G. Rowe, and C. Clifford, Sediment oxygen demand in an outwelling and upwelling area, *Tethys, 6*, 223-230, 1974.
Smith, K.L., Jr., G. Rowe, and J. Nichols, Benthic community respiration near Woods Hole sewage outfall, *Estuarine and Coastal Marine Science, 1*, 65-70, 1973.
Smith, K.L., Jr., G.R. Harbison, G.T. Rowe, and C.H. Clifford, Respiration and chemical composition of *Pleuroncodes planipes* (Decapoda: Galatheidae): Energetic significance in an upwelling system, *Journal of the Fisheries Research Board of Canada, 32*, 1607-1612, 1975.
Sørensen, J., B. Jørgensen, and N.P. Reusbech, A comparison of oxygen, nitrate, and sulfate respiration in coastal marine sediments, *Microbial Ecology, 5*, 105-115, 1979.
Staresinic, N., The vertical flux of particulate organic matter on the Peru coastal upwelling as measured with a free-drifting sediment trap, Ph.D. Thesis, Woods Hole Oceanographic Institution and Massachusetts Institute of Technology, 255 pp., 1978.
Steimle, F., and R. Stone, Abundance and distribution of offshore benthic fauna off southwestern Long Island, New York, NOAA Technical Report, National Oceanic and Atmospheric Administration NMFS SSRF-673, Seattle, Washington, 50 pp., 1973.
Thiel, H., Benthos in upwelling regions, in: *Upwelling Ecosystems*, R. Boje and M. Tomczak (eds.), Springer, Berlin, pp. 124-138, 1978.
Thomas, J.P., W.C. Phoel, F. Steimle, J. O'Reilly, and C. Evans, Seabed oxygen consumption - New York Bight apex, in: *Middle Atlantic Continental Shelf and New York Bight*, G. Gross (ed.), pp. 354-369, Special Symposia, 2, American Society of Limnology and Oceanography, 441 pp., 1976.
Visher, G.S., A simplified method for determining relative energies of deposition of stratigraphic units, *Journal of Sedimentary Petrology, 31*, 291-293, 1961.
Watson, S., Role of bacteria in an upwelling ecosystem, in: *Upwelling Ecosystems*, R. Boje and M. Tomczak (eds.), Springer, Berlin, pp. 139-154, 1978.
Wigley, R.L., and R.B. Theroux, Macrobenthic invertebrate fauna of the middle Atlantic Bight region: Part II. Faunal composition and quantitative distribution, U.S. Geological Survey Professional Paper, in press.
Zenkevitch, L.A., *Biology of the Seas of the USSR*, Wiley, New York, 955 pp., 1963.

PHYSICAL AND BIOLOGICAL STRUCTURE AND VARIABILITY IN AN UPWELLING CENTER OFF PERU NEAR 15°S DURING MARCH, 1977

K.H. Brink

School of Oceanography, Oregon State University, Corvallis, Oregon 97331

B.H. Jones *

Bigelow Laboratory for Ocean Sciences, West Boothbay Harbor, Maine 04575

J.C. Van Leer

Division of Physical Oceanography, Rosenstiel School of Marine and Atmospheric Sciences, University of Miami, Miami, Florida 33149

C.N.K. Mooers

Department of Oceanography, Naval Postgraduate School, Monterey, California 93940

D.W. Stuart

Department of Meteorology, Florida State University, Tallahassee, Florida 32306

M.R. Stevenson

Instituto de Pesquieras Espaciais./NEP
Caixa Postal 515, São José dos Campos - S.P. 12,200, BRAZIL

R.C. Dugdale

Institute of Marine and Coastal Studies, University of Southern California, University Park, Los Angeles, California 90007

G.W. Heburn

Department of Meteorology, Florida State University, Tallahassee, Florida 32306

* Present Address: Department of Biological Sciences, Allan Hancock Foundation, University of Southern California, Los Angeles, California 90007

Abstract. The upwelling region near 15°S off Peru was studied intensively during the period 12 to 30 March 1977. A surface cold area (apparent "plume") is a persistent feature of the region, and it has an associated biological signature. Horizontally, the cold water generally extended equatorward and offshore from the coast; vertically, the alongshore structure was 40% as strong at 40 m depth as at the surface, and imperceptible at 64 m. The size of the surface cold area increased as the local alongshore wind stress increased. Physical variability was also dependent

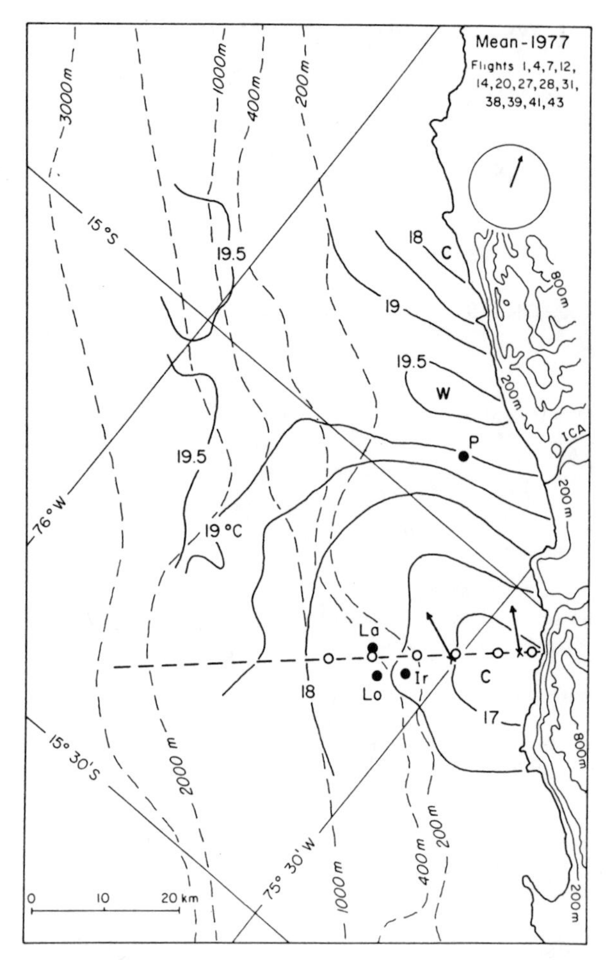

Fig. 1. Mean sea-surface temperature from 13 ART flights between March 21 and May 6, 1977. Plate III can be used to place this figure in a larger scale context. The heavy dashed line represents the C line, and the open circles denote the standard C stations: C-1 (closest to shore), out to C-6. The arrows represent the mean 4.5-m velocity vectors at PSS-Agave (closer to shore) and at PS-Mila. The length of the arrow is the distance that the flow would traverse in 12 h. The circle in the upper right hand corner contains the mean wind stress vector, and the radius of the circle is 1 dyne/cm^2. Symbols representing other current meter moorings: P = Parodia, La = Lagarta, Lo = Lobivia, Ir = Ironwood.

upon the passage of coastally trapped waves. Biological variability and structure were closely related to the physical regime. Physical and biological aspects of wind-driven upwelling events are described, and the biological dynamics of the upwelling center are discussed.

Introduction

The distribution of sea-surface temperature along the coast of Peru is far from uniform. As a region of persistent upwelling-favorable winds, it would be expected to have a band of cold, nutrient-rich water nearshore with progressively warmer, nutrient-depleted waters farther offshore. While this general pattern does obtain, there is comparably important alongshore variation in sea-surface temperature, with alternating warm and cold patches (Zuta, Rivera, and Bustamante, 1978). In particular, near 15°S, there is a cold patch previously described by Ryther *et al.* (1966), by Eber, Saur, and Sette (1968), and by Smith, Enfield, Hopkins, and Pillsbury (1971). The hydrographic and airborne radiometric measurements obtained throughout the CUEA (Coastal Upwelling Ecosystems Analysis) JOINT-II experiment of 1977 also confirm the existence of a surface cold spot, or horizontal plume, which generally had its origin at the same location on the coast (Fig. 1). The feature appeared to strengthen, deform, and decay as the local wind stress fluctuated in time. The coldest water of the patch, 16 to 18°C, was usually at the coast near 15° 05.5'S, 75°24'W, and spread about 50 km seaward and equatorward. Higher surface nutrient concentrations and biological productivity with low phytoplankton biomass are associated with the cold area (Walsh, Kelley, Dugdale, and Frost, 1971).

The upwelling center, representing both physical and biological phenomena, is perhaps best investigated by an interdisciplinary approach. Even for an approximate description of the physical regime, a combination of measurements is needed. The biological problem requires even greater breadth because nutrients, phytoplankton, and higher trophic levels interact in a highly variable, spatially complex physical environment. In the following, we shall attempt to describe the upwelling center through an integration of physical and biological observations. Ambiguities in description will arise because of the considerable temporal and spatial variability inherent to the system.

The Upwelling Center Study

The present study attempts a synthesis of several kinds of observations with the goal of jointly describing the physical, chemical, and biological regimes. The data were taken over a roughly 40 by 60-km region off the Peru coast near 15°S (Fig. 1), and the period of the study is an 18-day subset (12-30 March, 1977) of the 14-month JOINT-II program. This particular period was chosen for closer study because of the unusually intensive data acquisition at the time. Four ships were in the area, two conducting physical (drogue and synoptic-scale hydrographic) observations, and two primarily conducting biological, hydrographic, and chemical observations. Methods used in the drogue studies are described by Stevenson, Garvine, and Wyatt (1974) and by Stevenson (1974), while the synoptic-scale hydrography is described by Johnson, Koeb, and Mooers (1979). Details of the biological and large-

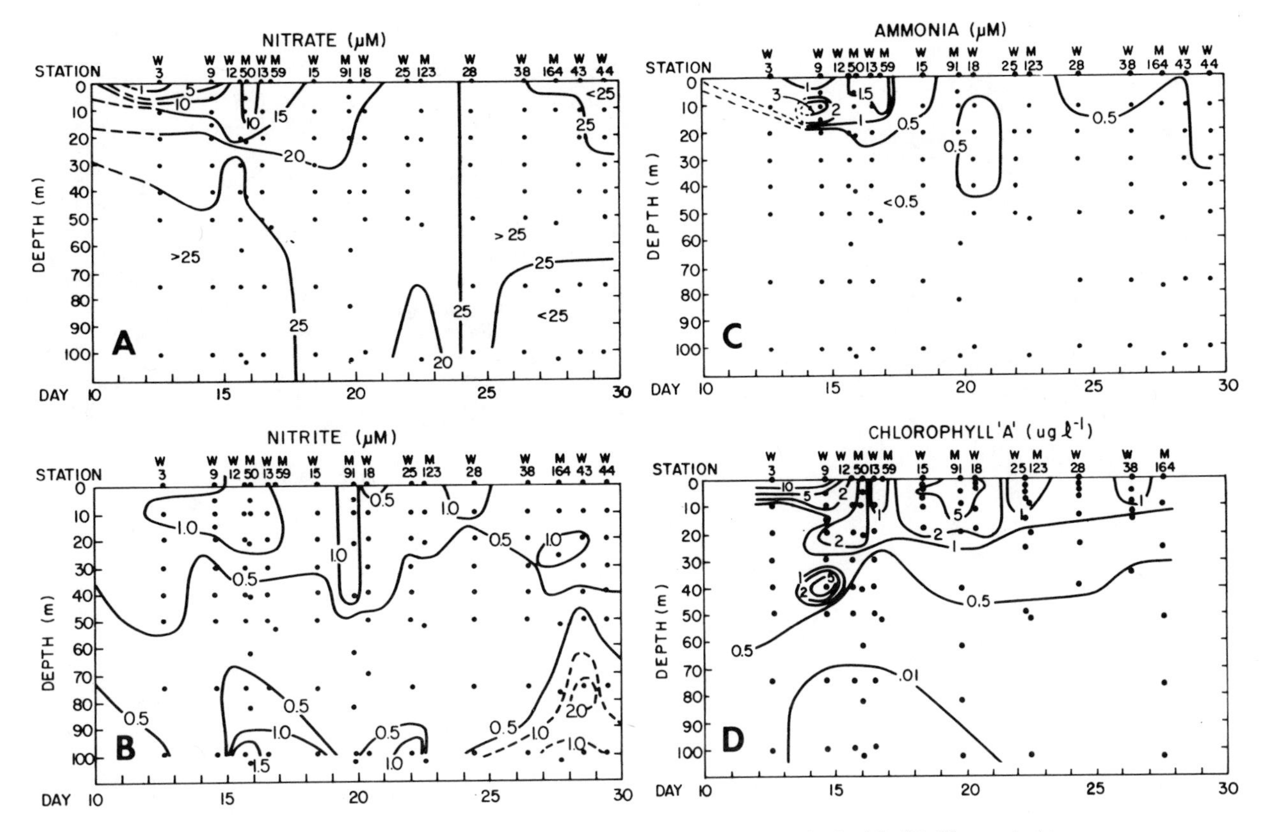

Fig. 2. Contour plots of chemical variables at the mid-shelf (C-3) position.

scale hydrographic studies can be found in Kogelschatz, MacIsaac, and Breitner (1980) and in Hafferty, Codispoti, and Huyer (1978). Also during the period, four airborne radiometric surface temperature (ART) charts were made at about noon local time (see Moody, 1979). In addition, fixed-point observations were made, including an array of 40 current meters, four Cyclosonde profiling current meters, and three meteorological buoys (Brink, Halpern, and Smith, 1980; Van Leer and Ross, 1979). Land-based observations included wind, solar radiation, and sea level (Moody, 1979; Brink *et al.*, 1980).

In the following, all wind and current time series are resolved into cross-shelf (positive for shoreward flow) and alongshore (positive for equatorward flow) components. The rotation angle of the coordinate system was 45° counterclockwise from true north, consistent with the general trend of the coastline. Over the shelf, this choice of coordinate system agrees to $\pm$ 2.5^{o} with the depth-averaged principal axis system used at each mooring by Brink *et al.* (1980). Throughout the following, only the wind stress record from mooring PSS, about 4 km from shore on the central mooring line, will be used. To synchronize with the times of aircraft flights, most time series data discussed are treated with a 12-h running mean, centered on 0500 and 1700 GMT. Following J.J. O'Brien (personal communication), the vertical velocity was estimated from current meter data at a position at 20 m depth roughly midway between the PSS current meter (Fig. 1) and the coast. The position was chosen because it is the estimated site of the most intense upwelling.

Most of the hydrographic data were observed along the C line, a line normal to the coast and passing through the area of low mean surface temperature. Stations along the line are denoted by C-1, C-2, etc., with the number increasing with distance from the coast. A time series of biological and chemical variables was sampled daily (Fig. 2) at the mid-shelf (C-3) position, near the most heavily instrumented current meter mooring.

The Mean Upwelling Center

Physical Observations

Throughout the March to May 1977 period, and in particular during our study period, the local alongshore winds were equatorward, hence favorable for upwelling (Fig. 3). Means of other physical variables over the 18-day period are also representative of longer term measurements (Brink *et al.*, 1980). The wind stress forced a pattern in cross-shelf velocity that was qualitatively con-

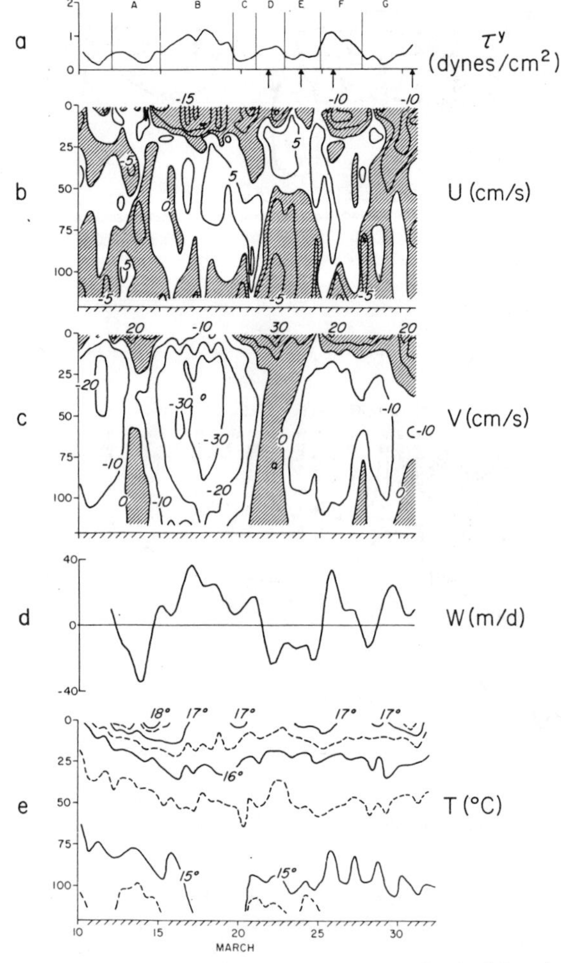

Fig. 3. Physical measurements at mid-shelf: a) alongshore wind stress at PSS, τ^y (dyne/cm^2); b) cross-shelf velocity (cm/s). Negative (shaded) values correspond to offshore flow; c) alongshore velocity (cm/s). Positive (shaded) values correspond to equatorward flow; d) estimated vertical velocity (m/d) at 20-m depth, 2 km from shore (J.J. O'Brien, personal communication). Positive values correspond to upward motion. e) Temperature (^{o}C). Panels b), 3), and e) were drawn from observations at depths 2.1, 4.6, 8.1, 12, 16, 20, 24, 39, 59, 80, 100, and 115 m.

sistent with Ekman upwelling dynamics at the mid-shelf C-line position (Fig. 4). Brink *et al.* (1980) demonstrated that the mean onshore transport is in approximate agreement with two-dimensional Ekman theory, but the offshore transport is about one-third that predicted by theory. The result implies a strong alongshore variability in the near-surface flow pattern associated with the upwelling circulation. The mean noon-time ART chart (Fig. 1) shows that there is considerable alongshore thermal structure, consistent with the presence of alongshore structure in the velocity field.

The mean (over 14 nightly CTD transects) temperature along the C-line section (Fig. 5a) has a relatively strong surface gradient (baroclinic zone) 10 to 25 km off the coast. The gradient marks the outer limits of the cool surface area evident in the ART charts. The companion salinity section (Fig. 5b) has a nearshore minimum apparently due to alongshore advection of the Ica River plume, as confirmed by visual observations of color from an aircraft. The low salinity water reduces nearshore density to the point where the mean σ_t = 25.6 isopycnal outcrops (Fig. 5c). The density structure below the upper 20 m is consistent, through a thermal wind bal-

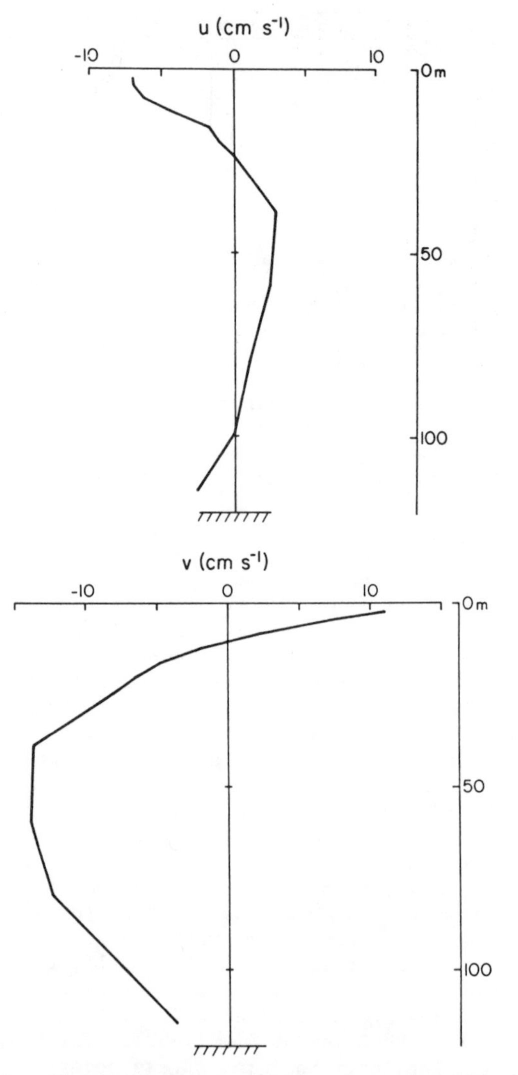

Fig. 4. Mean velocity components at the mid-shelf (PS-Mila) position, March 12 to 30, 1977. The cross-shelf component, u, is positive for onshore flow, and the alongshore component, v, is positive for equatorward flow.

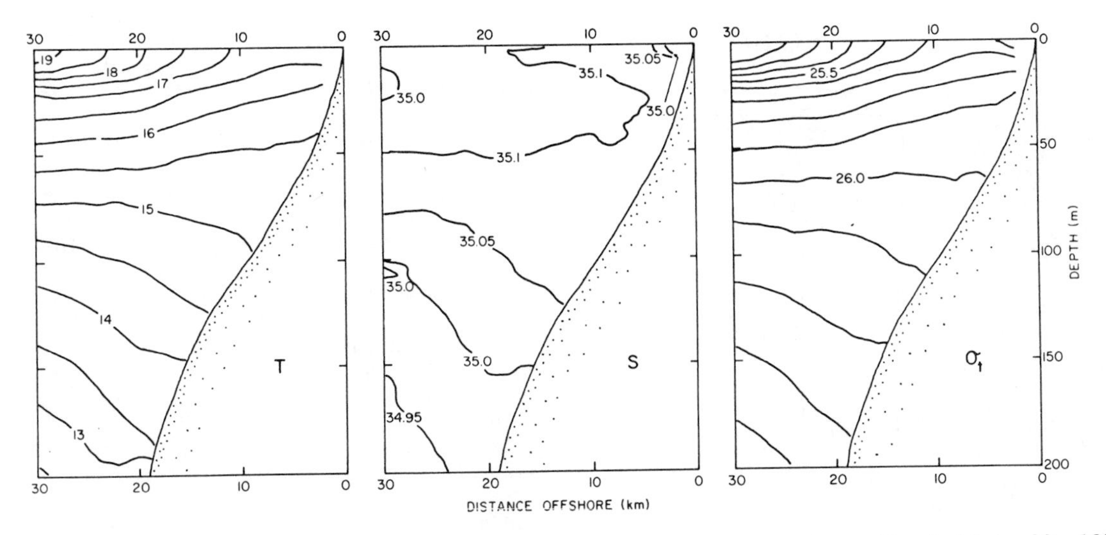

Fig. 5. Mean CTD hydrography along the C line from 14 nightly sections, March 16 to 30, 1977.

ance, with the observed mean alongshore currents, i.e., equatorward near-surface currents overlying a poleward undercurrent, which had its maximum mean speed at about mid-depth over the entire shelf.

Current measurements at fixed locations, cross-shelf hydrographic transects, and sea-surface temperature charts, however, give only a partial picture of the physical nature of the upwelling center. How deep did the structure observed at the surface penetrate? Because the mid-shelf mean onshore transport (below roughly 25 m) fits a two-dimensional model, the surface structure probably did not extend significantly below 25 m. Data derived from mid-shelf moored instrumentation (i.e., at Parodia and C-3) show that the mean alongshore temperature gradient, a measure of the structure's intensity, decreased considerably with depth; the mean gradients were 2.2×10^{-2} °C/km at 3 m, 0.8×10^{-2} °C/km at 24 m, and indistinguishable from zero at 64 m. The figures suggest that the alongshore structure disappeared by about 50-m depth on average, a notion supported by a CTD survey during our study period (Fig. 6).

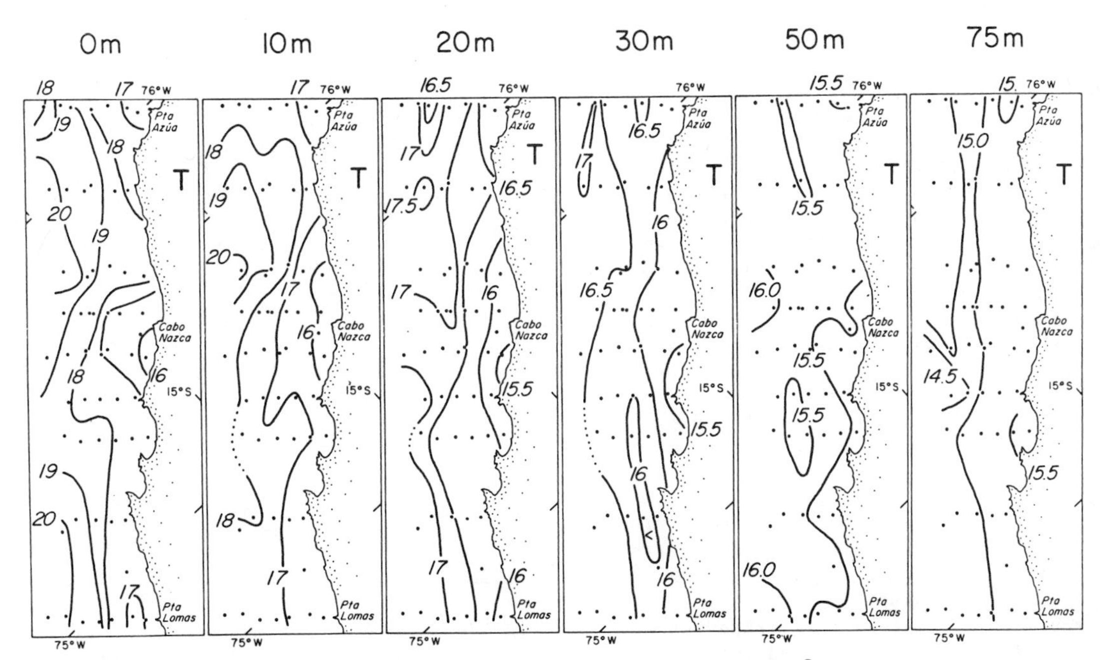

Fig. 6a. Ship-based mapping results, March 20-23, 1977. Temperature (°C) at depths 0, 10, 20, 30, 50, and 75 m (from Huyer and Gilbert, 1979).

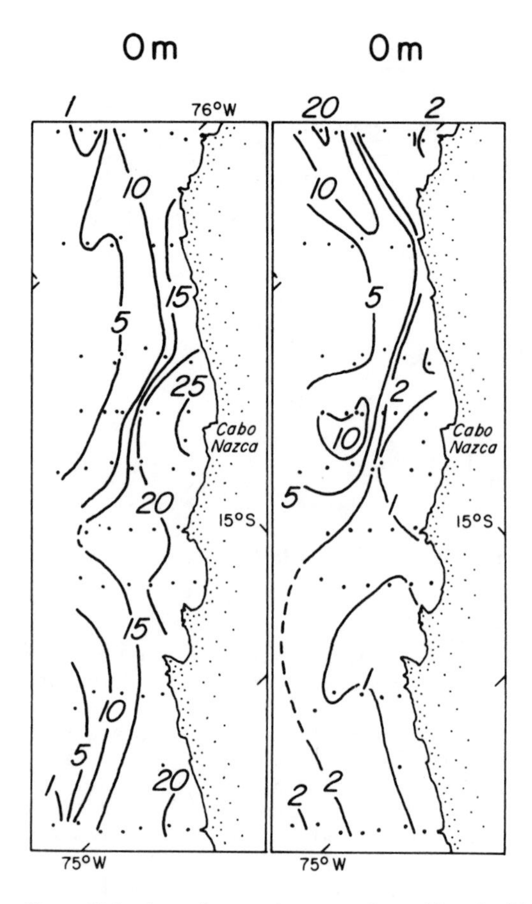

Fig. 6b. Ship-based mapping results, March 20-23, 1977. Surface nitrate (μM) and chlorophyll *a* (μg/1).

The overall structure and dynamics of the near-surface flow field are also poorly resolved. The mean wind is not uniform in space (Brink, Smith, and Halpern, 1978); it was strongest at position PSS (near shore on the C line), and weakened both farther offshore (C-3) and with alongshore distance from the C line. Unlike the winds, however, the mean 4.5-m velocities were stronger at C-3 and directed more nearly offshore than at PSS. Near-surface drogue data (Figs 7, 8, and 9) suggest a tendency for cross-isotherm (i.e., radially outward) flow, although the data are too few to be conclusive. To the extent that frontal dynamics (e.g., Mooers, Collins, and Smith, 1976) applied, a geostrophic component of flow was superimposed on the directly wind driven component. Order-of-magnitude estimates (Brink *et al.*, 1980) support this hypothesis for the mean flow. Frontal dynamics also imply the presence of a further small-scale frictional component in the flow. The mean surface temperature map (Fig. 1) shows no sharp fronts, as they tend to average out due to their variability in position and intensity. Individual maps, however, do occasionally show sharp frontal features (Figs 7 and 8), which also are clearly evident in some cross-shelf profiles of aircraft sea-surface temperature (e.g., Stuart and Bates, 1977).

Chemical and Biological Observations

Average C-line sections of chemical variables were computed using 14 transects occupied between March 5 and May 16, 1977 (Fig. 10). The longer time period was chosen to obtain a more stable mean than would have been obtained from the four chemical sections occupied during the interval discussed in this study. We feel that the shorter period is sufficiently representative that the use of the longer-term mean is reasonable (compare Figs 10, 11, 12, 13, 14).

The nitrite maximum-nitrate minimum was a persistent feature offshore, centered at about 200 m (Fig. 10a and 10b; see also Hafferty, Lowman, and Codispoti, 1979). The contour lines, 20 μM for

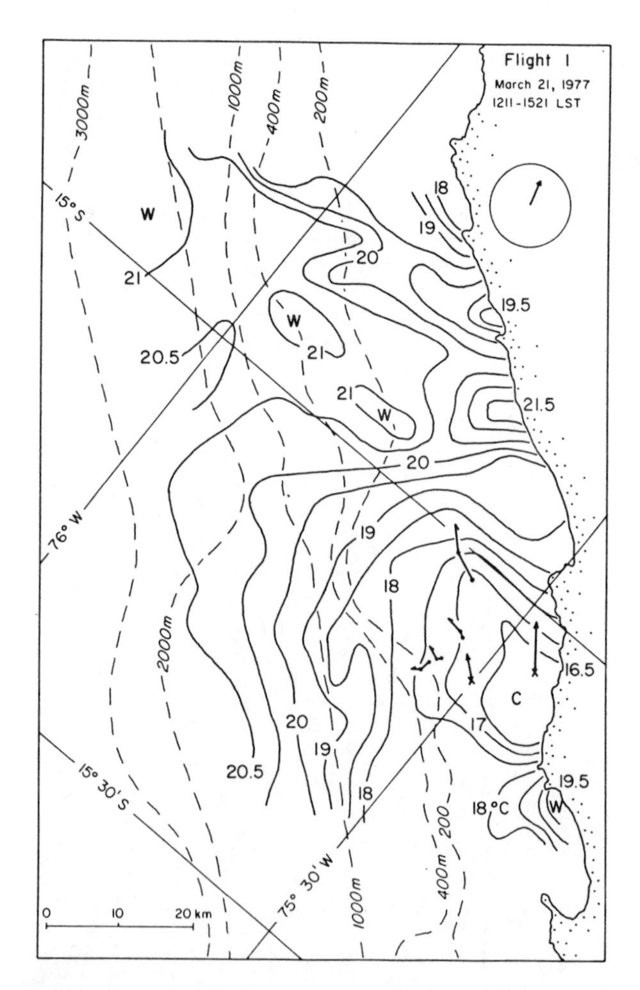

Fig. 7a. Sea-surface temperature (°C) from an aircraft flight near noon local time on March 21, 1977. Conventions for wind stress and velocity are given with Fig. 1.

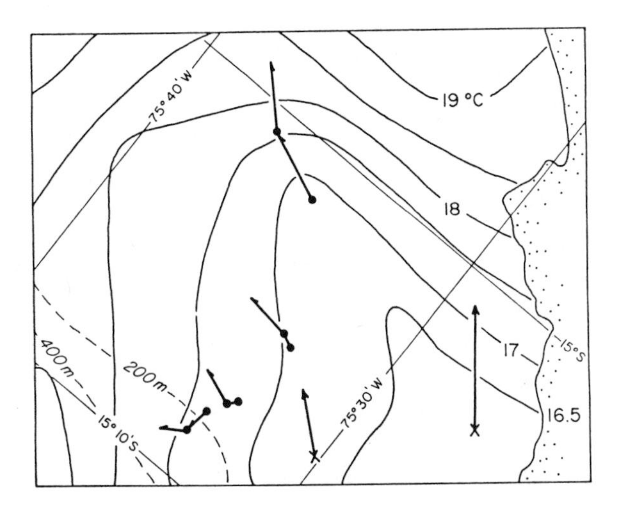

Fig. 7b. The displacement of 4-m drogues over 12 h centered on the aircraft flight near noon local time on March 21, 1977.

NO_3 and 1 μM for NO_2, both extend up over the shelf near the bottom indicating that some of the nitrite-rich — nitrate-poor water was frequently present on the shelf. A subsurface core of high nitrate (>24 μM) water was present at about 25 m near the coast and increased in depth to about 50 m at 30 km offshore (C-6). The mean surface nitrate concentration was essentially constant between 1 and 7 km offshore, this being the apparent region where most of the upwelling was occurring (Brink *et al.*, 1980). The surface nitrate concentration then decreased in the offshore direction but remained higher than 5 μM out to 50 km.

Silica (Fig. 10c), unlike nitrate, increased monotonically with depth. Near-bottom silicate concentrations of greater than 30 μM extended over the shelf to the most inshore station. Near-surface silica concentrations declined monotonically in the offshore direction. Nitrate and silica were about the same concentration nearshore, and silica was only slightly less concentrated than nitrate 30 km offshore.

Chlorophyll *a* (Fig. 10d) was low nearshore and increased to a maximum in the region of strong mean cross-shelf temperature gradient, about 18 km offshore. For a region known for its high primary productivity and the resultant high phytoplankton biomass, the values (∿3 μg/l) appear to be anomalously low. A year earlier in this area, concentrations had been 25 to 30 μg/l and even higher when dense *Gymnodinium splendens* blooms were present (Dugdale *et al.*, 1977; Jones, 1977). Also, during *Anton Bruun* cruise 15 in 1966 (Ryther *et al.*, 1966) and during the PISCO cruise in 1969 (Walsh *et al.*, 1971), near-surface chlorophyll *a* concentrations coincident with the 17.5°C isotherm were approximately 10 μg/l. Finally, Barber (1981) showed that, for reasons that are not understood, the area-averaged chlorophyll *a* levels in this vicinity were lower in 1977 than in 1966, 1976, or 1978.

The Variability

To explore the coupled physical and biological variability, events are discussed chronologically through the period of detailed observations. For convenience, the 18-day period is broken up into sub-sections defined primarily by the local alongshore wind stress, hence the strength of local coastal upwelling. The sub-sections are designated alphabetically (Fig. 3). A more general discussion with conclusions on time variability will follow the chronological description.

The major forms of physical variability can readily be isolated from the data. One is wind driven, which includes mixed-layer deepening and offshore Ekman transport, the clearest signatures of which are in the distribution of temperature and of onshore velocity (Brink *et al.*, 1980). The

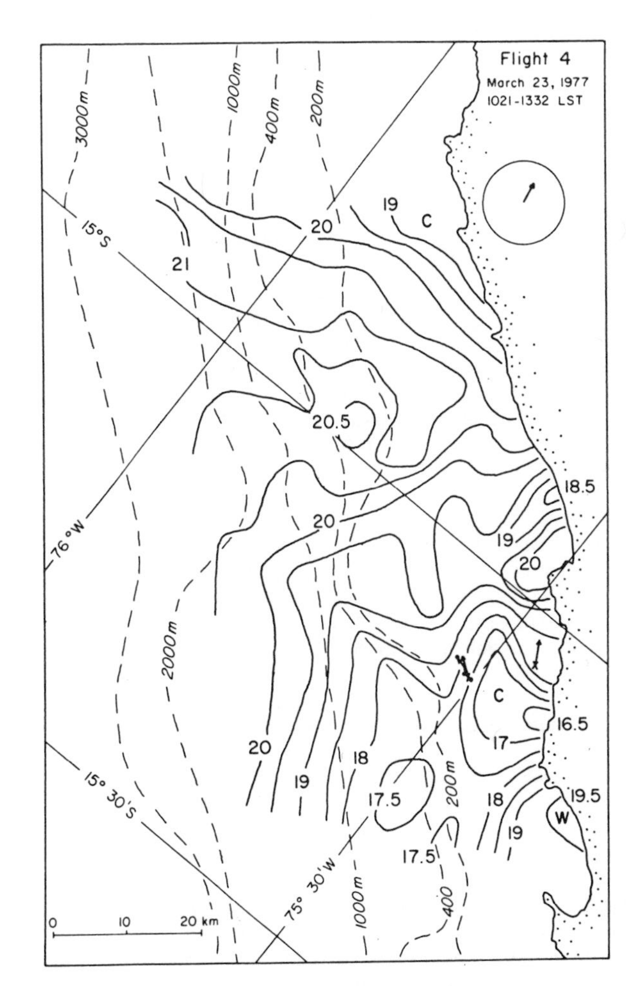

Fig. 8a. Sea-surface temperature (°C) from an aircraft flight near noon local time on March 23, 1977. Conventions are the same as in Fig. 7. Drogues were set to 4-m depth.

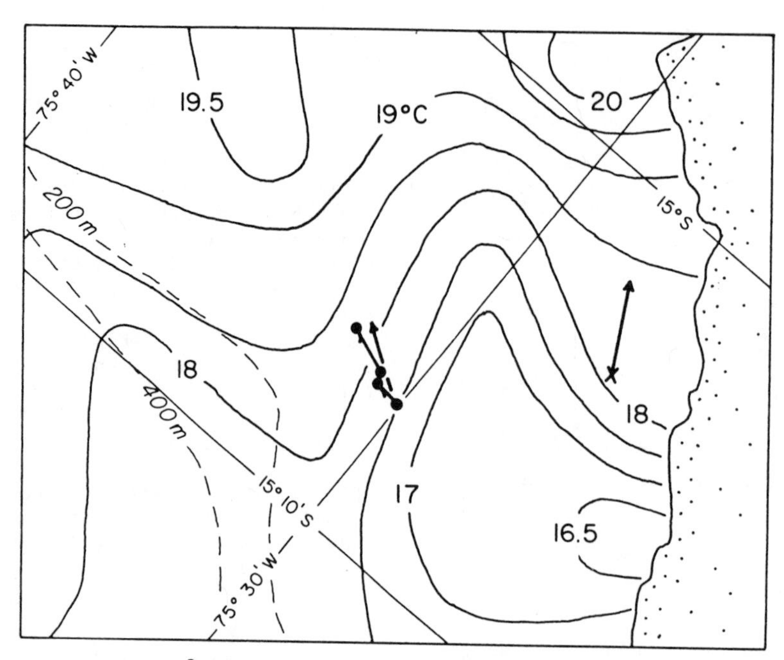

Fig. 8b. Sea-surface temperature (°C) from an aircraft flight near noon local time on March 23, 1977. Conventions are the same as in Fig. 7. Drogues were set to 4-m depth.

second major form of physical variability is related to free coastally trapped waves, which dominate the alongshore velocity fluctuations. The origin of the waves is not fully known, but they are baroclinic in nature (i.e., velocity fluctuations are associated with density fluctuations) and they propagate poleward at about 200 km/d (Smith, 1978; Brink *et al.*, 1980). In the following, the forms of physical variability will be described in conjunction with chemical and biological variability.

Period A: Weak Winds

During a period (March 12 to 14) of weak, though equatorward, winds, the estimated nearshore vertical velocities ranged from weakly upward to strongly downward (Fig. 3) and the surface mixed layer at mid-shelf was thin or nonexistent (Fig. 15). The onshore flow over the shelf was relatively disorganized. The mid-shelf near-surface water can be classified as "old" in the sense that phytoplankton biomass was relatively high (Fig. 2d) and nitrate (Fig. 2a) and other primary nutrients were depleted, presumably due to consumption by the phytoplankton. Near-surface nitrite (Fig. 2b) and ammonia (Fig. 2c) were high and presumably byproducts of phytoplankton consumption (Estrada and Wagensberg, 1977) and of zooplankton grazing (Smith and Whitledge, 1977). On March 15, at the end of the quiet phase, a hydrographic section showed a distinct maximum in chlorophyll *a* at the surface near the shelf break, about 20 km from shore (Fig. 11). Near-surface nitrate concentrations decreased gradually from relatively high values near shore to very low values more than 23 km offshore. An upwelling response was beginning, as evidenced by offshore near-surface flow, onshore currents below, and upwarped isotherms in the upper 50 m over the shelf.

Period B: Strong Upwelling

From 15 to 19 March, a strong wind-driven upwelling event was superimposed on a very intense coastally trapped wave. The onshore-offshore current structure (Figs 3 and 12) clearly showed the effects of the upwelling event: strong offshore flow in the upper 20 m, onshore flow greater than 5 cm/s over most of the remainder of the water column, and nearshore vertical velocities estimated to be as large as 36 m/day. The mid-shelf surface mixed layer had a deepening trend during this time (Fig. 15). The warm, nitrate-poor near-surface water had been advected farther offshore and replaced by waters richer in nitrate (Fig. 12). The mid-shelf waters (Fig. 2) were lower in both primary and secondary products of phytoplankton productivity: chlorophyll *a*, nitrite, and ammonia. The near-surface shelf-break chlorophyll *a* maximum remained at the same position as on March 15 (Fig. 12) but had grown considerably stronger.

Coincident with the wind event was the passage of a strong coastally trapped wave, which had passed Callao, near 12°S, two days earlier, producing a period of strong (velocities as large as 40 cm/s) poleward current (Brink *et al.*, 1978). The wave caused poleward flow throughout the mid-

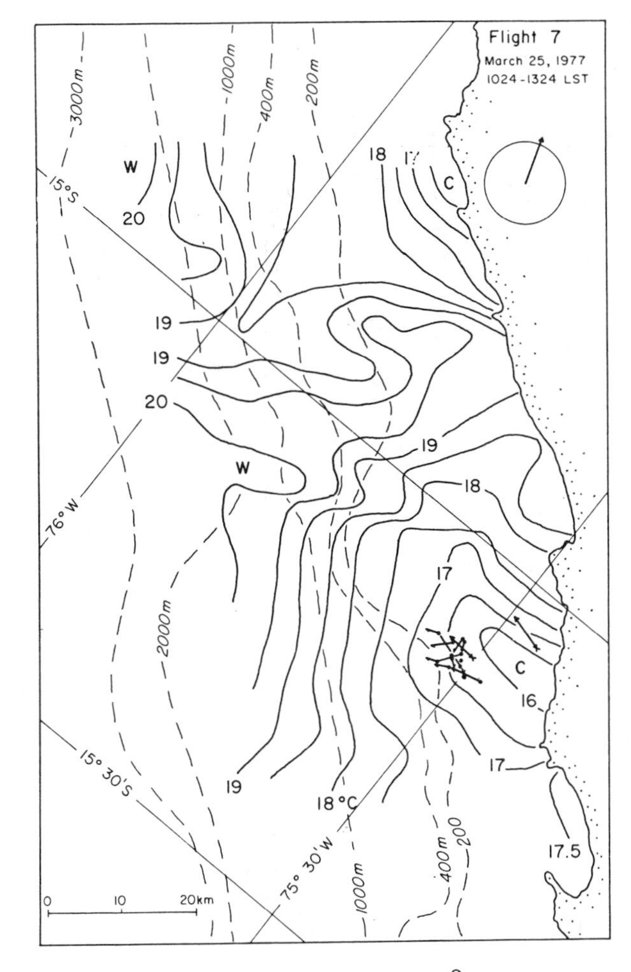

Fig. 9a. Sea-surface temperature (^{o}C) from an aircraft flight near noon local time on March 25, 1977. Conventions are as in Fig. 7. Drogues were set to 4-m depth.

shelf (C-3) water column (Fig. 3), despite the strong (greater than 1 dyne/cm^2) equatorward wind stress. Associated with the wave, through the thermal wind balance, isotherms in the depth range of 80 to 250 m tilted downward toward the coast (Fig. 12). For example, the 15° isotherm was displaced downward and seaward of C-3 for three days (Fig. 3). During the wave passage, nitrite values were relatively high as much as 50 m above the bottom (Fig. 2). Normally, enhanced nitrite was found only relatively close (within about 10 m) to the shelf bottom (Fig. 10a), apparently as a result of regeneration in the sediments or in the near-bottom waters. During the wave passage, unusually high speeds resulting in strong near-bottom shears apparently caused a significant increase of the bottom boundary layer thickness (Fig. 16). Thickening of the bottom mixed layer presumably allowed near-bottom water to penetrate higher into the water column.

This particular wave passage seems to have had considerable delayed biological consequences. Under mean conditions, an upwelled phytoplankter would be initially advected equatorward and offshore. As the individual sinks at a rate of, say, 10 m/d (Smayda, 1970) it would, after about two days, settle into a subsurface region of onshore and poleward currents. Thus, there is a tendency for phytoplankton to return to the general region of origin. Allowing for grazing, turbulent dispersion, and the time and spatial variability of the flow, it appears that at least some small fraction will generally be returned to the upwelling center, providing a phytoplankton inoculum to the upwelled water. This idea, treated two-dimensionally by Smith and Barber (1980), is supported by the high water-column averaged chlorophyl (1 μg/l) at position C-1. If the alongshore flow for some period is in the same direction throughout the water column, the reseeding mechanism can break down, because the settled phytoplankton would be upwelled at some considerable distance alongshore from their origin. As the coastally trapped wave caused poleward flow over the entire shelf, the phytoplankton stock normally associated with the upwelling center was apparently advected away. Thus, during the next upwelling event, waters from roughly 40 to 120 km equatorward would be upwelled, and they could have a considerably lower phytoplankton content (provided they came from a less productive region).

Biological observations following a 4-m (a depth well within the near-surface layer defined by PS onshore velocity; Fig. 3) drogue began on March 18 at C-3. The drogue and those launched at about the same time (at other depths) moved

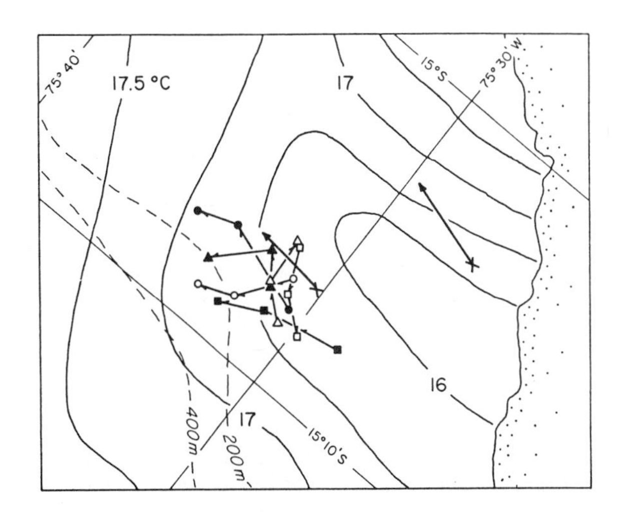

Fig. 9b. Sea-surface temperature (^{o}C) from an aircraft flight near noon local time on March 25, 1977. Conventions are as in Fig. 7. Drogues were set to 4-m depth.

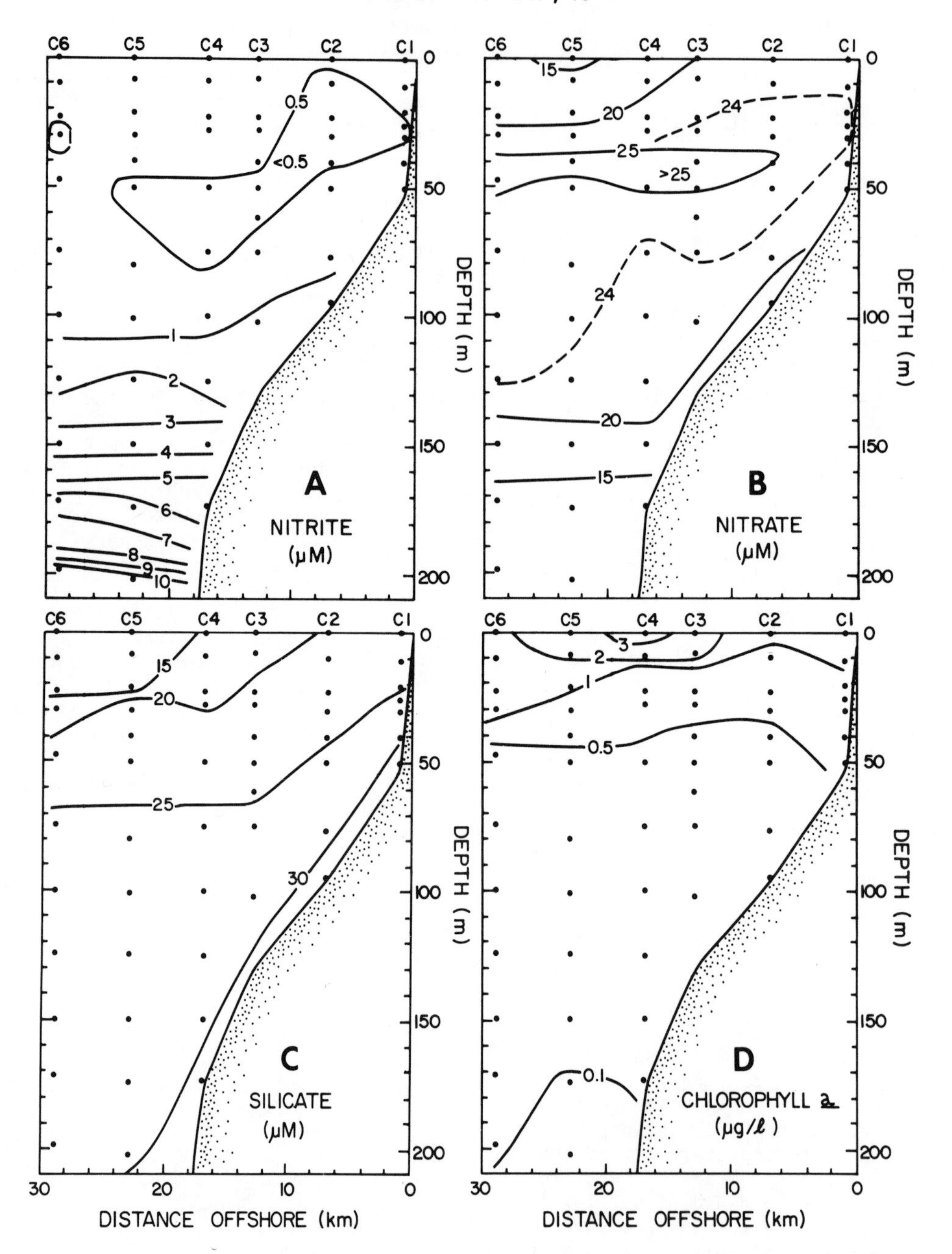

Fig. 10. Mean chemical C line, averaged over sections made between March 5 and May 6, 1977.

poleward initially (Fig. 17a), consistent with the dominance of the wave-associated poleward velocities throughout the mid-shelf water column (Fig. 3). On March 19, after the passage of the wave, the 4-m drogue moved generally offshore and equatorward, consistent with the return to dominance of wind-driven effects on alongshore currents at this depth. Near-surface biological measurements were made daily at the drogue's location and show a sequence of "aging" of freshly upwelled water (Fig. 17b): temperature, chlorophyll *a*, and nitrite increase monotonically, while nitrate, other nutrients, and nitrogen uptake rates decreased along the drogue's trajec-

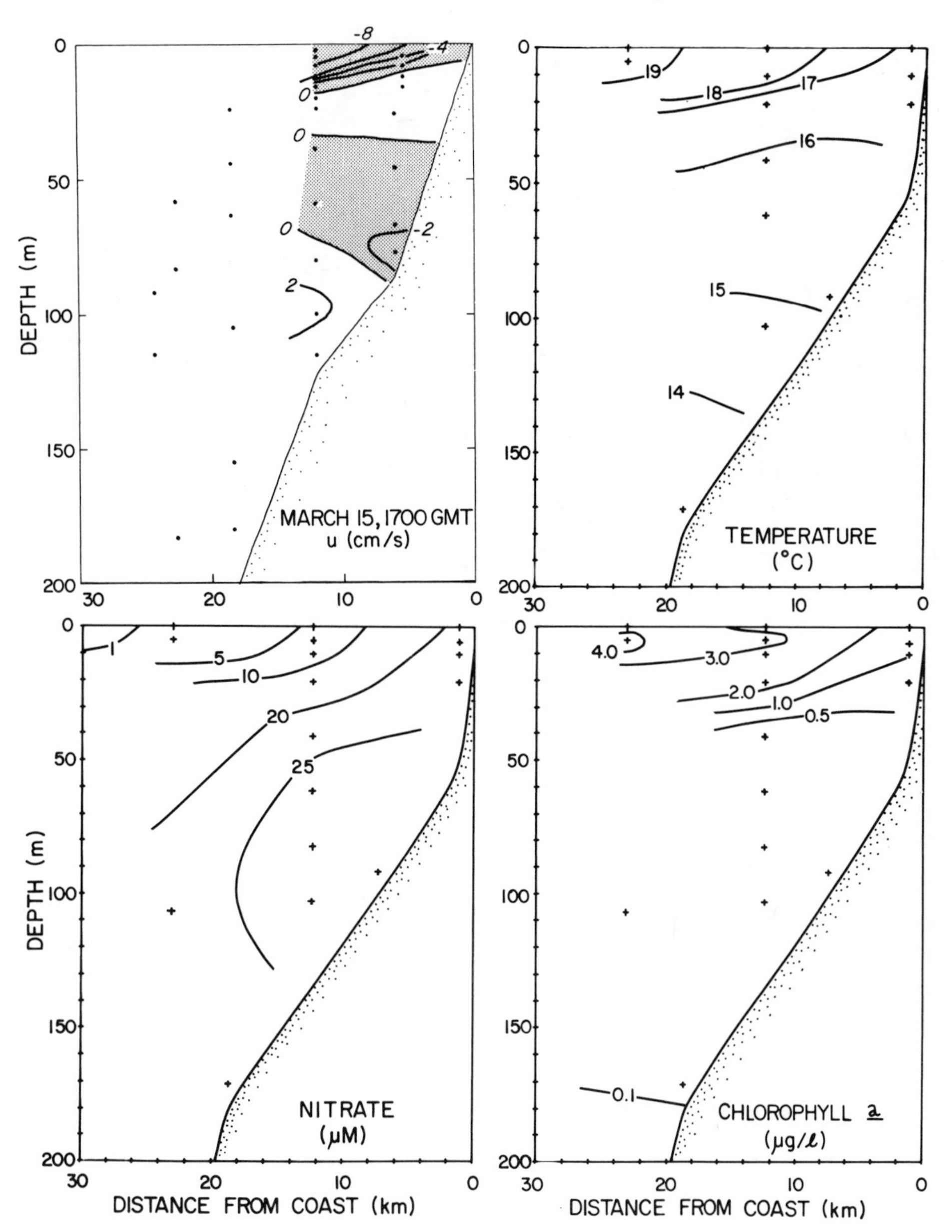

Fig. 11. Hydrographic section of March 15, 1977. Upper left, onshore velocity (cm/s). Negative values correspond to offshore flow.

tory. The events of March 15 to 19 suggest that although structures can change and new effects can be introduced by the coastally trapped wave's passage, upwelling processes continued despite the wave's presence, and immediate effects on the surface layer primary productivity were not apparent.

Period C: Weak Winds

During a brief period (March 19 and 20) of weak winds, the poleward alongshore flow associated with the coastally trapped wave decreased and the upwelling circulation remained detectable in the cross-shelf flow (Fig. 3). The mid-shelf

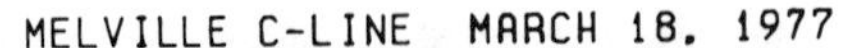

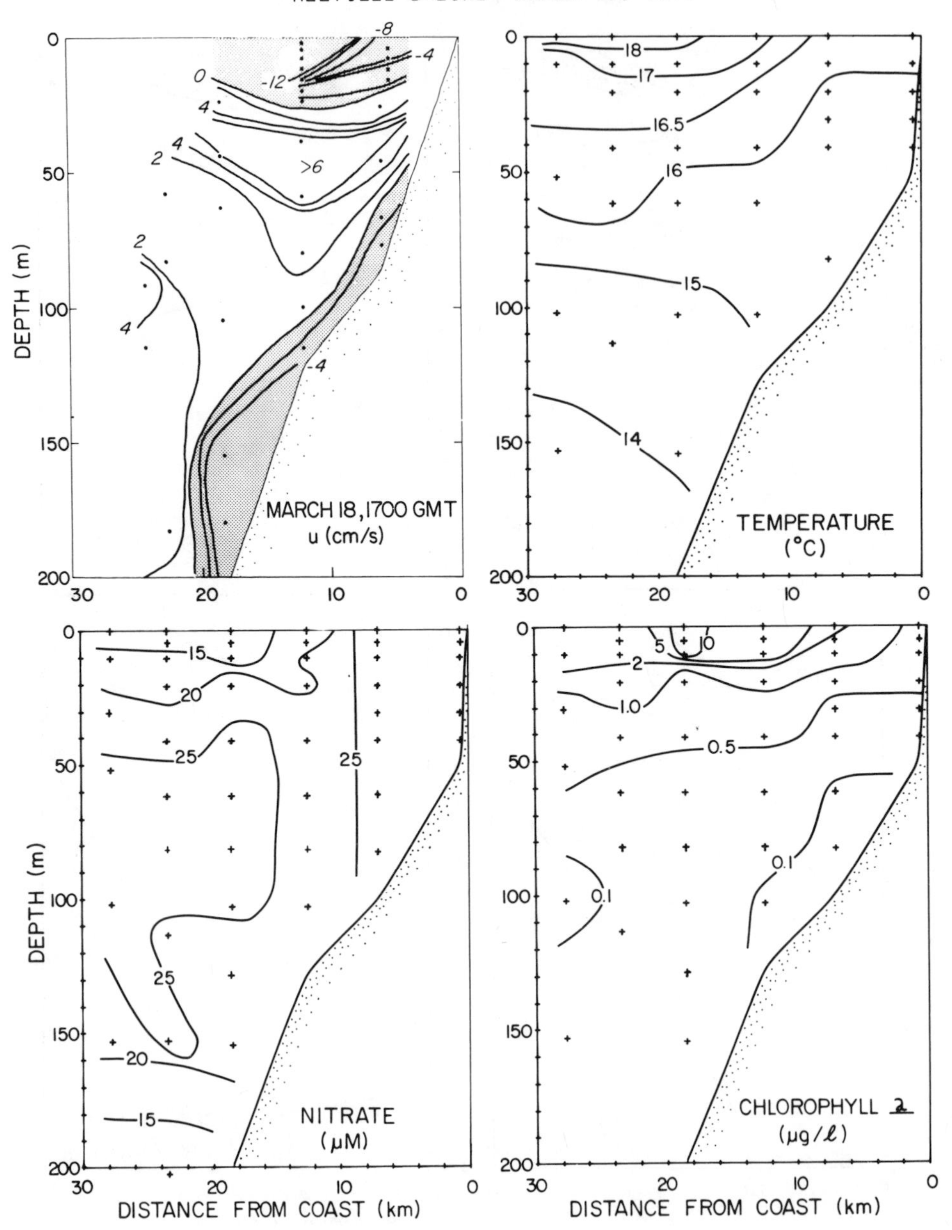

Fig. 12. Hydrographic section of March 18-19, 1977. Sections are the same as in Fig. 11.

near-surface waters warmed by an amount consistent with estimated surface heating (estimates show that solar radiation is generally the dominant surface heat flux), and the mid-shelf surface-mixed layer became thin or non-existent (Fig. 15). Shallow chlorophyll *a* also increased during the period (Fig. 2), indicating phytoplankton growth. This overall pattern appears to be typical of relaxation following an upwelling event.

Period D: Shallow Upwelling

Centered on March 21, a brief, weak wind event occurred. Coincident with the event and with the passage of a coastally trapped wave there was equatorward flow throughout the water column over the shelf. The flow structure could act to break down the phytoplankton reseeding as did the poleward flow during March 15 to 19. There appeared to be upwelling associated with the wind event,

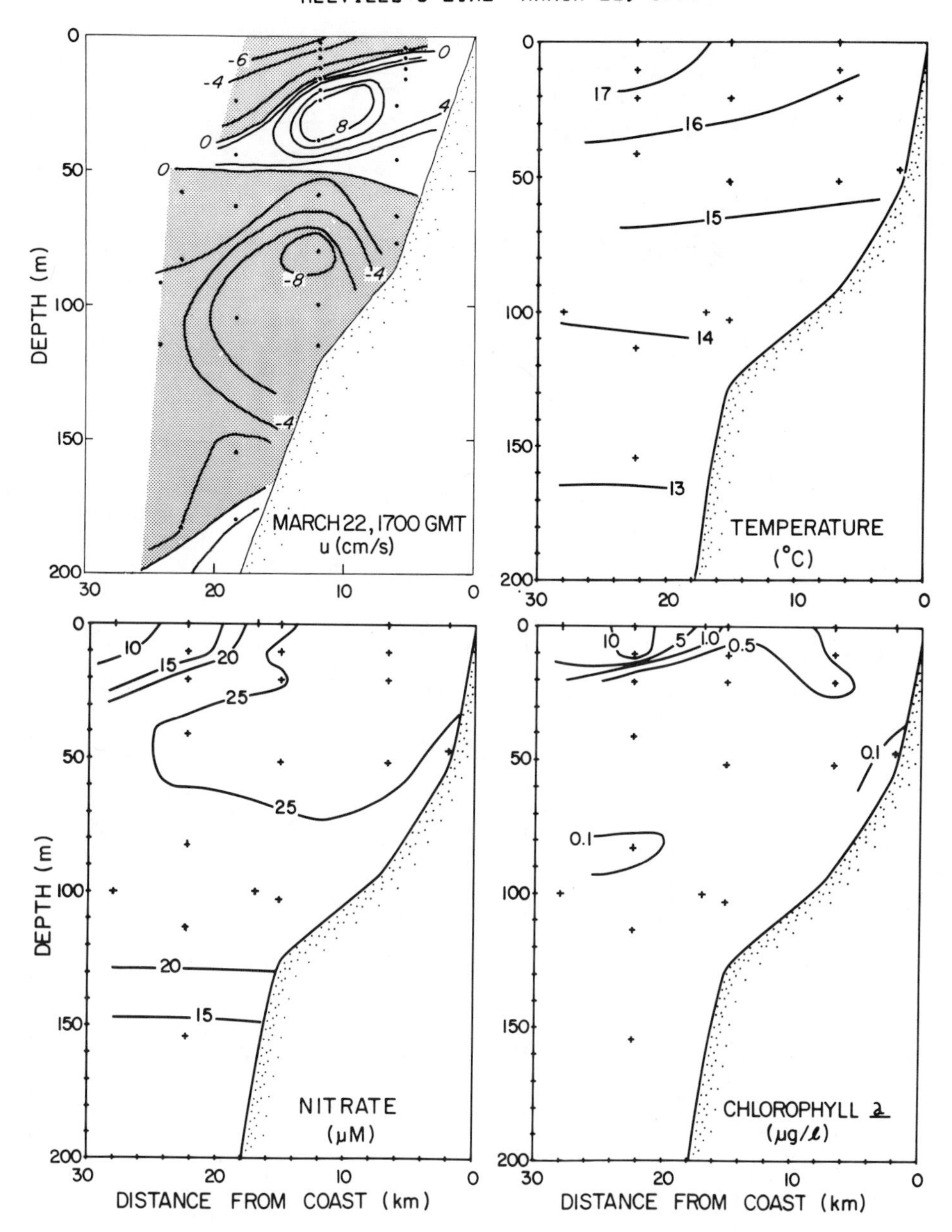

Fig. 13. Hydrographic section of March 22, 1977. Upper left, onshore velocity (cm/s). Sections as in Fig. 11.

with onshore flow between 15 and 50 m (Fig. 3). The mid-shelf surface mixed layer deepened to as much as 16 m during the event (Fig. 15). The near-surface flow at mid-shelf was strongly offshore in the upper 15 m, but nearshore, at PSS, the flow was onshore, so that estimated vertical velocities show the apparent upwelling only initially. The near-surface flow on March 21 was divergent and there was a large cold spot (<16.5°C) centered at the coast slightly south of the C line (Fig. 7). The structure was present to 30 to 50-m depth (Fig. 6a). The near-surface drogues at the time moved outward from the center of the cold area, suggesting a region of divergent flow. The axis of the cold area was directed equatorward from directly offshore, consistent with the near-surface northward flow. About 40 km north of the cold spot there was a distinct warm patch (>21.5°C). The strong alongshore variation in surface temperature is indicative of an accompanying alongshore structure in the upwelling circulation.

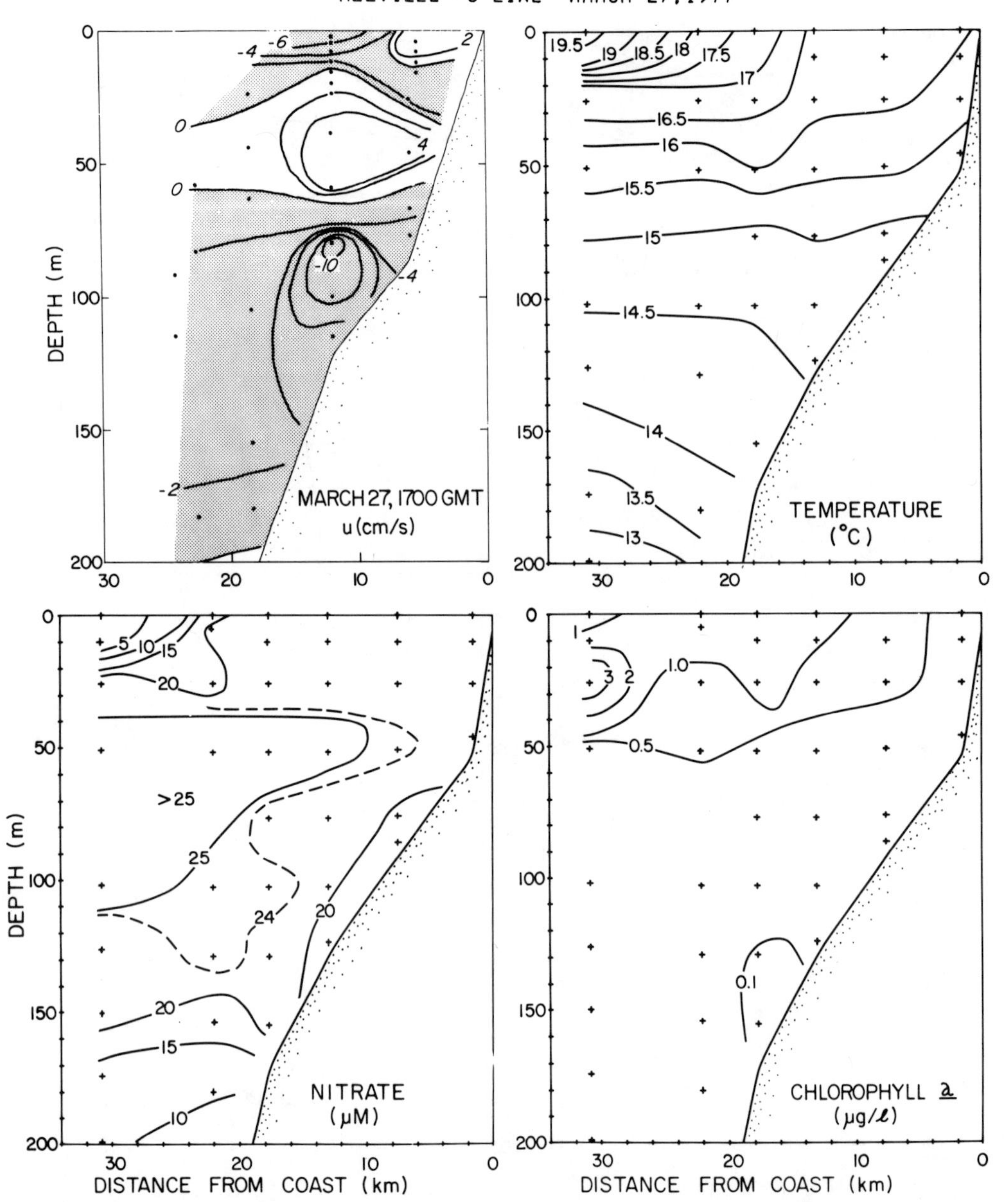

Fig. 14. Hydrographic section of March 27, 1977. Upper left, onshore velocity (cm/s). Sections are as in Fig. 11.

Although the wind event was weak and its apparent source water was of shallow (<50 m) origin, it did appear to have significant biological consequences. During period B, the phytoplankton recycling mechanism was apparently broken and the upwelling source water was not restocked. Water of unusually low phytoplankton content apparently surfaced. The immediate effect of the March 21 event was to decrease significantly near-surface chlorophyll *a* concentrations (Fig. 2). The mid-shelf phytoplankton population did not recover for the remainder of the observation period. The shelf-break chlorophyll *a* maximum remained, although the concentration was less than 1 μg/l throughout the water column within 15 km of the shore on the C line (Fig. 13). The surface chlorophyll *a* showed the same structure at the C line as did alongshore gradients associated with the surface cold area (Fig. 6b). Subsequently phytoplankton was depleted across the entire shelf and slope (Fig. 14). Near-surface nitrate increased slightly during the event, but

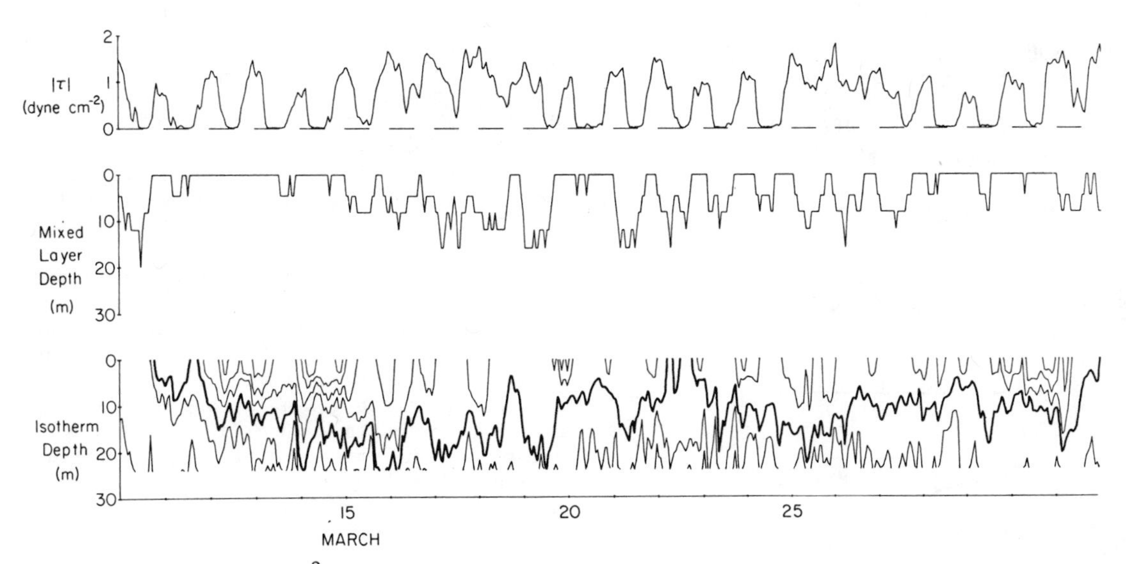

Fig. 15. Wind stress (dyne/cm^2), surface mixed layer depth (m), and isotherm depths (m) at the mid-shelf current meter mooring. Isotherm-depth contours are plotted at increments at 0.5^{o}C, with the 16.5^{o}C depth contour darkened. Wind and temperature data were treated with a filter having a half-power point of 2 h. A given depth is within the mixed layer if its temperature is within 0.1^{o}C of that at 2.1-m depth.

it did not subsequently decrease, apparently because the phytoplankton growth rate was low, as suggested by low specific uptake rates for nitrogen and silicon (Fig. 18).

The second set of drogue-following biological observations began on March 22, toward the end of the upwelling event. The 4-m drogue moved initially northward, and then offshore, fairly consistent with the flow expected in a wind-driven near-surface layer (Fig. 17a). The rates of phytoplankton productivity were initially low; the specific nitrate uptake rate was less than 0.03 h^{-1} for the first 4.5 days (Fig. 17b). During the first three days, chlorophyll *a*, interpolated to 4 m, along the drogue path increased to only 2.8 μg/l, and the associated declines of nitrate and silica were just 5 and 6 μg/l, respectively (Fig. 17b). These low growth and uptake rates again reflect the considerable decrease in biological productivity that may have been set up by the wave passage on March 15 to

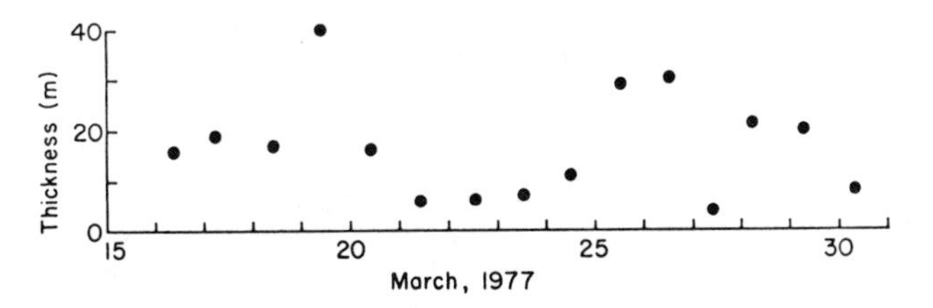

Fig. 16. Depth of the mid-shelf bottom mixed layer, derived from daily CTD observations. A given depth is within the bottom mixed layer if its temperature is within 0.1^{o}C of the deepest measurement.

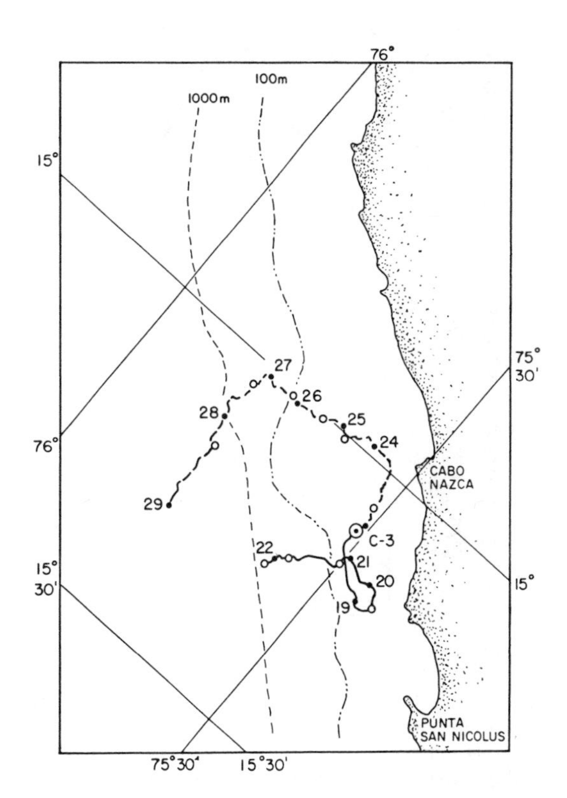

Fig. 17a. Results from the drogue-following biological observations. Drogue tracks, with solid dots representing positions at 0000 local time and open circles denoting positions of biological measurements. The first and second biologically dedicated drogue tracks are denoted by solid and dashed lines, respectively.

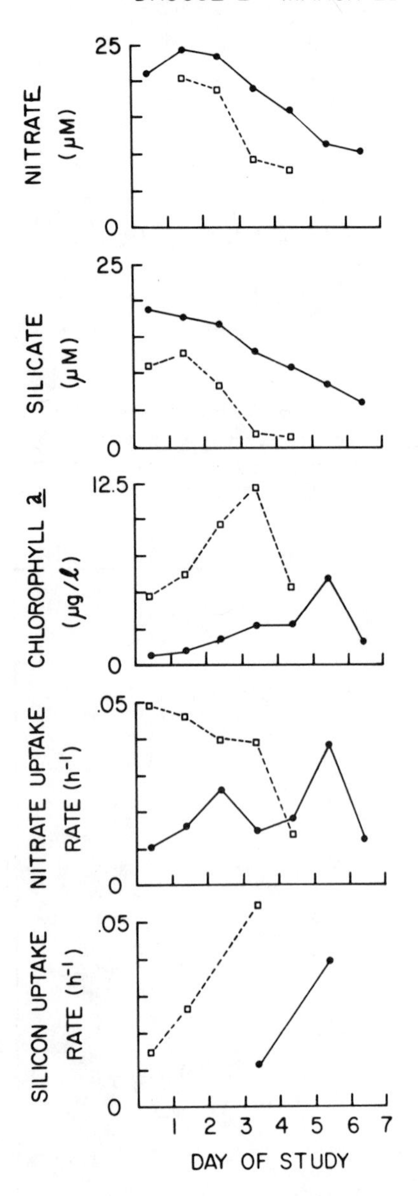

Fig. 17b. Results from the drogue-following biological observations. Time series of chemical and biological variables. Dashed line: March 18-22 drogue, solid line: March 23-28 drogue

19, as suggested above in conjunction with period B.

Period E: Weak Winds

The brief but biologically important event of March 21 was followed by a period of weak winds lasting from March 23 to 25. The onshore-offshore circulation over the shelf became relatively disorganized, with generally onshore flow above 50 m, offshore flow below 50 m, and an estimated downward vertical velocity near the coast (Fig. 3). The undercurrent was re-established, as indicated by the poleward flow below 15 m at mid-shelf. The surface mixed layer reached only moderate depths (Fig. 15). The sea-surface temperature pattern was relatively disorganized on March 23 (Fig. 8), and the central cold area was considerably smaller than on March 21. The drogues and current meters still had a tendency toward cross-isotherm flow. The physical system showed effects (e.g., no strong upwelling) ordinarily contributing to the "aging" of upwelled water, but the primary nutrients and nitrite at mid-shelf did not change significantly (Fig. 2) because of the very low phytoplankton population, hence low consumption.

Period F: Wind Event

A second strong wind event occurred during the period of March 25 to 27, but in the absence of a strong coastally trapped wave. The onshore flow pattern over the shelf suggested developed upwelling circulation and a vertical velocity estimate of as much as 33 m/day (Fig. 3), while the alongshore flow was strongly (peak > 20 cm/s) equatorward near the surface and poleward below 20 m. The 4-m drogue turned more offshore during the period (Fig. 17a). Surface temperatures were generally lower on March 25 than on March 23 (Figs 8 and 9) and a large cold area was centered just poleward of the C line. The drogue data again showed a tendency toward flow outward from the coldest part of the cold surface area. The near-surface flow at PSS turned onshore late in the event, which could be consistent with geostrophic (baroclinic) flow. Although the event

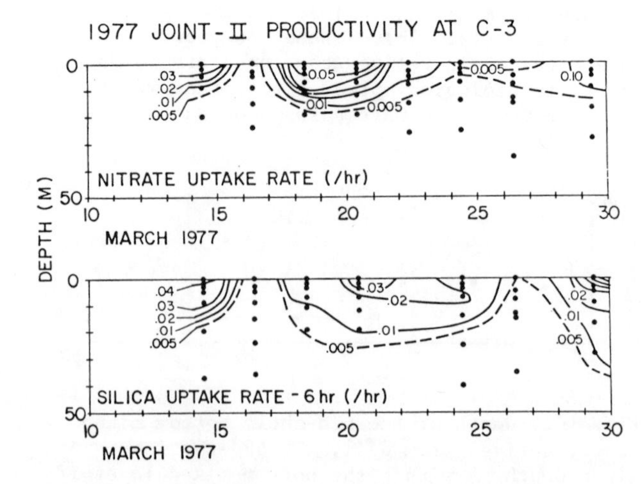

Fig. 18. Mid-shelf (C-3) time series of nitrogen and silicon specific uptake rates (h^{-1}).

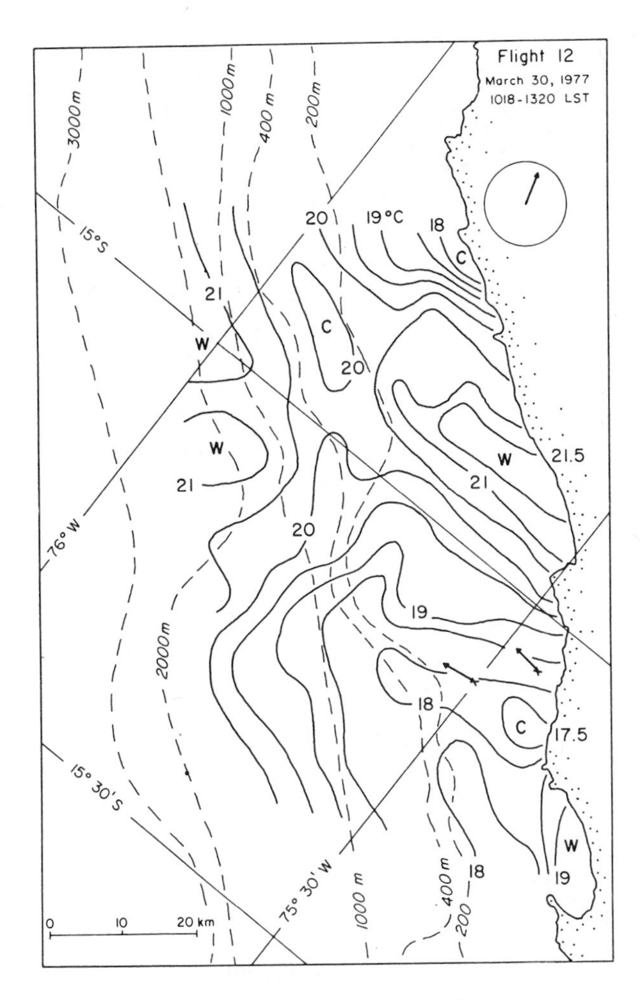

Fig. 19. Sea-surface temperature (°C) from an aircraft flight near noon local time on March 30, 1977. Conventions are as in Fig. 7.

was fairly strong and had deep source waters, the best near-surface expressions of upwelling were in the low surface temperatures (<16.0°C, Fig. 9). No biological or chemical consequences were observed at C-3, apparently because, in the absence of significant biological uptake, nutrient concentrations were nearly uniform with depth at mid-shelf (Fig. 2), i.e., the displaced "old" surface water had roughly the same nutrient content as the freshly upwelled water. The hydrographic section of March 27 (Fig. 14) supports this interpretation: nitrate concentrations were high (>20 μM) over most of the shelf, and chlorophyll *a* concentrations were low over the entire section. The temperature section, with its cold near-surface waters, and the cross-shelf velocity section do, however, show strong indications of upwelling.

Period G: Weak Winds

The intensive observation period ended with a phase of wind relaxation from March 27 to 30, during which the mid-shelf surface temperatures again increased (Fig. 3), the surface mixed layer was generally very shallow (Fig. 15), and the surface cold area contracted (Fig. 19). The mid-shelf onshore flow pattern became disorganized, estimated vertical velocities were variable, and the undercurrent remained strong (Fig. 3). Again, no biological "aging" of the near-surface water was evident at mid-shelf (Fig. 2); chlorophyll *a* remained low and nutrients high.

Conclusions on Temporal Variability

The period we studied may be unrepresentative. The coastally trapped wave of March 15 to 19 was the strongest that occurred during the March through May 1977 period, and the depletion of the phytoplankton stock was also unusual. Therefore, only tentative conclusions about the general patterns of variability can be drawn.

It seems that a normal sequence for upwelling, in the absence of very strong coastally trapped waves, would be the following: Within about one day (half an inertial period) of the onset of upwelling-favorable winds, upwelling of cool, nutrient-rich, phytoplankton-poor (but not totally depleted) water would be established. The vertical motion is concentrated in the nearshore area, so that the surface temperatures there become especially low, and (presumably) the nutrient concentrations high. The surface cold area would continue to enlarge as long as the upwelling continues or until it spreads to the extent that the dynamics change. During this growth phase, the surface temperature patterns are relatively regular, sharp horizontal gradients are unusual, and the near-surface flow is Ekman-like (Figs 9 and 19). The results of Brink *et al.* (1980) support these conclusions with respect to the currents. A clear statistical relation also exists between the size of the surface cold area and the alongshore wind stress: the area having sea-surface temperature less than 16.5°C on the ART charts has a correlation of 0.62 (13 samples, 99% confidence) with the alongshore wind stress at San Juan, averaged for the two days prior to the ART observation. After the wind relaxes, the near-surface waters become warmer with time, primarily because of solar heating and alongshore advection. During relaxation, the sea-surface temperature and near-surface flow fields become less regular, and frontal features occur (e.g., on the northern side of the surface cold area on March 23, see Fig. 8). The tendency for the organization of the surface thermal field to vary with wind strength has also been observed in other upwelling regions (e.g., Curtin, 1979 and Barton, Huyer, and Smith, 1977).

The biological "aging" of the surface water is largely due to the growth of phytoplankton as the water moves away from the upwelling center. As the water moves offshore, its phytoplankton content should increase, provided growth exceeds losses and that nutrient supplies do not severely limit growth. As the phytoplankton grow, they

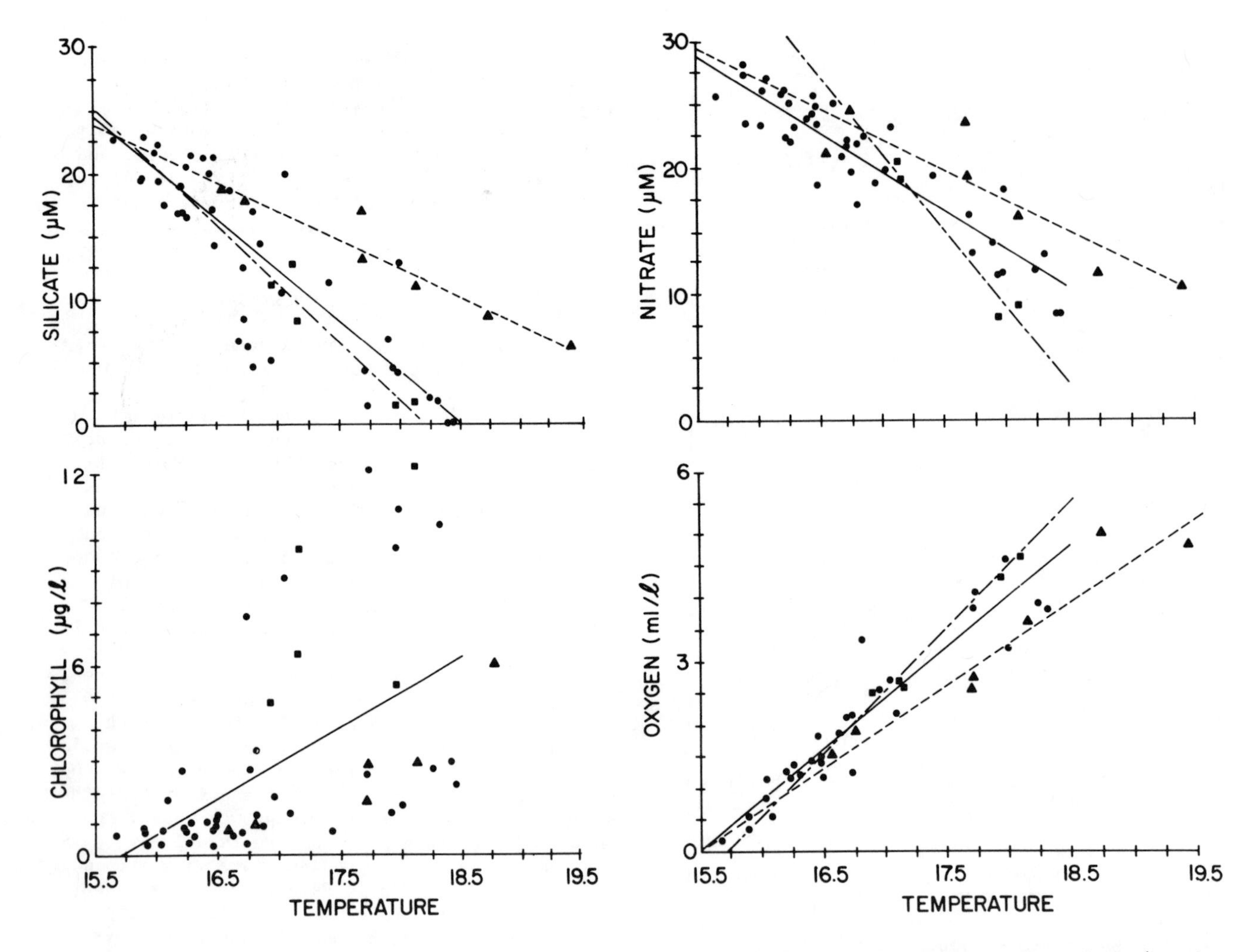

Fig. 20. Chemical variables as a function of temperature in the upper 25 m of the water column (March 12 to 30, 1977). Circles represent hydrographic data, and the solid line is the resulting linear regression equation. Squares (triangles) and the dash-dot (dashed) line are from the first (second) set of drogue-following biological observations.

consume primary nutrients in the near-surface layer, notably nitrate, phosphate, and silicate. Nitrite may increase due either to phytoplankton release or to bacterial nitrification of ammonia (Estrada and Wagensberg, 1977). Thus, as the upwelled water "ages", we expect increased phytoplankton biomass (chlorophyll *a*) and nitrite, and decreased nitrate and silicate concentrations. Losses to the phytoplankton are, however, expected. Diatoms, for example, tend to settle out of the euphotic zone, especially when nutrients become scarce. Grazing by both fish and zooplankton can also account for losses and leave increased ammonia concentrations as a by-product. Ammonia in the surface layer can usually be taken as a sign of grazing and thus of "aging" water. Horizontal advection out of the system is another potential means of removing phytoplankton. This sequence for "normal" upwelling has some support in our observations. The period March 12 to 16 seems to show the sequence fairly clearly, as did the first set of drogue-following biological observations. The physical component of this sequence appears to hold during most of the period.

The above sequence does not apply strictly throughout the March 12 to 30 period, however, because of both reduced phytoplankton concentrations and changes in the species composition, which may have been related to the passage of a coastally trapped wave. (Alternatively Smith and Barber, 1980, carried out some two-dimensional calculations that suggest that the species change of March 20 to 23 may have been due only to cross-shelf advection.) It is difficult to assess the biological impact of the waves. Over the shelf, a wave causes an alongshore velocity perturbation that is relatively depth-independent and causes large temperature fluctuations only below about

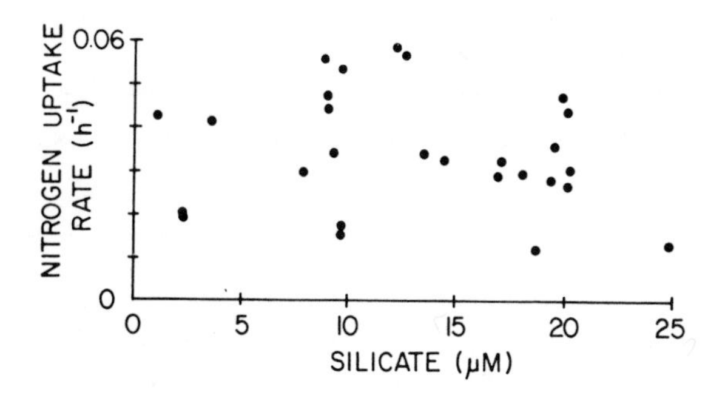

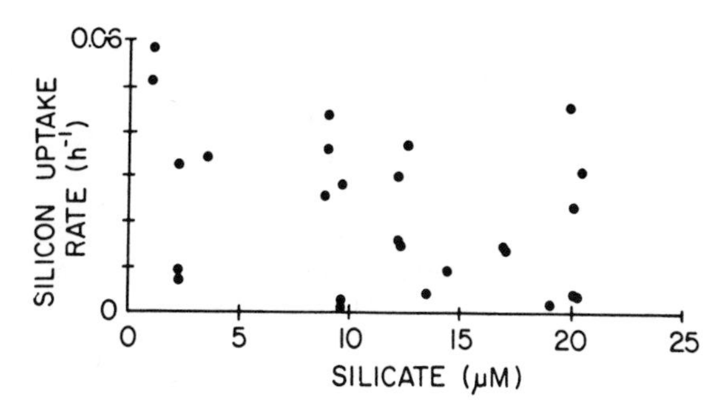

Fig. 21. Specific uptake rates at the 50 and 30% levels as a function of silica concentration (μM), March 12 to 30, 1977. a) nitrate uptake rate (h^{-1}); b) silica uptake rate (h^{-1}).

80-m depth (Brink *et al.*, 1980). Most of the upwelling source water seems to be of shallower origin (see, for example, Fig. 3), so that temperature-nutrient correlations would suggest only small perturbations in the chemical content of upwelled water associated with the waves. Apparently, then, the major direct effect that the waves could have on the biology is through alongshore advection, thus the breakdown of the hypothetical phytoplankton "reseeding" mechanism.

The Biological Surface Expression of the Upwelling Center

Maps based on observations from ships were occasionally drawn of the 15°S upwelling center during the JOINT-II observations, and these often showed chemical and biological structures analogous to that in sea-surface temperature (e.g., Fig. 6) (see also Kelley, Whitledge, and Dugdale, 1975; Boyd and Smith, 1980; Jones, 1977). The structures in the maps are not simple, and the correspondence of chemical and thermal features is not exact. Nevertheless, it is useful to explore the correlations of chemical variables with temperature.

Data from hydrographic sections and the drogue-following studies were used to construct plots of biologically-related variables vs. temperature in the upper 25 m (chosen to include the maximum depth of the surface wind-mixed layer and the euphotic zone). The observations from the drogue-following study were interpolated to 4 m, the depth of the drogue. Sea-surface temperature is taken to approximate the "age" of the upwelled water, as discussed above.

Primary nutrients decreased approximately linearly with increasing temperature. Silica (Fig. 20a) had a maximum of about 23 μM at 15.5°C and decreased to nearly zero at 18.5°C. Nitrate (Fig. 20b) and phosphate (not shown) also decreased linearly with temperature but did not intercept the temperature axis at or below 18.5°C. Thus, silicate was probably a limiting nutrient during the period, particularly at temperatures equal to or greater than 18.5°C. The depletion of silicate also suggests that diatoms were a major part of the local phytoplankton population. The distribution of chlorophyll *a* relative to temperature (Fig. 20c) is bimodal, unlike the primary nutrients or oxygen (Fig. 20d). Because of the good correlations of nutrients and oxygen with near-surface temperature, we expect that mean surface contour plots of these variables would show structure similar to that of the mean temperature plot (Fig. 1).

Differences in 4-m temperatures between the start and finish of each drogue study can be used to estimate nutrient depletion and oxygen generation with regression coefficients derived from Fig. 20. Using the Redfield, Ketchum, and Richards (1963) ratio of $\Delta O:\Delta C:\Delta N:\Delta P = -276:106:16:1$, estimates of the carbon fixed by the phytoplankton were made from the computed nutrient changes and compared with primary productivity integrated over the drogue periods using observations given by Barber *et al.* (1978) (Table 1). The direct estimates of carbon productivity, representing

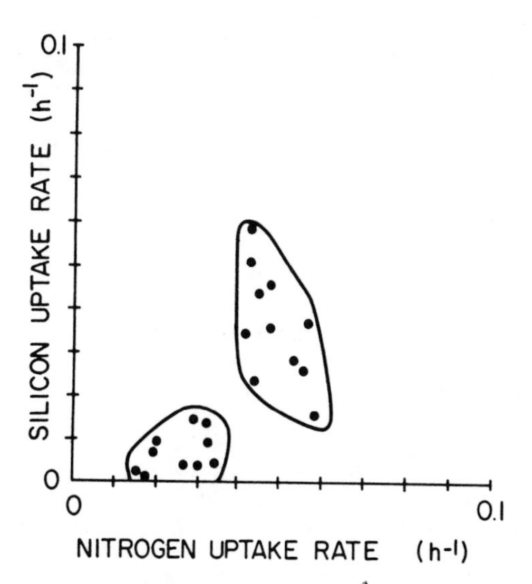

Fig. 22. Silica uptake rate (h^{-1}) vs. nitrate uptake rate (h^{-1}).

TABLE 1. Biological Rate Calculations

	Observed		Estimated from Regression		
	ΔT_{4m}	ΔC*	ΔO	ΔN	ΔP
Drogue Study 1:					
Change along drogue	1.03°C	54.9	177	-12.9	-0.64
ΔC based on Redfield Ratio			67.8	85.3	68.2
$\frac{\text{Measured } \Delta C}{\text{Potential } \Delta C}$ (%)			81	64	80
Drogue Study 2:					
Change along drogue	2.87°C	63.9	328	-13.8	-1.11
ΔC based on Redfield Ratio			126	91.4	118
$\frac{\text{Measured } \Delta C}{\text{Potential } \Delta C}$ (%)			51	70	54

* The measured carbon production is the sum of the depth-averaged euphotic zone simulated *in situ* measurements during the drogue-following studies (Barber *et al.*, 1978). All changes, except for temperature, are in μg-at/l.

averages for the entire euphotic zone, indicate that phytoplankton uptake accounted for 50 to 80% of the changes in chemical variables.

The nutrient-temperature regression results can be used to estimate phytoplankton process rates. The approximate mean travel time for surface transport between the coast and the 18.5°C isotherm was about three days. The mean surface chlorophyll *a* concentration at position C-1 was about 1 μg/l (Fig. 10d). From the nitrate-temperature regression, the decrease in nitrate concentration between the coast and the 18.5°C isotherm was about 19 μM. Using the rule of thumb that 1 μg/l of chlorophyll is produced per 1 μM of nitrogen consumed, then total estimated chlorophyll *a* production was 19 μg/l. From the chlorophyll *a*-temperature regression, an increase to 6.2 ± 2.8 μg/l at 18.5°C can be estimated at 99% confidence. Thus the estimated mean net growth rate is about 0.4 to 0.7/day. The gross growth rate estimated from the change in nitrate is 1.0/d. Assuming no losses due to mixing, a net loss rate of 0.3 to 0.6/day is estimated. We assume the loss is due to grazing and to sinking. Grazing by large copepods (*Calanus chilensis, Eucalanus inermis,* and *Centropages brachiatus*) accounted for less than 5% of the daily primary production during April, 1977 (Dagg *et al.*, 1980). Sediment trap measurements by Staresinic, Rowe, and Clifford (1980) and by Staresinic (1978) show that in 1977 about 9% of the daily surface primary productivity sank from the euphotic zone. Thus, there is an observed minimum loss rate of 14% compared to our estimated 30% to 60%. Our estimated growth rate (equivalent to 1.4 doublings/day) agrees with the carbon doubling rates at depths shallower than the 15% light penetration depth, as measured by simulated *in situ* uptake studies (Barber *et al.*, 1978). Thus, the observed (Fig. 6) or inferred (from regression) scales of surface chemical structures are fairly consistent with *in situ* rates.

Uptake rates of nitrogen and silicon show no simple dependence on the concentration of silica (Fig. 21), the apparent growth-limiting nutrient. A plot of nitrogen vs. silicon uptake rates (Fig. 22) shows that the observations fall into two groups. One group has high uptake rates of both nitrogen and silicon. The second group has low nitrogen uptake rates and lower silicon uptake rates. The observations in this low-rate group fall into two categories. The first includes measurements from period A, when it seems that the low uptake rates were due primarily to low nutrient concentrations. The other observations in the low-rate group were made after March 21, when nutrient concentrations were high (Fig. 2), but flagellates represented a relatively larger fraction of the phytoplankton species composition (Dolors Blasco, personal communication). Thus the changes in nutrient uptake rates during March 12 to 30 were apparently due both to nutrient limitation and to changes in the phytoplankton species composition. The changes in uptake rates will presumably cause the chemical signature of the upwelling center to vary in its intensity. Nonetheless, the tendency for the chemical reflection of the upwelling center to mimic the thermal structure should generally obtain.

Conclusions

Our study of the upwelling center near 15°S gives a partial description of its behavior.

While ambiguity remains concerning some aspects, a few general points are quite clear. The near-surface region especially displays strong temporal variability on the diurnal time scale and longer and is extremely responsive to variations in the wind and in surface heating. Within the upwelling center, there appears to be a considerable dependence of biological and chemical variables on the physical environment. Both the biological and the physical features are strongly three-dimensional and seem to have most of their structure concentrated in the upper 50 m or less of the water column. Although the physical "horizontal plume" is a shallow feature, its biological significance is assured because the phytoplankton are generally concentrated near the surface.

Questions remain concerning the near-surface physical and biological dynamics of the upwelling center. Is the edge of the surface cold area a deformed front, or is there generally a more gradual transition between the freshly upwelled waters and the ambient surface waters? Unambiguous intense fronts have not been thoroughly documented by sub-surface observations, although surface measurements (e.g., ART charts) occasionally show sharp fronts, usually on the lateral boundaries of the cold area. Closely related to the question of fronts is the nature of the near-surface flow. To the extent that the fronts are important, near-surface flow will tend to have a thermal wind balanced component parallel to the front and a frictional component perpendicular to the front. Our observations do not allow a clear evaluation of the importance of the two factors. Another question of considerable importance, which we have not addressed, is why does the surface cold area, i.e., the apparent plume, exist, and why is the location of its coastal origin so stable? The locally intensified alongshore component of the wind stress, discussed above, will certainly contribute to this phenomenon. Also, the numerical experiments reported by Preller and O'Brien (1980) suggest that upwelling is locally intensified because of interaction of the undercurrent with nearby irregularities in the continental shelf topography. The validity of the phytoplankton "reseeding" hypothesis must be resolved. The onshore flow at depth varies strongly in the alongshore direction (Brink *et al.*, 1980), so that estimates based on data from C-line current meters do not allow a sufficient test. Yet the substantial productivity of the upwelling center (relative to its immediate surroundings) and the significant temporal variations in nutrient uptake rates suggest that there is indeed a biological concentration mechanism related to the physical regime. Future studies of regions such as this will probably require a greater concentration of resources on near-surface observations. The numerous ambiguities of this study point out a need for physical, chemical, and biological observations to define the entire spatial extent of the cold area. Temporal intensity must be adequate to document, without ambiguity, the response of the structure to its forcing functions. These constraints will probably require at least daily mapping of the surface feature and careful monitoring of the wind forcing and sub-surface variability.

Finally, one might ask whether upwelling center phenomena as we have described them here are confined to the Peruvian and perhaps California (Traganza, Nestor, and McDonald, 1980) upwelling systems, or if they are more universal. In other words, is an upwelling center simply another manifestation of the more nearly alongshore uniform dynamics found, for example, off Oregon (Holladay and O'Brien, 1975)? It is our opinion that the upwelling center near 15°S is unusual only in its relatively short (∿30 km) alongshore scale, the persistent shallowness of the poleward undercurrent, and the importance of free, coastally trapped waves. These differences do not appear to make the upwelling center near 15°S particularly anomalous.

Acknowledgements. We would like to express our appreciation to all of those scientists who made their data available for this study: Drs. R.T. Barber, L.A. Codispoti, J.J. Goering, D. Halpern, A. Huyer, and R.L. Smith. Funding for the research was provided by the International Decade of Ocean Exploration (IDOE) office of the National Science Foundation (NSF) through grants OCE 78-03380, OCE 79-22547, OCE 75-22444, OCE 77-28354, OCE 77-27735, OCE 78-03042, OCE 77-27006, and OCE 78-00611.

References

Barber, R.T., Variations in biological productivity along the coast of Peru: the 1976 dinoflagellate bloom, *Science*, in press, 1981.

Barber, R.T., S.A. Huntsman, J.E. Kogelschatz, W.O. Smith, B.H. Jones, and J.C. Paul, Carbon, chlorophyll and light extinction from JOINT-II 1976 and 1977, CUEA Data Report 49, 476 pp., 1978.

Barton, E.D., A. Huyer, and R.L. Smith, Temporal variation observed in the hydrographic regime near Cabo Corveiro in the northwest African upwelling region, February to April 1974, *Deep-Sea Research, 24*, 7-23, 1977.

Boyd, C.M. and S.L. Smith, Interaction of plankton populations in the surface waters of the Peruvian upwelling region: A study in mapping, Abstracts of the IDOE International Symposium on Coastal Upwelling, University of Southern California, Los Angeles, February 4-8, 1980, American Geophysical Union, Washington, D.C. (abstract only), 1980.

Brink, K.H., D. Halpern, and R.L. Smith, Circulation in the Peruvian upwelling system near 15°S, *Journal of Geophysical Research, 85*, 4036-4048, 1980.

Brink, K.H., R.L. Smith, and D. Halpern, A compendium of time series measurements from moored instrumentation during the MAM 77 Phase of JOINT-II, CUEA Technical Report 45, 72 pp., 1978.

Curtin, T.B., Physical dynamics of the coastal upwelling frontal zone off Oregon, Ph.D. Dissertation, University of Miami, 338 pp., 1979.

Dagg, M., T. Cowles, T. Whitledge, S. Smith, S. Howe, and D. Judkins, Grazing and excretion by zooplankton in the Peru upwelling system during April 1977, *Deep-Sea Research, 27,* 43-59, 1980.

Dugdale, R.C., J.J. Goering, R.T. Barber, R.L. Smith, and T.T. Packard, Denitrification and hydrogen sulfide in the Peru upwelling region during 1976, *Deep-Sea Research, 24,* 601-608, 1977.

Eber, L.E., J.F.T. Saur, and O.E. Sette, Monthly mean charts: sea surface temperature North Pacific Ocean 1949-62, U.S. Bureau of Commercial Fisheries Circular, 258, 1968.

Estrada, M. and M. Wagensberg, Spectral analysis of spatial series of oceanographic parameters (fluorescence, temperature, concentrations of nitrite and of nitrate + nitrite), *Journal of Experimental Marine Biology and Ecology, 30,* 147-164, 1977.

Hafferty, A.J., L.A. Codispoti, and A. Huyer, JOINT-II R/V *Melville* Legs I, II, IV; R/V *Iselin* Leg II Bottle Data March 1977 - May 1977, Reference M78-48, University of Washington, Department of Oceanography, 779 pp., 1978.

Hafferty, A.J., D. Lowman, and L.A. Codispoti, JOINT-II *Melville* and *Iselin* bottle data sections March-May 1977, CUEA Technical Report, 38, 129 pp., 1979.

Holladay, C.G. and J.J. O'Brien, Mesoscale variability of sea surface temperatures, *Journal of Physical Oceanography, 5,* 761-772, 1975.

Huyer, A. and W.E. Gilbert, Vertical sections and mesoscale maps of temperature, salinity, and sigma-t off the coast of Peru, July to October 1976 and March to May 1977, School of Oceanography, Oregon State University, Data Report 79, 162 pp., 1979.

Johnson, W.R., P.I. Koeb, and C.N.K. Mooers, JOINT-II R/V *Columbus Iselin* Leg 1 CTD measurements off the coast of Peru March 1977, CUEA Data Report 56, 408 pp., 1979.

Jones, B.H., A spatial analysis of the autotrophic response to abiotic forcing in three upwelling ecosystems: Oregon, northwest Africa, and Peru, Ph.D. Dissertation, Duke University, 262 pp., 1977.

Kelley, J.C., T.E. Whitledge, and R.C. Dugdale, Results of sea surface mapping in the Peru upwelling system, *Limnology and Oceanography, 20,* 784-794, 1975.

Kogelschatz, J.E., J.J. MacIsaac, and N.F. Breitner, JOINT-II general productivity report R/V *Wecoma* March-May 1977, CUEA Data Report 50, 170 pp., 1980

Moody, G.L., Aircraft derived low level winds and upwelling off the Peruvian coast during March, April, and May 1977, CUEA Technical Report 56, 110 pp., 1979.

Mooers, C.N.K., C.A. Collins, and R.L. Smith, The dynamic structure of the frontal zone in the coastal upwelling region off Oregon, *Journal of Physical Oceanography,* **6**, 3-21, 1976.

Preller, R. and J.J. O'Brien, The influence of bottom topography on upwelling off Peru, *Journal of Physical Oceanography, 10,* 1377-1398, 1980.

Redfield, A.C., B.H. Ketchum, and F.A. Richards, The influence of organisms on the composition of sea water, in *The Sea,* M.N. Hill (ed.), Interscience Publishers, New York, Vol. II, Chap. 2, 1963.

Ryther, J.H., D.W. Menzel, E.M. Hulburt, C.J. Lorenzen, and N. Corwin, *Anton Bruun* Reports, Scientific results of the southeast Pacific expedition, Report No. 4, Texas A&M Press, 12 pp., 1966.

Smayda, T.J., The suspension and sinking of phytoplankton in the sea, *Oceanography and Marine Biology, Annual Review, 8,* 353-414, 1970.

Smith, R.L., Poleward propagating perturbations in currents and sea level along the Peru coast, *Journal of Geophysical Research, 83,* 6083-6092, 1978.

Smith, R.L., D.B. Enfield, T.S. Hopkins, and R.D. Pillsbury, Circulation in an upwelling ecosystem: the *Pisco* cruise, *Investigacion Pesquera, 35,* 9-24, 1971.

Smith, S.L. and T.E. Whitledge, The role of zooplankton in the regeneration of nitrogen in a coastal upwelling system off northwest Africa, *Deep-Sea Research, 24,* 49-56, 1977.

Smith, W.O., Jr., and R.T. Barber, Circulation, sinking, and vertical migration as determinants of phytoplankton success in the upwelling center at 15°S, *This Volume,* 1980.

Staresinic, N., The vertical flux of particulate organic matter in the Peru coastal upwelling with a free-drifting sediment trap, Ph.D. Dissertation, Woods Hole Oceanographic Institution and Massachusetts Institute of Technology, 255 pp., 1978.

Staresinic, N., G.T. Rowe, and C.H. Clifford, The vertical flux of particulate organic matter in the Peru coastal upwelling region, Abstracts of the IDOE International Symposium on Coastal Upwelling, University of Southern California, Los Angeles, February 4-8, 1980, American Geophysical Union, Washington, D.C. (Abstract only), 1980.

Stevenson, M.R., Measurement of currents in a coastal upwelling zone with parachute drogues, *Exposure, 2,* 8-11, 1974.

Stevenson, M.R., R. Garvine, and B. Wyatt, Lagrangian measurements in a coastal upwelling zone off Oregon, *Journal of Physical Oceanography, 4,* 321-336, 1974.

Stuart, D.W. and J.J. Bates, Aircraft sea surface temperature data--JOINT-II 1977, CUEA Data Report 42, 39 pp., 1977.

Traganza, E.D., D.A. Nestor, and A.K. McDonald,

Satellite observations of a nutrient upwelling off the coast of California, *Journal of Geophysical Research, 85,* 4101-4106, 1980.

Van Leer, J.C. and A.E. Ross, Velocity and temperature data observed by Cyclosonde during JOINT-II off the coast of Peru, Rosenstiel School of Marine and Atmospheric Sciences Data Report DR 7902, University of Miami, 34 pp., 1979.

Walsh, J.J., J.C. Kelley, R.C. Dugdale, and B.W. Frost, Gross features of the Peruvian upwelling system with special reference to possible diel variation, *Investigacion Pesquera, 35,* 25-42, 1971.

Zuta, S., T. Rivera, and A. Bustamante, Hydrologic aspects of the main upwelling areas off Peru, in *Upwelling Ecosystems,* pp. 235-257, R. Boje and M. Tomczak (eds.), Springer-Verlag, Berlin, 303 pp., 1978.

THE FORMATION OF PLANKTON PATCHES IN THE SOUTHERN BENGUELA CURRENT

L. Hutchings

Sea Fisheries Institute, Cape Town, South Africa

Abstract. The movement of patches of zooplankton and phytoplankton in the Southern Benguela Current was observed over a 6-day period during a complete wind cycle. During downwelling phytoplankton accumulated inshore, nutrients decreased, and stability increased in the upper layers. When upwelling began, eddy formation split the patch of phytoplankton in two, and zooplankton increased in one patch but remained low in the other. As upwelling continued the patches were displaced or dispersed along the plume at rates of 10 to 15 km/day. The patch of phytoplankton closest to the center of the plume, associated with low zooplankton levels, dispersed more rapidly than the patch to the south, where the zooplankton standing stock was some fivefold higher, suggesting that dispersion was the dominant factor affecting phytoplankton dynamics in the area. From 33 monthly transects, the distributions of temperature, chlorophyll *a*, and zooplankton along the major axis of the upwelling plume were examined with regard to short-term events prior to sampling. Plankton stocks in a more stable area, St. Helena Bay, were compared with those off the Cape Peninsula. The consequences of the variability associated with the active upwelling site are related to the distribution of pelagic fish.

Introduction

Perhaps the most characteristic feature of coastal upwelling zones is the intermittent nature of the supply of nutrients to the euphotic zone; "steady state" upwelling (Walsh, 1977) almost never persists for more than a few days before the circulation pattern is altered by the passage of wind events, shelf waves, current fluctuations, and other physical events. Also, the euphotic and the Ekman layers in these areas are generally shallow and subject to marked lateral movements, and the response time of diatoms to nutrient enrichment is short compared to the generation time of the planktonic and nektonic herbivores. The development of large populations of secondary producers is therefore likely to depend largely on the evolution of behavior patterns that will enable the consumers to utilize efficiently the dense but transient patches of phytoplankton and to remain within the upwelling zone itself. This implies that the trajectory of a water parcel after upwelling is of rather more importance than absolute concentrations of nutrients or rates of primary production. Water trajectories, countercurrents, and jet currents vary considerably between and even within different upwelling regions (Barber and Smith, 1980); off Oregon there appears to be considerable longshore homogeneity (Peterson, Miller, and Hutchinson, 1979) while off northwest Africa upwelling can be strongly influenced by bottom topography (Cruzado and Salat, 1980; Herbland, LeBorgne, and Voituriez, 1973). Off Peru, bottom topography and prominent coastal mountains and river valleys produce a series of plumes along the coast (Smith, Enfield, Hopkins, and Pillsbury, 1971). Wind regimes are of considerable importance in altering the near-surface flow patterns, and the constancy of upwelling winds will affect the consistency of water trajectories to a considerable extent. In the region of Cape Town, the Southern Benguela Current displays many complex features related to the proximity of warm Agulhas water, bathymetry, wind regime, and orography (Shannon, Nelson, and Jury, 1980). Very different conditions occur just 150 km north in St. Helena Bay (Fig. 1). This report examines the effect that wind reversals have on the plankton distribution in the particularly active upwelling region off Cape Town (Andrews and Hutchings, 1980) and contrasts the distribution of plankton in the area with that in St. Helena Bay, where pelagic fish availability is fourfold higher (Fig. 1) (Centurier-Harris and Crawford, 1974).

Methods

Data from three types of cruises were used to demonstrate the variability of plankton in the two areas:

(1) The first cruise was a two-ship survey of the upwelling center off the Cape Peninsula, where 35 stations in a 60 x 40 km area (Fig. 2) were repeatedly sampled at standard depths at roughly 36-h intervals. Zooplankton was sampled from close to the bottom to the surface

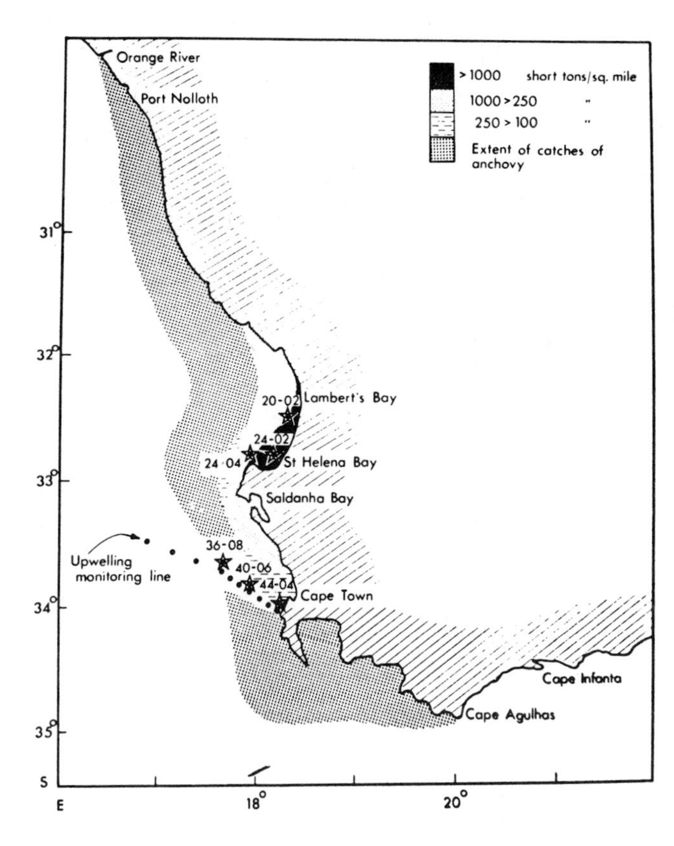

Fig. 1. The yield of anchovy per unit area in short tons per square mile during the period 1964 to 1972 (from Centurier-Harris and Crawford, 1974). Also indicated are the six stations selected for the comparison between St. Helena Bay (stas 20-02, 24-02, 24-04) and the Cape Peninsula (stas 44-04, 40-06, 36-08). The upwelling monitoring line of stations, sampled between October 1970 and March 1973, is also shown.

at each station with a WP-2 net (57-cm diameter, 200-μm mesh aperture) equipped with a depth-recording N10 flowmeter (Currie and Foxton, 1957). At 10 stations supplementary information on the depth distribution of zooplankton in the upper 50 meters was collected with a centrifugal pump (7.6-cm diameter; 550 l/min., 200-μm mesh filter) accompanied by closing net hauls with the WP-2 net. Zooplankton abundance was displayed as dry weight per square meter. Chlorophyll *a* concentrations in the upper 75 m were determined spectrophotometrically using the UNESCO (1966) formulae and were integrated to quantity per square meter for comparison with the zooplankton data.

(2) From October 1970 to March 1973, a transect of 7 to 10 stations running northwest from the Cape Peninsula (Fig. 1) was sampled once each month to observe trends along the mean position of the upwelling plume. Chlorophyll *a* and zooplankton were determined as before, but zooplankton hauls were collected in divided hauls, using the bottom of the upper mixed layer as the boundary between hauls.

(3) The third set of cruises was undertaken in 1977 and 1978 and was aimed principally at mapping the distribution of pelagic fish eggs and larvae. Chlorophyll *a* data were collected as before, while zooplankton was collected with 57-cm diameter Bongo nets of 300-μm mesh aperture hauled obliquely from close to the bottom to the surface down to a maximum depth of 100 m. Zooplankton abundance was recorded as displaced volume per 1000 m^3 filtered, the volume filtered being recorded during each haul with General Oceanics digital flowmeters. The samples provided the only suitable data for a comparison between St. Helena Bay and the Cape Peninsula region. Only six of the 120 stations occupied during each cruise have been used in the comparison (Fig. 1).

Results

Changes in plankton at the center of upwelling

During December 1969 it was possible to follow patches of plankton during a brief wind reversal. As the wind blew from the west on 12 and 13 December, 1969 (Fig. 2), the inshore surface water warmed rapidly and the frontal zone became less distinct. As the wind backed to the east and increased in speed, cold water reappeared in the south and rapidly spread offshore. Phytoplankton, indicated here as chlorophyll *a* in the upper 75 m (Fig. 3), became concentrated against the shore during westerly winds as downwelling occurred, and as the wind turned to blow from the east the patch appeared to split in two as the cold water advanced. As upwelling progressed, both patches were shifted to the northwest, with the northernmost patch, which was in the middle of the plume, being dispersed more rapidly than the southern patch. Changes in zooplankton were more obscure (Fig. 4), but the southern phytoplankton patch became associated with a zooplankton crop some five-fold higher than the northern patch, indicating that grazing was of little consequence in this situation. The patch of zooplankton was also displaced to the northwest as upwelling continued and retained its species composition (Andrews and Hutchings, 1980). Considering the station spacing and the fact that the patches of plankton were only observed at one or two stations on occasions, it was not possible to calculate precise rates of movement of the patches, but they appeared to move on the order of 10 to 15 km/day.

Examination of 18 years of wind observations shows that while wind reversals occur two to five times each month during the summer, they are

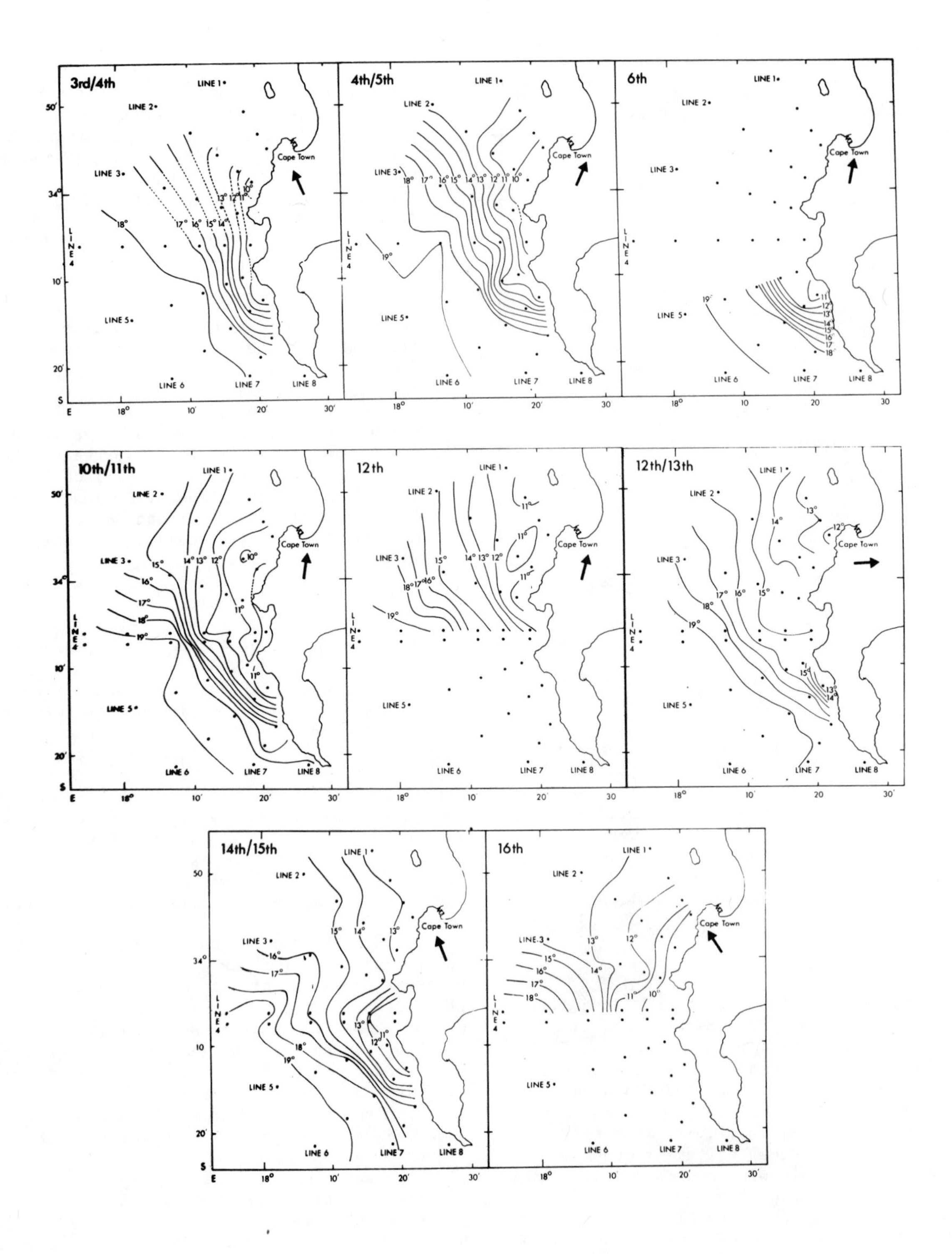

Fig. 2. The distribution of surface temperature (°C) 3-16 December, 1969. The arrows indicate the predominant wind direction.